The Biology of Pelagic Tunicates

Thalia democratica. Oozooid with an early stolon coiling out ventrally. The body is 5.5 mm across at the widest point. Photograph by A.C. Heron.

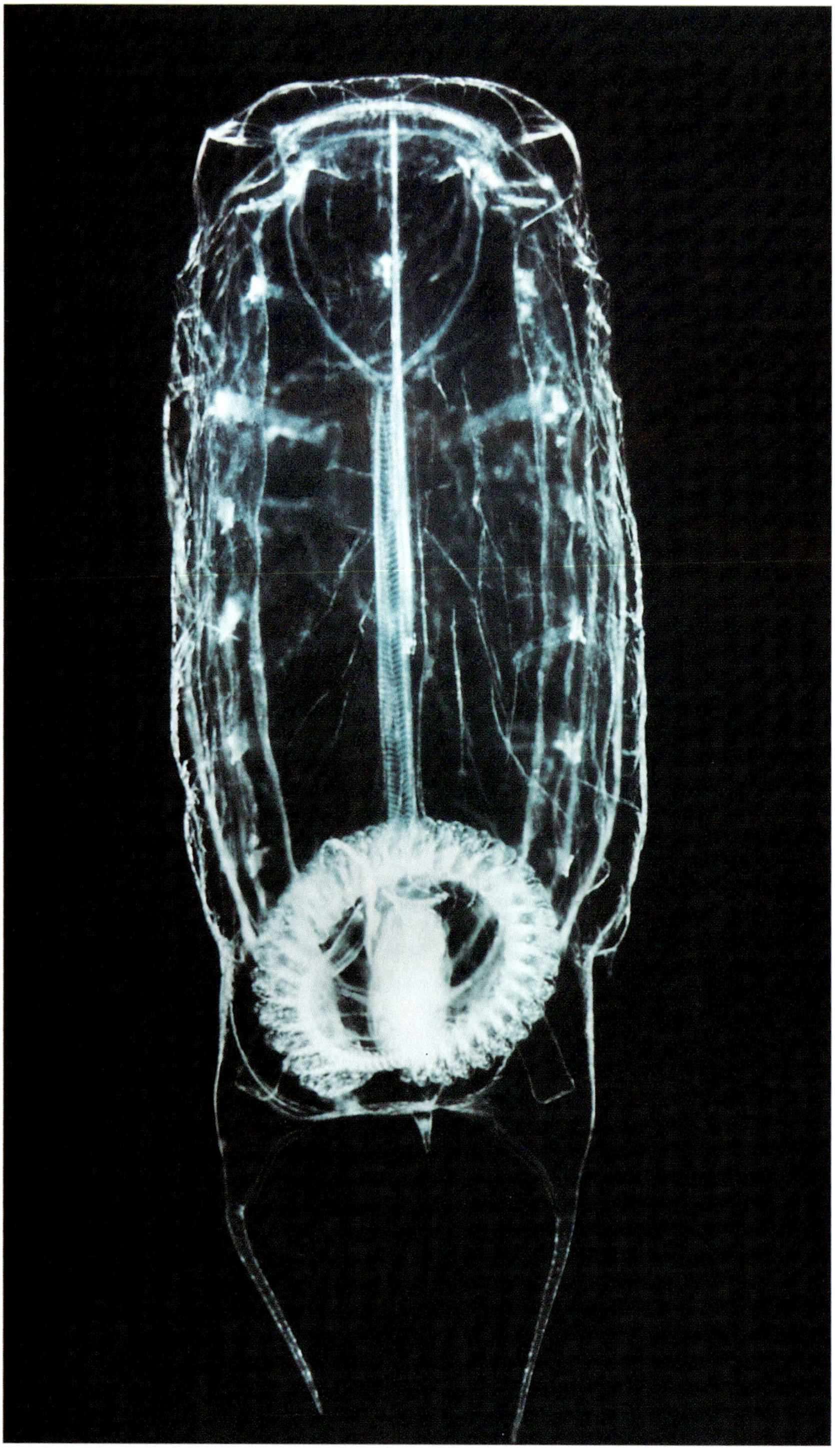

The Biology of Pelagic Tunicates

Edited by

Q. BONE
Marine Laboratory, Plymouth

Oxford New York Tokyo
OXFORD UNIVERSITY PRESS
1998

Oxford University Press, Great Clarendon Street, Oxford OX2 6DP
Oxford New York
Athens Auckland Bangkok Bogota Bombay
Buenos Aires Calcutta Cape Town Dar es Salaam
Delhi Florence Hong Kong Istanbul Karachi
Kuala Lumpur Madras Madrid Melbourne
Mexico City Nairobi Paris Singapore
Taipei Tokyo Toronto Warsaw
and associated companies in
Berlin Ibadan

Oxford is a trade mark of Oxford University Press

Published in the United States
by Oxford University Press, Inc., New York

A catalogue record for this book is available from the British Library

Library of Congress Cataloging in Publication Data

The biology of pelagic tunicates/edited by Q. Bone.
Includes bibliographical references and index.
1. Tunicata. I. Bone, Q.
QL613.B56 1998 596′.2–dc21 97–11329

ISBN 0 19 854024 8

Typeset by EXPO Holdings, Malaysia

Printed in Great Britain by
The Bath Press.

To the memory of Andy Heron (1940–1989)

This book is dedicated to the memory of Dr Andrew Heron, an unusually gentle and modest man, who was noted for his helpfulness to others, and for his enthusiasm for pelagic tunicates. He was the first to show, in a remarkable PhD thesis (later to appear as a number of seminal papers), what underlay the spectacular capacity of salps for bloom formation. He worked in the CSIRO Fisheries laboratory, first at Cronulla near Sydney, and then at the new Hobart laboratory. He was fortunate in possessing the most unusual combination of superior mathematical skills and a deep interest in natural history, and after a trip to Heron Island as a Sydney undergraduate, became a marine biologist whose best known and highly original work was on the population biology of the small salp *Thalia*.

This work, which resulted from field observations on a bloom of *Thalia*, soon became — and has remained — a classic of marine ecology, and as a glance at the bibliography of this book will show, he was interested in and contributed to functional anatomy as well.

He died on the 27th of November 1989, aged only 49, leaving his wife and colleague Shirley Jeffrey. Few scientists have made such an authoritative and original contribution to tunicate biology; he will be much missed by all who knew him.

Preface

This book arose from a fascination with pelagic tunicates that has if anything steadily increased over almost half a century that has elapsed since Sir Alister Hardy first introduced them to me in an undergraduate lecture at Oxford. Originally, the book was to be confined to salps, but it soon became obvious that all the pelagic tunicate groups should be included. By cajoling friends expert in these other groups to contribute, the book has taken longer to produce, but should prove to be much more interesting and useful.

I am grateful to the patient contributors who have waited remarkably uncomplainingly for the book to appear, to the Marine Biological Association and the Royal Society for supporting many visits to work on pelagic tunicates at the Villefranche Laboratory near Nice, where pelagic tunicates have been a speciality for 150 years, and above all, to my wife Susan, who has encouraged me to travel to work on pelagic tunicates despite having a large family to look after in my absences.

Plymouth Q.B.
March 1997

Contents

Contributors

V. Andersen Observatoire Océanologique, URA 2077, BP 28, 06234 Villefranche-sur-Mer Cedex France.

Q. Bone Marine Laboratory, Citadel Hill, Plymouth PL1 2PB, UK.

J.-C. Braconnot Observatoire Océanologique, URA 2077, BP 28, 06234 Villefranche-sur-Mer Cedex France.

R. Christen Observatoire Océanologique, URA 2077, BP 28, 06234 Villefranche-sur-Mer Cedex France.

D. Deibel Ocean Science Centre, Memorial University of Newfoundland, St John's Newfoundland and A1C 5S7 Canada.

R. Fenaux Observatoire Océanologique, URA 2077, BP 28, 06234 Villefranche-sur-Mer Cedex France.

Per R. Flood Bathybiologica, Gerhard Gransvei 58, N-5030 Landås, Bergen, Norway.

Charles P. Galt Department of Biological Sciences, California State University, Long Beach, CA 90840-3702, USA.

J. Godeaux Institut de Zoologie, Laboratoire de Biologie Générale, Université de Liège, Quai Van Beneden 22, B4020 Liège, Belgium.

G. Gorsky Observatoire Océanologique, URA 2077, BP 28, 06234 Villefranche-sur-Mer Cedex France.

G.R. Harbison Woods Hole Oceanographic Institute, Woods Hole, MA02543, USA.

L.R. Madin Woods Hole Oceanographic Institute, Woods Hole, MA02543, USA.

R.W.M. van Soest Institute of Taxonomic Zoology (Zoölogisch Museum), University of Amsterdam, PO Box 4766, 1009 AT Amsterdam, The Netherlands.

Introduction

Q. Bone

The phylum Tunicata is divided into three groups, which are usually considered to rank as classes. The largest in numbers of species is the sessile Ascidiacea, and this class is joined by the pelagic Appendicularia (Larvacea) and by the Thaliacea, the last class including the salps, doliolids and pyrosomas as sub-classes. All are marine, and although very different in form, all tunicates share a sufficient suite of obvious characters (one of these being the unusual 'cellulose' tunic which gives the phylum its name) to show that they are related. Salps were first described in the eighteenth century by Browne (1756) and Forsskål (1775), whilst *Pyrosoma* was first observed by Péron (1804), and its colonial nature recognized by Lesueur (1815). Both were allied with sessile ascidians in the Tunicata by Cuvier (1804), and then by Lamarck (1815) as the fourth class of the Mollusca though in the next year the salps and ascidians were separated from the Mollusca by Savigny (1816). Doliolids were first described a little later (Quoy and Gaimard, 1827, 1833, 1836, 1824–26), as were appendicularians (Chamisso and Eysenhardt, 1821), but the affinity of appendicularians with the other tunicate classes was not discerned until the middle of the century, when Huxley (1851) recognized them as adult tunicates, and Bronn (1862) gave them equal familial rank with salps, doliolids, and pyrosomas.

Although this established the essentially modern division of the phylum into the four groups of pelagic tunicates, and a single group of sessile ascidians, the rank of these various groups varied in different classifications until the modern division into three classes was agreed at the end of the century. However, the interrelationships between the different classes within the phylum, (and those within the Thaliacea) still remain the subject of speculation, though the preliminary molecular evidence discussed in one of the chapters of this book suggests that the current view of the Ascidia as the basal stock may be incorrect.

Sessile ascidians (sea squirts) are the most familiar and best-studied tunicates, partly because they are easily obtainable world-wide, (many species can be collected on the shore and live well in aquaria), and partly because the group has much interested zoologists since Kowalewsky (1867) showed that their tadpole larvae allied them to the Chordates. The more delicate pelagic forms are difficult to obtain and maintain in the laboratory, so that whilst their distribution, morphology and taxonomy have long been studied from fixed material collected in plankton net tows, their physiology, behaviour, and ecological impact have been less well known until recently. Indeed, rather few zoologists have had the opportunity to examine them alive. However, recent work by open-ocean divers from submersibles, and by workers at favoured shore laboratories where living pelagic tunicates can be obtained in good condition (and in some cases cultured), has meant that these fascinating and beautiful animals have become better known in the past decades. In particular, it has been realized that pelagic tunicates may have a very significant ecological impact, and are of considerable importance in the downward flux of carbon and nitrogen from the surface to the benthos. This recent work is scattered in the literature, and the excellent syntheses by Lohmann (1933, 1934), Neumann (1935) and Ihle (1935) in Kükenthal and Krumbach's *Handbuch der Zoologie* now require updating.

The present book provides an up-to-date account of the biology of the four groups of pelagic tunicates at a time when there is an increasing realization of the importance of the gelatinous filter feeding animals of the macroplankton in geochemical cycling. Quite apart from their significance in the economy of the sea however, the different pelagic tunicates are remarkable animals in their own right. Their life-cycles may be complex and bizarre (so much so that the original description of the salp life-cycle by the poet Chamisso (1819) was regarded as a flight of fantasy!), and as Berrill (1950) justly remarked of doliolids '... they have continued to excite interest by virtue of their transparent beauty, fantastic life-cycle, and general elusiveness'. In some salps, the reproductive rate, as Heron (1972) was the first to show, is apparently much in excess of most other multicellular organisms, recently, the same has also been found for appendicularians in the tropics (Hopwood and Roff, 1995). Again, they resemble hydrozoan cnidaria in their extensive use of excitable epithelia to transmit information, and exceed them in the unique way in which information is transferred from zooid to zooid along salp chains. Several pelagic tunicates luminesce (pyrosomas are amongst the mostly intensely luminescent of all animals), but the way their luminescent organs are controlled remains mysterious, and the luminescent system of the appendicularian house is both astonishing and unique. Perhaps the most remarkable of all aspects of pelagic tunicate biology however, is the unique appendicularian feeding mechanism, involving a large and most complicated filtering house which the animal secretes and inhabits; the most complex structure constructed by any animal.

In the chapters which follow, only relatively brief descriptions of the anatomy and reproduction of the different pelagic tunicate groups are given, for these have been well described in larger texts, beginning with the excellent *Traité de Zoologie Concrète* of Delage and Hérouard (1898), and the later extensive reviews of Lohmann, Ihle and Neumann. Berrill (1950) gives a briefer review and most recently Brien (1948) has given a well-illustrated account in volume XI of the Traité de Zoologie series. However, knowledge of the physiology and ecology of pelagic tunicates has greatly advanced since this last review, as has knowledge of their distribution, and for some groups, the classification has also been revised. Thus the major emphasis of this book is upon these topics, and in each chapter, as well as reviewing recent work, the authors have included unpublished work of their own, and special pains have been taken to indicate gaps in our present knowledge, and to suggest promising lines of further research.

As well as describing certain species in each pelagic tunicate group, dichotomous keys are given for the identification of all species in each.

ONE

Anatomy of Thaliacea

J. Godeaux (pyrosomas), Q. Bone and J.-C. Braconnot (salps and doliolids)

1.1. Introduction

The three thaliacean groups, pyrosomas, salps and doliolids, differ from each other not only in their anatomy, but also in their locomotion, buoyancy, sensory systems and embryonic development. They share, however, metagenetic life-cycles involving alternation of generations (although one generation is only transient in pyrosomas) and essentially similar stolonic budding. Traditionally grouped together, a grouping retained here for convenience, it now seems probable that the Thaliacea are polyphyletic, and thus that their grouping together is artificial (Chapter 16).

1.2. Pyrosomas

1.2.1 *General anatomy*

Pyrosomas are colonial animals, primarily warmwater forms, rather common in tropical and warm temperate waters except in the Red Sea. They are mainly caught in the epipelagic and upper mesopelagic layers, although specimens have occasionally been observed at greater depths (Trégouboff, 1956). In contrast to the other pelagic Tunicates, pyrosomas consist of permanent tubular hollow colonies; the blastozooids resulting from asexual propagation are essentially independent but remain embedded side by side in a common tunic or test, as in the aplousobranchiate Ascidians. These blastozooids are hermaphrodite and blastogenic and are responsible for the propagation of the species and the growth of the colony, whilst the oozooid is only a short-lived transient and blastogenic stage. The alternation of the two generations is thus rather different to that seen in salps and doliolids The test varies in consistency according to the species, being either firm and cartilaginous, as in *Pyrosoma atlanticum*, or soft, as in *Pyrostremma spinosum*, where the zooids are easily set free. Godeaux (1965) examined the test in *Pyrosoma atlanticum* where there is an outer thin (3–5 μm) cuticle, overlying an homogeneous test (finely fibrous at the EM level). Histochemically, the test contains acidic and basic polysaccharides in different regions, and hence stains metachromatically. Nothing is yet known of test secretion in pyrosomas, nor indeed, in any thaliacean. The colony itself is cylindrical or conical, either with a smooth outer surface or with protruding processes. It is closed at one end (the oldest part of the colony) and opens at the other at the common cloacal aperture (which sometimes has a kind of diaphragm). The size of the adult colony varies greatly in different species: *Pyrosoma aherniosum* is only a few centimetres long but *Pyrostremma spinosum* (Fig. 1.1) can reach lengths of 20 m or longer (excluding a tail many metres long), with the common cloacal aperture up to 2 m across (Griffin and Yaldwyn, 1970; Baker, 1971).

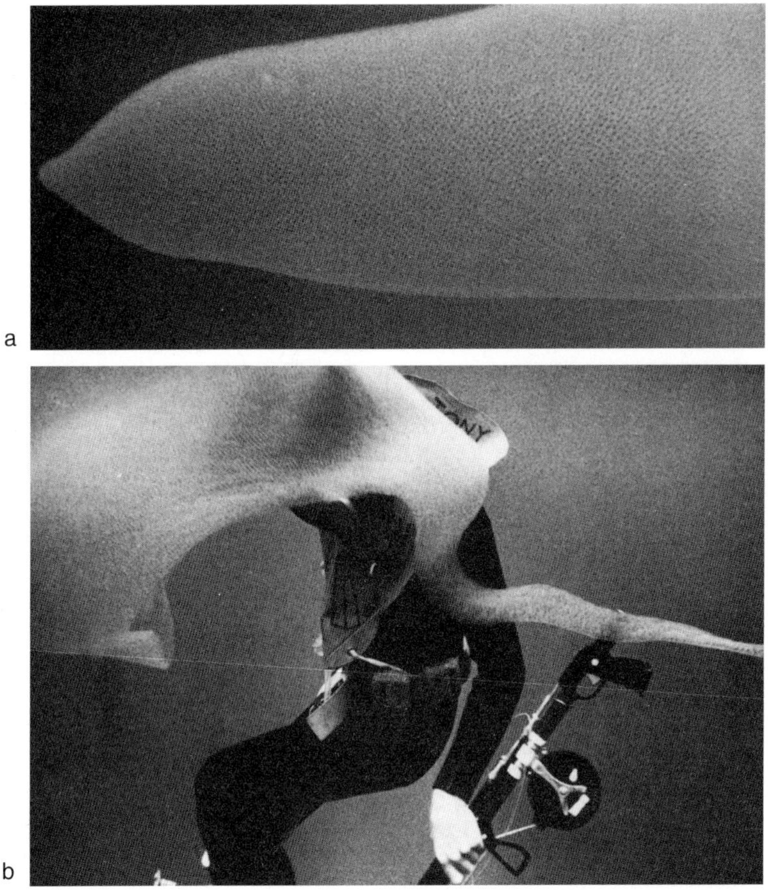

Fig. 1.1. (a and b) *Pyrostremma spinosum* of exceptional size. Photographs kindly provided by Dr. Baker.

1.2.2. *Blastozooid structure*

The blastozooid (Fig. 1.2) is essentially similar to a solitary ascidian, but the buccal and atrial siphons with their sphincters are located at opposite ends of the main axis of the animal. The buccal siphon opens at the outer surface of the colony, the atrial siphon into the common cloacal cavity. There may be a buccal vestibule between the siphon and the pharynx and/or a cloacal cavity in front of the atrial aperture (taxanomic characters). The pharyngeal cavity is ample and essentially similar to that of a sessile ascidian such as *Corella*, with a ventral, more or less incurved endostyle, several dorsal languets, and lateral walls pierced by numerous dorso-ventral gill slits or stigmata (corresponding to a single antero-posterior protostigma; Julin, 1912a). Numerous longitudinal bars cross the row of stigmata. As in sessile ascidians (Mackie *et al.*, 1974) and doliolids, the cilia of the branchial basket receive an inhibitory nervous innervation (section 4.2.1). On both sides of the pharynx the peribranchial chambers are fused posteriorly in a single cloacal cavity. Every blastozooid has its endostyle facing the anterior closed end of the colony. Above the first gill

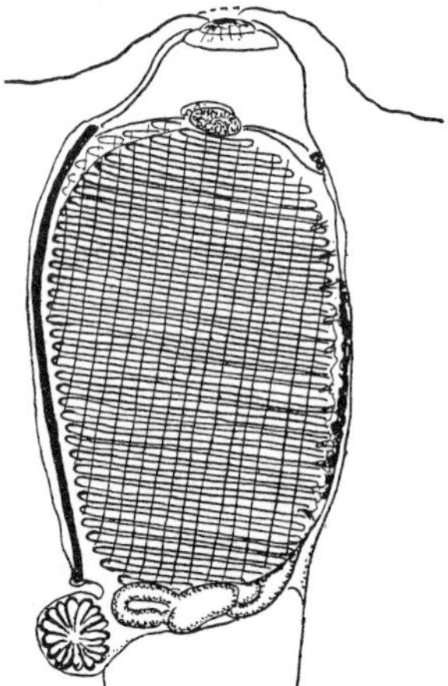

Fig. 1.2. *Pyrosoma atlanticum* blastozooid. Scale bar 500 μm. (After Metcalf, from Berrill 1950a.)

slits lies the neural complex with the eye, dorsal ganglion, and the ventral 'neural' gland connected to the pharyngeal cavity by a ciliary canal provided with a funnel. The gland has phagocytic properties (Godeaux, 1957–1958). Two peripharyngeal bands start at the front of the endostyle, run around the buccal aperture and fuse dorsally at the level of the funnel. Although feeding in pyrosomas has not been studied, the resemblance of the pharyngeal structures to those of sessile ascidians makes it clear that the endostyle secretes a mucous filter which is carried around the top of the branchial basket by the peripharyngeal bands, in the ascidian manner. A pair of luminous organs is present on both sides of the pharyngeal aperture. These are composed of cells containing curious striped sausage-like bodies that have long been supposed to be luminous bacteria (Pierantoni, 1921). However, this identification (on histological grounds and, more recently, ultrastructural grounds; Mackie and Bone, 1978) is still uncertain (Fig. 1.3). If they are indeed bacteria, they are unique in not emitting light continuously. The report of the presence of bacterial luciferase in *Pyrosoma* (Leisman *et al.*, 1980), which might seem to have settled the matter, may possibly have resulted from contamination (Campbell, personal communication). So the nature of the bodies in the cells of the luminous organs remains as mysterious as does the way light emission is controlled (section 4.2.1). Curiously, in some colonies of *P.spinosum*, Seymour Sewell (1953) was unable to detect luminous organs in the zooids and, as Seeliger (1895a) pointed out, Quoy and Gaimard (1827) observed a non-luminous pyrosoma (their *P.rufum*), as did Panceri (1872). Behind the pharynx there is a reduced abdomen in which lies the intestinal loop, the gonads, the stolon (attached to the hind end of the endostyle), and the cardiopericardium. As in all tunicates, the heart reverses periodically (see section 1.4.2), sending blood with a variety of cell types along sinuses and lacunae (since these are not lined by endothelium, they are not blood vessels). The arrangement is similar to that of sessile ascidians and salps, but not like that in doliolids and appendicularians. In some species, haemopoetic tissue is localized as two multilayered lines in the midline of the pharynx opposite to the endostyle, whilst in others, haemopoesis is performed by cell masses near the gut.

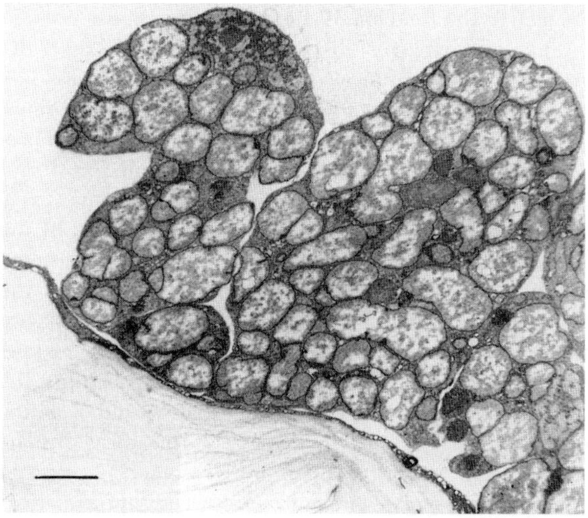

Fig. 1.3. *Pyrosoma atlanticum.* Supposed intracellular bacteria within cells of luminous organ of tetrazooid. Scale bar 5 μm.

Pyrosomas are hermaphrodite, an ovary with a large egg surrounded by follicle cells and a lobate testis are both derived from an ovotestis. Surprisingly perhaps, within a single colony, the oldest zooids next to the apex of the colony, are protandrous, the latest (and furthest from the apex) are protogynous, whilst in those intermediate, ovary and testis mature simultaneously. There is therefore the possibility for self-fertilization across the colony. Pyrosoma colonies move slowly forwards by virtue of the gentle flow out of the hinder aperture, which results from the activity of the branchial cilia of the zooids (Chapter 3).

1.2.3. *Development and blastogenesis*

A single and very yolky egg is produced with the cytoplasm reduced to a cap floating on the vitelline droplet. Meroblastic cleavage results in a disco-blastula, unusual among the tunicates. The early stages of development are best known in *Pyrosoma atlanticum*, thanks to the work of Julin (1912a), who compared them with those of an ascidian. The development follows the general tunicate pattern but is condensed and direct; the tadpole larva stage is missing.

Gastrulation takes place by the migration of endoblastic cells from the hind part of the disc, which are then assimilated to the dorsal lip of an enormous blastopore. The epiboly proceeds very slowly and is only just achieved when the first blastozooids are already formed. Most adult organs are completely absent in the embryo: buccal siphon, pharyngeal cavity (the endostyle is just outlined by two folds of its wall), peripharyngeal bands, the entire gut, and the muscles (Fig. 1.4). The embryo was called a cyathozooid by Huxley (1851) owing to its aspect of a goblet (κύαθος) resulting from the great extension of the cloacal siphon, after fusion of the two primitive peribranchial apertures (as in ascidians).

The only more or less differentiated organs present (Fig. 1.4a,d) are, in addition to the cloacal siphon (closed by a tunical plug), the beating heart, the peribranchial cavities (as tubes), and parts of the neural anlage (the ganglion never develops). The gill slits are reduced to a pair of openings equivalent to protostigmata, on both sides of the posterior neural cord. Thus development is obviously limited to the few structures involved in the production of the stolon.

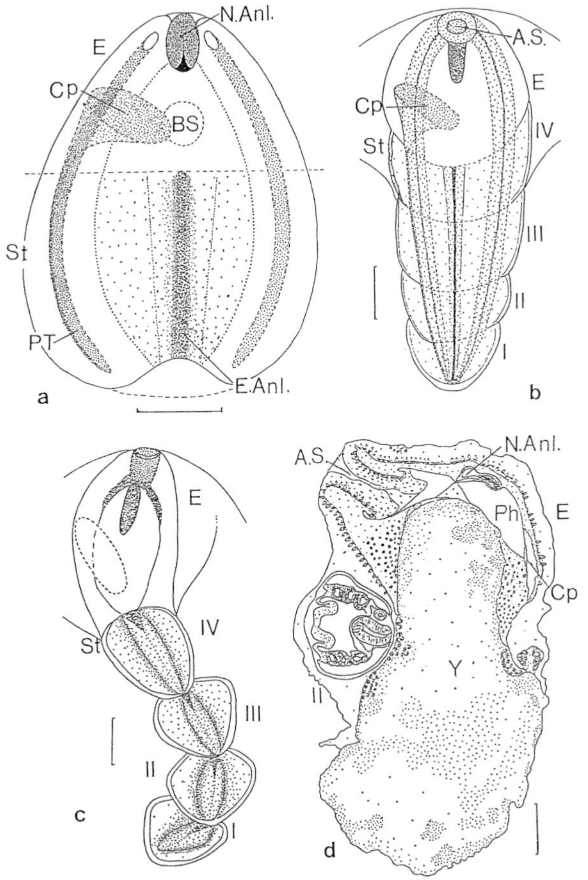

Fig. 1.4. Early developmental stages of *P. atlanticum* (scale bars 100 μm). (a) Embryo on the yolk droplet; (b and c) embryo and its stolon divided into the four primary blastozooids (I–IV), the future tetrazooid colony; (d) sagittal section of the embryo E and transversal section of the second blastozooid II. A.S., atrial siphon of the embryo; B.S., theoretical position of the buccal siphon of the embryo; Cp, cardiopericardium; E, embryo, E.Anl., endostyle anlage; N.Anl., nervous anlage; Ph, pharyngeal cavity of the embryo E; PT, peribranchial tube; St, stolon; Y, yolk droplet; I–IV, primary blastozooids.

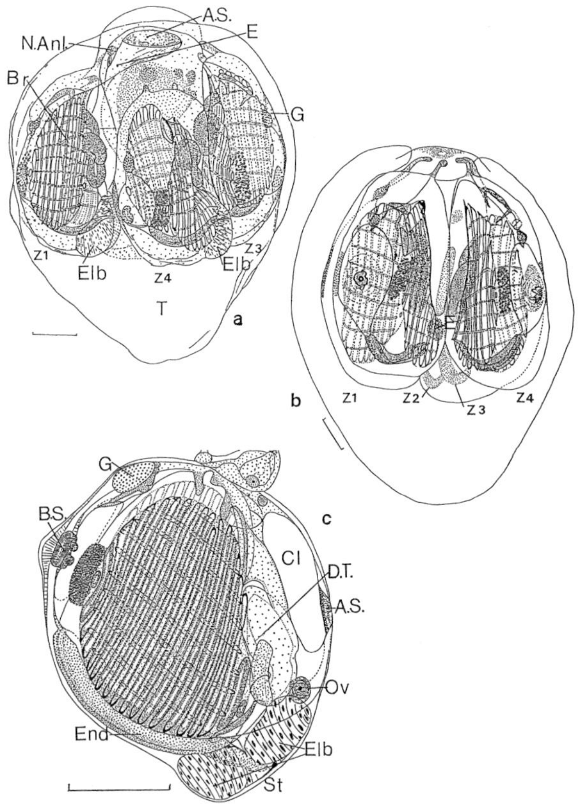

Fig. 1.5. Later developmental stages of *P. atlanticum*. (a and b) Young tetrazooid colonies (scale bar 100 μm); (c) young blastozooid (scale bar 25 mm). A.S., atrial siphon of the blastozooid; Br, branchial septum; B.S., buccal siphon of the blastozooid; Cl, cloacal cavity; E, embryo (cyathozooid); Elb, elaeoblast; End, endostyle; D.T., digestive tract; G, brain of the blastozooids; N.Anl., nervous anlage of the embryo; Ov, ovary; St, stolon; T, tunic; Z1–Z4 four first blastozooids. Scale bars 100 μm.

The endostylar anlage gives rise respectively to the pharynx and the intestinal loop of the blastozooid, the peribranchial tubes to the peribranchial cavities, and the cardiopericardiac outgrowth to the cardiopericardium. The free mesoblastic cells will give rise to the blood cells, the neural complex, and the genital strand (Fig. 1.5b,c). Pieces of this strand can be traced to the successive generations of blastozooids of *Pyrosoma atlanticum* while maturation of the oocytes slowly proceeds. In the small-sized colonies, such as *Pyrosoma aherniosum* or *Pyrosomella verticillata*, sexual reproduction occurs very early (when they are only a few centimetres long). The sexual cycle of *Pyrosoma atlanticum* (in the Ligurian Sea) has been described by Braconnot (1974b). Comparing the different species of pyrosomas, it seems that sexual reproduction is more and more delayed as the final size of the colony increases.

The ovoviviparity of pyrosomas means that the embryo is in a condition similar to that of the ascidian after fixation and metamorphosis and loss of the organs initially missing in the cyathozooid. As development proceeds, the endostylar anlage lengthens and transforms into a tube which pushes the ectodermal layer as a muff. It is accompanied by tubular outgrowths of the cardiopericardium and of the two peribranchial tubes; moreover mesoblastic cells can be seen free in the haemocoelian lacunae between the different anlagen (Fig. 1.5b,c). Progressively the stolon envelops the yolk droplet and divides by constrictions into four buds, the future primary blastozooids (or ascidiozooids), which progressively engulf the embryo. The fullgrown primary blastozooids constitute the aptly named tetrazooid colony (Fig. 1.5a,b). The embryo slowly disappears apart from its cloacal siphon and cavity which become the common cloacal cavity of the colony arising from the rapid budding of the first blastozooids and their descendants (Fig. 1.5c).

When fullgrown, the tetrazooid colony escapes from the follicular envelope and is expelled through the common cloacal cavity. This colony represents the starting point of the real colony of *Pyrosoma ambulata* as the primary blastozooids bud actively. The stolon of the successive blastozooid generations is composed of the same anlagen as the oozooid stolon and must be considered as a propagated piece of the latter. Thus, in fact, budding in pyrosoma is merely a strobilization. With the exception of the neural complex, the true nature of the cyathozooid tissues is preserved throughout the series of blastozooids.

In *Pyrosoma fixata*, (seemingly more primitive; Ivanova-Kazas, 1960, 1962), development is not condensed and produces an ascidian-like fullgrown embryo sitting on an enormous yolk droplet (Fig. 1.6a). All the organs are present, although not functional. The two siphons are visible. The branchial wall is pierced by some 10–15 vertical stigmata divided by longitudinal bars. The hinder part of the endostyle gives the central axis of a very long stolon displaying in cross section, the same organization

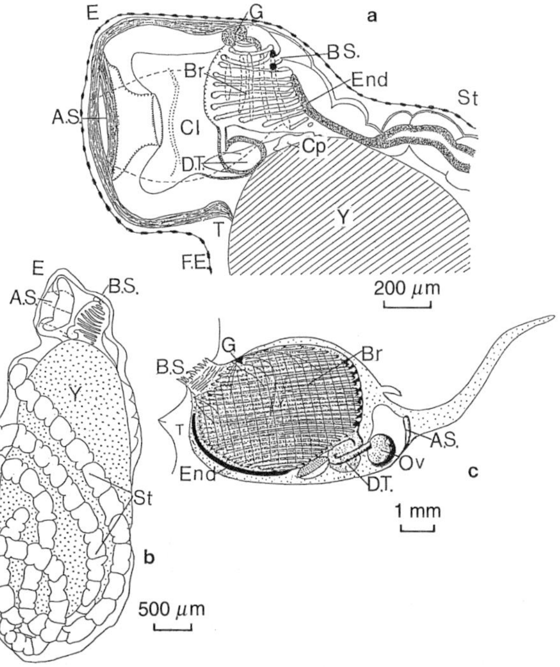

Fig. 1.6. *Pyrosoma fixata* embryos. Developmental stages of *Pyrostremma vitjasi* (after Ivanova-Kazas 1958–1959). (a) Fullgrown oozoid (cyathozooid); (b) cyathozooid and its stolon divided into a chain of buds; (c) fullgrown blastozooid. A.S., atrial siphon; Br, branchial septum; B.S., buccal siphon; Cl, cloacal cavity of the embryo; Cp, cardiopericardium; D.T., digestive tract; E, embryo; End, endostyle; F.E., follicular envelope; G, brain; Ov, ovary; St, stolon; T, tunic; Y, yolk droplet.

as in *Pyrosoma ambulata*. The stolon envelops the yolk like a scarf and divides by constrictions into many tens of buds, the future primary blastozooids (Fig. 1.6b). The further development of the colony is not known, but presumably the events are similar to those observed in *Pyrosoma atlanticum*, except that the secondary blastozooids do not migrate. The broad cloacal siphon and cavity of the embryo are the forerunners of the organs of a colony comprising many tens of thousands of zooids. It has to be stressed that sexual maturity is delayed, which explains the gaps in our knowledge of the life cycle of *Pyrosoma fixata* (Ivanova-Kazas, 1960, 1962).

As already noted, blastozooids of the two subfamilies are very similar but nevertheless exhibit different somatic characters of systematic importance.

In *Pyrosoma fixata* (Fig. 1.6c), a triangular spine is visible below the buccal aperture. The colony has no cloacal diaphragm. In *Pyrostremma spinosum* (Herdman, 1888), the colony bears a very long posterior whip-like tail; in *P.agassizi* (Ritter and Byxbee, 1905), the common cloacal aperture is provided with four fragile short quadrangular processes. In the *P.ambulata* group, the outer wall of the colony is either smooth [*Pyrosomella verticillata* (Neumann, 1909) and *P.operculata* (Neumann 1908)] or provided with more or less protruding projections containing the buccal cavity (*Pyrosoma atlanticum* Péron 1804, *P.aherniosum* Seeliger 1895, *P.ovatum* Neumann 1909, and *P.godeauxi* van Soest 1981).

In *P.fixata*, the stigmata are oblique with respect to the curved or even bent endostyle. The buccal aperture bears a crown of tentacles. The intestinal loop is brought down horizontally, with the anus facing the cloacal siphon, above which an appendix is present. The musculature is rather more evident than in other species: dorsal muscles radiating below the ganglion, ventral muscles arising from below the buccal siphon, cloacal vertical muscles at the level of the pharynx and connected with the next zooids.

In *P.ambulata*, the stigmata lie vertically with respect to the endostyle, which is straight or only slightly curved. There is a single ventral tentacle at the mouth. The vertical intestinal loop has the anus on the left side. The musculature is weak, with the cloacal muscles behind the gut and the dorsal musculature reduced to a few fibres. There is no cloacal appendix. A pair of haemopoetic organs lies in the dorsal blood lacuna; in *P.fixata* these organs are located near the gut.

1.3. Doliolids

1.3.1. *General anatomy*

Doliolids are small free-swimming pelagic forms, which are most abundant in the euphotic zone of warm continental shelf waters, though recent work by Ishikawa and Terazaki (1994) has shown that at least in one species, one of the stages in the complex reproductive cycle may be found down to 1000 m. The poorly known family Doliopsoididae (section 16.3.3) is apparently found below 400 m. As in other thaliaceans, the life cycle involves alternation of asexual oozooid and sexual blastozooid generations, though it is more complicated owing to the different kinds of buds from the oozooid stage, and in one species, there are two kinds of life cycle. The anatomy of individuals of the same species varies according to their position in the cycle, however, their basic structure is similar. This general basic structure is described for a single stage of one species, and then the differences from it in the different stages of its cycle are considered, before examining other species.

1.3.2. *The anatomy of the oozooid of* Doliolum denticulatum

The oozooid is barrel-shaped, 1.5–3.0 mm along the long axis, and almost perfectly transparent, so that details of its structure, such as the gills and heart and even individual external sensory cells and their axons, can readily be seen by normal and interference microscopy (Fig. 1.7). Each end of the barrel is provided with a dozen rounded flaps, which are normally extended to permit a flow of water into the anterior opening and out of the posterior as it filter feeds. In transverse section, the body is circular, the plane of bilateral symmetry passing through the spherical mediodorsal central nervous system and the ventral endostyle. Nine circular muscle bands encircle the body, like hoops on a barrel, giving doliolids their name (barrel: Latin *doliolum*: little cask). The muscle bands and other internal organs, such as the dorsal brain, ventral heart, and stolon, lie in the fluid-filled haemocoel space between the ectoderm and a similar endodermal layer, both made up of exceedingly thin (less than 0.3 μm thick) flattened polygonal cells. Although there are no defined lacunae or sinuses in the haemocoel, nevertheless there is a well-defined circulation path from either end of the heart, which like other tunicates, shows periodic reversals and seems to operate in the same way as that of salps (section 1.4.2). This path (Fig. 1.8) can be seen in living animals, because it is followed by occasional elongated cells free in the haemocoel first described by Uljanin (1884). There are no 'blood' cells as are found in salps (see section 1.4.2) The functional advantage of heart reversal in doliolids seems likely to be as in salps, to ensure an even distribution of metabolites and nutrients to all tissues from the gut region (section 1.4.2). The outer surface is covered with a delicately thin (0.25 μm) 'cellulose' tunic or test, which covers both sides of the anterior and posterior scalloped flap valves. Nothing is known of the chemical composition or mechanical properties of the doliolid test. In electron micrographs of doliolid material stained with uranyl acetate and lead citrate, inner and outer electron-opaque layers are separated by what appears as a finely fibrous medullary layer.

The flaps or valves of the anterior and posterior apertures bear ciliated sensory cells, and when touched, escape reactions are evoked. Similar sensory cells are scattered over the body surface, but are less abundant than on the valves. A statocyst (absent in the other stages of the life cycle) lies on the left side, between muscle bands III and IV.

The body cavity, or interior of the barrel, is divided into an anterior pharyngeal region and a posterior cloacal region by a thin oblique partition perforated by two rows of four ciliated gill slits. These are responsible for the feeding current of water flowing into the

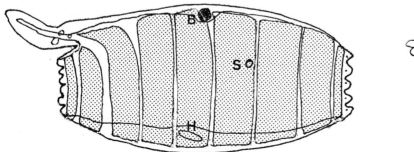

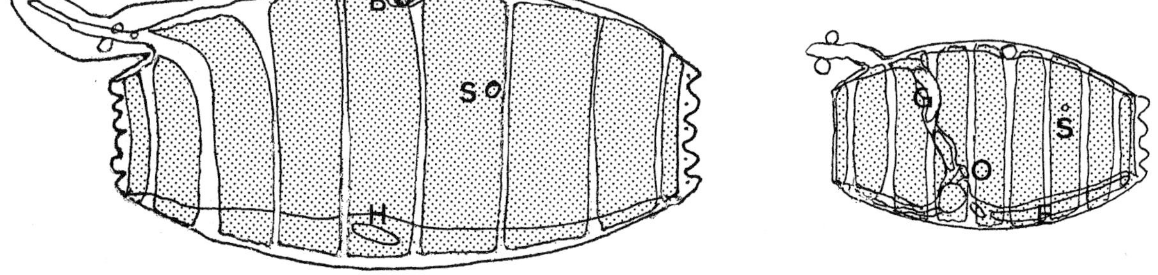

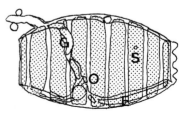

Fig. 1.7. *Doliolum denticulatum.* Left, living oozooid at stage where internal organs except heart (H), brain (B) and muscle bands (dotted) have degenerated and small buds are borne on posterior process. S, Statocyst. Redrawn from Grobben (1882). Right, younger oozoooid of *Doliolum gegenbauri* with gill apparatus (G), gut (O is to right of oesophageal opening) and endostyle (E). After Braconnot, (1970b).

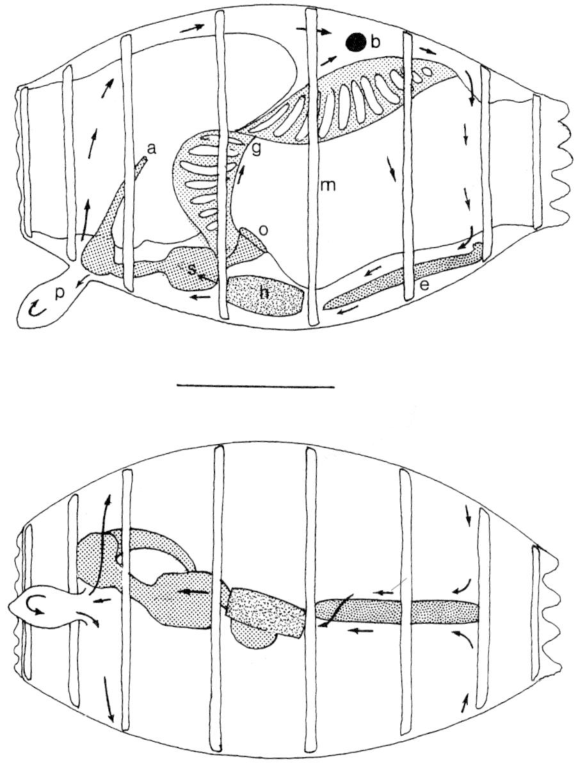

Fig. 1.8. *Doliolina mülleri.* Path of circulation of haemolymph. Scale bar 0.5 mm. From Bone, Braconnot, and Carré (1997a).

pharynx which is normally continuous if the animal is undisturbed, and the flow out of the cloacal aperture drives the animal slowly forwards and obliquely upwards. However, the gill cilia are under nervous control (Fedele, 1921; see section 4.3.2), and may be arrested at intervals for periods of a minute or more, when the animal slowly sinks. Much more rapid movements are brought about by abrupt contraction of the muscle bands evoked by touching the anterior or posterior valves (which are provided with mechanoreceptors). One or several contractions shoots the animal forward or backward by jet propulsion. The supposition that the elasticity of the thin test restores body shape after each contraction is incorrect. Although this has been referred to as elastic (e.g. Bone and Trueman, 1984), there is in fact no evidence for this assumption and since the 'haemolymph' fluid between ectodermal and endodermal layers forms an hydroskeleton (Chapter 4), the restoration of normal body shape after contraction is due mainly to the hydroskeleton resuming its original shape.

At the entry to the pharynx, peripharyngeal ciliated bands run upwards from the anterior end of the endostyle and encircle the pharyngeal cavity to end in a characteristic dorsal spiral or volute. This pericoronal arc plays a fundamental role in the deployment of the mucous feeding filter secreted by the endostyle, and it is the dorsal spiral arrangement that makes the operation of the doliolid feeding filter essentially different to that seen in salps (see Chapter 5). The structure of the endostyle is simpler than that of ascidians, consisting of five zones of cells (Fig. 1.9), and in addition to secretion of the mucous filter, a central protein-secreting layer of the endostyle probably secretes enzymes (Godeaux, 1971). The food collected by the filter passes into a short ciliated oesophagus, opening at the level of the gill slits, and thence into the stomach. From the stomach, a curved intestine ends at the anus. A branching pyloric gland begins at the duodenum and opens to the terminal region of the intestine via a pyloric canal. The function of this gland remains obscure, although Godeaux (1954) has observed that it concentrates basal dyes from the haemolymph and secretes them into the intestine. Apart from this, excretory organs are absent.

1.3.2. *The life cycle and the succession of stages*

The doliolid life cycle shows an alternation of generations; between the asexual *oozooid* stage and the sexual protogynous hermaphrodite *blastozooid* stage, that arises by budding from the stolon of the oozooid. But the complexity of the remarkable doliolid life cycle is far greater than in the other Thaliacea, for the blastozooid stage is represented by several types of individuals, and only one of these, the *gonozooid* has gonads, the others being asexual, with special roles in the life cycle.

1.3.2.1. *Development of the oozooid*

The oozooid develops from a free-living larva, which hatches from eggs shed into the sea after fertilization. This larva is the only 'chordate-like' stage in the doliolid life history, since some (but not all) doliolid larvae possess an elongate tail supported by a notaochord and oscillated by a series of three rows of striated muscle cells (Fig. 1.10). In this, the doliolid larva is similar to the tadpole larva of sessile ascidians, but in contrast to the ascidian tadpole larva, the movements of the tail are apparently entirely myogenic rather than under nervous control, and a dorsal nerve cord (as in the ascidian larva) never appears (Godeaux, 1955).

The egg undergoes the first divisions within the follicular envelope, but at gastrulation the embryonic cell mass becomes detached from the envelope and lies within a fluid that rapidly distends the envelope so that it becomes a very thin transparent covering, on a small part of which follicle cells persist for a time before degenerating. The fluid may possibly play a part in the buoyancy of the larva (section 3.3.3). The larval covering may become elongate (in caudate species where notochord and muscle cells are differentiated) or may

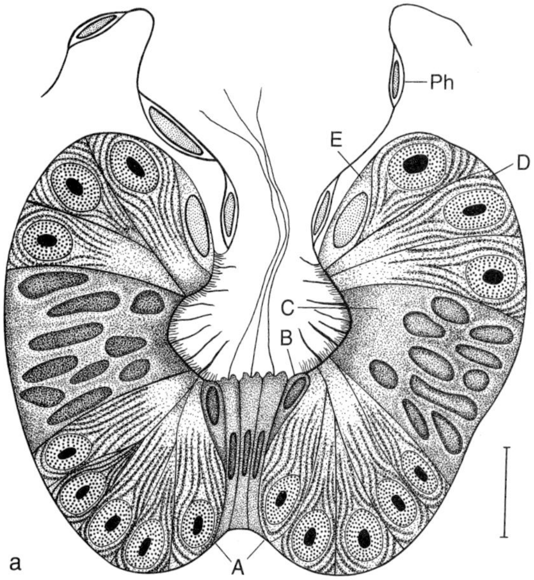

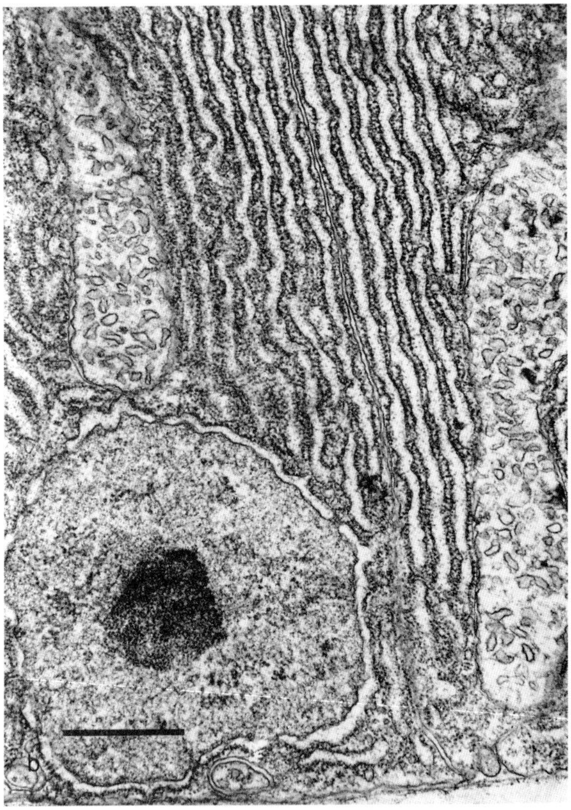

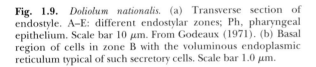

Fig. 1.9. *Doliolum nationalis.* (a) Transverse section of endostyle. A–E: different endostylar zones; Ph, pharyngeal epithelium. Scale bar 10 μm. From Godeaux (1971). (b) Basal region of cells in zone B with the voluminous endoplasmic reticulum typical of such secretory cells. Scale bar 1.0 μm.

remain more or less spherical in those where the tail is rudimentary, and where the notochord and muscles are simply represented by a cluster of undifferentiated cells. As the embryo develops, the anterior part (the cephalenteron) gradually takes on the adult barrel-shaped form, encircled by muscle bands, and the tail degenerates, although the covering does not change in shape. Finally the young oozooid quits the covering and begins to filter feed and grow. In culture, at 18°C, the young oozooid of *Dolioletta gegenbauri* is liberated 2 days after fertilization (Braconnot, 1970b).

1.3.2.2. *Later development of the oozooid*

In most species, the nine muscle bands of the oozooid gradually increase in width (by the development of new muscle fibres), until in the old oozooid the muscle bands touch or almost touch each other. In *Doliolum denticulatum* and *D.nationalis*, muscles II–VIII form a continuous cuff around the body, whilst in *D.mülleri* and *Dolioletta gengenbauri* they are separated by narrow intermuscular gaps. In some species, however, the muscle bands even in the old oozooids remain thin and widely separated. At the same time, as the oozooid ages, most of the internal organs disappear, so that although the brain, heart, and the stolon (linked to the heart) remain, the gills, gut, and endostyle disappear, and the oozooid can no longer feed. At this point, rather inappropriately termed the old nurse stage (Fig. 1.11), the posterior dorsal process or appendix of the oozooid lengthens; it is this process which bears the buds that will form the next blastozooid stage.

1.3.2.3. *The development of the blastozooid stages*

The stolon, of complex structure, lengthens and strobilizes to form a series of buds, which then extraordinarily enough, travel around the outer surface of the hinder end of the left side of the old nurse to pass to the base of the dorsal appendix. This migration involves amoeboid phorocyte cells, but nothing is known of the manner in which they are guided over the ectodermal cells to the appendix. Once they reach the base of the appendix, the buds become fixed to it, forming into three rows along its dorsal surface (Fig. 1.12a; see also Fig. 5.1b). Those in the lateral rows develop into gastrozooids or trophozooids, which since they are destined to nourish the colony have naturally as their chief feature a greatly enlarged gill apparatus. The gastrozooids have a very wide pharyngeal opening, leading to a reduced pharynx and no cloacal cavity, the gut is looped on itself, and the brain lies dorsally above the top of the gill bars

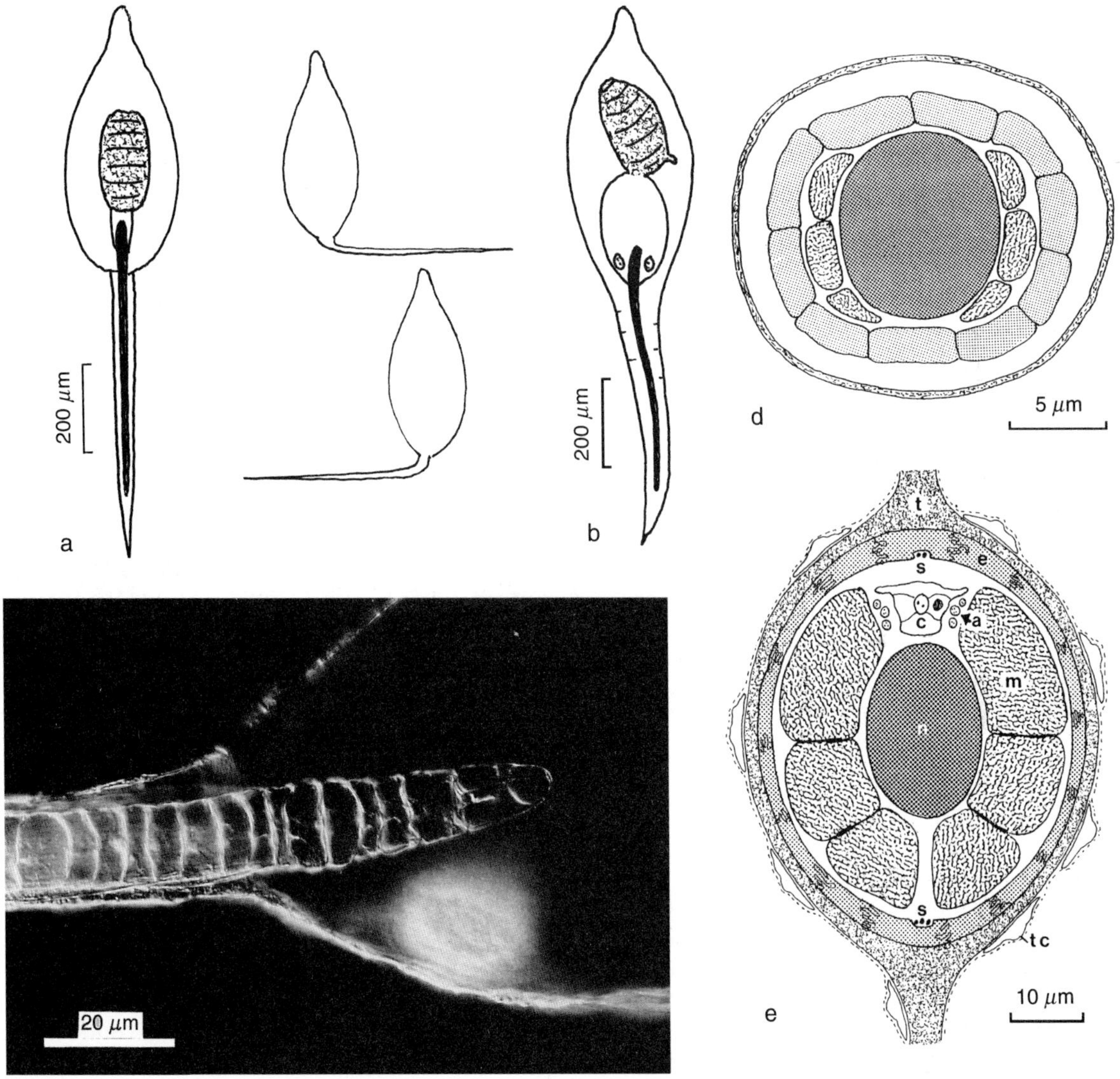

Fig. 1.10. Caudate doliolid larvae. (a) *Doliolum denticulatum*, on right, larval tail movement from base of tail, remainder held rigid. (b) *Doliolina mülleri*. (c) *Doliolina mülleri* notochord and base of tail. The white object below notochord tip is the mesoblast. Muscle striae faintly visible on left. (d) Cross-section of tail of doliolid larva. (e) Cross-section of tail of ascidian larva (*Ciona*), showing greater complexity and reduced nature of doliolid larva. Partly after Braconnot (1970b); (e) from Bone (1992).

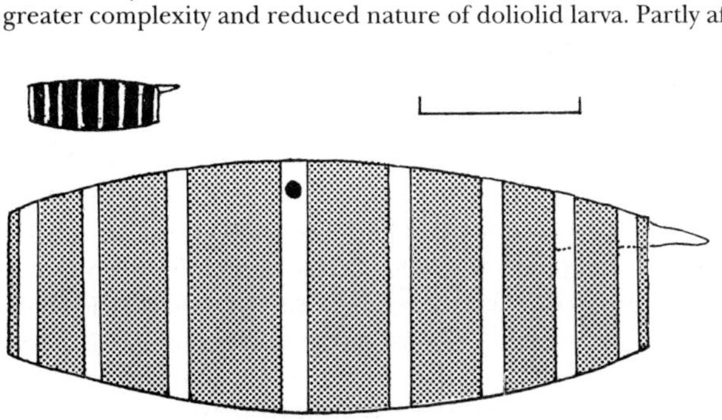

Fig. 1.11. Old nurses of *Doliolina mülleri* and *Dolioletta gegenbauri* (below). Note difference in size. Scale bar 1 cm.

(Fig. 1.12b). Ventrally, the gastrozooid is attached to the appendix via a mass of cells acting as a 'placenta'. Presumably, after digestion, food material collected by the gastrozooids passes from their haemolymph via this 'placental' attachment into the paired canals which run along the length of the appendix and so nourishes both the phorozooids and the old nurse itself. However, nothing is yet known of this process, nor of the fine structure of the 'placenta'.

The buds of the central row develop into phorozooids, attached to the appendix by a short ventral peduncle bearing a probud. The phorozooids are of

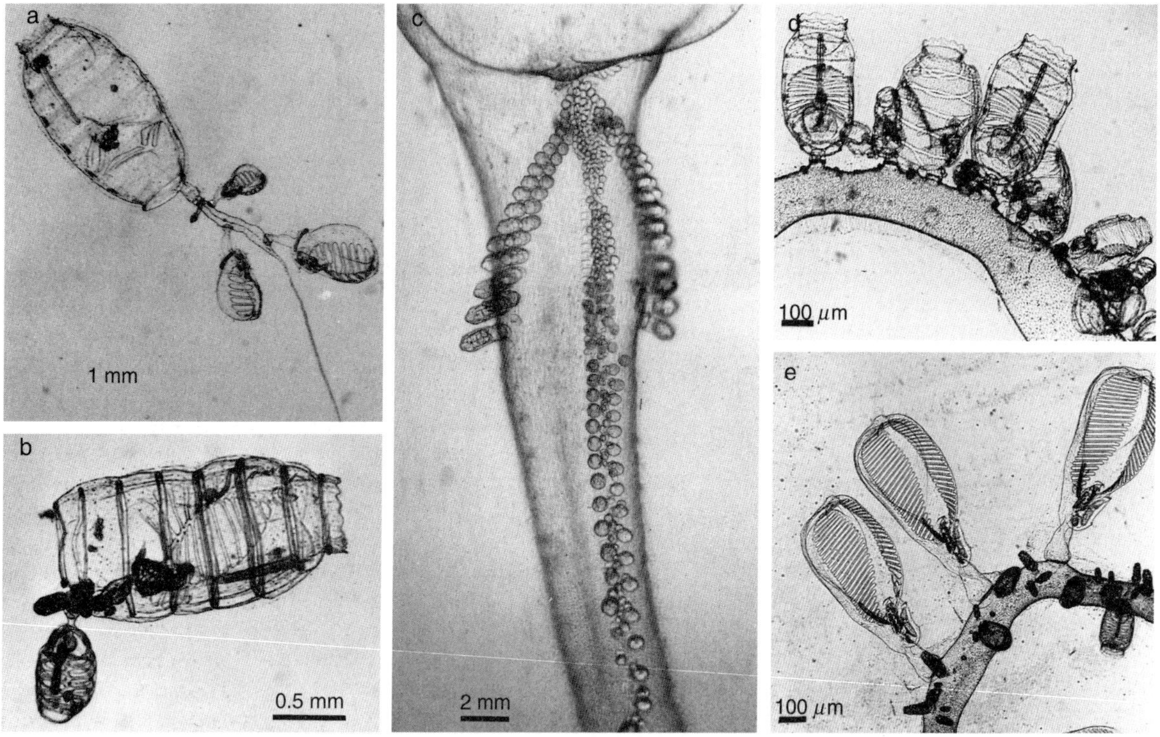

Fig. 1.12. (a) *Doliolina mülleri.* Late oozooid (early old nurse) with caudal peduncle bearing gastrozooids. (b) *Doliolum nationalis* phorozooid budding a second phorozooid on ventral peduncle. (c) *Dolioletta gegenbauri.* Base of caudal peduncle with its three rows of buds. (d) *Dolioletta gegenbauri.* row of phorozooid buds on caudal peduncle. (e) *Dolioletta gegenbauri.* Three gastrozooids and smaller phorozooid buds. From Braconnot (1970).

more typically doliolid form (Fig. 1.12c), barrel-shaped, encircled by eight muscle bands, and when detached from the appendix they are able to feed themselves. The probud borne by the ventral peduncle of the phorozooids divides and each gives rise to buds which develop into free-living hermaphrodite sexual gonozooids, very similar to the phorozooids apart from the presence of gonads. The ovary of the gonozooids produces several eggs (up to three) which once fertilized via a short oviduct, are liberated into the cloacal cavity, and thence into the sea. After this, the oviduct disappears and the testis matures, to release sperm to fertilize the eggs of younger gonozooids. The cycle then begins again, with the development of the eggs into larvae which become the next generation of oozooids.

The succession of stages described briefly above continues unceasingly, provided conditions are appropriate, leading to rapid increase in numbers. In culture at 18°C, the young oozooid of *Dolioletta gegenbauri* is liberated 2 days after the egg has been fertilized (Braconnot, 1970b), whilst in *Doliolum denticulatum* at 14°C development takes about 5 days (Braconnot, 1964).

1.3.3. *Diversity in the cycle: comparison of the stages in the cycle in different doliolid genera*

The complex cycle described above is found in all species, and sets doliolids apart from all other animals. Normally, it is possible to classify and identify any animal species by an ensemble of easily observed anatomical characters which define the species. But in doliolids, the diagnostic characters vary greatly according to the stage in the life cycle. For example, the oozooids of different species of the same genus are anatomically very similar, whilst they differ markedly from the corresponding blastozooids. Hence it is necessary to recognize the stage of a given individual before attempting to determine to which species it belongs. Detailed descriptions and diagnoses are given in the dichotomous keys in Chapter 17, but we may note here a few of the difficulties involved in attempting to make a comprehensive diagnosis for a particular species. All oozooids of all known species have nine muscle bands, whereas all barrel-shaped blastozooid stages have but eight. All oozooids have a flat gill with four ciliated slits, whilst all blastozooids have a curved (◡) or angled (⟨ ⟩) gill with at least five gill slits. Oozooids have a postero-dorsal process, whilst the blastozooids have a postero-ventral peduncle (in the gonozooids sometimes reduced and absent at maturity). No fewer than 16 doliolid species can be recognized from their gonozooids and phorozooids, but only four oozooids are known. The specific classification of doliolids began with the oozooid and phorozooid stages, and the attribution of the other stages corresponding to these has been difficult, and in some cases it has only been possible to refer the oozooids to generic level (as in the case of the cryptic species considered by Godeaux, 1961). Some of the stages of approximately 50% of doliolid species are still unknown.

1.3.3.1. *Phorozooids and gonozooids*

Phorozooids and gonozooids were early described by plankton workers, but the modern classification is based on the work of Garstang (1933), in which he gave very clear diagrams of the different species (see Chapter 16, Figs 16.5–7). Good figures of several species were however given in the last century [e.g. by Uljanin (1884) for *Dolioletta* and *Daliolina* and Grobben (1882) for *Doliolum denticulatum*]. Curiously, *Doliolum nationalis*, today the most abundant species in the sites where most doliolids have been described, has not been well figured since its original description by Borgert (1893, 1894). Remarkably, most of the specimens of this species in the Sea of Japan are readily observable in fresh plankton samples because they have a vermilion spot above the brain. This is external to the test and may perhaps be some kind of symbiont or parasite?

1.3.3.2. *Larval stages*

Doliolid larvae are divided into three morphological types: one (the anuran type) that is spherical and lacks a tail, and two elongate types which have tails, these distinguished by the presence or absence of a vesicle between the tail and cephalenteron. Only one of the two caudate larval types, however, can swim at all efficiently (see Fig. 1.10). Excellent early descriptions by Uljanin (1884) and Fol (1884) of the larvae of *Doliolina mülleri* raised from eggs of known adult gonozooids showed that it belonged to the elongate type with a median vesicle (sometimes called the elaeoblast). Since this larva shares both with the oozooid and blastozooid stages the characteristic gut bent back upon itself, its identification has never posed any problems. More recent work by Braconnot (1964, 1974a) has shown that the genus *Doliolum* has the other type of elongate larva, without a vesicle. The spherical larval type was very well described by Neumann (1906, 1913), who mistakenly attributed it to *Doliolum denticulatum*, and by Godeaux (1957–58), but later work by Braconnot (1968) showed that it belonged in fact to *Dolioletta gegenbauri*.

1.3.3.3. *The oozooids and the old nurses*

Most of the oozooids were described at the same time as the larvae by the same workers. That of *Doliolum nationalis* was only determined less than 20 years ago (Braconnot, 1977) by raising it from the larva. The great difficulty in distinguishing oozooids belonging to different genera (let alone species) is that they are remarkably similar, even if they arise from different types of larvae. Thus that of *Dolioletta* (developed from a spherical larva) only differs from that of *Doliolum* (developed from an elongate larva without a vesicle), in the length of the endostyle and the position of the stomach vis-à-vis the homologous muscle bands (see Chapter 16).

Similarly, for the old nurses, there has been much confusion in the older literature and only Grobben (1882) provided excellent figures (probably drawn with a camera lucida) of the width of the muscle bands in old nurses. Garstang (1933) later proposed the separation of the three known types of old nurse on this basis.

The striking old nurse of *Dolioletta* (Fig. 1.11) is notable for its size (up to 5 cm long without the appendix), and has been certainly identified from the phorozooids produced in culture (Braconnot, 1970b).

1.3.4. *Budding and colonial organization*

The process of budding from the ventral stolon of the old nurse and migration of the buds to the dorsal appendix was studied by several workers in the last century and in the early years of the present century (e.g. Grobben, 1882; Uljanin, 1884; Korotneff, 1904; Neumann, 1906, 1913. Neumann (1906) did not see fully the budding and young individuals on the dorsal appendix, for he missed the gastrozooids, although these had much earlier been figured by Gegenbaur (1856) and by Grobben (1882), and were shown in a clear diagram in the excellent text of the Traité de Zoologie Concrète by Delage and Hérouard (1898). Only two of the three types of old nurse have been cultured to view the life cycle in its entirety, for the third, where the muscle bands are so wide as to touch each other to form a sort of muff, and which belongs to *Doliolum*, is not only very rare, but has never been kept alive long enough to see the growth of its appendix. However, it probably has a similar cycle to that of *D.gegenbauri*. In the two types which have been cultured, there are differences in the relative length of the appendix. In *Dolioletta*, the large old nurse with its appendix forms a true colony of numerous individuals; gastrozooids along the two edges of the appendix, and phorozooids growing in a central row with the buds on their ventral processes (Fig. 1.12; see also Fig. 5.1b).

Such colonies are never fished entire, for they are too fragile, and fragment in plankton tows, but they have several times been observed from the submersible *Cyana* (during the *Migragel* cruises) off Villefranche-sur-Mer in the Western Mediterranean (Laval *et al.*, 1989, 1992). The colonies were all observed hanging immobile in a vertical position (the old nurse uppermost) with the gastrozooids well extended on either side of the appendix, which never exceeded three or four times the length of the old nurse itself, so that the whole colony was not more than 30 cm long. The much smaller *Doliolina* on the other hand, never has such long colonies; apparently the limit of three or four gastrozooids at a time on the appendix (seen in culture) is not greatly exceeded in the ocean. No doubt the phorozooids are continuously produced and liberated.

Since the development of the appendix and its buds has never been observed in the muff-like old nurse of *Doliolum*, the gastrozooid of this genus remains unknown, though probably similar to that of *D.gegenbauri*. The gastrozooids of the other two genera were well figured by earlier workers so that their species can be determined. In addition to their difference in size, the two types are distinguished by the number of gill slits (10–40 pairs in *Dolioletta*, 6–12 in *Doliolina*); by the presence in *Dolioletta* of foliaceous expansions around the peduncle lacking in *Doliolina*; and by the form of the digestive tract and the position of the anus: hence it is possible to identify the genus of isolated gastrozooids. In plankton tows gastrozooids are almost invariably separated from the parent old nurse since their peduncular attachment to the appendix is so fragile.

1.3.5. *The special cycle of* **Doliolum nationalis**

At certain periods, *Doliolum nationalis* is the most abundant species in the plankton of some regions of the Mediterranean. Indeed, it sometimes dominates the superficial mesoplankton to such an extent that all other competing species are eliminated. This situation, analogous to the blooms of salps, results from a modification of the life cycle permitting the multiplication of the phorozooid stage which can thus invade the superficial zone. Just as in salps, in doliolids sexual reproduction is incapable of this rapid multiplication, owing to the small numbers of eggs produced; hence such blooms can only be formed by budding. *In vitro* cultures have shown that isolated phorozooids can develop new individuals on their peduncles, which are then liberated. They are all asexual, and carry buds on their peduncles exactly as Borgert (1894) described, with a primordial X-shaped bud and smaller buds ready to produce a new generation of phorozooids. This 'short cycle' (Fig. 1.13) in which phorozooids bud new phorozooids with great rapidity, is evidently an extremely efficient way of invading the surroundings (Braconnot, 1967; Braconnot and Casanova, 1967). When a sexual individual of the species appears, in certain conditions rich in food particles, it is of an intermediate type, since it possesses both gonads *and* a ventral peduncle bearing buds: it is termed a *gonophorozooid* (Braconnot, 1967). The gonophorozooid is fertile, and in culture has produced larvae and subsequently oozooids, so that the species can be determined (Braconnot, 1974a, 1977).

The possibility of forming blooms to invade the surface layers by this species has probably been acquired since its separation from other species in the genus during its evolution; the most important consequence certainly is that *Doliolum nationalis* in some regions plays an important and fundamental ecological role in the plankton. In some regions at certain periods the blooms of this species are in direct competition with other species occupying the same niche (e.g. herbivorous copepods), by their massive grazing of the superficial phytoplankton population. Further, the consequent vast production of relatively dense faecal pellets, containing incompletely digested phytoplankton cells, gives the species an important role in the transfer of nutritive material to deeper layers and to the benthos. However, *D.nationalis* is not the only species that forms blooms; blooms of *Doliolетta gegenbauri* are common and play an important ecological role in the carbon and nitrogen cycles along the entire length of the Eastern North American continental shelf from Cape Canaveral in Florida to Cape Hatteras, North Carolina. It is not, therefore, entirely clear why the special cycle of *D.nationalis* has been evolved (but see p. 271).

1.4. Salps

1.4.1. *General anatomy*

Salps are tubular animals which swim continuously by rhythmic muscular contractions, these (in contrast to other thaliaceans) produce the current across the feeding filter. They are found in all seas down to 1500 m, and are often so abundant as to dominate the macroplankton (Chapter 7). Like the other Thaliacea, salps have a life cycle in which the aggregated sexual blastozooid generation alternates with the solitary asexual oozooid generation, but the cycle is simpler than in doliolids, since only two types of individuals occur. Apart from the presence or absence of gonads, the two types are anatomically sufficiently different that earlier authors (unaware of the alternation of generations in the

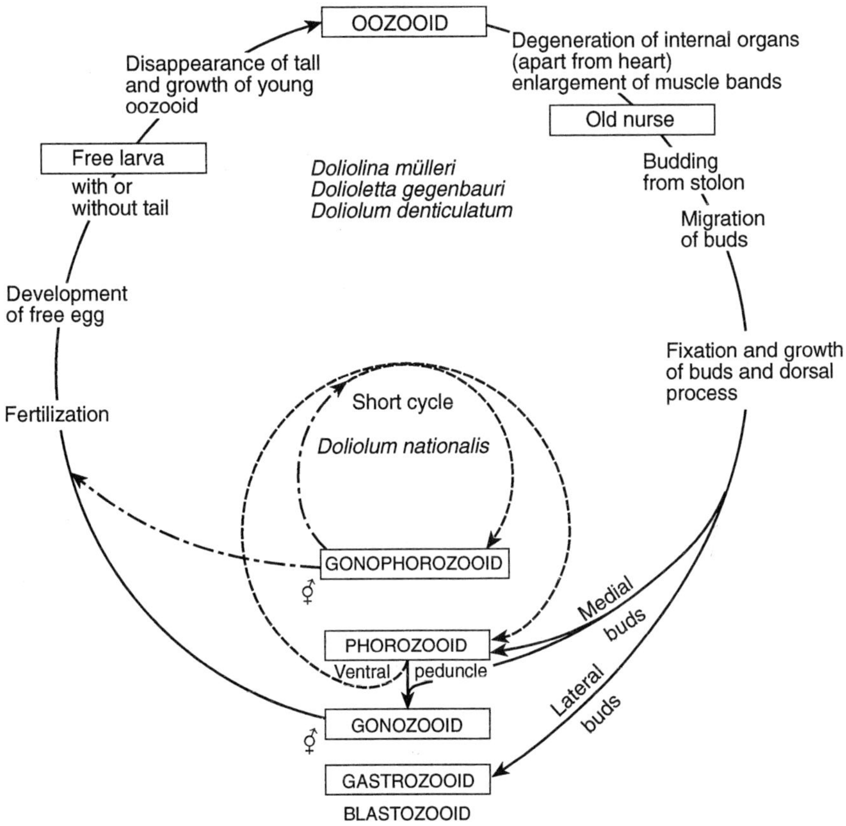

Fig. 1.13. The special short cycle of *Doliolum nationalis*. After Braconnot (1971).

life cycle) not unnaturally gave different specific names to the two stages of a single species (see Chapter 17). The oozoid and the blastozooid of the widespread species *Salpa fusiformis* will be described first, before considering the differences of other species from them.

1.4.2. *The oozooid of* Salpa fusiformis

The oozooid (Fig. 1.14) is essentially a hollow tube, 3–4 cm long, bilaterally symmetrical, with anterior inhalant and posterior exhalant apertures. The elastic outer transparent 'cellulose' tunic or test enclosing the animal is thicker ventrally than dorsally, and postero-ventrally encloses the opaque visceral mass forming a compact 'nucleus'. Very little is known of its composition, although as in *Iasis* (examined by Godeaux, 1965) and *Thalia democratica* (Bone, unpublished), metachromatic staining indicates regional differences. The wide oesophagus (opening at the base of the gill bar) leads to an expanded stomach with two caeca, the rectum curves from the stomach to open into the cloacal cavity. A pyloric gland with blind tubular caeca running over the hindgut opens into the gut at the duodenum, and perhaps excretes material accumulated from the haemolymph. Apart from the dark red or

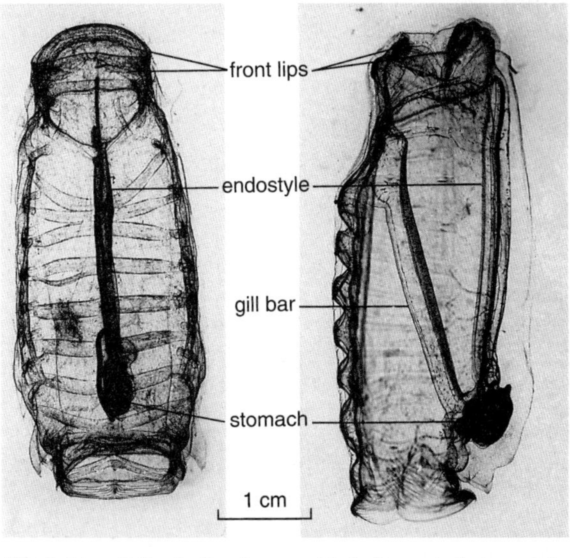

Fig. 1.14. *Salpa fusiformis* oozooid. Left: ventral view; right: lateral view. From Braconnot (1973).

brown-black pigment of the horseshoe-shaped eye above the dorsal circular brain (which is soon lost in formol-fixed specimens), the nucleus is the only coloured part of the oozooid, typically brown in *S.fusiformis*, reddish in *S.maxima*. Otherwise it is extremely transparent, so that the internal organs, like the heart, the ciliated gill bar, and such details as the nerve bundles from the brain crossing the broad flattened muscle bands, are easily seen in the living animal, and even in formol-fixed specimens.

Under the external test, the ectodermal and endodermal layers (between which lie the brain, heart, and muscle bands) consist of thin flattened polygonal cells only 0.2 μm or less thick (thicker around the brain, see Chapter 4). They enclose a large cavity, only partially divided into an anterior pharyngeal region and a posterior cloacal region, by the single median gill bar lying obliquely across the cavity. The gill bar is covered ventrally with transverse rows of cilia beating towards the oesophageal opening at its base. In principle, therefore, it may be said that there are two very large gill slits in the salp pharynx, differing from those of other tunicates in that they are not bordered by cilia. This difference in design is because the flow of water through the salp pharynx from which it filter feeds is due to muscular rather than ciliary action. The locomotor muscle bands, which produce the flow, and drive the animal forwards by jet propulsion, are made up of numbers of elongate flat rectangular cross-striated fibres (Chapter 3). The muscle bands occupy much of the dorsal surface (they are interrupted ventrally in *S.fusiformis*, but not in some other species); their number and disposition in different species are useful taxonomic characters (Chapter 17). Since the tunic is thinner dorsally, contraction of the muscle bands tends to flatten the oozooid dorsoventrally, the original shape being restored by the elasticity of the tunic as water is inhaled, mainly via the anterior aperture. In addition to the main locomotor muscle bands, there are also other thinner sphincters controlling the anterior and posterior apertures and the valves associated with them, normally permitting water to enter via the anterior aperture and be expelled via the posterior. Reverse locomotion, when the jet exits via the anterior aperture, is also possible (Chapter 3); naturally the relative timing of the contractions of the smaller muscle bands and the main locomotor bands has to be adjusted for forward or reverse swimming.

The ventral endostyle, as in other tunicates, is a deep U-shaped tube, more or less closed dorsally by the apposition of folds of the pharyngeal epithelium. That of *S.fusiformis* has not been examined, but Fol (1876) gave an excellent figure of the endostyle of *Pegea* (Fig. 1.15a) and Garstang and Platt (1928) examined the curious asymmetry of the dorsal ciliated marginal band in the endostyle of *Cyclosalpa*. Godeaux (1989) has noted the protein-secreting aspect of the glandular tract 2 (see Fig. 1.16) in *Thalia*, and (by analogy with his experimental demonstration that a wide spectrum of exoenzymes was secreted in the ascidian endostyle; (see also Fiala-Medioni and Pequignat, 1978) it seems very probable that as well as secreting the mucous feeding filter, the endostyle contributes enzymes to it. Presumably these begin to operate on the particles collected as soon as the

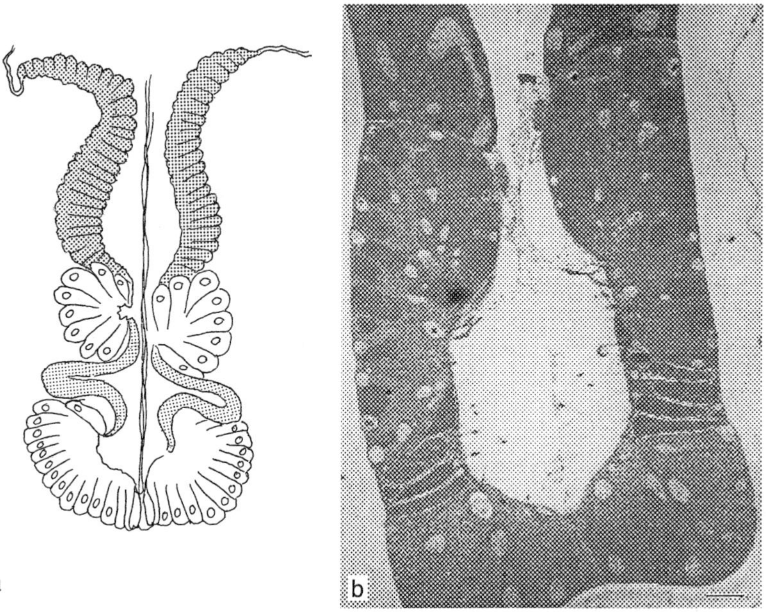

a b

Fig. 1.15. (a) *Pegea confoederata*. Transverse section of the endostyle. After Fol (1876). (b) *Thalia democratica*. Cross-section of ventral region of endostyle. After Goffinet and Godeaux (1992). Scale bar 20 μm.

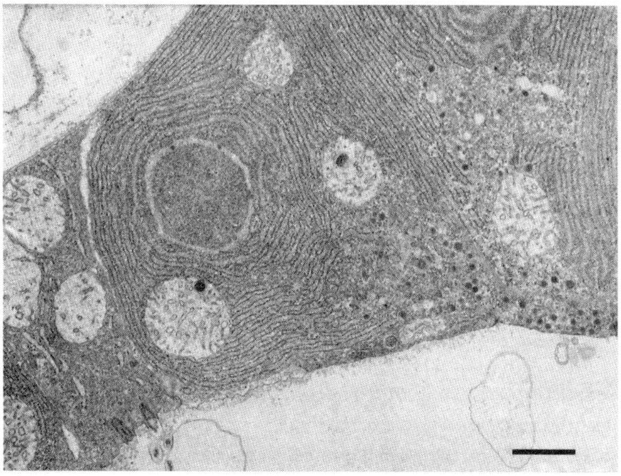

Fig. 1.16. *Thalia democratica.* Glandular cell of tract 2 showing rough endoplasmic reticulum. After Godeaux (1989). Scale bar 1.0 μm.

filter is carried upwards around the peripharyngeal ciliated bands at the entrance to the pharynx and deployed to form a large sac occupying most of the pharyngeal cavity (Chapter 5). The filter has not been studied in *S.fusiformis*, but is no doubt similar to the known in *Pegea confoederata* (Fig. 5.2d), which is a fine net with rectangular meshes, presumably manufactured by the endostyle in a similar manner to that suggested for the ascidian mucous filter by Hollie (1986). The endostyle receives a cholinergic secreto-motor innervation (Fig. 1.17), and secretion of the filter is inhibited by anaesthesia. The hinder end of the filter is rolled up and enters the oesophagus at the base of the gill bar. The peripharyngeal bands terminate at the upper end of the gill bar in a glandular cilated pit next to the brain.

In living oozooids, the blood circulation driven by the ventral (reversing) heart is clearly visible, for in contrast to doliolids, the blood contains numerous cells of a variety of types (Pérès, 1943). There are no true blood vessels lined by endothelium, but the blood circulates along well-defined paths through lacunar sinuses.

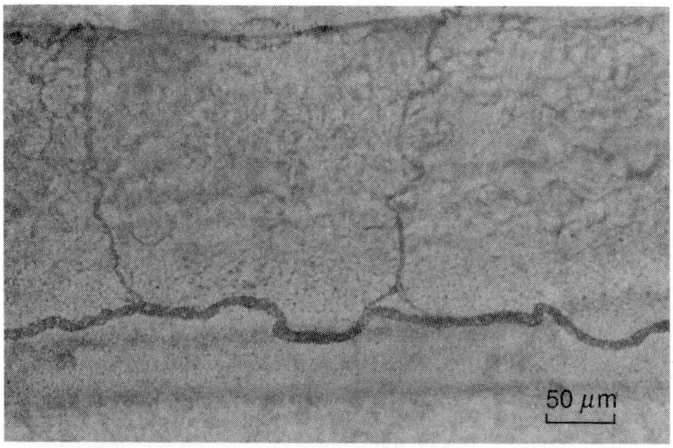

Fig. 1.17. *Salpa aspera.* Endostylar innervation after staining for cholinesterase.

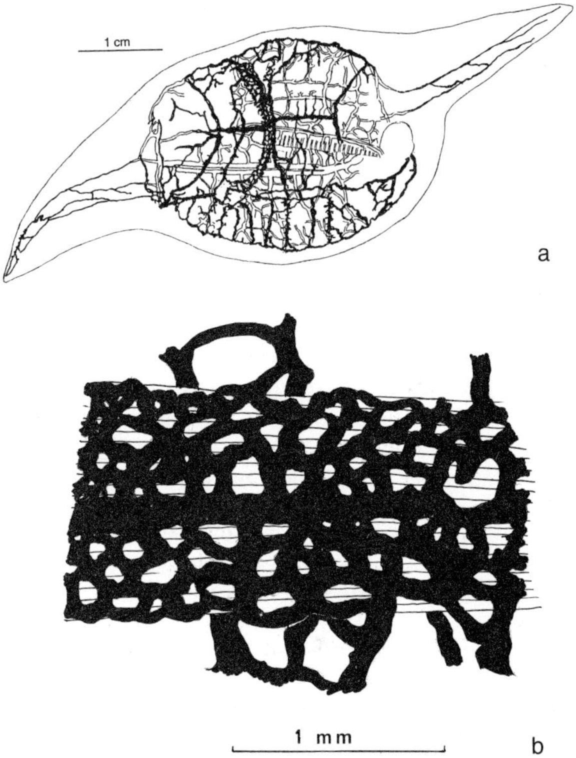

Fig. 1.18. *Salpa fusiformis.* (a) Blood sinuses of blastozooid injected with indian ink. (b) Detail of part of dorso-lateral part of system next to muscle band.

Figure 1.18 illustrates the system in the blastozooid of *S.fusiformis* as seen after injection of indian ink into the heart. Heron (1975) has given a schematic diagram (Fig. 1.19) of the circulation in the smaller *Thalia* in which it is the same as in *S.fusiformis*. From the anterior end of the heart blood passes into a rich lacunar

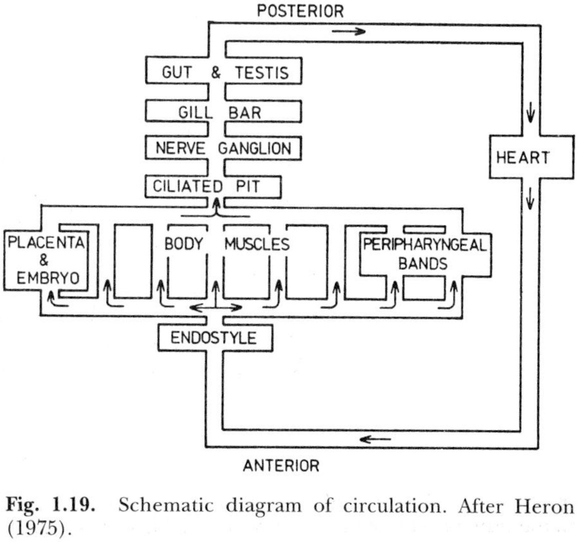

Fig. 1.19. Schematic diagram of circulation. After Heron (1975).

network around the endostyle, whence it ascends on either side of the body around the muscle bands and under the peripharyngeal bands to join a lacunar system around the brain at the top of the gill bar. Thence it passes down numerous lacunae down the gill bar to run backwards to the hinder muscles and through a lacunar plexus over the viscera to re-enter the heart. The circulation is in the opposite sense when the heart reverses. Heart reversal in *S.fusiformis* has been examined by Ebara (1954), who observed (at 11°C) 39 abvisceral heart beats, followed by a pause of 1.6 s, then 31 advisceral beats and a pause of 0.9 s, before abvisceral beating was resumed. These were mean values in a single individual where the heart rate was around 60 beats/min. The functional role of heart reversal was considered by Heron (1975), who suggested that in a linear circulation with periodic reversals, periodic accumulation and dispersal of blood cells in the gut region permits a more even distribution of metabolites and nutrients from the gut region. The heart itself is composed of a tube of striated myocytes, within a pericardium, and is apparently not innervated. Ebara (1954) has shown that brain ablation does not prevent heart reversal; the isolated heart beats and reverses normally. Contraction of the myocytes at one end of the heart passes towards the other, making a kind of rolling folded constriction, thus operating in a manner different to that known (except probably for *Doliolum*) in other animals, as Heron (1975) showed in *Thalia* (Fig. 1.20). Since the pharyngeal (endodermal) epithelium is very thin, it is not clear whether the circulatory system has in fact a respiratory role (rather than simply transferring metabolites), but the numerous blood lacunae in the gill bar, where the epithelium is particularly thin, and which is exposed to the main flow through the pharynx, suggest that these may be respiratory. At the hinder end of the oozooid, an excretory organ is formed by a tube connecting the blood with an expanded vesicle in the outside of the test, whence blood cells loaded with what are probably uric acid crystals are released. In the blastozooid, the tip of the gut complex forms a sac which migrates through the test to release its contents and then returns. These interesting excretory organs (Fig. 1.21) were described by Heron (1976) in *Thalia*, but were also observed in *S.fusiformis.* Next to the heart, there is a stolon, which strobilates to form the successive aggregated chains of blastozooids, the next generation. The frontispiece shows the beginning of stolon strobilization in the oozooid of *Thalia democratica.*

1.4.3. *The blastozooid of* **Salpa fusiformis**

The anatomy of the blastozooid (Fig. 1.22) is very like that of the oozooid in many respects, but it is asymmetrical, bearing anterior and posterior obliquely inclined conical processes (hence the specific name *fusiformis*), streamlining the body, which is more spherical than the tubular oozooid, with the anterior and posterior openings of the body cavity lying more dorsally rather than at either end. The conical prolongations of the body fit the individuals into the chain in such a way that alternate members of the chain are mirror images or enantiomorphs of each other. In the chain, they are attached by eight attachment plaques to four neighbouring zooids; at the two specialized connections between each individual, signals driving the escape responses of the

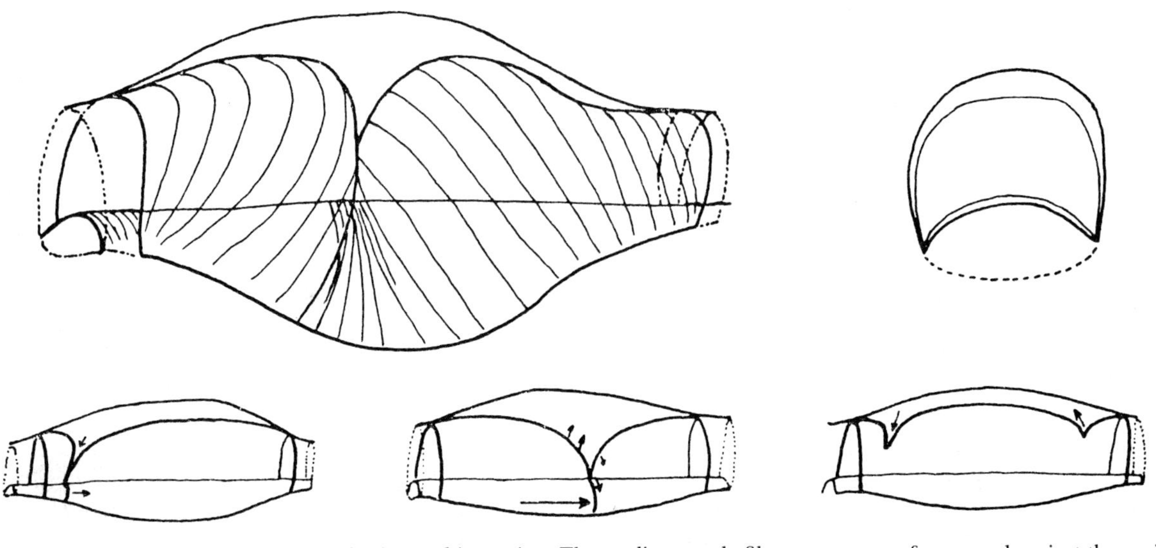

Fig. 1.20. The salp heart. Above, in side view and in section. The cardiac muscle fibres contract to form a seal against the pericardium, and as shown below, this seal travels along the heart to push the blood ahead of it. Below, the seal on the left at the beginning of heartbeat; middle in mid-beat, and right, the point of contraction moving to the left to begin the next beat. From Heron (1975).

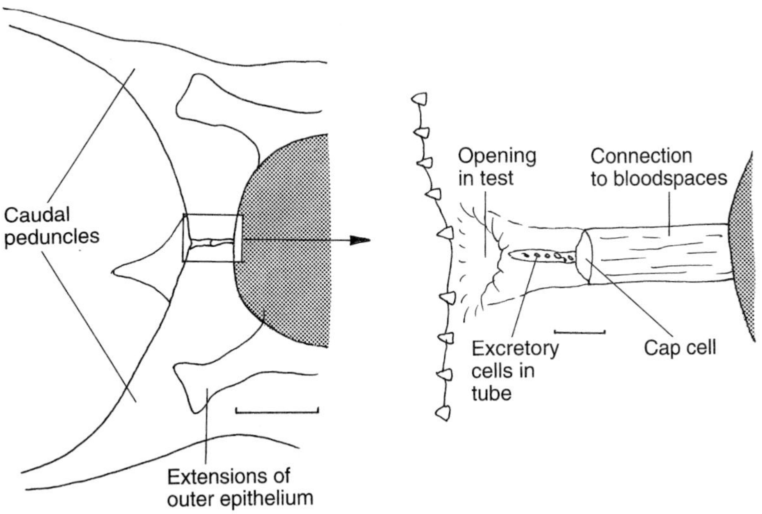

Fig. 1.21. *Thalia democratica.* Excretory organ of blastozooid. The portion within the rectangle on left enlarged on right. Scale bar 100 μm. After Heron (1976).

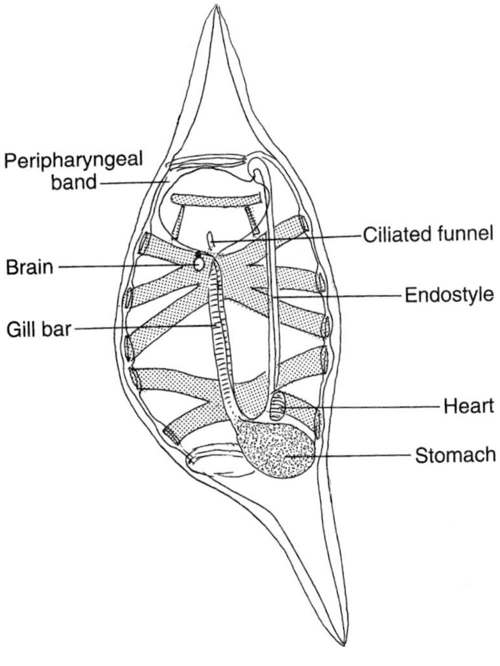

Fig. 1.22. *Salpa fusiformis.* Mature blastozooid isolated from chain after release of embryo. In comparison with the oozooid (Fig. 1.14), the blastozooid is more rounded and has two 'horns'. Overall length, including horns, 4.5 cm.

chain pass from one zooid to the next (section 4.4.4). If separated from the chain, however, individual blastozooids survive apparently without any inconvenience, though, as suggested in section 3.4.1, they can only swim rather more slowly than when in the chain, and therefore may be more vulnerable to predators when separated.

The dorsal eyes above the above the brain differ in arrangement from those of the oozooid, and the muscle bands differ in number and disposition to those of the oozooid, particularly around the posterior aperture. As in all salp species, the blastozooids of *S.fusiformis* have fewer muscle bands than the oozooids, and the musculature is weaker than in the oozooids, hence the blastozooids are less active. The stolon is absent. A lobate testis near the nucleus matures after the ovary, which lies on the right side postero-dorsally (the blastozooids are proterogynous), and after fertilization by sperm from an older blastozooid, the single egg divides within the follicle and forms an embryo attached to the maternal blastozooid by a placenta (Fig. 1.23). The embryo develops within the maternal cloacal region of the body

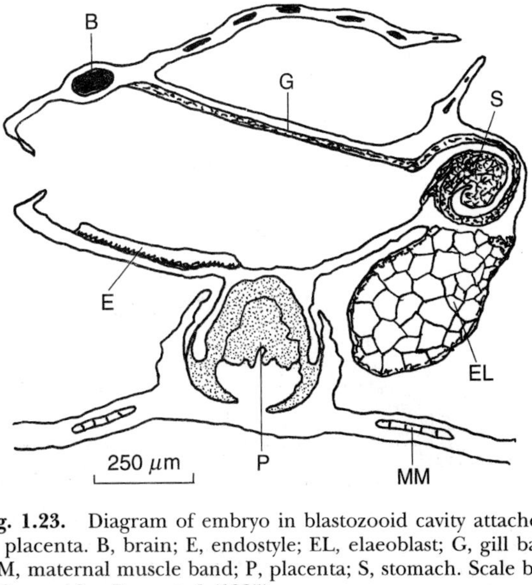

Fig. 1.23. Diagram of embryo in blastozooid cavity attached to placenta. B, brain; E, endostyle; EL, elaeoblast; G, gill bar; MM, maternal muscle band; P, placenta; S, stomach. Scale bar 250 μm. After Bone *et al.* (1985).

cavity, and finally swims off (taking the placenta with it) as a young oozooid to continue the cycle. The blastozooid lives on after release of the young oozooid, until the testis matures and sperm are shed into the body cavity and thence to the sea, to fertilize the eggs of younger blastozooids.

1.4.4. *The life cycle and its two stages*

The preceding brief anatomical description has not considered the importance of the alternation of generations in the salp life cycle. In fact, this dominates the life of salps, and has significant ecological implications. The blooms of some species at certain times of the year are only made possible as a consequence of the rapidity of blastozooid production by budding, and to a lesser degree by the remarkably rapid growth of the individuals, as Heron (1972a,b) showed in strikingly original papers on the population ecology of *Thalia*, where the life cycle may be completed in 46 h!

1.4.4.1. *The production of blastozooids from the oozooid stolon* Few authors have examined the growth of the free oozooid and the formation of blastozooid chains from the stolon. Brooks (1893) gave a good early account which was later extended in a detailed study by Brien (1928). The stolon appears in the embryo in a fold between the placenta and the elaeoblast, and consists of the different tissues which make up the animal. There is an outer ectodermal sheet, around an endodermal axis of endostylar origin, and cardiopericardial and mesodermal tissue. In the adult oozooid the stolon emerges behind the endostyle, in front of the nucleus, slightly to the left of the midline (Fig 1.24 and frontispiece). The

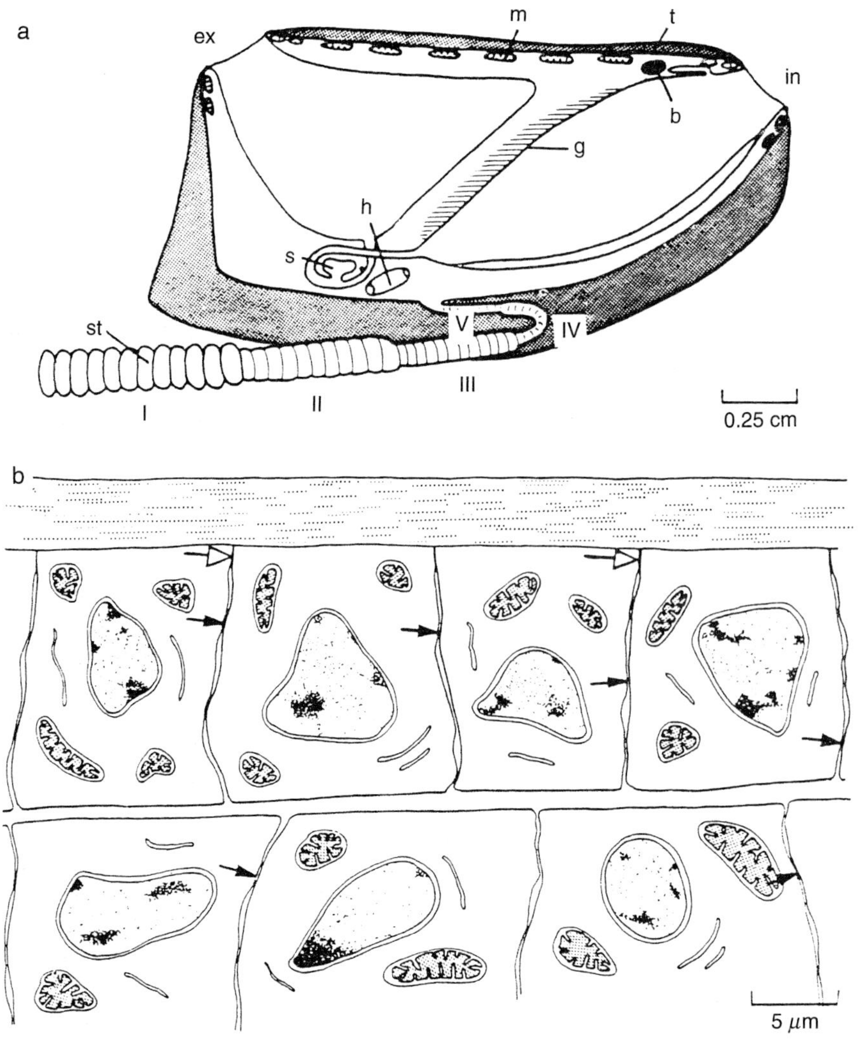

Fig. 1.24. *Salpa fusiformis.* Diagrammatic view of oozooid showing: (a) position and successive segments along stolon (I being the oldest); (b) section of young stolon showing cells connected by gap junctions (solid arrows) and tight junctions (open arrows). After Anderson (1979).

first portion passes forwards and downwards and then bends caudally, again in a straight line; this is the oldest portion which gives rise to the first blastozooid chain; there are two other successive chains in preparation on the stolon proximally. The stolon swells before strobilizing into numerous buds along a certain length to form the first distal segment of young blastozooids. It then lengthens to form a second and a third segment, each giving rise to a new set of identical blastozooids. At the distal end of each segment there is a weak link, composed of several abortive blastozooid buds, and each segment breaks off in turn as a free chain of young blastozooids (all of identical genetic constitution). Growth of the stolon is rapid, for in *S.fusiformis* in culture, the first chain of 100–150 blastozooids is liberated on the sixth day after the young oozooid leaves the maternal blastozooid, and only 2 days are required before the second chain is given off. In culture, up to five successive chains have been produced by the oozooid, but in favourable conditions in the sea, this number is probably exceeded. The young blastozooids remain in the chain as they feed and grow, maturing in 12 days, though the older chains may break up and their blastozooids continue as solitary individuals (Braconnot *et al.*, 1988).

1.4.4.2. *Sexual reproduction and embryonic development*

The eggs are fertilized in the young blastozooids (Leloup, 1929; Stier, 1938) by sperm from older blastozooids; so blastozooids with attached embryos are found before their testes mature. Curiously, although there is an 'oviduct' (Fig. 1.25), observations at the light microscope level were unable to resolve whether or not the 'oviduct' was solid, or if it had at any stage a lumen joining the body cavity with the space between the egg follicle and the oocyte. Recent ultrastructural observations by Holland and Miller (1994) have shown that in *Pegea socia* the 'oviduct' is indeed solid, and that the sperm (attracted to its junction with the wall of the body cavity) burrow through the oviducal cells to reach the oocyte! Fertilization takes place at the first meotic metaphase stage of the oocyte (Stier, 1938), and monospermy probably results from polyspermy block at the level of sperm–egg fusion (Holland and Miller, 1994). In *S.fusiformis* and in most species there is normally only a single embryo, but in one case Sutton (1960) found two embryos of equal age in a *S.fusiformis* blastozooid. In a few species (e.g. *Iasis zonaria, Metcalfina hexagona,* and *Thetys vagina*) the ovary contains several eggs, and several embryos develop successively. In the two related species *Ihlea ravocitzai* and *I.magalhanica*, although two embryos develop in each, Daponte and Esnal (1994) found that in the latter 8–12 oocytes develop in the ovary, and all but two degenerate, whilst in the former only two oocytes are present from the beginning.

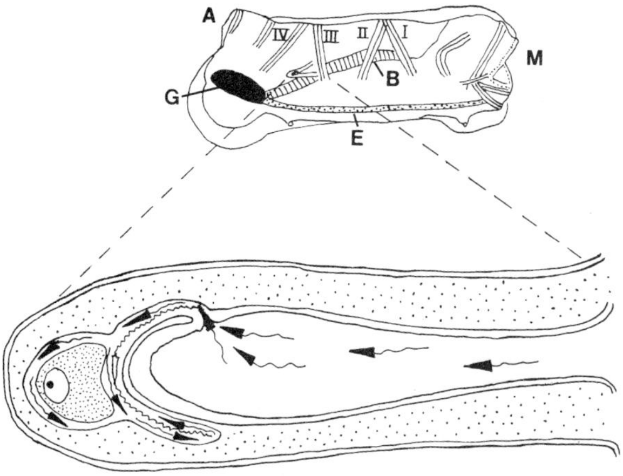

Fig. 1.25. *Pegea socia.* Above, general view of blastozooid showing position of oviduct enlarged below. A, exhalant aperture; B, muscle band; E, endostyle; G, stomach; M, inhalant aperture. Below, sperm swim into the oviduct and penetrate oviducal tissue to reach the egg. From Holland and Miller (1994).

Although salp embryology has been much studied at the light microscope level (e.g. Salensky, 1883, 1916; Brooks, 1893; Metcalf, 1918; and more recently, Brien, 1928; Stier, 1938; Berrill, 1950b), it has to be admitted that many points remain obscure. Brien (1928) noted, of earlier studies, 'Les interprétations les plus diverses, souvent contradictoires, ont été émises'. The most recent study by Sutton (1960) was later to point out that Brien's own work, and that of Berrill, had unfortunately altered this diversity of opinion but little! Whilst some of the difficulties of interpretation were a consequence of the absence of certain critical stages in these previous studies, the main problem has been that salp embryology is highly peculiar, unlike that of other tunicates or, indeed, any other animal. This is because, as Sutton (1960) remarks 'At an early stage the blastomeres become separated from each other by cells proliferated from the walls of the follicle within which the embryo begins its development'. At a later stage, the blastomeres move through spaces formed by the degeneration of the follicle cells (calymmocytes), and re-aggregate to continue the development of the embryo. Each cluster of blastomeres represents the anlage of an organ, re-aggregating progressively to form the body of the animal! It is hardly surprising that this strange pattern of development led to earlier misinterpretations of salp embryology, for instance Salensky (1883) neglected the blastomeres whilst describing the calymmocytes.

Depending on the position of the embryo within the maternal blastozooid, two kinds of embryonic development are known in the relatively few species (the commonest) so far studied in detail. In some, the thecogone salps (e.g. *Cyclosalpa pinnata, Salpa fusiformis,* and *S.maxima*), as the growing embryo breaks the follicle sac to become exposed to the maternal cloacal cavity, a circular fold of cloacal epithelium grows around it, eventually to entirely or almost entirely cover the embryo. In the gymnogone salps (e.g. *Thalia democratica* and *Iasis zonaria*), which have a more condensed embryonic development, the embryo remains uncovered. As yet, an insufficient range of species has been examined to determine whether thecogone and gymnogone development are simply the opposite ends of a continuous spectrum of salp embryogenesis.

Cleavage is total, despite the size of the egg (there is no yolk), and after the third cleavage the blastomeres form a morula attached to the follicle wall by a small group of follicular cells. During the 8-cell stage, the follicle cells proliferate rapidly and insinuate themselves between the blastomeres, covering the morula and enclosing each blastomere in a space surrounded by a single-layered sheath of flattened follicle cells (Fig. 1.26a and b).

According to Sutton (1960), each blastomere next breaks up into five or (most frequently) seven small fragments and one larger nucleated portion, the smaller fragments or daughter cells having a small nucleus which immediately begins to degenerate, and the fragments break down into a mass of detritus. A relatively normal division of the blastomeres forms the 16-cell stage, when more daughter cells are budded from the blastomeres, which are now more widely separated by calymmocytes. The blastomeres are not arranged in any particular pattern at this stage, but soon six or eight blastomeres migrate through the spaces remaining after calymmocyte degeneration, to the dorsal region of the embryo; these were interepreted as the presumptive neural and 'notochordal' cells. Unfortunately a promised paper in which Dr Sutton proposed to describe the structure, development, and subsequent degeneration of what she interpreted as the transient notochord never appeared, and the status of the 'notochord' remains debatable. Evidently, the whole process of early embryonic development in salps deserves re-investigation at the ultrastructural level. Later development is less bizarre, but complicated by the appearance of a placenta derived from a knob of calymmocytes (hence of maternal origin) which attaches the embryo to the maternal blastozooid. Although the placenta is of maternal origin, and is syncytial, since the boundaries between the placental cells break down, it also contains large cells from the embryonic blood space which pass into the placental roof (Fig. 1.27) and lose their boundaries to join in the syncytium (Bone *et al.*, 1985). As the embryo grows, a mass of mesodermal cells appears at the postero-ventral end, forming the elaeoblast. This greatly enlarges during later development, and its interior becomes vacuolated; it seems that it plays several important roles during development. First, it gives rise to the blood cells of the embryo, secondly, to the muscles, and lastly, as it begins to regress, it apparently acts as a source of nourishment. Although the embryo when fully developed as a yound oozooid (but still attached to the maternal blastozooid) is capable of nourishing itself by filter feeding with its own pharyngeal filter, when the blastozooid is feeding, the embryo lying behind the maternal filter only has already filtered water available to it, and hence cannot gain nourishment (Fig. 1.28). However, in some cases (e.g. *Thalia orientalis*) the young oozooid completely fills the cavity of the blastozooid and is as almost as large as the maternal blastozooid (Godeaux, personal observation). In *S. fusiformis*, development from fertilization to the liberation of the young oozooid only takes 5 days at 15°C.

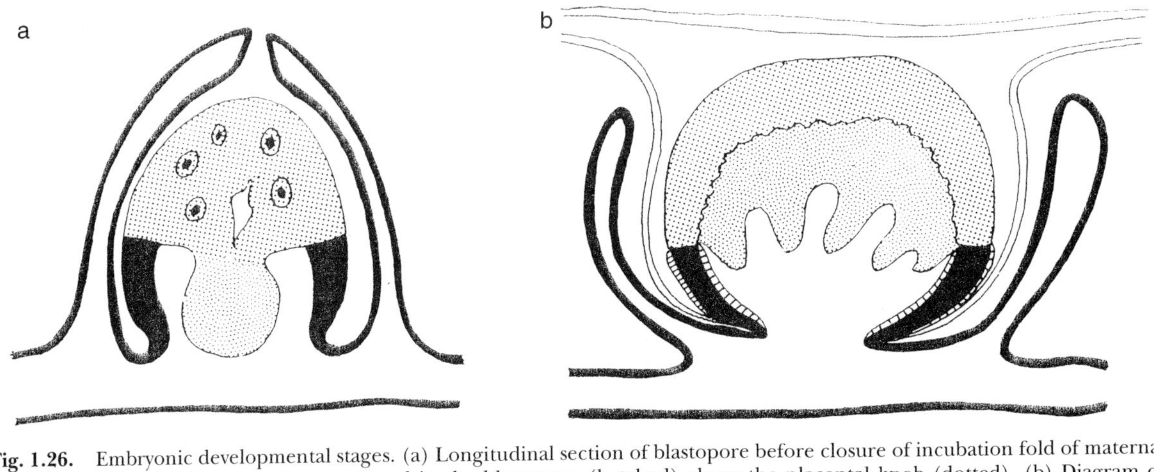

Fig. 1.26. Embryonic developmental stages. (a) Longitudinal section of blastopore before closure of incubation fold of maternal epithelium (black). Blastomeres are scattered in the blastopore (hatched) above the placental knob (dotted). (b) Diagram of mature placenta, where the placental knob hangs between thickened folds of maternal epithelium (black). After Sutton (1960).

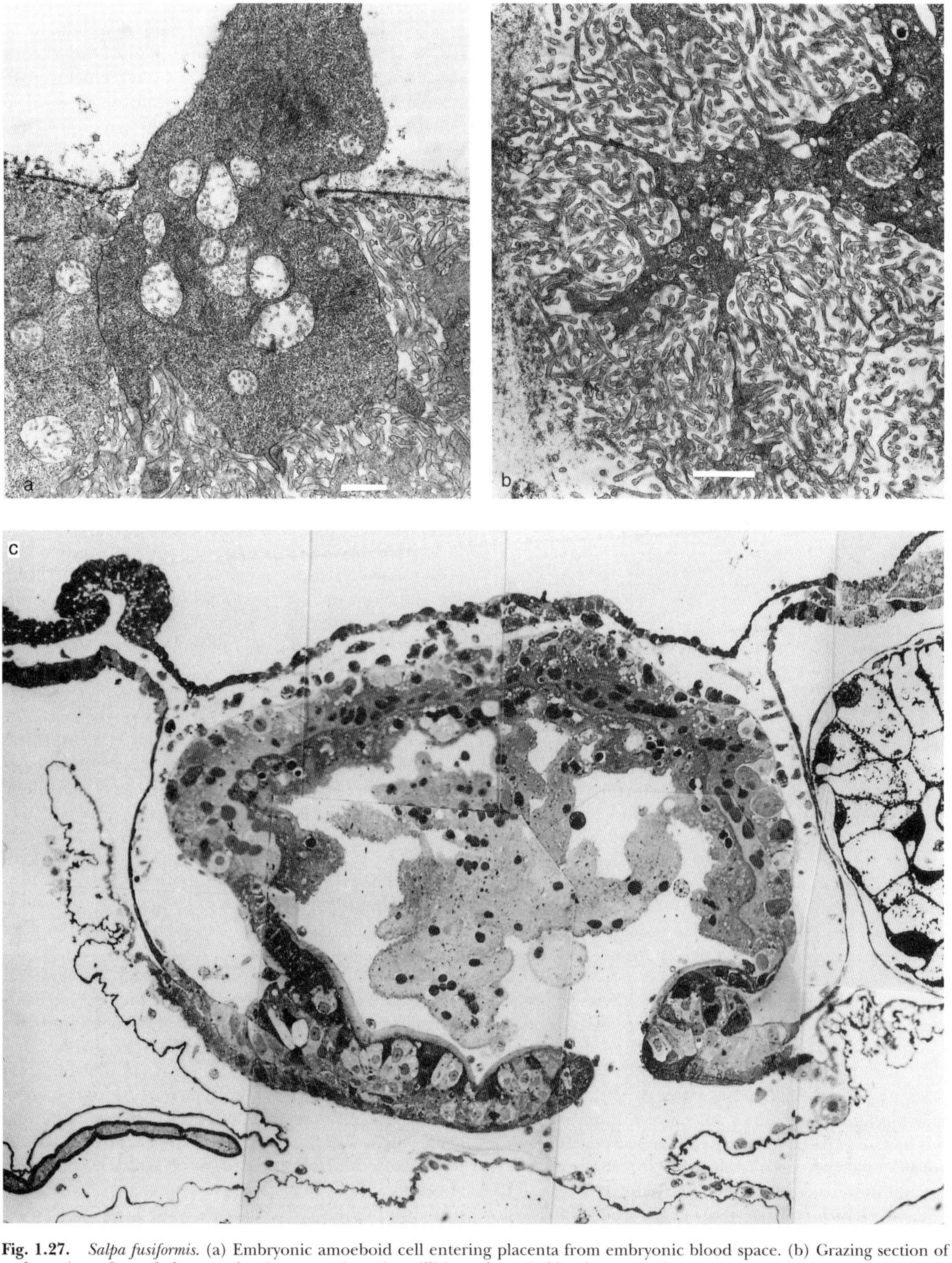

Fig. 1.27. *Salpa fusiformis.* (a) Embryonic amoeboid cell entering placenta from embryonic blood space. (b) Grazing section of embryonic surface of placenta showing extensive microvilli in embryonic blood space and numerous vesicles in cytoplasm. Scale bars 1 μm. (c) Sagittal section of placenta (almost 1 mm across) at similar stage to Fig. 1.26b. From Bone *et al.* (1985).

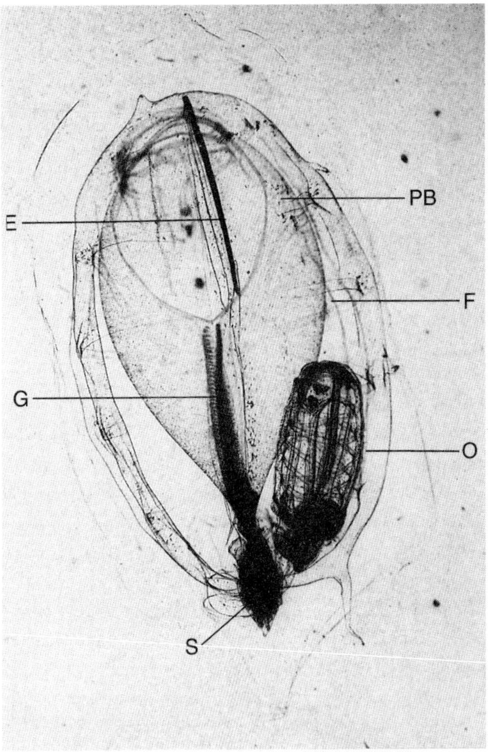

Fig. 1.28. *Thalia democratica.* Late embryo within maternal pharynx; note that when the mother deploys its feeding filter (F) as it is doing here, the embryo can only gain nourishment via the placenta. However, at this stage, the embryo is able to deploy its own filter and may perhaps do so when the mother does not. E, endostyle; G, gill bar; O, embryo oozooid; PB, peripharyngeal band; S, stomach.

1.4.5. *The morphology of other genera*

From the anatomical point of view, the different species of salps are basically similar to each other, as many be seen in detail in Chapter 17, where dichotomous keys and figures of the different species are provided. However, a notable difference between different genera is the form of the digestive tract. In most, it is arranged with its associated glands into a compact coloured nucleus, bluish violet in *Thalia*, more or less dark brown in *Salpa*, and orange-red in *Ihlea*, but in the oozooids of *Cyclosalpa* the gut is straight and elongated above the gill bar, the dorsal anus lying distinctly anterior to the oesophagus (see Chapter 16). Metcalf (1918) suggested that this linear disposition was more primitive than the nuclear pattern. Another difference between genera (of considerable taxonomic importance) is the number, organization and arrangement of the muscle bands. Some, such as *Iasis*, have wide muscle bands which completely encircle the body, whilst in others, such as *Pegea*, the muscle bands are much reduced, and lie dorsally rather than encircling the body. Allied with these differences in musculature are wide differences in the thickness and rigidity of the tunic, because the elasticity of the tunic is responsible for restoring body shape after the jet pulse. Thus genera with weak musculature like *Pegea* have relatively flaccid tunics, whilst that of *Iasis* is notably tough and rigid. The number of the locomotor bands varies widely, and in a few genera like *Ihlea*, they are edged by unusually wide muscle fibres (Chapter 3).

Metcalf (1918) suggested that a salp phylogeny could be based on the arrangement of the musculature, beginning with species in which the muscles completely encircle the body and leading by progressive reduction towards *Pegea*, where there are only four dorsal muscles making two characteristic Xs. As yet, insufficient molecular data (Chapter 16) are available to confirm Metcalf's proposed phylogeny, which was contested by Garstang and Platt (1928), who supposed that *Pegea, Thalia*, and *Thetys* were more primitive than *Salpa* or *Cyclosalpa*. As pointed out in Chapter 14, the analysis of salp biogeography, although providing some interesting hints, is most useful when considering evolutionary development at the species level.

Further differences between salp genera are seen in the varied processes of the tunic, differing between oozooids and blastozooids, as for example in *Thalia*, where the oozooid has two long caudal processes, absent in the blastozooids. Not all salps are completely transparent (apart from the nucleus), for example the blastozooids of *Ihlea* have scattered orange-red patches of ectodermal cells, the curious tentaculate *Traustedtia multitentaculata* has pigmented spots on its tentacles (perhaps, as Madin, 1990 suggested, mimicking another animal), and the ectoderm of the blastozooids of *Pegea* is yellowish or golden, particularly in the caudal processes.

Lastly, *Cyclosalpa* is set apart from all other genera by the presence of somewhat enigmatic 'light' organs forming more or less long interrupted lines or rows between the locomotor muscle bands. Several authors have doubted that they may be light organs, for they cannot be made to luminesce by the usual stimuli (e.g. H_2O_2; P.J. Herring, personal communication). However, it is certain that in *C.virgula* at least, they are luminescent in life, having been observed to glow a bright yellowish green when the salp was touched in the dark (Bone, personal observations). Julin (1912b) observed that these organs contained masses of blood cells of varying size and form, but amogst these were numerous cells which closely resembled (at the light microscope level) those he had observed in the luminous organs of *Pyrosoma*. In the *Cyclosalpa* specimens which could not be made to luminesce whatever the stimulus, no cells have been seen similar (at the ultrastructural level) to those of the *Pyrosoma* light organ. Unfortunately, the light organs of the specimen of *C.virgula* which was observed to luminesce were not well preserved. It would certainly be interesting to examine this situation further, and to see whether the salp follicle cells contain bacteria. Bacteria were associated with the light organ tissue in *C.virgula,* but were extracellular (Bone, personal observations).

1.5. Building and chain structure

Quite apart from the individual antomy of the oozooids and blastozooids, the morphology of the chains in which the linked blastozooids lie differs significantly in different species (Fig. 1.29). These differences in chain structures have important behavioural and even ecological consequences. Thus in *Pegea,* for example, where the-blastozooids in the chain, are attached so that their long axes lie at right angles to axis of the chain, forward locomotion of the chain is slow. In *Cyclosalpa,* the blastozooids are arranged similarly, but linked in wheels or circles of zooids. In contrast, where the zooids lie with their long axes parallel to the chain axis (as in *S.fusiformis*), the jet efflux is also parallel to the chain axis, and locomotion is much more rapid and efficient. Such chains may be several metres long (as in *S.maxima*) and are often streamlined by the way that the blastozooid tunics and their processes fit together, so enabling more effective swimming. Indeed, it is only in such species that significant vertical migrations have been observed, as in the nycthemeral migratory cycle of *S.fusiformis* (Chapter 7).

1.6. Concluding remarks

Of all groups in the sea, perhaps doliolids are the most elegantly beautiful, elusive, and intriguing, but all thaliaceans are beautiful and fascinating animals, and it is hardly surprising that anatomists since Huxley (1851) have been attracted· to examine their structure and development. Although much is now known of thaliacean anatomy and development, there are still very considerable gaps, and most studies are still based on light microscopy alone. It will be particularly intcresting

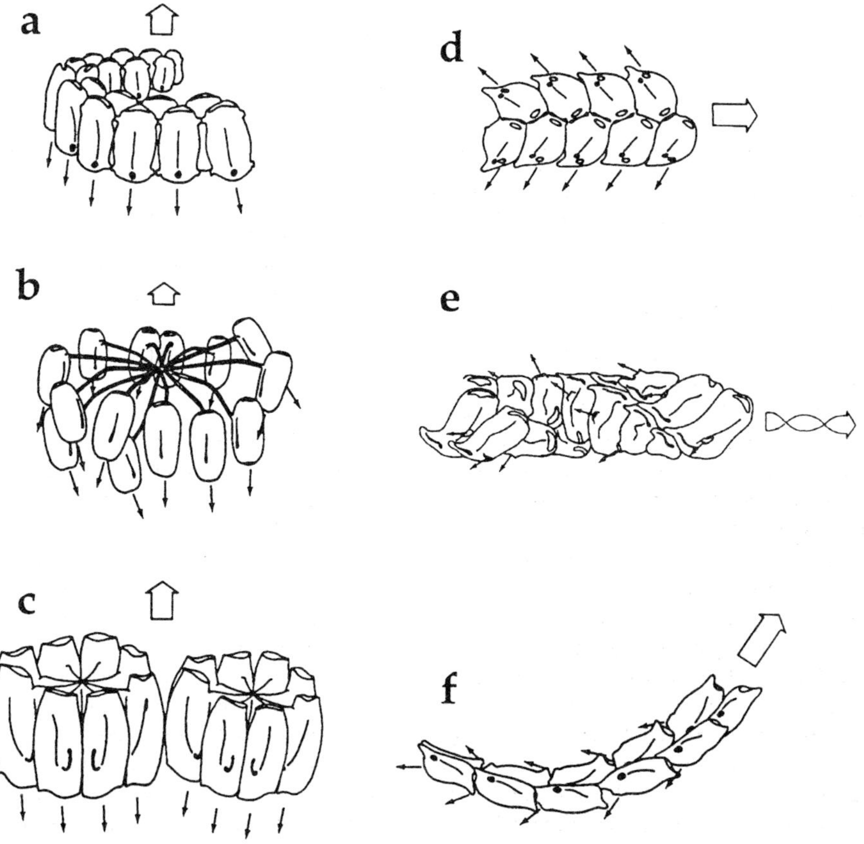

Fig. 1.29. Varieties of chain structure in different salp species. (a) Transverse chain, zooids at right angles to chain axis. (b) Cluster, zooids connected by large peduncle to cluster centre. (c) Whorl, zooids connected radially by short broad peduncle. (d) Helical chain twisted to form a double helix. (e) Oblique, axis of zooids at 60° to chain axis. (f) Linear. After Madin (1990).

to see ultrastructural studies of features that remain obscure at the light microscope level, such as, for example, the structure of the test in all groups , or the structure of the connections between gastrozooids and the other members of the doliolid colony borne on the caudal process of the old nurse. The ultrastructural studies by Holland on sperm and salp oviducal structure show how valuable such an approach is likely to prove. Biochemical studies of several aspects of thaliacean biology are also needed, for example, to unravel the remarkable movements of the buds in pyrosomas and doliolids.

TWO

Anatomy and functional morphology of the Appendicularia

R. Fenaux

2.1. General

Appendicularians are free-swimming pelagic tunicates with a trunk and chordate tail, found in all oceans, though most species are warmwater forms. Whilst the majority are found in the euphotic zone, an increasing number of mesopelagic and even bathypelagic species have been described recently. Most are only a few millimetres long, (including the tail), but some may be much larger, up to 8–9 cm. All appendicularians in the three families (Oikopleuridae, Fritillariidae, and Kowalevskiidae) secrete and inhabit a most complicated mucilaginous filtering house with many chambers and two levels of filters (Chapter 6). In the subfamily Bathochordaeinae, where tail length may be up to 70 mm, the house is remarkably large, more than 1 m in diameter (Barham, 1979; Hamner and Robison, 1992). Although in the Fritillariidae, the house was long described as a small filtration system that the animal does not inhabit, Flood (1994) has recently shown that fritillariid houses are homologous to those of the oikopleurids (Chapter 7). Fol's description and drawing of the house of Kowalevskiidae have recently been shown to be inaccurate, by observations from manned submersibles (Chapter 6). The house is secreted by regions of special secretory epithelia on the trunk, the oikoplasts (section 2.3.1.7 and Chapter 6).

The Appendicularia are thus very different in organization to the Thaliacea, and do not show alternation of generations, nor, as adults, (unlike any other tunicate) do they have an external cellulosic tunic. Instead, a glycocalyx consisting of chondronic mucopolysaccharides (Körner, 1952) covers the outer epithelium (if this is present, see p. 33). The first good descriptions of general anatomy are from Fol (1872); detailed anatomy and histology can be found in Salensky (1903, 1904a–c), Ihle (1906), Martini (1909a,b), Vernières (1933), Lohmann (1933) and Bogoraze and Tuzet (1969a, 1974).

Appendicularians consist of a trunk, normally from 1 to 8 mm long, to which is attached a muscular tail (supported by a notochord) usually several times longer than the trunk. The ratio between the trunk and tail is different in the three families, and may also vary between species.

A striking feature of appendicularians is the constancy of cell number. Soon after the tail shift (see Chapter 9), which occurs a little before the secretion of the first filter system, the number of the majority of the somatic cells is fixed. Although dividing cells in the gut of *Oikopleura* have been seen a considerable time after the tail shift, these were probably gregarine parasites (unpublished observations). Increase in size as the animal grows occurs by increase of somatic cell size (with nuclei becoming polyploid) and by the multiplication of germinal cells. Another obvious feature of appendicularians is the small cell number, as for example manifest in the nervous system (section 4.5). The haploid chromosome number of *Oikopleura albicans, O.fusiformis, O.longicauda* and *O.dioica* is eight (in *O.dioica* it may possibly be four). In the genus *Fritillaria, F.pellucida* has the low number of four chromosomes, exactly half that of *Oikopleura* (Fenaux, 1963; Colombera and Fenaux, 1973).

A part of the body is covered by a secretory epithelium which secretes the mucoid house with its highly specialized devices (different in each of the three families) for filter feeding (Chapter 6). A minor but noteworthy feature of at least some appendicularians is that they possess secondary sensory cells, otherwise only found (in invertebrates) in the ear of cephalopods. (Büdelmann, 1985). As perhaps might be expected from the small cell number of appendicularians and the

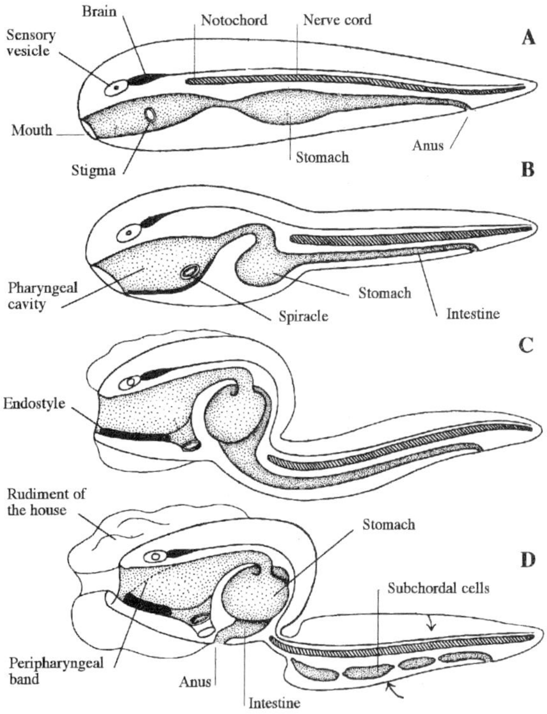

Fig. 2.1. Imaginary process of evolution from an hypothetical ancestral chordate. Modified after Tokioka and Caabro (1956). (a) Hypothetical ancestral chordate. (b) Differentiation of a trunk and a tail; notochord restricted to the tail. (c) Development of the endostyle, the spiracles and the oikoplastic layer. (d) Digestive tract restricted to the trunk. Some portions of the intestine remained in the tail as subchordal cells. The tail rotates through 90° towards the mouth. The last stage (not shown) would be the acquisition of gonads and expansion of the house from the secretion of the oikoplastic cells.

general 'miniaturization' of the animals, the life cycle can be short indeed, as little as 24 h in warm seas (Hopcroft and Roff, 1994).

The structure of appendicularians in general and of the family Oikopleuridae in particular is easiest to understand if we take a hypothetical ancestral form resembling the ascidian larva and follow (Fig. 2.1) the imaginary course of the appendicularian *bauplan* from such a form (Tokioka and Caabro, 1956). This does not, of course, imply that appendicularians have evolved from an ascidian larva, although this view has often been taken. It seems more probable in fact that appendicularians were the basal tunicate stock (see Chapter 16). Note that in the account which follows, the orientation is that the trunk is above, mouth and tail below. The tail is attached to the ventral side of the trunk, so that on the figures, the right of the animal is to the left.

2.2. The anatomy of the tail

The tail in all appendicularians is built on a common structural plan (Fig. 2.2). The epithelium covering it extends laterally, forming fin-like expansions on both sides. Thus the tail fins of Appendicularia differ from those of the ascidian tadpole larva where the fins consist of test material only. In the Fritillaridae, this epithelium is absent in the adult stage (except just at the tail base where it joins the trunk) and the tail is covered by what has been interpreted as a featureless basement membrane (Fol, 1872; Bone *et al.*, 1977). According to Fol (1872) the same structure can be observed in the Kowalewskiidae.

The central region of the tail is more or less inflated and as its elastic median axis contains the notochord with a rounded proximal extremity, tapering distally. The notochord is a thin-walled tube formed by flattened cells which secrete extracellular fluid substance containing sulphur-rich proteins into the cavity, so rendering it rigid (Olsson, 1965). The oikopleurid notochord thus forms a fluid skeleton similar to that of the ascidian tadpole, some doliolid larvae (Fig. 1.10), Agnatha and other fishes (but not amphioxus). The tail is flexed by a band of 10 flattened cross-striated muscle cells on each side (see section 3.5.2) operating against the notochord axis. Notochord and muscular bands are surrounded by the haemocoel formed by two blood sinuses. Between the left sinus and the notochord lies the nerve cord with several nerve ganglia. The first (and much the largest) is the caudal ganglion, which is linked to the Langerhans receptors situated on the latero-ventral part of the trunk (section 4.5.3).

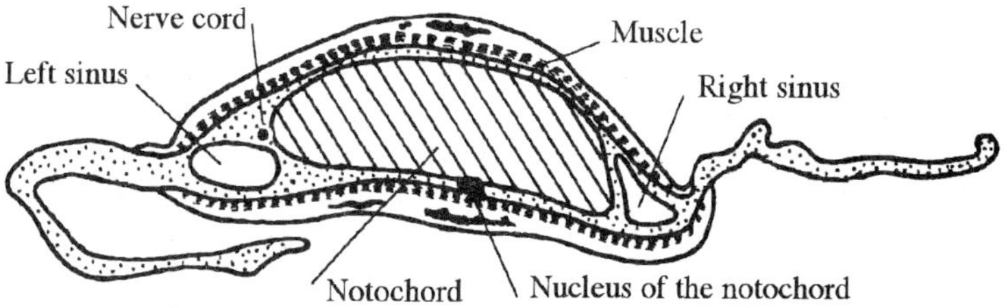

Fig. 2.2. Cross-section through the tail of an appendicularian. Modified after Seeliger (1895).

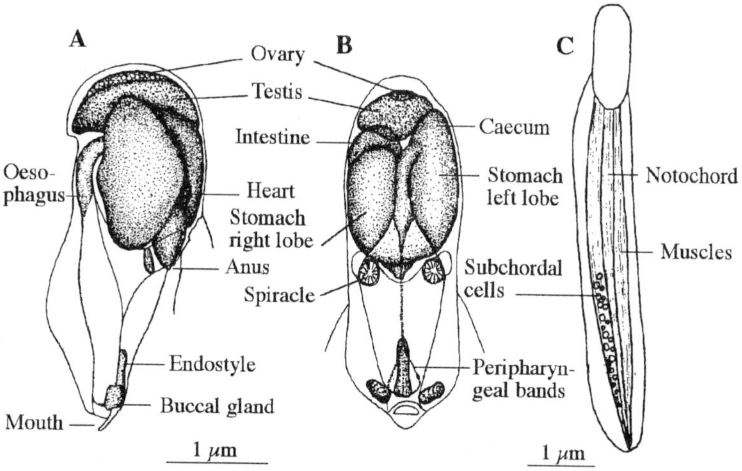

Fig. 2.3. Type of Oikopleuridae, *Oikopleura albicans*. After Fenaux (1967). (a) Left side view of the trunk. (b) Dorsal view of the trunk. (c) Complete animal.

In some oikopleurid species in the genera *Oikopleura* and *Stegosoma*, the tail contains subchordal cells lying on the left side only, between the lateral region of the muscles and the outer epithelium. The function and significance of these cells (whose shape and number are species specific) is not clear. According to Delsmann (1911), they represent vestiges of the ancestral caudal endoderm. Fredriksson and Olsson (1991) studied the ultrastructure subchordal cells of *Oikopleura dioica* and *O.albicans in vivo* and with light and electron microscopy, concluding that they were involved in protein synthesis. Considering their ontogeny and the resemblance in histology between these cells and the oral glands, they suggested that the two form a functional unit; the oral gland is involved in house luminescence. As yet, however, the subchordal cells have not been shown to be involved in bioluminescence. However, all species known to be bioluminescent possess subchordal cells (see section 12.5). The presence or absence of oral glands, correlated with the presence or absence of subchordal cells, is used in the classification to split the genus *Oikopleura* into two subgenera: *Vexillaria* and *Coecaria* (Lohmann, 1933). Other cells located on both sides of the fins in *Pelagopleura* and *Althoffia* are called amphichordal cells. Uni- or pluricellular cutaneous glands are also found in the tail of *Fritillaria pellucida, F.tenella* and *F.scillae.* They will be termed amphichordal *glands* to emphasize the difference between them and the amphichordal cells of *Fritillaria megachile* and *F.venusta.* The function of these glands is unknown, though Salensky (1904b,c) observed in *F.pellucida* not only liquid secretion but also solid concretions looking like the uric acid concretions in ascidians and salps (section 1.4.1), and supposed that the amphichordal glands were excretory organs.

2.3. Anatomy of the trunk

The anatomy of the trunk is very different in each of the three appendicularian families. One species of each of the genera *Oikopleura, Fritillaria,* and *Kowalewskia* will be described. Nevertheless, as the anatomy of the Oikopleuridae (Fig. 2.3) is much better known than of the Fritillariidae and the Kowalevskiidae, the description will be more detailed for this family and restricted to the important differences for the other two; differences occurring in the other species and genera may be found in Chapter 18.

2.3.1. *Oikopleuridae*

2.3.1.1. *Pharyngo-branchial region*

The trunk is more or less ovoid in shape, with an anterior mouth with a prominent lower lip. The pharyngeal cavity (Fig. 2.4) has a triangular prismatic form: ventrally more or less flat, with the other two faces joining medio-dorsally.

The endostyle (Fig. 2.5) lies ventrally in the pharynx, (beginning just behind the mouth) and is always straight, U-shaped in section. The edges of the U are formed by ciliated cells, whilst the walls are composed of glandular or ciliated cells in distinct longitudinal rows (Olsson 1963). The endostyle is prolonged backward, on the medioventral line, by a ciliated band reaching the oesophageal aperture. Two symmetrical ciliary bands, the peripharyngeal bands, arise at the front of the endostyle and turn obliquely to the median axis of the pharyngeal roof where they meet, a little before the entrance of the oesophagus. These bands carry the anterior borders of the mucous filter secreted by the

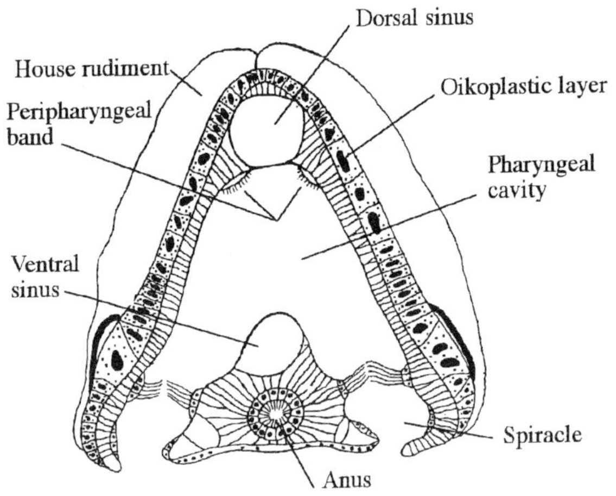

Fig. 2.4. Cross-section through the pharyngeal region of an *Oikopleura*. Modified after Seeliger (1895).

endostyle, and so a conical pharyngeal mucous filter (like that of salps) is formed, trapping the particles in the water entering the mouth. The production of the filter is under nervous control, and when it is not being secreted, particles in the inhalent flow simply pass out through the spiracles. Fredriksson *et al.* (1985, 1989) showed iodination and peroxidase activity in the central rows of the corridor cells of the endostyle. They may be considered as primitive forerunners of the vertebrate thyroid gland, though the significance of these biochemical features of the appendicularian endostyle is unknown. In some species, there is also an oral gland on each side of the anterior part of the endostyle. These glands are uni- or bicellular secretory glands, with a high level of polyploidy, and their 'secretion' (actually an extension of the cytoplasm, see section 12.5) passes externally. As Fredriksson

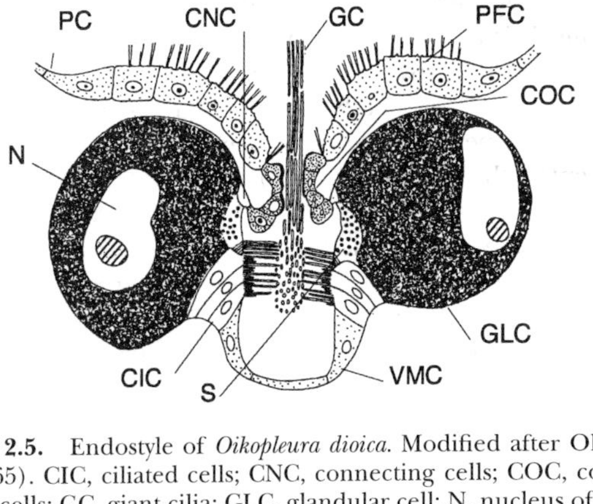

Fig. 2.5. Endostyle of *Oikopleura dioica*. Modified after Olson (1965). CIC, ciliated cells; CNC, connecting cells; COC, corridor cells; GC, giant cilia; GLC, glandular cell; N, nucleus of the glandular cell; PC, pharynx cells; PFC, pharynx floor cells; S, secretion, VMC, ventro-median cells.

and Olsson (1981) suggested, these glands secrete the luminescent particles which are the source of the bioluminescence in discarded houses (Chapter 13). Behind the endostyle on each side, there is a funnel-shaped duct, provided with a ciliary ring and opening to the exterior via a circular spiracular aperture. Ciliary beating around the spiracular ring induces water circulation in the pharyngeal cavity from the mouth to the spiracles or, in certain conditions, they reverse, and water flow is in the opposite direction (Fol, 1872; Galt and Mackie, 1961).

2.3.1.2. *Digestive tract*

The junction of the peripharyngeal bands and the medio-ventral ciliary band forms the oesophagus, so its aperture (the mouth of the oesophagus) is situated dorsally in the sagittal plane. The oesophagus is composed of short prismatic brush-bordered cells with cilia. Distally, it narrows a little and curves down to enter the left lobe of the stomach dorsally through the cardia. At this level, there are some rows of long cilia, directed toward the stomach cavity. The stomach has left and right lobes. The left is composed of short brush-bordered cells with some cilia, short excretory cells and a row of large glandular 'mucous' cells starting from the end of the oesophagus to the junction of the two lobes. In some species, the upper end of the left lobe extends, as a larger or smaller blind sac or caecum. The right lobe, which includes the median part, is composed of short brush-bordered cells containing lipid inclusions and showing alkaline phosphatase activity (Fenaux, 1963). There are two distinct parts in the intestine, the upper segment, almost vertical, behind and under the right lobe of the stomach with which it communicates widely. This region of the intestine is composed of small brush-bordered cells with strong cilia. This where the faeces are formed. The lower segment communicates with the upper by a narrow aperture which increases in size when the faeces pass. The intestine contains high, narrow brush-bordered cells showing alkaline phosphase activity. It turns to the left to the rectum on the median axis. Between intestine and rectum a ring of long cilia forms a kind of valve, directed now to the intestine' now to the rectum. The rectum is elongate, almost horizontal before the junction with the intestine, and is composed of the two categories of cells described in the stomach, but they are only half as high and lack alkaline phosphatase activity. The anus lies between the spiracles and consist of close ranked cells, forming a non-muscular sphincter. The contiguous walls show numerous folds which allow the anal aperture to enlarge as the faeces are voided.

The structure of the cells and the histochemical reactions indicate that part of the food ingested may be digested and assimilated during the short stay in the stomach, before the formation of the faecal pellets. A

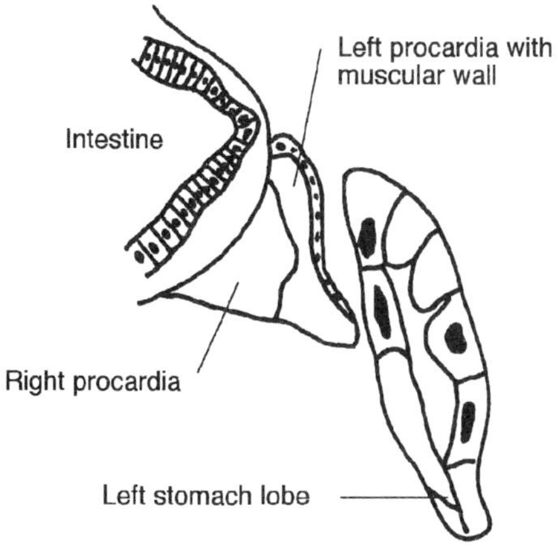

Fig. 2.6. Cross-section through the heart of *Oikopleura vanhoeffeni*. Modified after Salensky (1903).

second part of digestion takes place during the relatively long stay in the intestine. Finally, the absence of alkaline phosphatases in the rectum as well as the changes in direction of ciliary beat suggest that part of the food digested in the rectum may be assimilated in the intestine (Fenaux, 1963).

The anatomy of the digestive tract is used in the classification of the Oikopleuridae (Lohmann, 1896a, 1933, 1934). It varies from a single loop in the genus *Pelagopleura* to a complicated 'knot' in the genus *Oikopleura*.

2.3.1.3. *Heart and blood circulation*

The space between the ectoderm and internal organs is filled by the colloidal jelly of the haemocoel containing small fibres. This haemocoel is grooved by sinuses and lacunae where the haemolymph (without cells) circulates. The circulation is generated by the tail movements and by the heart. Owing to the fragility of the heart and its dimensions, the anatomy of the heart has been little investigated. The most comprehensive studies are by Salensky (1903, 1904). He described, for *Oikopleura rufescens* and *O.vanhoeffeni*, a heart consisting of two closed pouches (left and right procardial organs) lying behind the posterior wall of the stomach (Fig. 2.6). In the left, the left wall is composed of muscular cells and the right by a thin membrane. The right pouch is close to the dorsal face of the left and consists of a thin membrane. The cross-striated muscle cells are long and tapered; both sides being bordered by myofilaments, with a central mitochondrial core (Bogoraze and Tuzet, 1969a). The circulation is said to be produced between the stomach wall and the heart, as the heart undergoes undulating contractions. Haemolymph circulates in the trunk following the various sinuses. Fol (1872) described the main circulation as a flux which follows the ventral sinus to the endostyle, forks between the two pharyngeal bands and meets again in a dorsal sinus along the oesophagus to then surround the gut and the gonads. The circulation continues along the right sinus of the tail and returns to the trunk via the left sinus. This flow is inverse when the heart reverses. Some authors have denied the existence of heart reversals, but they certainly exist at least in those species such as *O.dioica* and *O.longicauda* which have been examined in most detail. The existing descriptions of the heart and its functioning are far from clear and new anatomical and physiological investigations are needed.

2.3.1.4. *Gonads*

All appendicularians are protandric hermaphrodites apart from *Oikopleura dioica* where, uniquely, the sexes are separate. The significance of this difference remains puzzling. The gonads lie in the posterior region of the trunk. In *Oikopleura*, there is generally a median ovary between two lateral testes; when they develop, they surround part of the viscera. Therefore, the number and shape of the gonads varies in different genera and even among species (Fenaux, 1963). The structure of the testes is always the same, two kinds of cells are mixed: germinal cells and accessory cells with big nuclei, generally interpreted as nutritive, which degenerate at maturity. The ovaries consist of oocytes mixed with accessory cells, but structurally the oocytes may be grouped in two types: (a) nude oocytes: *Oikopleura dioica, O.albicans* and *O.cophocerca*; (b) oocytes covered by a follicular layer: *Oikopleura longicauda* and in the genus *Megalocercus*.

The spermatozoa are released via a spermiduct and, some time later, the oocytes are released by rupture of the walls of the gonad and trunk, which causes the death of the animal.

2.3.1.5. *The nervous system and sense organs*

An antero-dorsal brain in the trunk with a vesicle containing a statocyte sends nerves to various lip receptors, to the ciliary rings of the spiracles, and to the oikoplastic epitheliums. It is linked to the nerve cord of the tail by a nerve cord from the posterior region of the brain which curves slightly to the right at the posterior part of the pharyngeal region, slopes behind the mid-region of the stomach, and so enters the base of the tail. In the tail, a large caudal ganglion near its base is linked to paired trunk Langerhans mechanoreceptors, and there are motoneurons more or less segmentally arranged along the remainder of the tail cord. The organization of the nervous system and the receptors are considered in detail in sections 4.5.1–3.

2.3.1.6. *Integument*

In the Oikopleuridae, the outer surface of the hinder part of the trunk and the tail is formed by a layer of large, thin polygonal epithelial cells which are mechanosensitive and, being linked by gap junctions,

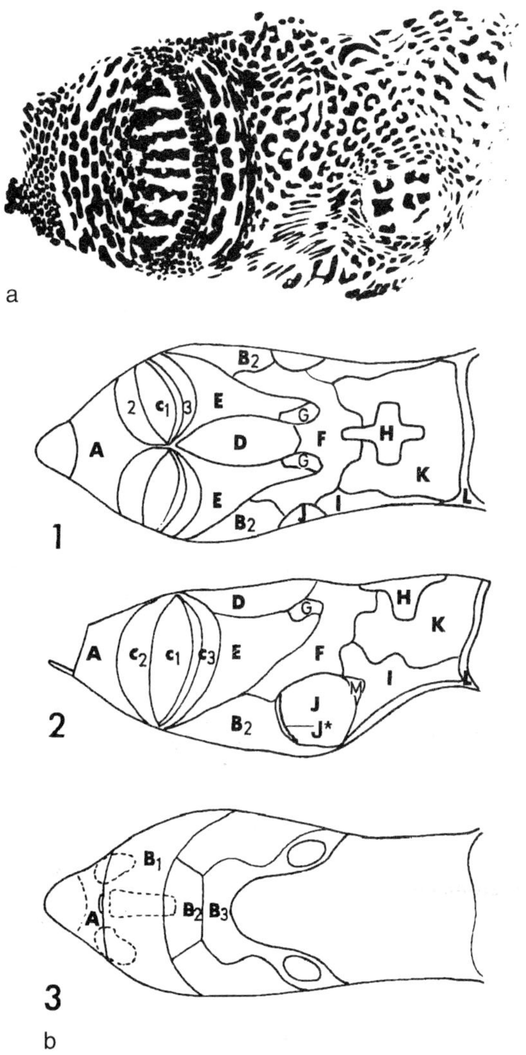

Fig. 2.7. (a) The lateral oikoplastic epithelium of *Oikopleura albicans* after Feulgen staining. After Fenaux (1971). (b) Delimitation of the different areas of the oikoplastic epithelium of *Oikopleura albicans*: 1, dorsal view; 2, lateral view; 3, ventral view. Modified after Fenaux (1971). A, oral field; B1, antero-ventral pharyngeal field; B2, medio-ventral field; B3, postero-ventral field; C1, giant cells of Fol's oikoplastic cluster; C2, anterior field of Fol's oikoplastic cluster; C3, posterior field of Fol's oikoplastic cluster (which form the feeding filter); D, anterior rosette; E, Martini's field; F, dorso-lateral field; G, small Leuckart field; H, posterior rosette; I, postero-lateral field; J, giant cells of Eisen's field; J*, tiny cells of Eisen's field (which form the inlet filter); K, postero-dorsal field; L, posterior edge; M, Ihle's field.

propagate action potentials (Galt and Mackie, 1971). These play an important role in the escape reactions of the animal (Bone and Mackie, 1975), for touching the animal anywhere (except upon the oikoplastic epithelium) evokes a burst of rapid swimming triggered by epithelial action potentials (see Bone, 1985). The integument therefore greatly extends the mechanosensitive field of the two bristle-bearing receptors of the trunk (see Chapter 4).

2.3.1.7. *The oikoplastic regions*

The anterior and mid-region of the trunk, from the mouth to the stomach dorsally and from the mouth to the spiracles ventrally, are covered by a glandular epithelium or oikoplastic layer which secretes the house (see Chapter 6). These glandular cells (Fig. 2.7a) have a mosaic appearance, and are specific in arrangement and number (Lohmann, 1899a; Bückmann, 1924; Lohmann and Bückmann, 1926, Fenaux, 1971a). The latter showed that polyploidy, which is the rule for all oikoplastic cells, increases with age and varies according to the region concerned. Growth is dependent upon the increase in cell size; thus whilst the cell number of the epithelium remains constant, the amount of DNA present in the nuclei increases by endomitotic processes, resulting in a polyploidy variable with the age of the animal and with the function of the cell (Fig. 2.8). The structure of the oikoplastic layer has been mainly studied by Lohmann (1899a, 1933), Martini (1909), Bückmann (in Lohmann and Bückmann, 1926) and

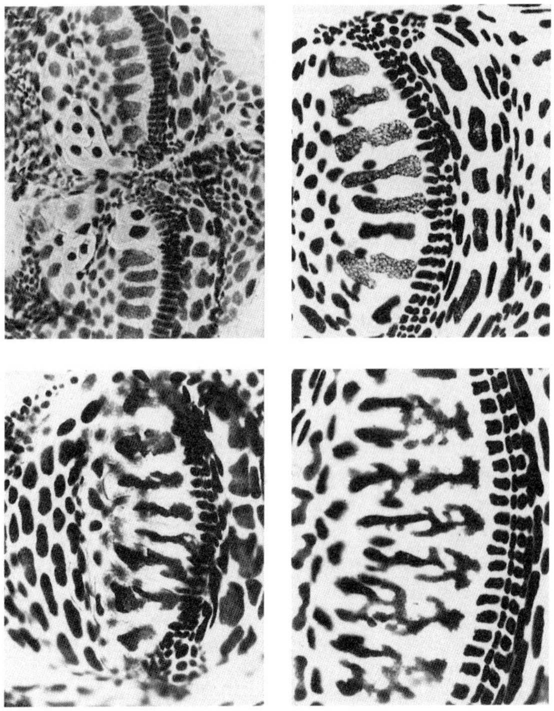

Fig. 2.8. Variation with age in shape and size of the nuclei of the oikoplastic layer of *Oikopleura albicans*. After Fenaux (1971). Magnification is identical for each picture which corresponds to specimens whose oikoplastic layers were (a) 400, (b) 550, (c) 650 and (d) 1,050 μm long.

Fenaux (1965, 1971a). Its shape and structure differs considerably between the subfamilies but varies also with the species. Details are presently known only for Oikopleurinae and the roles of the different areas (Fig. 2.7b) in the secretion of the house are not very well known, except that the fields of Fol and Eisen are involved in the secretion of the feeding and the inlet filters respectively (see section 6.3.3).

Due to the diversity of the cells which compose this epithelium and the precise and complex role played by its secretions, it deserves to be ranked as a true secretory organ rather than as a simple tissue (Fenaux, 1971a). Burighel *et al.* (1989) have shown that *Oikopleura dioica* has the unusual feature (in comparison with the other 'chordate' groups) that the apicolateral borders of the oikoplast epidermal cells are joined by tight junctions only, with no other associated junctions. In the normal epithelium, the cells have narrow tight junctions apically and subjacent intermediate junctions and macular gap junctions.

2.4. Fritillariidae

The trunk is generally cylindrical or flattened (Fig. 2.9). In front of the mouth there is a snout which often bears lobate lips. The endostyle is curved. Behind the endostyle, some large cells of the pharyngeal floor are sometimes differentiated as mucus-secreting cells (specially developed in *Fritillaria pellucida*; Martini, 1909a). The gill slits are short. The spiracular openings

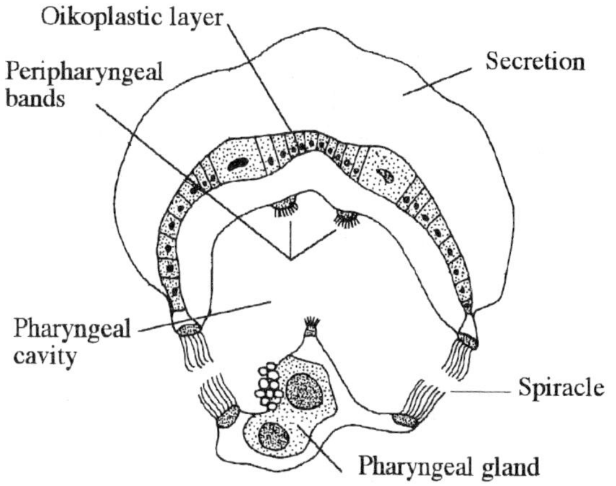

Fig. 2.10. Cross-section through the pharyngeal region of a *Fritillaria*. After Fenaux (1967).

are circular, oval, or slit-like according to species; they lie in the anterior part of the pharyngeal cavity (Fig. 2.10).

The following characters of *Fritillaria pellucida* are valid, with minor modifications, for the other species of the genus.

The oesophagus is horizontal, composed of ciliated brush-bordered cells. The cells of the cardia have longer cilia directed to the stomach. The stomach is made up of a small number of big cells with an internal brush border. The pylorus (Fig. 2.11) consists of three rings of cells, the third provided with long cilia directed now to the stomach, now to the intestine, forming a sort of

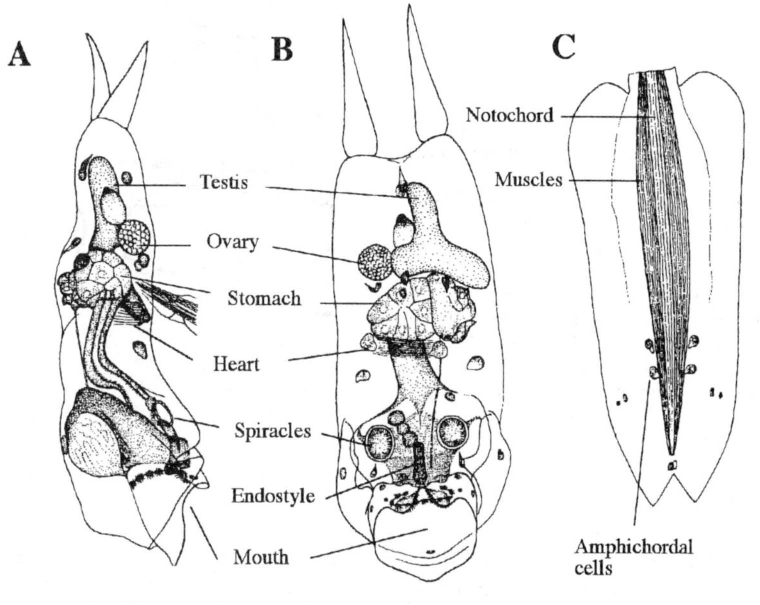

Fig. 2.9. Type of Fritillariidae, *Fritillaria pellucida typica* (modified after Tokioka and Caabro, 1956). (a) Left side view. (b) Ventral view. (c) Tail.

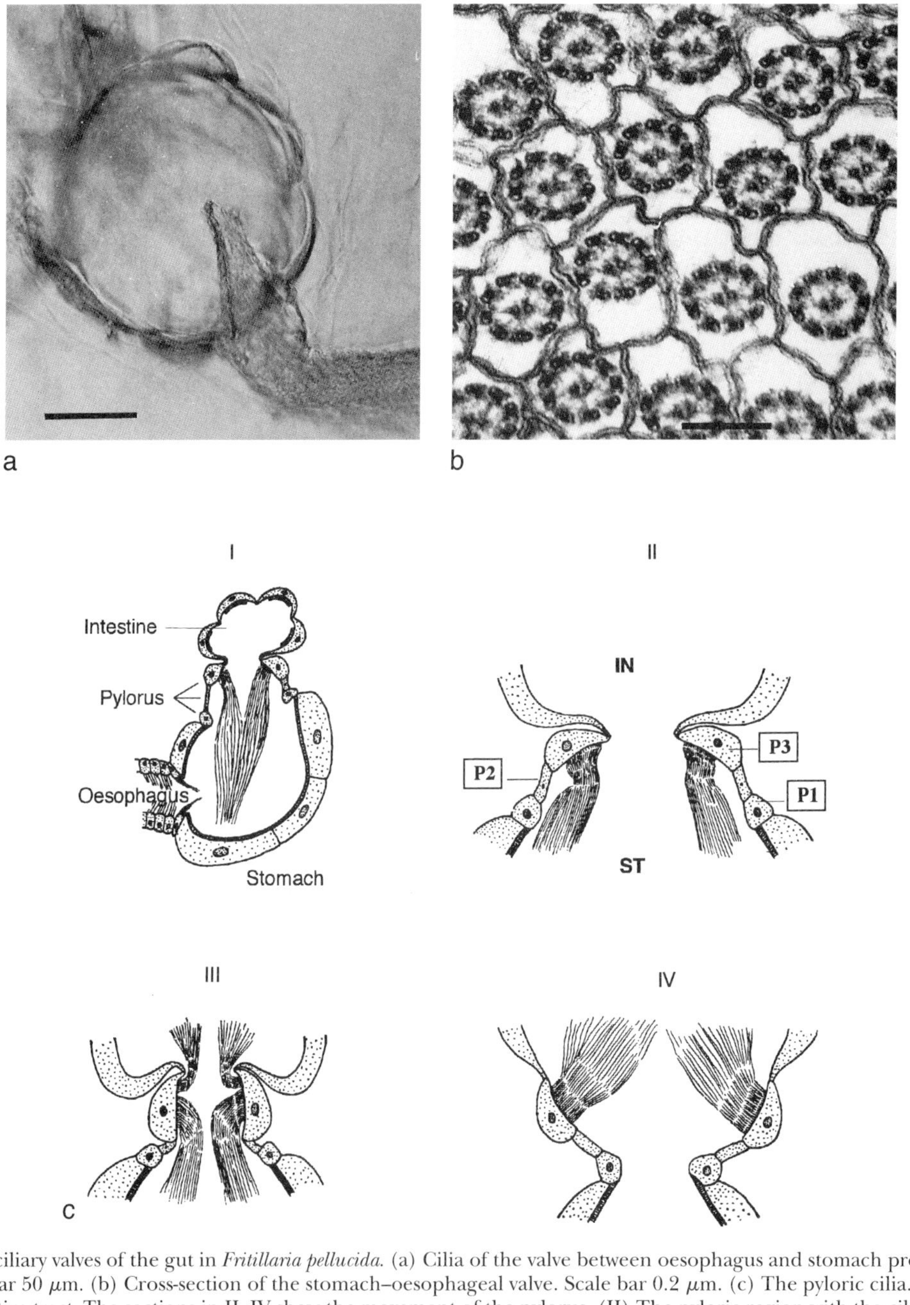

Fig. 2.11. The ciliary valves of the gut in *Fritillaria pellucida*. (a) Cilia of the valve between oesophagus and stomach projecting into stomach. Scale bar 50 μm. (b) Cross-section of the stomach–oesophageal valve. Scale bar 0.2 μm. (c) The pyloric cilia. (I) General view of the digestive tract. The sections in II–IV show the movement of the pylorus. (II) The pyloric region with the cilia within the stomach. (III) The cilia are beginning to move from the stomach to the intestine. (IV) The cilia are all in the intestine. Note that whilst in II the pyloric cells are in the lumen of the stomach cavity, in IV they are in the lumen of the intestine. P1, P2, P3, cells of the first, second and third pyloric rings; IN, intestine; ST, stomach. (a) and (b) from Bone *et al.* (1979), (c) after Fenaux (1961).

valve, moving the food from the former to the latter (Fenaux, 1961c). A globular short intestine starts from the right of the stomach, followed by a rectum formed by flat ciliated brush-bordered cells. The anus is located on the right side, anterior to the stomach.

The heart is clearly visible in living animals, lying transversely in front of the stomach and partially under the oesophagus. Whilst the circulation of the blood is similar to that described for the Oikopleuridae, the heart is reduced to the left pericardial sac, which is

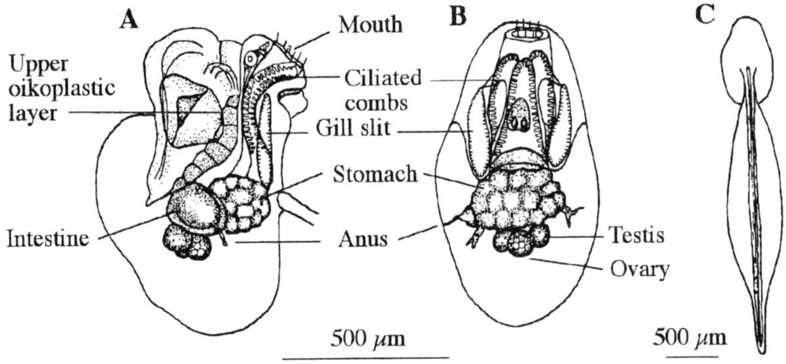

Fig. 2.12. Type of Kowalewskiidae, *Kowalewskia oceanica*. After Fenaux (1967). (a) Right side view. (b) Ventral view. (c) Complete animal. Note that in this species there are 2 testes, but a single ovary (see pp. 219).

perpendicular to the longitudinal axis of the trunk. Two big cells located at the left and right extremity anchor the six muscular cells forming the posterior upper wall. The remaining wall consists of thin membranes (Salensky, 1904b,c). As in *Oikopleura*, the structure and operation of the heart of fritillariids are not well understood, and ultrastructural investigations are needed.

Gonad shape is very variable (Fenaux, 1963). The structure of the testes is similar to that in the Oikopleuridae, but the ovaries (in the species where its structure is known: *F.pellucida, F.tenella,* and *A.sicula*) consist of a central nutritive syncitium with very large nuclei which disintegrate at maturity, and oocytes forming a cortex during the first stages of maturation, before expanding throughout the ovary.

As far as is known, the nervous and circulatory systems are basically built on the same plan as in the Oikopleuridae. Little is known of sense organs, though some are similar to those of oikopleurids (Bone *et al.*, 1987) (see section 4.5.1).

The portions of the trunk and tail which are not covered by the oikoplasts lack an external epithelium except in very young stages, and except just at the junction of the tail and trunk (Fol, 1872; Bone *et al.*, 1977a).

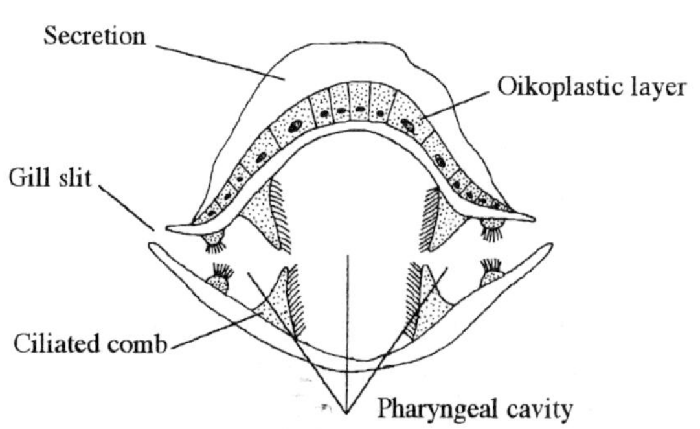

Fig. 2.13. Cross-section through the pharyngeal region of a *Kowalewskia*. After Fenaux (1967).

For this reason and although they have Langerhans receptors, Fritillariidae do not have epithelial action potentials (section 4.5.3). The dorsal oikoplastic layer covers the anterior part of the pharynx up to the mouth of the oesophagus and ventrally is limited to small rows behind the mouth.

Just under the surface of the trunk, in some species (e.g. *F.pellucida*), there are glands (usually unicellular) whose secretions pass to the surface of the body via a duct. Their function is unknown.

2.5. Kowalewskiidae

The trunk in Kowalewskiidae (Fig. 2.12) is globular posteriorly and anteriorly dorsoventrally compressed. The mouth is circular, without prominent lips but surrounded by bristles. The pharyngeal cavity (Fig. 2.13) has a very peculiar shape, and remarkably enough lacks an endostyle and peripharyngeal bands! It is divided longitudinally into three compartments by two rows of ciliated combs. On each side there are ventral and dorsal comb rows, whose ciliated 'teeth' lie on their internal sides, facing each other and almost meeting at their apices. The teeth are largest in the middle of the pharyngeal cavity, becoming smaller anteriorly and posteriorly. Fol (1872) described (from living animals) ciliature covering all the ventral part of the pharyngeal cavity, between the lower rows of the teeth, but flow between the rows of the the teeth must be driven by the ciliature of the teeth themselves. The spiracles are transformed into wide oval non-tubular gill slits. A short, flat, ciliated oesophagus, horizontal initially, curves up to the stomach. The stomach cardia are provided with a muff of long coalescent cilia which enters deeply into the gastric cavity. The stomach, as in Fritillariidae, is made up of a small number of large cells with an internal brush border. A short, but sometimes wide, ovoid intestine begins at the upper right of the stomach by a pylorus, with the same ciliary valve mechanism described in *Fritillaria*. The rectum is short and very contractile. It may be reduced to look like a ligament when empty. The anus emerges at the middle of the trunk on the right.

The gonads consist of an ovary and one or two testes, according to the species, and in structure are similar to that described in the Fritillariidae. A heart has never been observed by any author and this remarkable absence deserves further investigation.

The oikoplastic epithelium is limited to the dorsal part of the pharyngeal region and to a small area under the mouth. The upper oikoplastic zone consists of concentric rows around a very large cell with a large dense nucleus which makes a sort of dome in the middle (Fig. 2.14). Nothing has been published on this oikoplastic layer.

The tail is shaped like the leaf of a willow (*Salix*), with a very narrow musculature, less than twice the largest diameter of the notochord. Tail length varies with the species, between 4 and 10 times the length of the trunk.

Specimens of *Kowalewskia* are tiny, never found in abundance in plankton tows (but see section 15.3.1!), and rarely collected and preserved in good condition.

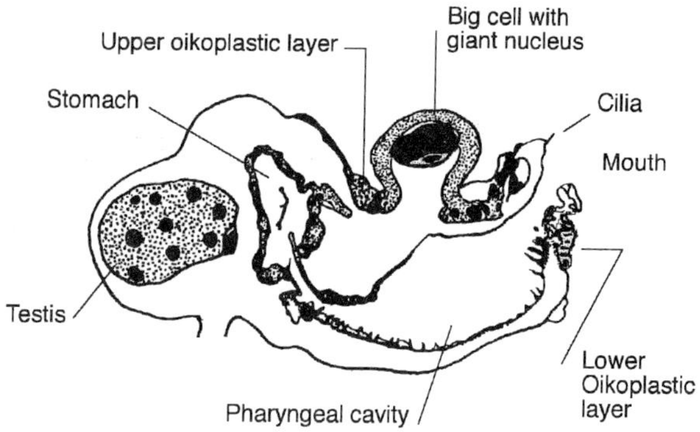

Fig. 2.14. Longitudinal section through the trunk of *Kowalewskia oceanica* (Fenaux, unpublished).

What information exists concerning morphology and histology needs to be confirmed and extended by new studies. Various features of *Kowalewskia* are shared with *Fritillaria*. In his classification Lohmann (1933) united the two genera in the Fritillariidae. However, the absence of endostyle and heart, and the very special shape of the oikoplastic layer and the house seem important enough to follow Lahille (1888, 1890) in placing the two genera in two separate families.

THREE

Locomotion, locomotor muscles, and buoyancy

Q. Bone

3.1. General

The four kinds of pelagic tunicates are rather different in their buoyancy and locomotor capabilities, and in the methods they use for generating thrust to move forwards. Pyrosomas, salps, and doliolids swim by continuous or intermittent jet propulsion, whilst appendicularians swim by oscillating a chordate-like tail. In all except pyrosomas (where cilia produce the water flow), locomotion is the result of the contraction of striated muscle fibres. The fine structure of these fibres has been examined, and there is some evidence for cholinergic innervation. In most, the ionic basis of the muscle potentials has been studied and in doliolids and salps something is known of the way that muscular contractions are graded. In both doliolids and salps, analyses have been made of their jet propulsion.

3.2. Pyrosomas

Pyrosomas move slowly forwards in consequence of the slow continuous jet which emerges from the aperture at the hinder end of the colony. This jet is produced by the ciliary activity of the gill bars of each zooid, which is under nervous control; mechanical stimuli such as tapping the colony evoke ciliary arrests and cessation of the propulsive jet, as well as luminescent responses. Directional swimming is not possible, for there is no adjustable sphincter around the jet aperture, as in salps or doliolids, so that it is not clear how the vertical migrations of some pyrosomas (e.g. *Pyrosoma atlanticum*) are brought about. No measurements have been made of the chamber pressure within the colony, nor of jet velocity or swimming speed, and until such measurements are made, all that can be said is that since the aperture is large (up to 2 m in some giant *Pyrostromella* colonies; see Chapter 1, Fig. 1.1) and the jet velocity presumably very low, thrust production is likely to be relatively very efficient (see below p. 44). Pyrosomas are close to neutral buoyancy, presumably by virtue of exclusion of heavy SO_4^{2-} from the test fluid (as Denton and Shaw, 1961, showed for salps), and it seems clear that the extensive vertical migrations undertaken by pyrosomas must be brought about mainly by ionic adjustment within the thick test of the colony (section 1.2.1). Surprising as it may perhaps seem, the exclusion of SO_4^{2-} for ascent and its re-entry to the test for descent might provide a sufficient buoyancy change to account for the observed rates of vertical migration. For example, Andersen and Sardou (1994) have recently shown that 51 mm colonies of *Pyrosoma atlanticum* show a migration amplitude of 760 m. Assuming that the movements up and down take 4 h (and 8 h is spent at the surface during the night) vertical movements of > 150 m h^{-1} would be involved.

Denton (personal communication) has calculated the changes of buoyancy that would be required for a sphere at different Reynolds numbers [Re = (length × velocity/kinematic viscosity) to give vertical movements at different rates. At $Re = 100$, for example, a change of buoyancy of only 0.01% would lead to a rate of vertical movement of 100 m h^{-1}. For a circular cylinder like *Pyrosoma atlanticum* 2 cm in diameter ascending and descending when lying horizontally (i.e. with the colony axis perpendicular to the direction of motion) at some 0.42 cm s^{-1} the Reynolds number is around 85. In fact, it is not known whether migration takes place with the axis of the colony horizontal or vertical. Since *Pyrosoma* is a cylinder and not a sphere, and has therefore a higher drag coefficient, a change in buoyancy of 0.03% would be required to give rise to the observed migration speed. No data are available for the SO_4^{2-} content of *Pyrosoma atlanticum*, but if it is assumed that this is similar to that of salps (54–87% of that of seawater gives neutral buoyancy; section 3.4.3), then a change of buoyancy of 0.03% would require a change in SO_4^{2-} content of ± 0.011% (see Denton and Shaw, 1961).

It is not yet known either whether pyrosomas do in fact utilize SO_4^{2-} exclusion (though by analogy with salps this seems very probable): it would be interesting to examine SO_4^{2-} content in *Pyrosoma* collected at the top of its vertical range at night and at depth during the day, to see if vertical migration does depend on changes in SO_4^{2-}. Locomotion by ciliary-driven jet propulsion is apparently so slow, that an ionic mechanism of this kind seems likely.

3.3. Doliolids

3.3.1. *Locomotion*

Like salps, doliolids swim by jet propulsion, but except for the rhythmically active large (5 cm) old nurse stages of *Dolioletta gegenbauri*, do so only when stimulated, otherwise remaining quiescent, filter feeding using the current provided by the cilia of the gill apertures. In the absence of a system of closed haemolymph sinuses (Chapter 1), and the requirement for metabolite exchange between the muscle bands of the old nurse and the feeding gastrozooids along the stolon, it is probable that the rhythmic bursts of muscular activity of the large old nurse stages serve to aid the heart to circulate the haemolymph in the body and stolon, as well as to maintain position in the water column (section 3.3.3).

If disturbed, for example by mechanical stimulation of the lips, the gonozooid and oozooid stages of doliolids show an extremely rapid escape reaction. The circular muscle bands around the barrelshaped body contract very rapidly, producing jet pulses driving the animal forwards at instantaneous velocities up to 50 Lengths (L)s^{-1} (Bone and Trueman, 1984). If stimulated anteriorly, they can close the anterior aperture and shoot backwards with equal rapidity. This rapid escape response usually consists of a single jet pulse, but if further stimulated the animals may produce several jet pulses in succession. A single jet pulse from an animal 0.4 cm long can drive it forward some 3 cm during 300 ms, so that it is an effective escape response.

Anterior and posterior apertures have scalloped valves (provided with sensory cells, Chapter 4), and mechanical stimulation of the valves at either end of the body evokes a rapid contraction of the muscle band furthest from that end, followed within some 5–7 ms by contraction of the other bands and a rapid jet pulse. Around 60% of the jet chamber volume is ejected during each jet pulse; maximum chamber pressures in a specimen 3.8 mm long were found to exceed 500 Pa. The positive ejection phase lasts some 80 ms and is followed by an inhalation phase lasting around 120 ms. Muscular contraction can be graded (as in salps) so that there is considerable variety in chamber pressures and in the instantaneous velocity curves resulting (Fig. 3.1).

Calculations from the pressure pulses (Bone and Trueman, 1984) indicate that a large doliolid making

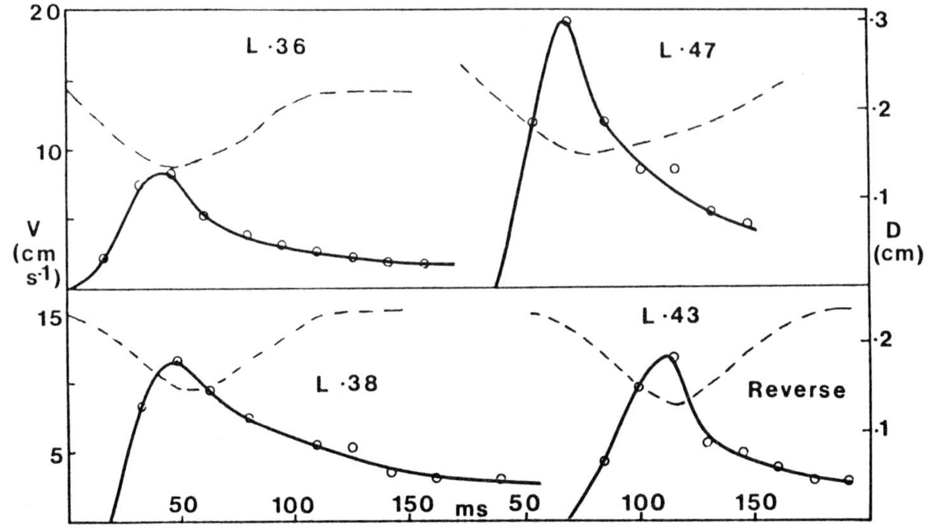

Fig. 3.1. *Doliolum.* Instantaneous velocity curves during escape responses evoked by posterior mechanical stimulation. Overall lengths of each specimen above. Dotted lines: changes in maximum body diameter. From Bone and Trueman (1984).

3 pulses s^{-1} with continued stimulation produced a power output of 1.3×10^{-5} W, the values for work g^{-1} muscle are similar to those of salps, around 8.7×10^{-2} J g^{-1} muscle. Maximum muscle stresses are much higher than those of salps, up to 15 N cm^{-2}, but the most striking difference between the two is that doliolid muscle contracts much more rapidly, normalized contraction velocity being 8.5 L s^{-1} at 18°C.

Doliolids contract the muscle band furthest from the end stimulated before the remainder, but the remainder contract simultaneously, so that the end band nearest the point of stimulation (encircling the jet aperture) contracts to decrease the aperture as the jet begins to emerge. For this reason, doliolids have a relatively smaller aperture than salps, and maximum jet velocities are higher, probably around 20–50 cm s^{-1}. Thus doliolid jet propulsion is likely to be a much more energetically costly process than that of salps, but since it is used as an escape response, energetic efficiency is of less consequence than a high escape velocity.

3.2.2. *Muscle fibres*

Doliolid muscle fibres differ from those of all other tunicates, for they are obliquely striated, and rectangular in section, cross-sections of the muscle bands resembling longitudinal sections of a loaf of sliced bread (Bone and Ryan, 1974), as seen in Fig. 3.2.

A central mitochondrial core is surrounded by a myofibrillar cortex in which the filament array spirals

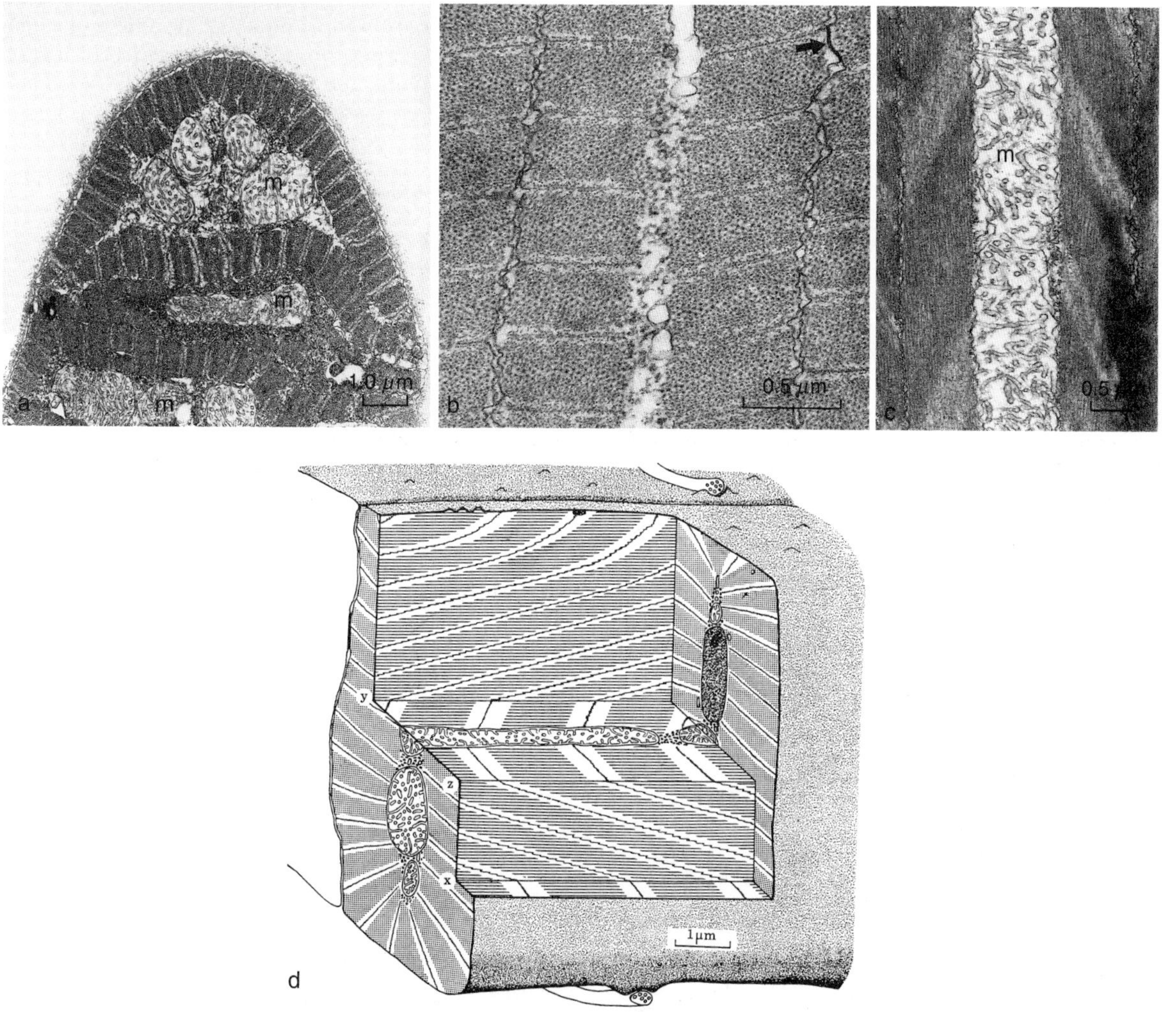

Fig. 3.2. *Doliolum.* (a) Cross-section of edge of muscle band showing parts of three fibres with central mitochondrial core (m). (b) Cross-section of part of muscle band at higher magnification; arrow indicates gap junction between two fibres. (c) Longitudinal section of muscle fibre showing oblique striation and central mitochondrial core. (d) Schematic diagram of myofibrillar array. All from Bone and Ryan (1975).

along the length of the fibre, resulting in a rather complex pattern (Fig. 3.2). Strikingly, no system of vesicles or subsarcolemmal cisternae exists which could be equivalent to a sarcoplasmic reticulum. Gap junctions occur between adjacent fibres, but dye injection and current injection experiments (Bone, 1989b) indicate that coupling extends only across a few fibres in the muscle band; the whole band does not form a single unit. As in salps, the muscle bands are innervated at a number of points along their length, punctate motor end plates normally lie on a small raised portion of the sarcolemma and lack subjunctional folds (Bone and Ryan, 1974). Again, acetylcholine iontophoresis evokes depolarizations, and acetylcholine is presumably the neuromuscular transmitter. Resting potentials of the fibres are around −75 mV; contraction is preceded by rapid spike potentials of varying amplitude (Fig. 3.3), varying according to the proximity of the recording electrode to the innervated site. Unlike salps, however, in most cases there is only a single potential for each contraction, except in the old nurses, where each contraction is preceded by a burst of summating spikes which may overshoot zero potential (Fig. 3.3d and e).

As might be expected, the activity of the anterior or posterior band which closes the end of the jet chamber opposite to that from which the jet issues is more prolonged than that of the other bands and here, electrical activity consists of a series of potentials (Fig. 3.3c). The spike potentials are probably carried by Ca^{2+} since they are abolished reversibly when placed in nominally 0 mM Ca^{2+} external solutions, as they are after such Ca^{2+}-blockers as Cd^{2+} or Mn^{2+}. In the absence of any sarcoplasmic reticulum, doliolid muscle fibres appear to provide an unique and unequivocal example of fibres in which the Ca^{2+} activation of the contractile apparatus is entirely due to Ca^{2+} influx during the spike potential (Bone, 1989b). Recent pharmacological confirmation that intracellular Ca^{2+} stores are not involved in contraction has come from the absence of any effect of caffein, or of > 10 μM ryanodine. In the absence of a sarcoplasmic reticulum, the Ca^{2+} entering to cause contraction is removed from the fibre by Na^{+}-Ca^{2+} exchange

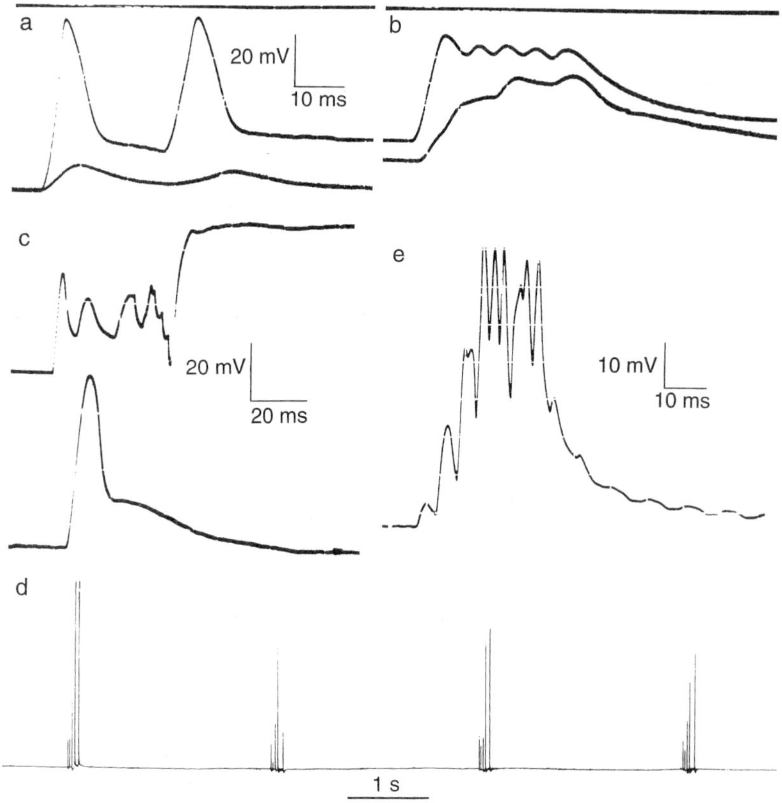

Fig. 3.3. *Doliolum.* (a) Intracellular records of spike potentials from different sites along same muscle band. (b) Same, showing multiple events during a short sustained contracture. (c) Simultaneous records from anterior muscle band (upper) and mid-body muscle band, showing multiple events associated with closure of anterior aperture during forward locomotion. (d) *Dolioletta* old nurse. Regular bursts of compound potentials. (e) As (d), at higher sweep speed, a single compound burst. From Bone and Trueman (1984) and Bone (1989).

across the sarcolemma; fibres stimulated electrically to contract in artificial seawater lacking Na^+ soon contract irreversibly as intracellular Ca^{2+} rises (Bone *et al.*, 1997a). Such a system would be unsuitable for muscle fibres that are more frequently active, but doliolids are mainly inactive, and even if stimulated usually make only one or a few contractions of the locomotor muscle bands.

3.3.3. *Buoyancy*

As Fedele (1923a) noted, when at rest, doliolids hang in the water with the anterior aperture obliquely upwards, so that the feeding current issues from the posterior aperture obliquely downwards. This oblique attitude presumably results from the densest part of the zooid, the gut, lying at the hinder end of the body. Doliolids are slightly denser than seawater, and maintain their position by the dynamic lift produced by the current of the gill cilia and by occasional 'spontaneous' upward single jet pulses (which Fedele termed the 'gallagiamento' reflex). The 'test' is very thin, (section 1.3.1.2) and hence cannot provide lift by exclusion of SO_4^{2-}; nothing is known of haemolymph composition, though it is apparently slightly hyperosmotic to seawater, since when doliolids are placed in solutions slightly more concentrated than seawater, normal body turgor is lost, and they become flaccid.

Doliolid eggs and larvae are found in the surface layers, and in view of Lambert and Lambert's (1978) observation that ascidian eggs float owing to the accumulation of ammonium ions in their follicle cells, it is possible that the same process may account for the buoyancy of doliolid eggs and larvae, the larvae having an inflated capsule (see section 1.3.2.1) conferring sufficient static lift to maintain position in the water column. If the capsule is ruptured, the larva sinks (personal observations). Fedele suggested that the intermittent rhythmic activity of the old nurse stages, which lose their gills and gut and bear a long stolon supporting the feeding and reproductive zooids (section 1.3.2.2), may be related to the loss of the lift produced by the feeding current in the other stages. Old nurse stages of *Dolioletta gegenbauri* are slightly denser than seawater and observations from submersibles show that they adopt a vertical head up attitude, so their intermittent short swimming bouts thus presumably maintain their position in the water column.

3.4. Salps

3.4.1. *Locomotion*

In contrast to pyrosomas, most species of salps are active and agile swimmers, and all propel themselves continuously by regular rhythmic contractions of the body muscle bands, which produce intermittent jet pulses from the posterior aperture during forward swimming. Salps can also reverse by sending the jet out of the anterior aperture. Although locomotion has only been studied in a few species, it seems essentially similar in all, though some salps are much more active than others. Unsurprisingly, since the form and muscular development of the two generations may be very different and since the blastozooids are normally linked in a chain for most of their lives, swimming speed and manoeuvrability may differ much in a single species between individuals of the two generations. Again, the swimming performance of the linked blastozooids in chains depends upon the way in which the zooids of the chain are arranged, so that, for example, the chains of *Salpa fusiformis* or *Iasis zonaria* (where the zooids overlap in the chain and lie along its long axis) swim much faster than those of *Pegea confoederata* (where they are attached laterally, the long axes of the zooids being at right angles to the axis of the chain; section 1.4.6).

A further complication is that as Harbison and Campenot (1979) showed, different species react differently to change of temperature. The frequency of locomotor contractions (which affects the speed at which salps swim), is highly temperature dependent in *Pegea socia, Cyclosalpa affinis,* and *Salpa maxima,* whilst *S.fusiformis* and *S.aspera* are relatively insensitive to change of temperature. As these authors point out, the open ocean is thermally stable, and insensitivity to temperature change characterizes the species which undergo vertical migrations and hence normally experience temperature changes as they do so. In normal forward swimming, chamber pressure and jet velocity can be varied by grading the contraction of the locomotor muscle bands, in addition to changes in contraction (pulse) frequency. These contractions are augmented by mechanical stimulation of the hinder end of the zooid which then accelerates into more rapid 'escape' swimming (see Fig. 4.18). In *Salpa fusiformis* for example (in which locomotion has been examined in most detail; Bone and Trueman, 1983), normal mean forward swimming speed of the undisturbed oozooids is around 2 cm s^{-1} (72 m h^{-1}), whilst if stimulated they can travel at over 6 cm s^{-1} (~240 m h^{-1}). The blastozooids swim rather more slowly, normally around 1.5 cm s^{-1} (~55 m h^{-1}), and up to 8 cm s^{-1} (~175 m h^{-1}) in forward 'escape' swimming. These values are for large individuals, examined in large containers soon after capture in plankton tows, representing maximum speeds of some 1.5–1.75 L s^{-1}. Similar values for *S.fusiformis* and *S.thompsoni* were obtained by Reinke (1987). Madin (1974a,b) measured swimming speeds of a number of species in the sea (noting that confinement in aquaria affected their performance), obtaining values between 1.0 and

Table 3.1 *In situ* swimming speeds and pulse frequencies of various salp species (After Madin, 1990; * indicates data from Nishikawa and Terazaki, 1994.)

Species	Zooid length (cm)	Mean speed (cm s^{-1})	Speed range (cm s^{-1})	Mean pulse rate (Hz)	Pulse rate range (Hz)
Oozooids					
Cyclosalpae polae	4–7	3.3	2.8–3.9	0.7	—
Pegea confoederata	3–6	4.1	1.7–5.6	1.1	0.9–1.5
Cyclosalpa affinis	5–7	4.8	2.5–5.9	0.9	0.8–1.1
Salpa maxima	4–10	6.2	1.1–10.0	1.4	0.7–1.8
S.cylindrica	3.5–6	6.0	2.5–13.9	1.9	0.7–2.5
**S.fusiformis*	1.65	4.5	4.3–4.7	2.5	2.4–2.7
**Thalia democratica*	—	—	—	1.4	0.9–2.1
Blastozooid chains					
Cyclosalpae polae	1–7	1.2	1.0–1.3	1.1	—
Pegea confoederata	1.5–6	2.0	0–4.0	2.0	1.1–2.8
Cyclosalpa affinis	2–6	3.6	1.6–4.6	2.0	1.7–2.2
Salpa maxima	1.5–8	3.9	2.2–10	1.3	0.9–1.8
S.cylindrica	0.8–1.2	9.0	2.2–15.3	2.7	2.0–3.1
**S.fusiformis*	15.6–20	4.7	3.4–5.8	2.5	2.0–3.1

13.9 cm s^{-1} (Table 3.1), representing 0.5–2.0 L s^{-1} for oozooids. Blastozooid chains (length not stated) swam at up to 15.3 cm s^{-1}.

Naturally, owing to the intermittent jet pulse rhythm, the zooids alternately accelerate and decelerate as they swim forwards undisturbed (Fig. 3.4).

Salps are also capable of reverse swimming when stimulated anteriorly (if they strike an obstacle for instance), when the posterior aperture is closed and the jet emerges from the anterior aperture; here too this avoidance reaction is more rapid than normal forward swimming. In *S.fusiformis*, the blastozooid chains cruise slowly forwards,

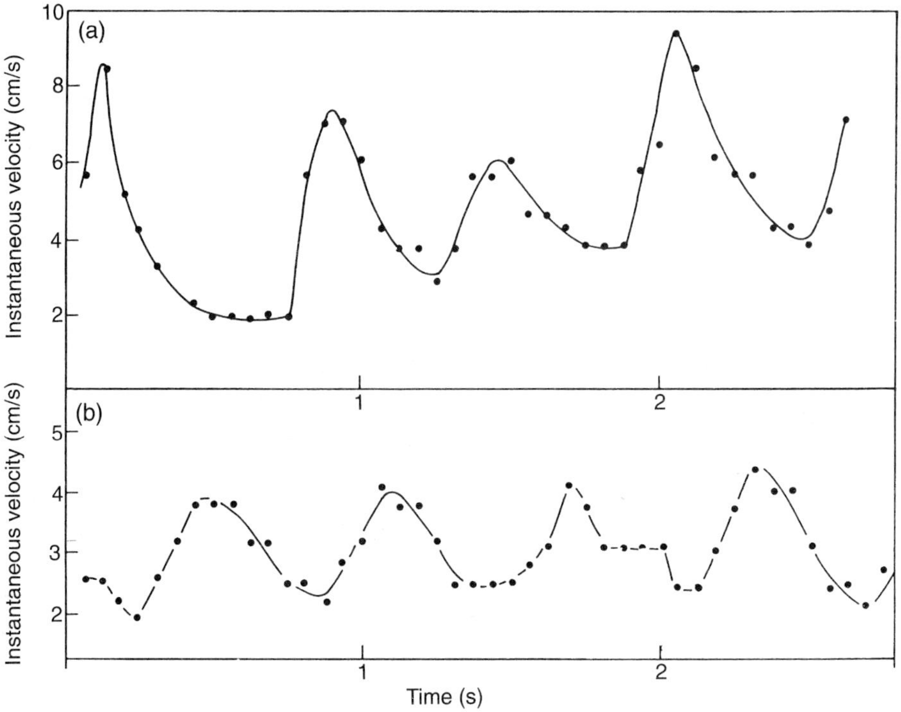

Fig. 3.4. *Salpa fusiformis.* Instantaneous velocity curves (from kinematic records) of (a) 3 cm chamber length oozooid; (b) 1.7 cm chamber length blastozooid, both during continuous swimming. From Bone and Trueman (1983).

as do isolated blastozooids and the oozooids, with the different zooids apparently operating independently, but remarkably, if either end of the chain is stimulated mechanically, all operate together to produce rapid escape swimming in either direction (see Fig. 4.21). If the chain is cut in two, the anterior part accelerates forwards, the posterior backwards. Evidently, some form of communication exists along the chain and this is provided by a most unusual system of alternating nervous and epithelial pathways which is described in Chapter 4.

Figure 3.5 shows the locomotor muscle bands of the two generations in *S.fusiformis*, and the form of the body in cross-section. The test is thickest ventrally, and since the locomotor muscle bands are not continuous around the body, but are dorsal, their contraction against the elastic test decreases the volume of the jet chamber dorsoventrally to produce the propulsive jet. At the anterior end there are flexible valves of test material, arranged to permit entry of water as the muscles relax, closing as they contract; these are provided with a series of small muscle bands. Some of these close the valves, whilst others can hold the valves open during contraction of the locomotor muscle bands so that the locomotor jet may issue from the anterior aperture when the salp is stimulated to reverse. At the hinder end, the rear aperture has a more complex series of thin muscle bands (particularly developed in the oozooid), which permit the jet to be controlled to some extent so that directional swimming occurs. Free-swimming oozooids can alter their direction with ease, sometimes looping the loop, presumably by small changes in the contraction of the different posterior muscles. Again, these muscles can also be employed to close the posterior aperture during reverse swimming. The timing of contraction of these various locomotor and anterior and posterior lip muscle bands in normal forward swimming is seen in Fig. 3.6, based on kinematic records from free swimming blastozooids, and on records of chamber pressure and electrical activity (such as in Fig. 3.4) from different muscles of a tethered blastozooid of *S.fusiformis*. The entire cycle occupies some 250 ms, in the oozooid between 180 and 300 ms. The muscles of

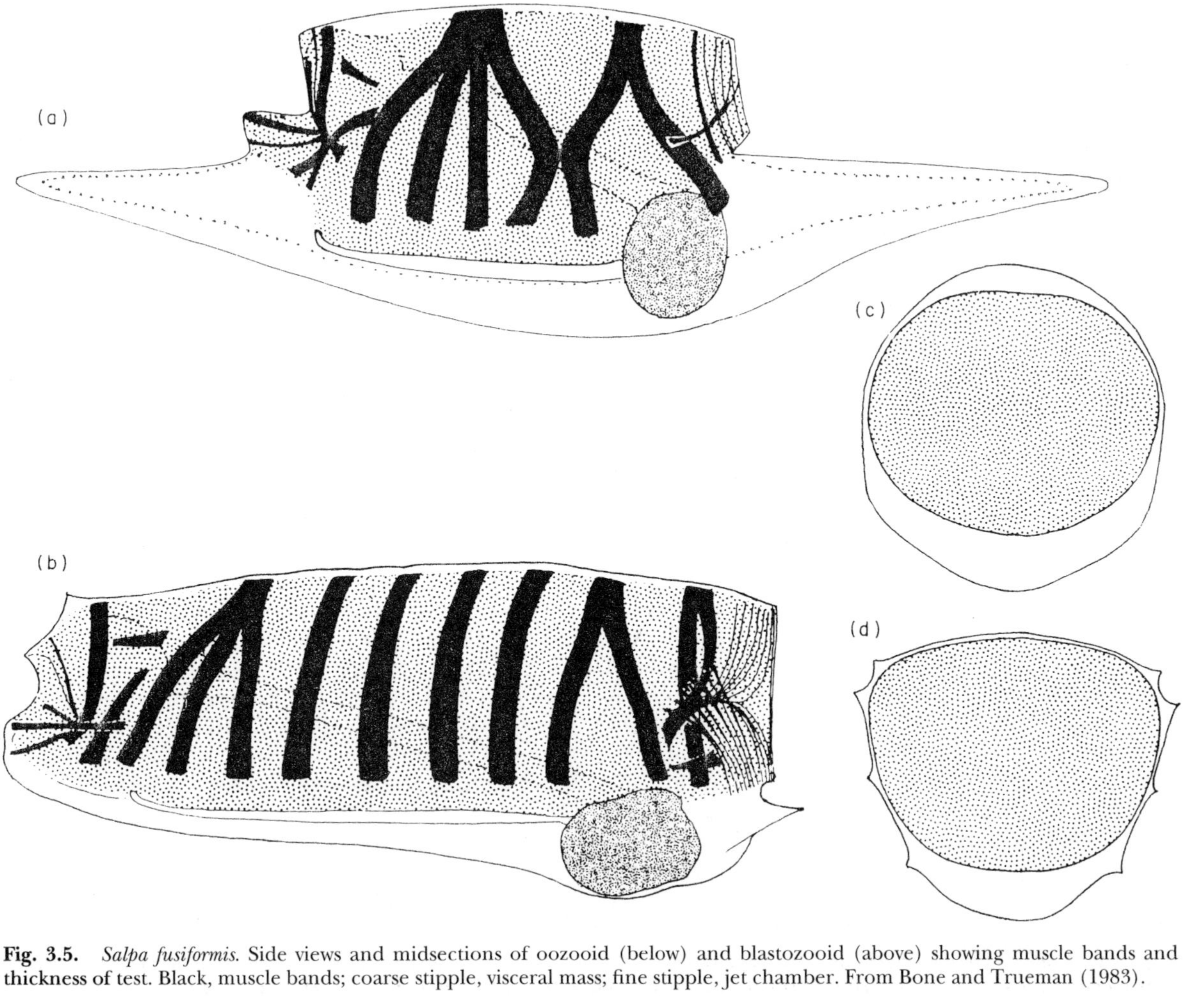

Fig. 3.5. *Salpa fusiformis.* Side views and midsections of oozooid (below) and blastozooid (above) showing muscle bands and thickness of test. Black, muscle bands; coarse stipple, visceral mass; fine stipple, jet chamber. From Bone and Trueman (1983).

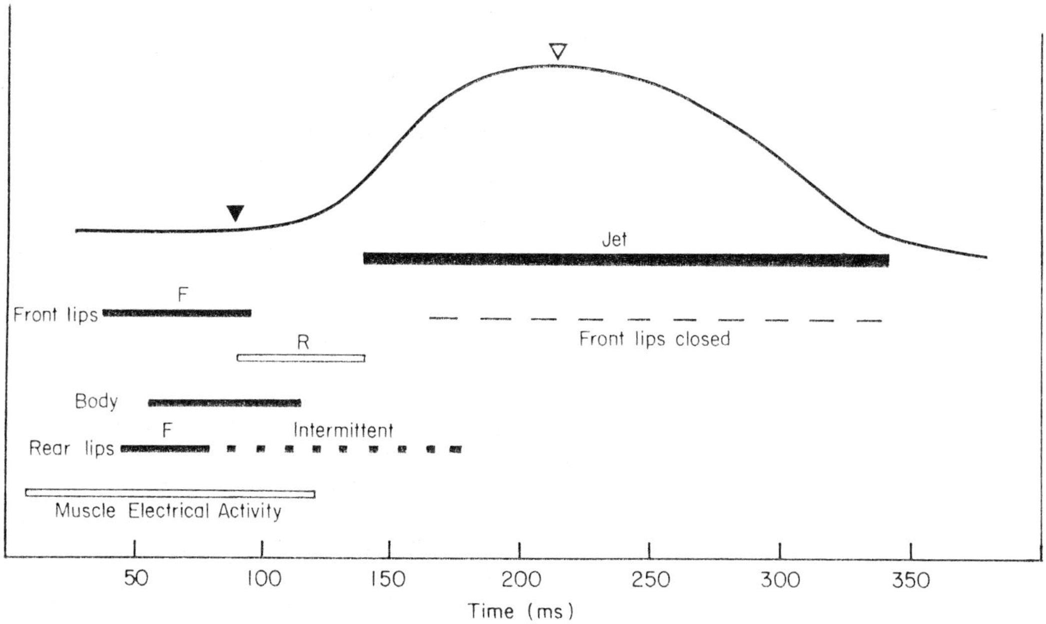

Fig. 3.6. Diagram illustrating events of propulsive part of jet cycle in blastozooid of *S.fusiformis* 1.5 cm chamber length. In forward swimming, the cycle begins with electrical activity (thinner black lines) of the front lip muscles, followed by that of the rear lips and the locomotor muscle bands of the body. The delay between activity of the locomotor muscular bands of the body and the beginning of contraction indicated by black triangle. Chamber pressure (top curve) then rises and the jet pulse begins (thick black line). Maximal tension of the locomotor muscle bands is developed at the peak of the pressure pulse (open triangle). Open bars: electrical activity of front and rear (R) lip muscles during *reverse* swimming. Combined from kinematic, pressure, muscle tension and electrical activity records. From Bone and Trueman (1983).

the front lips contract first, followed after 30 ms by more or less simultaneous contraction of the body muscle bands (Fig. 3.7).

During forward swimming the muscles around the posterior aperture show two contraction phases, the first probably aligning the jet, and the second closing the aperture so that the salp inhales for the next pulse mainly via the anterior aperture. During reverse swimming (evoked by mechanical stimulation of the anterior lips), the activity of the locomotor muscle bands is more prolonged (Fig. 3.7d) and the exhalant phase of the cycle lasts some 400 ms. In this case, the first event is the contraction of the posterior muscle bands.

In normal forward swimming, about 40% of the volume of the expanded jet chamber is expelled in each cycle, all via the posterior aperture (except for a small leak anteriorly before the front lips close completely). Refilling for the next jet pulse takes place via both apertures, though much more enters anteriorly; some 33% of the expanded volume, compared with only some 7% posteriorly. Refilling is brought about chiefly by the elasticity of the test, and to a lesser extent by the ram component due to the forward motion of the salp. Attempts to measure test resilience (Bone and Trueman, 1983), gave values of around 60%, but for various reasons this is certainly an underestimate. As Brooks (1893) remarked 'The test of salpa has never received the attention it merits'. Details of the structure and chemical composition of the test which underlie its elastic properties, such as those available for ascidians (e.g. van Daele and Goffinet, 1987) are not available for salps, nor indeed for pyrosomas and doliolids.

These jets produce cyclical thrust, and as a consequence forward velocity is cyclical (Fig. 3.4), as the salp alternately jets and coasts forwards inhaling water for the next jet pulse. Since the contractions of the locomotor muscle bands can be graded (see below), there is a wide variation in overall forward velocity, due to the variations in amplitude of the chamber pressures during the jet cycles. However, the larger salps appear to operate at lower chamber pressures than smaller individuals (Fig. 3.8).

From measurements of the (positive and negative) pressure pulses during the jet cycle in tethered salps, Bone and Trueman (1983) calculated the maximum work and power outputs for single jet cycles (Table 3.2). By examining the muscle bands in fixed specimens, muscle mass could be estimated (assuming muscle density to be 1.06 g cm^{-3}) hence the work g^{-1} of muscle was estimated; relative muscle mass varied greatly between different sizes of the same species and between species, the very active *Iasis* in particular having a large muscle mass (Table 3.2).

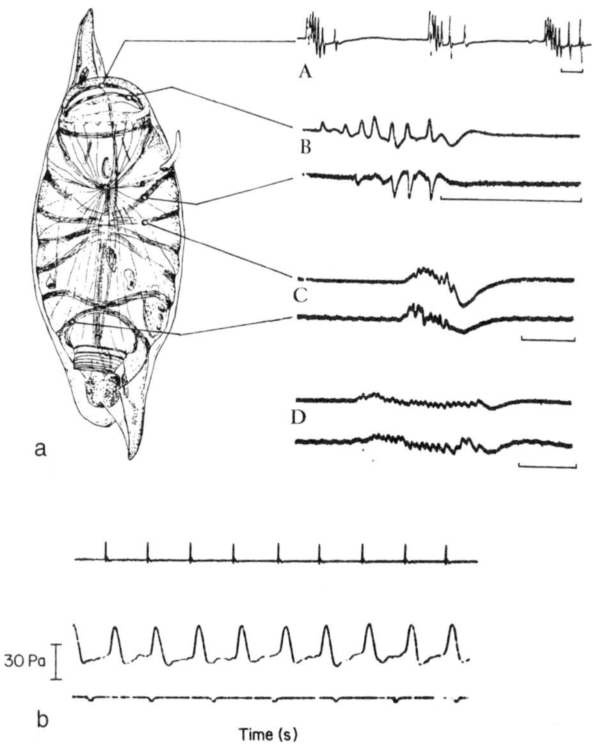

Fig. 3.7. *Salpa fusiformis.* Dorsal view of blastozooid (after Fedele, 1993b) showing extracellular (suction electrode) records from muscle bands. All forward swimming except D. Note delay between anterior lip activity and locomotor muscle bands and synchrony of different locomotor muscle bands. D shows longer activity (longer jet pulse) from same muscle bands as C but in reverse swimming. From Mackie and Bone (1977). (b) Regular cycles of jet pulses (lower) in tethered blastozooid recorded simultaneously with intracellular record from anterior lip muscle fibre. From Bone and Trueman (1983).

From the measurements on the muscle bands and from chamber pressures, muscle stresses during these (non-isometric) contractions were estimated to range from 3.87 N cm^{-2} in oozooids operating at 40 Pa chamber pressure to 10.26 N cm^{-2} for a small blastozooid at 100 Pa. These values for such slow muscle fibres are naturally much lower than those for the rapid muscle fibres of *Doliolum* (see section 3.2.1).

Another approach to the estimate of the work done during locomotion is to measure oxygen consumption as the salp swims in a respirometer chamber (Trueman *et al.*, 1984). By subtracting the basal rate of oxygen consumption (obtained by anaesthetizing the salp and again measuring oxygen consumption due to heart beat, ciliary activity, etc.), it is possible to calculate the work done per contraction (making some reasonable assumptions). Figure 3.9a shows the oxygen consumption of the locomotor muscles in *S.fusiformis* zooids of different sizes, and Fig. 3.9b the work performed during each contraction assuming 20% conversion of chemical to mechanical energy. The larger symbols denote work performed calculated from the pressure pulses during cruising and maximum performance; agreement between the hydrodynamic calculation and oxygen consumption values is reasonably satisfactory. The thrust generated by a propulsive jet (mu) is the product of the mass (m) ejected per second and the velocity (u) of the issuing jet. Evidently therefore, the same thrust could be generated by ejecting a large mass through a large aperture at low velocity as by ejecting a smaller mass at higher velocity through a smaller aperture. But since the power required to accelerate the fluid is the rate at which kinetic energy is given to it ($0.5\ mu^2\ s^{-1}$), it is more economical to eject a large mass at low velocity through a large aperture, as salps do. For continuous jet propulsion, the efficiency of momentum transfer to the water increases as the velocity of the issuing jet approaches the forward velocity of the machine, a measure of this being given by efficiency, $E = 2U/(2U + u)$, where U is the forward velocity and u is the jet velocity.

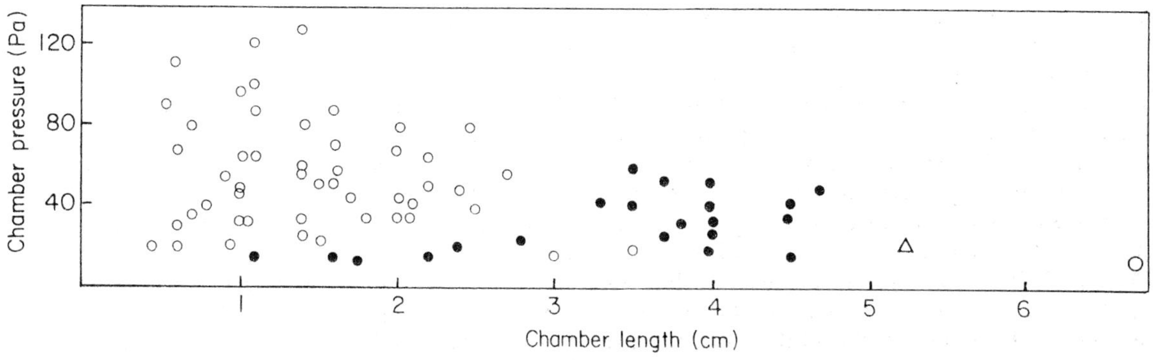

Fig. 3.8. Peak chamber pressure in salps of different sizes swimming at different velocities. Open circles, *S.fusiformis* blastozooids; filled circles, *S.fusiformis* oozooids; triangle, *Pegea confoederata* blastozooid; large open circle, *S.maxima* oozooid. From Bone and Trueman (1983).

Table 3.2 Locomotor muscle mass in different salps and work and power output for *S.fusiformis*. (From Bone and Trueman, 1983.)

Species	Chamber pressure (Pa)	Mean jet velocity, u (cm s^{-1})	Mean forward velocity, u (cm s^{-1})	$E = 2U / (2U + u)$
S.fusiformis				
Small blastozooid	100	22.8	3.8	0.25
Large blastozooid	80	29	4.8	0.25
Small oozooid	45	18.5	3.8	0.29
Large oozooid	60	27.9	6.6	0.27
Blastozooid chain, one active of 14		22.8	2.0	0.15
Same chain, all active◆		22.8	6.5 (estimated)	0.36
Siphonophores				
Chelophyes✧ anterior nectophore	400	71	16	0.31
Abylopsis✧ posterior nectophore	30	22	3	0.21
Squid				
*Alloteuthis**	20000	625	80	0.20

◆ Note that this assumes all zooids in chain operating at maximum performance, if they were not, and u was lower, efficiency would be much greater.
✧ From Bone and Trueman (1982).
* Trueman (unpublished).

Salp jet propulsion, like that of almost all animals, is oscillatory, but the same considerations apply, and some idea of their *relative* efficiency can be gained by applying this formula even though steady-state conditions obviously do not obtain. Table 3.3 shows that even where salps are operating at *maximum* chamber pressures, they compare favourably with most other designs.

Weihs (1977) pointed out that organisms like salps using pulsed jets for locomotion might gain significant thrust from the added mass entrained by the jet pulses as they rolled into vortices in the jet stream wake. His analysis showed that for short jet pulses producing thin disc-like vortex rings, extra thrust could be obtained when the ratio of the ring spacing to ring radius was less than 3. For *S.fusiformis* values for this ratio were 3.5–5.3 (Bone and Trueman, 1983). However, as Madin (1990) noted, if the vortex rings roll up into a spherical vortex, thrust gains are obtained at higher values for the ratio, and he found that the vortex rings did indeed roll up into nearly spherical vortices. Based on Weihs' analysis,

Table 3.3 Relative mechanical efficiency of various jet-propelled animals operating at maximum (burst) performance. (from Bone and Trueman, 1983.)

Species	Chamber length (cm) a	Muscle mass (g) b	Relative muscle mass (g) a/b	Work/cycle performed (J^{-2}/g)◆	Work/g muscle/ cycle performed (J^{-2}/g)◆	Power output at cycle frequency of 2 Hz (W × 10^{-5})◆
S.fusiformis						
Blastozooids	1.0	0.33	0.33	4.5	5.66	9.0
	2.1	1.26	0.6			
Oozooids	1.8	0.48	0.27			
	3.6	2.85	0.79	3.7	10.65	21
P.confoederata oozooid	3.6	0.82	0.23			
S.maxima blastozooid	19	12.12	0.64			
Oozooid	15	6.14	0.41			
Iasis zonaria blastozooid	2.1	4.4	2.13			

◆ At maximal performance.

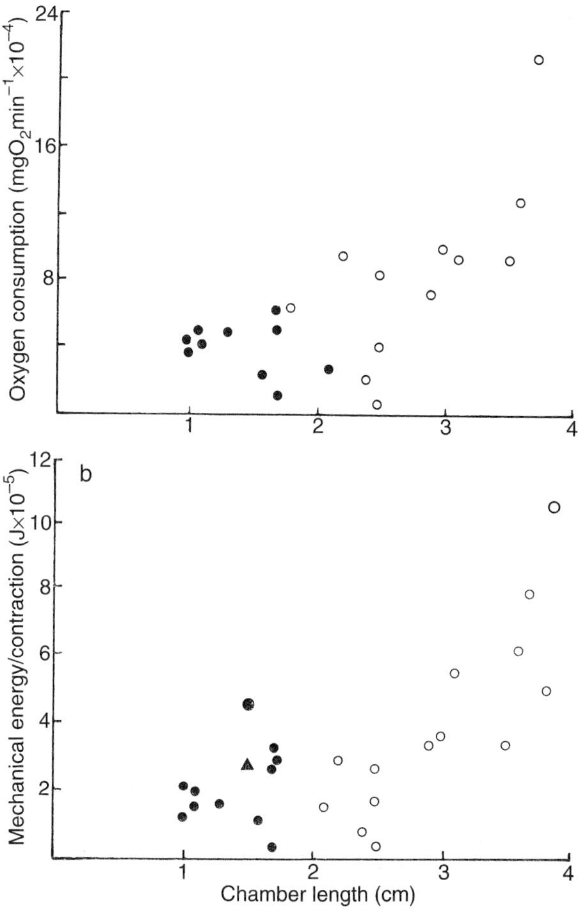

Fig. 3.9. *Salpa fusiformis.* (a) Oxygen consumption of locomotor muscle in blastozooids (filled circles) and oozoids (open circles) obtained by subtracting basal rate from total O_2 consumption. (b) Work performed during each contraction (assuming 20% conversion chemical to mechanical energy) by same individuals. Large circles, work/contraction during maximum performance calculated from pressure pulse; triangle, same for cruising performance. From Trueman *et al.* (1984).

Madin calculated a thrust advantage for salp pulsatile jets of some 30% for a wide range of salp species and sizes.

3.4.2. *Muscles*

The muscle fibres of salps are multinucleate and elongate, rectangular in section, forming flat bands or strips one fibre thick, lying between the inner and outer epithelium around the body and around the anterior and posterior apertures (see Fig. 3.5). They are invested with a dense plexus of blood lacunae (Fig. 1.17). In most salps, the locomotor muscle bands only partially encircle the body, ending ventro-laterally (hence the old terms desmo- or hemimyaria, contrasting with cyclomyaria for the doliolids with complete muscle bands), but in other salps such as *Ihlea* or *Cyclosalpa* the muscle bands join ventrally. The arrangement of the muscles in different salps is shown in Chapter 7. In the locomotor muscle bands of *Iasis* and *Thalia*, there are some 20–30 fibres in each band, about 100 μm wide in the larger *Iasis*, half this width in *Thalia* (Bone and Ryan, 1973).

In *Ihlea*, however, the marginal fibres of the bands are strikingly different to those forming the remainder of the bands (Fig. 3.10), for they are 10 times the width of the latter (in a large *Ihlea* oozooid the marginal fibres are 190 μm across, the central fibres 20 μm; Bone and Braconnot, in preparation). As Soest (1975) and Godeaux (1976) observed, in some species there may be clinal variations in the numbers of fibres in the muscle bands. Thus, in oozooids of *Ihlea punctata* from warm waters, the first locomotor muscle band has between 75–184 fibres, whereas in specimens from colder waters, the same band contains 200 and 256 fibres. Similarly, *S.fusiformis* from the eastern Mediterranean has a significantly smaller number of fibres in its muscle bands than in the western Mediterranean population. In contrast, although there is considerable variation in muscle fibre number in Mediterranean *Thalia democratica*, no cline is found.

There is an irregular junction across the band in the dorsal mid-line, and the muscle fibres are extremely elongate, for they run half the length of each muscle band. Adjacent fibres in the bands are closely apposed, but not linked by gap junctions (current injected into one fibre of the band is not seen in adjacent fibres). There is no transverse tubular system, but the myofibrillar cortex of the fibre is penetrated by an extensive sarcoplasmic reticulum at the Z-line level. An inner mitochondrial core occupies around 50% of the cross-sectional area of the fibre, and much glycogen is present; these are evidently fibres operating by aerobic glycolysis. Their structure is seen in Fig. 3.11. Toselli and Harbison (1977) have examined the development of muscle fibres in *Cyclosalpa affinis*, and conclude that they arise by the fusion of mononucleate myoblasts, a process giving rise to myelin figures that become associated with the mitochondria.

Intracellular records from rhythmically active fibres (Fig. 3.12a) show regular bursts of non-overshooting compound spike-like potentials from resting potentials around up to 70 mV; these vary in amplitude according to recording position. Close to the point where nerve bundles cross the muscle bands as they radiate from the brain the potentials are larger, whilst in between the innervated sites they are a good deal smaller; each fibre is multiply innervated and decremental potentials pass along the fibres from the innervated sites. Very similar bursts of compound summating potentials can be recorded following trains of stimuli to the nerve bundles containing the motor axons (see Fig. 3.12b).

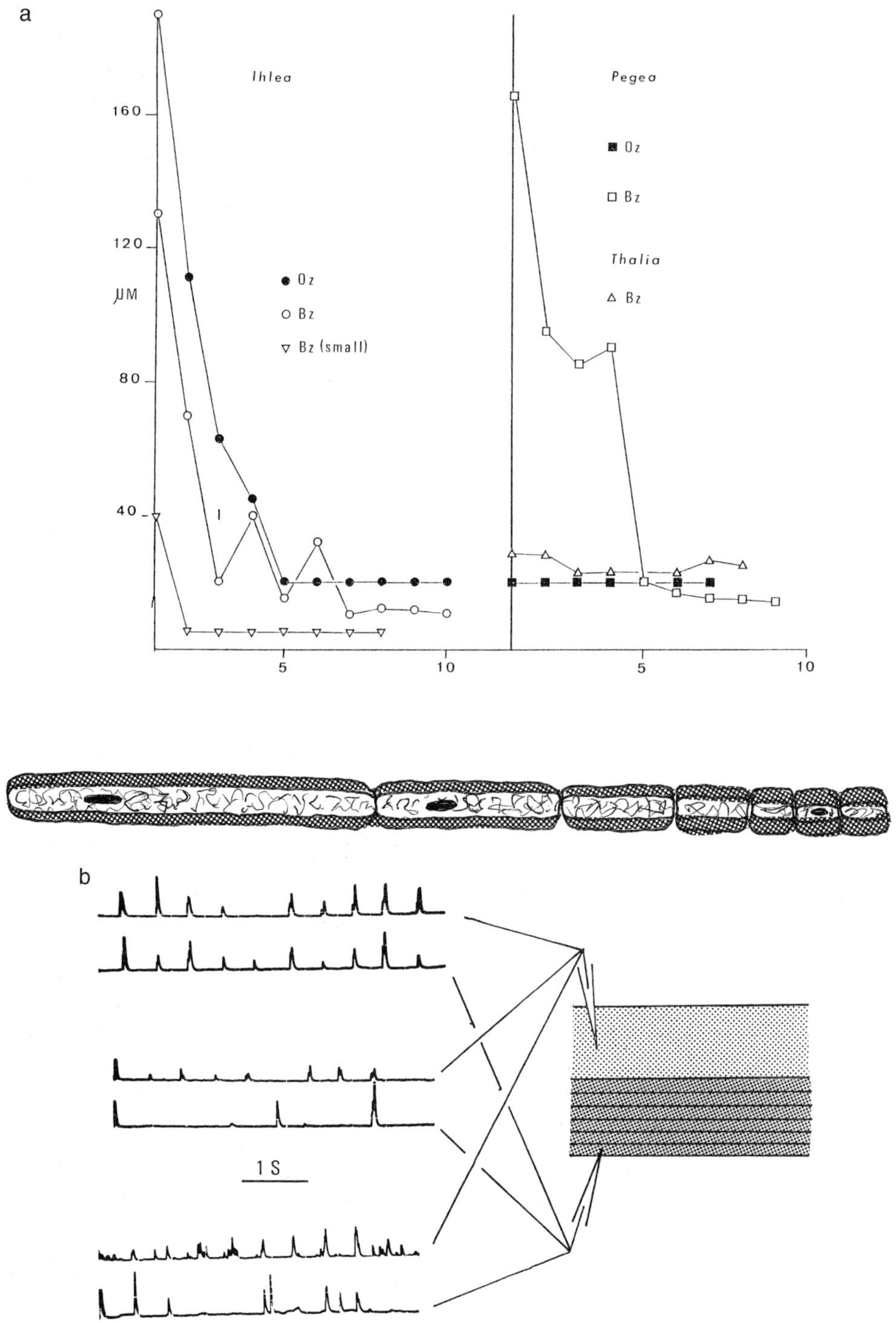

Fig. 3.10. *Ihlea.* (a) Widths of fibres edging locomotor muscle bands in *Ihlea* and other species. Ordinate, fibre width in μm; abscissa, number of fibre from edge of band. Oz, oozooids; bz, blastozooids. Below, semi-diagrammatic transverse section of edge of locomotor band in *Ihlea*, showing thinner central mitochondrial core in narrowed fibres towards middle of band on right. (b) Three sets of simultaneous intracellular records from the wide fibre edging band, and smaller 'central' fibres of an anterior lip muscle band. Note that in upper pair of traces, both fibre types are active together, whereas in those below the wide fibre is more often active than the narrow. Bone and Braconnot (in preparation).

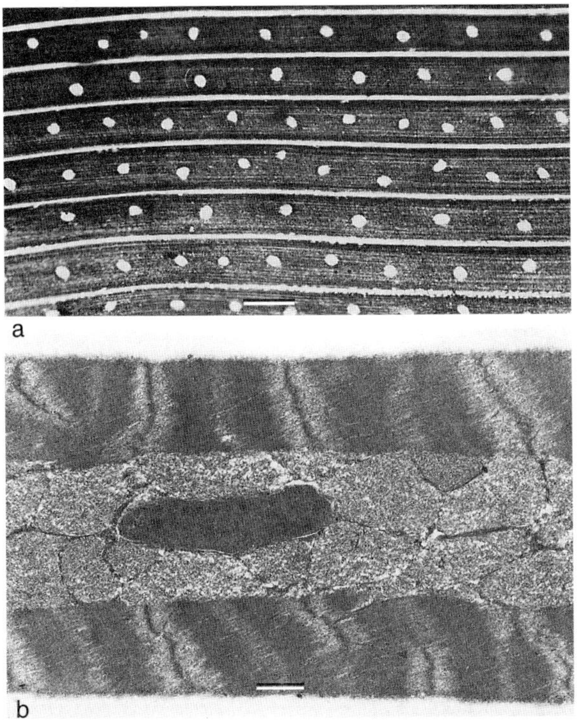

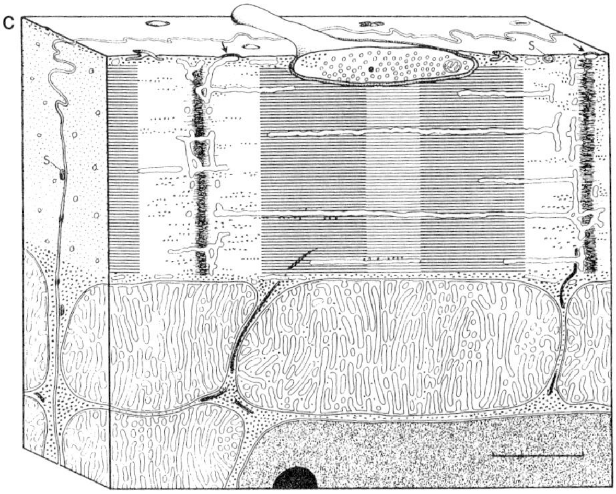

Fig. 3.11. *Iasis zonaria.* (a) Part of muscle band showing multinucleate fibres (nuclei unstained in this silver impregnation). Scale bar 100 μm. (b) Longitudinal section of salp muscle band showing mitochondrial core (with one of the nuclei of the fibre) surrounded by myofibrils. Scale bar 2 μm. (c) Schematic diagram of structure of muscle fibre. From Bone and Ryan (1973).

Since iontophoresis of acetylcholine near the fibre membrane evokes depolarizations, it seems probable that acetylcholine is the neuromuscular transmitter, as it apparently is in other tunicates, the nerve terminals (which lack subjunctional folds) contain electron-lucent vesicles similar to those seen at cholinergic junctions in other animals. They are found on both ectodermal and endodermal faces of the muscle bands (Fig. 3.13).

As yet, nothing is known directly of the mechanical properties of salp muscle fibres (though kinematic measurements indicate partially loaded contraction velocities of around 1.5 L s^{-1}), nor have their passive electrical properties been studied (which would be a relatively easy undertaking). It is clear, however, that they contrast in their mechanical properties with those of doliolids, which are rather different in design. The very wide marginal fibres in *Ihlea* can contract separately to the central fibres of the muscle bands, and since they have a relatively smaller myofibrillar content and larger mitochondrial content than the central fibres, it seems possible that in these slow moving salps they alone are responsible for normal cruising locomotion, whilst the central fibres are only employed during more vigorous escape swimming.

3.4.3. *Buoyancy*

Although Moseley (1892) long ago observed that dead salps sink (he measured the sinking rate of a salp in a jar of seawater, and determined that it descended 8 inches in 20 s), living salps achieve neutral or near neutral buoyancy by exclusion of SO_4^{2-} from the test. Denton and Shaw (1961) carried out measurements on a variety of gelatinous planktonic animals, and found that fluid extracted from the test of *Salpa maxima* contained about 85% of the SO_4^{2-} in seawater, giving a static lift of around 0.5 mg ml^{-1}, sufficient to confer neutral buoyancy. *Thalia democratica* excluded rather less from the test, but was also able to achieve neutral buoyancy. Bidigare and Biggs (1980) found that *Cyclosalpa pinnata* contained only about 54% of the SO_4^{2-} content of seawater, giving a greater lift than in *S. maxima*, around 1.5 mg ml^{-1}. The blood has only been analysed in *S.maxima* (Robertson, 1957), where the SO_4^{2-}, content is 65% of that in seawater. Blood volume in salps has not been estimated, but there are extensive blood lacunae, and hence the blood as well as the test may give significant lift. Reinke (1987) made continuous measurements of the underwater weight of *S.fusiformis* individuals and found variations up to 3 mg over 24 h, assumed to be brought about by ionic regulation.

Nothing is known of the SO_4^{2-} pumping mechanism, which presumably resides in the outer epithelial layer next to the test. *S.fusiformis, S.aspera,* and *S.thompsoni* are known to undertake daily vertical migrations of some 500–1000 m (Foxton, 1961; Wiebe *et al.*, 1979; Reinke, 1980), but whether adjustments in the amount of static lift provided by SO_4^{2-} exclusion are involved in these

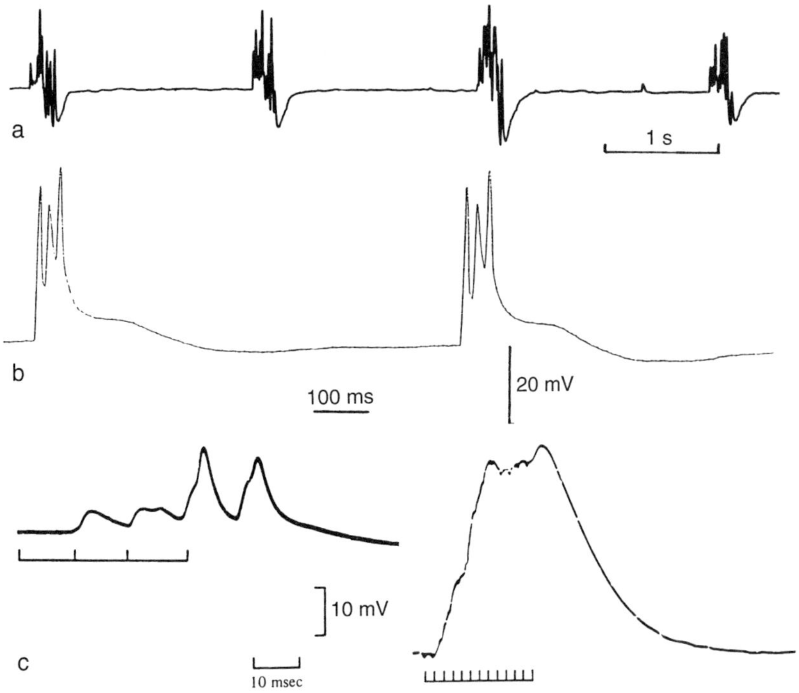

Fig. 3.12. (a) *Salpa fusiformis.* Extracellular suction electrode record of regular contractions of blastozooid locomotor muscle band. (b) *Ihlea.* Intracellular record of two compound muscle potentials from locomotor muscle of small blastozooid. (c) *Salpa fusiformis.* Same when nerve to muscle is stimulated at different frequencies (stimulus marker below each trace). Note that first stimuli do not evoke contractions. In part from Anderson *et al.* (1979).

migrations is unclear. Since sustained swimming speeds of 550 m h^{-1} have been measured for salps in the field (Madin, 1974a,b, 1990), it is unlikely that changes in buoyancy are significant in such daily migrations.

3.4. Appendicularians

3.4.1. *Locomotion and tail movements*

Oikopleurid appendicularians normally inhabit the feeding house which they secrete (Chapter 6), and use regular rhythmic bursts of tail movements to pump water through it; only if disturbed do they rapidly emerge to swim around frenetically in short bursts, before expanding and entering a new house. *Oikopleura dioica, O.labradoriensis, O.albicans* and *O.longicauda* are relatively dense, and outside the house sink rapidly trunk first between bursts. No doubt other smaller species are similar, but the swimming behaviour and buoyancy of the larger oikopleurids such as *Stegosoma,* and the giant abyssal forms such as *Bathochordaeus,* are unknown. Fritillariid appendicularians were until recently thought to differ from oikopleurids in being free-swimming throughout their lives, but it is now known that almost all the trunk and tail enter the feeding filter (section 6.4). *Fritillaria pellucida,* the only species that has been examined in any detail, is neutrally buoyant, in contrast to the common small oikopleurids.

No kinematic observations have been made on appendicularian movements, and what is known is based upon the activity of animals held by the tail tip with small flexible polyethylene suction electrodes, and on intracellular records from the caudal muscle cells in animals pinned out on Sylgard with *Opuntia* spines. Surprisingly perhaps, under these conditions, the burst frequency and tail beat frequency are similar to those of animals in their houses observed in large aquaria (Bone and Mackie, 1975). The impression gained from such observations is that appendicularians have but a small number of motor 'tapes', with which they drive stereotyped tail movements. *Oikopleura labradoriensis* and *O.dioica* (the two oikopleurid species examined in detail) show quite a limited repertoire of tail movements: (i) regular short bursts in which the tail oscillates at relatively low frequency (2–3 Hz), equivalent to the regular bursts of pumping behaviour within the house; (2) prolonged bouts of tail beats at 4–5 Hz which are apparently equivalent to the nodding movements made when the animals are expanding their house rudiments; and (3) regular short bursts of similar duration and interburst frequency to the pumping bursts, but in which the mean tail beat frequency is higher (5–6 Hz),

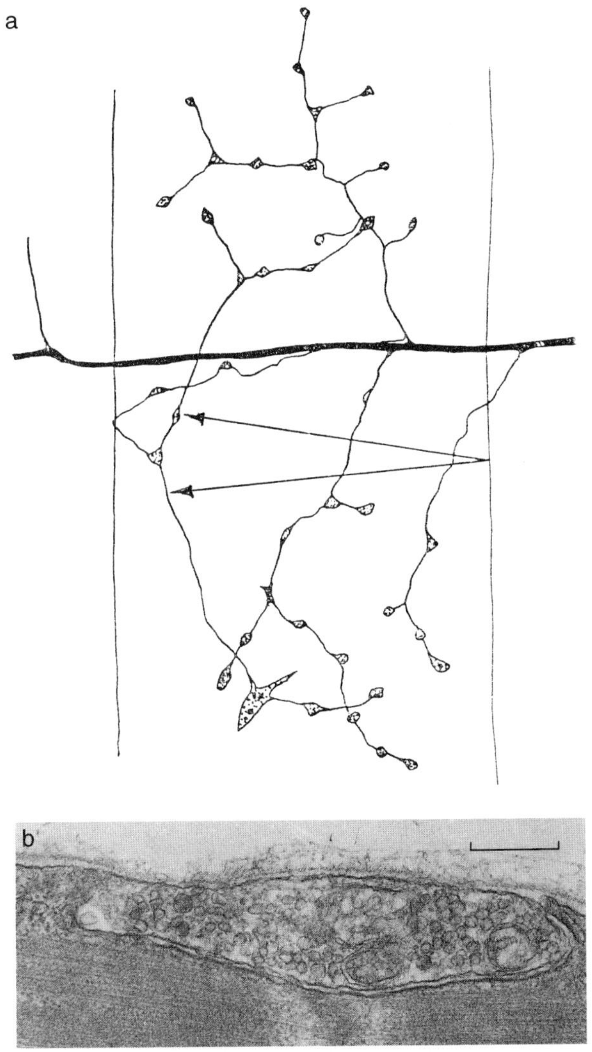

Fig. 3.13. (a) Pattern of nerve terminals (both intercalary and terminal) on a locomotor muscle band of *Thalia democratica*. Arrows indicate end-formations on other side of muscle band. (b) Motor terminal on locomotor muscle fibre of *Iasis zonaria*. From Bone (1959), and Bone and Ryan (1973). Scale bar 38 μm

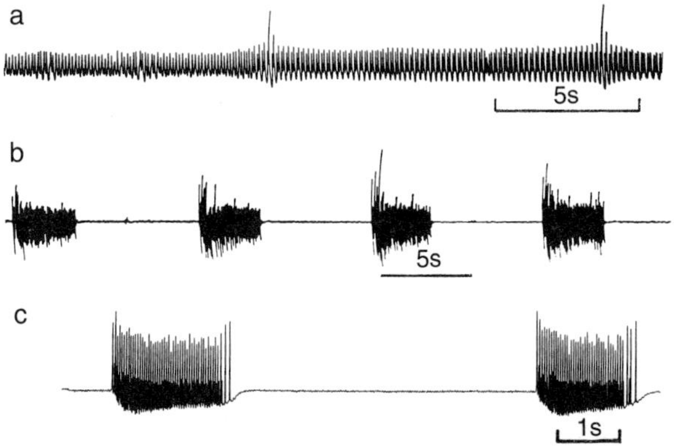

Fig. 3.14. *Oikopleura labradoriensis.* Different types of movement of *O.labradoriensis* and *O. dioica.* (a) House rudiment expansion behaviour. (b) 'Swimming' movements recorded with suction electrode on tail tip. (c) Swimming movements recorded by microelectrode in caudal muscle cell.

equivalent to the regular swimming bursts of free-swimming animals (Fig. 3.14). Presumably these regular spontaneous swimming bursts serve to maintain position in the water column, since the animals reverse their trunk first sinking attitude and swim upwards during the swimming bursts.

Mechanical stimuli to the tail and hinder part of the trunk evoke more rapid swimming than normal, whether they are given during swimming bursts or when the animal is at rest. This rapid escape response is evoked by propagated action potentials (outer skin impulses or OSPs) in the outer epithelial cells, which 'enter' the caudal ganglion at the base of the tail via axons to paired mechanoreceptors on the sides of the trunk. The system is described in Chapter 4, here we may simply note that it produces short bursts of escape swimming at tail beat frequencies up to 30 Hz, during which *O.dioica* swims at around 15 L s^{-1} (Fig. 3.15).

In *Fritillaria pellucida* the movements of the tail are rather more complex than in oikopleurid appendicularians (Bone *et al.*, 1979), for in addition to symmetrical swimming patterns, *F.pellucida* employs its tail in asymmetrical movements when it expands and contracts its house. When attached to suction electrodes, symmetrical movements occur in regular bursts, as in oikopleurids, at around 8–10 Hz. These movements correspond to the regular bursts of swimming shown by free-swimming animals, when Lohmann (1899b) found that they swam in spirals. These bursts of symmetrical

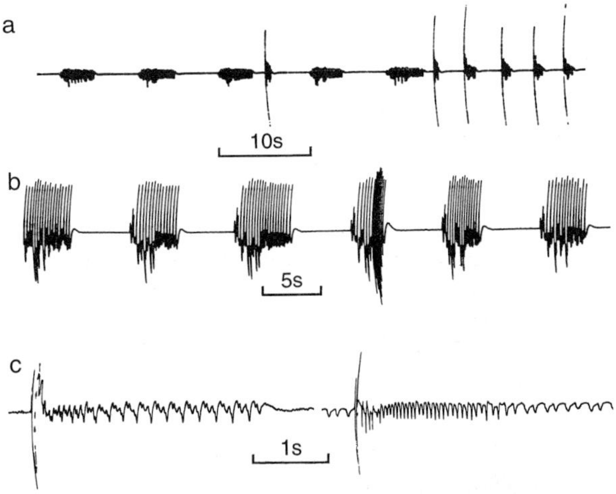

Fig. 3.15. *Oikopleura dioica.* (a) OSPs (large amplitude deflections) reset the swimming rhythm. (b) OSPs increase frequency within burst. (c) Two bursts evoked by OSPs. From Bone and Mackie (1975).

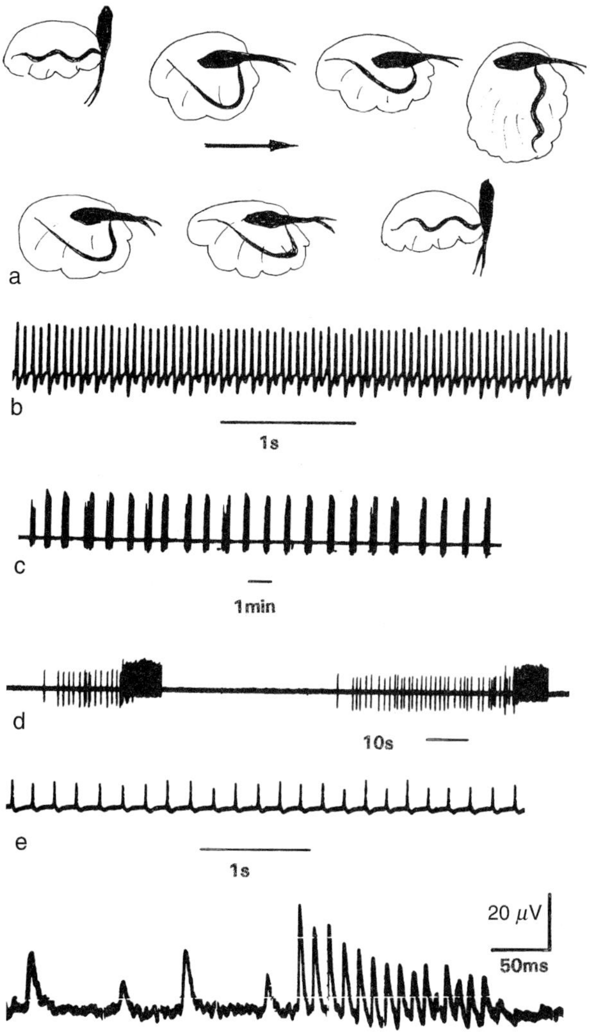

Fig. 3.16 *Fritillaria pellucida.* (a) Sequence of events in feeding cycle. From left to right: swimming; pumping up house; asymmetrical tail movements; filtering; unilateral flexions towards mouth, and swimming again. (b–d) Records of caudal muscle activity with tail inside suction electrode. (b) 'Escape' swimming at 10 Hz. (c) Regular bursts of symmetrical swimming at 5–6 Hz. (d) Asymmetric tail beats to mouth followed by symmetrical oscillations. (e) Low amplitude symmetrical oscillations (equivalent to drawing water through feeding filter) followed by unilateral 'wiping' contraction to mouth. From Bone *et al.* (1979); (a) modified after Flood, personal communication.

swimming are followed (in animals attached to electrodes) by a pause, and then by a series of asymmetrical tail movements towards the mouth (expanding the feeding house) and by regular symmetrical movements which draw water through it, but do not move the animal through the water; it remains suspended without changing position (Fol, 1872). There is then a second pause, and this is followed by 10–20 s of asymmetric tail movements towards the mouth, and a second burst of symmetrical swimming. The whole feeding process lasts about 2 min and is then repeated (Fig. 3.16).

Fritillariids have lost the outer epithelium (Bone *et al.*, 1977a), and hence do not have the same escape swimming mechanism involving action potentials in the epithelial cells, but they retain the paired mechanoreceptor cells (Chapter 4), and escape swimming is evoked (at around 10 Hz) by touching the anterior region of the tail where these receptor cells lie.

3.4.2. *Muscle cells*

In all appendicularians there are 10 large flattened muscle cells with branching nuclei on either side of the tail (Chapter 2). Each is interdigitated to its neighbours and coupled to them at several points by gap junctions. The muscle cells have an outer mitochondrial layer, and an inner myofibrillar layer penetrated by systems of vesicles (Bogoraze and Tuzet, 1969b; Bone *et al.*, 1977). In the two oikopleurid appendicularians examined (*O.labradoriensis* and *O.dioica*), there are invaginations of the sarcolemma at regular intervals, forming flattened tubules between the myofibrillar bundles; this system is equivalent to the transverse tubular (T) system of chordate fibres. A second internal tubular system, forming a fenestrated lattice between the myofibril bundles, is the sarcoplasmic reticulum, coupled to the T-tubules via regular end-feet, again as in chordate fibres, except that such couplings are found at all sarcomere levels, not just at the Z-line.

Fritillariid appendicularians (*F.pellucida* and *F.haplostoma* have been examined) and the Kowalevskiid *Kowalevskia tenuis* lack the T-tubule system and the sarcoplasmic reticulum is coupled directly to the inner face of the sarcolemma. Unusually, the sarcoplasmic reticulum tubules penetrate the mitochondrial zone of the fibres as well as ramifying in the myofibrillar zone. Presumably the absence of a T-tubule system in fritillariid appendicularians is a consequence of their smaller scale, for both groups contract their muscle cells equally rapidly. These arrangements are shown schematically in Fig. 3.17.

Although the muscle cells along one side of the tail are electrically coupled, each is innervated (Flood, 1973), and in oikopleurids, each receives two separate motor endings (see Fig. 4.31a). One type branches dichotomously in a corymbiform manner over the inner surface of the muscle cell, and arises from serially arranged motoneurons along the tail nerve cord (Chapter 4), whilst the other consists of much smaller finer fibres which branch to run anteriorly and posteriorly along the inner face of the muscle cell above the

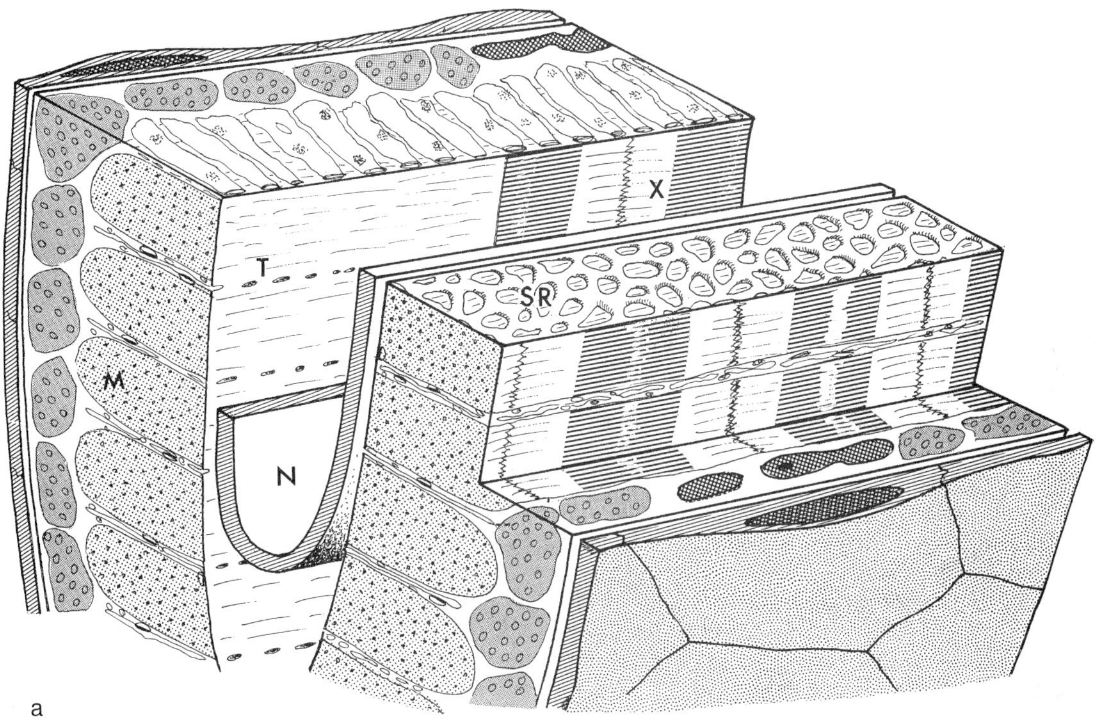

a

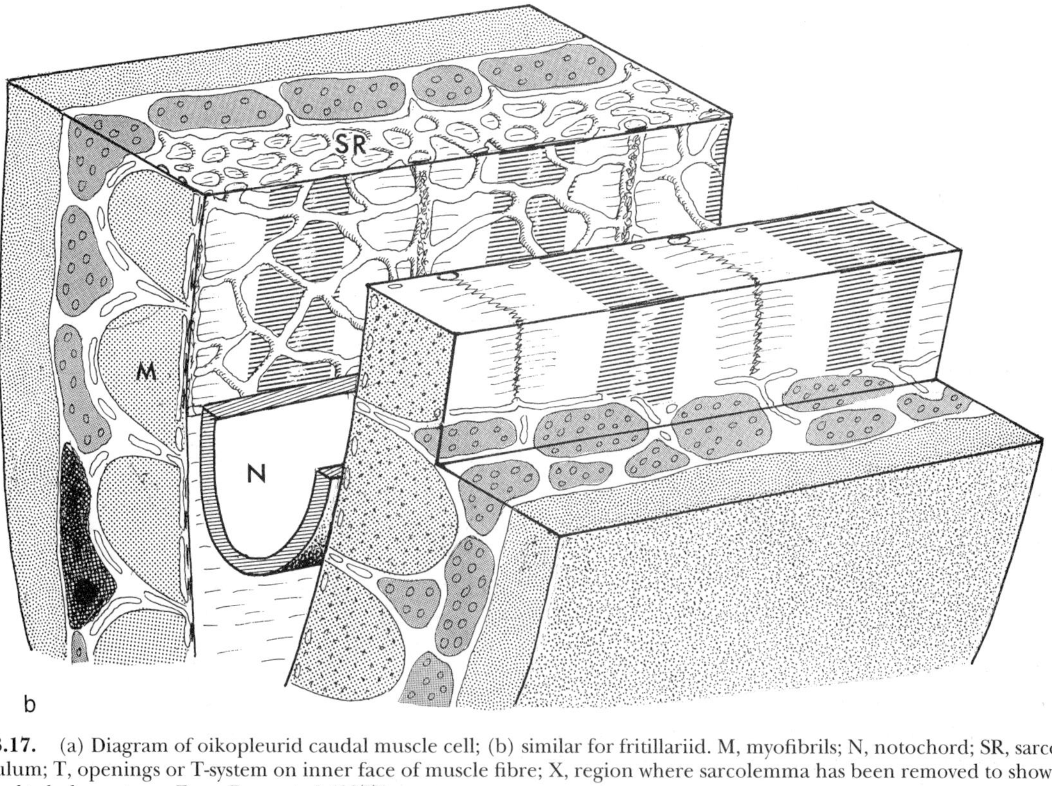

b

Fig. 3.17. (a) Diagram of oikopleurid caudal muscle cell; (b) similar for fritillariid. M, myofibrils; N, notochord; SR, sarcoplasmic reticulum; T, openings or T-system on inner face of muscle fibre; X, region where sarcolemma has been removed to show absence of basal tubular systems. From Bone *et al.* (1977).

level of the notochord. These smaller endings seem to be derived not from cells along the cord, but from somata in the caudal ganglion near its base. Flood has shown that the large corymbiform endings are probably cholinergic, since acetylcholinesterase can be demonstrated histochemically beneath them. Iontophoresis of acetylcholine near the sarcolemma evokes depolarizations (unpublished observations). Interestingly, however, Bollner *et al.* (1991) have recently demonstrated GABA-like immunoreactivity in what are probably the large motor endings, raising the question of possible aminergic modulation of the cholinergic responses. In *F.pellucida*, Fol (1872) figured branching nerve endings on the caudal muscle cells, somewhat simpler than the corymbiform endings of oikopleurids, and Flood (1973) observed nerve terminals (in *F.borealis*) rather smaller than those on the caudal muscle cells of *Oikopleura dioica*, but it remains to be determined whether the second type of oikopleurid motor ending occurs also in fritillariids.

Resting potentials of oikopleurid caudal muscle cells are around –65 mV, and contraction during normal swimming bursts is preceded by spike-like potentials up to 50 mV which do not overshoot resting potential (Fig. 3.18a–d). These potentials (and muscle contractions) are rapidly and reversibly abolished by such Ca^{2+}-blockers as Co^{2+} and hence are largely if not entirely carried by Ca^{2+} (unpublished observations). During escape swimming elicited by mechanical or electrical stimulation, the spike-like muscle potentials are larger than in normal swimming bursts, and may overshoot resting potential (Fig. 3.18e). Although it is tempting to suppose that this difference between muscle cell activity during normal and escape swimming may result from the activation of the two separate motor innervations, or perhaps by aminergic modulation, such speculations await experimental investigation.

As expected from the close electrical coupling of the cells, simultaneous records from two cells along one side of the tail show very similar potential changes (Fig. 3.18a), and it is presently unclear how the two separate innervations of each cell are employed in such a coupled system. Nothing is known of the mechanical properties of appendicularian muscle cells.

3.4.3. *Buoyancy*

Oikopleurid appendicularians are denser than seawater, but within the feeding house, some species at least are close to neutral buoyancy, and isolated houses sink very slowly. It is not known whether the larger deep sea

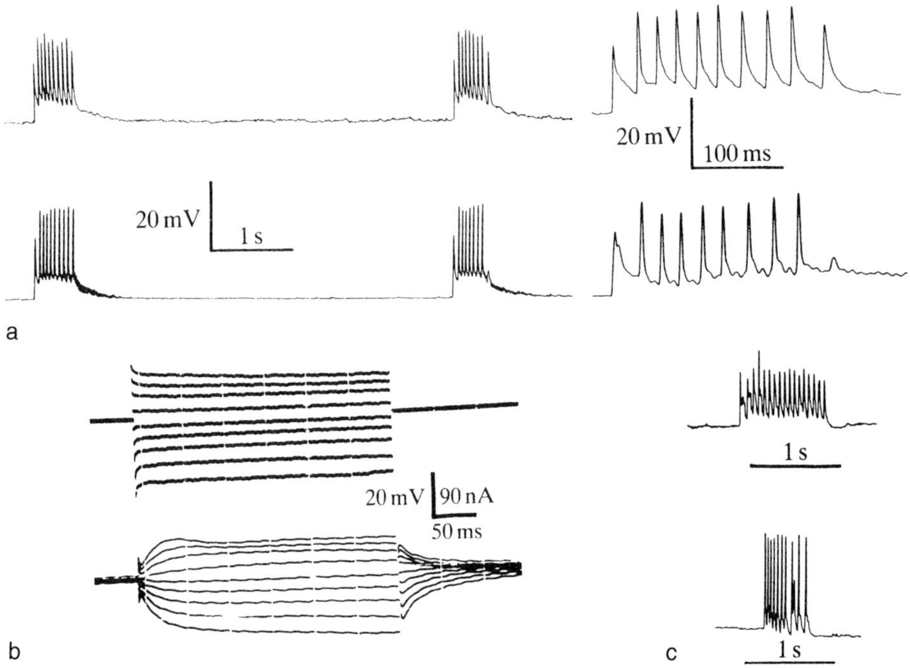

Fig. 3.18. *Oikopleura dioica.* (a) Simultaneous intracellular records from two different caudal muscle cells along tail showing close (but not exact) similarity (recorded at higher speed on right). (b) Current (upper trace) injection into one caudal muscle cell evokes membrane voltage changes in 2nd caudal muscle cell at other end of tail indicating close electrical coupling along tail. (c) Intracellular records from caudal muscle cell of a normal 'spontaneous' swimming burst (above), and an OSP-evoked burst. [(a) and (c) from Bone 1985.] From Bone (1985).

species are neutrally buoyant within the house, though this is suggested by observations from submersibles, and Garstang (1937) observed that a specimen of *Bathychordaeus stygius* swimming outside the house was 'obviously more or less buoyant'. *Fritillaria pellucida* in contrast appears to be neutrally buoyant, with the centres of buoyancy and gravity evidently coincident since it can remain hanging in the water in any attitude, showing the normal feeding rhythm. When placed in containers after plankton tows, its attachment to the surface of the seawater is due to the hydrophobic nature of the outer surface of the animal, which lacks an external epithelium (section 2.4), and not to positive buoyancy.

3.5. Conclusions

The locomotion of *Pyrosoma* colonies has not been studied experimentally, and this is unfortunate, since it is the only example known of an animal using continuous jet propulsion, and it would be interesting to compare the efficiency of this (ciliary) process with the intermittent muscular jet propulsion of salps, or indeed, squid. The necessary measurements of forward speed, jet chamber pressure, and jet efflux velocity would not be difficult to make for *Pyrosoma*. Although a good deal is known of the histology and physiology of the locomotor muscle fibres in salps, doliolids, and appendicularians, several important points remain to be clarified, and nothing is known of some aspects, such as the mechanical properties of the muscle fibres, in any of these groups. Pelagic tunicates are delicate animals, but it would certainly be feasible to examine the mechanical properties of their muscle fibres using the techniques devised for vertebrate single fibre preparations. As yet the passive electrical properties and the ionic bases of the spike potentials have not been examined, or have only been examined in a preliminary way in pelagic tunicate muscle cells, and these should be particularly interesting in *Doliolum*, which lacks both sarcoplasmic reticulum and T-tubule systems.

Probably the most interesting problem which remains to be cleared up is the way in which appendicularian caudal muscle cells are controlled. The present situation is that there appears to be a dual innervation of each muscle cell (despite the fact that they are closely coupled along the tail) and that one of these innervations appears to be cholinergic, though the terminals contain GABA-like material. Nothing is known of the transmitter at the other type of nerve terminal. It is puzzling too, how escape responses evoked by OSPs involve muscle spike potentials of greater amplitude than those during normal swimming bursts.

FOUR

Nervous system, sense organs, and excitable epithelia

Q. Bone

4.1. General

With the exception of the Appendicularia, all the pelagic tunicates have a single centralized brain from which mixed nerves radiate to innervate the muscles, gills, endostyle, and other organs; they also contain the axons of primary sensory cells of different kinds. In appendicularians, there is an anterior brain or cerebral ganglion in the trunk, but this is linked to a large caudal ganglion at the base of the tail, which controls the tail musculature, and in some cases the sensory cells are secondary sensory cells (section 4.5.3). The general morphology of the nervous system is described in Chapters 1 and 2. Unfortunately, as will become evident in this chapter, knowledge of most aspects of the histology and fine anatomy of the nervous systems of *Pyrosoma*, salps, and doliolids has lagged behind more recent work on the physiology of the system, and it is only in appendicularians that there has been recent advance in the fine structure of the nervous system and its circuitry.

A striking feature of most pelagic tunicates is that they possess excitable epithelia which propagate action potentials when mechanically stimulated. This epithelial sensory system (which in salps and appendicularians extends the receptor fields of the sensory cells lying in the epithelium), is also found in one doliolid stage, but not in *Pyrosoma* so far as is known. In salps and appendicularians the system plays an important role in locomotor behaviour, interacting with the nervous system and sense organs. Indeed, salps show the most complicated suite of excitable epithelia–nervous system interactions known in any animal, these are involved in the control of locomotion of blastozooid chains (section 4.4.4). In contrast, the function of the excitable epithelium in the doliolid 'old nurse' stage is unclear but does not seem to be linked with locomotor behaviour.

4.2. Pyrosoma

Little can be added to the classical descriptions of the nervous system of *Pyrosoma* referred to briefly in Chapter 1, since there has been no recent work on the system. Although it is probable that the central nervous system is very similar to that of sessile ascidians, this surmise is not very helpful, since the ascidian nervous system is itself poorly known. In ascidians, however, it is known that immunocytochemical techniques for a variety of neuropeptides stain different central neurons, and comparable studies on *Pyrosoma* (and indeed, all thaliaceans) would be of interest.

4.2.1. *Ciliary control and light emission*

Studies on the tetrazooid stage (Mackie and Bone, 1978) failed to reveal any trace of epithelial action potentials, but showed that the gill cilia received an inhibitory innervation; if the tetrazooid is disturbed by vibration all gill cilia are arrested simultaneously, producing a ciliary arrest potential (CAP) which can be recorded by suction electrodes on the outside of the test (Fig. 4.1a). A single arrest only interrupts the flow of water through the pharynx of the zooid momentarily, but stimuli evoking a series of CAPs cause sustained stoppage which may continue for many minutes. In the tetrazooid stage, and probably also in the adult

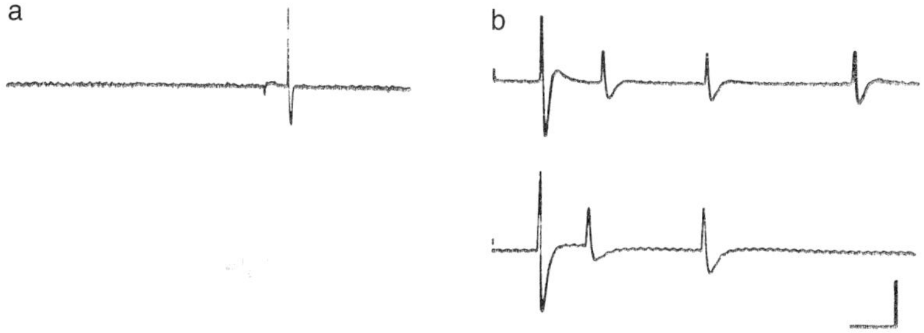

Fig. 4.1. *Pyrosoma atlanticum.* (a) Extracellular suction electrode record of ciliary arrest potential evoked by mechanical stimulation, preceded by small potential due to closure of siphonal muscle. (b) Ciliary arrest potential recorded by suction electrodes on two adjacent zooids of the tetrazooid stage. The first (larger) potential triggered by an electronic flash is succeeded by smaller potentials representing the firing of the ciliary arrest pacemaker. Scale bars: (a) 0.25s, 50 μV (b) 0.1s, 0.1mV

(Chapter 1), the ciliary arrests of adjacent zooids are not synchronized apart from the first, which is the response to a common stimulus (Fig. 4.1b). Presumably, ciliary arrests are driven by central arrest neurons similar to those described in the ascidian *Chelyosoma,* described by Arkett (1987).

Pyrosoma (as its name implies) has from the first been known to be brilliantly luminescent. Bennett (1833) gives a graphical account of a swarm of *Pyrosoma* off West Africa, where an 'extensive field of bright luminous matter emitted so powerful a light as to illuminate the sails, and to permit a book of small print to be read with facility near the windows of the stern cabins'. Kampa and Boden (1957) measured the intensity of light emitted following mechanical stimulation to be up to 4×10^{-6} cm^{-2} of receptor surface (at 1 m in the air). The light emitted by *Pyrosoma atlanticum* (Fig. 4.3a) is maximum at 492 nm (Swift *et al.*, 1977). Stimuli causing ciliary arrest also elicit light emission by a non-nervous pathway of uncertain nature; the light organs themselves are not innervated, nor linked to excitable epithelia (Mackie and Bone, 1978). Photic stimuli have long been known to evoke flashing in *Pyrosoma* colonies (e.g. Burghause, 1914), and in the tetrazooids examined by Mackie and Bone (1978), evoked ciliary arrests followed by light emission at the same interval (600 ms) as that following ciliary arrests evoked by mechanical stimulation (Fig. 4.3). The patterns of stimulated bioluminescence in colonies of *Pyrosoma atlanticum* and *Pyrosomella verticillata* have been examined by Bowlby *et al.* (1990). They found that owing to the different spacing of the zooids within the colony, the two species differed in their patterns of luminescence (Fig. 4.4), hence it is conceivable that colonies of a given species could recognize one another from this pattern. Photic stimulation between 350 and 600 nm evoked luminescence, maximum response (Fig. 4.2b) being with stimuli at 475 nm (i.e. close to the measured maximum of the light emitted by the zooids). Stimulus stength influenced the recruitment of flashing zooids, and by image intensification, these authors were able to show saltatory waves of luminescence across the colony resulting from photic triggering of the zooids by their neighbours.

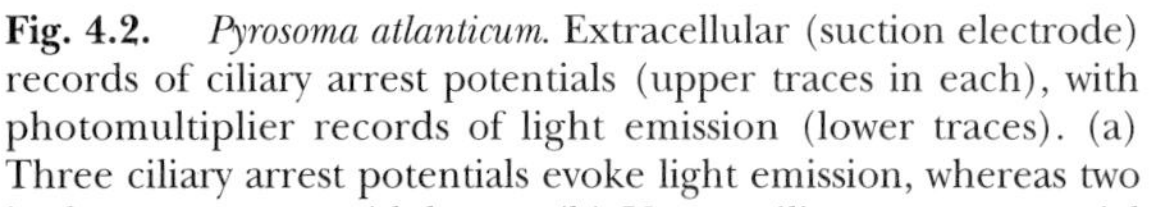

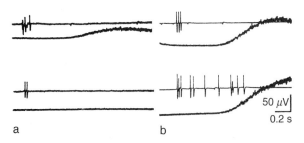

Fig. 4.2. *Pyrosoma atlanticum.* Extracellular (suction electrode) records of ciliary arrest potentials (upper traces in each), with photomultiplier records of light emission (lower traces). (a) Three ciliary arrest potentials evoke light emission, whereas two in the same tetrazooid do not. (b) Upper: ciliary arrest potential evoked by mechanical stimulation, followed by light emission. Lower: same, but ciliary arrest potentials evoked by photic stimulation, again followed by light emission. Note whichever the stimulus, light emission follows after some 600 ms. mV on scale bar refer to electrical record. From Mackie and Bone (1978).

Similar waves of bioluminescence are evoked by mechanical stimuli at a given point on the colony (Fig. 4.5), and these are quenched with an apparent refractory period for each zooid of around 18s. Although the nature of the bodies within the cells of the luminous organ remains uncertain (see section 1.2.1), the consensus is that they are probably luminous bacteria, albeit unique in luminescing only intermittently (when required to do so by the host).

4.2.2. *Eyes and sense organs*

Pyrosoma has a pigment-shrouded eye forming part of the brain, somewhat simpler than that of salps (see below), but neither the fine structure nor the physiol-

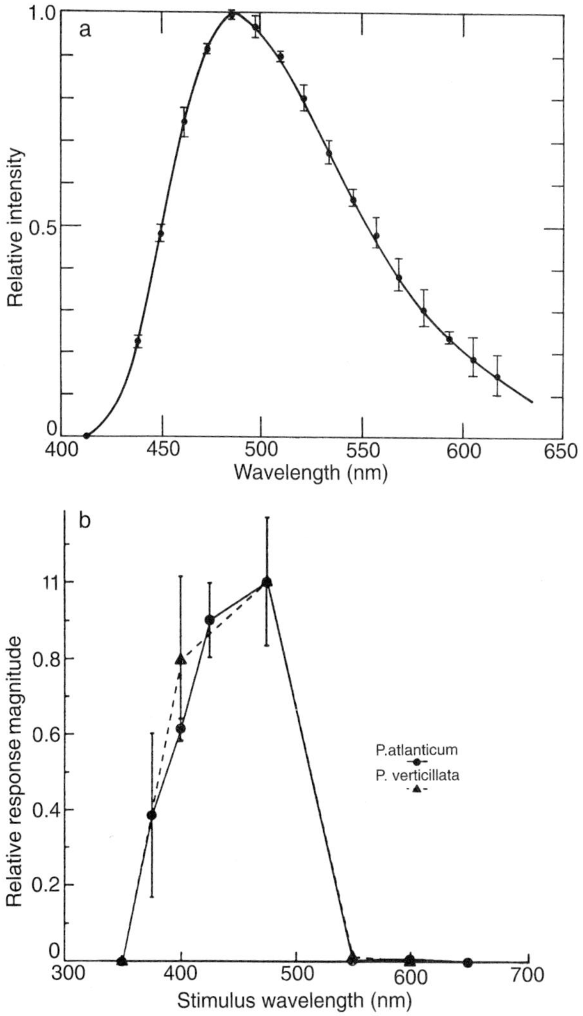

Fig. 4.3. (a) Emission spectrum of light from a colony of *Pyrosoma atlanticum*. From Swift *et al.* (1982). (b) Relative light emission responses of *P. atlanticum* and *P. verticillata* when stimulated by light of different wavelengths. Note that the most effective is close to the wavelength of maximum emission. From Bowlby *et al.* (1990) reproduced with permission.

ogy of the elongate receptors has been examined. Since light emission by one zooid of the colony excites others to luminesce, maximal receptor sensitivity is presumably close to the wavelength of the light emitted (490 nm), but this has not been tested.

Ciliated sensory cells with morphology typical of all pelagic tunicates are presumed mechanoreceptors (*Pyrosoma* is extremely sensitive to vibration), but their function has not been investigated experimentally, except for those around the inhalant sphincter (Mackie and Bone, 1978), where delicate touch evokes contraction of the sphincter, stronger stimulation sphincter contraction followed by ciliary arrest (Fig. 4.1b).

4.3. Doliolids

4.3.1. *Central nervous system*

As in *Pyrosoma*, the doliolid central nervous system has hardly been studied with modern methods. The fine anatomy of the brain is unknown, although the extreme rapidity of the escape movements and the small delays between the contraction of the lip muscle bands and the other muscle bands of the body suggests the presence of electrical synapses in the brain, not seen so far in the few ultrastructural observations made on doliolid brains. Much less is known of the doliolid nervous system and excitable epithelia than is known in salps. In a late oozooid of *Doliolum* 50–70 large neuron somata were found around the cortex of the brain (Bone, 1959); by analogy with salps these are probably motoneurons. They lie externally to a larger number of much smaller neurons. Although there is a small nerve passing to the viscera, it appears that the brain is mainly concerned with the control of the locomotor muscle bands (see next section).

4.3.2. *Peripheral nervous system*

The endostyle is innervated via nerve bundles passing out of the brain anteriorly and running to the front of the endostyle around the peripharyngeal bands, and Fedele (1923a) described (but did not figure) a visceral nervous system. An unpaired median posterior nerve passes from the brain to the fold supporting the gill apparatus, and is in connection with a small group of three neurons lying near the junction of the fold and the median partition separating the gill apertures of each side (personal observations). The connections of these neurons are not yet known, but they may provide the axons that run around the edges of the gill bars; the ciliated cells (which are coupled by gap junctions) are innervated only at the edges of the bars. The gill cilia show arrests triggered by vibration, but as in *Oikopleura* (Galt and Mackie, 1971; see section 4.5.1) regular patterns of 'spontaneous' synchronous ciliary arrests accompanied by CAPs seem to result from pacemakers intrinsic to the viscera (possibly the axons of the neurons innervating the bars), since they continue after ablation of the brain (Fedele, 1923a; Bone and Mackie, 1977). This is a notable difference to ascidians, where ciliary motor neurons lie in the brain (Arkett, 1987). Curiously, the occasional rapid contractions of the locomotor musculature which deform the gill bars do not alter the regular pattern of arrests. Fedele suggested that receptor cells for ciliary arrest lay near the oesophageal lips (where ciliated sensory cells are found; personal observations); possibly some of the visceral nerve fibres

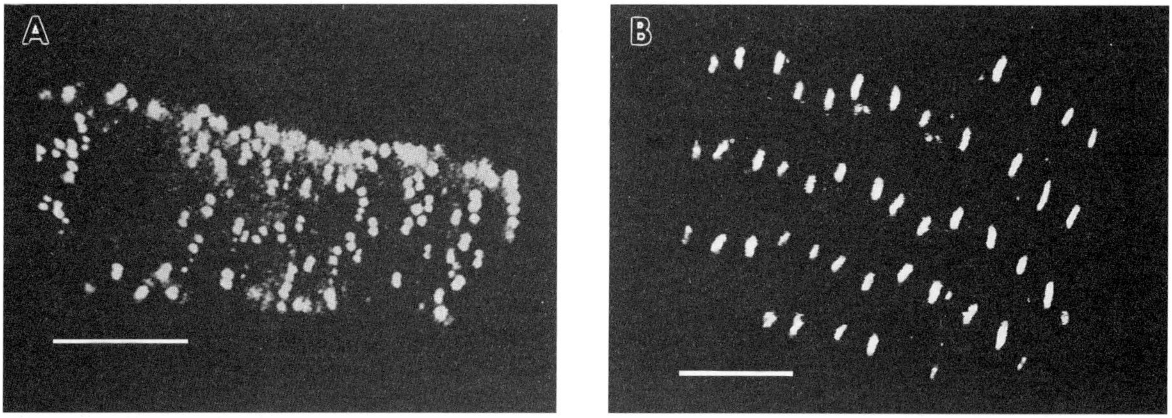

Fig. 4.4. Stimulated emission of light from colonies of *Pyrosoma atlanticum* (a), and *P.verticillata* (b). Note rows of double light sources in *P.verticillata, whereas in P.atlanticum* they are not resolved into twin sources since the zooids are closer together. From Bowlby *et al.* (1990) reproduced with permission.

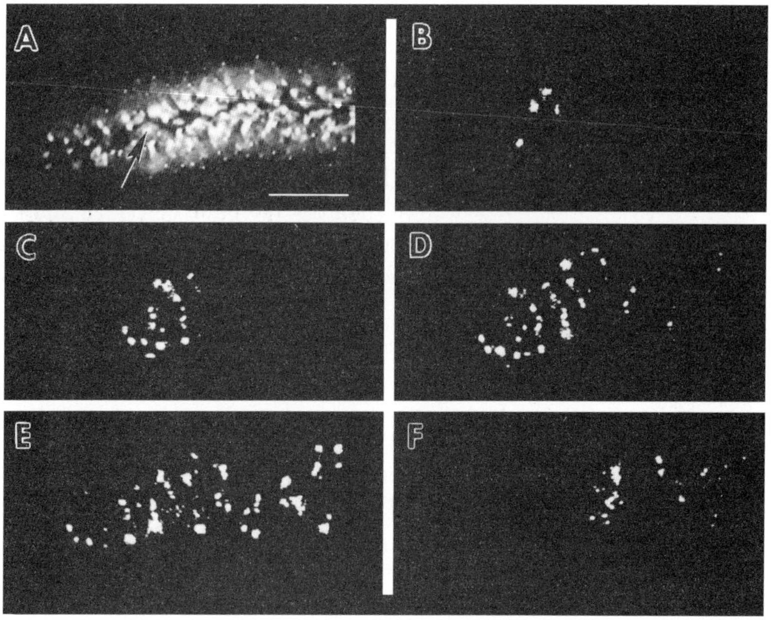

Fig. 4.5. Bioluminescent wave travelling across colony of *P.atlanticum.* (a) Colony seen in red light, arrow indicating stimulus site. (b) At time 0; (c) at 3 s; (d) at 6 s, wave spreading in both directions; (e) 9 s; (f) 15 s, response decaying. From Bowlby *et al.* (1990).

may be the axons of these cells. The patterns of CAPs in *Doliolum* are shown in Fig. 4.6.

4.3.3. *Sensory cells*

The greater transparency and smaller size of doliolids as compared with salps makes it possible to use Nomarski or phase microscopy to see sensory cells in living intact animals, and doliolids have a number of different sensory cell types, including an otocyst in the oozooid stage (section 1.3.1) although all stages lack eyes. All sensory cells so far as is known are of the usual tunicate ciliated type, although those in the external epithelium of the body surface (which may occur singly or in small groups) differ from other tunicates in being accompanied by 'supporting' cells, possibly differentiating sensory cells. At front and rear lips there are groups of sensory cells in a definite pattern (Fig. 4.7).

At the front lips, triads of sensory cells without accompanying supporting cells lie at the base of each scalloped extension of the lips; their cilia are aligned obliquely backwards when the lips lie open forwards. Fine nerve

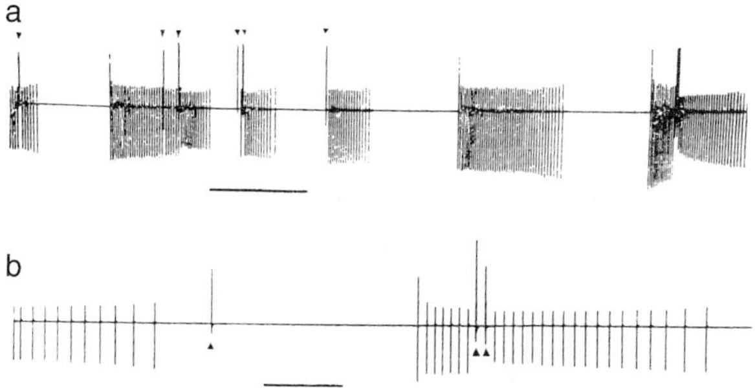

Fig. 4.6. *Doliolina mülleri.* Extracellular suction electrode records of ciliary arrest potentials (CAPs); electrode tip on outer surface of test. (a) Series of CAP bursts; note decline in frequency during burst. Scale bar 60 s. (b) Series of CAPs showing inhibition in second burst by muscle contractions (arrowed). Scale bar 10 s. From Bone and Mackie (1977).

fibres pass to the edges of the lips, and appear to terminate at the bases of the cells edging the lips. These bear short stout processes which are in constant irregular movement (like those of the olfactory mucosa of vertebrates; Pomerat, 1957); these interesting cells deserve further study, and may be the chemoreceptors known (by experiments with dissolved chemicals) to be present on the front lips. On the rear lips, the arrangement is different, for some of the sensory cells are accompanied by supporting cells, whilst others with shorter processes are not.

Special groups of 5–10 sensory cells lie at the base of the cadophore stalk of the oozooid, with their ciliary processes pointing away from the oozooid along the stalk. This cadophore organ, discovered by Uljanin (1884), is present in unhatched oozooids as well as in the old nurses; its function is unknown. Finally, there is an otocyst in the oozooid stage only, formed by an epidermal invagination; unfortunately its structure is not known in detail. As mentioned in section 3.3.3, both oozooids and gonozooids adopt the same oblique head up attitude at rest, hence the otocyst is more probably a vibration receptor than a gravity receptor.

As will be evident from this brief description, whilst the small size and extreme transparency of living doliolids makes them suitable objects for light microscopy of the nervous system, this is still poorly known, and little more has been added in this century to Uljanin's (1884) classical account. What is needed now (apart from further physiological studies) are ultrastructural investigations to resolve such questions as the status of the 'supporting' cells, the organization of the brain, and the structure of the otocyst. Examination of the otocyst would be interesting, to see whether it is similar in structure to that of appendicularians (section 4.5.1).

4.4. Salps

The salp nervous system and conducting excitable epithelia are linked to form a complex interacting system, for not only do the epithelial action potentials 'enter' the brain (probably via sensory cell axons) and alter behaviour, but motor axons from somata within the brain synapse with nearby epithelial cells and 'drive' epithelial action potentials, which then pass across the epithelium from the vicinity of the brain. This 'epithelio-motor' system is unique to salps, and is found in both oozooid and blastozooid stages, although its function is only understood in blastozooid chains. Unfortunately,

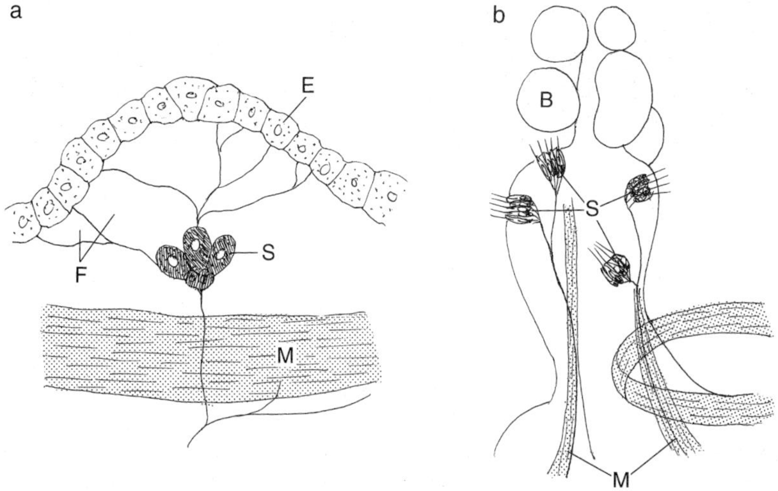

Fig. 4.7. The sensory cells of doliolids. (a) At the front lips. F: Fine fibres. It is still unclear whether these are free nerve endings amongst the epithelial cells (E), or whether (more likely) they are derived from the sensory cell triad (S). M: first muscle band. (b) Cadophore organ. B: buds, M: muscles of cadophore stalk, S: sensory cell groups. After Uljanin (1884) and Bone (1959).

knowledge of the histology of the salp peripheral and central nervous system is still fragmentary, and nothing is yet known of the circuitry within the brain.

4.4.1. *Central nervous system*

As seen in section 1.4.1, the circular brain lies dorsally, and from it mixed nerves radiate to supply the muscle bands and viscera, containing also the axons of peripheral ciliated sensory cells. The brain consists of a cellular cortex with a neuropil core; unfortunately nothing is yet known of the connections of the cortical neurons in the neuropil; the simple diagram in early work by Fedele (1933d) on the large *S. maxima* remains the only figure of the neurons within the salp brain (Fig. 4.8a and b).

Fedele (1933b,d) found that after destruction of the central core of the brain the normal swimming rhythm was not interrupted, but was no longer modifiable by sensory input; thus it appears that the rhythm is generated by neurons whose somata lie in the cortex, and that the core is the site of synaptic input from peripheral sensory cells. Similar observations have more recently been made by Mackie and Bone (1977). Both large and small neurons are found in the cortex, large cells staining supravitally with methylene blue around the posterior border of the brain (Fig. 4.8c) are presumed motoneurons, but their connections are lost within the central neuropil. Both tracer (e.g. horseradish peroxidase) and immunocytochemical studies on the brain neurons have yet to be carried out and should prove of much interest, since routine silver impregnation techniques have so far failed on the salp brain (personal observations). Intracellular records have, however, been obtained from brain motoneurons (Anderson, P.A.V. *et al.*, 1979), these are active in bursts of 8–15 spikes corresponding to the bursts recorded en passant in the nerves and to the bursts of muscle potentials seen preceding muscle band contraction (Fig. 4.9). Partial impalements of brain motoneurons show that their rhythmic activity is affected by skin impulse potentials (OSPs) in the outer epithelium.

Single epithelial action potentials do not appear to affect this rhythmic activity (though they may perhaps do so in unfatigued zooids in the sea), but two or more successive epithelial action potentials interrupt the rhythm and may increase firing rate after a short pause (Fig. 4.9d). Input from the eyes (changes in light intensity) alters or abolishes the regular rhythmic activity of presumed motoneurons (Fig. 4.9e). A second type of brain neuron has also been impaled, and these fire regularly at 1–2 Hz (with or without pre-potentials, i.e. some are driven and others fire spontaneously; Fig. 4.10a–c). It seems probable that such neurons are pacemaker neurons, although their rhythmic activity (1–1.7 Hz) is at a higher frequency than the locomotor rhythm.

4.4.2. *Peripheral nervous system*

The nerves passing to the muscle bands (section 3.4.2) contain small diameter axons (0.1–3.0 μm) and are not invested by sheath cells. Some of these axons are those of peripheral primary sensory cells, but these cannot be distinguished from the motor axons' although it is possible that the sensory axons are the largest fibres in the mixed nerves. Motor innervation of the muscle bands is intercalary and cholinergic (section 3.4.2).

Nerve bundles from the brain are also in connection with the endostyle (Fig. 1.17) and with the visceral

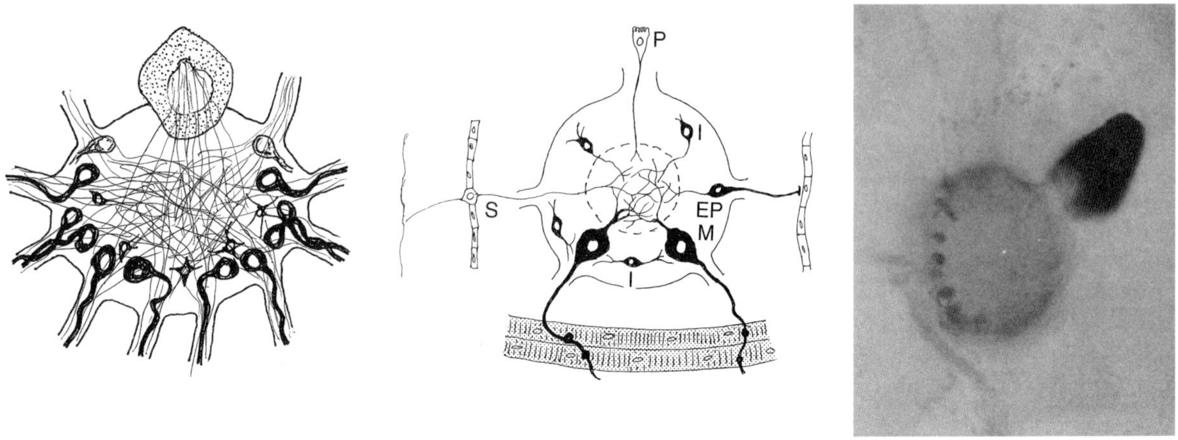

Fig. 4.8. Neuron types in salp brain. (a) *Salpa maxima.* Drawing from osmic–acetic preparation showing large motor cells and smaller 'associative' cells. (b) Interpretative scheme of preparations such as (a) with the addition of epithelio-motor cells. E, eye; EP, epithelio-motor cells; I, internuncial cells; M, motor cells; S, sensory cells. Both after Fedele (1933d). (c) *Thalia democratica.* Large motor cells in the brain stained supravitally with methylene blue.

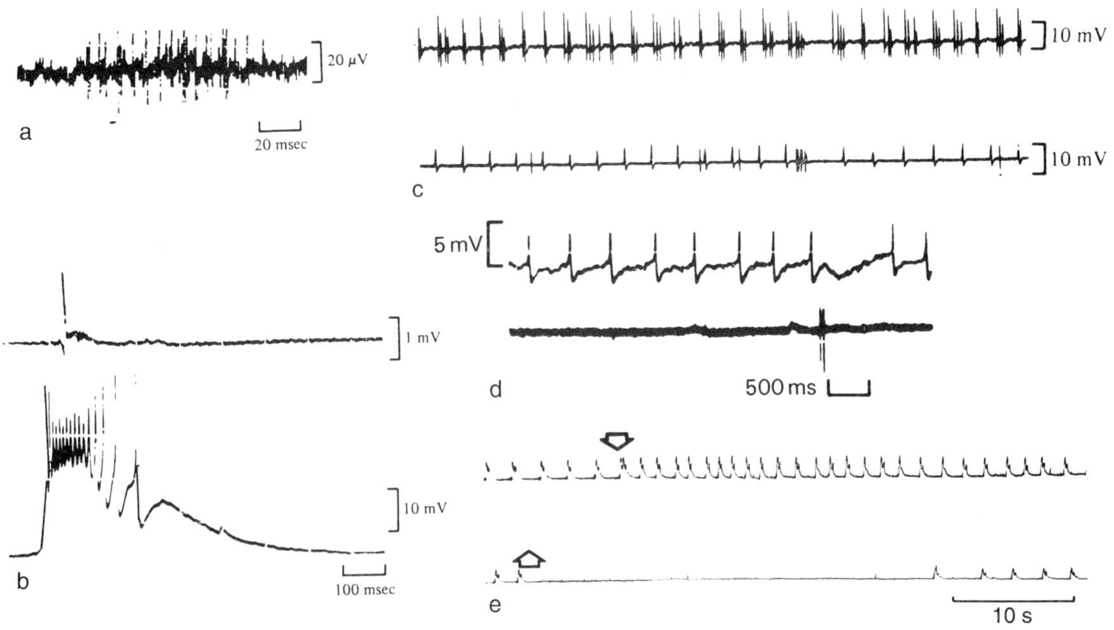

Fig. 4.9. *Salpa fusiformis*. Records from brain motoneurons and motor activity in nerve passing from brain. (a) *En passant* record from nerve bundle near muscle band. (b) Simultaneous intracellular record from brain motoneuron (lower) and muscle band (upper). (c) Similar record at lower recording speed (in this case the electrode tip lay close to but did not impale the motoneuron). (d) Two outer epithelial potentials (OSPs) (lower line) interrupt firing rhythm of partially impaled brain motoneuron. (e) Effects of light off (upwards arrow) and light on (downwards arrow) on firing frequency of presumed brain motoneuron. (a), (b) and (d) from Anderson *et al.* (1979); (e) from Mackie and Bone (1977).

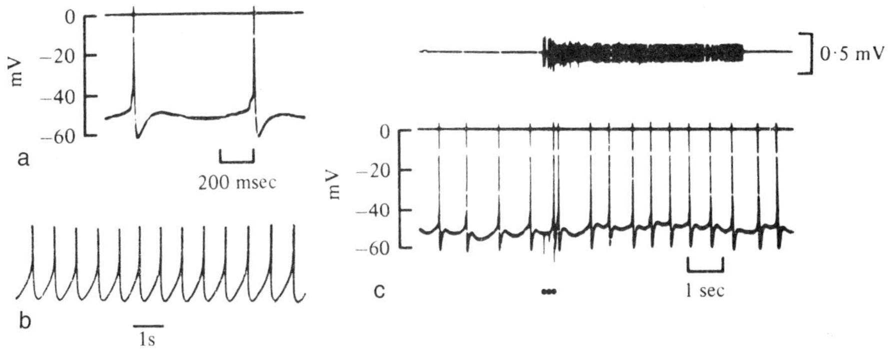

Fig. 4.10. *Salpa fusiformis*. Pacemaker brain neurons. (a and b) Intracellular records showing pre-potentials in (a) and their absence in another cell (b). (c) Effects of burst of outer epithelial potentials (OSPs) in upper trace on firing of pacemaker neuron (lower trace). (a) and (c) from Anderson *et al.* (1979); (b) from Mackie and Bone (1977).

plexus of neuron somata on the stomach (Fig. 4.11) described by Fedele (1933, 1938a). Fedele (1938a) claimed that the visceral plexus was very similar in arrangement to the vertebrate system, but this view is certainly premature, and later work (Bone, 1959), whilst confirming the presence of small multipolar neurons on the gut and base of the endostyle (much smaller than those regarded as neurons by Fedele), did not succeed in tracing their connections. Perhaps those on the stomach are secretomotor (the most likely function of the nerve fibres supplying the endostyle, since secretion of the feeding filter by the endostyle is blocked by MS222 anaesthesia). Recent immunocytochemical work on the ascidian *Corella* (Mackie, 1995) has shown a GnRH-like positive plexus containing neuron somata on the viscera; tests on salps (and doliolids) using the same GnRH antisera have failed to reveal any immunore activity (Mackie, personal communication and personal observations). The endostylar nerves are presumably cholinergic, since they stain with histochemical methods

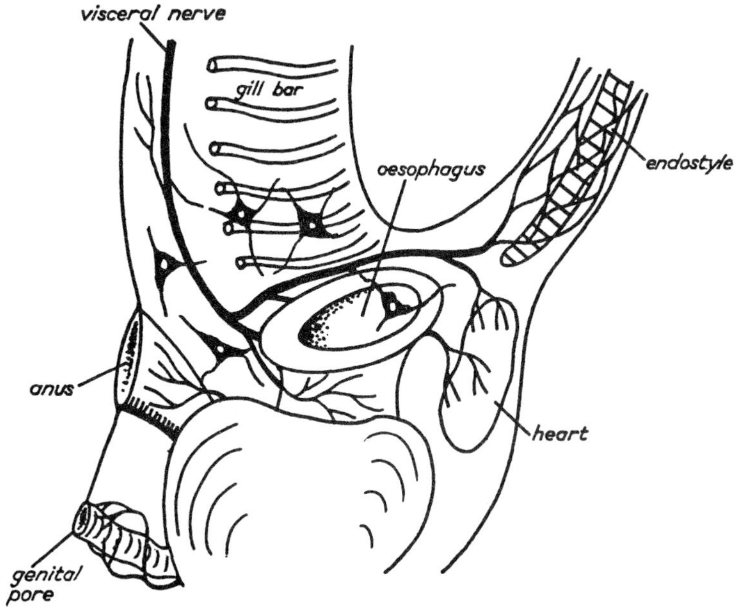

Fig. 4.11. Visceral plexus. After Fedele (1933b).

for cholinesterase (Fig. 1.17). Nerve fibres pass to the heart and course around the pericardium, but the heart muscle itself is probably not directly innervated.

In addition to motor axons to the muscle bands and possible secretomotor axons to the endostyle, motor axons from brain neurons pass to the outer epithelial cells around the brain, where they synapse at the bases of the epithelial cells (Fig. 4.12). These unique neuro-epithelial synapses evoke epithelial action potentials which spread over the outer epithelial sheet, as described in section 4.4.4.

4.4.3. *Sensory cells*

4.4.3.1. *The eye*

The eyes of salps, overlying the brain, are large conspicuous structures, the more so since their brownish black pigment is the most obvious feature in such transparent animals (see Fig. 1.14). Metcalf (1893) and Metcalf and Lentz-Johnston (1905) gave good morphological descriptions of the eyes of different species, and showed that the eyes differed much between species, particularly so in the blastozooids (Fig. 4.13), and in the same

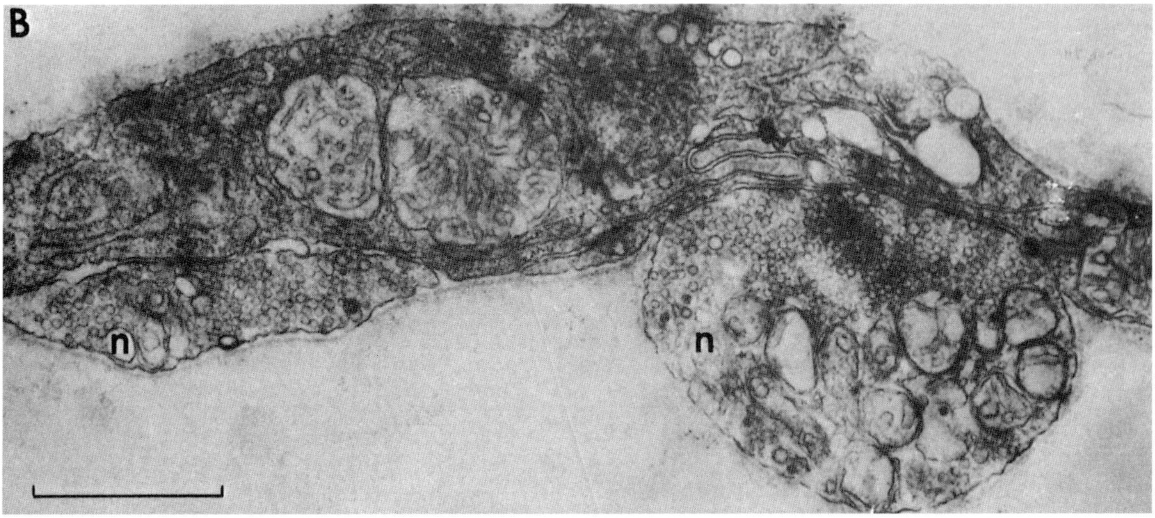

Fig. 4.12. *Salpa fusiformis.* Synapses of the axons of central neurons with cells of the outer epithelium around the brain. Scale bar 1 μm. From Bone (1982).

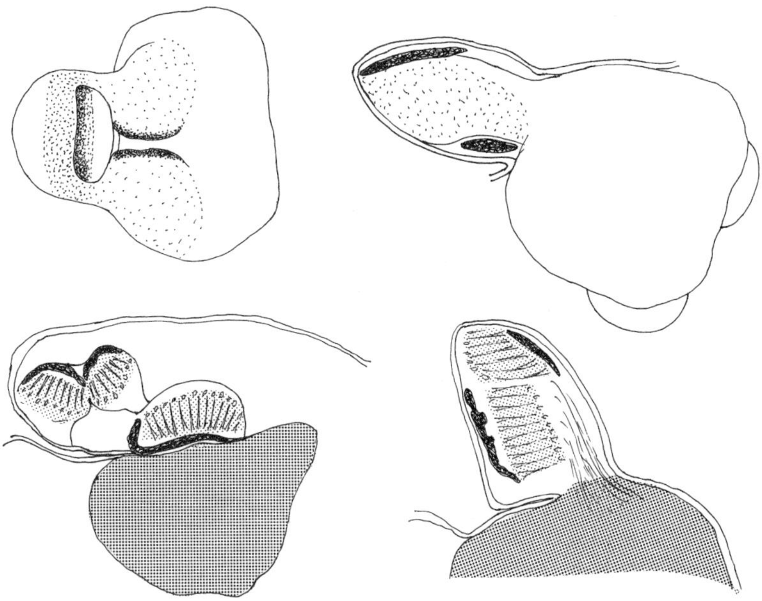

Fig. 4.13. Eye morphology in *Cyclosalpa pinnata* (left) and *Salpa fusiformis* (right). Dorsal views above and sections below. In sections: heavy stipple, brain; light stipple, receptor cells; black, screening pigment. After Metcalf and Lentz-Johnston (1905).

species, between oozooid and blastozooid. The photoreceptors (Fig. 4.14a) are unlike those described in ascidian larvae (those of *Pyrosoma* have not been studied), since the folded microvilli at the end of the cell furthest from incident light (the receptor cells are partially screened by pigment cells) are not derived from cilia. No lens is present, although some directional sensitivity is provided by different positions of the

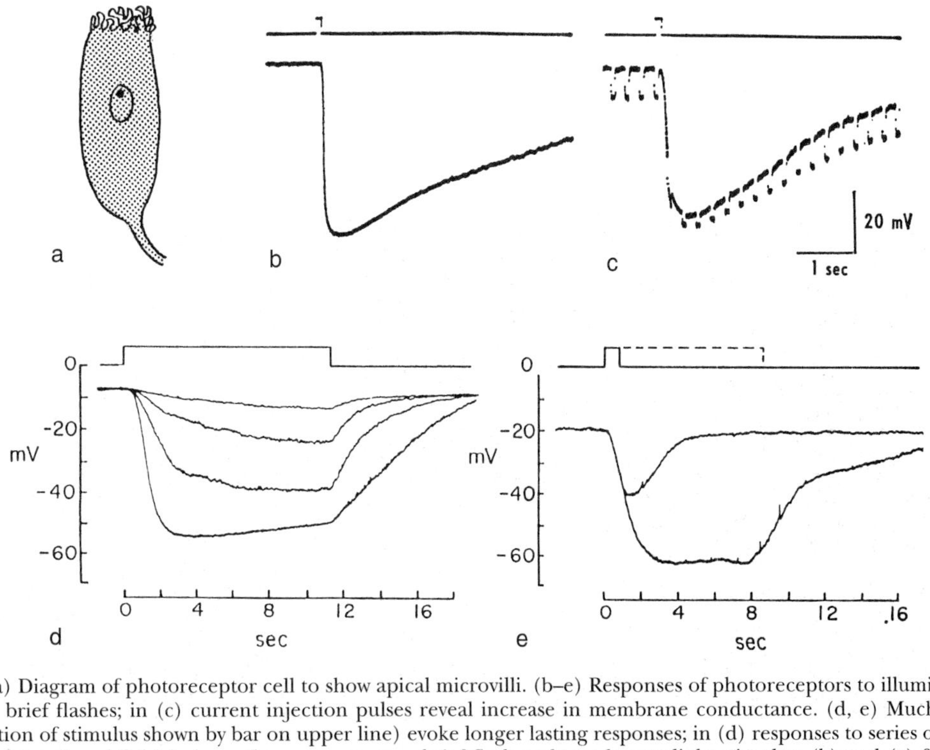

Fig. 4.14. (a) Diagram of photoreceptor cell to show apical microvilli. (b–e) Responses of photoreceptors to illumination. (b, c) Responses to brief flashes; in (c) current injection pulses reveal increase in membrane conductance. (d, e) Much longer light stimuli (duration of stimulus shown by bar on upper line) evoke longer lasting responses; in (d) responses to series of light stimuli at increasing intensity, whilst (e) shows the responses to a brief flash and to a longer light stimulus. (b) and (c) from Gorman, McReynolds and Barnes (1974); (d) and (e) from McReynolds and Gorman (1975).

pigment screen around different arrays of the photoreceptor cells. Since the eyes are above the brain and relatively large, they are accessible for recording, and Gorman *et al.* (1974) have shown that brief flashes of white light evoke large hyperpolarizations up to 70 mV from resting potentials around 10 mV, with an increase in membrane conductance (Fig. 4.14b and c). McReynolds and Gorman (1975) subsequently showed that in dim light, changes in illumination resulted in much slower longer lasting responses (Fig. 4.14d and e). The ionic basis of this change has not been studied, nor has the spectral sensitivity of the receptors. Changes in light intensity alter locomotor behaviour and alter the firing pattern of the rhythmically active brain motoneurons (see Fig. 4.9e), but the connections of the photoreceptors with the motoneurons are unknown.

4.4.3.2. *Other receptors*

The other sensory cells in salps are apparently all of the same morphological type, bearing long cilia arising from a deep invagination of the apical surface of the cell. Unlike those of doliolids (see p. 58), they are not accompanied by 'supporting' cells. With the exception of those of the attachment plaques of the blastozooids (see next section)' they have normal cilia with a 9+2 tubule array. Sensory axons pass to the brain from the basal region of the cells, partially invaginated into the bases of adjacent epithelial cells before they join the mixed nerves from the brain. Such sensory cells are most abundant on the rear and front lips (where they may be organized into small groups; Fedele, 1933a), but they also occur singly over most of the outer epithelium (Fig. 4.15b), where they are often rather flattened and partially overlain by adjacent epithelial cells. There is a special cupular organ in salps (Fig. 4.15a) which has a cushion of sensory cells at its base (Bolles Lee, 1891; Streiff, 1908), but apart from this organ, the eyes, and the junctional plaques of the blastozooids, there are no special sense organs known; the statocysts found in doliolids and in larvaceans are lacking in salps, which

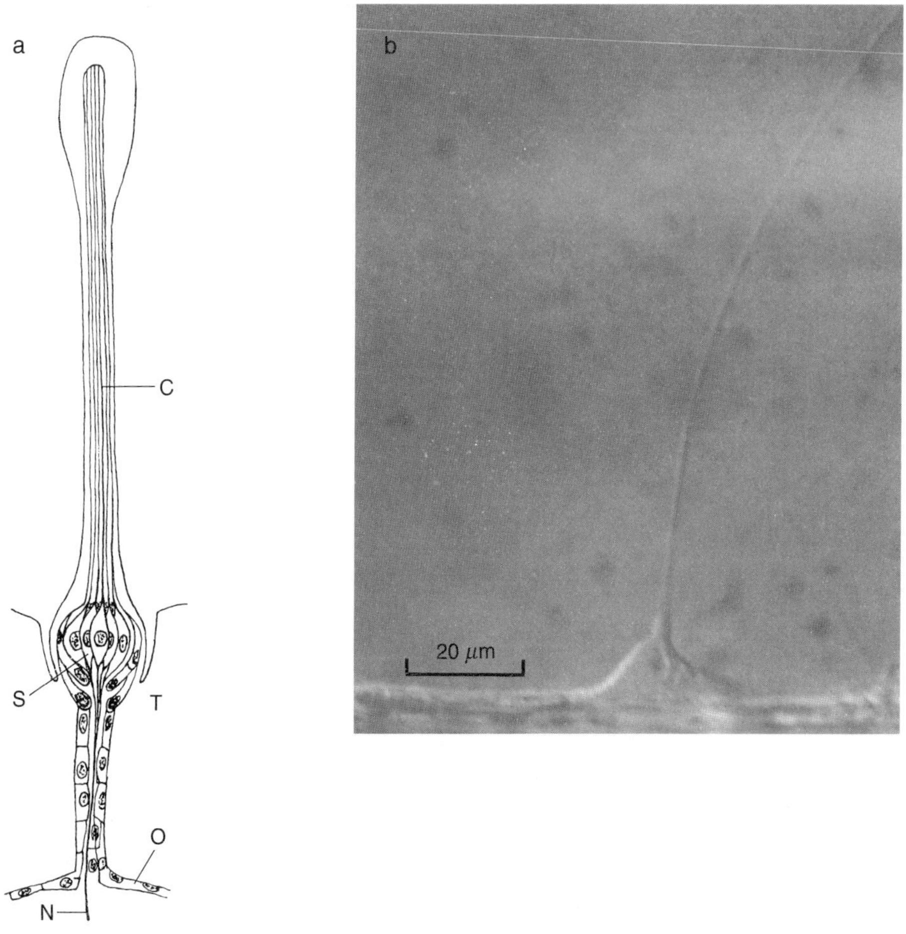

Fig. 4.15. Sensory cells. (a) The cupular organ. N, axon bundle from sensory cells; O, outer epithelium; S, sensory cell cluster; T, test. After Bolles Lee (1891). See Fig. 3.7 for position of the organ in a blastozooid. (b) Sensory cell with its process extending through test. Living, Nomarski interference contrast.

nevertheless normally swim horizontally, dorsal surface uppermost.

Little is known directly of the function of these sensory cells, although it seems clear that the majority must be mechanoreceptors, acting in conjunction with the skin pulse system of the mechanosensitive epithelia (section 4.4). Usually, mechanical stimulation of the surface of salps evokes skin impulses in the outer epithelium at the stimulus site, but on rare occasions delicate mechanical stimulation of the rear lips evokes skin impulses driven from the brain, rather than at the site of stimulation, presumably via direct mechanical stimulation of sensory cells, without evoking skin impulses. It would seem probable that the cupular organs respond to water movements (freshly collected salps are sensitive to nearby water movements), although gentle bending of the organ in salps collected in tow nets (and therefore possibly fatigued or damaged) does not lead to any obvious change in locomotor behaviour. Some of the sensory cells of the front lips are presumably chemosensitive, since salps show reversals when dilute chemicals such as the anaesthetic MS222 are applied to the front lips, but such chemoreceptor cells have not been identified morphologically.

4.4.4. *The epithelial sensory system*

4.4.4.1. *Physiological aspects*

Both the outer (ectodermal) and part of the inner (endodermal) epithelial layers of salps are mechanosensitive and propagate action potentials (skin impulses) equally in all directions from the point of stimulation (Mackie and Bone, 1977). The epithelia consist of flat polygonal cells 1–2 μm or less thick, coupled by gap junctions (Fig. 4.16); mechanical or electrical stimulation evokes rapid overshooting action potentials which can most simply be recorded extracellularly by fine polyethylene suction electrodes (even through the test in the case of the outer skin impulses!).

Outer skin impulses (OSPs) propagate at around 17 cm s^{-1}. Intracellular records of OSPs have been obtained from the outer epithelial layer around the brain (where the cells are thicker than elsewhere, 2.5–5 μm), these are seen in Fig. 4.17a, and are characterized by rapid rising and falling phases with no subsequent undershoot.

Although the ionic basis of OSPs has not been examined in mature oozooids or blastozooids, the dimensions of the epithelial cells in the stolon of oozooids (where the epithelial cells are cuboidal and 4–10 μm across) permit stable intracellular recording, and Anderson (1979) showed that tetrodotoxin (TTX), a Na^+-channel blocker, reversibly blocked OSPs, and that the Ca^{2+}-channel blocker Mn^{2+} diminished the magnitude of the OSP as well as lengthening the recovery phase of the potential. It therefore appears that OSPs in the stolon are largely carried by Na^+ (as in oikopleurid appendicularians; section 4.5.3) and that an inward Ca^{2+} current is also present. In the immature stolon (unlike the epithelium of mature zooids), successive stimuli at frequencies above 2 Hz produce a pronounced plateau following the rising phase of the second OSP (Fig. 4.17b and c), perhaps due to a poorly developed delayed rectification mechanism in this immature tissue, although other interpretations are possible.

The skin impulses of the inner epithelial layer (ISPs and ASPs; see Fig. 4.18a), do not travel over the whole of its area; the epithelium is instead divided into three separate zones, an anterior zone between the lips and the peripharyngeal bands, and left and right zones covering the ventral portion of the branchial chamber and each side of the gill bar (Fig. 4.18a). The upper part of the branchial chamber (except for the gill bar) is inexcitable, as is the posterior region behind the base of the gill bar. The inner epithelium does not extend into the anterior and posterior blastozooid horns, which are lined by outer epithelium only.

ASPs in the anterior region of the lips inhibit the regular rhythmic bursts of forward swimming (Fig. 4.18b),

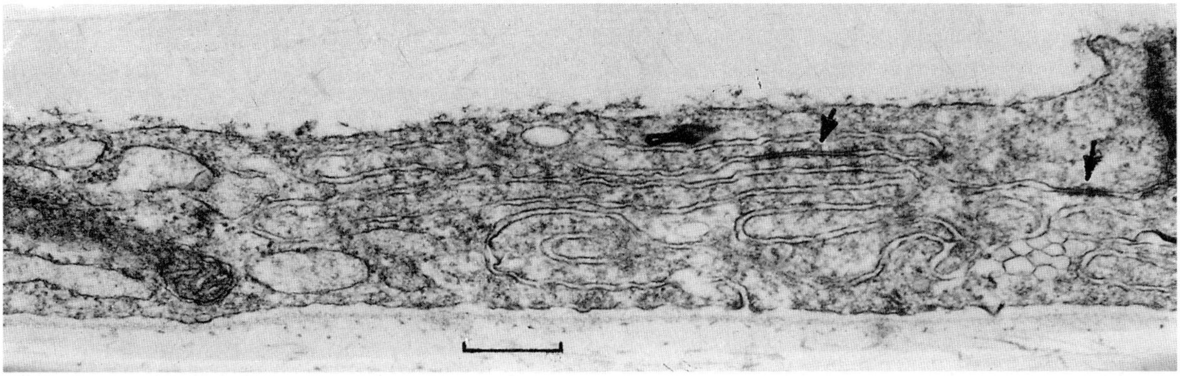

Fig. 4.16. *Salpa fusiformis.* Junction between outer epithelial cells showing gap junctions (arrow). Scale bar 1 μm.

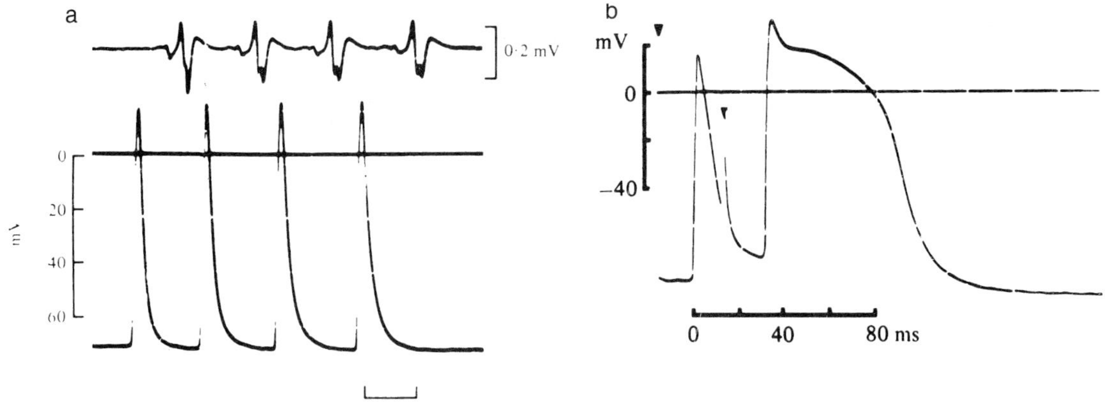

Fig. 4.17. *Salpa fusiformis.* (a) OSPs from mature blastozooid. Above, extracellular suction electrode record; below, intracellular record from epithelial cell near brain. Scale bar 20 ms. (b) Intracellular record of OSPs from immature stolon of oozooid. Note plateau on second potential (see text). From Anderson *et al.* (1979) and Anderson (1979).

and are presumably normally evoked by large objects inhaled into the front of the pharynx. It is not yet clear how the ASPs 'enter' the central nervous system to produce these changes in locomotor behaviour, but the most probable route is via the axons of the sensory cells in the inner epithelium of the front lip valves. This presupposes that gap junctions occur between sensory cells and adjacent epithelial cells either in the lips or (in the case of the OSPs) in the outer epithelium, but such junctions are yet to be demonstrated.

Although Fedele (1933b) stated that sensory cells are present in the inner epithelium, Mackie and Bone (1977) were unable to find them there, and ISPs in the other regions of the inner epithelium have no effect on locomotor activity. It is possible that ISPs may 'enter' the visceral nervous system via sensory cells around the oesophageal lips and alter the activity of the oesophageal cilia which draw the feeding filter into the stomach (Chapter 5). However, in doliolids, ISPs are present in the old nurse stage where the gut has been lost, whilst they are lacking in the other stages where it is present.

Unlike ISPs, OSPs in the outer epithelial layer underlying the test play an important role in locomotor behaviour. In both the solitary oozooids and in chains of blastozooids, the effects of OSPs evoked by mechanical stimulation vary according to the site of stimulation. If the oozooid is stimulated posteriorly, the frequency and amplitude of the regular bursts of muscle potentials greatly increases, chamber pressures are markedly increased, and the animal accelerates forwards. Conversely, if stimulated anteriorly, a series of OSPs leads to reversal with increased frequency and amplitude of muscle potentials and chamber pressure, and the animal swims backwards. These responses are seen in Fig. 4.18c. In either case, the OSPs can be recorded from any point on the outer epithelium, but presumably refractory periods in their link with the brain motoneurons ensure that the salp responds appropriately to the first OSP

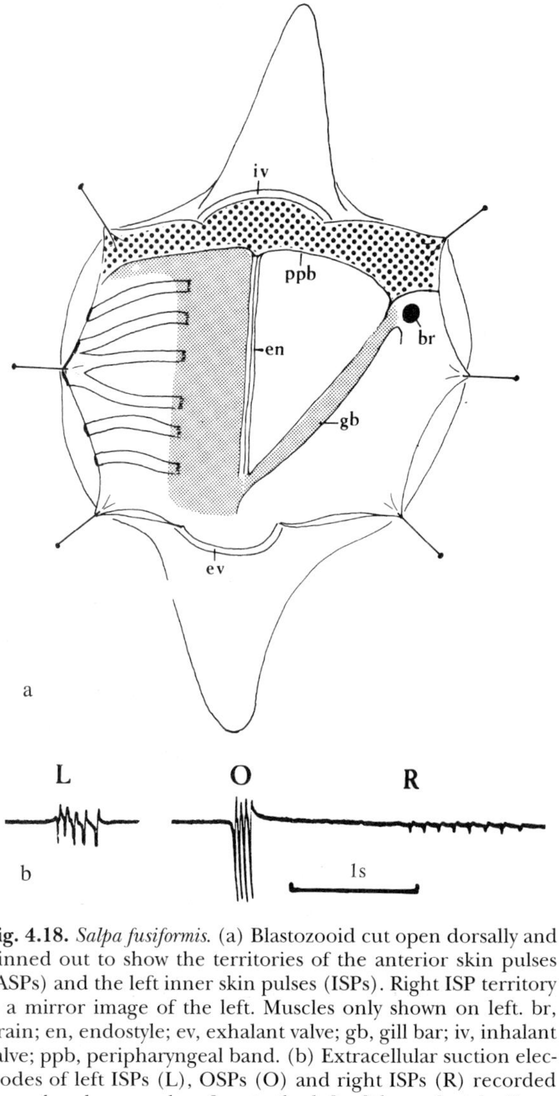

Fig. 4.18. *Salpa fusiformis.* (a) Blastozooid cut open dorsally and pinned out to show the territories of the anterior skin pulses (ASPs) and the left inner skin pulses (ISPs). Right ISP territory is a mirror image of the left. Muscles only shown on left. br, brain; en, endostyle; ev, exhalant valve; gb, gill bar; iv, inhalant valve; ppb, peripharyngeal band. (b) Extracellular suction electrodes of left ISPs (L), OSPs (O) and right ISPs (R) recorded from the pharyngeal surface to the left of the endostyle. From Mackie and Bone (1977).

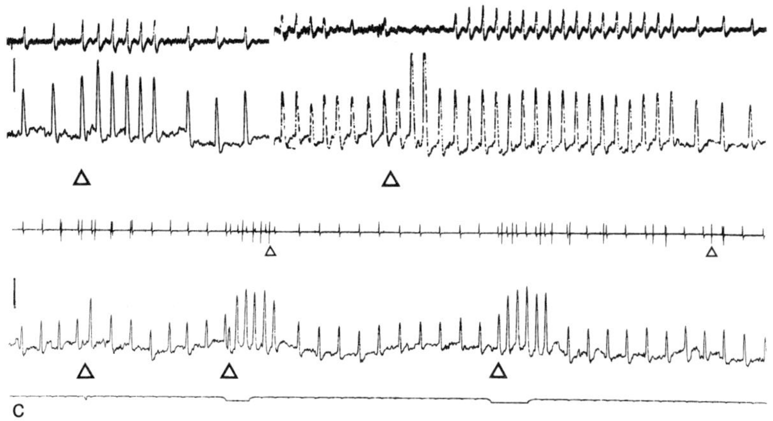

Fig. 4.18. (c) The effects of mechanical stimulation on the locomotor rhythm of blastozooids. Above: simultaneous tension (upper traces) and chamber pressure records from a blastozooid tethered posteriorly. Left, posterior stimulation evokes increased tension and chamber pressure as the locomotor muscles contract at higher frequency. Right, anterior stimulation during a period of rapid forward swimming evokes a reversal (when tension disappears), contraction and chamber pressures are augmented but there is no increase in frequency. The two reverse pulses are followed by a series of rapid forward pulses before the normal slower rhythm reappears. Below: simultaneous records of electrical activity of front lip muscle (upper) and chamber pressure (middle) in a small blastozooid, when the front lips were stimulated as indicated on lower line. Here, the reverse pulses evoked are at higher frequency than normal forward swimming, and OSPs are seen on the upper line (two are indicated by small triangles) when the front lips are stimulated mechanically at points shown by large triangles. Note that the activity from the front lip muscle diminishes during reversed swimming. Vertical bars: 20 Pa, time-marker: 20 s. From Bone and Trueman (1983).

'entering' the brain. Absolute refractory periods for electrically stimulated OSPs are around 7 ms, and they can fire at frequencies up to 5 Hz, during strong stimulation evoking an OSP burst. An OSP evoked by stimulation of the rear lips, for example, would activate a nearby sensory cell, and the brain would receive this information prior to activation of other sensory cells on the body. Thus 'knowing' the site of origin of the stimulus, the salp can take appropriate action (causing the zooid to accelerate forwards), as it would if the sensory cell itself was stimulated directly. Skin impulses evoked by mechanical stimulation in epithelial sheets also play an important role in escape behaviour in oikopleurid appendicularians (p. 77), as well as in a variety of groups, including Hydrozoa and some vertebrate embryos.

In animals with an excitable outer epithelium, this serves to extend the receptor field of sensory cells and so the system is well suited to an 'escape' response evoked by mechanical stimuli over a wide area of the animal. Salps are unique, however, in showing OSPs which are driven by central epitheliomotor neurons, and this remarkable feature (found in both oozooid and blastozooid stages) has an important role in the behaviour of the blastozooid chains. Such chains stimulated posteriorly accelerate forwards, if anteriorly, reverse, whilst if a chain is cut in half, the anterior portion swims off forwards, the rear portion reverses. These responses are controlled by OSPs and, remarkably, OSPs travel along the chain of separate blastozooids at around 12.5 cm s^{-1} (Fig. 4.19).

Since the outer epithelia of the adjacent zooids in the chain are not linked by gap junctions, the manner in which skin impulses are able to pass along the chain (see below) is complex, and involves driven OSPs. The axons of the brain epitheliomotor neurons synapse with the bases of the outer epithelial cells around the brain. Their terminals contain electron-lucent vesicles (Fig. 4.12), and since injection of acetylcholine (10^{-4} M) into the space between innner and outer epithelial sheets around the brain evokes trains of skin impulses, whilst injection of *d*-tubocurarine abolishes driven skin

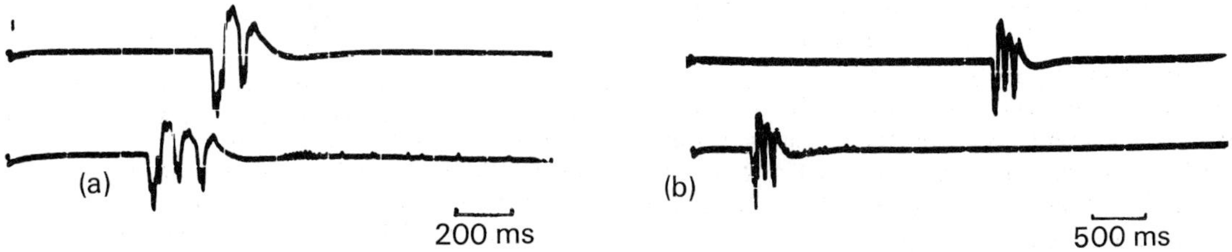

Fig. 4.19. *Pegea confoederata.* Suction electrode records of OSPs from blastozooid chain. (a) From two zooids separated by one zooid; (b) from two zooids with 20 zooids between them. From Mackie and Bone (1977).

impulses previously evoked by stimulation of the rear lips (without altering the skin impulse arising at the stimulus site), it seems probable that the synapse is cholinergic (Bone, 1982). Frequently, in both oozooids and blastozooids, mechanical stimulation evokes skin impulses arising at the stimulus site, followed by one or more 'driven' skin impulses arising from the innervated epithelial cells around the brain (Fig. 4.20).

The epithelial cells are thin and it is difficult to obtain stable impalements even in the thicker innervated cells near the brain, but successful records have been obtained showing excitatory presynaptic potentials preceding driven skin impulses (Fig. 4.20 bottom).

The function of this curiously complex system is not obvious in the solitary oozooid stage, but in the chains of blastozooids it is used to transmit escape signals from one zooid to another along the chain. As previously mentioned in section 3.4.1, individual zooids in blastozooid chains contract rhythmically but out of phase with each other, so that there is no obvious co-ordination between zooids. If stimulated mechanically, however, the activity of the zooids is co-ordinated to produce

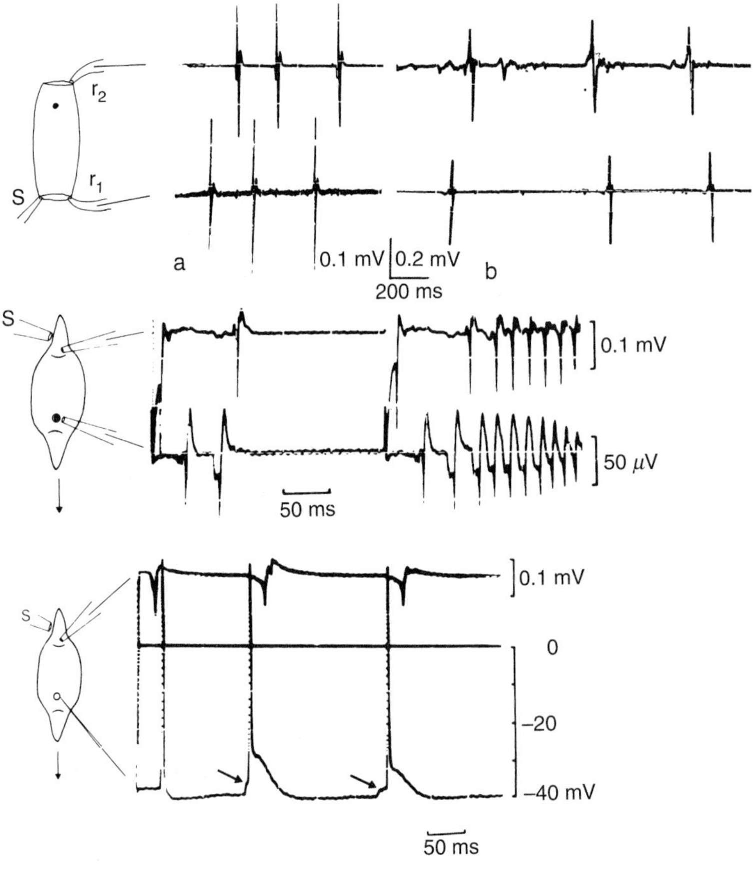

Fig. 4.20. *Salpa fusiformis.* 'Driven' OSPs. In both blastozooids and oozooids, stimulation first evokes OSPs which spread from the stimulus site, recorded first by a nearby electrode and secondly by an electrode near the brain. Then 'driven' OSPs are seen, arising near the brain, and recorded in the opposite sequence. Upper: oozooid. Recording extracellular suction electrodes R1 and R2 are attached to the front and rear lips; the solid circle is the anterior brain. A stimulating electrode S is attached to the rear lips. (a) 'Normal' OSPs arising at the stimulus site and recorded first by R1 (lower line) are recorded at an interval by R2. (b) The first two OSPs arise at the stimulus site, the third (recorded first by R1) is a 'driven' OSP arising near the brain. From Bone (1982). Middle: blastozooid. Similar arrangement as for oozooid. Upper trace, extracellular suction electrode close to stimulating electrode (S); lower trace, electrode close to brain. The first OSP is recorded first near the stimulus site, but the second is a 'driven' OSP and is recorded first near the brain as are those in the subsequent burst. Bottom: blastozooid. Upper trace extracellular suction electrode close to stimulus site, lower trace intracellular record from outer epithelial cell close to brain. The first OSP arises at the stimulus site and is followed by 'driven' OSPs with pre-potentials (arrows). The 'driven' OSPs are recorded first near the brain, and subsequently by the suction electrode (upper trace). From Anderson and Bone (1980).

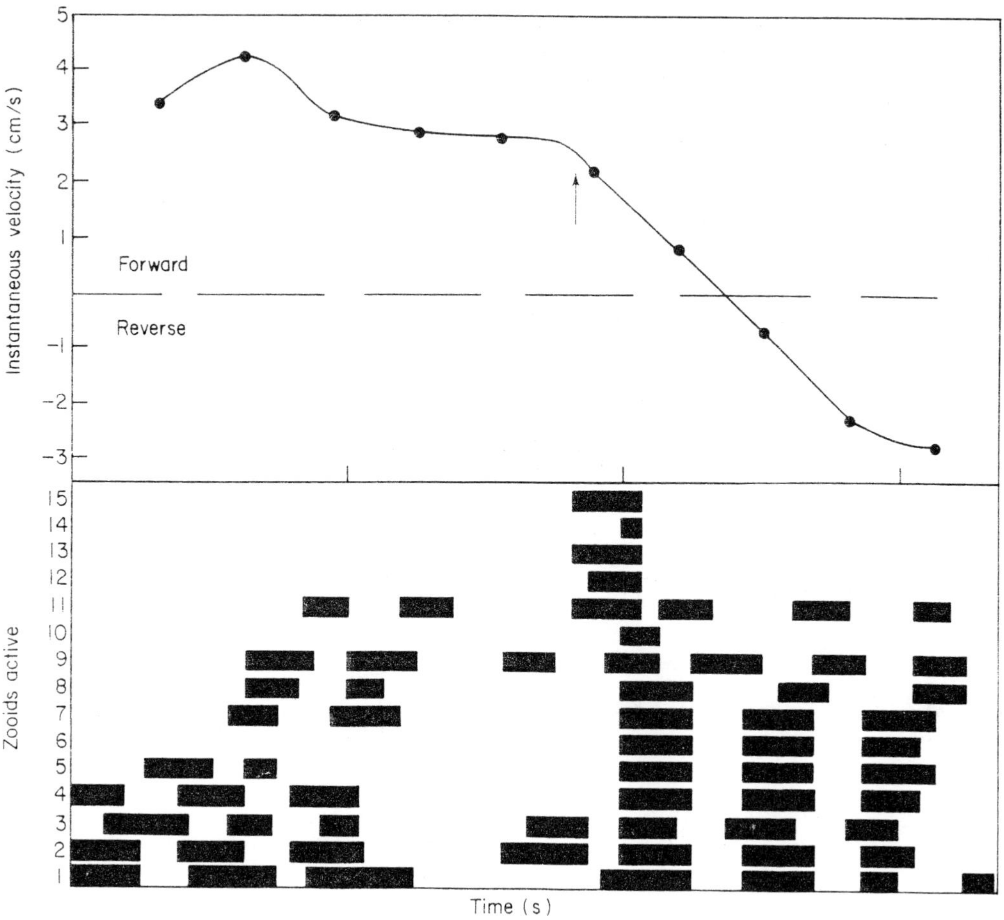

Fig. 4.21. *Salpa fusiformis.* Co-ordination of activity of the 15 members of a small chain of blastozooids when the chain is stimulated anteriorly to produce a reverse swimming pulse. Upper, record of instantaneous velocity; solid bars below indicate exhalant phase of jet cycle for each zooid. Note reverse pulses from all zooids including those previously inactive at the rear of chain. From Bone and Trueman (1983).

rapid swimming in the appropriate direction to remove the chain from the stimulus (Fig. 4.21).

4.4.4.2. *Morphology of the OSP pathway in the blastozooid chain*

In the chains of *Thalia democratica, Salpa fusiformis* and *Pegea confoederata* (the three species examined in detail; Bone *et al.*, 1980), each individual blastozooid is attached to its four neighbours by eight attachment plaques (section 1.4.3. and Fig. 4.22), i.e. to each neighbour by two plaques. At each plaque the outer epithelia of the adjacent zooids are 3–5 μm apart, separated by test material. The two plaques joining one zooid to another are not symmetrical, for towards one end of the plaque there are sensory cells on one side, and a sort of button of modified epithelial cells facing them on the other, and in one plaque the sensory cells lie on one side, in the other on the other side (Fig. 4.23).

On a single zooid therefore, there are four innervated plaques with sensory cells, and four with the button structures, linking the zooid to its four neighbours in the chain (Fig. 4.23). The cilia of the sensory cells are remarkably modified, for although they are of the normal 9+2 type as they leave the cell, they soon expand (Fig. 4.24a) and branch in the test material between the two zooids, to form a sort of bush of processes containing many tubules. These branches terminate in small expansions on the surface of the button cells of the other zooid, resembling synaptic structures (Fig. 4.24a). Under the button cell membrane at these junctions there are rows of electron-lucent vesicles. Morphologically, the junctions suggest that the button cells are

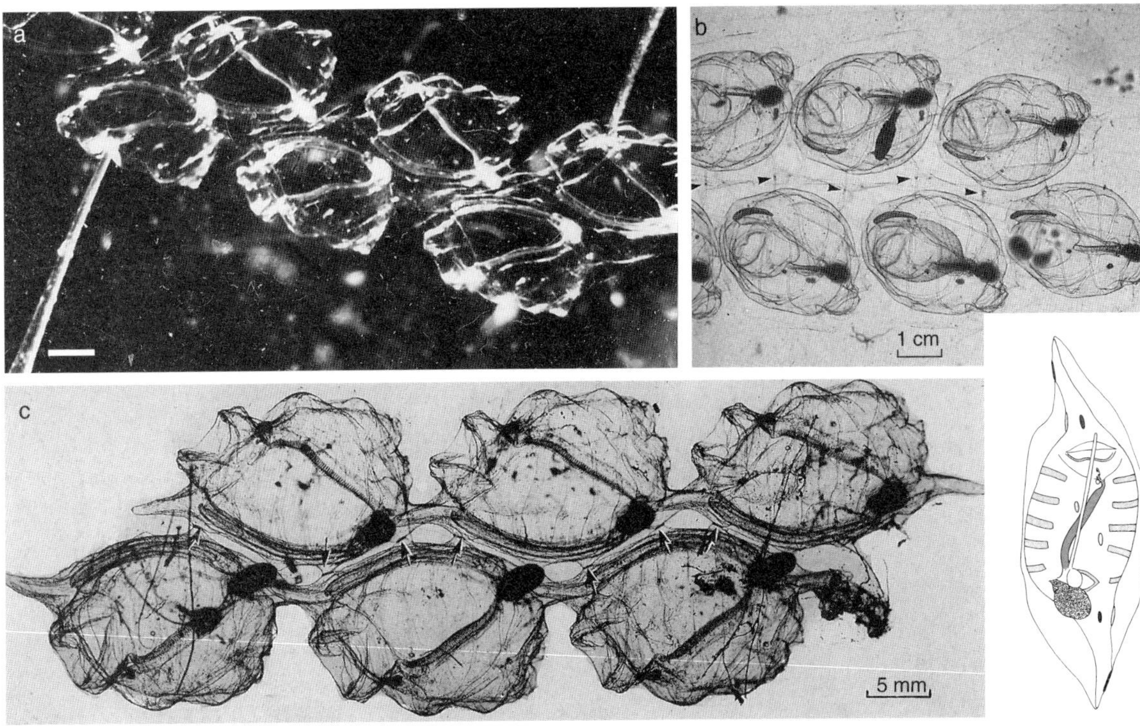

Fig. 4.22. (a, b) *Thalia democratica.* (a) Portion of living chain of young blastozooids attached to suction electrodes for recording passage of OSPs along chain. (b) Similar chain seen by transmitted light showing lateral attachments (arrowed) to neighbouring zooids. The blastozooid at bottom left has deployed its pharyngeal feeding filter. (c, d) *Salpa fusiformis.* (c) Fixed chain of young blastozooids, attachment plaques arrowed. Those between the horns of the adjacent zooids are not visible. (d) Position of the eight plaques on a blastozooid. The plaques shown solid are innervated. From Bone and Mackie (1977) and Bone *et al.* (1980).

presynaptic to the ciliary processes of the sensory cells, i.e. that the signal between zooids passes from the button cell side of the plaque to that with the sensory cells.

4.4.4.3. *Physiology of OSP system in blastozooid chains*
Physiological studies (Anderson and Bone, 1980) show that the signal indeed passes from the button cell (non-innervated) side of the plaque to the other, and that the morphological polarity of the plaques is reflected in their physiological polarity. When the blastozooid at one end of the chain is stimulated mechanically and a skin impulse evoked, this passes to the button cells of that zooid (which are coupled by gap junctions to the adjacent epithelial cells), and the sensory cells of the second zooid at the plaque are stimulated. These are not coupled to the adjacent epithelial cells; an OSP is not propagated across the epithelium from the sensory cells. Instead, via their axons, the stimulus enters the brain whence, epitheliomotor neurons drive an OSP from the cells around the brain across the zooid. This passes to the button cells of that zooid to fire the sensory cells in the plaque of the succeeding zooid in the chain (Fig. 4.25b). In this way, via alternating epitheliosensory and neuroepithelial junctions, OSPs are transmitted along the blastozooid chain, and escape reactions are rapidly transmitted along the chain (Figs 4.19 and 4.21). Although in anaesthetized chains of blastozooids it is quite difficult to tear the zooids apart at their attachment plaques, the zooids in unanaesthetized chains separate quite readily to swim off independently, and do so if damaging stimuli (evoking bursts of OSPs) are given to any zooid of the chain. Evidently, this separation has survival value when chains are attacked by such predators as the sunfish *Mola* or the stromateid *Tetraonurus*, both of which feed on salps (Chapter 12). The requirement for separation has perhaps precluded the development of a simpler system, for example, the direct apposition of the outer epithelia of adjacent zooids (which are all of identical genetic constitution) at the attachment plaques. The mechanism of separation remains obscure, and does not seem to involve muscular action; one possibility is that the actin filaments in the epithelial cells of the attachment plaques (personal observations) contract upon Ca^{2+} entry when bursts of skin impulses are propagated through the plaque, and so tear the blastozooids apart.

It is rather unexpected to find the unique salp system of 'driven' OSPs in the oozooid stage as well as in the blastozooids. In the solitary oozooid, one or more

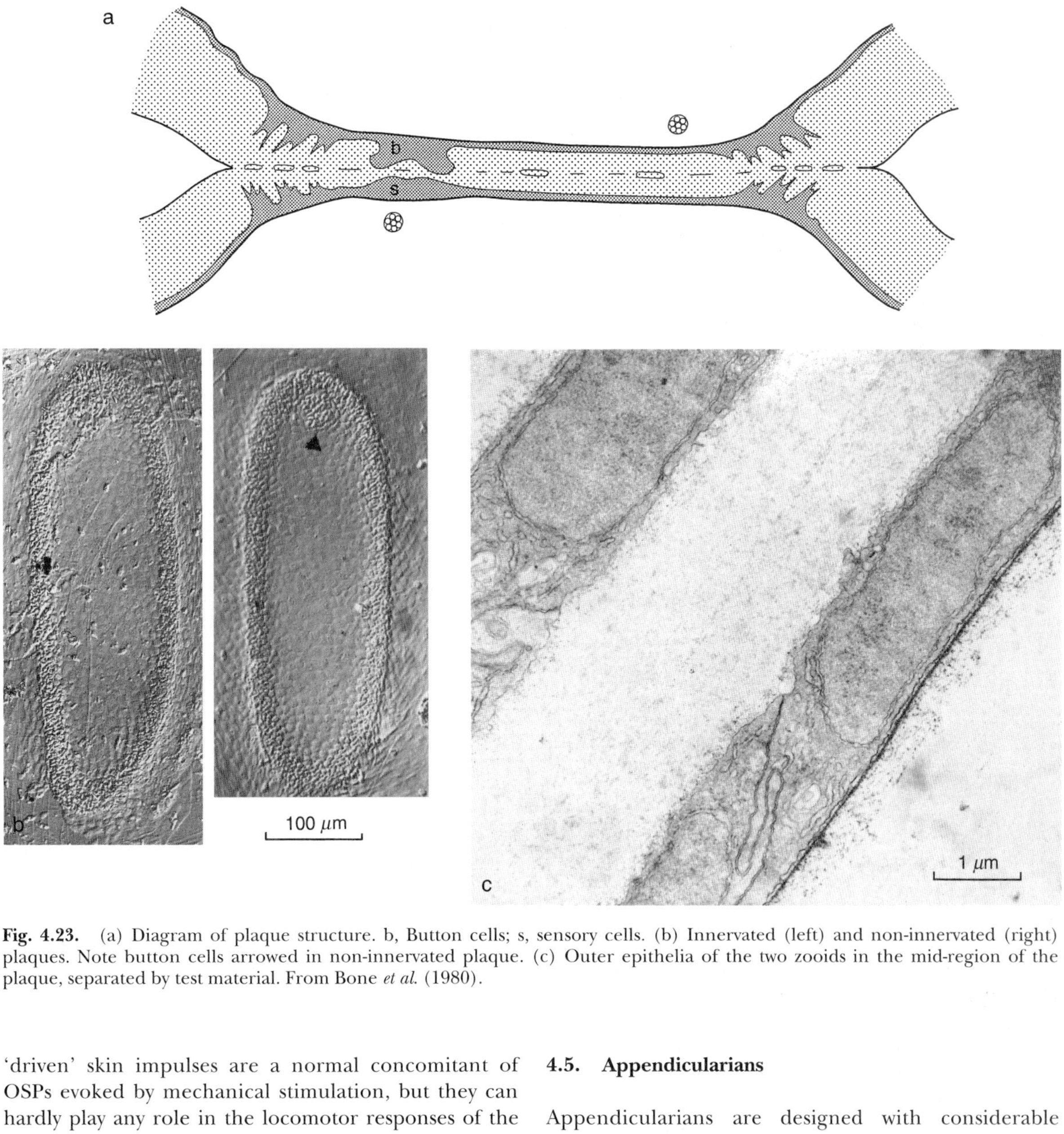

Fig. 4.23. (a) Diagram of plaque structure. b, Button cells; s, sensory cells. (b) Innervated (left) and non-innervated (right) plaques. Note button cells arrowed in non-innervated plaque. (c) Outer epithelia of the two zooids in the mid-region of the plaque, separated by test material. From Bone *et al.* (1980).

'driven' skin impulses are a normal concomitant of OSPs evoked by mechanical stimulation, but they can hardly play any role in the locomotor responses of the oozooid. Damaging stimuli, however, evoke bursts of 'driven' OSPs which propagate into the stolon (Fig. 4.25a), and the system may perhaps play a role in releasing the young blastozooid chain from the stolon.

There are evidently many aspects of the salp nervous system which remain to be investigated, most notably perhaps, the anatomy of the brain and the connections of the different neuron types within it, about which almost nothing is known, but there is also much to be done with the epithelial system; for example, the route of 'entry' of OSPs and ISPs to the brain needs to be confirmed, and the role of OSPs in the oozooids remains enigmatic.

4.5. Appendicularians

Appendicularians are designed with considerable economy of means. In oikopleurids for example, there are just two mechanoreceptors on either side of the trunk, whose receptor field is greatly extended by their link with the excitable epithelium covering the tail and posterior trunk. A single axon from one central neuron branches to supply both receptors (Holmberg, 1986). Again, in the oikopleurid brain, there are only some 70 cells of all types (Martini, 1909a), and as Olsson *et al.* (1990) remark, the brain of *Oikopleura dioica* shows an amazing miniaturization. This miniaturization has been carried even further in fritillariid larvaceans, where the brain contains only 52 cells (Martini, 1909b). In consequence, ultrastructural studies in *Oikopleura* have been

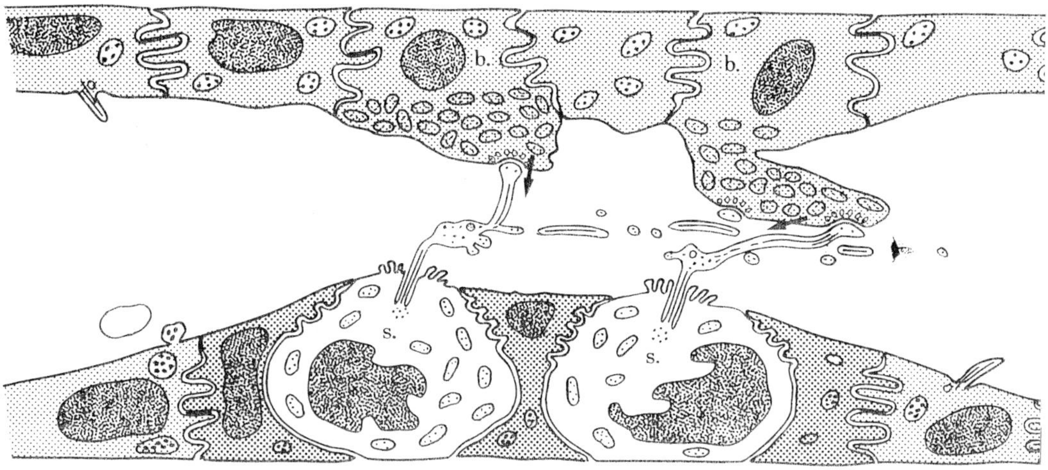

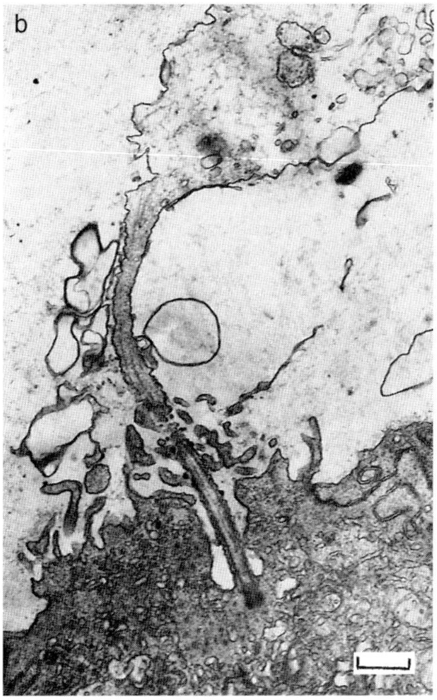

Fig. 4.24. (a) Schematic diagram showing connection of sensory cilia with button cells (b) of adjacent zooid at the plaque. (b) Expansion of cilium after passing from sensory cell. Scale bar 1 μm. From Bone *et al.* (1980).

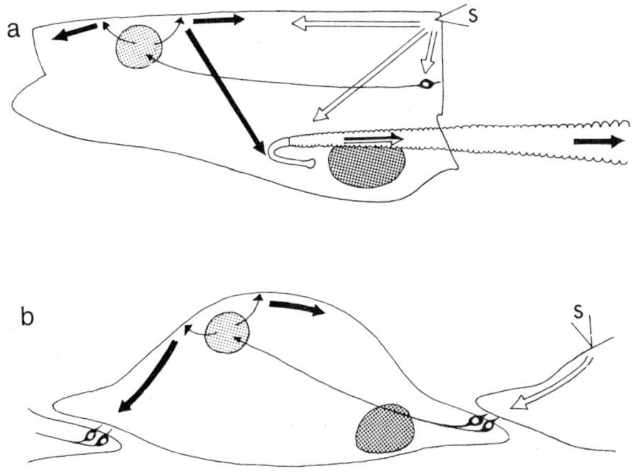

Fig. 4.25. The roles of 'driven' OSPs in oozooids (upper) and blastozooids (below). Driven OSPs are indicated by thick black arrows; OSPs initiated at the stimulus site S by open arrows; impulse propagation along nerve fibres by thin black arrows. The brain is lightly dotted, the stomach darker. In the oozooid impulses are propagated along the stolon and may lead to mature blastozooids breaking free. In the blastozooid OSP signals are propagated along the chain. After Bone (1982) and Anderson and Bone (1980).

able to reveal something of the structure and connections of neurons in the brain and caudal ganglia, whereas nerve impregnation techniques have so far failed on this small scale system.

4.5.1. *Brain and anterior sensory cells*

Knowledge of the brain and its associated nerves and receptors was largely due to the careful histological studies at the light microscope level by Martini (1909a,b), who extended the earlier work of Fol (1872). Martini's studies on *O.longicauda* and *F.pellucida* used conventional staining techniques to map the positions of the cell bodies in the brain, but were unable to reveal their processes and interconnections. Some of these connections in the brain have recently been worked out by the elegant ultrastructural study of Olsson *et al.* (1990), which has much advanced understanding of the brain. Using serial ultrathin sections, they were able not only to follow the three pairs of brain nerves to their terminations (Fig. 4.26a), but also to unravel something of the circuitry within the brain, as shown in Fig. 4.26b.

The most anterior paired nerve (nerve 1) has along its course a small swelling (containing four cells; Martini, 1909a), and consists peripherally of the axons of a group of 30 or so sensory cells lying in a group on the ventral surface of the anterior end of the trunk, under the lower lip (Fig. 4.26c). An ultrastructural study by Bollner *et al.* (1986) has shown that the long swollen cilia of these sensory cells project through the house rudiment after arising from deep invaginations in the cell apices. The sensory cilia are aligned in a row transverse to the long axis of the trunk, forming a ciliary fence similar to that seen in fritillariids, where, however, the axons linking them to the brain appear to be those of central neurons, i.e. as in *Fritillaria,* the lower lip receptors seem to be secondary sensory cells (Bone *et al.*, 1979).

The sensory axons synapse with some of the cells of the swelling on nerve 1 along their course; these cells

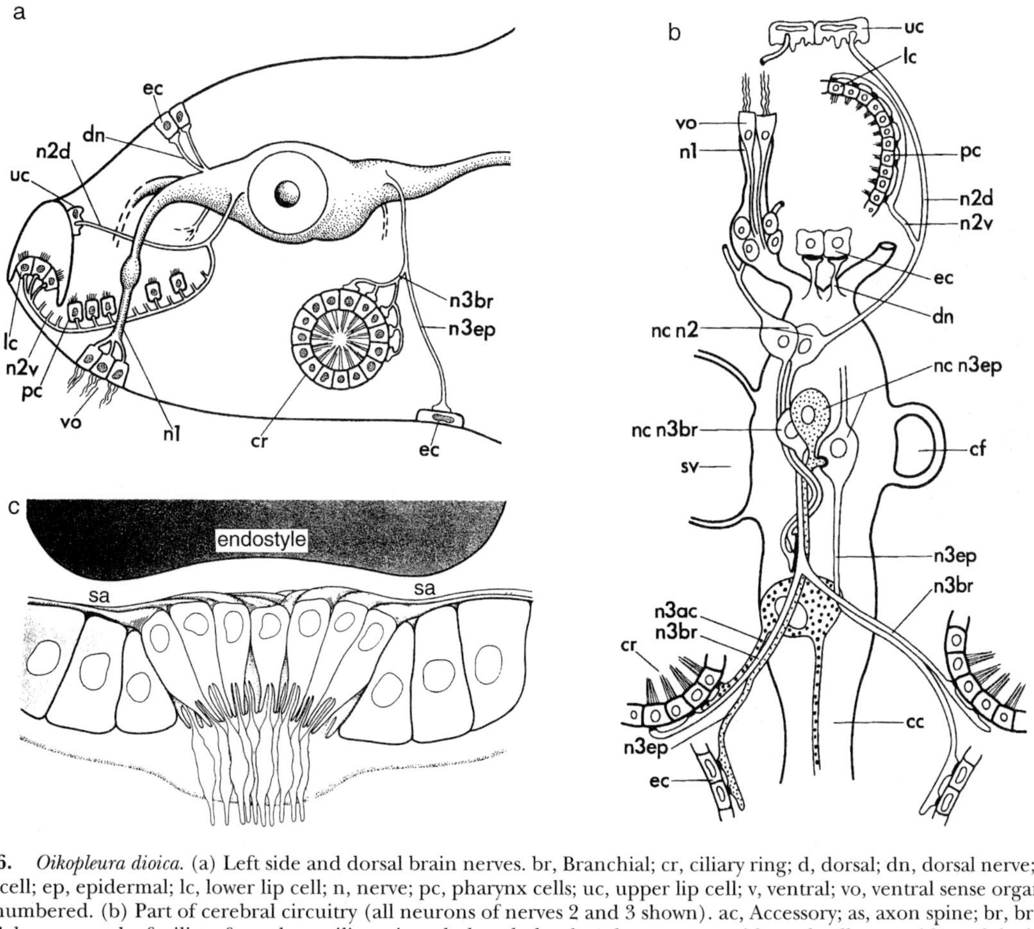

Fig. 4.26. *Oikopleura dioica.* (a) Left side and dorsal brain nerves. br, Branchial; cr, ciliary ring; d, dorsal; dn, dorsal nerve; ec, epidermal cell; ep, epidermal; lc, lower lip cell; n, nerve; pc, pharynx cells; uc, upper lip cell; v, ventral; vo, ventral sense organ. Brain nerves numbered. (b) Part of cerebral circuitry (all neurons of nerves 2 and 3 shown). ac, Accessory; as, axon spine; br, branchial; cc, caudal nerve cord; cf, ciliary funnel; cr, ciliary ring; d, dorsal; dn, dorsal nerve; ec, epidermal cell; ep, epidermal; lc, lower lip cell; n, nerve; nc, nucleus; pc, pharynx cells; sv, sensory vesicle (otocyst); uc, upper lip cell; v, ventral; vo, ventral sense organ. From Olsson *et al.* (1990). (c) Diagram of ventral sense organ. From Bollner *et al.* (1986).

Table 4.1 Synapses between cells of nerve 1 and sensory and central fibres (from Bollner *et al.*, 1986).

Synapse type	Type of swelling			
	1	2	3	4
Postsynaptic to sensory axon	13	7	7	
Presynaptic to sensory axon	3	3	2	
Postsynaptic to central fibre	4	4	2	
Presynaptic to central fibre	1	4	12	3
Reciprocal with sensory axon	1	1	1	

are postsynaptic to the sensory axons, and presynaptic to centrifugal nerve fibres passing from the brain. Bollner *et al.* (1986) recognized differences in the synaptic contacts of these four cells (Table 4.1), and therefore suggested that they had different functions. They pointed out that the system had some similarities with the craniate olfactory bulbs.

Recently, Bollner *et al.* (1991) have shown that some (perhaps all) of these cell types in the swellings are positive for GABA-like immunoreactivity, as are some of the cells within the brain (these have not been correlated with the different cells described by Ollson *et al.*, 1990).

The second nerve (nerve 2) consists of a single axon (1 μm in diameter at its origin from the brain) which branches to supply a specialized non-ciliated large cell in the epidermis of the upper lips, and the ventral branch supplies ciliated cells in the ventral region of the pharynx, as well as other ciliated cells of the lower lips (Fig. 4.26a). Typical chemical synapses were not seen at any of the appositions of this fibre with the cells innervated. The two neurons of the paired nerve 2 send processes caudally to synapse with the single neuron innervating both ciliary rings (Fig. 4.26b).

Nerve 3 is remarkable since the left nerve consists of three axons, whilst the right only contains two; this asymmetry is due to the asymmetry of the central neurons forming the nerves. The nerves arise from four neuron somata in the brain (Fig. 4.26b). The axon of the first monopolar neuron divides and one branch exits through each third nerve to innervate the ciliary rings, where several of the ciliary ring cells receive chemical synapses from the axon running along their bases. A second bipolar neuron sends an axon which synapses with several flattened epithelial cells in the region of the right Eisen's oikoplast (Chapter 6). The ciliary ring of the left side receives a second innervation by a process from a bipolar neuron, whose other process passes caudally. Finally, a third monopolar neuron sends its axon to synapse with epithelial cells in the region of the left Eisen's oikoplast. This last neuron is linked to the neuron supplying the right Eisen's oikoplast region by an axon spine which pushes into the soma of this neuron, apparently forming an electrical synapse with it (Fig. 4.26b). In addition to the three nerves described above, Ollson *et al.* (1990) found a few axons passing out of the dorsal anteriormost region of the brain to synapse with epithelial cells in the region of Fol's oikoplasts.

The brain circuitry and peripheral connexions which these authors have established are clearly linked with the control of the ciliary rings whose activity draws food particles into the pharyngeal mucous filter, and presumably also the epidermal innervation is linked with the secretion of the feeding house by the oikoplasts.

Early observations by Fol (1872) (who remarked that the direction of water flow through the ciliary rings depended upon the will of the animal) had shown that the cilia of both ciliary rings show a pattern of synchronous reversals (which result in rejection of particles in the flow into the pharynx). Galt and Mackie (1971) showed that the synchronous reversals are associated with relatively large ciliary reversal potentials (CRPs). These are much increased in frequency when the lower lip receptors are stimulated; in other cases, flow is continuous and CRPs are absent unless particles in the inhalent flow touch the lip receptors. (Figs 4.27a and b and 4.28). In *Fritillaria pellucida* (as in *Pyrosoma* and doliolids), although similar patterns of potentials (Fig. 4.27c) can be recorded from the cilia of the ciliary rings, these are not signal reversals, but simply short ciliary *arrests*, and hence are ciliary arrest potentials (CAPs). Treatment with *d*-tubocurarine (10^{-6}) abolishes the periodic arrests in *F.pellucida* (Bone *et al.*, 1979), so it appears that innervation of the ciliary rings is cholinergic.

As seen in Fig 4.26b, in *O.dioica* this reflex involves two neurons only. The second asymmetric ciliary ring innervation in the left nerve 3 remains enigmatic. In view of the innervation from somata in the brain, it is curious that removal of the brain and anterior part of the trunk does not abolish the pattern of CRPs from the ciliary rings (Galt and Mackie, 1971), suggesting that pacemaker activity resides either in the ciliated cells of the rings themselves, or in the distal fragments of the severed axons of nerve 3.

Synchrony of CRPs after removal of the brain (and the asymmetric innervation of the two ciliary rings) suggests that the two are linked by a non-nervous pathway, perhaps by an excitable epithelium, in which passive

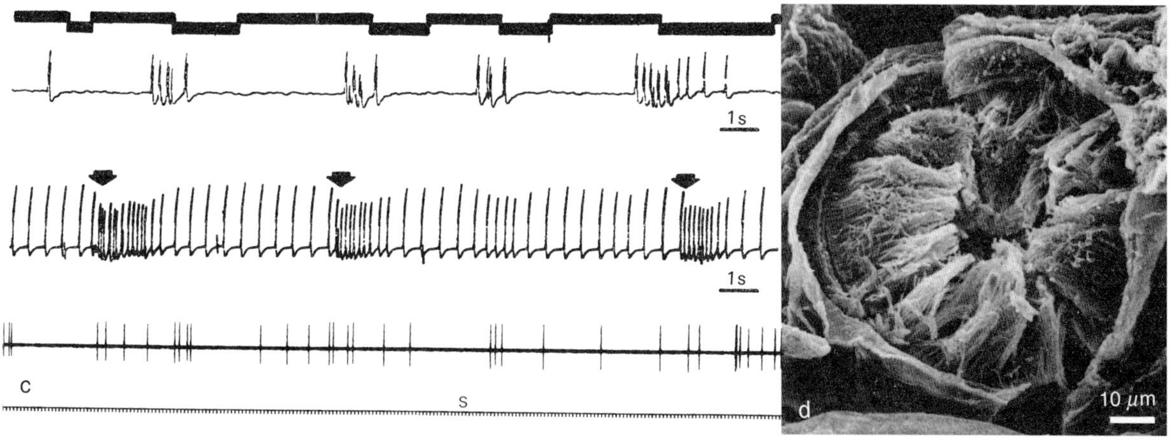

Fig. 4.27. *Oikopleura dioica.* (a) Extracellular suction electrode record of CRPs evoked when particles in inhalant stream strike lip receptors (recorded on upper line with delay due to observer's time lag and inertia of water mass in pharynx). (b) Similar record showing acceleration of regular CRP pattern by mechanical stimulation of lower lips (arrows). (c) *Fritillaria pellucida.* Spontaneous CAP pattern. (d) *F.pellucida.* Scanning electron micrograph of ciliary ring showing long cilia. (a) and (b) from Galt and Mackie (1971); (c) and (d) from Bone *et al.* (1979).

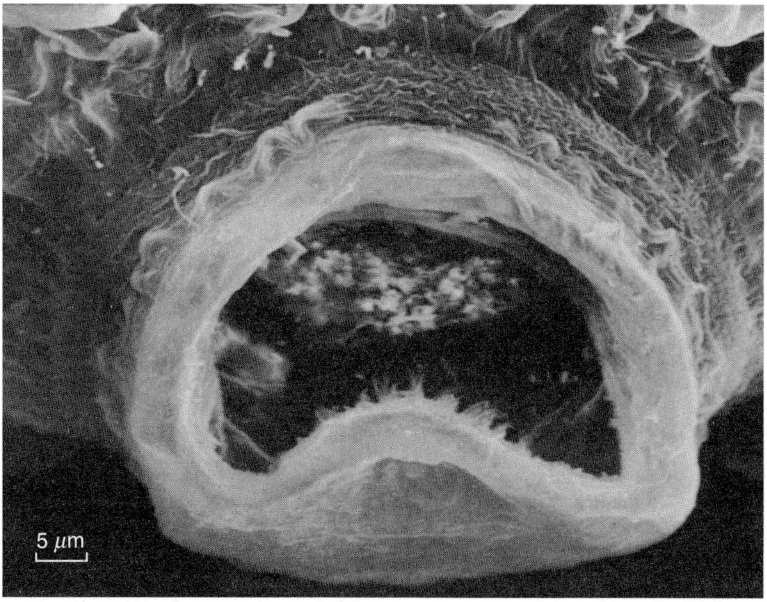

Fig. 4.28. *Oikopleural dioica.* Scanning micrograph looking into mouth showing lower lip receptors.

current spread rather than propagated action potentials might suffice for co-ordination.

Although the brain is small, it is not impossible that future intracellular studies may enable these functions to be examined in more detail; though Galt and Mackie (1971) obtained records from the ciliary rings using microelectrodes, it is not certain that these were truly intracellular records, and intracellular recording has not been attempted from the brain. Similarly, further immunocytochemical work should reveal putative neurotransmitters in the circuitry, which could then be examined pharmacologically.

In addition to the different receptors of the pharynx and lips, there is a statocyte on the left side of the brain (Fig. 4.26a), containing a statolith with concentric secretion layers (Fig. 4.29). The central connections of the sensory cells whose cilia touch the statocyte in the cerebral vesicle (Holmberg, 1984) are unfortunately still unknown; furthermore, it is not known whether they respond to vibration, acceleration, or to gravity. Since

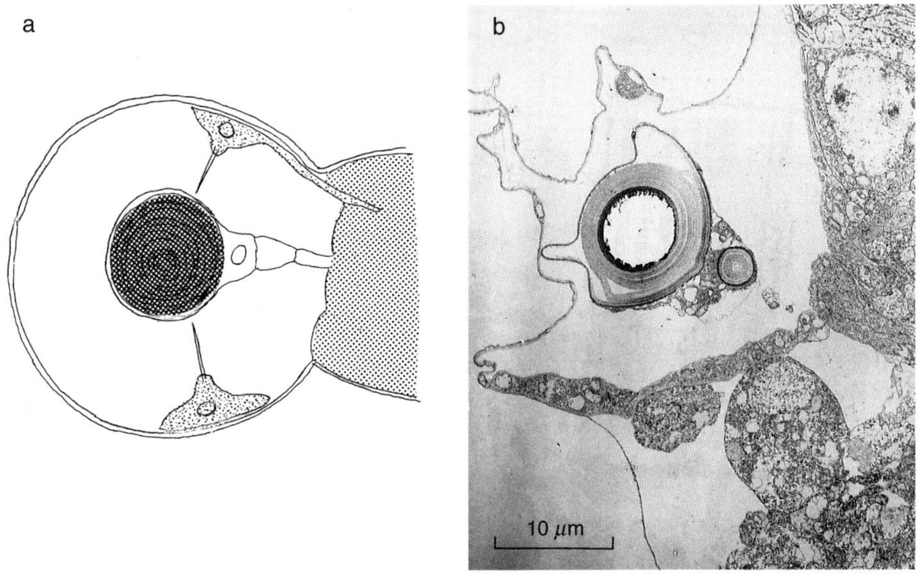

Fig. 4.29. (a) *Oikopleura dioica.* Diagram of statolith. (b) *O.labradoriensis.* Section of statolith (central portion torn out of section) showing concentric structure and secondary statolith formation. (a) Modified from Holmberg (1984); (b) photograph by C.L. Singla.

F.pellucida can apparently adopt any attitude in the water when feeding, it seems probable that the statocyte is not a gravity receptor.

4.5.2. *Ciliated brain duct*

A ciliated brain duct, opening into the buccal cavity on the right of the brain, develops from the brain vesicle during development (Delsman, 1912). Holmberg (1982) has examined the ultrastructure of this enigmatic structure in *Oikopleura dioica* (Fig. 4.30), showing that it opens at its tip into the haemocoel, and that its wall appears to receive synapses from brain neurons. The opening into the buccal cavity is surrounded by long immobile cilia. Since this organ arises in development from an outgrowth on the right side of the cerebral vesicle, it is an extension of the brain. A neural gland is absent in appendicularians, but otherwise the brain duct is similar to the ciliated funnel portion of the neural gland complex of ascidians and salps. The function of the ciliated duct is unclear, but Holmberg (1982) suggested that it might have a dual role: as an excretory organ, and, in view of the apparently cyclical appearance of amorphous material at the duct tip, and in secreting this material into the haemocoel. Earlier authors (see Lohmann, 1933) had suggested that the ciliated duct was a chemoreceptor, but Holmberg found no evidence for this view.

4.5.3. *Caudal ganglion*

Nerves pass from the brain to a small caudal ganglion at the base of the tail (Chapter 2), from which the nerve cord of the tail arises. Along the nerve cord there are groups of neuron somata of two different sizes, which supply the caudal muscle cells. Although Martini (1909a) (and previously Damas, 1904) supposed that there were both motor and sensory fibres issuing from the cells of the cord, both cell types are probably motor, and the muscle fibres receive a dual innervation (section 3.5.2). The groups of ganglion cells along the cord vary somewhat in position in different specimens (a detailed account of individual variability in *O.longicauda* is given by Martini, 1909a). Cell numbers in the caudal ganglion and along the cord have been examined by Martini (1909a,b). In the caudal ganglion of *O.longicauda* there are some 36 neurons and other cells; in *F.pellucida* only 21, whilst along the cord in the former there are 54 cells.

Immunocytochemical studies by Bollner *et al.* (1991) have shown the presence of GABA-like immunoreactivity in some somata and in the neuropil of the caudal ganglion, but the connections of the different neurons are not known, except for the single dorsal cell whose axon branches to pass to the paired Langerhans receptors on either side of the trunk (Holmberg, 1986). The axon forms the two recurrent caudal nerves first described by Fol (1872), which are the largest fibres (5 μm) found in any tunicate. Apart from the neuron supplying the

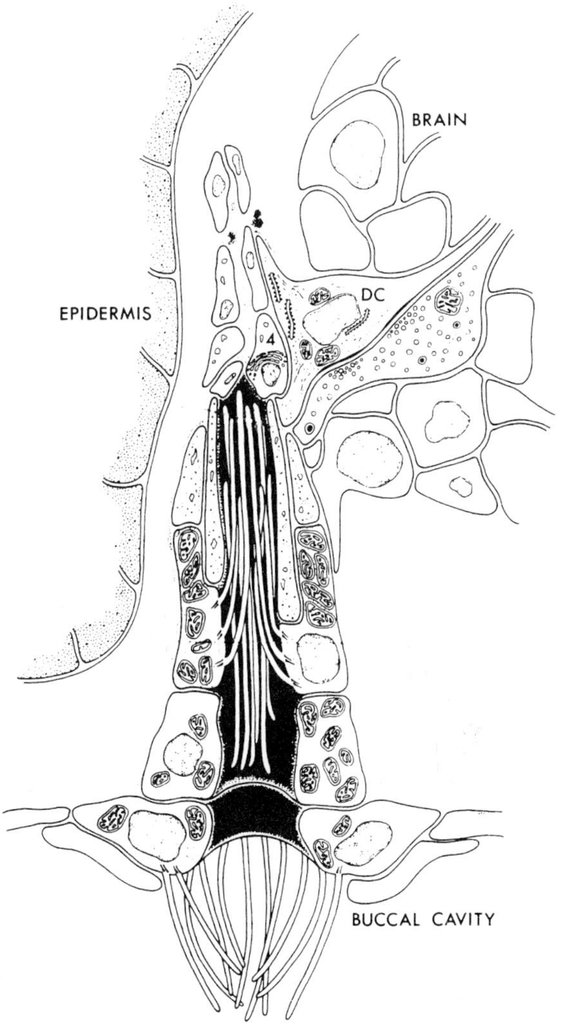

Fig. 4.30. *Oikopleura dioica*. Ciliated brain duct. DC, duct cell receiving synapses from central neuron. From Holmberg (1982).

Langerhans receptors (which are secondary sensory cells; see below), the other identified cell in the caudal ganglion is a large dorsal ependymal cell which secretes Reissner's fibre into the central canal of the cord along the tail (Holmberg and Olsson, 1984).

4.5.4. *Langerhans receptor cells and the epithelial sensory system*

The axon from the caudal ganglion terminates in gap junctions with the base of the Langerhans receptor cell and with the base of an adjacent epithelial cell (Fig. 4.31a–d). There are no synaptic vesicles in the axon terminals, and the synapses therefore appear to be electrical. The receptor itself is a cell bearing a large stiff multitubular bristle (a modified cilium) up to 150 μm long. The anatomical arrangement thus is that the OSPs can either pass direct to the Langerhans axon via its gap junction connection with the base of the accessory cell, or indirectly, via the Langerhans receptor cells themselves to which the adjacent epithelial cells are coupled by gap junctions. Stimulation of the receptor cell alone does not evoke OSPs (Fig. 4.3b). Evidently, the outer epithelium acts as an extension of the receptor field of the Langerhans receptors, whether or not these are stimulated by OSPs. Direct stimulation of the bristle evokes bursts of escape swimming (Bone and Ryan, 1979), but these are more easily evoked by mechanical stimulation of the excitable outer epithelial cells (Bone and Mackie, 1975; Bone, 1985), which propagate rapid overshooting action potentials (OSPs) over the tail and posterior part of the trunk (Fig. 4.32).

Although the epithelial cells are only around 1.4–2.8 μm thick, it is not difficult to obtain stable intracellular impalements, which show that the OSPs propagate at around 40 cm s^{-1}, and that graded generator or receptor potentials precede OSPs evoked by graded mechanical stimuli (Fig. 4.32b and c). The surprising ease with which these thin cells can be impaled has permitted a two-electrode study of their passive membrane properties (Table 4.2).

Table 4.2 Passive electrical properties of appendicularian epithelia compared with those of the excitable epithelium of the hydrozoan medusa *Euphysa*.

Tissue	Space constant (λ) (μm)	Membrane resistivity (R_m) (kΩ cm^{-2})	Specific internal resistivity (R_i) (Ω cm^{-1})	Mean thickness (μm)	Mean restin potential (mV)
O.dioica, outer epithelium	922	4.3	82.7	1.64	82.6
O.longicauda, outer epithelium	3350	35.6	104.5	3.3	80.2
Euphysa, exumbrellar ectodem*	1300	23	196	1.4	46

* From Josephson and Schwab (1979); appendicularian data from Bone (1985).

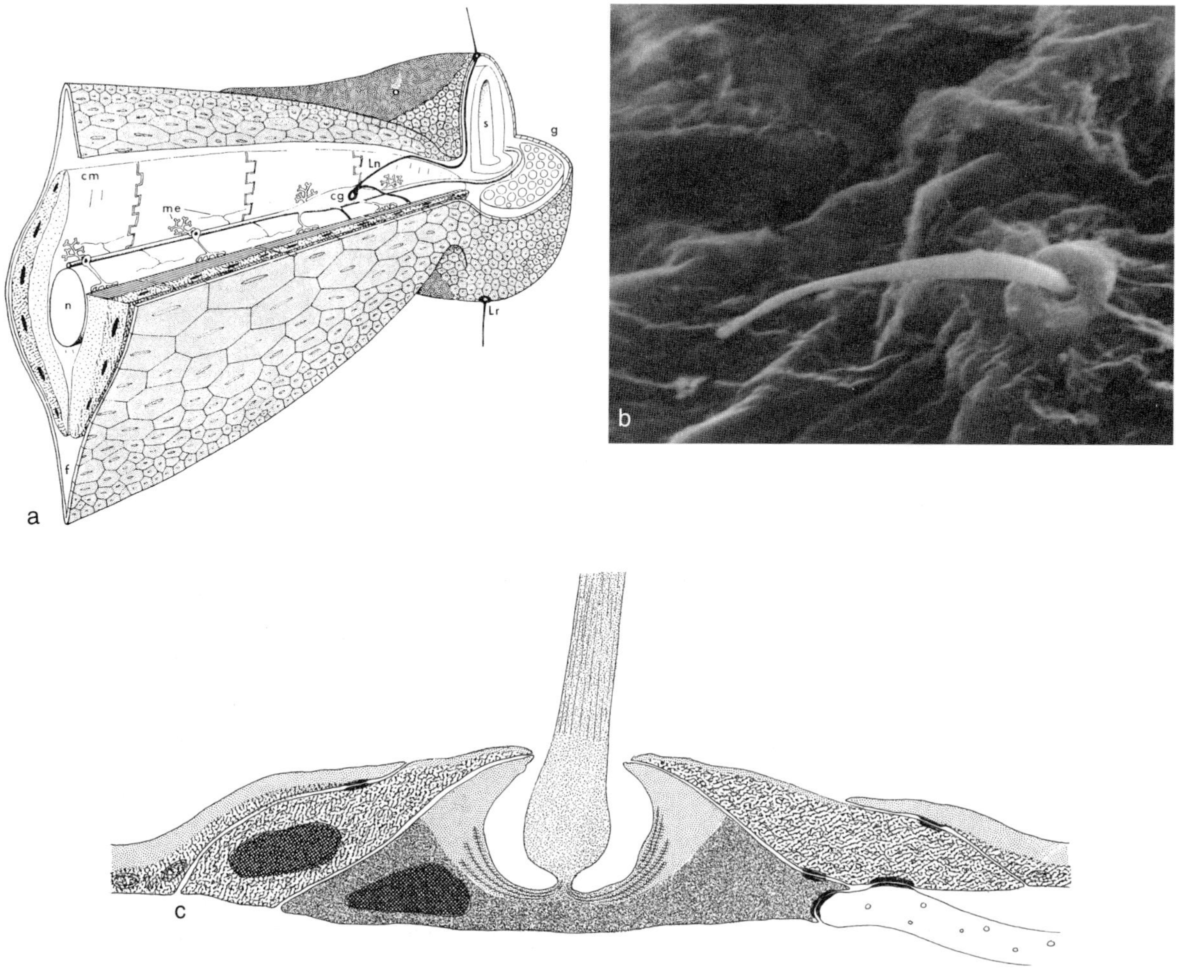

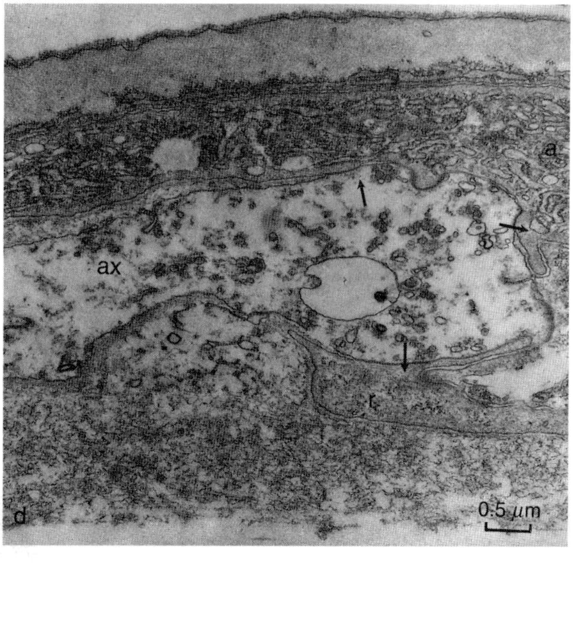

Fig. 4.31. The Langerhans receptor of *Oikopleura*. (a) Stereogram showing position of receptor cells (Lr) on either side of trunk, linked to Langerhans neuron in the caudal ganglion (cg) by its axon (Ln). cm, caudal muscle cell; g, gonad; me, motor endings on muscle; n, notochord; o, oikoplastic epithelium. From Bone and Mackie (1975) modified in the light of Holmberg (1986). (b) Scanning micrograph of surface of trunk showing Langerhans receptor cilium. From Bone and Mackie (1975). (c) Schematic section across receptor showing Langerhans axon on right contacting both receptor and accessory cells. (d) Gap junctions (arrowed) between Langerhans axon (ax) and accessory (a) and receptor (r) cells. From Bone and Ryan (1979).

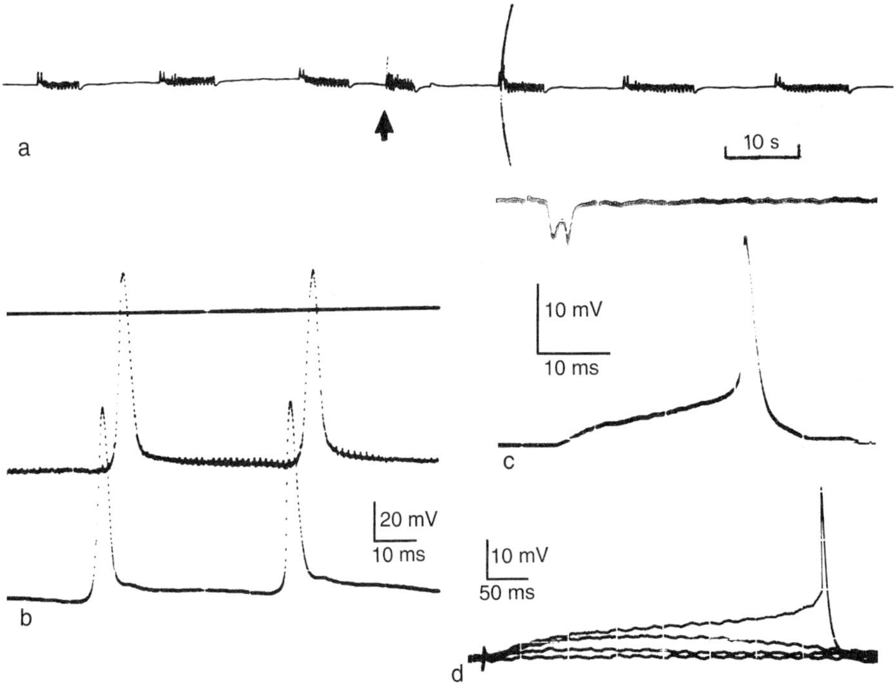

Fig. 4.32. *Oikopleura labradoriensis.* (a) Suction electrode records of regular swimming bursts interrupted by out-of-sequence escape swimming bursts either evoked by an OSP (large potential), or by direct stimulation of the Langerhans receptor bristle (arrow). (b) *O.albicans.* Dual electrode intracellular record of OSPs from electrodes 3.2 mm apart. (c) OSP arising from receptor potential evoked by mechanical stimulation of same epithelial cell as impaled by recording electrode at marker on upper trace. (d) *O.dioica.* Intracellular record of graded receptor potentials leading to an OSP, evoked by a series of graded mechanical stimuli. (a) from Bone and Ryan (1979), (b) from Bone and Mackie (1975), and (c) from Bone (1985).

The ionic basis of the oikopleurid epithelial action potentials has not been definitely established, but preliminary data (Bone, unpublished) suggests that (as in salps, section 4.4.4.1) they are carried mainly if not entirely by Na^+, and that repolarization involves a voltage-dependent K^+ channel. Recent whole-cell patch clamp observations on oikopleurid epithelial cells by Dr I. Inoue (personal communication) have shown that membrane deformation evokes very rapid currents that are almost certainly carried by Na^+.

Skin impulses (or direct stimulation of the Langerhans receptor) lead to rapid escape swimming (Chapter 3) at higher frequency than during the normal rhythmic swimming bursts (see Fig. 3.19). Presumably, in the house, the Langerhans receptor bristles are in contact with the house material and they are involved in the rapid escape from the house when it is touched, as Seeliger (1895) suggested. Fritillariid larvaceans have lost the outer epithelium (and hence lack the skin impulse system; Bone *et al.*, 1977a) and the recurrent caudal nerves (the axon of the Langerhans neuron in the caudal ganglion) terminate near the base of the tail, where stimulation evokes escape responses (Chapter 3). In some fritillariids, there is a single cell at the base of the notch at the end of the tail, which may be a sensory neuron.

4.5. Concluding remarks

Salps and Oikopleura show a remarkable development of skin impulse (or epithelial action potential) systems that are otherwise known in tunicates only from a single doliolid stage and from a single ascidian tadpole larva (Mackie and Bone, 1976). These systems interact with the nervous system in escape behaviour. Only in Oikopleura is the route of 'entry' of skin impulses to the nervous system known unequivocally, and only in salps are epitheliomotor neurons found which drive the skin impulse system. In both groups, small number of motoneurons control locomotor movements, but the process of reduction in cell number in larvaceans has led to a most striking decrease in size of the nervous system, which has had the consequence that ultrastructural studies have been able to work out important central circuitry in oikopleurids, although fritillariids (where cell numbers are even further reduced) remain to be examined. Much clearly remains to be explored, and in salps and doliolids, the way in which reversal depends upon the change in timing between the body locomotor muscle bands and those of the anterior or posterior lips offers attractive material for the study of reversable delay circuits. Very little is known of the transmitters used in the central and peripheral nervous systems; here immunocytochemical techniques seem promising, and have begun to provide interesting results in *Oikopleura*.

FIVE

Feeding and energetics of Thaliacea

L.P. Madin and D. Deibel

5.1. Filter feeding

Pelagic tunicates are true filter feeders. While copepods and other crustaceans are now known to have complex behavioural mechanisms for detecting, entraining, and grasping food particles, tunicates actually pass seawater through a mucous sieve and ingest the retained particles. The feeding mechanisms of the Thaliacea can be interpreted as a series of modifications of the ascidian branchial basket (Madin, 1990). However, there are distinct differences in the feeding mechanisms of thaliaceans and appendicularians, and care must be used when attempting to generalize rates among them.

The rates at which thaliaceans remove particles from the water when they filter feed are determined by a variety of methods. There is still some confusion in the literature over the exact meaning of 'filtration rate' and 'clearance rate' as applied to zooplankton. The following definitions of Omori and Ikeda (1984) will be used here: *filtration rate* refers to the total volume of water passing the filter apparatus per unit time; *clearance rate* refers to the volume of water from which particles are completely removed. The two are equal only if all suspended particles are removed with 100% efficiency. Experiments which measure the depletion of food particles from the surrounding water or their accumulation in the gut are properly determinations of clearance rate, because they calculate the apparent volume of water from which a certain class of particles has been removed. Measurements or calculations that determine the total volume of water pumped through a filter without regard to particle content are properly filtration rates. Because most feeding measurements are of clearance rate, this term will chiefly be used in this chapter.

5.1.1. *Feeding mechanisms and behaviour*

5.1.1.1. *Pyrosomas*

Pyrosomas are the least studied of the pelagic tunicates with respect to energetics or physiology. These colonial organisms are fairly commonly collected in trawl nets and can be quite large (see Fig 1.1), but rarely survive long in the laboratory. The colonies are made up of tens to thousands of minute, ascidian-like individuals (section 1.2.1), arranged in the common wall of the colony with their oral siphons on the outside and the atrial siphons opening to the lumen of the colony. The feeding mechanism as described by Fenaux (1968a) is essentially like that of ascidians. The pharyngeal cavity is occupied by a ciliated branchial basket, separated from the body wall by two lateral peribranchial spaces which connect to the smaller atrial cavity. Mucus produced by the ventral endostyle is carried dorsally over the surface of the basket by the cilia. Water is pumped through the branchial basket by ciliary action, and particles are retained on the mucus-covered surfaces. The collected food particles are carried along with the mucous sheet, which rolls into a cord and moves posteriorly to the oesophagus. Fenaux (1968a) observed that the mucus was formed either into a single food strand along the dorsal mid-line, or sometimes into two lateral cords, which were then rolled into one as they approached the oesophagus at the end of the basket. There appeared to be some variability in both the velocity of the water current through the basket, and in the quantity of mucus secreted, and at times part of the mucus formed a net suspended in the branchial basket. The conditions of observation may have affected the feeding behaviour, and there has been no further descriptive or experimental work on the mechanism.

5.1.1.2. *Doliolids*

The doliolid feeding mechanism was first described by Fol (1876) and Fedele (1921, 1923a), and is discussed in reviews by Neumann (1935), Jørgensen (1966a,b) and Fenaux (1968a). There have been only four modern studies of feeding by doliolids, one of feeding behaviour (Deibel and Paffenhöfer, 1988) and three of feeding rates (Deibel, 1982b; Crocker *et al.*, 1991; Tebeau and Madin, 1994). The brief description that follows is based on these earlier descriptions and on recent video studies of feeding doliolids by Bone *et al.* (1997). The last authors observed that the manipulation of the filter within the pharynx was akin to that in ascidians and entirely different to that in salps.

Doliolids are small, barrel-shaped macrozooplankton surrounded by a thin tunic containing hoop-like muscle bands (Fig. 5.1a). The ciliated pharyngeal wall which is typical of ascidians is reduced in doliolids, limited to a curved gill septum which separates the pharyngeal and atrial cavities. The pharyngeal cavities of all stages are similar in design and extent, except for that of the trophozooids, which have an expansive anterior chamber (Fig. 5.1b). The morphology of the different zooids of the doliolids and of the colony stage is discussed in Chapter 1.

Fol (1872) was the first to use carmine suspensions to show that living doliolids pump water through the branchial and atrial apertures, which have cuspate valves. The endostyle in the floor of the pharyngeal cavity produces mucous secretions which are drawn around the entrance to the pharynx by the peripharyngeal bands. Dorsally, the left and right peripharyngeal bands unite in a curious spiral or volute, which imparts a rotation to the filter, as well as providing a thin rotating central core (leading to the oesophageal opening) around which the expanded conical filter is wrapped. Fenaux (1968a) gives a good photograph of this core and its origin from the dorsal volute. Microcine-

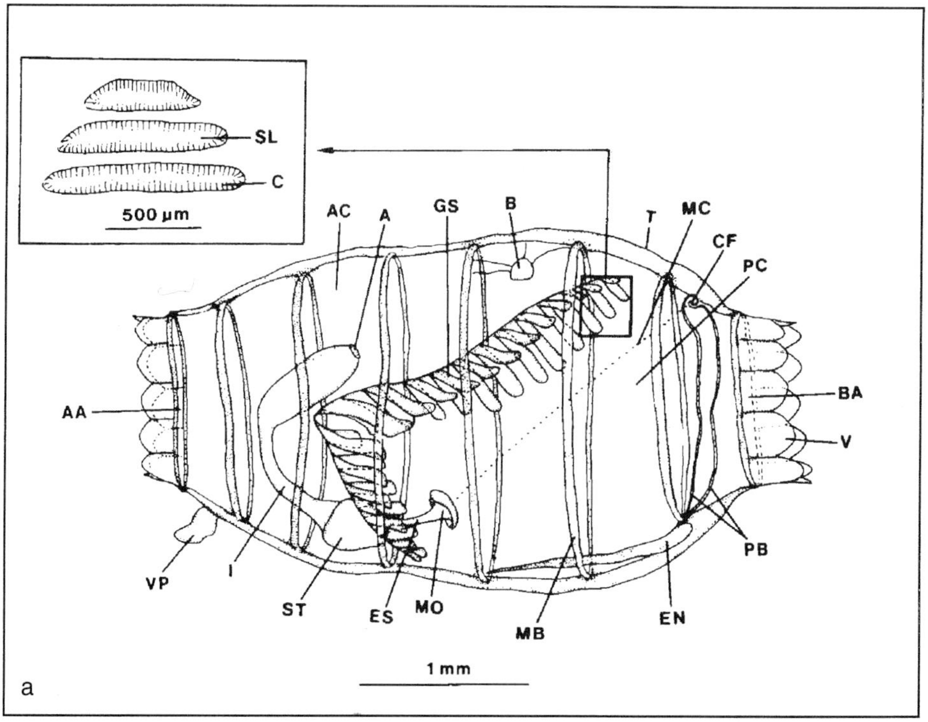

Fig. 5.1. (a) Line drawing of the phorozooid of *Doliolum nationalis* in lateral view; inset is a close-up view of several gill slits showing cilia. A, anus; AA, atrial aperture; AC, atrial cavity; B, brain; BA, branchial aperture; C, cilia; CF, ciliated funnel; EN, endostyle; ES, oesophagus; GS, gill septum; I, intestine; MB, muscle band; MC, mucous cord; MO, mouth; PB, peripharyngeal band; PC, pharyngeal cavity; PH, phorozooids; SL, gill slits; ST, stomach; T, tunic; TR, trophozooids; V, valve; VP, ventral peduncle. Drawings by S.H. Lee, adapted from Deibel and Paffenhöfer (1988).

Fig. 5.1 (b) Part of a colony of *Doliolletta gegenbauri* showing trophozooids and phorozooids. The largest trophozooids are 4–5 mm long. Photograph by L.P. Madin.

b

Fig. 5.2 Mucous filter nets of salps. (a) Aggregates of *Pegea confoederata* with their nets made visible with red carmine particles. (b) Solitary *Cyclosalpa polae*, anterior to left, net made visible with carmine. (c) *Pegea confoederata* solitary, carmine on filter net, green dye in water passing through filter and out of the atrial siphon. (d) SEM of fixed filter net of *P. confoederata*; scale bar 1.0 μm. Arrows point to 'errors' in construction. Photos (a)–(c) by L.P. Madin; (d) from Bone *et al.* (1991).

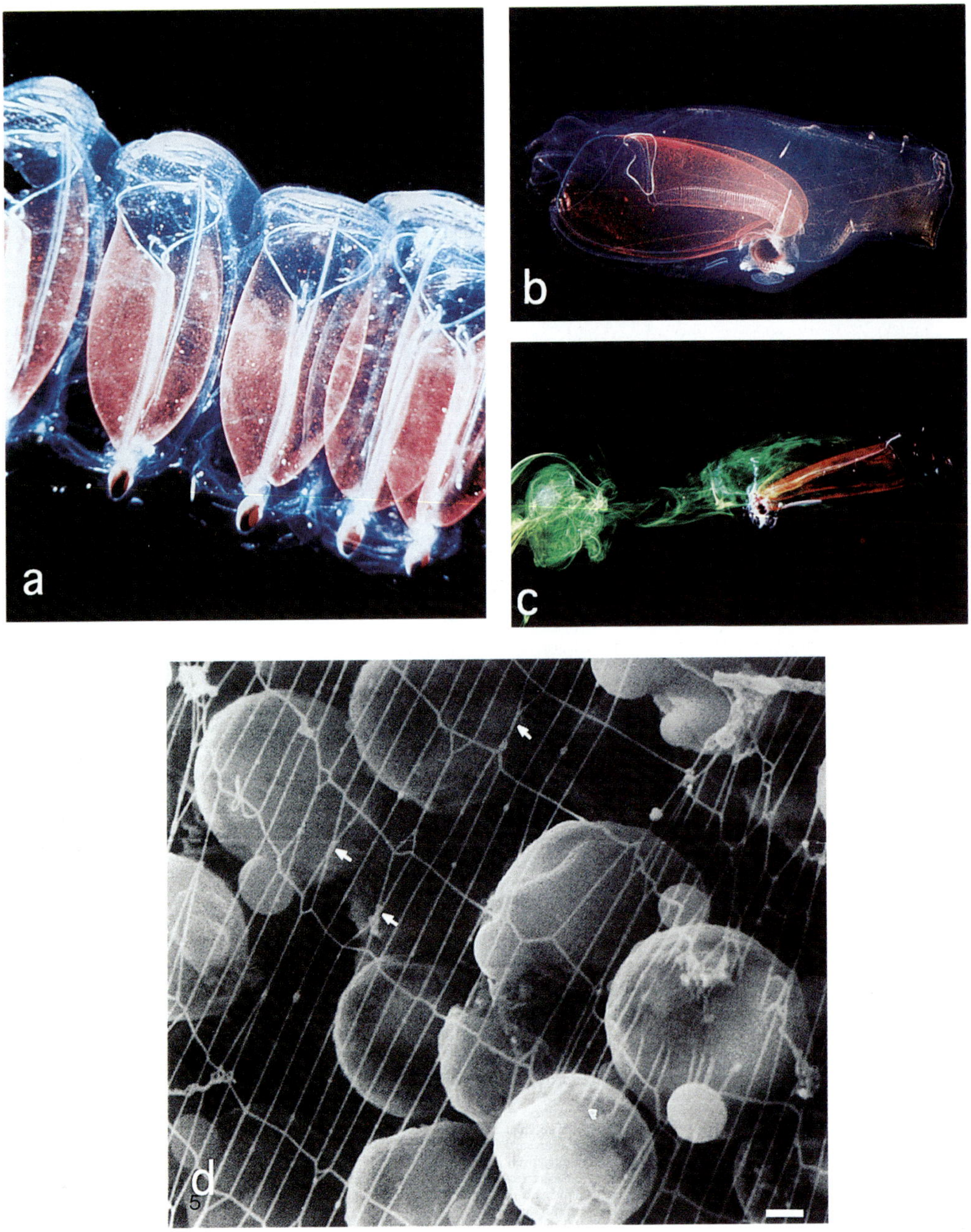

matographic observations by Deibel and Paffenhöfer (1988) indicate that, when undisturbed, doliolids always have a complete pharyngeal mucous filter in place, just as salps do. This suggests that previous observations of mucous threads streaming backwards from the peripharyngeal bands (as figured by Fol, 1872) were artifacts of containment in aquaria. The gastrointestinal tract is simple, consisting of a short oesophagus, globular stomach, and short intestine.

Doliolids pump water using ostial cilia (Fig. 5.1a inset; Fenaux, 1968a; Deibel and Paffenhöfer, 1988), not via muscle contractions as do salps. Earlier views that both ciliary and muscle power are used by doliolids to create feeding currents are incorrect. The cilia are 130 nm in diameter and beat at about 11 Hz, thus operating at Reynolds numbers in the range of 10^{-5} (Deibel and Paffenhöfer, 1988). The ciliary beat frequency of 11 Hz is similar to that of *Oikopleura* spp. (10–15 Hz; Galt and Mackie, 1971) and is in the middle of the range of values for bivalves, ctenophores, and protozoans (2–20 Hz; Sleigh, 1974). The Reynolds number of 10^{-5} is similar to that of ciliated protozoans and bivalves but is larger than that of the choanocytes of sponges (10^{-7}) and smaller than that of the setae of copepods (10^{-2}; Jørgensen, 1983).

Inflowing particles are trapped by the pharyngeal filter, which is rotated by the action of the cilia of the dorsal volute, and wound around the cord from volute to oesophagus, and ingested. In this way, the particles are actually trapped between two layers of the filter (Bone *et al.*, 1997). The mean velocity at which *Doliolum nationalis* ingests its mucous cord is 36 ± 11 μm s^{-1} (Deibel and Paffenhöfer, 1988). This is much slower than the estimated impact velocity of particles at the surface of the pharyngeal filter of 900 μm s^{-1} (Deibel and Paffenhöfer, 1988). Recent video-assisted microscopic studies of *Oikopleura vanhoeffeni* have indicated somewhat lower impact velocities, in the neighborhood of 300 μm s^{-1} (Acuña, Deibel and Morris, unpublished observations). Although doliolids are able to pump water at either 'slow' or 'fast' rates, perhaps in response to food particle concentration (Fedele, 1921), these different filtration rates may be an artifact of containment in aquaria. However, even if all the ostial cilia are active, and water flows through the pharynx, the doliolid does not necessarily deploy its mucous filter and feed. Remarkably, doliolids can feed apparently normally after the brain has been removed (Bone *et al.*, 1997)! As Fedele (1921, 1933a) observed, control of the feeding process (which involves activity of ostial, oesophageal, and peripharyngeal band cilia, as well as secretion of the filter by the endostyle) is largely autonomous, controlled by a visceral nervous system.

If a food particle that is larger than the diameter of the oesophagus is trapped on the pharyngeal filter and blocks the oesophageal opening, the ciliated band within the oesophagus described by Fedele (1933) is arrested, and the rolled up filter (which is elastic) is pulled out toward the ciliated volute, whose cilia are also arrested. Feeding may then begin again, and the rolled up filter again ingested, when if the opening is blocked once more, the oesophageal cilia are again arrested, and the process is repeated. Several such cycles may result in the particle being crushed and wound into a mucous 'cocoon', which is eventually ingested, or if ingestion is impossible, one or more strong contractions ofthe muscle bands drive water out of the anterior aperture, rupturing the pharyngeal filter and dislodging the bolus (Fedele, 1921; Deibel and Paffenhöfer, 1988). These contractions are rapid (section 3.3.1), and the jet of water exits from the pharynx at velocities up to 20–50 cm s^{-1}, an effective 'cough' (Bone and Trueman, 1984).

Arrest of the ostial cilia driving the flow through the filter occurs each time filter entry to the oesophagus halts, whether in response to noxious food particles or during escape contractions (Galt and Mackie, 1971; Deibel and Paffenhöfer, 1988). After remaining at rest for several seconds, the cilia resume their normal beat pattern. Several investigators have suggested that ciliary arrest is triggered by sensory cells at the opening of the oesophagus (Fedele, 1921; Braconnot, 1971c; Bone and Mackie, 1977). This suggests that doliolids may suspend filtering for short periods in response to chemical cues from food particles.

5.1.1.3. *Salps*

Salps exhibit the most extreme modification from the ascidian type, and achieve the highest rates, per individual, of water clearance and transport. The reduction of the ascidian branchial basket to a single gill bar (section 1.4.1), and the use of circumferential muscle bands instead of cilia to pump water through the pharynx and atrium, permits large volumes of water to be passed through the pharyngeal filter, at velocities high enough to create rapid jet locomotion (Chapter 4).

Early descriptions of the feeding mechanism of salps (Brooks, 1876; Fol, 1876; Fedele, 1933a,b; Carlisle, 1950) were clearly hampered by laboratory artifacts, and are now known to be partly or wholly incorrect. These descriptions have since found their way into review articles (Ihle, 1938; Fenaux, 1968a); their inclusion here is meant to illustrate the kinds of misconceptions than can arise when salps are observed under abnormal conditions.

Brooks (1876) observed carmine particles to be caught on the inner walls of the pharynx, and thought they were then conveyed by cilia backwards along the endostyle to the oesophagus. Fol (1876) did observe in *Thalia democratica* the formation of mucous sheets between the gill bar and endostyle, but his interpretation was still of isolated mucous strings drawn together by the cilia of the gill bar and moved into the oesophagus, an idea which was to persist for nearly a hundred years.

More extensive laboratory observations, mainly on *T.democratica* were made by Fedele (1933a,b), who described different modes of feeding, depending on the amount of mucus secreted by the endostyle. He attributed the extent of mucus production to such factors as sensory perception of the presence of suspended particles, or degree of intestinal fullness. A low level of mucus production produced 'slow feeding', in which strands of mucus proceeded from the anterior end of the endostyle up to the front of the gill bar, and then back along it into the oesophagus. Other mucous strands moved from the posterior end of the endostyle along the ventral floor of the pharynx and up into the oesophagus. Suspended particles of carmine or indian ink were seen to be caught in these strands and ingested. At a higher level of mucus production, Fedele saw sheets of mucus suspended between the endostyle and gill bar, and the occasional formation of a conical mucous net suspended from the peripharyngeal bands and the gill bar. He thought that the 'slow' ingestion was the normal feeding mechanism, punctuated by production of the conical mucous net at short intervals, perhaps in response to food availability.

These laboratory observations were repeated by Carlisle (1950), who saw similar 'slow' and 'fast' feeding mechanisms in *Salpa maxima*. He observed 'slow' feeding to be accompanied by rapid, incomplete contractions of the body muscles and incomplete closure of the oral and atrial siphons, which produced oscillating currents in the pharynx. Particles caught in these currents eventually stuck to mucous strands from the peripharyngeal bands and were moved to the oesophagus. During 'fast' feeding, the body muscles contracted slowly and more fully, and the siphons closed completely, to produce a strong unidirectional current. Carlisle described the conical mucous net produced from the peripharyngeal bands, but, like Fol, though it formed because individual mucous strands were twisted together by cilia of the gill bar. He illustrates the mucous net occupying only the anterior portion of the pharynx (Carlisle, 1950, fig. 3).

The first description of feeding *in situ* was given by Madin (1974a,b), who used a carmine suspension while SCUBA diving to visualize feeding by free-swimming salps. It is clear from his observations of six species, and from subsequent studies (Harbison and Gilmer, 1976; Madin, 1990; Bone *et al.*, 1991) that salps normally feed only with a full pharyngeal mucous net. 'Slow feeding' with isolated threads of mucus is an artifact of laboratory conditions, and does not occur in nature.

The pharyngeal net is formed from mucus secreted by the endostyle and transported by ciliated tracts up to the peripharyngeal bands. These form a hoop around the inner pharyngeal wall just behind the mouth. The baglike net fills the entire pharyngeal cavity, attached anteriorly at the peripharyngeal bands, dorsally along the gill bar, and entering the oesophagus at the posterior. Neither conical nor funnel-shaped, the net expands to conform to the dimensions of the pharynx, though without contacting the body wall (Fig. 5.2a–c, see also Fig. 1.28). New mucus is continually released from the peripharyngeal bands so that the net is constantly renewed from the front as it is rolled up and ingested at the back. The time required to completely renew the nets of free-swimming salps ranged from 10–300 s depending on the size (Madin, 1974b).

Under laboratory conditions of high particle concentration or restricted mobility, the pharyngeal net may remain in place for only a few minutes (Bone *et al.*, 1991), but in undisturbed animals *in situ* the filter appears to be in place almost all the time (Madin, 1974b, but see below). The net is often broken off when salps reverse their swimming direction, although the remnants may still be ingested (Bone *et al.*, 1991). Very high particle concentrations may clog the net, causing it to break off. Sometimes the salps reverse their locomotion with enough force to blow the offending net back out of the oral siphon (Madin, 1990). Continuous high concentrations of particles may result in formation of a bolus in the oesophagus that effectively blocks ingestion (Harbison *et al.*, 1986). Particle concentrations in oceanic waters are usually low enough that the pharyngeal net can function continuously, but high particulate loading in near shore waters may be a factor which excludes most species of salps (Harbison *et al.*, 1986).

5.1.2. *Particle retention and natural diet*

5.1.2.1. *Doliolids*

Nothing is known about the pore size of the pharyngeal filter of doliolids, or if pore size differs among the different life history stages. Particle capture by any suspension feeder is a complex function of pore size, fibre diameter, water velocity, and particle size, shape, and charge. As with salps, predictions of filtering performance based on ultrastructural observations may not agree with observed retention spectra of small particles. Although the ultrastructure of doliolid filters has never been examined, there are laboratory and field observations which provide some indication of their particle retention capabilities. Doliolids do appear to retain very small particles efficiently. Crocker *et al.* (1991) found little difference in the clearance rate of radiolabelled diatoms ($\sim 8 \times 100$ μm) and bacteria (0.2–5 μm in diameter) by *Dolioletta gegenbauri*. Working with the same species, Deibel (1982b) found no difference in the clearance rates of *Isochrysis galbana* (5 μm diameter) and *Peridinium trochoideum* (17 μm diameter), but Tebeau and Madin (1994) found that retention efficiency for 1.0 μm plastic beads relative to 2.5 μm beads was 30% for trophozooids and 56% for phorozooids. These laboratory results are in agreement with field observations made by Paffenhöfer *et al.* (in press), which showed that doliolid abundance was inversely correlated with the concentration of nanoflagellates 2–6 μm in diameter. Deibel (1985b) also found that areas with high concentrations

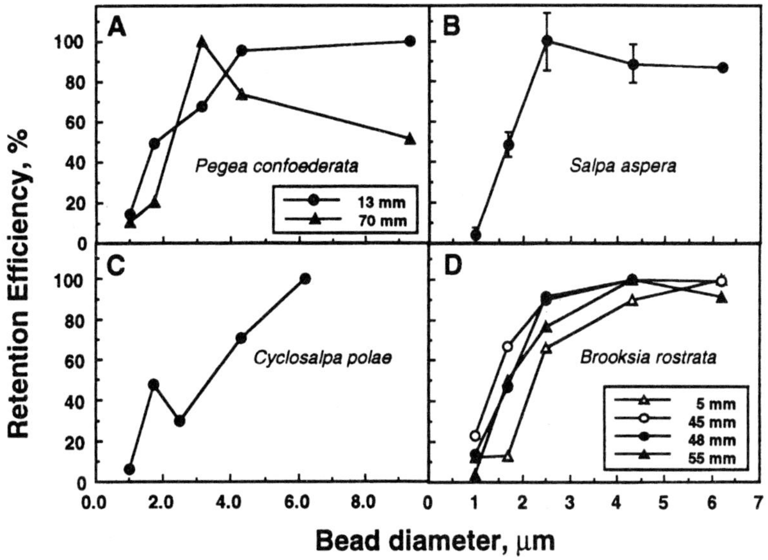

Fig. 5.3. Particle retention efficiency for four species of salps. Curves in (A) and (D) for different size individuals. From Kremer and Madin (1992).

of doliolids had very low concentrations of particles from 0.2–0.5 μm in diameter. This suggests that doliolids may retain small nanoplankton and perhaps even picoplankton with high efficiency. They are thus apparently significantly different from salps, which have low retention efficiency for particles < 2 μm (Fig. 5.3).

If the mucous filter of doliolids is similar to that of salps and appendicularians, then the size composition of their natural diets should be similar as well (Pomeroy and Deibel, 1980; Bruland and Silver, 1981; Deibel and Turner, 1985). Although appendicularians (see p. 141) are known to ingest colloidal DOM (Flood *et al.*, 1992), we have no idea of the capability of doliolids to ingest these submicrometre particles. Fedele (1921) reported some detrital particles, nanoplankton, and bacteria in the guts of doliolids, but primarily larger phytoplankton, including *Rhizosolenia* spp., *Chaetoceros* spp., and other diatoms, dinoflagellates, and coccolithophores.

5.1.2.2. *Salps*

The salp pharyngeal net is another remarkable example of a tunicate self assembling mucous structure. Like the filters of ascidians or appendicularians, it apparently combines high porosity with effective retention of small particles. The actual structure of the net has been difficult to see because of its delicacy and tendency to collapse into a wad. Silver and Bruland (1981) published TEMs of what they considered remnants of the pharyngeal net in faecal pellets (probably of *S.fusiformis*). The filter had a regular rectangular mesh measuring 1.9 × 0.2 μm. Bone *et al.* (1991) were able to fix the pharyngeal nets of actively feeding *Pegea confoederata*, and found a similar rectangular mesh, but with larger pores, measuring 3.31 × 0.57 μm (Fig. 5.1d). Discrepancies in these measured sizes may be due to shrinkage of the mucus during preparation for microscopy. Bone *et al.* (1991) estimated a 20% shrinkage loss for their preparation, and applied a 65% loss to the data of Silver and Bruland (1981), based on different preparation methods. Corrected in this way the meshes of the salp pharyngeal filter would be approximately 4–5 μm high and 0.3–0.7 μm wide. This is smaller than the mesh of the pharyngeal filter of the appendicularian *Oikopleura vanhoeffeni* (6.4 × 3.3 μm average; Deibel and Powell, 1987a) but has a comparable porosity (91% open area; Bone *et al.*, 1991). As noted on p. 161, however, there is some doubt about the dimensions for *O.vanhoeffeni.* Collection of particles smaller than the mesh pore size due to direct interception by single mucous fibres has recently been observed in appendicularians (Acuña *et al.*, 1996), but to date there is no experimental evidence that this mechanism also functions in salps.

The structural characteristics of the pharyngeal filter might be expected to determine the sizes of particles that it retains. However, empirical determinations of particle retention by salps, either by examination of faeces (Madin, 1974b; Silver and Bruland, 1981; Reinke, 1987) or experiments with particles of known size (Harbison and McAlister, 1979; Caron *et al.*, 1989; Kremer and Madin, 1992), indicate that cells less than 1.0 μm are caught only with low efficiency, while in most species particles larger than about 3–4 μm are retained with 100% efficiency (Fig. 5.3). Some of the data suggest that there are differences in retention threshold among different species (Kremer and Madin, 1992; Madin and Purcell, 1992), or that small individuals retain smaller particles than large individuals of the same species (Harbison and McAlister, 1979; Kremer and Madin, 1992). The general conclusion of these experimental studies, however, is that there is a sharp increase in retention efficiency around 2.0 μm, and that cells smaller than this suffer relatively little grazing pressure from salps. Ingestion of very large objects is probably limited by the strength of the pharyngeal net or the size of the oesophagus, but particles up to about 1.0 mm (nauplii, foraminifera, veligers, radiolaria, etc.) are ingested (Madin, 1974b).

The observed retention characteristics are not entirely consistent with the mesh dimensions actually measured for two species, although correction for apparent shrinkage narrows the gap between expected and observed particle retention (Bone *et al.*, 1991). Silver and Bruland (1981) suggested that the meshes might be enlarged by water pressure in feeding salps, a speculation echoed by Madin (1990), who calculated a pressure drop of 3.3 mm H_2O across the net of *Salpa maxima.* Bone *et al.* (1991) calculated a pressure drop of 2.1 mm H_2O across the net of *Pegea confoederata* by a different means, but discounted any significant stretching of the filter meshes. It should be borne in mind that these pressure drops are high in comparison to those of ascidians (see Riisgård, 1989) and that to achieve high porosity, because the fibres of the mesh need to be strong (relatively large diameter) to withstand the pressure drop, the salp filter may not have as fine a mesh as that of doliolids or pyrosomas.

There is no selection of particles on a qualitative basis (i.e. live vs dead material, cell type, or condition), except for the rejection, by reversing water flow, of clogged nets or noxious substances (Fedele, 1933a,b; Madin, 1990). Gut contents and faeces contain a broad range of both living and detrital particles (Fedele, 1933a,b; Yount, 1958; Madin, 1974b), and studies which have compared gut contents with available particulate matter (Silver, 1975; Silver and Bruland, 1981; Madin and Purcell, 1992) found no qualitative differences in their composition. The apparent overlap in diet among salp species has led several authors (Metcalf, 1918; Yount, 1958) to regard salps as 'ecological equivalents' lacking the trophic niche specialization seen in many other planktonic herbivores. While this appears to be the case, other physiological and behavioural characteristics may define niche boundaries among sympatric species (e.g. Harbison and Campenot, 1979; Cetta *et al.*, 1986; Madin, 1990).

In some cases, salps may ingest other zooplankton as a significant part of the diet. Hopkins (1985) and Lancraft *et al.* (1991) reported *Euphausia superba* debris in salp guts, but this may have been detrital material. It has been suggested (Huntley *et al.*, 1989; Nishikawa *et al.*, 1995) that *Salpa thompsoni* in the Antarctic might consume eggs or larvae of krill, and that this mechanism, more than competition for food, is responsible for the paucity of krill in years when salps are abundant. It is probable that small, weakly swimming crustaceans or larvae would be caught in the mucous net and ingested, but this kind of 'predation' has not been verified experimentally.

5.1.3. *Measurement of feeding*

5.1.3.1. *Pyrosomas*

There has been only a single study of feeding rate in *Pyrosoma.* Drits *et al.* (1992) collected large numbers of *P.atlanticum* in the southeast Atlantic, and estimated their grazing effects by measuring gut pigment content and defaecation rate. Gut pigment in the zooids and faecal pellets was measured directly by extraction and quantification of fluorescence. Defaecation rates were determined by microscopic observations of colonies during the first 15 min after collection. A gut turnover time of 0.75 h was calculated from these data, and Drits *et al.* estimated that the mean quantity of pigment ingested by *Pyrosoma* 50–65 mm in length was 4.13 μg colony^{-1} h^{-1} during the 10 h period at night that the pyrosomas were present in the surface (1–10 m) layer.

5.1.3.2. *Doliolids*

More attempts have been made to measure feeding rates for doliolids than for pyrosomas. Water pumping rates (i.e. filtration rates) have been determined for *Doliolum nationalis* from high speed video records of particle trajectories (Deibel and Paffenhöfer, 1988). For individuals 2.0–3.3 mm long, the rates ranged from 60 to 140 ml day^{-1}, which is a maximum rate and does not take into account time spent not feeding (i.e. time during ciliary arrest). Using carmine and indian ink suspensions, Fedele (1921) estimated the filtration rate of *D.mülleri* to be ~130 ml day^{-1}, very close to the rate determined by Deibel and Paffenhöfer (1988) for *D.nationalis.*

Three techniques have been used to determine the particle clearance rate of doliolids: (1) uptake of radiolabelled diatoms and bacteria (Crocker *et al.*, 1991); (2) particle removal from experimental bottles over time, determined using a Coulter counter (Deibel, 1982b); and (3) uptake of plastic beads during short-term laboratory or *in situ* incubations (Tebeau and Madin, 1994). Results of these studies suggest a wide range of clearance rates for *Dolioletta gegenbauri*: 1–355 ml day^{-1} for gonozooids and phorozooids, 14–697 ml day^{-1} for oozooids, and 5–235 ml day^{-1} for trophozooids. However, Deibel (1982b) and Tebeau and Madin (1994) worked with smaller animals on average than did Crocker *et al.* (1991). Thus, to make a meaningful comparison of clearance rates among the studies, regression equations from the primary publications were used to convert all rates to units of ml day^{-1} for a 'standard' gonozooid 4 mm long and a 'standard' oozooid 3 mm long. For the standard gonozooid, radioisotope uptake gave a clearance rate of 29 ml day^{-1} (Crocker *et al.*, 1991), 12% of the rate determined by particle removal experiments (i.e. 241 ml day^{-1}; Deibel, 1982b). For the standard oozooid, particle removal experiments lead to a clearance rate of 156 ml day^{-1} (Deibel, 1982b), nearly equivalent to the mean rate of 225 ml day^{-1} determined by radioisotope uptake for much larger oozooids, 10–30 mm long (Crocker *et al.*, 1991). Extrapolation of *in situ* bead uptake results predict a rate for 4 mm phorozooids of 181 ml day^{-1}, and of 152 ml day^{-1} for 3 mm trophozooids (Tebeau and Madin, 1994).

This simple re-analysis of the clearance rate data points to a fundamental disagreement in rates determined by particle removal vs radioisotope uptake. Similarly, clearance rates of the salp *Thalia democratica* were greater when determined by particle removal vs radioisotope uptake (see Table 5.1). The doliolid clearance rate of 241 ml day^{-1} for a standard gonozooid 4 mm long determined by particle removal (Deibel, 1982b) is similar to the filtration rate of 160 ml day^{-1} for an animal 3 mm long determined directly by microcinematography (Deibel and Paffenhöfer, 1988). This suggests that radioisotope uptake may underestimate the clearance rate of thaliaceans. A possible explanation for this underestimate is that Crocker *et al.* (1991) ran their radioisotope uptake experiments for 1–5 h, much longer than the probable gut passage time of doliolids, which Tebeau and Madin found to be ≤15 min. This could mean that much of the isotope consumed by the doliolids in their experiments was defaecated and thus lost to their further analyses (K. Crocker, personal communication). Most importantly, this kind of disagreement emphasizes the need for intercomparison of techniques for measuring clearance rates.

5.1.3.3. *Salps*

Feeding rates of salps have been measured using several laboratory and *in situ* methods; a comparison of different techniques for different species has recently been made by Madin and Kremer (1995). Handling stresses can severely affect feeding rates, and early efforts (Pavlova *et al.*, 1971; Silver, 1971) produced low or erratic results. The first satisfactory laboratory data were from Harbison and Gilmer (1976) for *Pegea confoederata*, a species that adapts well to the laboratory. These authors kept salps in tanks with cultured diatoms, and monitored the depletion of cells in the water over time using a Coulter counter. Harbison and Gilmer reported two feeding rates, an overall rate (F_{overall}) based on starting and ending concentrations, and a high rate (F_{high}) based on the highest short term rates between successive counts (Fig. 5.4). They regarded the F_{high} as a better estimate of continuous feeding rates in the field, since F_{overall} integrates all interruptions of feeding due to confinement artifacts. Container effects can be a serious drawback to particle clearance experiments for species that are active swimmers. Additional data for other species have since been obtained by Harbison and McAlister (1979), Deibel (1982b, 1985a), Andersen (1985, 1986) and Harbison *et al.* (1986) using Coulter counter methods. Huntley *et al.* (1989) conducted similar experiments but measured changes in chlorophyll concentration instead of particle counts; others (Reinke, 1987; Caron *et al.*, 1989) made microscopic counts of various cells (cyanobacteria and flagellates) used in feeding experiments.

Since all the water pumped through the body of a feeding salp as it moves fowards passes through the pharyngeal net, it is theoretically possible to estimate filtration rate from data on swimming. The volume of water pumped through the salp should represent the maximum possible clearance rate. This approach eliminates problems of artificial diet or particle counting, but depends on a knowledge of hydrodynamic parameters for freely swimming animals. Andersen (1985) used data from Bone and Trueman (1983) on pumping rate and volume ejected to estimate the volume of water pumped by swimming *Salpa fusiformis*. These values were as much as ten times higher than clearance rates determined from particle removal, leading Andersen to conclude that the nets of *S.fusiformis* had capture efficiencies of only 6–32%. Size retention data (Harbison and McAlister, 1979; Kremer and Madin, 1992) suggests that capture efficiency of the cells used by Andersen should have been much higher than this. Similarly, Reinke (1987) used measurements of pressure and thrust to calculate velocity and volume of water pumped through the pharynx of *Salpa* spp., also obtaining values higher than clearance rates based on removal of algal cells. Species of *Salpa* are active swimmers and it is likely that feeding in laboratory clearance experiments is intermittent due to contact with the container walls, causing the clearance rates to be much lower than filtration calculated from continuous water transport. Another calculation of volume pumped by *Pegea* was made by Bone *et al.* (1991) who measured the internal volume of the pharyngeal filter and multiplied it by the pulsation frequency. In the case of this slow swimming species, the estimate came close to values obtained by Harbison and Gilmer (1976) for similar salps.

As with direct feeding experiments, locomotory parameters measured from salps confined in small containers may be atypical. The parameters of volume pumped that Andersen (1985) used for *S.fusiformis* were considered to be 'escape behaviour' by Bone and Trueman (1983), and probably higher than for normal swimming. On the other hand, the pulsation frequency observed by Bone *et al.* (1991) for *Pegea* was eight times slower than frequencies observed for the same species *in situ* (Madin, 1990), suggesting a different response of this species to confinement.

Two *in situ* methods have been used to measure filtering rates of salps. Mullin (1983) collected *Thalia democratica* in 5 l containers while diving, introduced radioactively labelled bacteria and phytoplankton, and incubated the chamber underwater. At the end of the experiment, radioactivity in the food particles and the bodies of the salps was measured, and clearance rates were calculated from the ratio of the two measurements. Although minimizing handling, this approach does require containment and provision of an artificial diet.

The second technique uses fluorometric measurement of plant pigments in gut contents or faecal pellets of salps to determine relative gut fullness or feeding rates. This has the advantage that the salps need not be contained or given an unusual diet. Animals feeding freely on naturally occurring particulates are collected from the field with plankton nets or by divers and analysed immediately for gut pigment content. Early applications of the technique were only qualitative (Nemoto and Saijo, 1968), or simply related gut contents to ambient concentrations of phytoplankton (Jansa, 1977). Determination of ingestion or clearance rates with this approach requires additional information on gut passage time, and on interference with pigment measurements from pigmented tissues of the salp. Quantitative application of the method has been used by Madin and Cetta (1984) and Madin and Purcell (1992) to determine *in situ* clearance rates for several oceanic species. In most cases, feeding rates calculated from gut content methods are higher than those determined in laboratory incubations of the same species (Fig. 5.4a and b).

The use of plant pigments as an index of gut fullness or feeding rate in planktonic herbivores has several pitfalls. Pigment fluorescence does not appear to be a conservative tracer (i.e. does not always retain its original fluorescent yield), and hence measurement can underestimate actual ingestion. In copepods, these pigment losses are typically about 30% (Dam and Peterson, 1988; Downs, 1989), but can be higher or variable, depending on conditions (Lopez *et al.*, 1988). In salps, losses of about 34% (Madin, unpublished) and 50% (Madin and Purcell, 1992) have been found. This loss can be estimated and corrected for by comparison with a truly inert label material (e.g. silica) in the gut, or from pigment budget experiments. The loss may be variable with species and diet, and so has to be determined for each investigation using this approach.

Gut passage time must also be determined in order to calculate feeding rates. Although the simplest approach has been to hold animals without food and monitor gut fullness or faecal production over time, these rates are often unnaturally slow in the absence of normal ingestion. Salps that normally feed continuously and have short digestion times often become moribund during such incubations. An alternative that works for salps (where there is no mixing in the gut) is to feed them coloured particles and monitor their appearance in the faeces (Madin and Cetta, 1984). Because some kind of confinement is necessary to measure gut passage times, there is often variation by a factor of two in times for similar individuals, and the results are likely to overestimate passage times.

If one assumes a steady-state between ingestion and defaecation, ingestion rates can be calculated from the production of faecal pellets by salps feeding on natural particle mixes. Kremer and Madin (unpublished results) maintained salps collected by divers in large tanks of unfiltered seawater, which was changed at frequent intervals to provide nearly constant ambient levels of natural particles. Plant pigments in the faeces were measured, and the cumulative faecal output plotted over

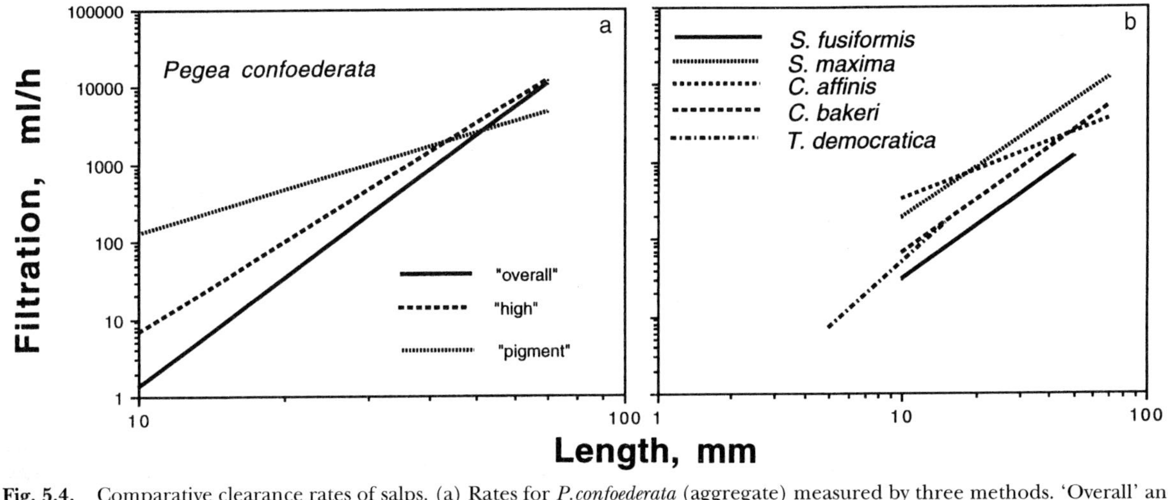

Fig. 5.4. Comparative clearance rates of salps. (a) Rates for *P.confoederata* (aggregate) measured by three methods. 'Overall' and 'high' from Harbison and Gilmer (1976), 'pigment' from Madin and Cetta (1984). (b) *Salpa fusiformis* (solitary) and *C.affinis* (solitary) rates from particle clearance experiments (Andersen, 1984; Harbison and McAlister, 1979); *S.maxima* (aggregate) from gut pigment (Madin and Cetta, 1984); *C.Bakeri* (both) from diatom ingestion (Madin and Purcell, 1992); *T.democratica* (both) from *in situ* grazing of radiolabelled algae (Mullin, 1983).

time After correction for degradative loss of pigments, the slope of this function represents ingestion rate. Clearance rate is then calculated from the concentration of pigment in the water.

Each of these methods has advantages and disadvantages; some probably work for some species and not others. The closest approximation to realistic feeding rates of salps in nature will probably be derived from a comparison of methods, constrained further by energetic and locomotory requirements that put bounds on the minimum and maximum feeding rates possible (Madin and Kremer, 1995).

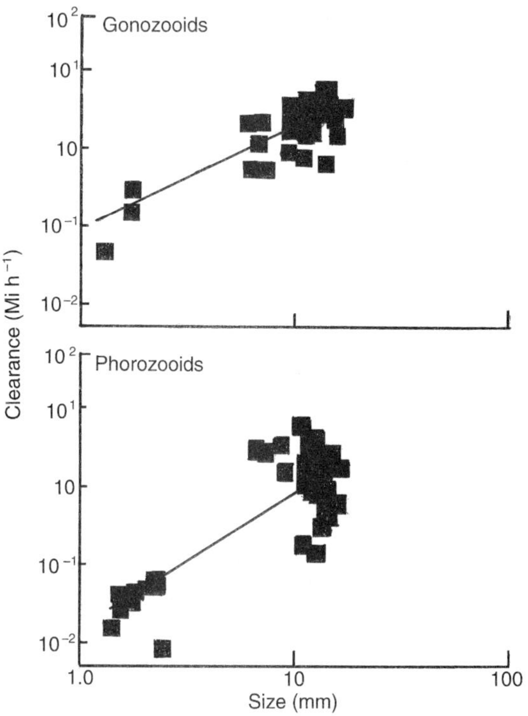

Fig. 5.5. Clearance rate of *Dolioletta gegenbauri* versus body size. Uptake of ^{14}C labelled diatoms; uptake of ^{3}H-labelled bacteria. The linear regression equations are for gonozooids, $\log Y = 1.34 \times \log L - 0.73$ ($r^2 = 0.66$, $P < 0.05$), and for phorozooids, $\log Y = 1.77 \times \log L - 1.53$ ($r = 0.72$, $P < 0.05$). Adapted from Crocker *et al.* (1991).

5.1.4. *Comparative feeding rates*

5.1.4.1. *Pyrosomas*

Drits *et al.* (1992) calculated a clearance rate for a 55 mm long colony of 5.5 l h^{-1}, comparable to individual salps of similar length. A colony of that length was estimated to contain 1200 zooids, so that the individual filtering rate would be approximately 4.6 ml zooid^{-1} h^{-1}, comparable to individual doliolids (or copepods). These rates indicate that the population they sampled near the surface could consume 53% of the phytoplankton standing stock per night under 'swarm' conditions (9.5 colonies m^{-3}), but only 4% during 'non-swarm' conditions (1.9 colonies m^{-3}). Below 10 m depth, the impact of the pyrosomas was much less.

5.1.4.2. *Doliolids*

Clearance rates for *Dolioletta gegenbauri* were found to increase exponentially (Deibel, 1982b; Crocker *et al.*, 1991) or linearly (Tebeau and Madin, 1994) with increasing body size (Fig. 5.5; Table 5.1). On a weight-specific basis, clearance rates ranged from 3–33 ml μgC^{-1} day^{-1} (Deibel, 1982b), and 0.7–113 ml mg dry weight^{-1} day^{-1} (Crocker *et al.*, 1991). These results are here expressed in the original units because conversion factors from dry weight to carbon content have not yet been determined for doliolids. However, dry weight is not a reliable measure of gelatinous zooplankton mass (Madin *et al.*, 1981; Schneider, 1990) and the rates per unit carbon are likely more meaningful.

Although the feeding rates in Table 5.1 are expressed per individual zooid, it is important to remember that doliolids occur not only as single oozooids, gonozooids, or phorozooids, but also as a polymorphic colony consisting of a single non-feeding oozoid ('old nurse') trailing a chain of attached trophozooids and phorozooids that may number in the thousands. The phorozooids themselves bud a cluster of gonozooids which remain attached for a while. Crocker *et al.* (1991) found that ca. 80% of the total clearance of particles by phorozooids was due to juvenile gonozooids on the ventral peduncle, and a similar proportion of clearance by oozooids was due to the trophozooids on the dorsal cadophore. Tebeau and Madin (1994) measured the clearance rates of intact colonies and confirmed that the trophozooids are responsible for most of the feeding activity.

Nothing is known about the diel cycle of feeding activity by doliolids or of the functional response of feeding activity to food particle concentration or quality.

5.1.4.3. *Salps*

It is apparent from all methods used that salps are capable of filtering much larger volumes of water, on an individual basis, than almost any other grazer in the zooplankton. Data from these studies show that clearance rates can be several l salp^{-1} h^{-1}, increasing exponentially with the length of the salp. This relationship is to be expected, since internal volume of the salp will increase as a power function of length. When clearance rate is expressed as a function of dry weight or, preferably, carbon weight, values are similar to those of other grazers, like copepods, owing to the high water and salt content of salps (Madin *et al.*, 1981). Weight-specific clearance rates of a variety of zooplankters do not vary greatly; for a 1200-fold range in body size, weight-

Table 5.1 Clearance rates of salps and doliolids.

Species Generation	Method	*a*	*b*	*n*	r^2	Weight-specific (ml mg^{-1} h^{-1})	Example (ml h^{-1})	Reference
Cyclosalpa affinis								
agg	A	0.00262	3.34	4		266	122	Harbison & Gilmer, 1976
agg	A	2.28	1.35	6		383	176	Harbison & McAlister, 1979
sol	A	4.88	1.73	4		4153	1279	Harbison & Gilmer, 1976
sol	A	19.51	1.22	11		3214	990	Harbison & McAlister, 1979
Cyclosalpa bakeri								
both	G	0.275	2.09	162	0.81	161	230	Madin & Purcell, 1992
both	H	0.361	2.25	15	0.85	354	505	Madin & Purcell, 1992
Cyclosalpa floridana								
agg	A	0.477	1.87	10		664	75	Harbison & McAlister, 1979
sol	A	0.951	1.93	9		1566	177	Harbison & McAlister, 1979
Cyclosalpa polae								
sol	A	1.02	1.75	13		363	285	Harbison & McAlister, 1979
Pegea confoederata								
agg	A	0.000929	3.85	23	0.74	626	224	Harbison & Gilmer, 1976
agg	G	1.87	1.84	24	0.62	1951	698	Madin & Cetta, 1984
sol	A	0.0573	2.62	9	0.66	338	263	Harbison & Gilmer, 1976
Salpa cylindrica								
sol	G	13.87	1.44	74	0.08	4064	1429	Madin & Cetta, 1984
Salpa fusiformis								
agg	B	0.0145	3.32	19	0.67	173	635	Andersen, 1985
sol	B	0.162	2.27	16	0.58	234	241	Andersen, 1985
Salpa maxima								
agg	A	10.55	1.40	8		1235	956	Harbison & Gilmer, 1976
agg	G	1.34	2.15	28	0.72	1753	1357	Madin & Cetta, 1984

Table 5.1 *(Continued)*

Species Generation	Method	a	b	n	r^2	Weight-specific (ml mg^{-1} h^{-1})	Example (ml h^{-1})	Reference
Salpa thompsoni								
agg	C	89.95	(W)0.71	7	0.96	70	169	Huntley *et al.*, 1989
agg	D	10.1	0.97	54	0.20	95	229	Reinke, 1987
Thalia democratica								
agg	B	0.14*	2.76	25	0.85	801	81	Deibel, 1982b
both	E	0.0624	2.75	54		250	35	Mullin, 1983
both	F	0.0726	2.83	41		350	49	Mullin, 1983
Dolioletta gegenbauri								
gnz	E, F	0.186	1.34		0.65		1.2	Crocker *et al.*, 1991
phz	E, F	0.03	1.77		0.72		0.3	Crocker *et al.*, 1991
gnz†	B	0.37	2.38	23	0.58		10.0	Deibel, 1982b
ooz	B	0.14	3.60	18	0.88		21.0	Deibel, 1982b
phz	I	-0.52‡	2.02‡	52	0.23		7.6	Tebeau & Madin, 1994
trz	I	-1.65‡	2.66‡	42	0.22		9.0	Tebeau & Madin, 1994

Coefficients (a, b) given for equations relating clearance rate to body length as $C = aL^b$, where C = clearance rate (ml $individual^{-1}$ h^{-1}), L = body length (mm) (except *Salpa thompsoni*, mgC weight). Weight-specific clearance rates are given in ml mgC $weight^{-1}$ h^{-1} for individuals 25 mm long, except *C.floridana* (15 mm), *Thalia* (10 mm), and *D.gegenbauri* (4 mm); example values of clearance rates are for individuals of these same sizes. Length to carbon weight conversions from Table 5.3.

Key to methods: A, particle clearance experiment using cultured algae and Coulter counter, F_{high} rates; B, as above, $F_{overall}$ rates; C, as A, but chlorophyll as measurements instead of Coulter counter; D, as B, but microscopic counts instead of Coulter counter; E, *in situ* chamber with radiolabelled bacteria as food; F, as E, but with radiolabelled algae; G, *in situ* gut pigment content; H, *in situ* diatoms in gut contents; I, uptake of plastic beads.

* Length-based regression recalculated from data in Deibel (1982a,b). Values of n and R^2 from original weight-based regression.

† Includes some phorozooids.

‡ Coefficients of linear regression, $a + b \times$ length.

specific clearance rates change only 25-fold, from 83–1041 ml^{-1} h^{-1} (Deibel, 1982b; Alldredge, 1984). This suggests that there are fundamental rules of metabolism in marine zooplankton that span gelatinous and crustacean taxa.

Representative clearance rates for all salp and doliolid species studied to date are given in Table 5.1, with sample values calculated for individuals 25 mm long. The original data for this table were reported in a variety of units and expressions, which have been converted here to power functions relating clearance to body length. The table serves to illustrate the considerable variation in rates, not only among species, but between different methods applied to the same species (Madin and Kremer, 1995). The variability makes assignment of characteristic values to particular species rather imprecise, but species of the genus *Salpa* have some of the highest rates, and the cyclosalps some of the lowest.

Weight-specific rates are also given in Table 5.1, calculated for individuals of the same length. Although weight-specific physiological functions, like filtration rate, are inversely related to body size in most organisms (e.g. copepods and ascidians), some studies of salp feeding suggests that these rates remain constant (Andersen, 1985) or even increase (Harbison and Gilmer, 1976; Deibel, 1982b; Mullin, 1983) in larger individuals. If this is the case, it might be attributed to the differences in exponents relating filtration and weight to body length (Mullin, 1983), or to the use of muscular rather than ciliary pumping mechanisms (Harbison and Gilmer, 1976). The variability of existing feeding data makes such generalizations rather speculative.

Clearance rate is a useful parameter for comparison of feeding among species, or for calculations of ecological effects, because it represents inherent capabilities of the salps. Ingestion depends on clearance rate and the concentration of available food. In nature, food availability will vary spatially and temporally, sometimes over quite small scales; in experiments, food level is often arbitrary. For this reason, ingestion rates are not summarized here, although some are included in the section below on energy budgets.

The erroneous distinction between 'slow' and 'fast' feeding (Fedele, 1933a; Carlisle, 1950) led to the hypothesis that salps regulated their feeding effort in response to gut fullness and/or the type or concentration of food particles in the water. Salps possess a variety of innervated structures around the oral lips and on the inner pharynx wall that are considered sensory in nature (Madin, 1995; see also Chapter 4). Fedele (1933a,b) had evidence from laboratory experiments of the chemical sensitivity of salps. He thought it might allow salps to select water masses and avoid unpleasant stimuli, but did not regulate feeding in response to chemical properties of the food. Short-term control of feeding activity he attributed instead to mechanoreception. Tactile receptors were thought to sense relative food concentration from the collisions of suspended particles, and regulate the production of mucus to switch from 'slow' to 'fast' feeding.

Since salps are now known to feed only with an expanded pharyngeal filter net, the only ways they could modulate feeding rates are to change the rate at which they pump water through the net (also affecting swimming speed), or to vary the amount of time the net is in place. The frequency of muscle contraction is known to vary with temperature in some species (Harbison and Campenot, 1979), but there is no evidence to date for variation related to food availability. Madin (1974 a, b) found pharyngeal nets in place an average of 91% of the time in six species observed underwater, but also observed that repeated labelling of the nets with carmine particles interrupted feeding. He concluded that salps fed continuously in oligotrophic environments (at least during daylight), but could interrupt production of the mucous net if they encountered high particle densities. Irregular production of the mucous net is common in laboratory experiments, but has generally been interpreted as artifact (Harbison and Gilmer, 1976; Bone *et al.*, 1991). Because of the sensitivity of salps to laboratory confinement and the difficulty of visualizing the pharyngeal net, it has not been possible to determine experimentally a correlation between presence of the net and food concentration.

Only a few feeding studies have varied food availability, and the effect of food concentration on clearance rates remains equivocal. Deibel (1982b) found no relationship between algae concentration and clearance rate in *Thalia democratica*, but in a later study with natural seston as food, noted a 3-fold decline in clearance rate as particle concentration increased by a factor of 18 to about 1.2 p.p.m. (Deibel, 1985a). At concentrations around 5 p.p.m. (5×10^6 μm^3 ml^{-1}), Deibel observed that muscle contractions of the salps became shallow and irregular, probably causing reduced clearance efficiency. Andersen (1985) reported an exponential decline, also by a factor of about three, in clearance rate of *Salpa fusiformis* over a 32-fold range of particle concentrations, with a maximum of about 5 p.p.m. Particle concentrations around 5 p.p.m. almost always produced clogging and diminished ingestion in *Pegea confoederata* (Harbison *et al.*, 1986). It is possible that the comparable high particle concentrations used by Andersen (1985) diminished apparent clearance in *S.fusiformis* by the same mechanism. Variability of the results makes it difficult to see clear functional responses in most of these cases, but particle concentrations above 5 p.p.m. seem to have an adverse effect on feeding.

The role of temperature in regulating feeding rate was also investigated by Andersen (1986), who found that rates for *S.fusiformis* increased by a factor of about six from 13 to 19°C, and then diminished at higher temperatures. Low clearance rates for the Antarctic species *S.thompsoni* at 1.0°C, relative to other species at 16–29°C, have been attributed mainly to the difference in ambient temperature (Reinke, 1987; Huntley *et al.*, 1989). Slowing or cessation of muscle contraction by some species as they are cooled would also affect filtering rates. Harbison and Campenot (1979) reported values of Q_{10} based on changes in swimming pulsations ranging from about 2 to over 20 in non-migratory species of salps and generally < 2 in vertical migrators.

There is now some evidence of periodicity of feeding on diel time scales. Purcell and Madin (1991) found that *Cyclosalpa bakeri* undertook diel vertical migrations over a distance of about 30 m, and ceased feeding in the surface waters at night. This result is contrary to the prediction of most theories for vertical migration. There was evidence that the salps were spawning while aggregated at the surface, and Purcell and Madin suggested that the cessation of feeding prevented the pharyngeal net from filtering out sperm before they reached the oviduct located in the atrial wall. It has not yet been determined whether this kind of feeding periodicity occurs in other salps, although several other species are vertical migrators (Franqueville, 1971; Wiebe *et al.*, 1979; Madin *et al.*, 1996; see also Chapter 7).

5.2. Assimilation and defaecation

5.2.1. *Doliolids*

The assimilation capabilities of doliolids have not yet been investigated, nor has faecal pellet production been quantified. In addition, the degree to which they digest and assimilate material from the ingested mucous cord is unknown. Ascidian guts have been found to contain amylase, saccharase, lipase, and a protease, but primarily enzymes for digesting polysaccharides (Barrington, 1965). This fits with the O:N ratio of doliolids (see below) and with the absence of storage lipids in tunicates generally (Deibel *et al.*, 1992). The endostyle of ascidians also produces several digestive enzymes, and presumably extracellular digestion begins as soon as the particles are trapped in the filter (Godeaux, 1989). Although Fedele (1921) suggested that the pharyngeal mucous filter may possess extracellular digestion capabilities, the occurrence of these enzymes in doliolids has not been examined. Gut passage time ranges from 4 (Fedele, 1921) to 15 min (Tebeau and Madin, 1994), and faecal pellets range in volume from 4×10^4 to 1.4×10^8 μm^3 (Fedele, 1921; Deibel, 1990). Pellets at the large end of this size range are likely to sink sufficiently fast to leave the mixed layer intact (Pomeroy *et al.*, 1984; Deibel, 1990) and should contribute to near-bottom food webs.

5.2.2. *Salps*

Qualitative observation of gut contents and faecal material suggests that digestion by salps is incomplete. Madin (1974a and b) reported cells with intact chloroplasts in gut contents, and others have found that diatoms (Harbison and Gilmer, 1976; Pomeroy and Deibel, 1980), thecate dinoflagellates (Silver and Bruland, 1981), and cyanobacteria (Caron *et al.*, 1989) can pass through the salp's intestine unharmed. Experiments using the ratio method (Conover, 1966) of calculating assimilation efficiency showed that the percent assimilated ranged from 28 to 39% for *Salpa fusiformis* fed a diatom (*Phaeodactylum*), but from 39 to 81% for salps fed the flagellate *Hymenomonas* (Andersen, 1986), perhaps reflecting relative digestibility of the algae. Assimilation efficiency did not vary significantly as a function of food concentration over a 10-fold range for either alga.

An analogous ratio method was used by Madin and Purcell (1992) with *Cyclosalpa bakeri*, using chlorophyll and its degradation pigments in the faeces as the unassimilated portion. They obtained values of 61% C and 71% N, but this method is likely to overestimate efficiencies because some loss of the pigments does occur in the gut. The available data suggest (as might be expected!) that salps eating natural diets have assimilation efficiencies similar to many other gelatinous and crustacean zooplankton (Alldredge, 1984).

Like ingestion, defaecation rates depend partly on the concentration of available food. Because of the significance of salp faecal pellets for the transport of small particulate material to the deep ocean, several investigators have measured faecal production rates (Wiebe *et al.*, 1979; Madin, 1982; Huntley *et al.*, 1989). Defaecation rates have also been used as an index of feeding (see above). Examples of these rates for several species are given in Table 5.2; generally defaecation amounts to between 10 and 30% of body C day^{-1} (Madin, 1982; Huntley *et al.*, 1989; Kremer and Madin, in preparation). Carbon:nitrogen ratios of salp faecal pellets range from about 10 to 20 (Madin, 1982; Caron *et al.*, 1989). These values could indicate relatively higher assimilation of nitrogen than of carbon from ingested food, but may also reflect the presence in the faeces of carbon-rich mucus from the feeding net.

Faecal pellets of most larger salp species (see Fig 7.1) are compact and fairly dense (Silver and Bruland, 1981; Madin, 1982), and probably maintain their integrity for some time after release. Caron *et al.* (1989) found that faecal pellets incubated for 10 days supported substantial microbial and protozoan populations, but remained physically intact and retained an average of 78% of their original carbon content, and 69% of their original nitrogen. The dense pellets have sinking rates as high as 2400 m day^{-1} (Caron *et al.*, 1989), and presumably can contribute significantly to fluxes out of the euphotic

Table 5.2 Weight-specific rates of defaecation by salps μg faecal C × mg body C^{-1} h^{-1}.

Species	Generation	Defecation (mean ± SD)	Reference
Cyclosalpa affinis	agg	9.9 ± 4.4	Madin, 1982
Cyclosalpa pinnata	agg	3.7 ± 1.9	Madin, 1982
Pegea bicaudata	agg	6.5 ± 1.4	Madin, 1982
Pegea confoederata	sol	27.7 ± 8.3	Madin, 1982
Pegea socia	agg	9.9 ± 1.0	Madin, 1982
Salpa cylindrica	sol	12.2 ± 8.0	Madin, 1982
Salpa maxima	agg	13.0 ±7.7	Madin, 1982
Salpa thompsoni	agg	4.25*	Huntley *et al.*, 1989

*Calculated from daily rate.

zone at times of high salp abundance. A more recent study by Yoon *et al.* (1996) of the faecal pellets of *Pegea confoederata* examined the changes in the faecal pellets held in suspension for 10 days. They obtained essentially similar results; however, sinking rates were measured under different conditions of turbulence, varying between 2706 and 3646 m day^{-1}. The pellets never remained longer in the 0–200 m layer than 100 min. In contrast, the faecal pellets of the smaller *Thalia democratica* and aggregate *Salpa cylindrica* are much more flocculent, and are degraded almost completely within about 4 days (Pomeroy and Deibel, 1980; Pomeroy *et al.*, 1984). The more flocculent faeces of small salps are degraded and recycled largely within the mixed layer, where they appear to be a source of marine snow (see Chapter 10).

5.3. Metabolism

5.3.1. *Pyrosomas*

No work has yet been done on the metabolism of pyrosomas.

5.3.2. *Doliolids*

Specialized excretory organs in doliolids are unknown (Barrington, 1965). The only published data on respiration and excretion of doliolids are from Biggs (1977), and these weight-specific values are given in Tables 5.4 and 5.5. The weight-specific rates are quite high in comparison with other species in the tables, or with other gelatinous zooplankton measured by Biggs (1977), as is the mean O:N ratio of 45 ± 24. The high O:N ratio suggests that doliolids do not have a protein-based metabolism, and since pelagic tunicates probably do not store much high energy lipid either (Deibel *et al.*, 1992), they presumably rely primarily on carbohydrates for metabolic energy. This is supported by the finding that the tunicate endostyle produces large quantities of enzymes that digest polysaccharides (Barrington, 1965), and by the presence of much glycogen in the muscle fibres (Bone, unpublished results).

5.3.3. *Salps*

Several investigators have measured rates of oxygen consumption by salps, using different methods. As with feeding rates, metabolic measurements are susceptible to artifacts from the collection or confinement of the salps. In many of the earlier studies (Rajogopal, 1962; Silver, 1971; Nival *et al.*, 1972; Mayzaud and Dallot, 1973; Ikeda and Mitchell, 1982) salps used for respiration incubations were collected with plankton nets, separated from the catch and held in laboratory aquaria for varying periods of time before oxygen measurements began. These results are variable, and probably unreliable. Other authors have used diver-collected salps in an effort to minimize stress due to sampling and handling (Biggs, 1977; Cetta *et al.*, 1986; Madin and Purcell, 1992). Even so, almost all data on oxygen consumption by salps is based on batch incubations in fairly small containers, and therefore suspect to some degree. The best results are probably those from short-term incubations at ambient temperatures and food levels.

Although metabolic rates of salps and other gelatinous animals often are lower on a length or dry weight basis than for non-gelatinous species, they are generally quite similar when expressed on a carbon weight basis (Ikeda and Mitchell, 1982; Cetta *et al.*, 1986; Schneider, 1990). This is because the high salt content of thaliacean tissues makes dry weight an inaccurate measure of biomass. Energetic parameters for gelatinous organisms should be expressed as functions of carbon weight wherever possible. Carbon and nitrogen weights of salps as a function of live length have been determined by several authors (Cetta *et al.*, 1986; Heron *et al.*, 1988; Madin *et al.*, 1981; Madin and Purcell, 1992). Regression equations for these relationships in 18 species are summarized in Table 5.3.

Table 5.3 Regression equations for carbon weight to live length for 18 species of salps.

Species Generation	Length Range	*a*	*b*	*n*	r^2	Reference
Brooksia rostrata						
agg	3–15	0.0011	2.04	12	0.91	Madin & Kremer, unpublished
sol	24–50	0.0146	1.46	5	0.88	Madin & Kremer, unpublished
Cyclosalpa affinis						
agg	7–54	0.0008	1.98	19	0.95	Madin *et al.*, 1981
sol	9–153	0.0109	1.54	16	0.92	Madin *et al.*, 1981
Cyclosalpa bakeri						
both	10–100	0.0051	1.75	47	0.83	Madin and Purcell, 1992
Cyclosalpa floridana						
agg	9–22	0.0162	0.72	9	0.56	Madin *et al.*, 1981
sol	6–22	0.0033	1.48	9	0.94	Madin *et al.*, 1981
Cyclosalpa polae						
agg	14–38	0.0017	1.72	6	0.99	Madin *et al.*, 1981
sol	7–75	0.0025	1.79	8	0.97	Madin *et al.*, 1981
Iasis zonaria						
agg	11–50	0.001	2.26	9	0.94	Madin & Kremer, unpublished
sol	26–60	0.0006	2.25	5	0.88	Madin & Kremer, unpublished
Ihlea asymmetrica						
agg	6–25	0.0041	1.53	10	0.65	Madin *et al.*, 1981
sol	8–33	0.0328	0.98	10	0.67	Madin *et al.*, 1981
Ihlea punctata						
agg	6–39	0.0005	2.38	9	0.94	Madin & Kremer, unpublished
Pegea bicaudata						
agg	16–72	0.0002	2.43	23	0.92	Madin *et al.*, 1981
sol	5–75	0.0131	1.74	8	0.96	Madin *et al.*, 1981
Pegea confoederata						
agg	8–90	0.0008	1.91	23	0.95	Madin *et al.*, 1981
sol	3–90	0.0282	1.03	26	0.86	Madin *et al.*, 1981
Pegea socia						
agg	29–91	0.0005	2.22	18	0.96	Madin *et al.*, 1981
Ritteriella retracta						
sol	30–72	0.0002	2.60	7	0.86	Madin & Kremer, unpublished
Salpa aspera						
agg	10–30	0.0005	2.47	24	0.89	Madin & Kremer, unpublished
sol	15–75	0.003	1.81	13	0.91	Madin & Kremer, unpublished
Salpa cylindrica						
agg	2–5	0.0016	1.67	6	0.53	Madin & Kremer, unpublished
sol	7–61	0.0014	1.92	19	0.91	Madin *et al.*, 1988
Salpa fusiformis						
agg	2–20	0.0005	2.78	12	0.98	Cetta *et al.*, 1986 Kremer, unpublished
sol	10–55	0.0014	2.05	20	0.94	Cetta *et al.*, 1986 Madin & Kremer, unpublished
Salpa maxima						
agg	9–93	0.0001	2.06	30	0.93	Madin *et al.*, 1981
Salpa thompsoni						
agg	n.d.*	0.0013	2.33	49	0.88	Huntley *et al.*, 1989
Thalia democratica						
agg	0.5–12	0.0014	2.04	22	0.91	Heron *et al.*, 1988
sol	1.7–12	0.0029	1.59	17	0.71	Heron *et al.*, 1988

Coefficients of the equation $C = aL^b$ are given, where C = carbon weight (mg), L = length (mm). Length ranges are for salps used for carbon determinations; values for a rounded to four decimal places.

*n.d., not determined.

Comparison of rates of oxygen consumption for different species is complicated because of differences in methods and formats for expressing results. In Table 5.4, respiration data for 12 species are expressed as functions of body weight and as weight-specific values. Examples are also calculated for salps of the same length, as previously for clearance rates. These data are not adjusted for temperature differences, but values for coldwater species (e.g. *Salpa thompsoni*) are approximately equivalent if corrected by a Q_{10} factor of about 2 (Andersen and Nival, 1986; Harbison and Campenot, 1979; Madin and Purcell, 1992). Remaining variability in weight-specific rates (Table 5.4) probably reflects both interspecific differences and those due to methods. Higher respiration rates were usually obtained with salps collected by divers and subjected to minimal handling. Cetta *et al.* (1986) compared rates of respiration and excretion of some individuals incubated directly in the jars in which they were collected by divers, and others held in filtered water for a few hours and then transferred to incubation vessels. They also compared rates from initial and subsequent incubations (4–6 h later) of the same individuals. In both cases, the freshly collected and least handled salps had rates up to twice as high as the others.

Table 5.4 Oxygen consumption by salps and doliolids.

Species Generation	Temperature	*a*	*b*	*n*	r^2	Weight-specific	Example	Reference
Cyclosalpa affinis								
agg	21–28	1.23	1.16	10	0.92	1.09	0.50	Cetta *et al.*, 1986
Cyclosalpa bakeri								
both	11	0.164	1.22	27	0.92	0.18	0.25	Madin & Purcell, 1992
Cyclosalpa pinnata								
agg	21–28	0.603	1.30	9	0.99	0.57	0.47	Cetta *et al.*, 1986
Cyclosalpa polae								
sol	21–28	0.371	1.63	6	0.77	0.32	0.25	Cetta *et al.*, 1986
Ihlea racovitzai								
not given	–1.0					0.09		Ikeda & Mitchell, 1982
Pegea confoederata								
agg	23–29	0.893	1.06	21	0.88	0.89	0.30	Biggs, 1977
sol	23–29					1.58		Biggs, 1977
agg	21–28	1.41	1.23	23	0.73	1.11	0.40	Cetta *et al.*, 1986
Pegea socia								
agg	21–18	0.80	1.07	7	0.93	0.75	0.45	Cetta *et al.*, 1986
Salpa cylindrica								
sol	23–29	1.82	0.71	13	0.88	2.45	0.84	Biggs, 1977
agg	23–29					1.74		Biggs, 1977
sol	21–28	3.63	1.06	19	0.93	3.40	1.14	Cetta *et al.*, 1986
Salpa fusiformis								
agg	13.5–19.5	0.426	0.68	15	0.90	0.28	1.03	Cetta *et al.*, 1986
sol	13.5–19.5	0.707	1.15	10	0.95	0.71	0.74	Cetta *et al.*, 1986
agg	15							
Salpa maxima								
agg	21–28	0.977	1.22	16	0.88	0.92	0.71	Cetta *et al.*, 1986
agg	23–29					0.63		Biggs, 1977
sol	23–29					1.25		Biggs, 1977
Salpa thompsoni								
agg*	–1.0	0.095	1.02	16	0.95	0.10	0.23	Ikeda & Mitchell, 1982
Thalia democratica								
agg	23–29					0.71		Biggs, 1977
sol	23–29					1.03		Biggs, 1977
Unidentified doliolids								
Unknown	23–29					2.4		Biggs, 1977

Coefficients of the equation $R = aW^b$ are given, where R = respiration rate (μmol O_2 individual^{-1} h^{-1}, W = weight (mg carbon or protein). Weight-specific rates are in μmol O_2 mg C (or protein) weight^{-1} h^{-1}. Examples values in μmol individual^{-1} h^{-1} are calculated for salps 25 mm long, except *Thalia* (10 mm). Where weight-specific rates are not given in the original source, they have been calculated for 25 or 10 mm individuals. Length to carbon weight conversions from Table 5.3. Original data have been converted to common units, but not adjusted for temperature differences.

*Mainly *S. thompsoni*, mixed with some *Ihlea racovitzai*.

Metabolic rates of organisms are commonly expressed as power functions of animal size or weight, $R = aW^b$. When the value of b is 1.0, respiration increases as a linear function of body weight. It has long been a paradigm of physiology that the metabolic rates of organisms ranging from protists to large mammals are related to body mass as a power function with an exponent somewhere between 0.67 and 0.88 (Withers, 1992), so that weight-specific respiration decreases as body weight increases. Various explanations of this 'universal' relationship have been suggested, but none satisfactorily encompasses the range of ectotherms and endotherms having similar values of b. One hypothesis suggests that exponents near 0.67 reflect the fact that exchange processes like respiration are proportional to surface area, and surface area is related to weight$^{2/3}$ (Withers, 1992).

Values of b for salps (Table 5.4) range from about 0.7 to 1.63, with a mean of 1.12, indicating that weight-specific respiration is usually independent of or increases with size, in contrast to most other organisms. The explanation for this is not obvious at present, but may be related to the geometry of salp bodies or to their swimming activity. Salps are essentially hollow cylinders, with the metabolically active tissue arranged in a nearly two-dimensional layer on the inner surface of the tunic, rather than in a three dimensional solid bounded by an outer epithelium. This would confer a high surface to volume ratio for animals of their size. Cnidarians and ctenophores also have thin layers of tissue over a substrate of non-metabolic mesoglea, and values of b close to or greater than 1 have been reported for some of these organisms (Biggs, 1977; Kremer, 1977; Kremer *et al.*, 1986). Because salps normally swim continuously, respiration experiments measure 'active' rather than 'basal' metabolism. A comparison of the respiration rate of swimming vs anaesthetized *S.fusiformis* (Trueman *et al.*, 1984) suggests that swimming accounts for well over half of total oxygen demand. The work provided by the muscles for swimming must both overcome the elasticity of the test and expel the water contained within the body chamber (Bone and Trueman, 1983). Test thickness tends to increase as salps grow larger, and the internal volume of the salp is an approximate cube function of length. Thus work for swimming, and metabolic demand of the muscles, might be expected to increase as a power function > 1 as body size increases. However, these possibilities remain conjectural until specific investigations are made.

The main nitrogenous excretory product of salps is ammonium, and this has been measured by several investigators in incubation experiments similar to those used to measure respiration (Andersen and Nival, 1986; Biggs, 1977; Cetta *et al.*, 1986). All the caveats mentioned above regarding collection, handling, and container size apply to excretion measurements as well as respiration. Rates of excretion of ammonium nitrogen for nine species of salps and unidentified doliolids are summarized in Table 5.5, converted to common units. The effect of temperature on excretion rate was studied by Andersen and Nival (1986) for *S.fusiformis.* They found an exponential increase with temperature, and calculated Q_{10} values of 2.3 for aggregates and 2.1 for solitaries.

While ammonium is the principal nitrogenous excretory product, salps also excrete other compounds. Cetta *et al.* (1986) found that urea accounted for up to half the total excretion in *Traustedtia multitentaculata*, but was low or absent in other species. Non-ammonium primary amines constituted about 12% of ammonium excretion in *Cyclosalpa bakeri* (Madin and Purcell, 1992), and non-ammonium nitrogen averaged 24% of total excretion in *C.affinis, P.confoederata, S.cylindrica,* and *S.maxima* (Cetta *et al.*, 1986). Production of uric acid as an excretory product has been reported for *Helicosalpa virgula* (Todaro, 1902), similar to benthic ascidians which sequester the crystals in the tunic (Barrington, 1965). Heron (1976) has described a structure (see p. 9) in *Thalia democratica* that removes crystalline particles from the bloodstream and excretes them through a pore in the test. He surmised that these crystals were uric acid resulting from purine metabolism, and that they were excreted so that their presence in the body would not make the salps more visible to predators. Heron observed similar organs in *S.fusiformis,* but there has been no further anatomical or physiological study of the excretion of solid nitrogenous wastes.

The only study to date of phosphate excretion is that of Ikeda and Mitchell (1982) for *Salpa thompsoni* and *Ihlea racovitzai.* Phosphate was about half the ammonium excretion of the former and about equal to the ammonium excretion of the latter.

5.4. Growth

5.4.1. *Measurement of growth rates*

Several methods have been used to attempt to measure salp growth rates. The simplest, in principle, is to maintain salps in the laboratory under favourable conditions and measure their growth as changes in length either directly or from photographs (Braconnot *et al.*, 1988). This approach has been used for several species (e.g. Braconnot *et al.*, 1988; Deibel, 1982a; Heron, 1972a; Madin and Purcell, 1992; Madin, unpublished results), but results have been variable owing to the difficulty of culturing salps in the laboratory under conditions which allow survival and growth comparable to that of field populations. Differences in container size, food supply, temperature, and initial condition of the animals appear to

Table 5.5 Excretion of ammonium nitrogen by salps and doliolids.

Species Generation	Temperature	*a*	*b*	*n*	r^2	Weight-specific	Example	Reference
Cyclosalpa affinis								
agg	21–28	0.102	1.09	10	0.94	0.096	0.044	Cetta *et al.*, 1986
Cyclosalpa bakeri								
both	11	0.027	1.01	27	0.85	0.03	0.04	Madin & Purcell, 1992
Cyclosalpa pinnata								
sol	21–28	0.055	1.33	10	0.96	0.051	0.042	Cetta *et al.*, 1986
Cyclosalpa polae								
sol	21–28	0.036	1.69	6	0.93	0.031	0.024	Cetta *et al.*, 1986
Pegea confoederata								
agg	26 ± 3	0.042	1.22	16	0.79	0.034	0.012	Biggs, 1977
agg	21–28	0.074	1.56	18	0.66	0.042	0.015	Cetta *et al.*, 1986
Salpa cylindrica								
sol	26 ± 3	0.141	0.64	9	0.51	0.209	0.070	Biggs, 1977
sol	21–28	0.290	0.38	17	0.21	0.569	0.190	Cetta *et al.*, 1986
Salpa fusiformis								
agg	16	0.100	0.94	11	0.87	0.092	0.339	Andersen & Nival, 1986
agg	14–20	0.043	0.59	14	0.35	0.025	0.092	Cetta *et al.*, 1986
sol	16	0.083	0.64	9	0.47	0.083	0.085	Andersen & Nival, 1986
sol	14–28	0.098	0.65	7	0.76	0.097	0.100	Cetta *et al.*, 1986
Salpa maxima								
agg	21–28	0.081	1.10	11	0.95	0.079	0.061	Cetta *et al.*, 1986
Salpa thompsoni								
agg	−1.0	0.448	0.92	16	0.78	0.417	1.01	Ikeda & Mitchell, 1982
Unidentified doliolids								
Unknown	23–29			3		2.2		Biggs, 1977

Coefficients of the equation $E = aW^b$ are given, where E = excretion rate (μmol NH_4^+ individual^{-1} h^{-1}), W = weight (mg carbon or protein). Weight-specific rates are in μmol NH_4^+ mg C (or protein) weight^{-1} h^{-1}. Sample values are calculated as in Table 5.4. Where weight-specific rates are not given in the original source, they have been calculated for 25 or 10 mm individuals. Length to carbon weight conversions from Table 5.3. Original data have been converted to common units, but not adjusted for temperature differences.

have significant effects on measured growth rates. For example, rates of growth for *Thalia democratica* obtained by Heron (1972a), were about 10 times higher than rates measured by Deibel (1982a) for the same species, a contrast attributed by Deibel and by Heron and Benham (1984) to differences in food availability and general condition. Container size in particular needs to be appropriate for the size and species of salp. For *Thalia*, Deibel (1982a) used 2.5 l bottles on a rotating plankton wheel, but for the larger and more active *S.fusiformis* Braconnot *et al.* (1988) used containers of up to 700 litres capacity. Data from Madin (unpublished results) was mainly from salps maintained in static aquaria of 4–20 litres capacity, but Madin and Purcell (1992) used a flow-through aquarium ('salpostat') that provided a continuous supply of surface seawater through 2.8 litre vessels.

Although Braconnot *et al.* (1988) and Madin and Purcell (1992) were able to maintain salps for up to 2 weeks in aquaria, survival times are usually much shorter, and growth rates can decrease rapidly during confinement. Heron (1972a) found that growth rates of *T.democratica* dropped sharply after only 5 h in captivity. Since it has rarely been possible to maintain individual salps through their entire lifespan, estimates of growth rate and generation time have sometimes been based on incremental rates of increase measured for salps of different sizes. Madin and Purcell (1992) used data from 87 individuals of *C.bakeri* that grew for varying lengths of time in aquaria to calculate the exponential growth rate, *g*, for salps of different sizes. The values were quite variable, but it was assumed that experimental conditions would only depress normal growth rates, and that the highest observed rates were most realistic. An exponential curve was fitted both to the entire data set and to the highest values to generate high and low bounds for instantaneous growth rate at any size. From these estimates, a growth curve for *C.bakeri* was obtained predicting generation times of 14 days at the high growth rate and 24 days using all data.

A similar approach was used with data for six other species (Madin, unpublished results). Only the highest value of *g* for each size category (1 mm size intervals) was plotted against size at the beginning of the growth increment (Fig. 5.6a). A best-fit equation was then fitted to the data to estimate growth rates as a function of size, which could then be used to generate a growth curve as in Madin and Purcell (1992) and estimate growth times to reproductive maturity and to maximum size (Fig. 5.6b).

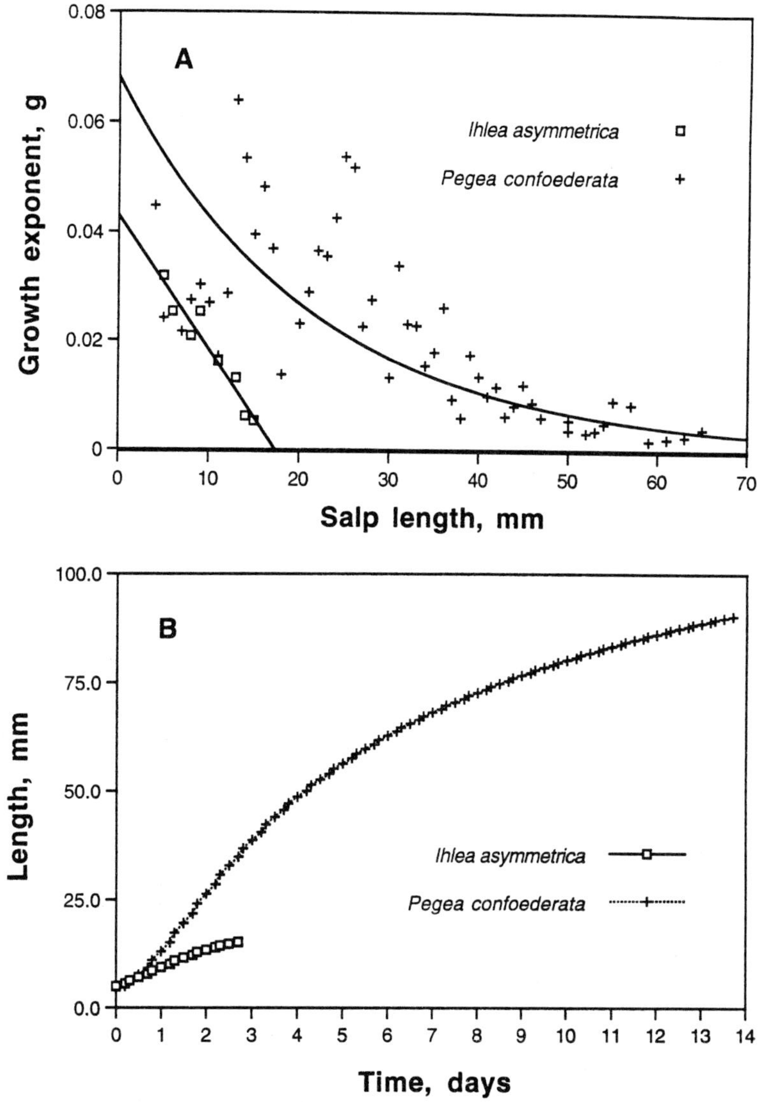

Fig. 5.6. Growth curves for *Ihlea asymmetrica* and *Pegea confoederata*. (A) Coefficient of instantaneous growth, *g*, determined from laboratory incubations, highest values for each size are plotted against salp length; for Ihlea $g = -0.002 \times \text{length} + 0.043$, ($r^2 = 0.93$), for *Pegea* $g = 0.0068 \times 10^{-0.02} \times \text{length}$ ($r^2 = 0.72$). (B) Growth curves from birth to maximum size, calculated from exponential growth equation using values of *g* from (a). Data from Madin (unpublished).

An alternative to laboratory culture is to estimate growth rate by cohort analysis of field samples, as commonly done for other zooplankton (e.g. Omori and Ikeda, 1984). Heron (1972a,b) sampled a developing population of *Thalia democratica* at intervals of 2 h, and calculated growth rate from the changes in size frequency peaks of solitary and aggregate salps. Rates of up to 20% length increase h^{-1} were found, and supported by very similar rates obtained from growth experiments in the laboratory. In a re-analysis of these and other data, Heron and Benham (1984) refined their method of cohort analysis and calculated rates for the aggregate generation of populations sampled in the winter, spring, and summer off Australia. The rather astonishing maximum growth rates they reported, comparable to some protists and bacteria, lent considerable support to the characterization of *T.democratica* as a colonizing species, capable of quickly dominating the zooplankton. A similar approach was used by Tsuda and Nemoto (1992).

In a subsequent paper, Heron and Benham (1985) developed a method which related growth rate to biometric characters of salps in a population. They identified five parameters of the life history of *T.democratica* that could be determined from preserved samples, and that were strongly correlated with one another. These parameters were also significantly correlated with growth rates previously determined for some of the same populations by cohort analysis (Heron and

Benham, 1984). In fact, a multiple linear regression on two parameters, number of aggregate buds per chain and ratio of aggregate offspring to their parents, accounted for 97% of the variation in growth rate; inclusion of three other parameters in the equation made no significant improvement in the correlation. Heron and Benham did not make any new estimates of growth rate based on this approach, but the method was subsequently used to estimate growth of *T.democratica* in the South Pacific (Le Borgne and Moll, 1986) and Kuroshio Current (Tsuda and Nemoto, 1992).

This method has the advantage of being applied to preserved samples, but the equations and coefficients determined by Heron and Benham for *Thalia democratica* are not directly applicable to other species. The appropriate life history parameters have to be determined empirically for each species, and correlated with measurements of growth rate made by some other method, before the method can be used with preserved samples. This has not yet been done for any species besides *T.democratica.*

A fourth alternative approach to estimating salp growth rates also avoids the necessity of rearing and measuring live animals. It uses a comparison of the elemental ratios (carbon:nitrogen:phosphorus) of the food supply and body tissues to calculate production rates, by weight, of these constituents of the salps (LeBorgne, 1978). These changes in the biomass represent individual growth rates, which can be multiplied by standing stock to estimate population growth (LeBorgne and Moll, 1986).

5.4.2. *Comparative growth rates of salps and doliolids*

Growth rate data from the literature and some previously unpublished work by Madin are summarized in Tables 5.6 and 5.7. The first table lists growth rates as percentage increase in length or carbon weight per hour, the units used in early work by Heron (1972a,b) and adopted by later authors for comparative purposes. Change in length is the actual measurement made in these experiments, and increase in weight is calculated from a known relationship between length and weight (Table 5.3). An exception is the method of Le Borgne and Moll (1986), which calculates weight changes directly from the C:N:P ratio. It is clear from Table 5.6 that the very high rates initially reported by Heron (1972a) are also found by other investigators, but only for *Thalia*, indicating that this small species grows and matures significantly faster than other salps studied. For most species listed, Table 5.6 gives a range of percentage increases based on growth increments measured on salps of different initial size. In all cases, the highest growth rates are found in the youngest salps, and diminish by a factor of 10 or more as salps near full size. Growth rates for most species seem to fall in the range 10–0.1% length increase h^{-1} over the life of the salp. These rates are quite comparable to increases reported for other gelatinous animals (Alldredge, 1984).

Growth measurements can also be used to calculate coefficients for various equations describing the growth curve, which can then be compared among species. Table 5.7 summarizes growth data in this form. Different equations (exponential or von Bertalanffy) have been used by different authors to describe the growth patterns they observed. The coefficient(s) of these equations (usually *g*, the exponent) have been measured or estimated in various ways. For example, Madin and Purcell (1992) and Madin (unpublished results) plotted the high values of *g* against body length at the beginning of each growth increment, and chose a best-fit equation to describe *g* as a function of length. Examples of such plots are given in Figure 5.6a. The growth equations given in Table 5.7 also permit creation of a model growth curve (Fig. 5.6b) and estimation of the time required for growth to a certain size. Estimated times required to grow to reproductive size and maximum observed size, determined in this way, are also given in Table 5.7. Since these times apply only to one generation, growth times to reproductive size must be summed for both generations in order to estimate generation time.

5.5. Energetics

5.5.1. *Doliolids*

At food concentrations of 30–146 μgC l^{-1}, ingestion rates of *Dolioletta gegenbauri* ranged from 1–27 μgC individual^{-1} day^{-1}, with a mean carbon-specific daily ration greater than 100% day^{-1} (Deibel, 1982b). These daily rations are similar to those of some salps (Table 5.8), appendicularia (Paffenhöfer, 1976), and copepods (Paffenhöfer and Harris, 1976; Paffenhöfer and Knowles, 1978), but they are much lower than the indirect estimate by Fedele (1921) for *Doliolum mulleri* of ~500% of dry weight day^{-1}.

The available respiration data coupled with carbon-specific clearance rates of ca. 20 ml μg C^{-1} day^{-1} for *D.gegenbauri* (Deibel, 1982b), equates to 15.4 litres of water filtered for each millilitre of oxygen consumed. This is in the middle of the range of most marine organisms (10–20 litres filtered ml^{-1} respired; Barrington, 1965). Converting these respiration rates to CO_2 respired allows us to estimate the metabolic demand, or amount of carbon required for respiratory maintenance of the organism per unit body carbon. Using $RQ = 0.9$ (Cetta *et al.*, 1986), this figure is ca. 1.2 μgCμgC^{-1} day^{-1} for doliolids, or 120% day^{-1}. This is almost identical to the daily ration of *D.gegenbauri* estimated from laboratory feeding studies of 132% day^{-1} (Deibel, 1982b), without accounting for losses due to an assimilation of less than 100%, and to excretion and growth. This suggests that in nature the daily rations of doliolids must be higher than

Table 5.6 Growth rates of salps expressed as percent increase in length (L) or carbon weight (W) per hour for nine species, as determined by various methods.

Species Generation	Size Range	Method	T	% L h^{1}	% W h^{-1}	Note	Reference
Cyclosalpa affinis							
sol	12–70	A	18–25	5.1–0.2	8.2–0.2	1	Madin, unpublished
Cyclosalpa bakeri							
agg	10–65	A	11	1.2	n.a.*	2, 4	Madin & Purcell, 1992
sol	15–100	A	11	0.6	n.a.	2, 4	Madin & Purcell, 1992
Ihlea asymmetrica							
sol	5–15	A	18	3.5–0.6	3.4–0.5	1	Madin, unpublished
Pegea bicaudata							
sol	17–93	A	18–25	4.9–0.1	10.4–0.1	1	Madin, unpublished
Pegea confoederata							
sol	4–65	A	18–25	7.1–0.2	7.3–0.2	1	Madin, unpublished
Pegea socia							
sol	35–100	A	18–25	4.2–0.2	n.a.	1	Madin, unpublished
Salpa fusiformis							
agg	2–22	A	15	1.5–0.4	n.a.	3	Braconnot *et al.*, 1988
sol	20–40	A	15	1.7–0.2	n.a.	3	Braconnot *et al.*, 1988
Salpa maxima							
sol	19–150	A	18–25	3.9–0.1	9.0–0.3	1	Madin, unpublished
Thalia democratica							
agg	1–6	A	20	0.9–0.3	n.a.	3	Deibel, 1982a
		A	n.a.	14	n.a.	4	Heron, 1972
		B	16–22	20–10	n.a.	5	Heron & Benham, 1984
	1–12	B	19–23	8.0	n.a.	6	Tsuda & Nemoto, 1992
	1–12	C	19–23	15.8	n.a.	6	Tsuda & Nemoto, 1992
both		C	30	28–25	n.a.	7	Le Borgne & Moll, 1986
		D	30	n.a.	36–34	8	Le Borgne & Moll, 1986
sol		A	n.a.	20	n.a.	4	Heron, 1972

See notes for methods and description of results. Size range (mm) is for salps used for growth data; T is experimental or environmental temperature (°C). Values rounded to two significant figures. Numbers of salps (N) and growth measurements used (n) for data from Madin (unpublished) are given for the same species in Table 5.7.
Methods: A, direct measurements in laboratory culture; B, cohort analysis from field data; C, ratio of life history parameters; D, C:N:P ratio.
Notes: 1, range from minimum to maximum initial size; 2, mean rate to maximum size; 3, range of different experiments; 4, maximum rate; 5, range of populations from three seasons; 6, mean for population; 7, range of multiple samples from one population; 8, range of calculations based on C,N,P.
*n.a., not available.

Deibel's laboratory estimate of 132% day^{-1}, which was determined at relatively low food concentrations (i.e. < 150 $\mu gC\ l^{-1}$; see section 4.2). As food concentrations in the field are likely to be greater than 150 $\mu gC\ l^{-1}$ (Deibel, 1985a) and ingestion rates and the daily ration are food concentration dependent (Deibel, 1982b), the true daily ration in the field may well exceed 132% day^{-1}.

At a mean ingestion rate of 15 μgC individual $l^{-1}\ day^{-1}$ (Deibel, 1982b), swarm populations of doliolids (i.e. 3000 individuals m^{-3}) should ingest 45 mgC $m^{-3}\ day^{-1}$, or ca. 20% of the standing stock of POC in Georgia coastal waters (Deibel, 1985b). This is comparable to 10% day^{-1} calculated for *D.gegenbauri* off the coast of California, where they occurred at concentrations of ca. 800 individuals m^{-3} (Crocker *et al.*, 1991). This grazing capacity amounts to 38% of peak primary production during shelf-break phytoplankton blooms and doliolid swarms (i.e. 120 mgC $m^{-3}\ day^{-1}$), and 188% of the daily primary production during non-bloom intervals of both phytoplankton and doliolids (i.e. 24 mgC $m^{-3}\ day^{-1}$; Yoder *et al.*, 1983). This suggests strongly that doliolid swarms require phytoplankton blooms to achieve the observed high concentration of zooids, and that those populations can have a significant grazing impact on the phytoplankton stocks.

5.5.2. *Salps*

Based on laboratory data, several authors have calculated daily rations, or expressed other energetic parameters, as a function of body carbon. Determination of a full carbon and nitrogen budget has apparently been done only for *Cyclosalpa bakeri* (Madin and Purcell, 1992), but budgets have been estimated based on partial data for several other species (Table 5.8). The values in the table are given on an hourly basis for comparison with rates given

Table 5.7 Growth equations and coefficients for 10 salps and doliolids.

Species Generation	Growth equation	Growth coefficient N, n, r^2	Time to grow (days) Repro	Max	Note	Reference
Cyclosalpa affinis						
sol	$L_t = L_{t-1}e^{gt}$	$g = 0.059 \times 10^{(-0.017L)}$ $N = 33$ $n = 44$ $r^2 = 0.45$	4.8	60	1	Madin, unpublished
Cyclosalpa bakeri						
agg	$L_t = L_{t-1}e^{gt}$	$g = 0.028 \times 10^{(-0.017L)}$	8.0	12	1	Madin & Purcell, 1992
sol		$N = 14$ $n = 14$ $r^2 = 0.89$	6.0	30		
Ihlea asymmetrica						
sol	$L_t = L_{t-1}e^{gt}$	$g = 0.043–0.0025L$ $N = 5$ $n = 8$ $r^2 = 0.93$	n.a.	2.7	1	Madin, unpublished
Pegea bicaudata						
sol	$L_t = L_{t-1}e^{gt}$	$g = 0.098–0.0051 \times \log L$ $N = 9$ $n = 30$ $r^2 = 0.71$	2.2	5.0	1	Madin, unpublished
Pegea confoederata						
sol	$L_t = L_{t-1}e^{gt}$	$g = 0.068 \times 10^{(-0.02L)}$ $N = 44$ $n = 52$ $r^2 = 0.72$	4.2	14	1	Madin, unpublished
Pegea socia						
sol	$L_t = L_{t-1}e^{gt}$	$g = 27.335 \times L^{-2.016}$ $N = 9$ $n = 25$ $r^2 = 0.51$	4.2	15	1	Madin, unpublished
Salpa fusiformis						
agg	$L = A(1-e^{-kt})^B + L_0$	$A = 11.73$, $k = 0.24$, $B = 4.27$, $L_0 = 2.91$	8.0	10	2	Braconnot *et al.*, 1988
	$L = At + B$	$A = 1.33$, $B = 1.53$			3	Braconnot *et al.*, 1988
sol	$L = A(1-e^{-k(t-t_0)})$	$A = 36.47$, $k = 0.21$, $t_0 = 2.14$	6.0	8.0	3	Braconnot *et al.*, 1988
Salpa maxima						
sol	$L_t = L_{t-1}e^{gt}$	$g = 0.618 \times L^{-1.705}$ $N = 16$ $n = 42$ $r^2 = 0.40$	11	15	1	Madin, unpublished
Thalia democratica						
agg	$L_t = L_{t-1}e^{gt}$	$g = 0.0065$	ca.11	n.a.	4	Deibel, 1982a
	$L_t = L_{t-1}(1 + Gt)$	$G = 0.2219L^{-0.6875}$	n.a.	n.a.	5	Heron & Benham, 1984
		$G = 0.5143L^{-0.866}$	n.a.	n.a.	6	Heron & Benham, 1984
		$G = 0.3759L^{-0.853}$	n.a.	n.a.	7	Heron & Benham, 1984
Dolioletta gegenbauri						
gonozooid	$L_t = L_{t-1}e^{gt}$	$g = 0.007$	n.a.	n.a.	8	Deibel, 1982a

Predicted times (days) for growth to first reproductive size (repro) and to maximum reported (max) size are based on the growth equation given or on direct observations. When times for growth to reproductive size are known for both solitary and aggregate forms of a species, generation time can be estimated as their sum. See Figure 5.6 for examples of growth curves. Entries beneath growth coefficient equations: N = number of individual salps used; n = number of growth-increment measurements; r^2 = correlation coefficient for fitted curve; n.a. = not available.

Notes: 1, equation for g is best fit (r^2) to highest values for each size class; 2, mean of two individuals, growth times from direct observation; 3, mean of three individuals, growth times from direct observation; 4, mean value from two experiments for growth to maximum size achieved; 5, mean of winter populations; 6, mean of spring populations; 7, mean of summer populations; 8, mean value from one experiment for growth to maximum size achieved.

in other tables. The data available so far suggest that energetic requirements of salps are generally similar, on a carbon weight basis, to those of other kinds of zooplankton, but can vary among different salp species. For example, daily ration and metabolic rates for the relatively slow swimming *C.bakeri* and *P.confoederata* are lower than for the more active *S.cylindrica* and *S.fusiformis.* Temperature effects are presumably responsible for the low energetic demand of the Antarctic *S.thompsoni.*

Salps are notorious for rapid growth of individuals and populations (Tables 5.6 and 5.7). The allocation of energetic resources to these functions will vary depending on the size and reproductive state of the salp. Young salps and oozooids (solitaries) producing chains may be expected to invest the most in growth, while older male aggregates probably expend the least. Information on the energetics of growth is still limited, but in the case of *C.bakeri*, growth and reproduction of a 50 mm solitary producing an average of 5 aggregates day^{-1} (Madin and Purcell, 1992), amounted to almost 30% of body C day^{-1}. Even larger investments in reproduction would be expected from species of *Salpa* which produce multiple chains containing hundreds of individuals (Braconnot *et al.*, 1988; Foxton, 1966).

Table 5.8 Energetic parameters for salps and doliolids.

Species Generation	Length (mm)	C ingestion	C egestion	C respiration	N excretion	C growth	C reproduction	Reference
Cyclosalpa bakeri								
both	50	3.8	1.5	0.25	0.25	0.38	0.83	Madin & Purcell, 1992
Pegea confoederata								
agg	40	2.2*	0.62	1.5	0.52			Cetta *et al.*, 1986
Salpa cylindrica								
sol	40	5.3*	1.2	4.1	2.3			Cetta *et al.*, 1986
Salpa fusiformis								
agg	16	4.5						Andersen, 1985
sol	28	4.9						Anderson, 1985
Salpa maxima								
sol	40	2.4*	1.3	1.1	0.54			Cetta *et al.*, 1986
Salpa thompsoni								
sol	21	1.0*	0.43	0.62†	0.29†			Huntley *et al.*, 1989
Thalia democratica								
agg	3–4	2.5						Deibel, 1982b
agg	4–7	2.0						Deibel, 1985a
agg	8	8.3						Mullin, 1983
Dolioletta gegenbauri								
gnz		5.5						Deibel, 1982b

Data are expressed as % body C h^{-1}, except N excretion as % body N h^{-1}, uncorrected for temperature differences. Values are direct measurements except as noted.
* Ingestion calculated as sum of egestion plus respiration, assuming no growth or reproduction.
† Values calculated as function of body weight and temperature.

SIX

The appendicularian house

Per R. Flood and D. Deibel

6.1. History

The 'house' building capacity was early considered an important feature of the appendicularians, and the name 'Oikopleura' was coined by Mertens (1830, p. 210) soon after their original description by Chamisso and Eysenhardt (1821). *Οικος* (Oikos) is Greek for 'house' and *πλευρα* (pleura) is Greek for 'rib' or the side of an animal, but came early to be applied to the lining membrane of the chest wall (Skinner, 1961). Thus, Mertens' term 'Oikopleura' most likely points to the respiratory function which he erroneously assigned to the houses of these animals. However, it took almost 30 years before Mertens' (1830) observation of a house surrounding an appendicularian was verified by Allman (1859). He suggested the house to be a 'definitely shaped secretion, destined to act as a nidamental covering for the ova'. Moss (1870) gave some structural notes on inflating houses and presented interesting chemical data on the house material of an unidentified oikopleurid species, but did not comment on its function. Fol (1872) gave the first detailed description of the external houses in all three main families of appendicularians. His drawings still represent the most detailed published accounts of those of the Fritillariidae (26 species) and Kowalewskiidae (two species). A single photograph of a *Kowalewskia* house (Alldredge, 1976c) and a behavioural study of *Fritillaria pellucida* (Bone *et al.*, 1979) add little to this knowledge. For the Oikopleuridae (about 29 species) on the other hand, Fol's (1872) description of the house of *Oikopleura albicans* (at that time named *O.cophocerca*) was soon supplemented by the study by Eisen (1874), probably on *O.dioica*. Whereas Fol (1872, p. 495) suggested that the house served some protective function against predators, Eisen (1874, p. 6) was the first to vaguely suggest it served 'till at införa näringsämnen till munöppningen' (to introduce nutritional substances to the mouth opening). However, undoubtedly the detailed structural studies of Lohmann (1898, 1899c, 1909a, 1933, 1934), mostly on *O.albicans*, were needed before any precise ideas could be established on the functional significance of the house with respect to feeding ecology. The great complexity and extreme fragility of appendicularian houses has been a challenge for numerous subsequent investigators and improved understanding of the precise structure and mode of operation of the house has only slowly emerged.

6.2. General considerations

For appendicularians, with their recurved and twisted tails, several terms of anatomical orientation, like dorsal – ventral, oral – caudal, are ambigous and easily misunderstood. In addition, for several species the mouth of the animal points backward compared to the direction in which the animal and house complex moves through the sea. The use of anterior and posterior may therefore also cause some confusion. It has been proposed (Flood, 1991a) to regard the house as a vessel or ship, and to describe the position of its structural components relative to the way in which this vessel normally orients itself and moves through the water mass (Alldredge, 1976b). This works well for the better known houses of small Oikopleuridae (Flood, 1991a), and a comparable 'reference' orientation should be adopted for other appendicularian houses. In this reference orientation the animal should be suspended within the house in its sagittal plane; trunk up with mouth pointing aft or up, gonad pointing forward, and tail pointing down and backward.

In brief outline, appendicularian houses are complex filtration devices which allow the animals to feed on a highly concentrated suspension of particles extracted from the water pumped through the houses. The muscular activity of the animal's tail, lodged within a close fitting tail chamber of the house, represents the driving force for this pumping action. The tail pump sucks water into the house, often through bilateral coarse-meshed inlet filters. These are assumed to prevent large and potentially harmful particles from gaining access to the interior of the house. Downstream of the tail, water is forced through extensive food concentrating filters of extremely small and regular pore size before it can leave the house. Living and dead particles, down to about 0.2 μm in diameter, are efficiently trapped in these filters and a food collecting tube, spanning the gap between the filters and the mouth of the animal, allows the animal to drain the particles into its pharynx. Here, however, the particles have to be *recaptured* by an endostylar pharyngeal filter, as in all other tunicates. Further details of the feeding of appendicularians are given in Chapter 8. Today the houses of several oikopleurid species are fairly well known, and these will therefore be considered below under several headings.* A better understanding is also emerging for the houses of some Fritillariidae (Flood, unpublished observation). Since these are quite different from oikopleurid houses, they are treated separately. The houses of the third appendicularian family, the Kowalevskiidae (and those of several genera within the Oikopleuridae and Fritillariidae) are obviously and profoundly different from the types mentioned above. These houses are unfortunately poorly known, and will be treated only summarily at the end of this survey.

6.3. The oikopleurid house

6.3.1. *Anatomy of the oikopleurid house*

The size of oikopleurid houses varies according to species; from that of *O.dioica*, approximately 4 mm in diameter, to the houses of *O.vanhoeffeni*, 6–7 cm long. With few exceptions, the house diameter is about twice as large as the total length of the animal (Table 6.2). The external shape of the houses varies as well, from the almost globular house of *O.dioica* to rather elongated and more complex shapes in several larger species. Our most detailed accounts of the appearance of the house of *O.albicans* are still those of Lohmann (1898, 1933, 1934). Alldredge (1977) gave the most detailed accounts of the houses of *O.cornutogastra, O.fusiformis, O.intermedia, O.longicauda, O.rufescens, Megalocercus huxleyi,* and *Stegosoma magnum.* Flood (1983) and Fenaux (1986a) gave details of *O.dioica*, Deibel (1986), Flood *et al.* (1990) and Flood (1994) of *O.vanhoeffeni*, and Flood (1991a) of *O.labradoriensis* (Fig. 6.1a). Detailed knowledge also exists of the house of a newly discovered mesopelagic species, named *Oikopleura villafrancae* by Fenaux (1992) (Flood, Galt and Youngbluth, in preparation).

Based on a detailed study of the contributions mentioned above, and with due regard to both technical, documentational, and interpretational aspects, the present authors agree with most previous authors, that in principle all oikopleurid houses contain the same water passages, chambers, valves, and filters. Thus, they all are built on the same architectural plan and function in the same general way. This is not to say that they all look similar. Obviously, pronounced species-specific differences exist. However, these differences are mostly related to the presence or absence of specific appendages or crevices to some chambers, or to the dimensions or precise geometrical shapes of chambers and filters. It is beyond the scope of this presentation to review and discuss these species–specific differences. We will rather consider uncertainties related to the functional principles of the oikopleurid house. However, since the different nomenclatures used by previous authors may cause considerable confusion, a tabular comparison of the terms used, and those recommended, is given in Table 6.1.

In general terms (and with reference to Fig. 6.1), the oikopleurid house may be said to be spheroid and to possess two inlet openings with funnels leading to a midline trunk and tail chamber in which the animal is suspended from its walls only by its trunk. Regular periods of undulatory movements of the tail in a close-fitting tail chamber act like a peristaltic pump to suck water into the house through the bilateral inlet funnels, which normally is spanned by a coarse-meshed inlet filter. The frequency of tail beat during these bursts of pumping is low for the larger species; in *O.labradoriensis* 2.5–3.0 Hz (Bone and Mackie, 1975) and in *O.vanhoeffeni* 1–2.5 Hz (Morris and Deibel, 1993), whilst it is higher in the smaller species (*O.dioica* ≈10 Hz; Flood, personal observation). Temperature may also influence tail beat frequency significantly.

The tail pump also forces water from the downstream end of the tail chamber through bilateral supply passages to the lateral border of a multilayered food concentrating trap. This trap consists of two filtering screens held together by a complex set of suspensory filaments. The entire trap is fairly rigidly suspended across the posterior

*The houses of the genera *Bathochordaeus* and *Pelagopleura* are treated separately later in this chapter under the heading 'Other appendicularian houses', even if they belong to the family of Oikopleuridae according to Buckmann and Kapp (1975).

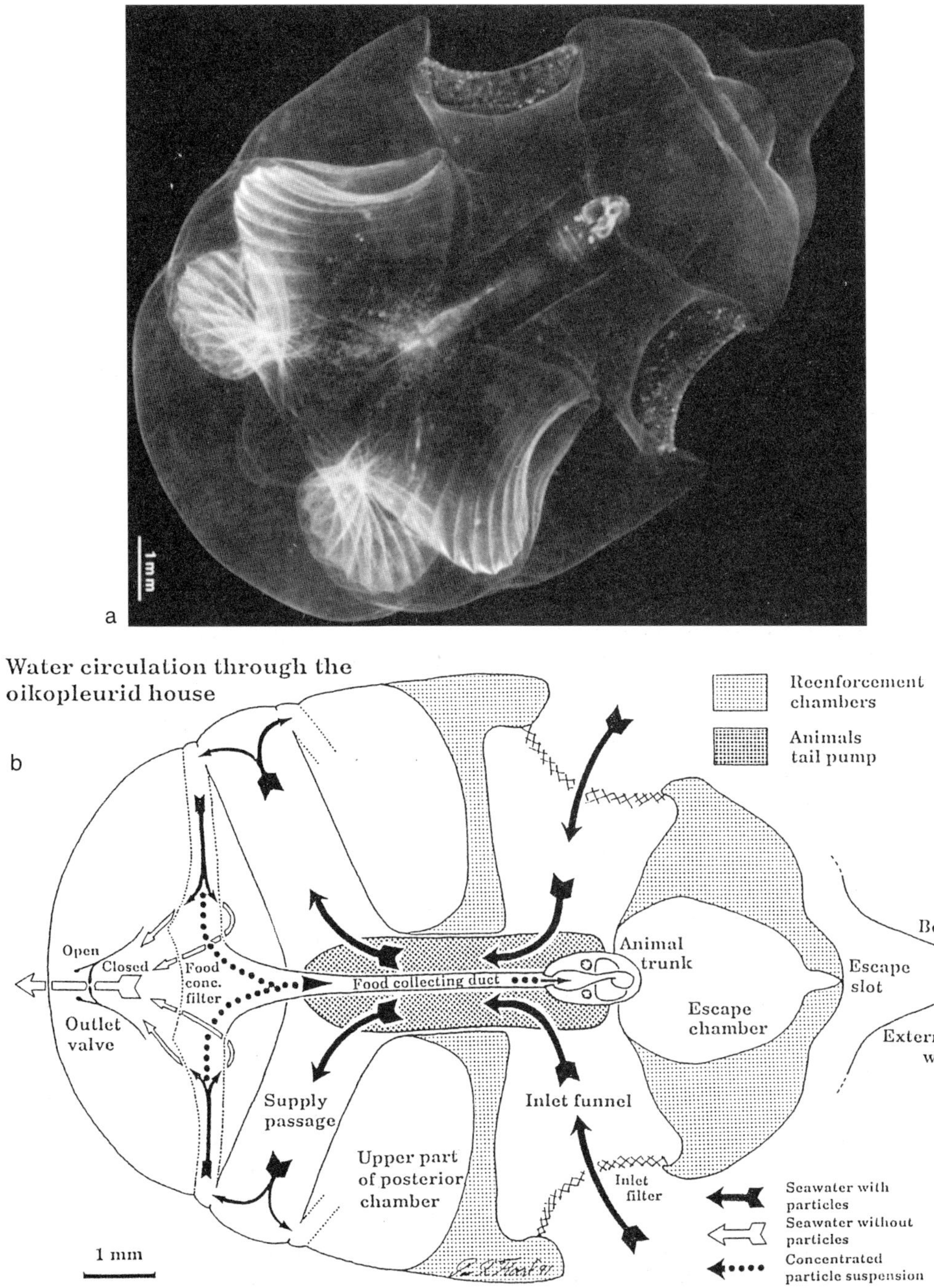

Fig. 6.1. Dark field macro-photograph (a) and explanatory diagram (b) of *O.labradoriensis* in its house, as seen from above after staining with carmine particles and melanin granules from *Sepia officinalis* ink. Black arrows indicate water flow through the house before, white arrows after, its passage through the food concentrating filter. Dotted arrows indicate the flow of trapped particles towards the mouth of the animal. Note: The external wall of the house is only visualized in the bow area.

half of the house as a recurved and corrugated set of wings. Water enters between the upper and lower filtering membranes and the tangential flow forces particles to accumulate near the mid-line where a food collecting tube connects the anterior margin of the trap to the mouth of the animal. Water that seeps through the filtering membranes accumulates in a large posterior chamber above and below the filter wings until the

Table 6.1 Recommended and previously used terms for structures associated with appendicularian houses.

Recommended terms and their first user	Synonyms used by different authors or explanation (=) of term
House-producing epithelium	
Oikoblastic cells[1]	Oikoplasten[2]
Oikoblastic epithelium[14]	Oikoplasten-Epithel
Oikoblastic cell groups/fields	Oikoplasten-Feldern[15,16]
Oikoplasts[2]	= oikoblastic cell groups
Non-expanded house	
House rudiment[8]	Haus Anlage[3], proto-house[4]
Bioluminescent bags[P]	Körperchen[4], Häutungskorpern[5], House etchings[6], inclusion bodies[8]
Lumisome[P]	= granules present in bioluminescent bags
House exterior	
External (outer) house wall[11]	Outer house membrane[7]
Bow[11]	Schnabelwurtzel[3], anterior keel[9]
Keel[10]	= Wedge-shaped bottom of house[10]
Posterior process[P]	Posterior keel[9], keel[11]
Lateral flaps[11]	Sails postero-lateral to inlet opening
Inlet opening	Einstromungs-Offnung[3], incurrent opening[7]
Inlet filter[13]	Einstromungs-Gitterfenster[3], incurrent filter[7]
Escape slot[14]	Flucht-Pforte[3], exit port[7], escape passage[11]
Outlet valve	Ausstromungs-Offnung[3], exit port[9], exit valve[13] Exit sphincter[10], excurrent passage[7]
Outlet spout	Erectile (outlet) appendage[10], exit spout[11]
Trailing threads	Schleppfaden[3]
House interior	
Internal house wall[11]	= internal aspect of peripheral house wall[11], house wall[9]
Inlet funnel[13]	Einstromungs-Trichter[3]
Trunk chamber	Rumpf-Kammer[3], anterior chamber[10]
Escape chamber[11]	Flucht-Kammer[3], anterior chamber[10]
Tail chamber	Schwanz-Kammer, tail passage[7]
ruffles[11]	quergerillt[3], streaked[10]
Supply passage[13]	Zwischenflugel-Kammer Sinus[18], tail passage[7], trap canal[10]
Cushion chambers[11]	
upper[14]	= above escape chamber
anterior[13]	= lateral to escape chamber
lateral[13]	= lateral flaps
inferior[14]	Reinforcement chamber[10]
Food concentrating trap[12]	Reusenwerk[3], trap[10]
Food conc. filters[13]	Fangnetz[3], feeding filter[7] Fol'sfilters[19]
Filter ridge[14]	Reusengang[3], corrugated fold[7] scalloped
Entrance slots[14]	Eintritts-Offnungen, trap inlet[10]
Upper filter[14]	Dorsal layer[17], external sheet[10]
Intermediate suspensory screen[14], middle layer[9]	
Lower filter[14]	Ventral layer[17], internal sheet[10]
Food concentrating filter junction	Median channel[7], middle junction[9]
Food collecting tube[13,14]	Mundrohr[3], buccal tube[7]
Posterior chamber[13]	
upper compartment of[13]	Rucken-Kammer[3], dorsal chamber[10]
lower compartment of[13]	Exit chamber[10]
Animal interior	
Feeding filter	Pharyngeal filter, endostylar filter

[1]Klaatsch, 1895; [2]Lohmann, 1896a; [3]Lohmann, 1896b; [4]Lohmann, 1905; [5]Lohmann, 1914; [6]Bückmann and Kapp, 1975; [7]Alldredge, 1976; [8]Galt, 1972; [9]Deibel, 1986; [10]Fenaux, 1986a; [11]Flood *et al.*, 1990; [12]Flood, 1981; [13]Flood, 1983; [14]Flood, 1991b; [15]Bückmann, 1923; [16]Lohmann and Bückmann, 1926; [17]Jørgensen, 1966a; [18]Lohmann, 1899a; [19]Fenaux, 1976; [P]present work.

pressure is sufficient to force open a complex water outlet valve projecting from the house as a short appendix.

Whereas Kümmel's (1956) transmission electron microscope study of the house of *O.dioica* added little to our understanding, Fjørdingstad (in Jörgensen, 1966a, fig. 1.47, 1966b figs 16 and 17), also studying *O.dioica*, presented the first evidence for regularly meshed food concentrating filters with a pore size in the 0.1 μm range. However, the extent of these exceedingly fine submicrometre filters inside the oikopleurid house, and their significance for the food collecting capacity of the house, was not fully appreciated until the scanning electron microscope studies of Flood (1978). Later, several studies of filters mounted directly on plastic- and carbon-coated grids for transmission electron microscopy (Flood, 1981, 1983; Flood *et al.*, 1983; Deibel *et al.*, 1985; Deibel and Powell, 1987b; Flood, 1991a) have enabled us to present precise filter dimensions for houses from a number of oikopleurid species (Fig. 6.2e and Table 6.2). It has also become clear (Flood, 1991a) that the upper and lower filtering membranes of *Oikopleura labradoriensis* reveal easily recognizable structural and dimensional differences. Similar differences are documented in Fig. 6.2a–d for *O.dioica* and are also known to exist for *O.vanhoeffeni* and *O.albicans* (Flood, unpublished observations).

The mesh size of the food concentrating filters appears to be nearly constant, irrespective of species and size of the houses. The mesh width, which is crucial for the straining capacity, ranges only between 0.15 and 0.24 μm (Table 6.2). Further, the meshes are always rectangular, with a length to width ratio ranging between 3.8 and 6 or 8. From a constructional point of view this design seems to represent the most economical way of making a small pore filter with the least expenditure of matter (Wallace and Malas, 1976). The filaments making up the meshes are also very thin, from 5 to 50 nm in diameter (Flood, 1981, 1991a; Deibel *et al.*, 1985), giving the filters the extremely high open area fraction (or porosity) of > 90%. Based on this porosity and the low velocity of water across these filters, ~2.5 μm s^{-1} according to Flood (1991a), it has also been calculated that the boundary layer around individual filaments is 30% of the pore width (Morris and Deibel, 1993). In spite of Reynolds numbers < 1, indicating laminar flow dominated by viscous forces, flow through these filters is therefore possible on theoretical grounds.

The mesh width of the inlet filters shows much larger species differences, mean values of 12.7 to 8.1 μm in the 11 species examined so far, with a tendency for larger houses to have wider pores (Table 6.2). It has also been shown that within one species (*Oikopleura vanhoeffeni*) individuals of increasing body size have inlet filters of increasing mesh width. A range from 35 to 155 μm was found by Deibel (1986). Again the porosity is very high, 90%, but the length to width ratio ranges from 1 (Alldredge, 1977) to 5.8 (Flood, 1991a). Comparing these parameters for the inlet and food concentrating filters, it may be suggested that the pore size of the latter is determined by the molecular properties of the cross-linked filaments, whereas the inlet filter pore size apparently is determined by the oikoblastic cells that secrete the anlage to the filter. Further, it may be stated that all the appendicularians examined have a comparable lower cut-off value for the size of particles they will concentrate and ingest, whereas the upper cut-off value differs among species. At present we can only speculate on the functional significance of the highly variable pore sizes of inlet filters in distinct size classes and species of oikopleurid appendicularians. The differences may represent an adaptation to distinct habitats, allowing the animals to graze upon distinct particle size populations, or put the other way, to prevent harmful organisms from gaining access to the interior of the houses. Acuña (1994) suggested that the vertical distribution of *O.longicauda* off northern Spain was linked to the absence of an inlet filter in this species, enabling it to feed on larger particles than other oikopleurids.

6.3.2. *Function of the oikopleurid house*

The detailed water circulation pattern through the oikopleurid house has been a matter of dispute until recently. Differences of opinion are mainly related to the inlet funnels, food concentrating filters, water outlet valve, and to the suspension of the animal relative to the chambers of the house. Lohmann (1899c) described a flap internal to the inlet filters in houses that prevented effective backwashing of these filters of the *O.albicans.* In *O.dioica*, Fenaux (1986a) described a membrane lateral to the inlet filter that closed off the greater part of the funnel, except for a middle slit with a mobile flap. Beyond individual filaments or reflections from the external walls of the houses we have been unable to see comparable structures in *O.vanhoeffeni* (Flood *et al.*, 1990), *O.labradoriensis* (Flood, 1991a), and *O.dioica* (Flood, unpublished observations). Lohmann (1899a, 1933, 1934), studying *O.albicans*, and Körner (1952) and Jörgensen (1966a,b), studying *O.dioica*, believed the food concentrating trap to be triple-layered, forming a lower set of influx and an upper set of efflux channels delineated by prominent ridges. Water was supposed to enter the influx channels laterally, and particles to be strained from the water only at the mid-line of the trap where the flow shifted from medially through the lower channels to laterally through the upper channels. Jörgensen (1966a,b) modified this hypothesis to include a porous middle layer where water could be strained from the lower to the upper channel over a larger area. Alldredge (1977), chiefly studying *Stegosoma magnum* and *Megalocercus huxleyi*, denied both the entrance of water at the lateral aspect of the filter ridges and the

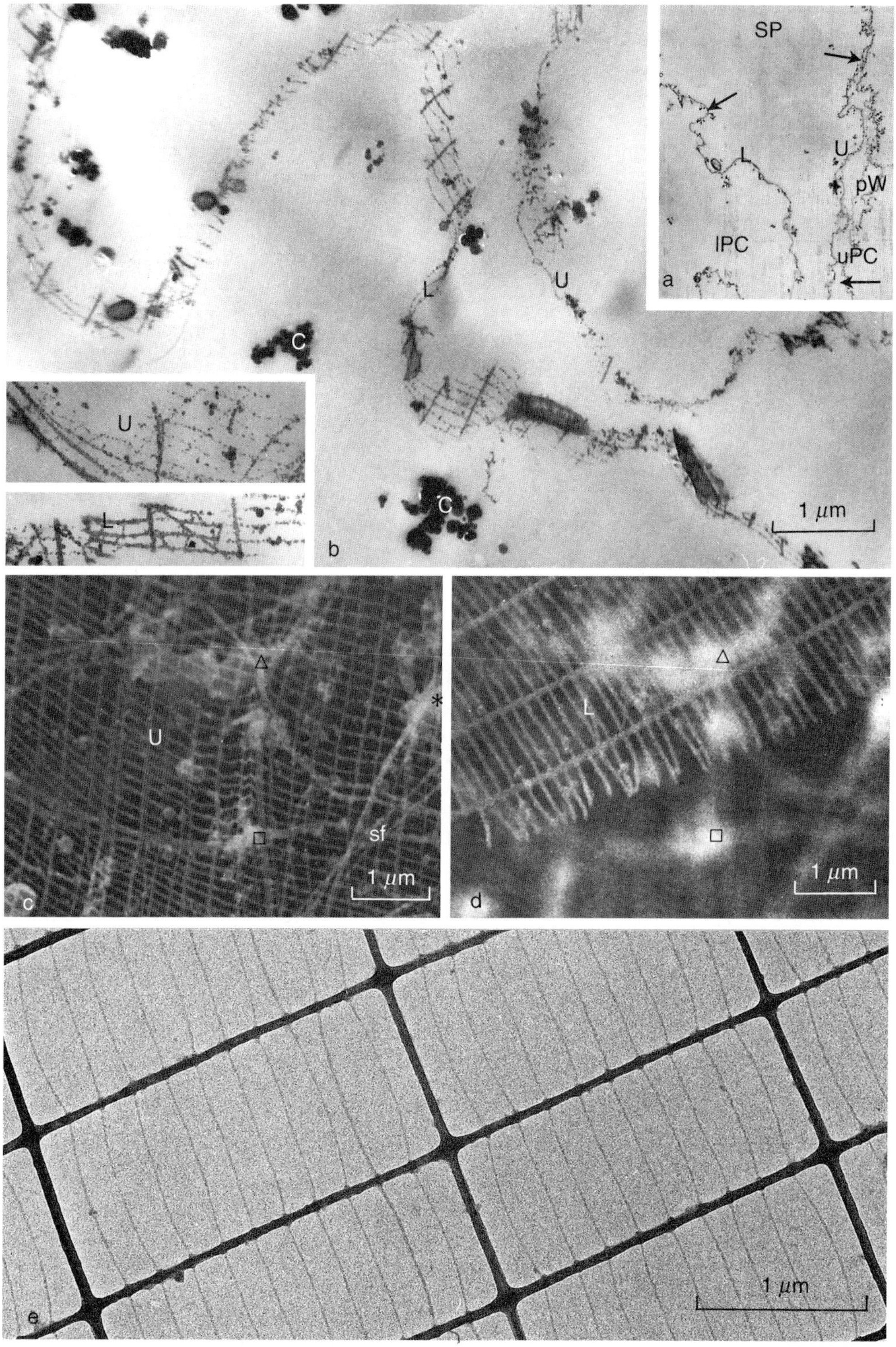

Fig. 6.2. (a) Light micrograph of a 1 μm thick section through a potassium permanganate fixed and plastic embedded house of *O.dioica*. The area shown represents the junction (arrows) between a supply passage (SP) and the upper (U) and lower (L) membranes of the food concentrating filter. The internal aspect of the peripheral house wall (pw) and the upper (uPC) and lower (lPC) parts of the posterior chamber are also seen. Magnification ×200. (b) Transmission electron micrograph of ultrathin section from the same material as shown in (a). The upper (U) and lower (L) filters of the food concentrating trap are included, together with a variety of trapped particles, including colloidal carbon particles (C) used to visualize the house *in vivo*. Magnification ×10 000. (Insets) Details of the upper (U) and lower (L) filters. Magnification ×15 000. (c and d) Scanning electron micrographs revealing respectively the difference in ultrastructure of the upper (U), the lower (L) food concentrating filter and the presence of intermediary suspensory filaments (sf) in the food concentrating trap of *O.dioica*. The two images represent identical fields of view (corresponding points are indicated by various symbols) at different focal settings. Magnification ×12 000. (e) Transmission electron micrograph of the lower food concentrating filter of the house of *O.albicans* mounted directly on a carbon and Formvar-coated grid and stained by uranyl acetate and lead hydroxide. Magnification ×26 000.

Table 6.2 Numerical data on selected appendicularian species and their houses.

Genus Species	*Animal size* Trunk length* (mm)	Tail length (mm)	House diameter (mean ± SD) (mm)	*Inlet filter mesh* Width (mean ± SD) (μm)	Length (mean ± SD) (μm)	Length: width ratio	*Food concentrating filter mesh* Width (mean ± SD) (nm)	Length (mean ± SD) (nm)	Length: width ratio		Open area fraction (%)	Filtration (F) or clearance (C) rate (ml h^{-1})		House renewal rate (houses day^{-1})
Oikopleura														
albicans	3.0[1]	8.0[1]	17.5[1]	34.5	127	3.7 : 1[2]	190 ± 10	920 ± 60	4.8 : 1[3]	(L)	≈90[3]			
dioica	1.3[1]	3.9[1]	5.0[2]	300	100	3.3 : 1[4]	150 ± 20	980 ± 220	6.5 : 1[3]	(U)	≈90[3]	8.5[5]–10[6]	(C)	4.1–10.6[21]
fusiformis	1.1[1]	4.4[1]	10.2 ± 2.7[7]	13 ± 5.1	13 ± 5.2	1 : 1[8]						72[8]	(F)	≈6[19]
cornutogastra	1.3[8]		29.4 ± 9.0[8]	35 ± 6.1	36 ± 5.0	1 : 1[8]								
intermedia	2.4[8]		37.0 ± 4.7[7]	34 ± 5.8	38 ± 5.6	1.1 : 1[8]								
labradoriensis	3.6[9]	14.4[9]	18.0[10]	12.7 ± 2.1	74 ± 12	5.8 : 1[10]	180 ± 30	690 ± 200	3.8 : 1[10]	(U)	95[10]	35[10]	(F)	2.32 ± 1.03[18]
							240 ± 30	1430 ± 170	6 : 1[10]	(L)	95[10]			
longicauda	1.1[8]		6.2 ± 1.4[7]		absent		150 ± 20	610 ± 130	4 : 1[11]	(L)	85[11]	36[8]	(F)	5.3 ± 2.9[20]
rufescens	1.8[1]		10.2 ± 0.6[7]	25 ± 8.7	82 ± 10.8	3.3 : 1[8]						144–480[8]	(F)	
vanhöffeni	6.5[18]	32.8[18]	70.0[13]	81 ± 34	163 ± 65	2 : 1[12]	220 ± 40	1040 ± 170	4.7 : 1[14]	(L)	91[14]	182[15]	(C)	1.7 ± 0.78[18]
villafrancae	5.0[16]	23.0[16]	50.0[16]	75.5 ± 3.7[16]			150 ± 20[16]			(L)	≈90	1722[16]	(F)	
Megalocercus														
huxleyi	3.5[8]		25.0 ± 5.5[7]	54 ± 9.0	170 ± 25	3.1 : 1[8]						594–1188[8]		
Stegosoma														
magnum	3.0[8]		29.7 ± 5.0[7]	25 ± 7.2	29 ± 5.1	1.2 : 2[8]						860–1477[8]	(F)	
Fritillaria														
borealis	1.3[17]	3.0[17]	2.5[17]	(< 30)[‡17]			(< 0.45)[‡17]					12[17]	(F)	

[1]Fol, 1872; [2]Lohmann, 1899a; [3]Flood, 1981; [4]Fenaux, 1986; [5]Alldredge, 1981; [6]Fenaux and Malara, 1990; [7]Alldredge, 1976b; [8]Alldredge, 1977; [9]Shiga, 1976; [10]Flood, 1991a; [11]Deibel and Powell, 1987a; [12]Deibel, 1986; [13]Flood, 1991; [14]Deibel *et al.*, 1985; [15]Knoechel and Steel-Flynn 1989; [16]Flood *et al.*, 1995; [17]Flood, 1994; [18]Riehl, 1992; [19]Lohmann, 1909; [20]Taguchi, 1982.
[21] Fenaux, 1985a. * Including gonad.
‡ Based on trapped particle sizes.
L = lower filter screen; U = upper filter screen.

bilaminar flow of water in opposite directions through the food concentrating trap. She proposed instead, a unidirectional flow from the anterior and posterior margins of the trap, obliquely across several filter ridges towards a median channel. From here all the water pumped was directed posteriorly towards an exit passage, whereas particles were 'sucked out of the feeding filter into the median channel and up into the mouth by the ciliary action of spiracles on the appendicularian's trunk' (cited from Alldredge 1977, p. 183). Recognizing the need for fine structural studies of the food concentrating filters, Alldredge (1977, p. 183) was rather vague in her explanation of how particles were separated from the water in this process: 'As water travels through the feeding filter, food particles are collected by the membranes within the filter'. Flood (1983) found water to enter the food concentrating trap of *O.dioica* laterally between two exceedingly fine-meshed screens. Water seeped through these screens to a dorsal and ventral portion of a posterior chamber, whence it could leave the house through a pressure regulated outlet valve. Particles withheld by the screens slowly approached the mid-line, to be sucked into the mouth of the animal at intervals. Fenaux (1986a), studying the same species, stated that water only (or mainly), seeped through the lower membrane of the trap. He also described orifices with valves between the upper and lower compartment of the posterior chamber (his dorsal and exit chambers) lateral to the food collecting tube. In *O.vanhoeffeni*, Deibel (1986) again concluded that water was forced through both the upper and lower layers of the trap. He established our current view that water entered at both anterior and posterior margins and at the lateral border of the food collecting trap, pointing out that a middle or third layer in the trap was too coarse-meshed to have any straining capacity for the food particles in question. Flood (1991a) gave further details of the construction and function of the three layers of the trap and emphasized the difference in ultrastructure of the two filtering screens and the importance of particle agglutination caused by partial collapse of the food collecting trap during the brief resting phases between pumping bursts of the animal. Due to the tangential flow of water over the food concentrating filters, the food particles gather near the mid-line of the food concentrating trap. From here food particles are periodically sucked into the mouth of the animal through a food collecting tube connecting the anterior medial margin of the trap and the mouth of the animal. However, in the pharynx the food particles have to be recaptured on a mucous filter secreted by the endostyle, in a process similar to that of all other tunicates (see Deibel and Powell, 1987a; Deibel and Lee, 1992; Bedo *et al.*, 1993; Chapter 8).

Oikopleurids somehow monitor the particles that enter their mouth or pharynx and are capable of rejecting or bypassing particles they find superfluous or disagreeable. The so-called mechanoreceptors of the lips or the ciliated funnel in the pharynx, combined with reversal of the spiracular ciliary beat (see section 4.5.1), have been suggested to be responsible for such rejection (Fenaux, 1986a), but this has not been confirmed by experiment. The upper (Deibel, 1986; Fenaux, 1986a) or lower lip (Flood, 1991a) may detach from the food collecting tube and so cause particles to flow directly into the posterior chamber and out of the house, without passing through the animal's pharynx or gut. Alternatively, the secretion of a pharyngeal mucous filter may be interrupted and the spiracles activated to drain particles through the pharynx, bypassing the gut (Flood, 1991a; Morris and Deibel, 1993).

The connections of the spiracles and anus with the chambers of the house have been debated. Fenaux (1986a) described both spiracles and anus in connection with the posterior chamber (his exit chamber) and thus both faecal pellets and particles which had bypassed the food concentrating trap were allowed to escape directly from the house. Deibel (1986) described the same openings as connected to the tail chamber alone, thus allowing for recycling of faecal pellets and bypassed particles. Lastly, Flood (1991a) described the anus as connected to the posterior chamber and the spiracles to the tail chamber. With these connections, food particles bypassed or missed by the pharyngeal filter (but not faecal pellets) could be recycled through the food concentrating trap. Although Deibel and Powell (1987a) have suggested that the pharyngeal filter is rather coarse-meshed compared to the food concentrating filter, some doubt remains about this; hence it remains unclear whether recycling of missed food particles for further coagulation in the food concentrating trap may be an important aspect of the function of the oikopleurid house (Flood, 1991a).

Water escaping from the food concentrating filter leaves the house through an outlet spout equipped with a complex pressure regulated valve (Flood, 1983, 1991a; Fenaux, 1986a; Flood *et al.*, 1990). Although some disagreement exists concerning the precise operation of this valve, all agree that it functions as a water jet propulsion system for the house. The speed of propulsion evidently depends on the pumping activity of the animal. Fenaux (1986a) related this to the abundance of nutritive particles in the environment and stated that a high abundance led to slow pumping in *O.dioica*, allowing the animal to remain in the favourable environment, whereas low particle abundance led to high pumping activity and forward movement of the house until more favourable nutritive conditions were encountered. However, Flood (1991a) suggested the exact opposite for *O.labradoriensis*, where low pumping activity led to linear cruising of the house and high pumping activity caused a somersaulting behaviour. The linear movement may be advantageous in searching for more nutritious environments, and the somersaults may enable the animal to prolong its exploitation of a nutritive environment.

The introduction of washed *Sepia* ink (colloidal melanin particles in the 0.1–0.2 μm range) as a convenient stain for structures of the oikopleurid houses, both upstream and downstream of the food concentrating filters (Flood *et al.*, 1990), clarified some of the remaining details concerning the functional anatomy of the oikopleurid house. It should, however, be remembered that we still lack a reliable and simple technique for visualization of the real external walls of these houses (cf. Flood *et al.*, 1990). The bioluminescent inclusion bodies (Galt and Sykes, 1983) resting on the outer surface of the house (Galt and Flood, 1984; Flood 1991a) are usually separated from the stained internal walls by a considerable gap, illustrating this separation (Flood, unpublished observations). The most detailed accounts on the water circulation through, and particle straining capacity of, the houses of oikopleurid species are probably those of Flood (1991a) and Morris and Deibel (1993), dealing with *O.labradoriensis* (Fig. 6.1b) and *O.vanhoeffeni* respectively. In these papers a hydrodynamic rationale for the function of the houses is also presented. Based on such considerations it seems clear that all oikopleurid houses examined so far are able to trap and agglutinate a significant fraction of the colloidal 'DOM' (dissolved organic matter) in the sea (Flood *et al.*, 1992). Further, the animals, through the use of their houses, feed on a 200- to > 1000-fold concentrated suspension of such material (Flood, 1991a; Morris and Deibel, 1993).

6.3.3. *Secretion of the oikopleurid house*

The oikopleurid house is produced as an anlage or rudiment by the secretory activity of the oikoblastic (Klaatsch, 1895) cells that cover the anterior trunk (section 2.3) as a single-layered oikoplastic* epithelium (Lohmann, 1896a,b), seen in Fig. 6.3e. The cells probably stop dividing as soon as the young animal starts to build its first house (Martini, 1909a; Fenaux, 1971a). However, they vary enormously in size and shape, and the nuclei (see p. 29 and Fig. 2.7) can become a 1000-fold polyploid (Fenaux, 1971a). Light microscopical studies of the secretory process itself were reported by Lohmann (1899a,b) and Körner (1952), but none of these studies are satisfactory in the light of present day knowledge of cell biology. Studies of this secretory process at the ultrastructural level are now in progress (PRF).

The oikoblastic cells, with their variable size and shape, are gathered in specific assemblages; the oikoplasts or oikoplastic fields named after previous investigators by Lohmann (1899a), Bückmann (1923) and Lohmann and Bückmann (1926). Although these authors stressed that the regions named were purely descriptive and not related to their role in the house-building process, the two most prominent, Fol's and Eisen's oikoplasts, are obviously related to the production (respectively) of the food concentrating filters and inlet filters of the fully expanded, functional house (cf. Figs 6.3e and 6.4). Lohmann (1933, 1934, pp. 32–50) discussed the relation between house structures and oikoplasts in detail and found that the number of small 'trap cells' (*Reusen-zellen*), present in three rows immediately posterior to a row of giant cells in Fol's oikoplast, corresponded exactly to the number of folds or ridges in each wing of the food concentrating filter. The structural appearance of the secretory product of these cells strengthened his belief that they produced the particle trap of the house. However, the increased resolving power of transmission electron microscopy has shown that the three rows of trap cells only produce the system of suspensory filaments associated with the two filter membranes, and the intermediate suspensory screen of the food concentrating filter. The upper and lower filter membranes themselves are produced respectively by Bückmann's (1923) *post-* and *prae-oval zellen* (Flood, unpublished observations). The lower filter membrane is secreted from the preoval cells as two superimposed mats of parallel filaments which in each mat are at 90° to each other (cf. Lohmann, 1933, 1934, p. 48). During the house inflation process these mats stretch out, first to thin films, and then to individual filaments which apparently cross-link at regularly spaced and chemically well defined 'glueing' points along their length. Accordingly, rather than being produced directly as net structures by a cellular 'loom', these filters are produced by a poorly understood self-assembly process of previously secreted extracellular filaments (Flood, personal observations; see also section 6.3.6).

The Eisen oikoplast, its six or seven giant cells and a row of about 15 *Neben-zellen* on their oral side, produces the anlage to the inlet filters of the house in a somewhat similar process: The *Neben-zellen* produce pairs of parallel filaments that stretch across the giant cells behind. The

*Since 'blast' means germ or something capable of forming and 'plast' means formed or molded or something of definite shape (Skinner 1961), it has been suggested that the term oikoblastic (=house-forming) should be associated with these cells and the epithelium they constitute (cf. Flood 1991a). This is in full agreement with accepted priority rules and with the use of the 'blast' term in modern cell biology. However, most authors still use the term oikoplast associated with these cells and epithelia. In our view the term 'oikoplast' should only be used to designate the definitely shaped groups of oikoblastic cells, so characteristically found in specific areas of the oikoblastic epithelium (e.g. Fol's oikoplast, Eisen's oikoplast).

giant cells, in turn, produce a second set of zig-zag shaped filaments below the parallel ones. As the house inflates, the inlet filter is unfolded to reveal its elongated meshes, where the parallel filaments produced by the *Neben-zellen* constitute the transverse strands, and the zig-zag shaped filaments of the giant cells stretch out to form the longitudinal strands (Lohmann, 1899a, 1933, 1934).

The disposition of so-called granular inclusion bodies (Galt and Sykes, 1983), or *Häutungskörper* of Lohmann (1914), on the house rudiments of some oikopleurids (Fig. 6.3b–d) represents the third and last case where components of the house rudiment have been directly related to the oikoblastic cell pattern. These bodies form species-specific patterns on the dorso-lateral rudiment surface of all oikopleurids that possess oral glands and subchordal cells (the subfamily Vexillaria) and are related to the bioluminescent properties of the same species (Galt, 1978; Galt and Sykes, 1983; Galt *et al.*, 1985) (cf. Chapter 13). Lohmann (1896b), studying *O.labradoriensis*, and Körner (1952), studying *O.dioica*, assigned specific oikoblastic cells as producers of these bodies. However, for neither of the two species mentioned, nor for any other species we have examined, have we found any such direct correspondence between oikoblastic cell pattern and inclusion body pattern. We rather support the interpretation of Fredriksson and Olsson (1981) that the inclusion bodies originate as a

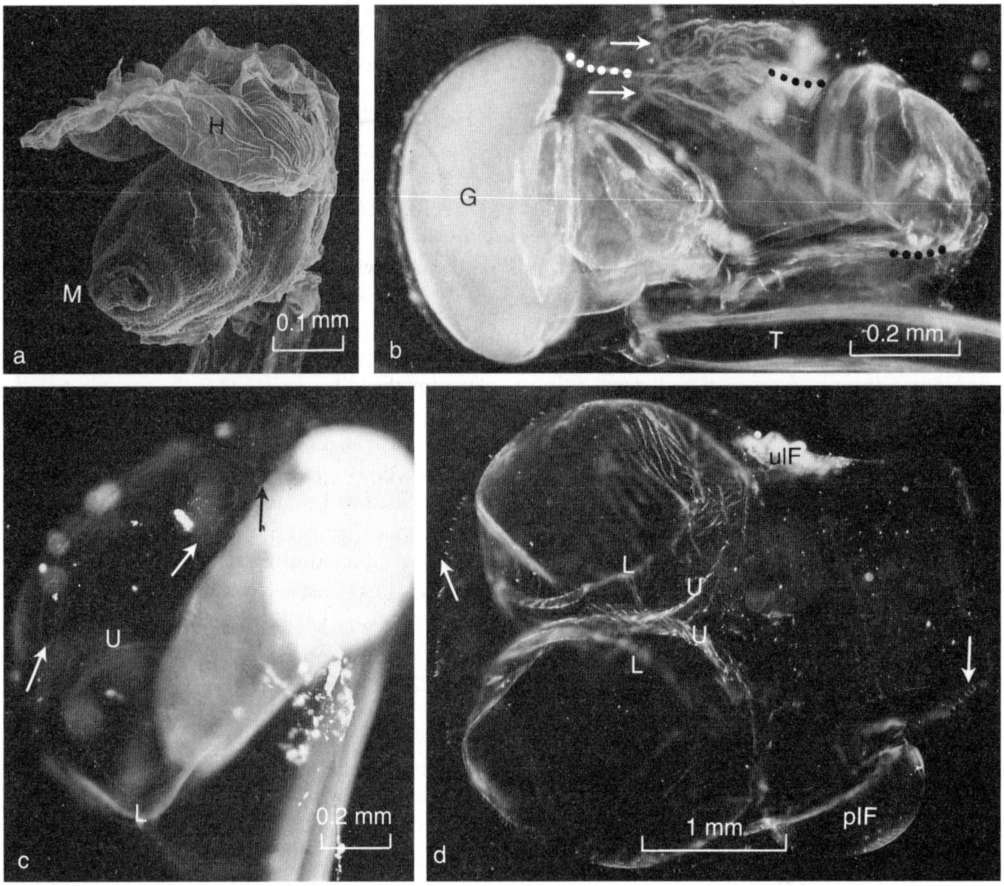

Fig. 6.3. (a) Scanning electron micrograph of the trunk of *O.longicauda* demonstrating the pronounced acellular hood (H), probably derived from the external limiting membranes of previously expanded houses. M: mouth. Magnification ×200. (b) Dark field light macrograph of the trunk of *O.labradoriensis* surrounded by two acellular house rudiments. Note the two partially superimposed patterns of inclusion bodies (arrows) in the rudiments. Dotted lines point to the underlying surface of the oikoblastic epithelium. Side view, fixed and unstained specimen. G, gonad; T, tail. Magnification ×100. (c) Dark field light macrograph of partially inflated (stage 2) house rudiment attached to the trunk of *O.labradoriensis*. Note the spreading out of inclusion bodies at the dorso-lateral surface of the house and the location of the anlage for the upper (U) and lower (L) food concentrating filters. Magnification ×50. (d) Dark field light macrograph of a partially inflated house of *O.labradoriensis* abandoned and fixed soon after the tail of the animal started to pump water into the house (early stage 3). Note the position of the anlagen for the upper (U) and lower (L) food concentrating filters, the unexpanded (uIF) and partially expanded (pIF) inlet filters and how the inclusion bodies stand up like tiny rods (arrows) on the external surface of the house. Magnification ×20.

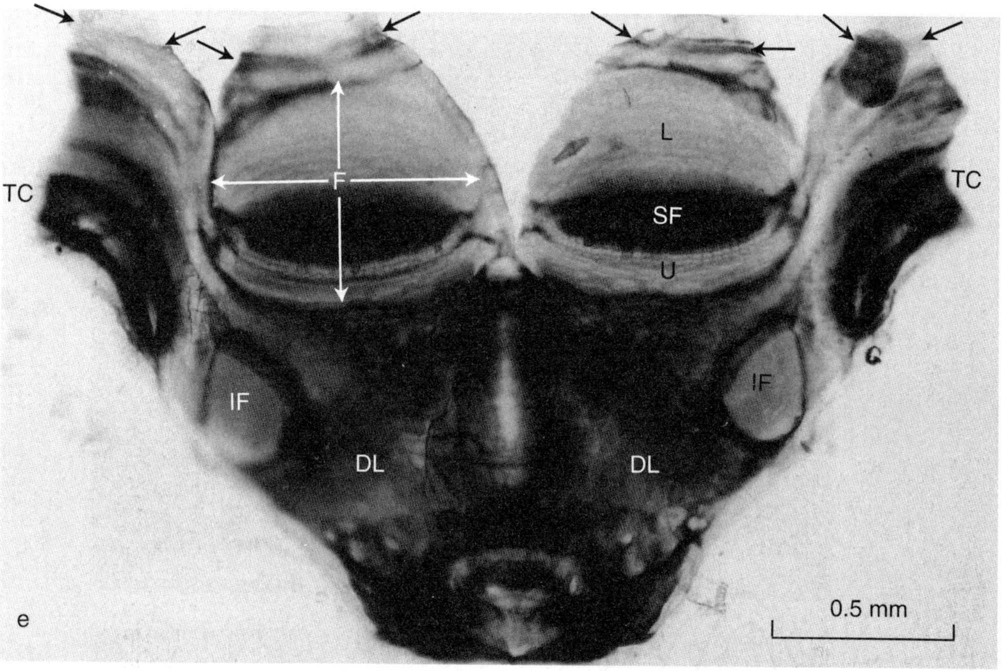

Fig. 6.3. (e) Bright field light micrograph of unfinished house rudiment and the attached, underlying oikoblastic epithelium as removed from the trunk of *O.labradoriensis*. The specimen was aldehyde fixed and stained in a dilute solution (~0.01%) of toluidine blue before it was cut completely open along the midventral line, cut partially open at the anterior mid-dorsal and lateral edges, unfolded and mounted for microscopy. Note the weak stainability of the anlagen for the upper (U) and lower (L) food concentrating filters, corresponding to Fol's oikoplasts (F), and the inlet filters (IF), corresponding to Eisen's oikoplasts. In contrast, the dorsolateral surfaces of the house rudiment (DL) and the suspensory filament system of the food concentrating trap (SF) are strongly stained, as are a circum-oral ring (arrows) and the anlage for the tail chamber (TC) on the ventral surface of the rudiment.

secretory product of the oral glands. However, at its origin this secretion appears to be non-bioluminescent and has an ultrastructure completely different from that of the mature inclusion bodies (Flood, unpublished observations). Fredriksson and Olsson (1981) offer no satisfactory explanation of how the oral gland secretion could be transported within the acellular rudiment to the remote position of the dorso-lateral pattern of inclusion bodies. Time lapse video microscopy of *O.labradoriensis* by one of us (PRF) has shown the oral gland 'secretion' to be in fact a cytoplasmic extension of the oral gland that spreads over most of the dorso-lateral surface of the rudiment just below its external limiting membrane. Independently of this process, the oikoblast cells responsible for the formation of the dorso-lateral region of the rudiment have incorporated into their acellular (but highly structured) secretion an almost invisible template for the species-specific pattern of inclusion bodies. Each template spot appears to consist of an acellular area of slightly higher density than in the surrounding rudiment material. Minute granules contained in the cytoplasmic process of the oral gland settle in these template spots and undergo a pronounced maturation process, enlarging until they are recognizable as the typical bioluminescent 'organelles' or 'lumisomes' of a mature inclusion body. Oral gland material that is unable to settle within the template spots of the rudiment fails to undergo a comparable differentiation and the entire cytoplasmic process start to degenerate before the rudiment is fully formed (Flood and Andersen, unpublished results, Chapter 13). In other words, the bioluminescent inclusion bodies have a dual origin; their location, size, and shape as a species-specific pattern is determined by the underlying oikoblastic epithelium, but are not distributed in a pattern respecting their cell boundaries. Their substructure and content of bioluminescent 'organelles' or 'lumisomes' is determined by the oral glands (Flood and Andersen, unpublished results).

Apart from Fol's and Eisen's oikoplasts it is difficult to see any obvious relation between the intricate pattern of cell boundaries of the oikoblastic epithelium and the structural organization of the overlying rudiment. However, such a relation must exist. It also seems obvious that most of the cells must contribute to the production of several distinct components of the rudiments in a sequential manner. In addition to the external and internal walls and septa of the house and the potential channels and chambers they delineate, most

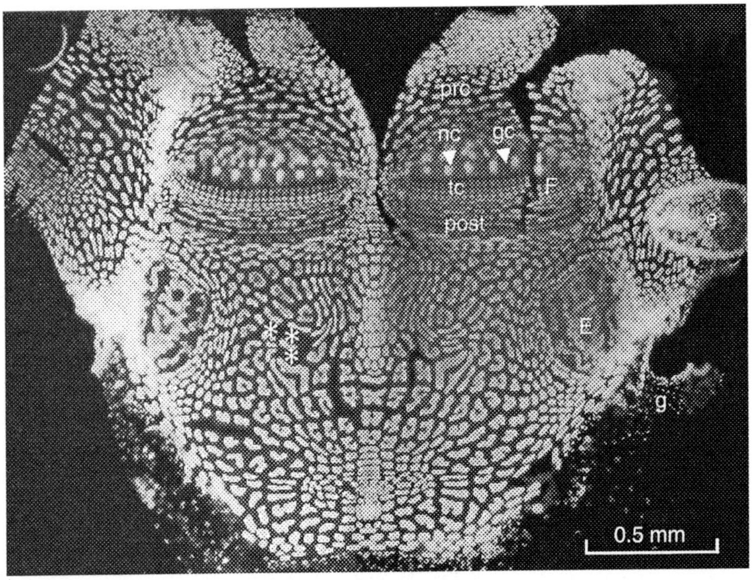

a

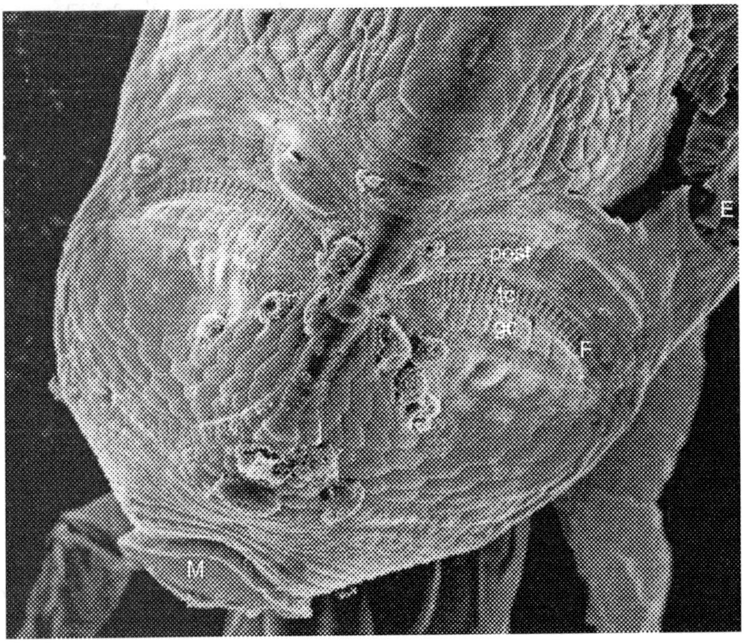

b

Fig. 6.4. (a) Oikoblastic epithelium of formaldehyde-fixed *Oikopleura labradoriensis*, microdissected, unfolded and mounted for microscopy like the house rudiment of Fig. 6.3e. Stained by the DNA specific fluorochrome DAPI and observed in an epifluorescence microscope under UV excitation, (revealing the cell nuclei as white dots against a black background). Eisens (E) and Fols (F) oikoplasts represent prominent landmarks. In the Fols oikoplasts three rows of trap cells (tc) behind one row of eight giant cells (gc) and six interposed 'Neben' cells (nc) constitute an oval area bounded by a large field of pre-oval cells (pre) and a more narrow band of post-oval cells (post). Three oikoblastic cell nuclei have been lost from the left dorsolateral surface (asterisks) during the preparation. Note the near perfect bilateral symmetry of the cellular pattern and minor differences in the shape of some nuclei in corresponding cells on the two sides. Note also the minute size of the cell nuclei of the gonadal epidermis (g) as compared to the much larger and variable size of the oikoblastic cell nuclei. The endostyle (e) is attached to the specimen. (b) Scanning electron micrograph of anterior dorsal aspect of the trunk of *O.labradoriensis* fixed after removal of innermost house rudiment revealing 'cobblestone' pattern of the (slightly shrunken) oikoblastic cells. Photograph by Q. Bone. Note mouth (m), the bilateral Fol's (F) and the left Eisens' (E) oikoplasts. Further labelling as in (a) Magnification ×50.

oikoblastic cells must also contribute to the external limiting membrane of the rudiment, which is shed or partially detached during the house expansion process (Flood, unpublished results). Accordingly, the oikoblastic epithelium offers an unique opportunity for the study of the spatial and temporal coordination of cellular activity in the production of a highly structured extracellular secretory product.

Körner (1952, fig. 44) presented detailed maps relating oikoblastic and rudiment areas to areas of the expanded house in *O.dioica*. However, from the appearance of his house drawings, it may be questioned if he ever saw the house of this species. So far, we have been able to relate the anterior and posterior oikoplastic rosettes of Bückmann (1923) to the production, respectively, of the posterior 'keel' and the 'escape passage' of *O.vanhoeffeni*, as described by Flood *et al.* (1990) (Flood, unpublished results). Further, the complex outlet valve of the house appears to be derived from the circumoral oikoblastic cells and the tail chamber from a ventral, posterior, and crescent-shaped cell group (Fig. 6.3e) (Flood, unpublished results).

Most oikoblastic fields are discernible to a variable degree in all oikopleurid species, and this variation probably reflects the differences in appearance of the houses of different species. One of several exceptions to this rule is the apparent total lack of Eisen's oikoplast (and accordingly the inlet filter) in *O.longicauda* (Lohmann, 1896b, 1898, 1909a). I have, however, repeatedly seen rudimentary Eisen oikoplasts in this species (Flood, unpublished results). In the expanded houses of *O.longicauda*, the filter is not visible, and 100 μm plastic beads are readily drawn into the interior of the house, even in small individuals (Bone, Gorsky and Fenaux, unpublished observations). It remains to be seen, however, if some strands representing the inlet filter are present in the unexpanded rudiments, and ruptured as the rudiment is expanded. Epipelagic oikopleurids captured undisturbed from the sea, still pumping water through their filter houses, usually have one or two fully synthesized and one partly synthesized house rudiments attached to their trunks (Flood, unpublished results). However, up to seven superimposed rudiment have been reported on an animal incapable of inflating new houses (Fenaux, 1985).

6.3.4. *Inflation frequency of oikopleurid houses*

Oikopleura dioica is known to escape from its old house, and expand a new one every 4 h (Lohmann 1909a). Päffenhofer (1973), who first succeeded in keeping *O.dioica* in culture, reported a house production rate of 5.1 $\pm$ 0.7 (mean $\pm$ SD) houses day^{-1} at 13°C. Fenaux (1985) reported a linear increase in the hourly house production rate for the same species from 0.17 $\pm$ 0.04 houses h^{-1} at 14°C to 0.44 $\pm$ 0.04 houses h^{-1} at 22°C These numbers are equivalent to daily house production rates of 4.1 and 10.6. Daily production rates of 5.3 $\pm$ 2.9 (Table 6.1) were found for *O.longicauda* at Hawaii by Taguchi (1982). Larger species and species living at low temperatures keep their houses for longer periods. At temperatures between –0.6 and 5.8°C, Riehl (1992) found renewal rates of 1.7 $\pm$ 0.78 houses day^{-1} for *O.vanhoeffeni* and 2.32 $\pm$ 1.03 houses day^{-1} for *O.labradoriensis* (Table 6.1).

Rather than resulting from clogging of the filters or damage to the house, as Lohmann (1909a) and Alldredge (1976b) suggested, house renewal seems to depend on the rudiment secretion rate and on the availability of particulate and dissolved organic matter in the ambient water (Fenaux, 1985). Obviously, the rudiment production rate must keep pace with the house inflation or renewal rate.

6.3.5. *The inflation process*

As soon as an oikopleurid appendicularian has escaped from its old house, it will try to expand a new one. Provided that a fully synthesized house rudiment is available on the trunk, this process takes about a minute. Several stages can be distinguished in the process (Fig. 6.5). First, the rudiment lifts away from the next rudiment below it by a peculiar swelling process lasting for some seconds. This stage 1 process is probably under nervous control and may involve an obscure secretory activity of the oikoblastic epithelium, bursting or shedding of the external limiting membrane of the rudiment and a subsequent activation and swelling of preformed material, rather than muscular movements of the animal's tail (Flood, 1994). Thus the prominent 'hood' structure (Lohmann, 1896b) found on well preserved specimens of *O.longicauda* (Fig. 6.3a) probably represents remnants of a burst external rudiment limiting membrane (Flood, personal observation). In other species, like *Oikopleura labradoriensis* and *O.dioica*, this external protective layer is shed and may be retrieved as barrel-shaped, extremely transparent flaccid structures from the aquarium tanks (Flood, unpublished observation).

Secondly, the animal will use its tail muscles to nod the trunk backwards and forwards (see section 3.5.1), probably to force water underneath the outermost rudiment to cause some expansion of the house (Fig. 6.3c). This stage 2 may last as little as 20 s, or can be protracted if no rudiment is ready for expansion (Fenaux and Hirel, 1972; Galt, 1972; Flood, unpublished observation). Alldredge (1976b) discerned two substages of different motor behaviour at this stage.

During stage 3 (Fig. 6.3d) the tail of the animal slips inside the slightly expanded rudiment, initially to be markedly curled up inside the tail chamber of the

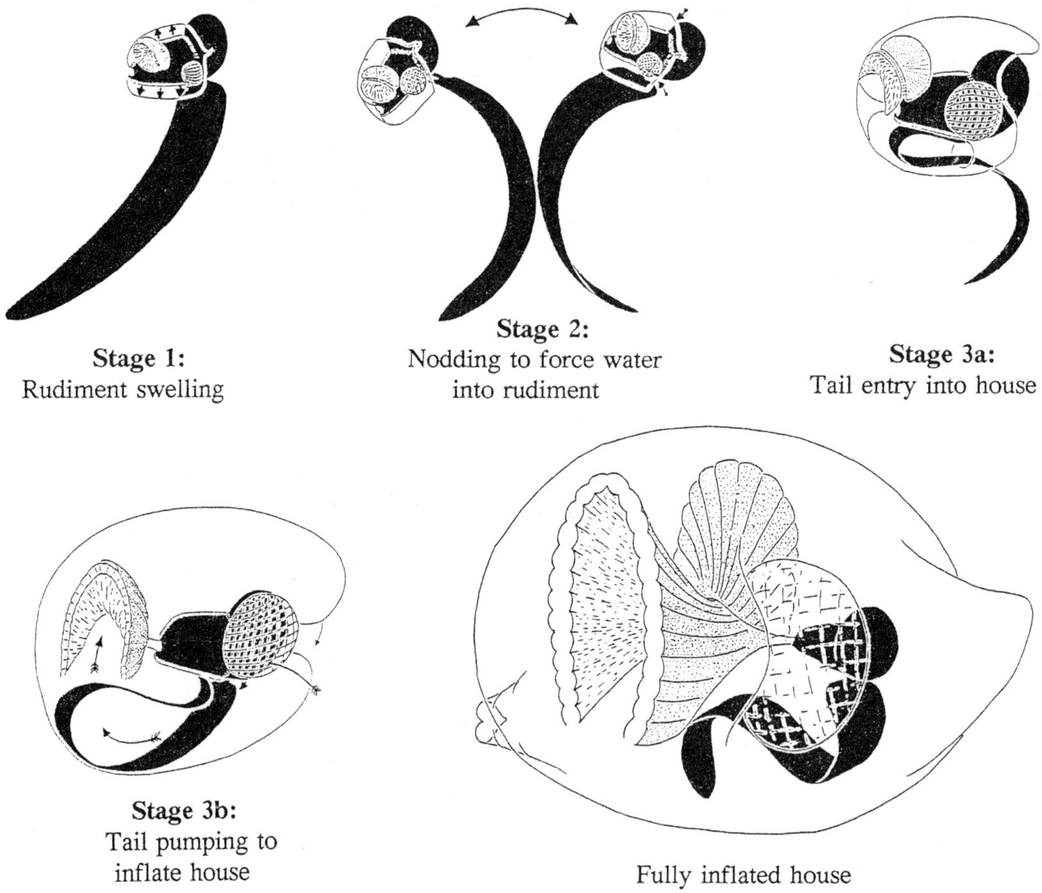

Fig. 6.5. Diagrammatic representation of successive stages of house inflation in *O.labradoriensis*. For further explanation see text. From Flood (1994).

forming house, but it soon straightens as its undulatory movements steadily force water into the house until it reaches its fully expanded size, with a diameter around twice the total length of the animal (Fig. 6.1a). Its volume will then be some 300 times the volume contained within the unexpanded rudiment. Since the house rudiment is secreted only from the oikoblastic epithelium on the front of the trunk anterior to the gonads, during the expansion process the posterior dorsal and ventral regions of the house rudiment must join and eventually fuse somewhere below the inlet filters and above the tail chamber (Fenaux and Hirel, 1972; Galt, 1972).

The mechanisms responsible for stopping house expansion at a specific point are obscure. Reaching the elastic limit of the structure does not seem to be involved since the elastic recoil of abandoned houses is limited to < 10% (Flood, unpublished results). Similarly, mechanical stretching and straightening of previously coiled or undulant filaments seems irrelevant since most fibres in the rudiments are quite straight (and perhaps less straight in the expanded house!). Chemical cross-linking (curing or hardening) of the house substance during expansion, or a sudden drop in hydrostatic pressure when the house reaches the size at which the exit valve opens, seem the most likely basis for the size limits of oikopleurid houses.

6.3.6. *Chemical composition*

Lipids appear to be unlikely candidates for any of the house structures identified so far by electron microscopy, and proteins and carbohydrates would seem to be the two relevant classes of biological macromolecules to be considered.

Chemical analysis of oikopleurid houses has been greatly hampered by their fragility, low weight, and the ease with which they are contaminated by filtered material. So far, most elemental carbon and nitrogen analysis have been performed on material filtered onto pre-ashed glass fibre filters. However, such filters will absorb substantial amounts of both particulate and dissolved organics from the water used to transfer the samples

(Johnson and Wangersky, 1985). Unless filters used to filter exactly the same volume of comparable water, rather than just pre-ashed filters, are used as blanks, erroneously high values will be obtained. Unfortunately, such details are rarely reported in the literature. Alldredge (1976a) found mean values of 10 μg C and 0.04 μg N (giving a molar C:N ratio of 214) in 12 'newly secreted' houses of *O.rufescens*. If this high C:N ratio is correct, it indicates that amino acids and amino sugars are minor constituents of the house. Deibel (1986) reported a mean of 20.7 μg C in nine 'newly secreted, particle-free' houses of *O.vanhoeffeni* (no N values were given). Riehl (1992) found means of 8.4 μg C (n = 8) and 2.6 μg N (n = 11) in 'not yet functional' houses of *O.vanhoeffeni*. This gives a molar C:N ratio of 2.8, which is in the low range for proteins, indicating that other nitrogen sources than proteins may be present. However, for another house, expanded in 0.2 μm filtered seawater, Riehl (1992) reported 11.7 μg C and no detectable N. Körner (1952) applied some histochemical tests to *O.dioica* houses and rudiments (apparently mostly on histological sections through the trunks with attached rudiments). He concluded that the house material was made up of a very uniform substance quite distinct from the tunic of ascidians; a polysaccharide with esterified amino sugars.

Unpublished results by one of us (PRF) on the houses of *O.dioica, O.labradoriensis* and *O.vanhoeffeni* indicate first of all a prominent difference in the stainability of non-expanded rudiments (Fig. 6.3e) as opposed to expanded houses. Rudiments stain heavily and houses faintly by many histological techniques. This difference cannot simply be explained by stretching and thinning of the stainable material alone. Loss of non-structural organic material during the house expansion process, in particular acidic polysaccharides stainable by Alcian blue at low pH, probably takes place (cf. Lohmann, 1933, 1934 p. 47). Secondly, there is a profound difference in the stainability (and accordingly the chemical composition) of the inlet and food concentrating filters as opposed to the rest of the rudiments. The entire rudiment, except the filter areas, stains prominently and metachromatically with toluidine blue at high pH (Fig. 6.3e). Alcian blue staining at pH 1, indicative of sulphated mucosubstances (Kiernan, 1981), stains most of the rudiment weakly, parts of the food concentrating trap moderately, and the inlet filters prominently. At pH 2.5 the same technique (now indicative of acidic mucosubstances in general; Kiernan, 1981) stains the same structures as mentioned above, but also several irregular patches throughout the rudiment, without any obvious relation to its structural components. The protein-specific dye Coomassie blue (Brilliant Blue G) stains the food concentrating filter anlage more strongly than the inlet filters. The rest of the rudiment is very weakly stained, except for the bioluminescent inclusion bodies, which are the most prominently stained structures by this technique (Flood, unpublished results).

Microchemical analyses of house rudiments with minimal organic, but considerable (maybe 80%) salt contamination resulted in no trace of amino sugars. This analysis, kindly performed in B. Larsen's laboratory at the Institute of Biotechnology, University of Trondheim, showed galactose, glucose and mannose to be the main sugars. However, taken together these sugars made up only 6% of the total sample dry weight. Amino acid analyses of comparable samples, kindly performed in B. Walther's laboratory, Department of Biochemistry, University of Bergen, showed that amino acids again accounted for only 6% of the sample dry weight. Close to 25% of these amino acids (on a molar basis) were acidic (aspartic and glutamic acids) compared to about 5% of basic amino acids (arginine and lysine). An additional 37% represented other hydrophilic amino acids (mostly glycine, threonine, and serine), hence the hydrophobic amino acids were about 33% of the total. Amino acids known to cross-link to sugars were asparagine (13%), serine (9%), and threonine (8%). Sulphur-containing amino acids constituted less than 1% of the total amino acid content on a molar basis, yet splitting of disulphide bonds by mercaptoethanol produced a pronounced increase in the amount of peptides entering polyacrylamide gels during electrophoresis (Flood, unpublished observation). Tyrosine residues constituted 3–4% of total amino acids and, based on a weak blue fluorescence at high pH upon excitation by ultraviolet light, appeared to be concentrated in the food concentrating filter areas of the rudiments (Flood, unpublished observation). Based on the variable saccharide stainability and the presence of a weak peroxidase activity in the same areas, detected by the diaminobenzidine reaction (Flood, unpublished observation), perhaps dityrosine bridges between weakly or non-glycosylated proteins are involved in the cross-linking of filaments in the food concentrating filter meshes. Maybe a process analogous to the hardening of the fertilization membrane in sea urchin eggs (Foerder and Shapiro, 1977) is responsible.

Apart from the filters, the bulk of the house (i.e. most of the wall material), rather than being composed of polymerized amino sugars as proposed by Körner (1952), is most likely composed of proteins heavily glycosylated by galactose, glucose, and mannose. Acidic sugars may also be present in some structures, as indicated by the Alcian blue staining results mentioned above. However, much of the acidic polysaccharides detected histochemically appear to be non-structural, and are likely to be washed out of the houses during their expansion process. The rinsing of rudiments in distilled water prior to drying for microchemical analysis may also cause loss of such substances.

6.4. The fritillariid 'house'

Fol (1872, p. 474) described an extremely delicate mucous bag surrounding only the mouth of *Fritillaria megachile*. When the animal's tail was beating this hollow vesicle was inflated, whereas when the tail was at rest, it collapsed to be barely seen by the naked eye. An opening opposite to the attachment of the mouth allowed water pumped by the tail to enter the vesicle. As long as the animal was attached to this mucous bag it remained suspended in the sea without moving, even when the tail was undulating rapidly. Similar observations and additional notes on the motor behaviour of *F.pellucida* were reported by Bone *et al.* (1979).

A recent video microscopic study of *F.borealis* (Flood, unpublished observation) much extends and corrects previous observations (Fig. 6.6). First, most of the animal, except the posterior half of the trunk, is definitely lodged *inside* a house. This house may be considered homologous to the oikopleurid house in being produced by an oikoblastic epithelium. However, it is constructed according to quite different architectural and functional principles. Water enters directly into a tail chamber and appears to be pumped through a coarse-meshed pre-filter on the downstream side of the tail before it enters the fine-meshed food concentrating trap. This trap appears to consist of two filtering membranes as in oikopleurids. However, one of these membranes is also the external, scalloped wall of the house through which much of the filtered water escapes directly to the exterior. The other filtering membrane delineates a scalloped, interior, mid-line chamber which probably also serves as a water outlet. No pressure regulated outlet valve has been noted so far in this type of house, although the animal may keep the house in a non-pumping, semi-inflated state by pressing the tip of the tail against the house structures in front of its mouth.

The entire house, and in particular its food concentrating trap, is very elastic and collapses around the animal and its tail as soon as the tail ceases beating. The food concentrating trap, which usually contains numerous opaque particles, then occupies most of the space between the pharynx and the tail of the animal, and it seems reasonable that this is the only part of the house visible to the previous investigators in *F.megachile* (Fol, 1872) and *F.pellucida* (Bone *et al.*, 1979). The coarse-meshed pre-filter is efficiently backwashed during these brief resting phases and a patch of coarse particles is expelled from the house behind the animal. During the resting phase the tail also slides inside a fin-like prolongation of the tail chamber, which thereby becomes very difficult to recognize as a separate structure. This fin-like prolongation acts like a paddle during the first

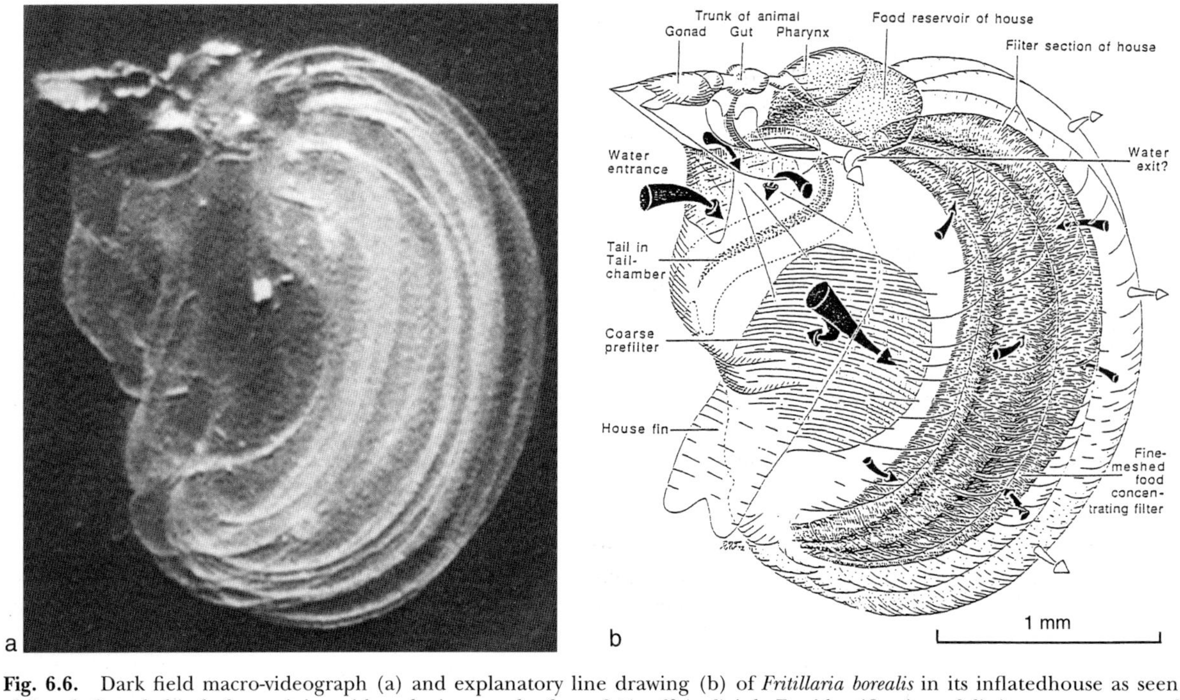

Fig. 6.6. Dark field macro-videograph (a) and explanatory line drawing (b) of *Fritillaria borealis* in its inflatedhouse as seen obliquely from behind after staining with melanin granules from *Sepia officinalis* ink. For identification of distinct components of the animal and house refer to the line drawing. Arrows point to the direction of water flow through the house. Black arrows to water before filtration through the food concentrating trap, outlined arrows to water after filtration.

tail strokes when the animal resumes its pumping activity. This is enough to move the animal and its collapsed house through the sea a few millimetres before the tail chamber and house are inflated again to a level where the tail slides out of the fin and undulates freely in the tail chamber again. In this way, the animal can move away from the coarse backwashed particles and from water depleted of food after each pumping cycle.

From experiments with dilute and filtered *Sepia* ink (cf. Flood *et al.*, 1990, 1992) it has been shown that *F.borealis* can feed efficiently on particles < 0.45 μm in diameter. However, the structure of the filters responsible for this straining capacity has yet to be visualized; owing to their extreme elasticity, attempts to do so have failed so far. A 'giant' fritillariid, probably identical to *F.magna* as described by Lohmann (1896b) and distinct from *F.aberrans* as redescribed by Tokioka (1958), has also been videoed within its house at about 150 m depth in Monterey canyon from the remotely controlled submersible vehicle of the Monterey Bay Aquarium Research Institute. This house looked rather similar to the house of *F.borealis* and was approximately 4 cm in diameter (Silver and Flood, unpublished results).

6.5. Other appendicularian houses

The house of *Kowalewskia* was drawn in great detail by Fol (1872) (Fig. 6.7) and photographed in the field by King (in Alldredge, 1976c). This house is very different from both oikopleurid and fritillarid houses and looks like a flattened rotational ellipsoid with 24–28 evenly spaced and vertical, radial ridges somewhat below their external surface. A single, centrally located water inlet opening spans the central 1/4 diameter of the house. This opening appears to be spanned by a coarse-meshed and fragile inlet filter (Flood, unpublished observation) and *in situ* observations from a manned submersible indicate that the house is usually located with the opening pointing downwards in the water column (Flood, unpublished observations). The water circulation pattern through these houses, as well as their filter parameters, are unknown. Fol (1874) also gave the only description and drawing of the house of *Appendicularia sicula*. Again, further details of functional importance are lacking.

The house of *Bathochordaeus* sp. has been studied for some years by *in situ* photography and video recording from manned and remotely controlled submersibles (Barham, 1979; Robison, in Gore, 1990; Youngbluth *et al.*, 1990; Hamner and Robison, 1992; Monterey Bay Aquarium, 1996). Hamner and Robison (1992), who have published the only detailed description of *Bathochordaeus* houses, make a clear distinction between the feeding filters (= food concentrating filters), the tail

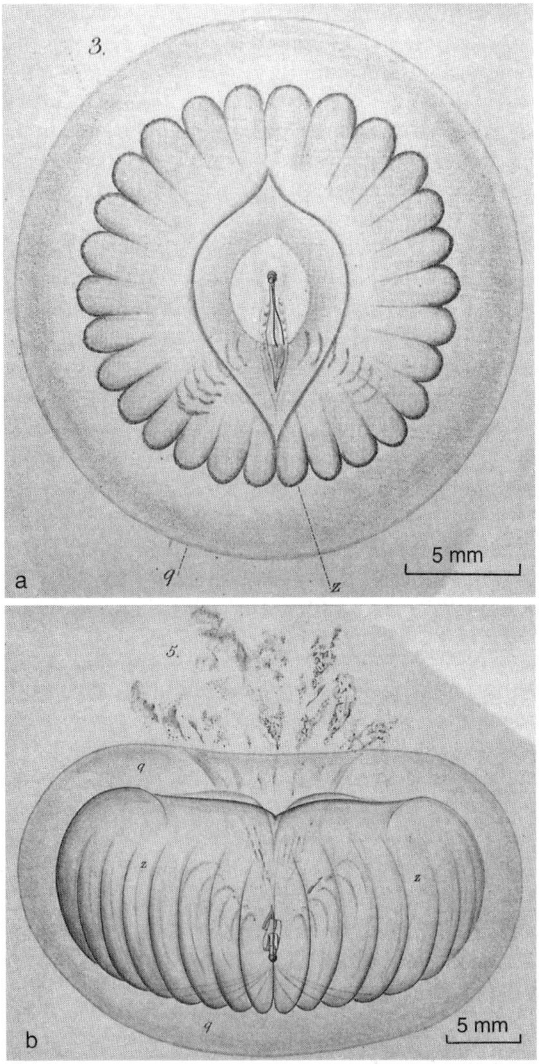

Fig. 6.7. Reproductions of the line drawings presented in Fol (1872) of *Kowalewskia tenuis* in its house as seen from below (a) and the side (b), q, coquille (house); z, three-dimensional internal cavity.

chamber (= supply passages), and the buccal tube (= food collecting tube), which they called the 'core' structures of the house, and what they termed the outer structure of the house. They claim that *Bathochordaeus* houses are not homologous with the typical oikopleurid house, that they lack the external house seen in oikopleurids, and that instead they secrete mucous continuously from the spiracle region of the body (or the oikoplast epithelium) to gradually envelop the animal's body and the inner filter in a curved mucous sheet. Further, that this mucous envelope continues to grow and expand, ballooning above and around the animal and the feeding filter, reaching a diameter usually less than 1 m, but occasionally over 2 m, during a period of about 30 days. Using Alldredge's (1977) nomenclature they state the feeding filters (= food concentrating filters) and tail chamber (= lower, fused part of the supply passages) to have a diameter about 2.3 times the total body length of the animals, which was measured as 39.2 ± 11.9 mm (mean ± SD). A valve at the transition between the lateral arms and the central cavity of the tail chamber (= supply passages) flickers open and shut in synchrony with the tail beat frequency. Water and particles move into the ribbed flutes of the feeding filter which converge and terminate in short left and right buccal tubes which attach in turn to the mouth.

From the evidence presented by Hamner and Robison (1992), *in situ* video recordings provided by M. Youngbluth, photographic diapositive slides supplied by G.R. Dietzman and L. Madin, and from several specimens of *Bathochordaeus* sp. supplied by M. Silver and her group from the Monterey canyon to one of us (PRF), a clear homology may be seen between the 'core' structure of the house of *Bathychordaeus* and the typical oikopleurid house described in section 6.3.3. Although there are manifest differences in the orientation and shape of the completely divided bilateral food concentrating trap, the similarities in its fluted construction and its links to supply passages and food collecting tubes (Fig. 6.8), makes the homology obvious. The lateral walls of a real tail chamber must be closely fitting along the entire length of the tail to explain its pumping efficiency. This chamber has yet to be identified in the *Bathychordaeus* house. The chamber termed 'tail chamber' by Hamner and Robison (1992) must be the downstream prolongation of this pumping chamber as it is far too wide and distal to the tail to generate any effective momentum to be given to the contained water.

A homology between the external mucous envelope of the *Bathychordaeus* house and the external walls of oikopleurid houses is also suggested. There certainly are difficult problems involved in visualizing these walls and the spaces between them and the 'core' chambers of the house (Flood *et al.*, 1990; Flood, 1991a). It is true that the distance between the 'core' chambers and the exterior is enormous in *Bathychordaeus* compared to most oikopleurid houses, but it seems impossible to us to explain how the animal could pump water in one direction (into the food concentrating trap) and to a flaccid mucous web in another direction (to the exterior of the house) as suggested by Hamner and Robinson (1992). Instead we propose the following interpretation:

The enormous space between the 'core' chambers and the exterior of the house contains several mucous layers or an extremely dilute, gelatinized substance, so weak that pieces of it easily are torn off in turbulent water or in collisions with other animals. New surfaces will then be exposed to the exterior, but due to their high transparency, these will only gradually become visible as particles start to accumulate on them. In other words, it is believed all constituent parts of the *Bathochordaeus* house, including the external mucous envelope, are produced as integral parts of the house rudiment, secreted as a unit at intervals by the oikoblastic epithelium, and expanded together with the food concentrating trap. The notion that a lack of Eisen's oikoplasts in *Bathochordaeus* should explain the lack of an outer house (Hamner and Robison, 1992, p. 1302) is wrong. Eisen's oikoplast has never been shown to produce anything but the inlet filter (Lohmann, 1933, 1934). From a comparison of sediment trap data with the particle densities found on still inhabited and abandoned *Bathochordaeus* houses (M. Silver *et al.* personal communication) we believe that houses are renewed at less than 1 day intervals, rather than at the 1 month intervals suggested by Hamner and Robison (1992). The particle straining capacity of the food concentrating filters of the *Bathochordaeus* house is uncertain. Within such houses net structures with very regular square meshes, about 1.4 μm across have been identified by G. Matsumoto (personal communication). However, since the morphology of these nets is so different from that of all other appendicularian food concentrating filters examined so far, more evidence is needed before it is concluded that they represent the food straining capacity of these animals.

Pelagopleura sp. were on several occasions seen within their houses at 900 m depth during the MEDAPP-91 submersible cruise (Gorsky and Youngbluth, unpublished observations). These houses appeared globular with a diameter ranging between 1 and 3 cm. They contained bilateral food concentrating filters, which after preparation for light microscopy, appeared similar to those of oikopleurid houses examined previously (Flood, unpublished results). Inlet filters with a pore width of some 140 μm were found in two successive houses from one specimen, which was unfortunately lost before it could be identified. In six other *Pelagopleura* sp. houses, no inlet filters could be found (Flood, unpublished results).

6.6. Ecological impact (see also Chapter 8)

The rate of flow through some appendicularian houses has been determined for several oikopleurids and one fritillariid (Table 6.1). Depending on the size of the species and their houses, volumes of 12–1500 ml h^{-1} were filtered through a single house (Alldredge, 1977; Flood, 1991b; Morris and Deibel, 1993) Particle clearance rates are of the same order (Päffenhofer, 1976; Gorsky, 1980; King *et al.*, 1980; Alldredge, 1981; Deibel,

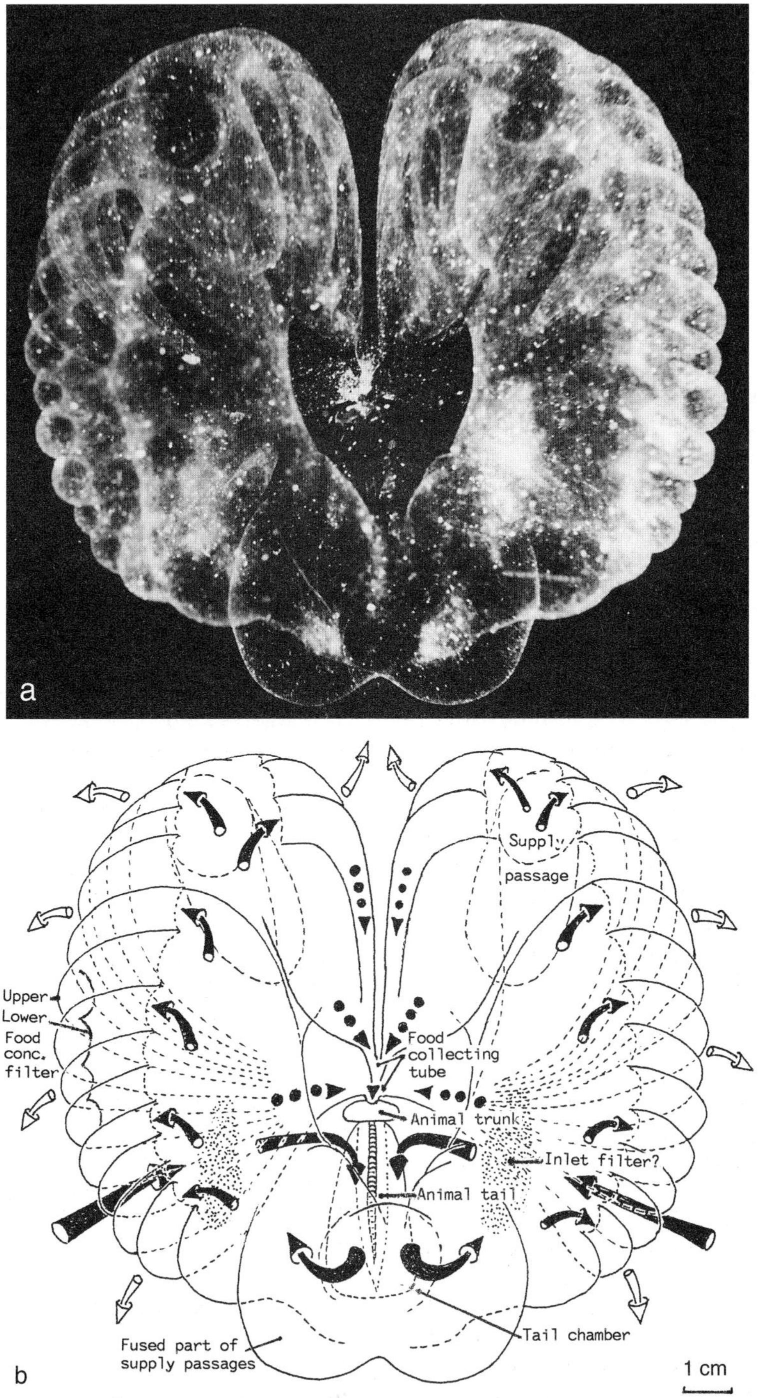

Fig. 6.8. *In situ* flash photograph made by a scuba diver of the 'core' structures (~15 cm wide) of a giant appendicularian house seen from above (a) and a corresponding explanatory line drawing (b). Black arrows indicate tentative flow of water towards the food concentrating filters, white arrows the tentative flow from these filters and dotted arrows the tentative flow of trapped particles towards the mouth of the animal. The reflections marked 'inlet filters?' are rather speculative. Although this particular specimen remains unidentified it was probably *Bathochordaeus stygius*. Courtesy of G.R. Dietzman, from *Oceanus* dive no. 165 near Bermuda, April 1985, 20.5°C surface temperature. Specimen photographed at ~25 m depth.

1988; Knoechel and Steel-Flynn, 1989; Fenaux and Malara, 1990; Deibel and Lee, 1992). Using commonly reported densities of appendicularians, anything from < 1% to > 60% of the water mass in which they are living may be cleared of particles on a daily basis (Alldredge, 1981; Deibel, 1988; Knoechel and Steel-Flynn, 1989). Under bloom conditions, with up to 25 600 oikopleurids m^{-3} (Seki, 1973), or 240 000 fritillariids m^{-3} (Gran, 1929), the entire water mass may be filtered more than once each day!

Appendicularian houses gather large amounts of particulate and colloidal material as long as the animals pump water through them. As much as 30% of this material may remain trapped in the houses, rather than being processed to the mouth of the animal (Gorsky, 1980). When abandoned, such loaded houses gradually collapse and sink through the water column. Their sinking rate appears to depend on their size and degree of collapse. For the giant houses of *Bathochordaeus* sp., sinking rates of 818 ± 199 m day^{-1} have been reported (Hamner and Robison, 1992). The sinking houses represent an important source of organic carbon for other organisms (Alldredge, 1976a; Gorsky *et al.*, 1984, 1989; Alldredge and Youngbluth, 1985; Youngbluth *et al.*, 1986, 1990; Alldredge and Silver, 1988; Davoll and Youngbluth, 1990; Fisher *et al.*, 1991; Silver and Gowing, 1991).

6.7. Preservation of appendicularian houses

A technique has been developed for the preservation of oikopleurid houses as thin films on paper, membrane filters, or glass substrates (Flood, 1991a). This technique has recently been improved and used with some success on houses of *Pelagopleura* sp. and *Kowalewskia* sp. (Flood *et al.*, 1993) and is promising for future advance in our knowledge of these and similar delicate structures, including the large houses of *Bathochordaeus*.

6.8. Acknowledgements

The authors wish to thank Miss Wenche Grønås, MS, Department of Biochemistry, University of Bergen, for the amino acid analyses reported above. Our sincere thanks are extended to G.R. Dietzman, L. Madin, M. Silver, and M. Youngbluth who generously supplied one of us (PRF) with data and material on *Bathochordaeus* sp. The constructive criticism of the draft to this chapter, largely completed while PRF was Professor of Zoology at the University of Bergen by R. Fenaux and C.P. Galt, is also acknowledged. This work was supported by the Norwegian Research Council.

earlier records of the times of appearance of swarms of *T.democratica* in various regions, from the Hebrides to the southeastern Australian coast.

Among the pyrosomas, *Pyrosoma atlanticum* is the commonest and most widespread species, occurring between 50° N and 50° S in all oceans (Van Soest, 1981).

Table 7.1 shows some abundance maxima of salps and pyrosomas reported in the literature. These swarms are usually restricted to the superficial layers (the upper 100 m). Concentrations as high as 1000 individuals m^{-3} were observed for *Salpa fusiformis* in the Gulf of Guinea (Roger, 1982a) and for *Thalia democratica* off the Florida coast (Paffenhöfer and Lee, 1987). *Salpa thompsoni*, the commonest salp species of the Southern Ocean (Casareto and Nemoto, 1986), is also known to form swarms in these waters. Huntley *et al.* (1989) reported a maximum biomass of 671 mgC m^{-2} in the 0–200 m water column for this species, which is comparable to the value of 909 mgC m^{-2} (65 individuals m^{-3}) for the *Salpa aspera* swarm observed by Wiebe *et al.* (1979). More recently, Nishikawa *et al.* (1995) found 559.9 mgC m^{-2} as the austral summer biomass maxima of a swarm of *Salpa thompsoni* (with *Ihlea racovitzai*) near the South Shetland Islands. Most recently, Siegel and Harm (1996) have observed, in the northern part of the Bransfield Strait, up to 263 *S.thompsoni* 1000 m^{-3} i.e. 1134 g 1000 m^{-3}, a biomass dominance of up to 98.9%. Salp swarms have been reported for only a few species, notably *Salpa fusiformis, S.aspera, S. thompsoni*, and *Thalia democratica*, though Seymour Sewell (1953) reports *Pegea confoederata* swarms with up to 50 individuals m^{-2} at the surface in the Gulf of Aden in autumn 1933.

Reports on the abundance maxima of pyrosomas are few. The maximum density of 41 individuals m^{-3} (Drits *et al.*, 1992) appears very low compared to the values for salps. But in the case of *Pyrosoma atlanticum* these numbers are of colonies of many zooids, rather than for single individuals (blastozooid or oozooid) in the case of salps. In general, the reported abundance values suggest that salps and pyrosomids form smaller swarms in the Mediterranean than elsewhere, which could be related to the relatively oligotrophic nature of the Mediterranean.

7.3. Body composition

The high water content of living salps and pyrosomas, and their high ash contents, are typical of gelatinous species (Clarke *et al.*, 1992). Several studies provide carbon and nitrogen body contents and the C:N ratios of different salp species and of pyrosomids (Table 7.2). Carbon weight of salps ranges from 3.2 to 11.9% of dry weight (DW), the relatively high value of 18% measured for *Thalia democratica* by Heron *et al.* (1988) being excepted. Nitrogen weight values are in the range 0.6–2.8% DW. There are much fewer data on pyrosomids; only two values of 9.4 and 11.3% DW for carbon content and one value of 2.8% DW for nitrogen content. Average C:N ratios (by weight) of both salps and pyrosomids are in the range 3.7–5.3 (16 species), the value of 6.2 reported by Roger (1982b) appearing relatively high. The salps and pyrosomids have very low C and N body contents, compared to non-gelatinous zooplankton, as a result of their very high ash contents. For example, copepods and euphausiids have, respectively, C contents of 35.2–47.6 and 31.1–47.4% DW and N contents of 8.16–12.5 and

Table 7.1 Abundance maxima of salps and pyrosomids.

Species	Location	Depth (m)	Concentration (no. m^{-3})	Reference
Salpa fusiformis	Gulf of Guinea	surface	1000	Roger, 1982a
	Gulf of Guinea	surface	165	Le Borgne, 1983
	Off western Ireland	0–100	700	Bathmann, 1988
	NW Mediterranean	0–200	19	Nival *et al.*, 1985
	NW Mediterranean	100	7.5	Morris *et al.*, 1988
Thalia democratica	Off Florida coast	0–30	1000	Paffenhöfer and Lee, 1987
	Off Florida coast	0–12	60	Atkinson *et al.*, 1978
	Off Southern California	0–70	275	Berner, 1967
	Off Ivory coast	surface	100	Binet, 1976
	SE Australian shelf	50	44	Heron and Benham, 1984
Salpa aspera	NW Atlantic	100	65	Wiebe *et al.*, 1979
Pyrosoma atlanticum	NE Atlantic	0–10	41	Drits *et al.*, 1992
	NE Atlantic	0–300	2	Goy, 1977
	NW Mediterranean	0–75	0.2	Andersen and Sardou, 1994

Table 7.2 Body composition and defaecation rates of salps and pyrosomids.

Species/genus	C as % DW	N as % DW	P as % DW	C:N ratio (by weight)	Defaecation rate (μgC mg body C^{-1} h^{-1})
Salps	11.6[2], 6.3[8]	0.7–2.2[2], 1.4[8]	0.08–0.19[2], 0.10[8]	6.2[8]	
Cyclosalpa affinis				agg. 4.7; sol. 4.1[4]	agg. 5.0–16.5 (9.9)[7]
Cyclosalpa bakeri	3.2 ± 2.6[16]	0.6 ± 0.[illegible][16]		5.3 ± 1.7[16]	15.1[16]
Cyclosalpa floridana				agg. 5.2; sol. 4.9[4]	
Cyclosalpa pinnata				agg. 4.5; sol. 4.1[4]	agg. 1.7–8.2 (3.7)[7]
Cyclosalpa polae				agg. 4.6; sol. 4.3[4]	
Ihlea asymetrica				agg. 4.2; sol. 3.7[4]	
Ihlea punctata	6.4 ± 1.1[11]	1.7 ± 0.3[11]		3.7 ± 0.2[11]	
Ihlea racovitzai	10.1[5]	2.8[(5)]	0.16[5]		
Pegea bicaudata				agg. 4.4; sol. 4.1[4]	agg. 5.0–9.5 (6.5)[7]
Pegea confoederata	agg. 7.2[4]	agg. 1.[illegible][4]		agg. 4.8; sol. 4.7[4]	agg. 2.6–27.5[7]; sol. 19.6–37.3 (27.7)[7]
Pegea socia				agg. 4.6; sol. 4.6[4]	agg. 6.0–14.6 (9.9)[7]
Salpa aspera					9.4[3]
Salpa cylindrica				agg. 4.1; sol. 4.4[4]	sol. 2.4–27.2 (12.2)[7]
Salpa fusiformis	3.9[9], 8.2[6], 11.9 ± 2.0[14], 7.8[1]	1.0[9], 2.1[6], 2.7 ± 0.5[14]	0.195[6]	4.0[9], 4.6[6]	12.4[9]
Salpa maxima	agg. 10.7[4]	agg. 2.5[4]		agg. 4.6; sol. 4.4[4]	agg. 3.9–31.4 (13.0)[7]
Salpa thompsoni	4.2–4.7[10], 4.[illegible][5], 3.69[13]	1.0[10], 1.2[5], 0.84[13]	0.09[5]	4.2–4.7[10], 4.24[13]	4.2[13]
Thalia democratica	18.0[12]			3.81[12], agg. 4.0; sol. 4.4[4]	
Pyrosoma sp.	9.4[1]				
Pyrosoma atlanticum	11.3 ± 1.1[11]	2.8 ± 0.1[11]		4.0 ± 0.2[11]	24.5–43.5*[15]

Average defaecation rates are in brackets. Superscripts correspond to literature references. DW, dry weight.

[1] Curl,1962; [2] Beers, 1966; [3] Wiebe *et al.*, 1979; [4] Madin *et al.*, 1981 (only mean given); [5] Ikeda and Mitchell, 1982; [6] Le Borgne, 1982; [7] Madin, 1982; [8] Roger, 1982b; [9] Small *et al.*, 1983; [10] Ikeda and Bruce, 1986; [11] Gorsky *et al.*, 1988; [12] Heron *et al.*, 1988; [13] Huntley *et al.*, 1989; [14] Clarke *et al.*, 1992; [15] Drits *et al.*, 1992; [16] Madin and Purcell, 1992.

* Rates given in μgC individual^{-1} h^{-1} converted by using the relations DW (mg) = 0.111 Length (mm)^{190} (Sardou, personal communication), and C = 11.3% DW (Gorsky *et al.*, 1988).

8.5–11.6% DW (Beers, 1966; Ikeda and Mitchell, 1982; Ikeda and Bruce, 1986; Gorsky *et al.*, 1988). The phosphorus body content of salps is less known than that of C and N. It ranges between 0.08 and 0.195% DW, and C and N content is also lower than in non-gelatinous zoo-plankton, e.g. 0.53–0.87% DW in copepods and 0.75–1.60% DW in euphausiids (Beers, 1966; Ikeda and Mitchell, 1982). Nevertheless, for a predator that can process tissue rich in water and mineral ash, the organic matter in gelatinous species does contain normal amounts of energy (Clarke *et al.*, 1992), and many animals feed partially or exclusively on salps and pyrosomas (see Chapter 12).

7.4. Feeding and grazing impact

7.4.1. *Feeding*

Salps filter water at very high rates (e.g. Harbison and Gilmer, 1976; Harbison and McAlister, 1979; Andersen, 1985; Madin and Purcell, 1992) relative to other planktonic organisms, as discussed in Chapter 5. Here the discussion will therefore be focused on the diet, filtration and assimilation efficiencies and grazing impact of salps and, when possible, of pyrosomas.

Salps are non-selective feeders which can graze particles less than 1 μm. For example, the mucous net of *Pegea confoederata* blastozooids can retain particles at least as small as 0.7 μm (Harbison and Gilmer, 1976), bacteria are grazed by *Thalia democratica* (Mullin, 1983), and salp faecal material contains particulate matter ranging from less than 1 μm to about 1 mm (Madin, 1974b). But larger particles are retained with higher efficiencies. Particles with diameters >4.6 μm are always retained with 100% efficiency, and particles with diameters <1.8 μm are never retained completely by *Cyclosalpa affinis, C.floridana,* and *C.polae* (Harbison and McAlister, 1979); *Cyclosalpa bakeri* does not efficiently retain particles < 5 μm (Madin and Purcell, 1992). Four other salp species and *Cyclosalpa polae* have been shown to retain plastic beads > 2.5 μm diameter with high (> 60%) efficiency and 1.0 μm beads with low (< 15%) efficiency (Kremer and Madin, 1992).

As non-selective feeders, salps are known to feed also on particles other than phytoplankton cells. Krill larvae are often found in the guts of *Salpa thompsoni* in the Southern Ocean (Huntley *et al.*, 1989). Salp faecal material also contains small faecal pellets, probably produced by small crustaceans or larval forms (Silver and Bruland, 1981b), showing evidence of coprophagy in salps.

The diet of pyrosomids is much less studied. These organisms are reported to feed only on coccolithophores (Dhandapani, 1981). But in a more recent study, they appeared to feed on a large range of particles: their faecal pellet content consists mainly of unidentified small phytoplankton cells and coccoliths, but also of centric diatoms and silico-flagellates, and fragments of small crustaceans and particles looking like the faecal pellets of copepods (Drits *et al.*, 1992).

7.4.2. *Grazing impact*

From experiments on filtration rate, one *Pegea confoederata* blastozooid 50 mm long is estimated to have the grazing impact of at least 450 large calanoid copepods (Harbison and Gilmer, 1976). Salp feeding is also reported to accelerate the termination of the phytoplankton spring bloom. On the continental slope off western Ireland, in 1984, feeding by *Salpa fusiformis* occurring in very dense swarms in the upper layers ended the diatom spring bloom prior to nutrient depletion in surface waters and, thus, prior to mass sedimentation of algal cells (Bathmann, 1988).

Several attempts have been made to quantify the grazing impact of salp populations. In a coastal area of the northwestern Mediterranean, ingestion of phytoplankton by the *Salpa fusiformis* population of the 0–200 m water column would represent up to 35% of the total primary production if salps ingest the phytoplankton of the 0–75 m layer, and up to 74% if salps feed during the night on the phytoplankton of the 0–10 m layer (Nival *et al.*, 1985). Estimated carbon demands to sustain the growth rate of an aggregated population of *Thalia democratica* in the northwestern Pacific are as high as 0.63 and 0.83 gC m^{-2} day^{-1} (Tsuda and Nemoto, 1992). Comparison with primary production, 1 g C m^{-2} day^{-1}, measured at the same site, shows that a considerable fraction of the primary production would be consumed by the salp population. Moreover, these estimates of carbon demand are much less than grazing rate, as they do not include carbon loss by respiration nor assimilation efficiency during digestion.

The final example to illustrate the grazing impact of salps concerns a third oceanic area. Off the Antarctic Peninsula, in March 1984, Huntley *et al.* (1989) estimated that the respiratory requirements of the *Salpa thompsoni* populations would have required approximately 10% of the daily primary production at most locations, and could conceivably have required 100% of primary production at three stations. Moreover, the comparatively low biomass of krill larvae might be attributed not only in part to competitive removal of food by salps, but also to direct predation: salps were estimated to remove approximately 10–30% of the krill larvae in the upper 200 m. Nishikawa *et al.* (1995) likewise observed near the South Shetland Islands in the austral summer of 1990–1991 that salp and krill distributions were segregated; krill being in low numbers in offshore waters where salps were most abundant. Their calculations suggested that at most, 9% of daily primary production would have been removed by salps, hence that competitive pressure on krill was not then important and, as Huntley *et al.* (1989) had found in the same area previously, direct predation by salps on krill eggs and small individuals was significant. Finally, in a coastal shallow (60 m) bay on the NE shore of New Zealand, Zeldis *et al.* (1995) studied zooplankton and phytoplankton samples over three years, one of which was a salp-poor year. At many locations, salp grazing rates overtook phytoplankton population growth, but the deepening of the chl *a* maximum that followed the salp blooms of the previous years was seen even in the salp-poor year. Using their observations and simulations, Zeldis *et al.* (1995) concluded that the major effects of salp grazing in the coastal environment were to deepen phytoplankton distributions and reduce biomass, rather than to completely remove phytoplankton biomass from the euphotic zone, as can occur in slope and oceanic waters.

The filtration rate of pyrosomids has not been measured until very recently. The most complete study, if not the only one, which gives some idea of the nutritional physiology of *Pyrosoma atlanticum* is the recent study of Drits *et al.* (1992). From measurements on faecal pellet production and content, and on gut turnover time, and assuming that pyrosomas fed during 10 h in the surface layer, Drits *et al.* calculated that the grazing impact of *Pyrosoma atlanticum* while swarming was 53% of the phytoplankton standing stock in the 0–10 m layer, whilst the non-swarming was only 4%. The grazing pressure on the phytoplankton of the whole euphotic layer (0–50 m) was considerably lower (swarm, 12.5%; non-swarm, 0.96%), as the pyrosomids were concentrated in the upper 10 m.

7.5 Assimilation, gut transit time, and defaecation

Much of the material in salp pellets appears to be undigested, and cells with intact chloroplasts are common (Madin, 1974b; Silver and Bruland, 1981), whilst *Pegea confoederata* passes many diatoms through the gut intact (Harbison *et al.*, 1986). The assimilation efficiency of *Salpa fusiformis* varies between 28 and 39% (mean 32%) for salps feeding on a diatom and between 39 and 81% (mean 64%) for salps eating a flagellate (Andersen, 1986). Values for *Cyclosalpa bakeri* are 61% for C and 71% for N (Madin and Purcell, 1992). As pointed out by Andersen (1986), these values are in the range of those determined for other herbivorous zooplankton such as copepods, but are lower than values found for omnivorous or carnivorous plankton such as euphausiids and chaetognaths.

The gut fluorescence experiments of Madin and Cetta (1984) indicate that gut clearance times of *Pegea confoederata* and *Salpa maxima* are positively correlated with salp size, clearance times of 3.2–8.3 h and 3.2–12.3 h corresponding respectively to size ranges of 27–65 mm for *P. confoederata* and 32–77 mm for *S. maxima*. However, in *Salpa cylindrica*, in which the clearance time is between 1.4 and 5.2 h, it is not correlated with salp size. The average gut turnover time of *Pyrosoma atlanticum* is 0.75 h for 55 mm colonies (Drits *et al.*, 1992).

Specific defaecation rates for oozooids and blastozooids of seven species of oceanic salps, measured by Madin (1982), range from 1.7 to 37.3 μgC mg body C^{-1} h^{-1} (Table 7.2), with mean values from 3.7 (*Cyclosalpa pinnata* blastozooids) to 27.7 μgC mg body C^{-1} h^{-1} (*Pegea confoederata* oozooids). The other average values reported in the literature fall in this range, varying between 4.2 and 15.1 μgC mg body C^{-1} h^{-1}. On a C-specific basis salp faecal pellet production is much higher (12.38 μgC mg body C^{-1} h^{-1}) than the rate for various crustaceans (0.89–1.36 μgC mg body C^{-1} h^{-1}), because of the low body C content of salps (Small *et al.*, 1983). Defaecation rates of 43.5 μgC mg body C^{-1} h^{-1} were measured for *Pyrosoma atlanticum* colonies of 18 and 55 mm length, respectively, (Drits *et al.*, 1992). These values are near the upper limit or even higher than those reported for salps, and require confirmation as they come from a single study, compared to the data for salps. Meanwhile, these differences cannot be attributed to the C content of the animals, as the C content of pyrosomas (11.3% DW) is equal to or exceeds the C content of salps (3.2–11.9% DW). Therefore, it appears that the specific defaecation rate of *Pyrosoma atlanticum* is particularly high. According to Drits *et al.* (1992), nine colonies of 50–65 mm m^{-3} produced 1.6×10^4 pellets h^{-1} which equals the faecal pellet production of 3000–5000 *Calanus* m^{-3} (herbivorous copepods) at their usual defaecation rate of 0.5 pellets individual^{-1} h^{-1}.

7.6. Faecal pellets

7.6.1. *Composition (C,N,P)*

Salp faecal pellets appear as rectangular packets covered by a delicate peritrophic (?) membrane (Fig. 7.1), sometimes attached end to end. Pellets of the eight salp species studied by Caron *et al.* (1989) ranged in size from approximately 0.3 × 1.0 mm (for small *Pegea bicaudata*) to 3 × 4.0 mm (for large *Salpa maxima*). The faecal pellets of *Pyrosoma atlanticum* are drop-shaped with an average size of 0.26 × 0.20 × 0.10 mm and a volume of 0.005 mm^3 (Drits *et al.*, 1992).

A few studies report the carbon and nitrogen contents of faecal pellets produced by salps and pyrosomids (Table 7.3). The average C and N contents of salp pellets are respectively in the range 24.4–27.0% DW and 1.7–4.2% DW (Madin, 1982; Small *et al.*, 1983;

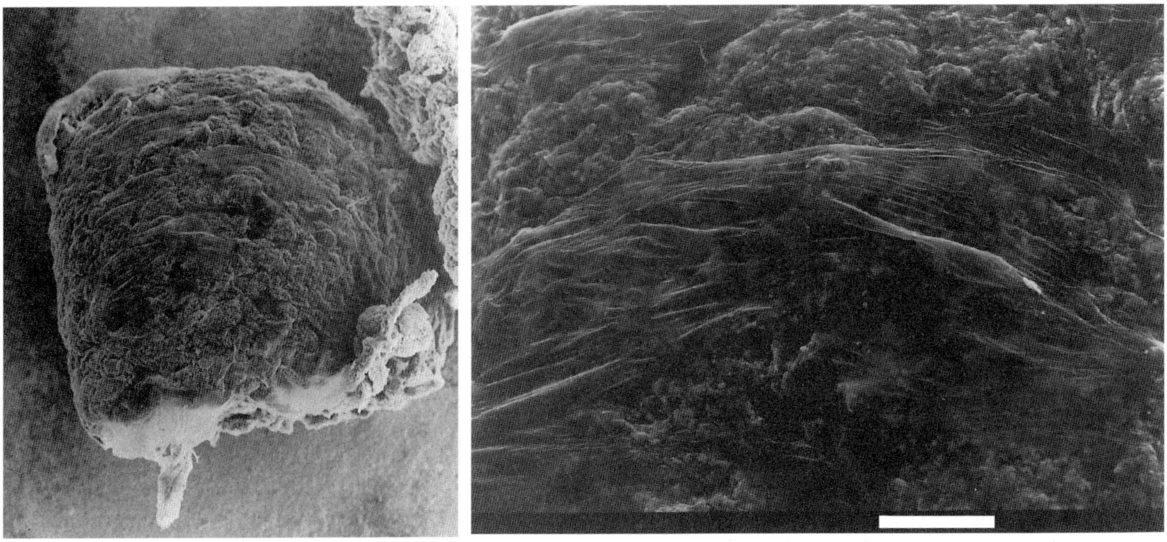

Fig. 7.1. Faecal pellet of *P. confoederata* showing rectangular shape and (at higher magnification on right) thin membrane covering pellet. Scale bar 100 μm. Scanning micrographs by Q. Bone.

Table 7.3 Composition and sinking rates of faecal pellets of salps and pyrosomids.

Organism	Length (mm)	C as % DW	N as % DW	C:N (by weight)	Sinking rate (m day^{-1})
Brooksia rostrata	47–53			14.5[5]	1040[5]
Cyclosalpa affinis	22–55			7.3–20.6[5]	650–1290[5]
	agg.			10.5 ± 1.1[2]	
Cyclosalpa bakeri				8.0 ± 2.1[7]	
Cyclosalpa pinnata	agg.			11.5 ± 1.5[2]	320–950 (588)[2]
Ihlea punctata	20–25			10.2[5]	530[5]
Pegea bicaudata	17–47			12.8–22.8[5]	650–1500[5]
	agg.			11.3 ± 2.4[2]	
Pegea confoederata	12–36			10.0–11.5[5]	300–1040[5]
	agg.			11.9 ± 2.7[2]	
	sol.			10.0 ± 0.8[2]	
Pegea socia	17–70			17.1–24.1[5]	460–1670[5]
	agg.			12.5 ± 1.0[2]	588–1218 (850)[2]
	sol.				1797–2238 (2022)[2]
Salpa cylindrica	30–40			8.9[5]	490[5]
	sol.			12.4 ± 5.6[2]	
Salpa fusiformis		27.0 ± 6.7[3]	4.2 ± 1.3[3]	6.4[3]	
		24.5 ± 10.0[4]	3.1[4]	9.3[4]	
				6.3–7.2[1]	450–2700[1]
Salpa maxima	22–85			12.8–18.1[5]	950–2470[5]
	agg.	24.4[2]	1.7[2]	11.1 ± 2.9[2]	small: 588–1642 (1024)[2]
	agg.				large: 1210–1987 (1702)[2]
Pyrosoma atlanticum	49–62	22[6]			70 ± 19[6]

DW, dry weight.
[1] Bruland and Silver, 1981; [2] Madin, 1982; [3] Small *et al.*, 1983; [4] Bathmann, 1988; [5] Caron *et al.*, 1989; [6] Drits *et al.*, 1992; [7] Madin and Purcell, 1992.

Bathmann, 1988); *Pyrosoma atlanticum* faecal pellets present a similar C weight of 22% DW (Drits *et al.*, 1992). Faecal pellet C and N concentrations are therefore higher than body contents, mean body C content ranging from 3.2 to 11.9% DW and N content from 0.6 to 2.8% DW, as previously noted (cf. Table 7.2). This contrasts with non-gelatinous zooplankton, for which the C and N concentrations of faecal pellets are lower than the concentrations in the bodies of the organisms which produce them (e.g. Small *et al.*, 1983). Phosphorus content appears somewhat higher in faecal material (0.38% DW for *Salpa fusiformis* pellets; Le Borgne, 1982) than in the organisms (0.08–0.195% DW for different salps species). Faecal pellets of salps, and presumably those of pyrosomids, are therefore rich in organic matter.

Literature values for C:N ratios of faecal pellets are more numerous than data on C and N contents, but are quite variable. The results of Caron *et al.* (1989) indicate C:N ratios (by weight) ranging from 7.3 to 24.1 in eight oceanic species, no trends being observable in this ratio with species or the size of the salps. Other studies report average values from 6.3 to 12.5 (nine salp species). Meanwhile C:N ratios of faecal pellets are always greater than the C:N ratios of the animal themselves (3.7–5.3, cf. Table 7.2). This phenomenon appears as a general rule in various zooplankton types (Small *et al.*, 1983). No data are available on the N content, and therefore C:N ratio, of pyrosomas faecal pellets.

7.6.2. *Sedimentation rates*

The sinking velocities of salp faecal pellets are very high, ranging from 300 to 2700 m day^{-1} (cf. Table 7.3) in 10 species (Bruland and Silver, 1981; Madin, 1982; Caron *et al.*, 1989; Yoon 1995; Yoon *et al.*, 1995). The highest sedimentation rates noted for copepod pellets are 12–225 m day^{-1} (adults of several species; Small *et al.*, 1979), 180–220 m day^{-1} (*Calanus finmarchicus*; Honjo and Roman, 1978), and for euphausiid pellets, 126–862 m day^{-1} (Fowler and Small, 1972). The maximal sinking rate for *Salpa fusiformis* pellets of 2700 m day^{-1} is therefore three times faster than the highest rate noted for euphausiid pellets, and an order of magnitude higher than those for copepods, Salps, which can also be coprophagic, therefore play an important role in accelerating the descent of smaller pellets which, with their reduced sinking rates, are not expected to reach deep water. In contrast to salps, the sinking rate of pyrosomid faecal pellets is in the range of those of copepod pellets, the unique estimate being 70 ± 19 m day^{-1} (Drits *et al.*, 1992). The only available value on the sedimentation rate of dead organisms, between 240 and 480 m day^{-1}, concerns measurements on preserved *Salpa aspera* (Wiebe *et al.*, 1979).

7.6.3. *Degradation*

The experiments of Pomeroy *et al.* (1984), Caron *et al.* (1989) and Drits *et al.* (1992) are the three main studies which address the degradation of faecal pellets of salps or pyrosomas. The decomposition of faecal pellets of *T. democratica* and *S. cylindrica* in the laboratory seems to be very rapid (Pomeroy *et al.*, 1984). During the first day of experiments there is a rapid increase in bacteria within the faecal pellets, then on day 2 protozoans enter the faecal particles and rapidly reduce the population of bacteria. Over 2–3 days faecal pellets become amorphous particles of small size. According to Pomeroy *et al.* (1984), these particles have a maximum residence time of 2–3 days and are a potential food source for coprophagy only for about 36 h on average. If not ingested, they are rapidly consumed and respired through the action of bacteria and protozoa. A recent study (Yoon *et al.*, 1995) has confirmed and extended these observations, using faecal pellets of *Pegea confoederata.*

In contrast, 10 day decomposition studies on pellets of eight salp species (*Salpa cylindrica* and seven other oceanic species) lead to different conclusions (Caron *et al.*, 1989). In fact, although bacterial activity in the pellets increased, the faecal pellets generally remained intact throughout the 10 day study. The average losses of ash-free dry weight and carbon and nitrogen content are small compared to their initial values. Overall, the sinking rates for all groups after 10 days of degradation averaged 80% of the initial rates. Caron *et al.* (1989) conclude that microbial degradation of large salp faecal pellets would not prevent the vertical flux to the deep ocean of a significant fraction of the particulate organic material contained in the pellets.

Several explanations were proposed by Caron *et al.* (1989) to explain the differences between their results and those of Pomeroy *et al.* (1984): (1) the species studied by Pomeroy *et al.* are mainly coastal species of salps, smaller (except the solitary form of *Salpa cylindrica*) than the larger species of oceanic environments which produce larger and denser faecal pellets; (2) the slow rates of decomposition observed by Caron *et al.* may be due, in part, to the exclusion of protozoa from the pellets, while protozoa are shown to be important for the physical disruption of the less cohesive faeces of the neritic tunicates studied by Pomeroy *et al.*; and (3) the consistency of the faecal material produced by salps in neretic and oceanic environments may differ as a result of the nature of the particulate food available.

Degradation experiments performed on *Pyrosoma atlanticum* faecal pellets (Drits *et al.*, 1992) gave comparable results to those observed with coastal salp species, the size of *P. atlanticum* pellets being small compared to those of oceanic salps. After 45 h at 23°C the faecal pellets contained many bacteria and a few protozoans, their membrane was practically disrupted and they lost 72 and 59% of their pigment and carbon content. At 8°C, faecal pellet pigment remained unchanged during 45 h; after 11 days the pigment showed an 83% decrease, the organic carbon was the same as the initial value, but the faecal pellets were pale and loose with disrupted membrane. The sinking rate slightly decreased during 45 h at 23°C (initial, 70 m day^{-1}; 45 h, 54 m day^{-1}). After 11 days of incubation at 8°C the sinking rate was significantly lower (49.4) compared with that of freshly collected pellets.

7.6.4. *Elemental and radionuclide composition*

Several studies have addressed the elemental and radionuclide composition of salp faecal pellets; three of them (Krishnaswami *et al.*, 1985; Caron *et al.*, 1989; Roméo *et al.*, 1992) provide an overview. Of the various elements analysed, calcium is the most abundant, making 8.2–12.6% of the DW of faecal pellets in four species (Krishnaswami *et al.*, 1985; Caron *et al.*, 1989). In the faecal pellets of *Cyclosalpa affinis, Pegea confoederata,* and *Salpa maxima,* magnesium, strontium, and aluminium are also important elements with respectively 1.08, 0.95 and 0.30% of the mean DW of faecal pellets (Caron *et al.*, 1989). In *Thalia democratica* pellets (Krishnaswami *et al.*, 1985) aluminium is also abundant (2.85% of DW), but not strontium (0.04%). The authors of these two studies indicate that calcium and strontium presumably originate from the coccolithophorids and planktonic sarcodina (foraminifera and acantharia) in the diet of the salps, while some of the magnesium may derive from chlorophyll in ingested phytoplankton. Aluminium apparently comes from the filtration and defaecation of suspended lithogenic particles and from aluminium absorbed to particles and microorganisms. Particularly noteworthy with *Thalia democratica* faeces is also the fact that they contain 1.18% Fe and 4 μg g^{-1} ^{232}Th (Krishnaswami *et al.*, 1985).

When nuclide concentration is measured both in the body of the salp and in its faecal pellets, it is evident that for most nuclides there is considerable enrichment in the faecal material of salps. For *Thalia democratica,* except for P and Sr, faeces/body ratios for the elements and radionuclides (Ca, Fe, Al, Zn, Cu, Mn, ^{234}Th, ^{228}Th, ^{232}Th, ^{238}U, ^{210}Pb and ^{210}Po) are > 1, with, for example, ratios of 31 for A1, 22 for ^{210}Pb, and 16 for ^{234}Th (Krishnaswami *et al.*, 1985). Faeces/salp ratios are also high for Cu and Zn for *Salpa maxima,* and for Cd, Cu, and Zn for *Pegea confoederata* (Roméo *et al.*, 1992).

7.7. Diel vertical migrations

7.7.1. *Existence of diel vertical migration*

Table 7.4 summarizes the daytime depths, night-time depths, and existence of diel vertical migration (DVM) or its range recorded in the literature for the commonest species of salps and pyrosomas. Part of these observations result from detailed studies of vertical distributions, while others, less accurate, have been made during the sampling of organisms for different purposes. So, for some species, the existence of DVM is simply presumed by the absence of organisms from the surface by day and their presence at night. Six salp species appear as weak or non-migrants (migration amplitude lower than 50 m) and occur in the surface layer or the upper 150 m: *Cyclosalpa bakeri, Ihlea punctata, Thalia democratica*, Pegea confoederata, Salpa cylindrica,* and *S. maxima.* Note that for these three latter species, no detailed studies have been made on their vertical distributions. The occurrence of migratory behaviour in *Pegea socia* is based on its absence from the surface during daylight hours and its abundance at the surface at night. In this case also, further studies are needed on its migration amplitude. Madin *et al.* (1996) examined the distribution and vertical migration of salps off Bermuda, finding that different species were found in surface waters at particular times during the day, suggesting different schedules for their diel vertical migrations.

Extensive (strong) migrators belong to the genus *Salpa. Salpa aspera* performs migration of at least 800 m amplitude (Wiebe *et al.*, 1979); *Salpa thompsoni* is also a relatively strong migrator, with migration amplitude of about 200 m (Casareto and Nemoto, 1986). Although more studied, the migration of *Salpa fusiformis,* still remains a debated subject. According to Franqueville (1971), *S. fusiformis* undergoes a large migration in the northwestern Mediterranean, from 300–800 m by day to the surface at night. But further studies show that this behaviour is not so clear cut. Two studies in the north western Mediterranean reached different conclusions. During most of the eight submersible dives they performed, Laval *et al.* (1992) observed a surface population (0–200 m) and a deep population (400–600 m) both by day and night, and concluded that the bulk of the population did not move up and down on a diel basis. In the study of Sardou *et al.* (1996), the behaviour of *S. fusiformis* appears more complex. From January to March, a small population is concentrated in the 0–225 m layer by day and caught throughout the water column at night. In contrast, the greatest population of April and May occurs essentially below 600 m both during the day and at night, but its upper limit of occurrence appears to be shallower at night (225 compared to 400 m), suggesting an upward migration of a small part of the population. Sampling down to 600 m in the northwestern Pacific also indicates *S. fusiformis* as a weak or non-migrant (Tsuda and Nemoto, 1992). In the northeastern Pacific, existence of a migratory behaviour is deduced only from absence or occurrence in surface waters (Bruland and Silver, 1981; Silver and Bruland, 1981). The pulsation rates of *S. fusiformis* and *S. aspera* are largely insensitive to changes in temperature, which is

* Most recently, Gibbons (1997) has confirmed the absence of DVM for *Thalia* on the Agulhas bank.

Table 7.4 Daytime depths, night-time depths, and the existence of diel vertical migration or its range for the most common species of salps and pyrosomids.

Species	Daytime depth (m)	Night-time depth (m)	Migration	Area	Reference
Cyclosalpa bakeri	30–60	<30	30 m	Subarctic Pacific	Purcell and Madin, 1991*
	Surface	Surface	No	NW Pacific	Tsuda and Nemoto, 1992*
Ihlea punctata	0–100	0–100	No	NW Mediterranean	Laval *et al.*, 1992*
	0–150	0–150	No/weak	NW Mediterranean	Andersen (unpublished)*
Pegea confoederata	Surface	Surface	No	NW Atlantic	Caron *et al.*, 1989
Salpa cylindrica	Surface	Surface	No	NW Atlantic	Caron *et al.*, 1989
Salpa maxima	Surface	Surface	No	NW Atlantic	Caron *et al.*, 1989
Thalia democratica	0–20	0–20	No	NW Pacific	Tsuda and Nemoto, 1992*
	0–75	0–75	No	NW Mediterranean	Sardou *et al.*, 1996*
Pegea socia	Absent from surface	Surface	Yes	NE Pacific	Silver and Bruland, 1981
	Absent from surface	Surface	Yes	NE Pacific	Bruland and Silver, 1981
Salpa aspera	400–800	Surface	Large	NW Atlantic	Wiebe *et al.*, 1979*
Salpa fusiformis	0–200 and 400–600	0–200 and 400–600	No (p)	NW Mediterranean	Laval *et al.*, 1992*
	300–800	Up to surface	Large	NW Mediterranean	Franqueville, 1971*
	0–225[†]	0–800[†]	Reverse?	" (January–March)	Sardou *et al.*, 1996*
	400–>800[†]	[illegible]25–≥800[†]	No (p)	" (April–May)	
	agg. 20–75	[illegible]0–75	No	NW Pacific	Tsuda and Nemoto, 1992*
	sol. 75	[illegible]0	25		
	Absent from surface	Surface	Yes	NE Pacific	Silver and Bruland, 1981
	Absent from surface	Surface	Yes	NE Pacific	Bruland and Silver, 1981
Salpa thompsoni	200- 300	0–75	215*bt*bt	Southern Ocean	Casareto and Nemoto, 1986*
Pyrosoma atlanticum	450- 700	0–75	515	NW Mediterranean	Andersen *et al.*, 1992*
	500–900	0–150	625*bt*bt	NW Mediterranean	Franqueville, 1971*
	700–800	0–200	650*bt*bt	NE Atlantic	Roe *et al.*, 1987*
	500–900	0–100	650*bt*bt	NE Atlantic	Angel, 1989b*

Day- and night-time depths correspond to the layers where the abundance is maximum.
[†] Detailed study of vertical distributions.
[†] Estimated from middle of day-and night-time depths.
[‡] Estimated from middle of day- and night-time depths.
p, most of the population does not migrate.

compatible with occurrence of an extensive migration (Harbison and Campenot, 1979).

Pyrosoma atlanticum is well known as a strong migrator, with mean migration amplitude of the population in the range 515–650 m (Franqueville, 1971; Roe *et al.*, 1987; Angel, 1989b; Andersen *et al.*, 1992). Even the small colonies migrate, the amplitude of the migration ranging from 90 m for 3 mm length colonies to 760 m for 51 mm length colonies (Andersen and Sardou, 1994). As noted in Chapter 3, it is possible that these extensive vertical migrations involve buoyancy changes as well as active swimming, since pyrosomas can only swim slowly by means of the jet efflux from the colony produced by the gill cilia.

7.7.2. *Migration speeds*

Salpa aspera, S.thompsoni, and probably *S.fusiformis* can be considered as large diel migrators. In the laboratory, *S.fusiformis* shows swimming speeds of 54–173 m h^{-1} for aggregates and 47–238 m h^{-1} for solitaries (Bone and Trueman, 1983). Swimming speeds observed by Madin (see Table 3.1, Chapter 3) in *S.cylindrica, S.maxima, Cyclosalpa affinis* and *Pegea confoederata* range from 144 to 252 m h^{-1} (Cetta *et al.*, 1986; Madin, 1990). Unfortunately the swimming speeds of the migrating species *S.aspera* and *S.thompsoni* have not been measured, although *S.aspera* is reported as capable of high speeds in the laboratory (Wiebe *et al.*, 1979). If we assume that the ascent or descent of the organisms would take each 4 h at a maximum, for migration amplitudes of 200, 300, 400, and 800 m, migration speeds of respectively 50, 75, 100, and 200 m h^{-1} are needed. These estimated speeds fall in the range of those observed, but further studies are obviously needed, particularly on pyrosomas.

7.7.3. *Effect on downward particulate flux*

By feeding in the superficial layers and defaecating at depth, vertical migrants could accelerate the vertical flux of particulate matter. This control of the downward flux would be important provided that, essentially: (1) the organism undergoes a deep migration and its migration rate is commensurate with or faster than the settling rates of its faecal pellets; and (2) the gut retention time is long enough for the animal to reach maximum depth before defaecation, but short enough to ensure that the organism will not transport ingested material back up in the water column during its next ascent (Angel, 1984, 1989a).

Faecal pellets of *S.fusiformis* sink at rates between 19 and 113 m h^{-1}. If the *in situ* swimming speed of this species is in the range 47–238 m h^{-1}, migration could enhance the high sinking rate of the faecal pellets produced by this organism. For the two other migrating salp species no data are available on the sinking rate of their pellets. However, all species considered, swimming speeds of 47–252 m h^{-1} appear somewhat higher than the observed faecal pellet sinking rates of 13–113 m h^{-1}. Pellets of *Pyrosoma atlanticum* sink comparatively slowly, at about 3 m h^{-1}. At present, no data exist on the swimming speed of pyrosomas. To undergo a large migration of 600 m in 4 h, the necessary swimming speed would be 150 m h^{-1}, which is much higher than the pellet sinking rates.

Data on the time that migrating salps and pyrosomas spend at night in the surface layers are lacking. If we assume (reasonably) that this time is 8 h, the gut clearance time values for salp (1.4–12.3 h) would be long enough to have part or all of the food ingested in the surface layers defaecated at greater depths. Note that these clearance time measurements concern non-migrating species. In the case of pyrosomas, the gut turnover time is comparatively low (0.75 h). But as the sinking rate of their faecal pellets is low and swimming speed (hypothetically!) high, a small part of the food ingested in the surface layers would be defaecated at greater depths where the temperature is lower and, therefore, degradation of the faecal pellets reduced. This attempt to estimate the influence of the vertical migration on the downward flux of matter obviously emphasizes the need for further studies, on locomotor and digestive physiology as well as on vertical distribution.

7.8. Downward flux

7.8.1. *The flux of particulate organic matter*

Salp faecal pellets often account for a large part of the matter in sediment traps. In the northeastern Pacific, Iseki (1981) found a predominance of salp faecal pellets in a trap at 200 m, as did Matsueda *et al.* (1986) in a trap at 740 m. Dunbar and Berger (1981) also reported that tabular pellets resembling salp faecal pellets accounted for at least 55% by weight of the total material trapped at 341 m in the Santa Barbara basin. Moreover, during the cruise 'CEROP I' in the California Current, salp pellets accounted for 90% of the matter collected at 58 and 580 m (Coale and Bruland, 1985). In the shallow waters (40 m) of the Bay of Villefranche, Fernex *et al.* (1996) were able to show that maximum ammonification rate in the surficial sediments directly followed blooms of *Thalia democratica* in the waters above.

Estimated values of faecal fluxes produced by salps cover a large range (Table 7.5): 0.01–0.07 mgC m^{-2} day^{-1} for a biomass of 0.02–0.15 mgC m^{-2} (Caron *et al.*, 1989) to 8.5–137 mgC m^{-2} day^{-1} for a large biomass of 909 mgC m^{-2} (Wiebe *et al.*, 1979) from population data, and as high as 576 mgC m^{-2} day^{-1} from sediment trap measurements, but for unknown salp biomass (Morris *et al.*, 1988). Estimated flux from pyrosomid populations reaches 561 mgC m^{-2} day^{-1} for a biomass of 1246 mgC m^{-2} (Drits *et al.*, 1992). From the values reported in Table 7.5, faecal carbon fluxes per unit of biomass produced by pyrosomids appear therefore higher than those reported for salps.

Fluxes of organic carbon due to salp pellets tend to decrease with the increasing depth of the sediment trap. The study of Iseki (1981) indicates a decrease in the flux of 36% between 200 and 900 m, and that of Matsueda *et al.* (1986) a decrease of 22% from 740 to 940 m and of 71% from 740 to 1440 m. The concentration of organic C in the sinking particles also decreases with the increase in depth. According to Caron *et al.* (1989), the flux losses between sediment trap depths could not be due only to microbial degradation of pellets while sinking, but other mechanisms such as horizontal advection, midwater coprophagy, or interception by other particles must account for these losses. The significance of such processes is presently unknown.

The high values of the vertical flux of organic carbon, as well as the high C content in the sinking salp faeces (Matsueda *et al.*, 1986), are advantageous for the supply of organic matter to the deep-sea benthos. Note that the following estimates assume that the faecal matter produced reaches the bottom without any loss. In the central North Atlantic, the faecal carbon produced by a sparse population of *Cyclosalpa pinnata* would amount to 6% of the local benthic carbon demand (Madin, 1982; cf. Table 7.5). The swarm of *Salpa aspera* observed in the northwestern Atlantic would supply from their faecal pellet production more than 100% of the daily metabolic needs of the deep benthic infauna (lower estimate considered, cf. Table 7.5). When the flux of dead salp tests (3.6 mgC m^{-2} day^{-1}) is taken into account, this estimate is revised upwards to 180% (Wiebe *et al.*, 1979). No such values are yet available for pyrosomas. The main part of pyrosomid pellets most likely does not leave the upper water column (Drits *et al.*, 1992). But in death these large animals might have considerable ecological influence: the arrival of a corpse of *Pyrosoma* on the sea bed (5540 m) and its ingestion by a starfish and munidian decapod is reported by Angel (1989b; from Lampitt, personal communication). A patch of several decaying pyrosomas on the bottom was described by Monniot and Monniot (1966) as a new benthic species, but according to Van Soest (1981) these specimens were *P. atlanticum*.

7.8.2. *Role of salp faecal pellets in the transport of elements*

The production, sinking, and decomposition of large biogenic particles are important factors controlling the distribution of trace elements within the oceans (see the review by Fowler and Knauer, 1986). The large faecal pellets of oceanic salps which sink at very high rates are particularly significant in this respect. Moreover, they are enriched with a large number of elements and radionuclides, particularly calcium, aluminium, and thorium. The elements can originate from direct filtration of lithogenic particles or from ingestion of organic particles and microorganisms containing these elements, and also upon sinking, the salp pellets can adsorb and 'scavenge' many elements from the dissolved phase.

When small coccolithophores are abundant, salp pellets contain such cells and the derived coccoliths, while pteropod faecal pellets in the same area consist chiefly of larger phytoplankton, especially diatoms. Changes in species composition of herbivores at a given geographic location can therefore change the chemistry of materials entering deep waters (Silver and Bruland, 1981). In the oligotrophic waters of the Mediterranean, salps ingesting the quite common coccolithophores may contribute significantly to the carbonate flux from surface to deep waters (Krishnaswami *et al.*, 1985).

In the Gulf Stream, the observed vertical distribution of particulate aluminium, characterized by a minimum concentration at 100 m, is consistent with a process of primary input from the atmosphere and first order removal in the upper 100 m of the water column; filter feeding by salps might be largely responsible for calculated removal rates (Wallace *et al.*, 1981). Sinking of salp pellets enriched in aluminium might also explain the relatively high ratio of aluminium flux to organic flux in the deep North Atlantic (Caron *et al.*, 1989).

Coale and Bruland (1985) used the extent of radioactive disequilibrium between the particle reactive daughter, ^{234}Th, and its soluble parent, ^{238}U, to quantify removal rates for thorium in the water column of the California Current. Their study shows that the mean life of the particulate ^{234}Th in the surface waters ranges from 2.4 to 18 days, the low value of 2.4 days coinciding with a period of an intense salp swarm. Therefore, particle residence times seem to be governed by the rate of zooplankton grazing and the types of zooplankton present, and salps are believed to be effective agents in removing thorium from the upper water column. Furthermore, where comparisons of fluxes could be made for the same elements and radionuclides (e.g. Sr, Fe, Mn, ^{210}Pb, ^{210}Po, ^{232}Th, and ^{238}U), fluxes through salps range from 3 to 35 times greater than corresponding fluxes through euphausiids from the same waters in the Mediterranean (Krishnaswami *et al.*, 1985).

7.9. Models

The occurrence of salps in high numbers and their role in the downward flux of matter has evoked the develop-

Table 7.5 Estimates of faecal production.

Species	Location	Depth (m)	Biomass (No. m^{-2})	(mgC m^{-2})	Faecal flux (mgC m^{-2} day^{-1})	Reference
(i) Estimates from population data						
Salpa cylindrica	NW Atlantic	25	0.21–2.21	0.02–0.14	0.01–0.04	Caron *et al.*, 1989
Salpa cylindrica,		50	0.06–0.14	0.05–0.15	0.01–0.07	
Salpa maxima, Pegea confoederata,						
Cyclosalpa pinnata	N Atlantic	25	0.2	1.6	0.14	Madin, 1982
Salpa aspera	NW Atlantic	100	6500	909	8.5–137	Wiebe *et al.*, 1979
Salpa thompsoni	Antarctic Peninsula	200		46–671	5–67	Huntley *et al.*, 1989
Pyrosoma atlanticum	SE Atlantic	10	51.5; 410	1308; 1246*	305; 561	Drits *et al.*, 1992
Pyrosoma atlanticum	NW Mediterranean	75	14	45	10	Andersen and Sardou, 1994
(ii) Estimates from sediment trap data						
Salpa fusiformis	NW Mediterranean	100	750	n.d.	576	Morris *et al.*, 1988
		100	12.5	n.d.	18	
Salpa spp.	NE Pacific	200	n.d.	n.d.	10.5	Iseki, 1981
		900	n.d.	n.d.	6.7	
Salps	NE Pacific	740	n.d.	n.d.	23	Matsueda *et al.*, 1986
		940			18	
		1440			6.7	
		3440			6.7	
		4240			8.7	

Depth is the sampling depth for population data, or trap depth for sediment trap data; n.d., not determined

* From the relations DW (mg) = 0.111 Length $(mm)^{1.90}$ (Sardou, personal communication) and C = 11.3% DW (Gorsky *et al.*, 1988).

ment of two types of models, stochastic and simulation models. A stochastic model for ordered categorical time series has been developed from weekly sampling of *Thalia democratica* in the Bay of Villefranche-sur-Mer (northwestern Mediterranean) from November 1966 to December 1990 (Ménard *et al.*, 1993). It shows that significant information is contained in the abundances of the two previous weeks, this fortnightly period corresponding to the maximal generation time of *T.democratica* reported in the literature. It also provides evidence for a significant influence of temperature and density on the emergence of the blooms; blooms tend to occur when hydrological variables exhibit rapid changes (increase of temperature, decrease of the density of surface waters, and so stability of the water column).

Similar interest in the development of salp swarms resulted in a model of population dynamics of *Salpa fusiformis* (Andersen and Nival, 1986a). Reproduction, transfer, and mortality rates are based on physiological functions (feeding and excretion) and the influence of the changing environment is taken into account via temperature and food concentration. This model, applied to the coastal shallow waters of Villefranche-sur-Mer (Mediterranean), reproduces the annual distribution of *S.fusiforms* (field data of 1983), the prevalence of the blastozooids compared to the oozooids and the major contribution of the young blastozooids, released in long chains, to the abundance of the population. Simulation of a simplified version of this model (Braconnot *et al.*, 1988) predicts that: (1) increase in the total salp abundance by two orders of magnitude would occur every 22 days, (2) 1 salp 1000 m^{-3} could give rise to 10 salps m^{-3} in 1 month's time.

An ecosystem model comprising salps and copepods as herbivorous zooplankton (Andersen and Nival, 1988) underlines the major role played by salps in the vertical flux of matter. After a 40-day simulation, it predicts that material in a trap set at 200 m depth consists of 47% salp faecal pellets and 18% copepod faecal pellets, although salp and copepod biomasses are similar. Simulated fluxes at 200 m (up to 72 mg C m^{-2} day^{-1} and a mean of 45 over 40 days) are in good agreement with those estimated from field data (cf. Table 7.5).

More recently, a third type of model, object-oriented, has been developed which simulates the development of a tunicate bloom and the colonization of space by its members (Laval, 1995, 1996). Using initial parameter settings based on values found for *S. fusiformis* (Braconnot *et al.* 1988) interestingly variable results have been obtained, and the model is being further developed to take account of the physiology of salps (Laval, personal communication).

7.10. Conclusion

This review points out some of the studies which have to be performed in future to obtain better estimates of the importance of salps and pyrosomas in biogeochemical cycles.

(i) The physiology of pyrosomas is far less known than that of salps, although these colonial organisms can occur in large swarms. Measurements of their filtering rate and particle retention efficiency are particularly needed to assess their grazing impact.

(ii) Detailed analyses of the pyrosoma faecal pellet composition, organic constituents as well as trace elements, and of their assimilation efficiency will contribute to better estimates of their importance in the vertical flux of matter.

(iii) More efforts should be made to quantify the influence of the diel vertical migration on the downward fluxes of matter. The main processes to characterize are: the gut transit time; the day and night vertical distributions; the chronology of migrations (time spent in surface layers and at depth); the migration speeds.

(iv) Corpses of salps and pyrosomids could be of particular importance to the ecology of the deep sea benthos when these organisms occur in large swarms. The sedimentation and decomposition rates of the dead animals should therefore be examined.

(v) At the same time, the development and improvement of models of population dynamics or ecosystem processes need to be continued, for example the introduction of diel vertical migration in ecosystem models.

Acknowledgements

The author's work was supported by the CNRS (Centre National de la Recherche Scientifique) funds through URA 2077 'Ecologie du Plancton Marin' (Sciences de l'Univers).

EIGHT

Feeding and metabolism of Appendicularia

D. Deibel

8.1. Introduction

Lohmann (1898, 1899, 1909a, 1933) was the first investigator to recognize the true function of the appendicularian house, which is to concentrate and collect particulate matter from the sea for feeding. He examined the gut contents of *Oikopleura albicans* and in this way identified an entirely new size class of plankton in the sea, which he called 'nanoplankton' (Lohmann, 1909a).

The structure of the house is described in detail in Chapter 6; it is remarkably complex and its operation is only well understood in detail in a few oikopleurid species, and less well in a single fritillariid. Essentially, there are in oikopleurids usually three filters involved in food collection; their structure determining the size class of particles a given species can feed upon. The inhalent flow produced by oscillation of the tail draws water into the house via an external relatively coarse filter, then a food collecting filter or trap concentrates particulate material, which is drawn along a short tube to the mouth by the action of spiracular cilia. After entering the pharynx, the particles are trapped on a mucous filter and they enter the oesophagus.

8.2. *Particle size selection and the natural diet*

Appendicularians historically have been considered to consume only very small particles, from ~1 to 20 μm in size (Lohmann, 1909a; Gerber and Marshall, 1974). Most of the particles Lohmann (1909a) observed in the gut contents of *Oikopleura albicans* were small naked flagellates and bacteria, although he also reported some small diatoms in the gut contents, including spores of chain-forming species. Large diatoms and diatom chains were mostly absent.

Recent evidence points to the wide size spectrum of naturally occurring particles captured by some oikopleurids, including not only nanoplankton but also large diatoms and armoured dinoflagellates, approaching 100 μm in least dimension (Deibel and Turner, 1985; Urban *et al.*, 1992, 1993a). Deibel and Turner (1985) showed that *O.vanhoeffeni* seems to consume most of the particles in its environment in proportion to their relative abundance, which results in a seasonally fluctuating diet in Newfoundland coastal waters, from diatom dominated in the spring to nano- and picoplankton-dominated in the summer and autumn. Thus the magnitude of the 'appendicularian shunt' of the microbial loop may vary seasonally (Urban *et al.*, 1992)

There is further evidence from field studies of interesting contrasts in food particle selection between appendicularians and planktonic copepods (Redden, 1994). In Newfoundland coastal waters, storms sometimes result in nearly unialgal blooms of the prymnesiophyte *Pyramimonas* spp. These blooms reveal a distinctive pigment signature using HPLC, with a strong peak of chlorophyll *b* (Fig. 8.1). In the marine environment, chlorophyll *b* is restricted to the green algae and is not found in diatoms. Copepods collected from these waters showed complete avoidance of *Pyramimonas*, with little chlorophyll *b* or any degradation products of chlorophyll *b* in their guts (Fig. 8.2a). *Oikopleura vanhoeffeni* from the same samples showed active ingestion of *Pyramimonas*, producing faecal pellets filled with chlorophyll *b* and its characteristic breakdown products (Fig. 8.2b). Thus, sinking oikopleurid houses and faecal pellets may be an avenue for the vertical transport of small, naked flagellates that would otherwise not leave the upper mixed layer. The magnitude of this vertical flux may be significant, as recent sediment trap studies from the Barents Sea show that appendicularian houses

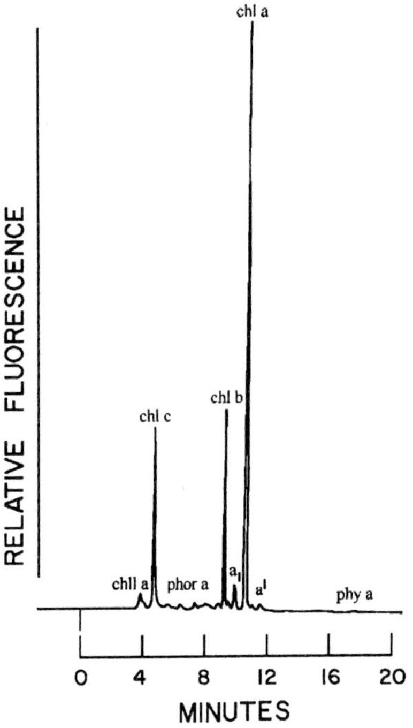

Fig. 8.1. HPLC chromatograph of phytopigments in seston < 15 μm in size, from the water column of Conception Bay, Newfoundland. Chl's *a*, *b* and *c* are shown. chll *a* = chlorophillide *a*. phor a = phaeophorbide *a*. a_1 and a^1 = chl *a* allomers. phy a = phaeophytin *a*. Adapted from Redden (1994).

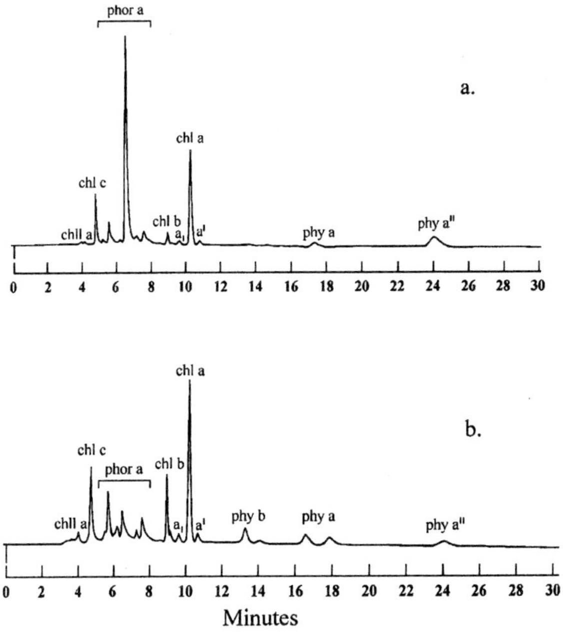

Fig. 8.2. HPLC chromatographs of phytopigment breakdown products in macrozooplankton from Conception Bay, Newfoundland. phy a^{11} = phaeophytin *a* isomer. phy b = phaeophytin *b*. All other labels are as in Fig. 8.1. (a) Sample from the gut of the copepod *Calanus finmarchicus*. (b) Sample of faecal pellets from the pelagic tunicate *Oikopleura vanhoeffeni*. Adapted with permission from Redden (1994).

and faecal pellets dominate the zooplanktonic fraction of the vertical flux during most times of the year (Gonzalez *et al.*, 1994; Zeller *et al.*, 1994; Bauerfeind *et al.*, unpublished).

In contrast to much of the above fieldwork, laboratory studies of particle size selection by appendicularians has focused on their retention of the smallest particles that exist in the sea. Deibel and Lee (1992) determined the functional particle size selection of *O.vanhoeffeni* by feeding them a series of sizes of fluorescent latex microbeads. Their results showed that particles of 3 μm diameter were retained with 88% efficiency and those 0.6 μm in diameter were retained with an efficiency of 44% (Fig. 8.3). This suggests that the functional mean pore width of the pharyngeal filter of *O.vanhoeffeni* is ~0.7 μm, in keeping with recent studies of the ultrastructure of the pharyngeal filters of salps (pore width 0.7 μm; Bone *et al.*, 1991) and ascidians (pore width ranging from 0.2 to 0.5 μm; Flood and Fiala-Medioni, 1981). This empirical determination of the functional mean pore width is at variance with ultrastructural predictions, which suggest that the particle size retained with 50% efficiency by *O.vanhoeffeni* should be 3.3 μm (Deibel and Powell, 1987). How can this be?

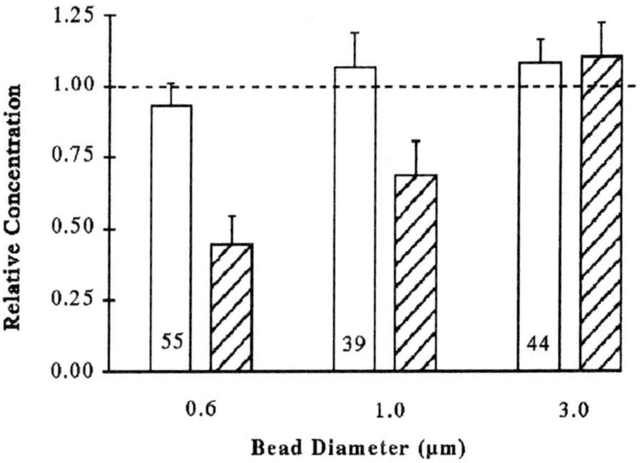

Fig. 8.3. Histogram of the mean volume concentration of several bead sizes relative to that of 7 μm beads. The open histograms are the relative volume concentrations available in the food suspension (ca. 1:1:1:1 by volume), and the hatched bars are the relative volume concentrations in the guts of *Oikopleura vanhoeffeni*. The number at the bottom of the bars is 'n' for both the open and hatched bars. The error bars are the upper 95% confidence limit. The relative volume concentration in the gut is approximately equal to the retention efficiency. Adapted from Deibel and Lee (1992).

Many explanations have been given in the past to explain differences between microscopic measurements of pore dimensions of filters and their empirically determined size retention spectra (Flood, 1981). In general, however, most of these rationalizations involve the issue of the introduction of structural artifacts when fixing tissue for microscopy. However, recent video-assisted microscopy of living appendicularians suggests another, more satisfying, possibility. Oikopleurids use their food concentrating filters to produce a concentrated particle suspension of 100–1000 times ambient (see above). It is likely that submicrometre particles form aggregates at these extreme concentrations, producing larger particles that can be sieved by the pharyngeal filter (Morris and Deibel, 1993). Also, small particles that pass through the pharyngeal filter are not discharged from the house but are passed back into the tail chamber to be recycled through the food concentrating filter (Deibel and Powell, 1987; Bedo *et al.*, 1993). This should result in further concentration and aggregation of small particles. Support for this hypothesis comes from the observations of Bedo *et al.* (1993), in which they report an increase in the proportion of 0.2 μm beads retained in the filters of the house over time — exactly what one would expect if the beads were being recycled and given more time to form aggregates. Finally, the pharyngeal filter may not function simply as a sieve, but may also capture particles by direct interception onto individual fibres. Recent video analyses of *O.vanhoeffeni* have shown that estimates of the magnitude of direct interception are almost exactly what is needed to account for the experimentally determined 3-fold overcapture of 0.6 μm beads discussed above (Acuña *et al.*, submitted for publication). This suggests that fixation artifacts no longer need be invoked to explain disagreements between ultrastructural predictions and empirical observations of particle capture by mucous filters.

Recent experimental evidence suggests that oikopleurid appendicularians may play a significant role in the consumption of submicrometre colloidal and 'dissolved' material in the sea. Flood *et al.* (1992) found *Oikopleura vanhoeffeni* to retain colloidal melanin (particle size 0.17 μm) from the ink sac of the cuttlefish *Sepia officinalis* with an efficiency of about 10%. Bedo *et al.* (1993) found *O.dioica* to retain 0.2 μm plastic spheres with efficiencies of 34–62% those of 0.75 μm spheres. Even though these retention efficiencies are rather low, projected ingestion rates of colloidal organic carbon are significant, primarily because of new information showing that the concentration of carbon in colloidal size fractions in the sea is more than an order of magnitude higher than that of particulate carbon (Kepkay, 1994).

All of the above studies of the ingestion of submicrometre particles are based on the use of surrogates of naturally occurring material. We know nothing about the capability of oikopleurids to consume naturally occurring colloidal material and about the degree to which they are able to digest these particles. This is a major impediment to furthering our understanding of the ecological role of appendicularians in the sea. Recent work by Gowing (1994) has revealed virus-like particles in the guts of oikopleurids from several of the world's oceans. This is the first corroborative evidence from the field that the laboratory experiments of Bedo *et al.* (1993) and Flood *et al.* (1992) represent the capability of appendicularians to capture naturally occurring small particles.

The considerations above suggest that oikopleurids may serve to 'short circuit' the microbial loop, directly transferring energy from very small particles to much larger predators of oikopleurids, including larval and adult fish (Gadomski and Boehlert, 1984; Keats *et al.*, 1987). This shunt of the linear biomass spectrum (*sensu* Sheldon *et al.*, 1972) is significant, similar to the particle size relationship between baleen whales and the krill on which they feed (see Chapter 10). As we shall see below, this shunt may also operate to transport microbial cells and colloidal material to great depths in the form of sinking faecal material and houses.

8.3. Filtration, clearance, and ingestion rates

The feeding rates of oikopleurids were first determined by Paffenhöfer (1975). Before his experiments are discussed, some definitions are required. 'Filtration rate' will refer to the bulk flow of water pumped through the house and will have dimensions of volume time^{-1}. Filtration rate will not take into account time spent not pumping water, i.e. time spent in tail or ciliary arrest. 'Clearance rate' will refer to the volume of water swept clear of a particular particle size and also will have dimensions of volume time^{-1}. The filtration rate times the particle retention efficiency equals the clearance rate. Also, since clearance rate is generally determined empirically over several tens of minutes to hours, it includes time spent not feeding. For oikopleurids, clearance rate is generally determined as the particles swept clear by the house and the animal combined. This is an important distinction, since not all of the particles removed from suspension are ingested (Gorsky, 1980). The 'ingestion rate' refers to the number or mass of cells consumed and may have dimensions of cells time^{-1} or of biovolume time^{-1}, i.e. μm^3 h^{-1} or μg carbon h^{-1}. The 'daily ration' is the ratio of mass of particles ingested to the mass of the individual oikopleurid and will have units of % day^{-1}. The ambiguous terms 'grazing rate' and 'feeding rate' will not be used in this chapter (see also Chapter 5).

Filtration rates for a variety of appendicularians vary from 1,200–35,400 ml day^{-1}, depending on the species (Table 8.1). Flood (1991b) used stroboscopic photographs to trace particles flowing through the house of *O.labradoriensis*. He found flow rates through the tail chamber of 0.84 ml min^{-1}, which equates to a filtration rate of 50 ml h^{-1} (trunk length 1.2 mm, temperature 11–12°C). Morris and Deibel (1993) used video analyses

Table 8.1 A comparison of the filtration rates of several oikopleurid appendicularians.

Species	Trunk length (mm)	Volume of house passageways (ml)	Filtration rate (ml d^{-1})	Reference
Stegosoma magnum	NA	1.6	20,640–35,448	Alldredge, 1977
Megalocercus huxleyi	NA	0.33	14,256–28,512	Alldredge, 1977
Oikopleura vanhoeffeni	1.5–5.5	NA	1,440–25,920	Morris and Deibel, 1992
Oikopleura rufescens	NA	0.04	3,456–11,520	Alldredge, 1977
Oikopleura cornutogastra	NA	0.13	4,248–5,832	Alldredge, 1977
Oikopleura fusiformis	NA	0.02	1,728	Alldredge, 1977
Oikopleura labradoriensis	ca. 1.2	NA	1,210	Flood, 1991

These rates are for the total pumping capacity of the animal and do not take into account time spent not feeding. Species are shown in rank order of maximum filtration rate.
Alldredge (1977) was done at 25°C by visual tracking of inspired dye. Flood (1991) was done at 11–12°C by microscopic tracking of inspired particles using a stroboscope. Morris and Deibel (1993) was done at 0–5°C by videomicroscopic tracking of inspired plastic beads.

to determine flow rates inside the house of *O.vanhoeffeni* at 0–5°C, finding values of 25 ml h^{-1} for a standard animal 1.2 mm long: within a factor of two of the rate determined by Flood (1991b) at temperatures about 10°C higher. The filtration rate of *O.vanhoeffeni* increased 20-fold over a trunk length increase of 1.5–5.5 mm (Morris and Deibel, 1993). Very low water temperatures do not seem to be a hindrance to the filtration of *O.vanhoeffeni*, as the smallest individuals of 1.5 mm trunk length have almost the same filtration rate as do *O.labradoriensis* of 1.2 mm trunk length, pumping at 11–12°C (Table 8.1). The increased viscosity at these low temperatures may explain the large inlet filter pores of *O.vanhoeffeni*, the largest pore size yet measured (Table 6.1, Chapter 6), which lives in the coldest water temperatures. Larger pores at higher viscosities would reduce drag and hence the energy required to pump water through the inlet filters.

Clearance rates have been determined for oikopleurids using a variety of techniques over a wide range of water temperatures (i.e. −1.6 to 24°C, see Fig. 8.4). Given these differences, there is surprisingly little scatter in the data, with about a 5-fold range for *O.dioica* of a given body size (total range overall body sizes 1–320 ml day^{-1}) and a 5- to 7-fold range for *Stegosoma magnum* and *O.vanhoeffeni* for a given body size (total range overall body sizes 60–14 000 ml day^{-1}). For *O.dioica*, the values of Alldredge (1981) are the highest and the values of Gorsky (1980) and Bedo *et al.* (1993) are the lowest. In general, the values for *S.magnum* are higher than are

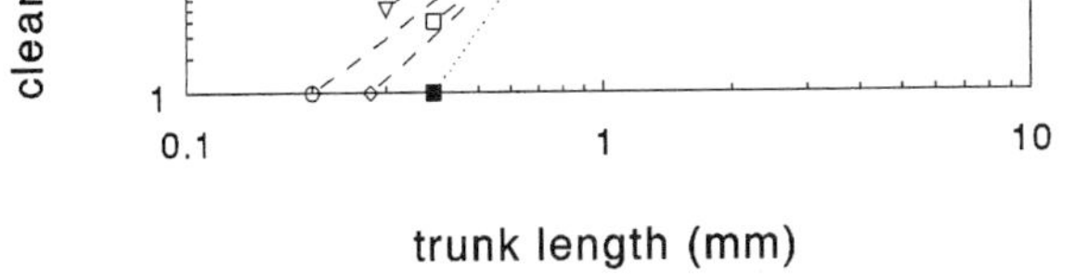

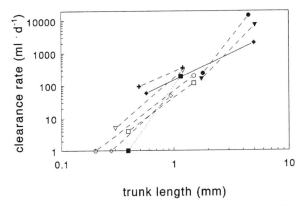

Fig. 8.4. Log–log plot of the clearance rate of several oikopleurid appendicularian species vs. body size. All rates represent the particles cleared by the house and the animal together, unless otherwise indicated. ▽ *Oikopleura dioica* at 12°C (Paffenhöfer, 1975). ○ *O.dioica* at 13.5°C, particles ingested by animals only (King *et al.*, 1980). □ *O.dioica* at 13.5°C (Bedo *et al.*, 1993). ◇ *O.dioica* at 17°C (Gorsky, 1980). ■ *O.dioica* at 18°C (Fenaux and Malara, 1990). + *O.dioica* at 23–24°C (Alldredge, 1981). ● *Stegosoma magnum* at 23–24°C (Alldredge, 1981). ▼ *O.vanhoeffeni* at −1.6 to +4.5°C (Deibel, 1988). ◆ *O.vanhoeffeni* at −1.2 to +11.5°C (Knoechel and Steel-Flynn, 1989).

those of *O.vanhoeffeni* of similar body size (Fig. 8.4), while the clearance rates of both species are from 1 to 2 orders of magnitude lower than the filtration rates (compare Fig. 8.4 to values in Table 8.1). This certainly reflects the fact that oikopleurids do not pump continuously while feeding (Deibel, 1988; Flood, 1991) and that particle retention is probably not 100% efficient.

In general, clearance rate increases exponentially with increasing oikopleurid body size (Table 8.2). Power curve slopes range from 1.2 to 4.0 for *O.dioica*, from 1.9 to 6.6 for *S.magnum* and from 1.6 to 1.8 for *O.vanhoeffeni* (Table 8.2).

Variation is generally high except for *O.dioica* (44%; Alldredge, 1981) and *O.vanhoeffeni* (49–56%; Deibel, 1988; Knoechel and Steel-Flynn, 1989). This relatively high scatter for *O.vanhoeffeni* may be explained by the fact that both investigators made use of wild-captured animals collected on many days throughout the year. Most of the remaining investigators made use of animals in culture (e.g. almost all of the *O.dioica* studies) or of wild-captured animals collected on several days within a single week (i.e. Alldredge, 1981). There is no apparent reason for the large scatter in the slopes for *S.magnum*, as the experiments were done with wild-captured animals collected on 3 days within one week (Alldredge, 1981).

Ingestion rates of oikopleurids are difficult to determine empirically, because some of the particles removed from suspension remain stuck in the house and are not ingested (Gorsky, 1980) and because oikopleurids may retain colloidal particles (Flood *et al.*, 1992) smaller than are retained by the GF/C or GF/F glassfibre filters typically used in zooplankton studies (Paffenhöfer, 1975). Given these limitations, there have been few attempts to estimate daily ingestion rates. Paffenhöfer (1975) found *O.dioica* to ingest around 100–200% of its body weight each day at a particle concentration of 65 μg C l^{-1}, and ~200–400% day^{-1} at 105 μg C l^{-1}. This is much higher than the daily ration of *O.dioica* estimated from the ingestion of radiolabeled bacterioplankton, viz. 60% day^{-1} at a food concentration of 60 μg C l^{-1} (Sorokin, 1973). It is similar to the value found by Gorsky (1980) of 350% day^{-1} while *O.dioica* was feeding on *Isochrysis galbana* at ~1 × 10^4 cells ml^{-1}. Approximately 30–60% of the particles removed from suspension remain stuck in the house (Gorsky, 1980; Crocker, Deibel and Rivkin, unpublished).

These values for the ingestion rate of *O.dioica* are much higher than the estimates for *O.vanhoeffeni*: a median value of 64% day^{-1} based on a particle concentration range of 29–85 μg C l^{-1} (Deibel, 1988). The range of daily ration for *O.vanhoeffeni* is quite high however, up to 980% day^{-1} (Deibel, 1988). Recent estimates of the ingestion rates of *O.vanhoeffeni* living in the Northeast Water Polynya, north east Greenland

Table 8.2 A comparison of the regression parameters for the clearance rate of several oikopleurid appendicularians versus body size.

Species	Temperature (°)	Predictor	Units of predictor	Units of clearance rate	Slope	Intercept	r^2	Reference
O. dioica	12	Ash-free dry weight	μg	ml day^{-1}	1.19[a]	10	0.96	Paffenhöfer, 1975
					1.24[a]	6.3	0.94	
O. dioica	13.5	Trunk length	μm	ml day^{-1}	3.08	4×10^{-8}	0.94	King *et al.*, 1980
O. dioica	23–24	Trunk length	mm	ml h^{-1}	1.62	0.10	0.44	Alldredge, 1981
O. dioica	13.5	Trunk length	μm	ml day^{-1}	1.97[b]	5.8×10^{-5}	0.83	Bedo *et al.*, 1993
					2.32[c]	4.7×10^{-6}	0.79	
O. dioica	17	Trunk length	mm	ml day^{-1}	3.39	126	0.84	Gorsky, 1980
O. dioica	18	Trunk length	μm	ml day^{-1}	3.95	2.3×10^{-10}	0.80	Fenaux and Malara, 1990
S. magnum	23–24	Trunk length	mm	ml h^{-1}	1.89[d]	32.4	0.94	Alldredge, 1981
					4.71[d]	0.36	0.86	
					6.60[d]	0.006	0.71	
O. vanhoeffeni	1.6–4.5	Trunk length	mm	ml h^{-1}	1.79	12.6	0.49	Deibel, 1988
O. vanhoeffeni	1.2–11.5	Tail length	mm	ml day^{-1}	1.58	22.6	0.56	Knoechel and Steel-Flynn, 1989

The parameters shown are the slope and intercept of the power curve, $y = aX^b$, determined after log-log transformation. All investigators determined the clearance of particles by the animal and house together except for King *et al.* (1980), which are data for uptake of particles by the animal only (i.e. equivalent to the ingestion rate). The grand mean (± standard deviation) of all 13 slopes is 2.72 ± 1.59.

[a]Determined at two different food concentrations.
[b]Feeding on *Synechococcus* spp.
[c]Feeding on *Isochrysis galbana*.
[d]Determined on 22, 23 and 26 July.

(78–80° N), have been made by the gut pigment technique (Deibel and Acuña, 1994). A 'standard' animal 1 mm long contained 7 ng of total pigments. Given a gut evacuation rate of 0.86 ng h^{-1} (Acuña and Deibel, 1994) and continuous feeding over 24 h (Redden, 1994), the ingestion rate is 144 ng pigment day^{-1}. Madin and Purcell (1992) have suggested that about half of the ingested pigment is lost during gut passage in salps. Assuming that this factor applies to appendicularians, this gives an ingestion rate of 288 ng pigment day^{-1}. Assuming further a carbon:chlorophyll ratio of 40:1 for the ingested phytoplankton, a daily ingestion of phytoplankton carbon of 11.4 μg, or about 50% of body carbon day^{-1} is obtained. This value is close to the estimate of Deibel (1988) of 64% day^{-1} (see above). If phytoplankton carbon is 20% of total seston carbon, then the total daily ration of *O.vanhoeffeni* in the Northeast Water Polynya would be about 250% day^{-1}.

Oikopleura dioica displays a peculiar functional relationship between ingestion rate, food concentration and body size (King, 1982). Although ingestion rate increases with increasing food concentration as would be expected, large individuals ingest more food per unit body mass than do small individuals (Fig. 8.5). This interesting relationship allows *O.dioica* to maintain an exponential growth rate for its entire life cycle (King, 1982) and must be due to underlying relationships between body size, body mass and clearance rate that are somewhat atypical with respect to similar allometric relationships for copepods (for example see Table 8.2, above). Also, King (1982) demonstrated that *O.dioica* does not display saturation of ingestion up to a seston concentration of 125 μg C l^{-1}, equivalent to a chlorophyll *a* concentration of 2–3 μg l^{-1}. As this has been the

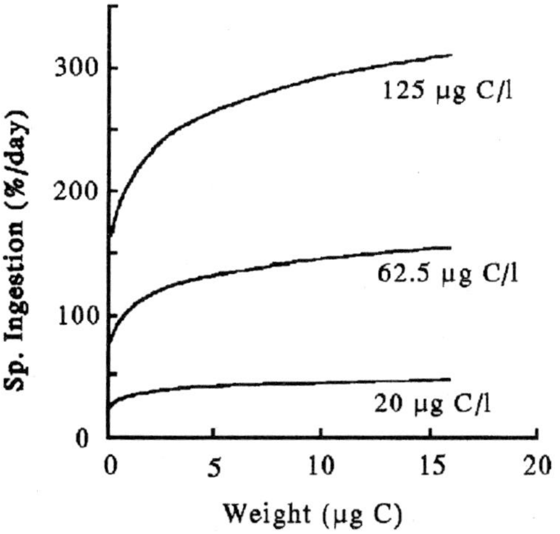

Fig. 8.5. Weight-specific ingestion rate vs. body weight of *Oikopleura dioica* at three different food concentrations. The temperature was 14°C. Adapted with permission from King (1982).

only examination of the functional response-feeding relationship for any pelagic tunicate, we are still quite ignorant of the relationships between food concentration and ingestion rate.

8.4. Modulation and periodicity of feeding

Oikopleurids have the capability to accept or reject particles by regulating the size of the mouth opening in a continuous fashion, rejecting a variable portion of the flow received from the food tube (Deibel, 1986; Fenaux, 1986a; Flood, 1991b). The rejected particles are pumped into the dorsal chamber and out of the house. Note that there is no discrimination between individual particles in this rejection mechanism and that it permits continual monitoring of the incoming particle field as to quality and quantity. Why they should reject some or all of the particles collected by the food concentrating filter is not clear. The beat of the spiracular cilia which draws particles into the pharynx via the buccal tube is under nervous control and can be reversed (Galt and Mackie, 1971). Reversal is triggered by contact of large or noxious particles with the lip or mouth area. Presumably the size (mass) and/or chemical composition of the particles could be sampled by the sensory cells on the lower lips (section 4.5.1).

Oikopleurids also have possibilities for modulating their feeding behaviour over a range of time scales from minutes (episodic arrest of tail beat) to days (i.e. diel feeding rhythm). Field studies have shown that *O.vanhoeffeni* only spends about 50% of its time pumping water through the house (Deibel, 1988; Knoechel and Steel-Flynn, 1989). Deibel (1988) found that the concentration of ingestible food particles accounted for around 25% of the total variation in time spent feeding by *O.vanhoeffeni*, much more than could be accounted for by temperature (7%) or trunk length (1%). At present, neither filter clogging nor satiation can be eliminated as possible causes. Diel studies of gut pigment content in the field indicate that *O.vanhoeffeni* feeds at similar rates in the day and at night (Redden, 1994). This is in marked contrast to most of the copepods that live along with *O.vanhoeffeni* in Newfoundland coastal waters, which show a marked diel feeding rhythm (Redden, 1994). Of course, so far as known, appendicularians lack any light receptors.

8.5. Assimilation and defaecation rates

There has been only a single study of the assimilation capacity of *O.dioica* (Gorsky, 1980). Depending on the food available, assimilation efficiency ranged from 17% (*Isochrysis galbana*) to 88% (*Thalassiosira pseudonana* and

Platymonas sueica). Assimilation of naturally occurring particles was 80%. Thus, the assimilation efficiency of *O.dioica* is similar to that of copepods, which is surprising considering that oikopleurids neither chew nor grind, having no mouthparts with which to pulverize their prey, and that their gut passage time is relatively short, of the order of tens of minutes to a maximum of about 1 h (Acuña *et al.*, 1994). We have no idea of microzooplankton or colloid assimilation by oikopleurids. This is an obvious and important gap which needs rectification.

There have been only two studies from which we can estimate a daily defaecation rate. These studies make for an interesting comparison, because one was made in the laboratory (Gorsky, 1980) and the other in the field, in the Northeast Greenland Polynya (Deibel and Acuña, 1994). Gorsky (1980) found that *O.dioica* fed *Isochrysis galbana* at 1.1×10^4 cells ml^{-1} produced 3 ± 1 pellets every h^{-1}. Since each pellet contained an average of 53.1 ng C, the hourly carbon egestion rate was 0.16 μg C h^{-1}. Assuming that *O.dioica* feeds continuously, this hourly rate is equivalent to 3.2 μg C day^{-1}, or three times the body weight of 1.2 μg C egested each day as faeces.

The daily defaecation rates found by Deibel and Acuña (1994) for *O.vanhoeffeni* off northeast Greenland differ from these laboratory results. Feeding at chlorophyll *a* concentrations < 1 $\mu g\ l^{-1}$, a 'standard' *O.vanhoeffeni* 1 mm long produced 2 faecal pellets h^{-1}, or 40 day^{-1} assuming continuous feeding (Redden, 1994). Each pellet contained an average of 3.5 ng total chloropigments. If we assume a C:chloropigment ratio of 40, then each pellet contained 140 ng C. Multiplied by the daily faecal pellet production rate of 40 day^{-1}, the daily carbon egestion rate (i.e. of phytoplankton carbon only) was 5.6 μg C, or 25% of body carbon day^{-1}, assuming an average body size of 22.4 μg C. This is about one order of magnitude lower than the estimate for *O.dioica* fed laboratory phytoplankton at much higher water temperatures (see Gorsky, 1980, above). This disagreement could be due to real differences between species, but note that the estimate for *O.vanhoeffeni* included phytoplankton carbon only. If detrital carbon made up 50% of the diet of *O.vanhoeffeni* for example, then the daily carbon defaecation would be doubled, bringing it closer to the laboratory estimate of Gorsky (1980). A second difference is the water temperature, which differed by 20°C. At 0°C, *O.vanhoeffeni* may well have a much slower gut evacuation rate than *O.dioica* living in warmer waters.

Recent results indicate significant transport of faeces and houses of appendicularians to considerable depths in the Barents Sea (Gonzalez *et al.*, 1994; Zeller *et al.*, 1994; Bauerfeind *et al.*, unpublished). This suggests that either sinking velocities are high or that the decay rates of appendicularian houses and faeces are very slow. Reports from the tropics suggest high sinking velocities. Taguchi (1982) found that 66% of the total volume of all zooplankton faeces in Kaneohe Bay, Hawaii, was produced by *O.longicauda*, despite the fact that it comprised only 1% of the total net zooplankton abundance. What is known of the sinking velocities of appendicularian faeces and houses?

The faecal pellets of oikopleurids range in size from < 300 μm long and 1×10^4 μm^3 for *O.dioica* (Gorsky, 1980) and *O.longicauda* (Taguchi, 1982) to > 500 μm long and $1–5 \times 10^7$ μm^3 for *O.vanhoeffeni* (Urban *et al.*, 1993b). Urban (1992) determined sinking velocities of *O.vanhoeffeni* faeces using a SETCOL device. The pellets sank with $R_e > 1$, but at much lower velocities than was predicted by Stokesian dynamics. The velocities ranged from 86–518 m day^{-1} over a pellet density range of 1.15–1.29 g cm^{-3} (sinking experiments were done at 0°C and 32% salinity. Gorsky (1980) found that the sinking velocity of houses of *O.dioica* increased with increasing house size and was an exponential function of water temperature between 4 and 24°C, ranging from ~30–160 m day^{-1} (Fig. 8.6a). Faecal pellets sank at about the same velocities (Fig. 8.6a), which were much lower than were the sinking velocities of faecal pellets produced by *O.vanhoeffeni* (see above). Sinking velocities decreased linearly with increasing water density (Fig. 8.6b).

Taguchi (1982) found in laboratory tests that the faecal pellets of *O.longicauda* sank at velocities intermediate between those of *O.dioica* and *O.vanhoeffeni*, i.e. 103–368 m day^{-1} ($\bar{X} \pm 95\%$ CL = 189 ± 31 m day^{-1}). However, in the field, sediment trap data indicated much slower sinking velocities than those determined under quiescent conditions in the laboratory, i.e. 7 m day^{-1} (Taguchi, 1982). These results may be indicative of the general problem of attempting to apply laboratory determined sinking velocities to field conditions.

The density of faecal pellets is evidently extremely important in determining their sinking velocity. Urban *et al.* (1993b) determined the densities of faecal pellets of *O.vanhoeffeni* using isosmotic density gradient centrifugation. They found the densities to range from 1.09 to 1.32 g cm^3 and to vary seasonally, with (paradoxically), the lowest densities during the spring diatom bloom and the highest densities in the summer and winter, when microbial food dominated the diet. They concluded that this seasonal difference in density was due to the less efficient packing of diatoms in the pellets in spring (12% open area) versus winter (3.5% open area). This suggests that the compactability of faecal pellet contents is more important in determining their density than is the primary density of the con-

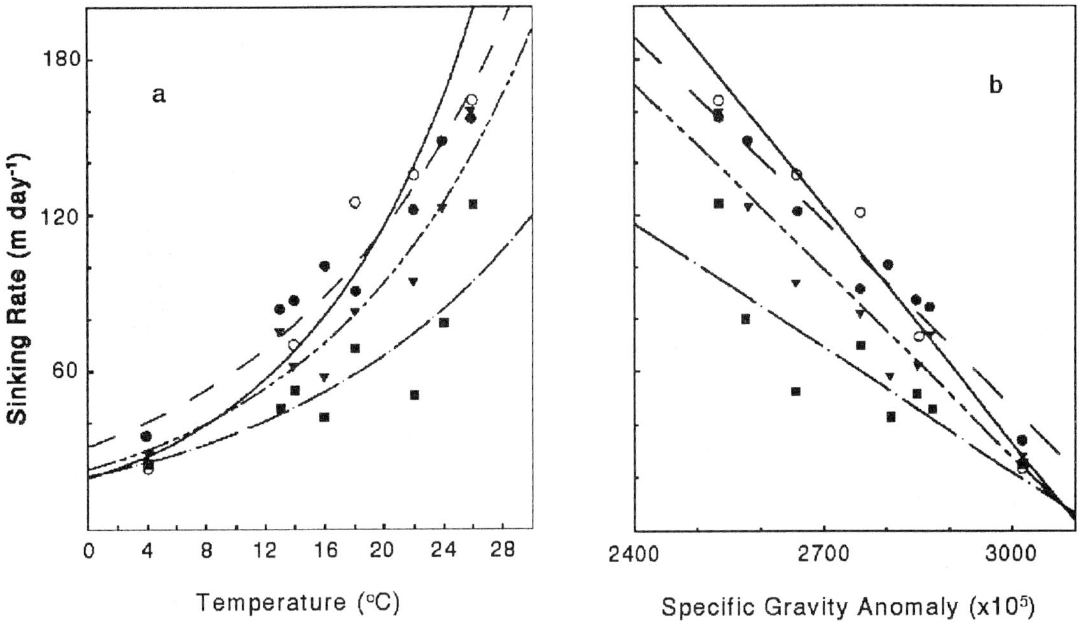

Fig. 8.6. Sinking rate of faecal pellets (○) and houses (solid symbols) of *Oikopleura dioica* vs. temperature 'a' and seawater density 'b' expressed as the specific gravity anomaly. The symbols represent 3 size classes of oikopleurid, ● is 800 μm trunk length, ▼ is 600 μm trunk length and ■ is 400 μm trunk length. The various lines in panels 'a' and 'b' represent the least-squares, best-fit regressions. Adapted with permission from Gorsky *et al.* (1984).

tents. It also suggests much greater compaction of naturally occurring particles than was found by Gorsky (1980) for laboratory grown phytoplankton (i.e. 33% open area).

8.6. Respiration rates

Respiration rates have been determined for *O.dioica* by Gorsky *et al.* (1987) in incubations of 4–8 h in water screened through 30 μm mesh. Rates increased both with increasing body size and increasing temperature, ranging from 0.021 to 0.18 $\mu l\ O_2\ h^{-1}$ at 15°C, from 0.032 to 0.30 $\mu l\ O_2\ h^{-1}$ at 20°C and from 0.039 to 0.35 $\mu l\ O_2\ h^{-1}$ at 24°C. This corresponds to a dry weight specific respiration rate of 23–54 $\mu l\ O_2\ mg^{-1}\ h^{-1}$ at 15°C, 34–79 $\mu l\ O_2\ mg^{-1}\ h^{-1}$ at 20°C and 43–130 $\mu l\ O_2\ mg^{-1}\ h^{-1}$ at 24°C. These weight-specific rates are much higher than are those of a range of other tunicates and copepods (4–11 $\mu l\ O_2$ mg dry weight^{-1} h^{-1}; Gorsky, 1980). Assuming a respiratory quotient of 0.8, these respiration rates are equivalent to 141% of body C day^{-1} at 15°C, 257% day^{-1} at 20°C and 332% day^{-1} at 24°C. In some cases, these values exceed the daily ration at these temperatures (see above). This obviously suggests that oikopleurids may depend on other, smaller, particles for food energy (e.g. colloids; Flood *et al.*, 1990, 1992), but there is also the possibility that the rates are overestimates, perhaps due to difficulties in determination of the dry weight of gelatinous zooplankton. More studies are needed to determine if these high rates are typical.

The power curve relationship between respiration rate and dry weight had an exponent (i.e. a slope) of 0.52–0.71, depending on temperature (Gorsky *et al.*, 1987). Smaller animals had a higher weight-specific respiration rate than did larger animals, with the slope of the power curve relating weight specific respiration rate to dry weight of –0.32 to –0.38, depending on temperature. The Q_{10} of respiration between 15 and 23°C was 3.51.

For comparison, Gorsky *et al.* (1984) also determined the respiration rate of *O.longicauda* at 20°C using similar techniques to those above for *O.dioica.* Dry weight-specific rates ranged from 20 to 31 $\mu l\ O_2\ mg^{-1}\ h^{-1}$, somewhat lower than those shown above for *O.dioica* (i.e. 34–79). The above excretion and respiration rates for *O.dioica* result in an atomic O:N ratio ranging from 15 to 20 depending on temperature and body size, suggestive of protein-based metabolism (Gorsky *et al.*, 1987). Colloids may be protein rich (Parrish, personal communication) due to the adsorption of dissolved organic matter (DOM). These O:N ratios are consistent with the assimilation of protein-rich food by oikopleurid appendicularians.

8.7. Excretion rates

Excretion rates have been determined in 0.2 μm filtered sea water for pre-fed *O.dioica* from laboratory cultures over 2–8 h incubation intervals at a range of temperatures (Gorsky *et al.*, 1987). Rates ranged from 0.92×10^{-4} to 16×10^{-4} μM NH_4-N h^{-1} at 15°C, from 1×10^{-4} to 18×10^{-4} μM NH_4-N h^{-1} at 20°C and from 1.9×10^{-4} to 29×10^{-4} μM NH_4-N h^{-1} at 24°C. Highest excretion rates were found for the largest animals, with the slope of the power curve relating excretion rate to dry weight of 0.78–0.99 at the various temperatures. Weight-specific excretion rates were highest for the smallest animals, with slopes of the power curve relating weight-specific excretion rate to dry weight ranging from –0.28 to –0.35. This is the only comprehensive study of the excretion of an oikopleurid appendicularian, and it is obvious that many more studies are required before the nitrogen metabolism of this group of abundant animals will be fully understood.

8.8. Energy budget

There has only been a single attempt to formulate a complete energy budget for an oikopleurid appendicularian. Gorsky (1980) found that the growth, development and fecundity of *O.dioica* was dependent on the nature of the diet and on the food concentration. A mixture of *Isochrysis galbana, Thalassiosira pseudonana* and *Platymonas sueica* at a concentration of 3.4×10^5 μm^3 ml^{-1} was the optimal diet and provided for exponential growth throughout the life cycle, which took 6.2 days at 13°C and 3.8 days at 23°C. A food concentration of 2.7×10^6 μm^3 ml^{-1} was the lower threshold below which the gonads did not mature. At food concentrations near and below the threshold, *O.dioica* directed all available energy into gonad maturation, at the expense of the growth and maintenance of somatic tissues.

Large bag enclosures provide an empirical approach midway between a beaker and a bay. King (1982) examined growth and production rates of *O.dioica* in the CEPEX bags in Saanich Inlet, British Columbia, Canada. In the bags at 14°C and 125 μg C l^{-1} of food, the generation time was 7–10 days. This is identical to the generation time observed by Paffenhöfer (1973), of 9.5 days at 13°C. Over the entire life cycle from egg to reproductive stage, the weight of *O.dioica* doubled each day. This is similar to the results of Paffenhöfer (1975), showing instantaneous growth coefficients (based on dry weight) of 0.57–1.09 day^{-1}. However, it is much lower than recent estimates of the population growth of *O.dioica* at 28°C in Kingston Harbor, Jamaica, which showed a remarkable increase in population biomass of > 1000% day^{-1} (Hopcroft and Roff, 1994). Investigations are currently underway in Jamaica to see if individual ingestion rates are sufficiently high to account for such explosive population growth (Hopcroft, personal communication). Egg production and juvenile recruitment in the CEPEX bags were very tightly coupled to food supply (King, 1982), as the laboratory experiments of Gorsky (1980) would suggest. Maturation, egg production and growth of juveniles followed blooms in nanoplankton and bacterioplankton. Given the generation times in the bags (King, 1982) and in the laboratory (Gorsky, 1980), it is likely that most field studies of oikopleurid growth and generation time (e.g. Wyatt, 1973b) are in error, because the sampling interval was too long relative to the generation time (Paffenhöfer, 1975). *Oikopleura dioica* is a colonizing species, having instantaneous rates of population increase (*r*) of 0.44–1.2 day^{-1} (King, 1982). Oikopleurids do not store lipids (Deibel *et al.*, 1992) nor do they have diapause eggs or specialized overwintering stages (King, 1981). Thus, how oikopleurids maintain seed populations through times of low food concentrations is not known. It is possible that at these times they rely on the ingestion of picoplankton (Deibel and Lee, 1992; Bedo *et al.*, 1993) or colloids (Flood *et al.*, 1992). Gorsky (1980) found that the scope for growth of *O.dioica* depended on the nature and concentration of the food supply (Table 8.2). When *Isochrysis galbana* was offered at a concentration of 10^4 cells ml^{-1} (i.e. 145 μg C l^{-1}), the scope for growth was negative. However, when naturally occurring particles that passed through a 50 μm sieve were offered as food, the scope for growth was 1 μg C day^{-1}, giving a coefficient of daily exponential growth of 0.61, close to a doubling of body mass each day. The low scope for growth when the animals were fed the *I.galbana* cells was attributed by Gorsky (1980) to a low assimilation efficiency (17%). The assimilation of naturally occurring particles was much higher (80%). Gorsky (1980) did not take into account energy lost via the excretion of non-solid waste products. Thus, the scope for growth given naturally occurring particles may be somewhat lower than 1 μg C day^{-1}.

8.9. Summary

Oikopleurid appendicularians possess many unique adaptations for living in a nutritionally dilute medium. They use their external, gelatinous 'house' as a tangential flow filter to concentrate ambient food particles up to 1000-fold prior to their collection by the pharyngeal filter. The pharyngeal filter captures particles the size of colloids and viruses up to diatoms, by sieving and also by direct interception of particles onto individual filter fibres. Thus, they are able to mediate a shunt of very small, submicron particles directly into their bodies, houses and faecal pellets, making them available relatively directly to their predators and packaging them for transport to the benthos.

Large oikopleurids are capable of pumping impressive volumes of water, up to 20 litres individual^{-1} day^{-1}, meaning that they can have considerable grazing impact on their prey populations. This feeding activity is only weakly modulated, with no apparent saturation of ingestion up to at least 125 μg C l^{-1}, with daily rations in terms of carbon of several hundred % day^{-1}.

Appendicularians produce a great deal of mucous material and faecal pellets which are available for vertical export. Assimilation rates appear to be similar to those of copepods (i.e. 80%) and faecal pellet production rates seem to be from 50% to several hundred % of body carbon each day. Coupling these faecal pellet production rates with sinking velocities of more than 100 m day^{-1} means that appendicularian faeces can be a significant component of the particulate carbon, nitrogen and pigment vertical flux.

The metabolism of appendicularians is poorly understood. Laboratory energy budgets indicate high respiratory rates and low or negative scope for growth. These experiments are at variance, however, with field observations of very high individual and population growth rates. This suggests that we are far from understanding the mechanisms used by appendicularians to make a living in the sea.

NINE

Life history of the Appendicularia

R. Fenaux

9.1. Introduction

The first serious attempt to study appendicularian development was that of Delsman (1911, 1912), which still remains the only description of cleavage and embryonic development. The chronology of development in cultured *Oikopleura dioica* was given by Galt (1972), Paffenhöfer (1973) and Fenaux (1976). Fenaux and Gorsky (1983) decribed the development and growth of *O.longicauda*, the only hermaphroditic oikopleurid to be cultured so far; they found a similar general pattern with a few differences in detail. Most of the information given here concerns *O.dioica*. A comprehensive review of the reproductive biology of appendicularians can be found in Galt and Fenaux (1990b).

The life of an appendicularian can be divided into four distinct periods. Each period is associated with important changes, as shown in Fig. 9.1.

After fertilisation, which occurs in the sea, early development gives rise to a larva consisting of two parts, the trunk and the tail. At hatching, the tail has practically all the components of the adult. However, in the trunk, only the cerebral ganglion and statocyst can be distinguished. The organization of the trunk and the modification and growth of the tail continue until the tail twists or shifts. This very rapid development is soon followed by the construction of the first house. At this stage, most of the somatic cells reach their definitive number. Further growth is only by an increase in the volume of the somatic cells (associated in many cases

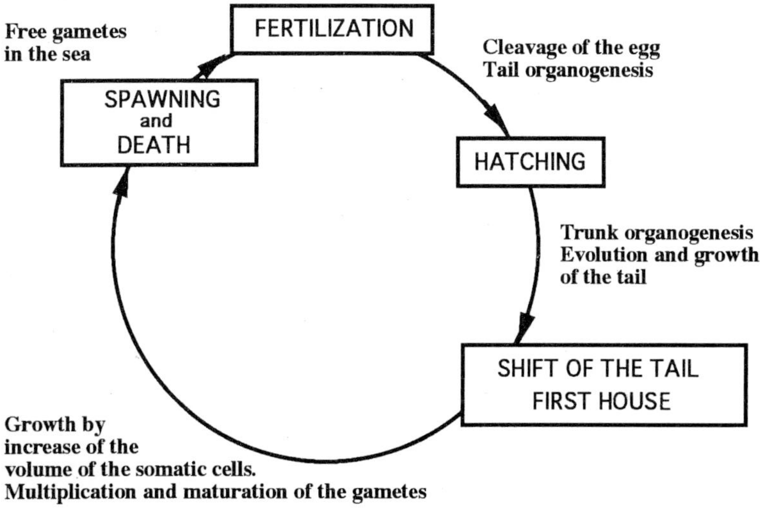

Fig. 9.1. Life cycle of *O.dioica*. Modified after Fenaux (1976).

with a high degree of polyploidy) and by multiplication and maturation of the gametes. Spawning is followed by the death of the animal.

In culture, *Oikopleura* no longer secretes houses when the gonads become close to maturity. Examination of the trunk shows that the cells of the oikoplastic layer and gut are almost empty, suggesting metabolite transfer from them to the gonads (Fenaux and Gorsky, 1983). Fenaux (1985) suggested that the absence of the house reduces the possibility of self-fertilization in hermaphroditic species (and ensures that no larvae resulting from fertilization within the house remain trapped in it). In the sea, mature specimens without houses of *Oikopleura longicauda* and *Stegosoma magnum* have been observed in dense windrows, apparently forming reproductive aggregations (Alldredge, 1982).

Oikopleura dioica is the only dioecious appendicularian. Why this should be so remains an interesting enigma for those zoologists interested in the evolution of life histories! All other tunicates are hermaphrodites. The sex ratio has been estimated on the first generation (1104 animals), obtained from nine pairs of the appendicularian *O.dioica*. A χ^2 test showed that the hypothesis of a 1:1 sex ratio cannot be rejected, either on the progeny from each couple, or on the pooled data. (Fenaux, 1986).

9.2. Spawning

Observations on cultured *O.dioica* and *O.longicauda* (Fenaux, 1976; Fenaux and Gorsky, 1983) showed that spawning occurred most frequently during the night and at dawn. In *O.longicauda*, Fenaux and Gorsky (1983) observed sperm release some 30–60 min before eggs were spawned. Sperm are released through a spermiduct located in the dorsal side of the genital cavity. The duct is short and simple in *O.dioica* but longer and more complex in *O.longicauda* (Fenaux, 1976; Fenaux and Galt, 1990). The spermatozoon of *O.dioica* (Fig. 9.2), whose fine structure was described by Flood and Afzelius (1978), consists of a round head 1 μm in diameter containing the acrosome, the nucleus and the base of the axoneme, which terminates in a simple centriole. This is followed by a mid-piece (3 μm long) with a single U-shaped mitochondrion and a flagellum 25 μm long. This is considered to be the least derived of the sperm of any pelagic tunicate (Holland, 1990).

Oocytes at the stage of first meiotic metaphase (Fenaux, 1976) are released by rupture of the walls of the ovary and genital cavity, resulting in the death of the animal. The number of oocytes produced varies with the size of the female. For example, in *O.dioica* with a trunk length of 800 μm, 21 eggs were produced. A female with a trunk length of 920 μm produced 94 eggs and another with a trunk measuring 1000 μm gave rise to 118 eggs. A large female whose trunk measured 1100 μm produced 187 eggs. In the North Sea, Päffenhofer (1973) found that between 120 and 360 eggs were produced.

The diameter of appendicularian eggs varies between 44 and 130 μm, according to the species (Galt and Fenaux, 1990). The diameter of the eggs produced by the same female was very similar. In *O.dioica* it varied from 97 to 107 μm. (in Villefranche-sur-Mer). These dimensions are larger than those published from the Baltic Sea (88 μm) and the North Pacific coast (80–85 μm). Some are enveloped by a layer of follicle cells, as for instance *O.longicauda* and *Megalocercus abyssorum*. Others are naked, like *O.dioica*, where Holland *et al.* (1988) have shown that each egg is surrounded by a thin (100 nm) vitelline layer covering plasma membrane and sparse cortical granules. The cytoplasm does not contain lipid droplets or yolk granules but shows multivesicular bodies.

9.3. Fertilization

Fertilization may occur immediately after spawning or (at 14°C) at any time within 24 h according to Galt (1972). Miller and King (1983) showed that the spermatozoa of *O.dioica* are chemically attracted to the eggs. In the laboratory, fertilization has been observed for *O.dioica* (Galt, 1972; Fenaux, 1976), *O.longicauda* (Fenaux and Gorsky, 1983) and *O.cophocerca* (Fenaux, unpublished). Galt (1972) and Fenaux, (1976) pointed

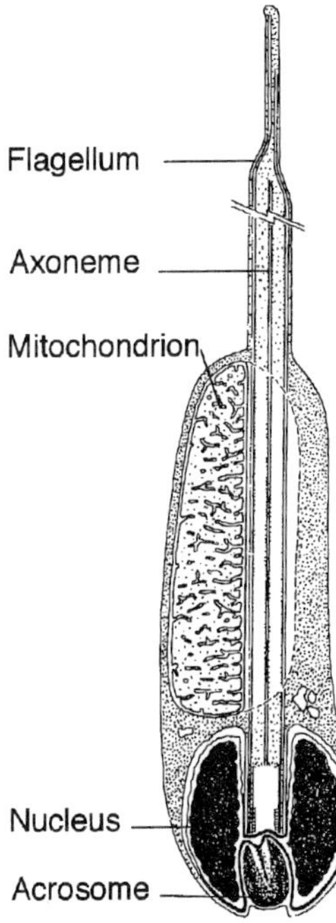

Fig. 9.2. Structure of the spermatozoon of *O.dioica*. Modified after Flood and Afzelius, 1978.

out that overfertilization by a dense suspension of sperm causes abnormal embryonic development. The penetration of the spermatozoon triggers a cortical reaction that causes the release of the contents of the cortical granules, which then migrate to the inner surface of the vitelline layer where they remain intact through several cleavages (Holland *et al.*, 1988).

The acrosome reaction of *O.dioica* has one feature not previously described for the acrosome of any animal: the anterior portion of the nuclear envelope evaginates together with the acrosome membrane and may help to stiffen the acrosomal tubule (Holland *et al.*, 1988). Two or three minutes after fertilization, the egg produces two polar bodies (Fig. 9.4, 1). The percentage of successful fertilizations in relation to the total number of eggs varies from 85 to 95%.

9.4. Embryonic development

Embryonic development is very rapid. From fertilization to hatching in *O.dioica* takes less than 3 h at 22°C and about 12 h at 7°C.

Little is known of cell lineages and that only for *O.dioica*; what is known derives entirely from Delsman's observations (1911, 1912) (Fig. 9.3). New investigations

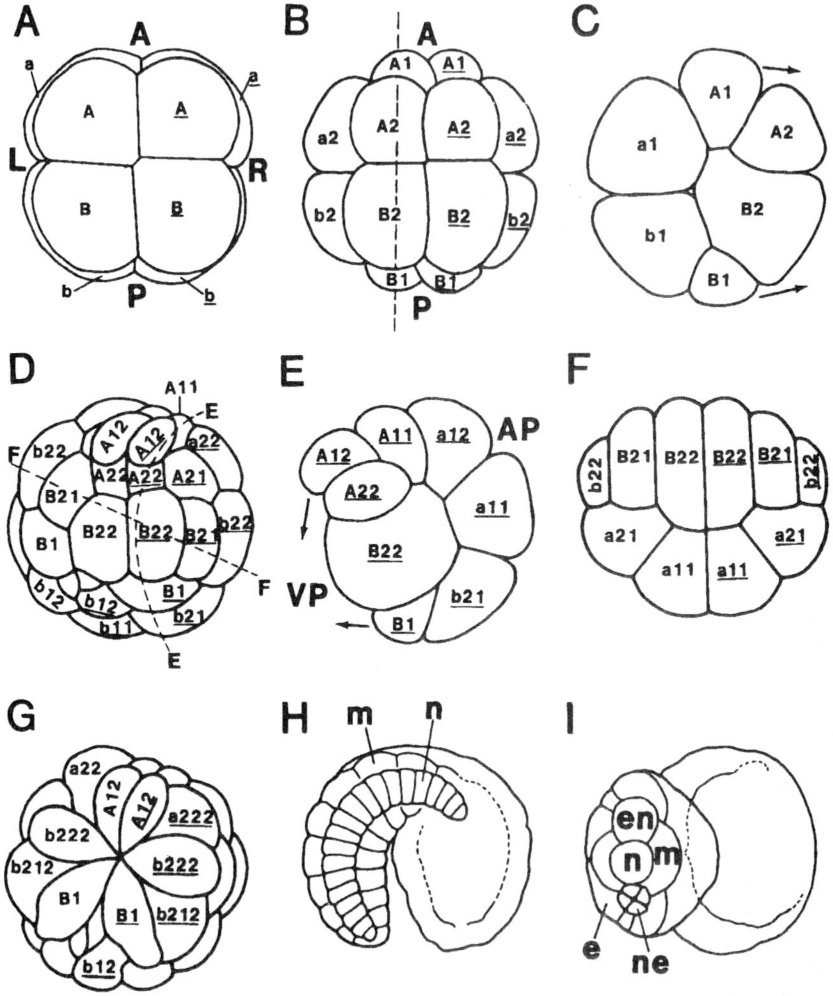

Fig. 9.3. Cell lineage in the embryonic development of *O.dioica*. (a) 8-cell stage, from vegetal pole. (b) 16-cell stage, from vegetal pole. Section plane of (c) marked by dashed line. (c) 16-cell stage, optical section through plane indicated in (b). Arrows indicate growth of the cells during gastrulation. (d) 32-cell stage ('Placula'). Dashed lines mark approximate planes of (e) and (f). (e) 32-cell stage. Optical section through plane E–E in (d). (f) 32-cell stage. Optical section through plane F–F of (d). (g) 32- to 44-cell stage, from vegetal pole. Gastrulation complete. (h, i) Optical sections of tail at tailbud stage, in lateral (h) and frontal (i) views shortly before hatching. A, Anterior; P, posterior; L, left; R, right; AP, animal pole; VP, vegetal pole; e, epidermal cell; en, caudal endoderm; m, muscle cell; n, notochord; ne, caudal nerve. After Galt and Fenaux (1990), redrawn from Delsman (1910).

with modern methods are needed to confirm and extend Delsman's earlier work; it will be particularly interesting to examine the expression of genes related to trunk organization and nervous system segmentation.

9.5. Cleavage

The first division into two cells of equal size occurs 15 min after fertilization, (Fig. 9.4, 2). The second cleavage 20 min after fertilization is also meridional but per-

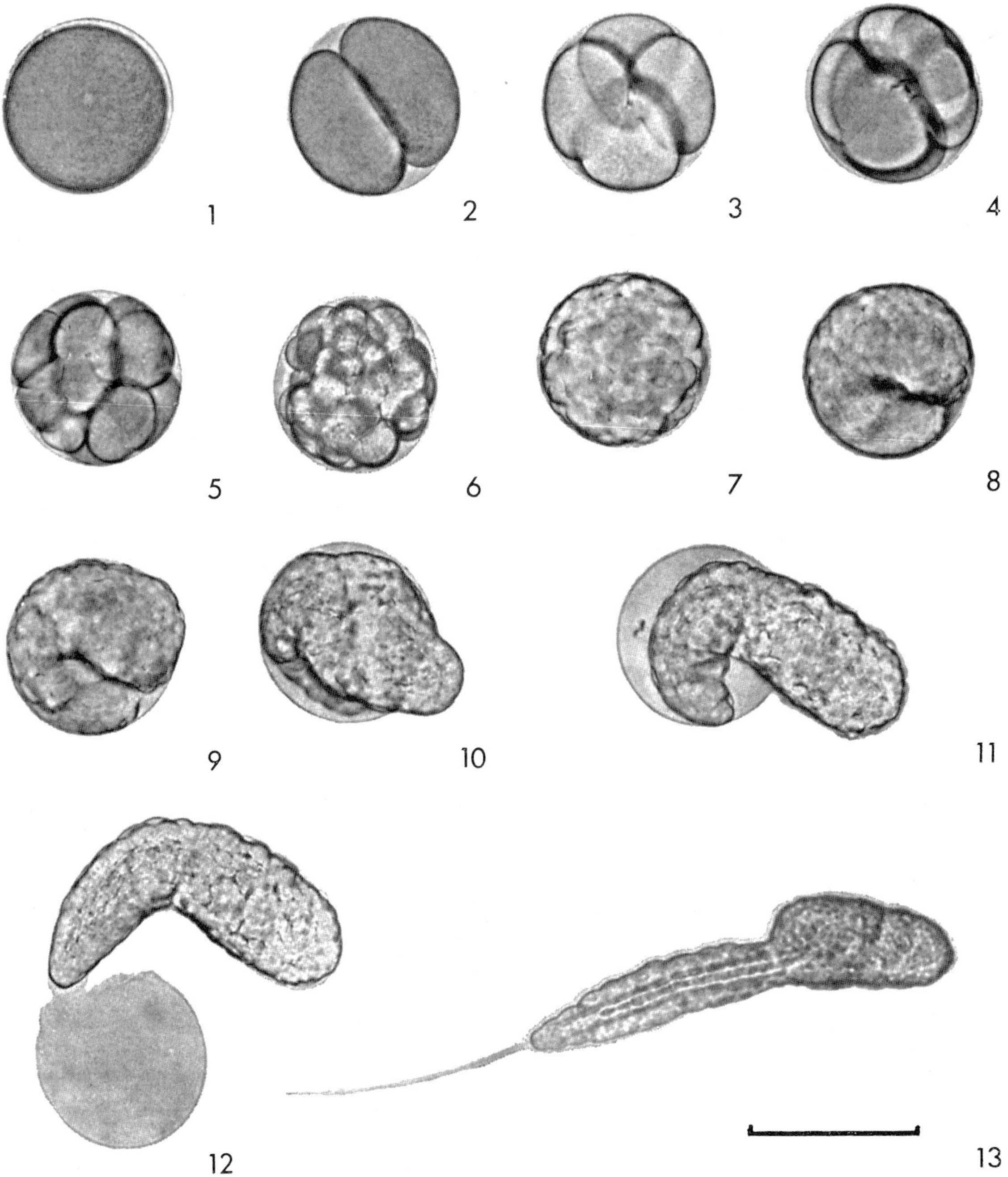

Fig. 9.4. Early development and hatching of *O.dioica*; 1, fertilized egg; 2, stage 2; 3, stage 4; 4, stage 8; 5, stage 16; 6, stage 32; 7, gastrula; 8, embryo; 9–12, hatching; 13, young animal 30 min after hatching. Scale bar 100 μm. After Fenaux (1976).

pendicular and gives rise to 4 cells (Fig. 9.4, 3). The third division, 5 min later, is parallel to the equatorial plane and gives rise to two groups of 4 cells of different sizes (Fig. 9.4, 3). The smallest are the vegetative cells, presumptive endoderm and mesoderm, and the largest include the animal cells and presumptive ectoderm. *Note that this is the reverse of the situation in ascidians.* The fourth division into 16 cells (Fig. 9.4, 5) occurs 32 min after fertilization.

With the fifth division, gastrulation begins, and further segmentation gives rise to a spherical gastrula which becomes a little flattened 80 min after fertilization (Fig. 9.4, 7). The differentiation of the trunk and tail starts 40 min later (Fig. 9.4, 8), then the notochord becomes clearly visible in optical section (tailbud stage), and very soon hatching occurs.

9.5. Hatching

After about 150 min (at 22°C), the portion in front of the trunk produces a swelling which grows, ruptures and extrudes the contents of the egg membrane. The entire body comes out slowly through the aperture (Fig. 9.4, 9). Hatching generally lasts a few minutes.

The entire animal is covered by a thin acellular membrane which forms a process at the distal end of the tail; this structure becoming clearly visible after hatching and remains until stage III (see below). It is unlikely that this membrane is equivalent to the test of other tunicates. A somewhat similar thin acellular membrane covers the doliolid larva (section 1.3.2.1), where it is clearly distinct from the test. In solitary ascidian tadpole larvae, there is a bi-layered larval tunic, the outer layer forming the tail fin, the inner the base of the juvenile test after metamorphosis (Gianguzza and Dolcemascolo, 1984). Since some doliolid larvae have tail fins, in addition to the thin outer covering, there is no reason to equate the outer membrane with any part of the test. Nothing is known of its composition, in either doliolids or appendicularians. In *O.dioica*, the egg membrane keeps its spherical shape even after the animal leaves it (Fig. 9.4, 12). Among the species like *O.longicauda*, where the eggs are provided with follicle cells, the egg membrane shrinks after hatching.

9.6. Trunk organogenesis

9.6.1. *Stage I* (Fig. 9.5, 1)

Just after hatching, the trunk and tail are almost the same size (about 90 μm). The animal is at first curved, but after a few minutes it becomes almost straight ('streckform'; Lohmann, 1933, 1934) The tail is an extension of the trunk. Ventrally, the tail is a direct prolongation of the trunk, but dorsally the narrowing of the tail distinguishes the two parts of the animal. Practically no changes are visible in the trunk but occasionally part of the cerebral gland and the statocyst can be seen. The first obvious change is a rapid increase in the size of the tail, which increases in length by 45% in the first 15 min and by more than 200% after the first hour. Occasional vibrations of the tail are seen although no chordal and muscular cells are yet visible.

9.6.2. *Stage II* (Fig. 9.5, 2–3)

An hour after hatching, the organs in the trunk begin to appear. The cerebral ganglion develops. First appearance of the gut lumen and of the body cavity occurs. An epithelium is visible, and the proximal end of the tail (which now shows rapid feeble movements) moves a little ventrally. The animal is still covered by the outer membrane.

9.6.3. *Stage III* (Fig. 9.5, 4)

Two hours after hatching, the organs are becoming easily recognizable. The mouth begins to appear. The epithelium of the trunk is well differentiated. Often the outer membrane disappears from the trunk region but still persists on the tail.

The animal swims about rather awkwardly using rapid tail movements. When in a large container it swims freely in a characteristic circling fashion.

9.6.4. *Stage IV* (Fig. 9.5; Fig. 9.6, 3–5)

During the third hour, the mouth is well formed and open. A buccal gland is seen at each side of the endostyle. The movements of the cilia of the gut and of the spiracles become visible. The heart beat is also clearly seen. The epithelium of the trunk is differentiated into a thin posterior area and a thick anterior oikoplastic area. The tail shows fusion of notochordal lacunae and becomes dorso-ventrally flattened; lateral fins appear on both sides. At this stage the tail is still directed posteriorly, but the animal swims strongly and its movements are now in the form of open curves.

9.6.5. *Stage V* (Fig. 9.6; Fig. 9.6, 4)

About 4 h after hatching, the mouth and the spiracles are completely open, and an inhalent current starts through the mouth. All the organs show a continuous

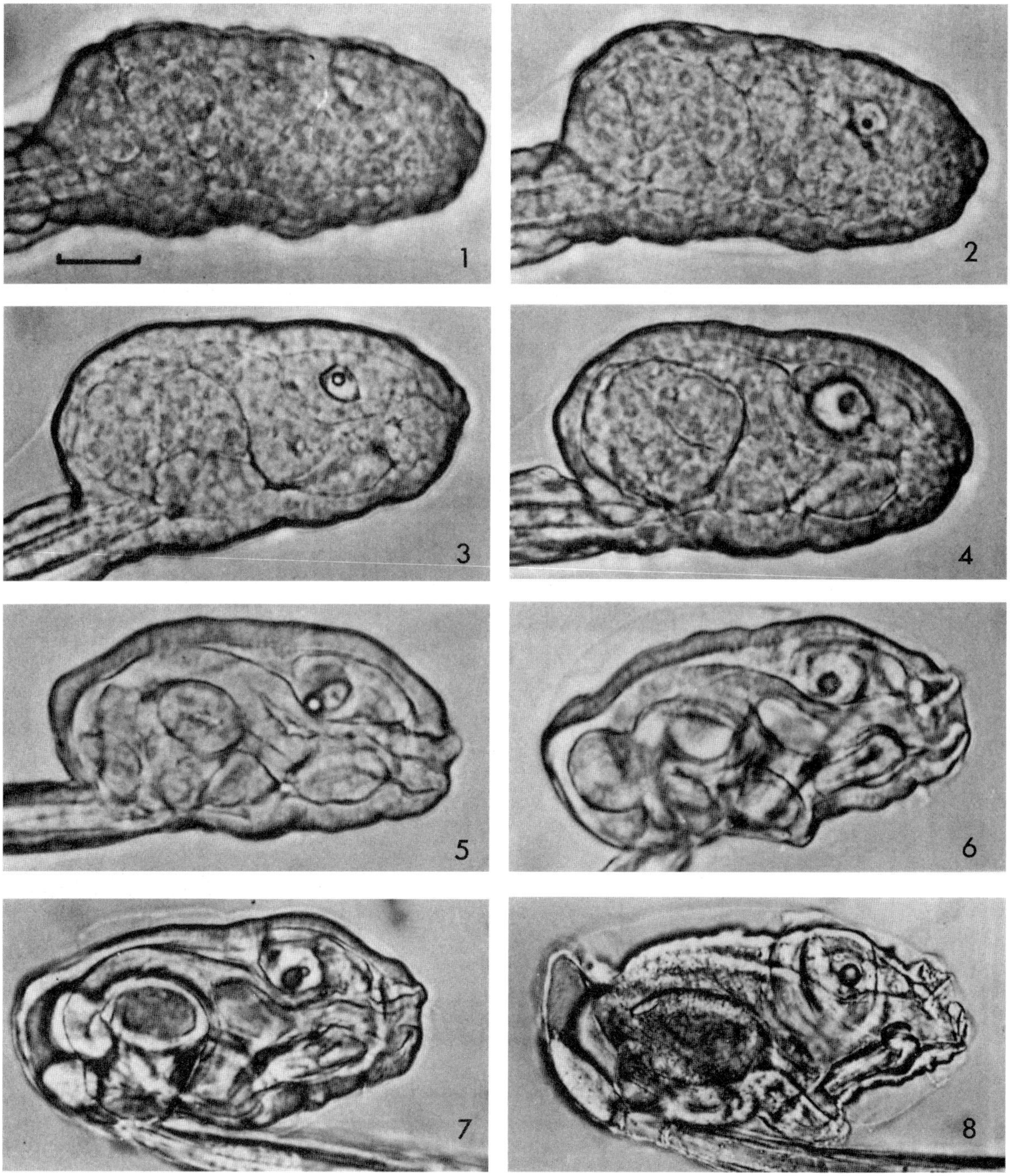

Fig. 9.5. *Oikopleura dioica*, organogenesis of the trunk. Scale 20 μm. After Fenaux (1976). Note: 1–7 same (larger) scale than 8.

No.	1	2	3	4	5	6	7	8
Stage	1	2	2′	3	4	5	Shift	After shift
L (μm) trunk	95	100	105	105	110	110	115	220
L (μm) tail	140	200	220	250	310	370	380	680

lumen. The oikoplastic region starts to secrete the mucus substance which will build the house. The tail now makes an obtuse angle (12°) with the trunk and the small process at its tip disappears. Although the point of attachment of the tail moves forwards ventrally, its distal extremity is still opposite the mouth. The swimming movements are by now very vigorous and the swimming path is almost rectilinear.

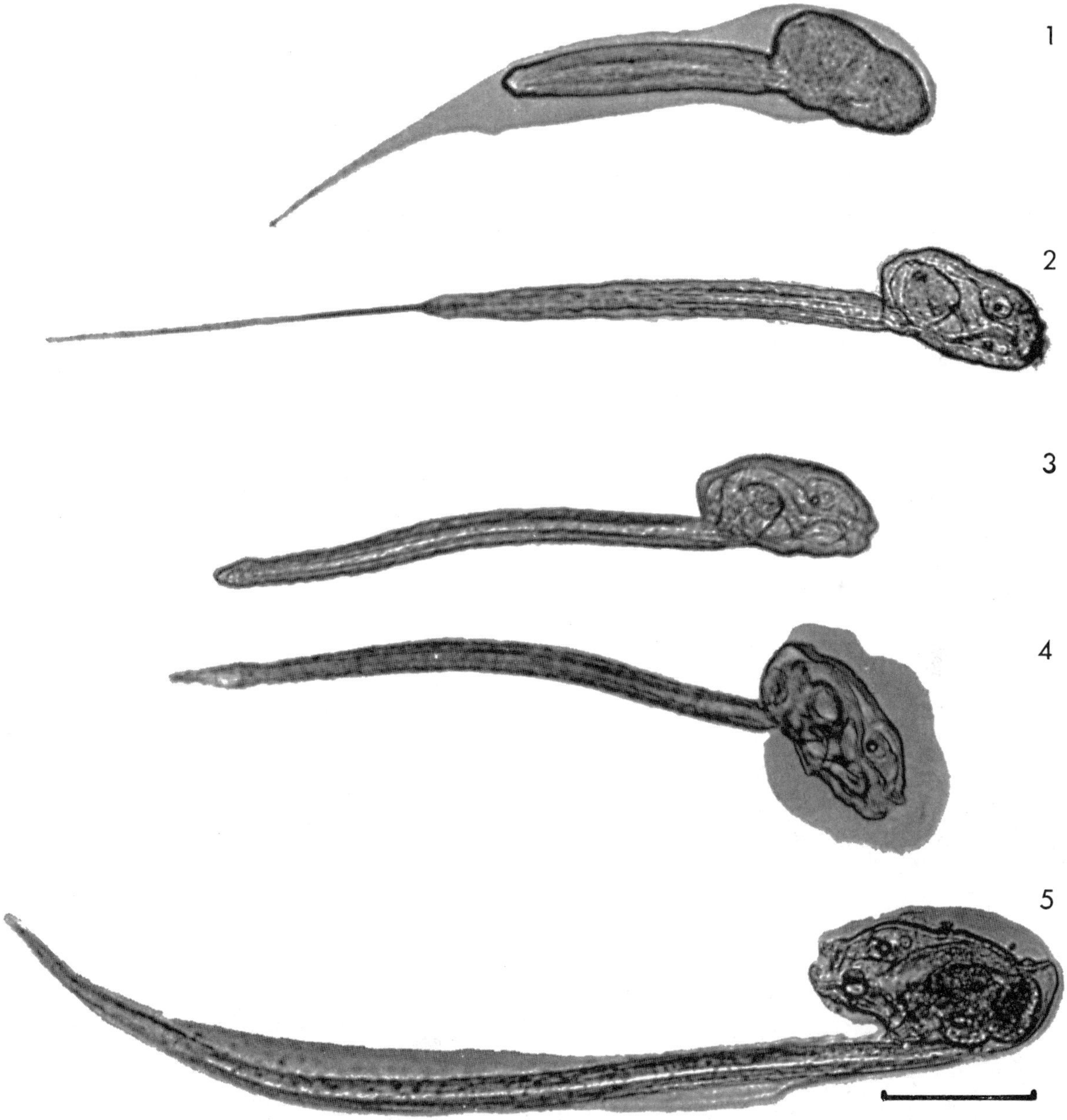

Fig. 9.6. Development of *O.dioica* after hatching. Scale bar 100 μm. After Fenaux (1976b). (I) Stage I. The swelling of the acellular cuticle membrane covering the body was obtained artificially. (2) Stage III. The caudal appendage is still present but the cuticle is partially detached from the body. A part of the cumpled cuticle is visible in front of the trunk. (3) Stage IV. The cuticle and caudal appendage have now disappeared. (4) The tail insertion moves antero-ventrally. The secretion of the first house begins. The tail twist will occur in a few seconds. (5) Appearance of the animal out of its house a few hours after the tail twist.

At 22°C, this stage ends about 8 h after fertilization, and at 14°C after 14–18 h. In each case, the tail is about 370 μm long.

9.7. Tail shift and formation of the first house

The change in orientation of the tail is a most significant event in appendicularian life history. After a brief period of intense movement which lasts for a few seconds, the tail suddenly twists in such a way that its distal extremity falls on the same side as the mouth (termed '*Knickform*' by Lohmann, 1933) (Fig. 9.5, 7 Fig. 9.6, 4–5). With this, the swimming pattern of the animal changes into the 'looping' trajectories described in detail by Fenaux and Hirel (1972).

Soon after the tail shift, the oikoplasts begin to secrete what appears as a mucous covering. The animal then enters into a period of characteristic rapid movement, during which the tail tip describes a circle centred on the trunk, while the secretion swells and separates from the trunk to form a sleeve (house rudiment) which encloses most of the trunk. This swelling probably involves a special secretion of some oikoplastic cells, rather than being the result of the tail movements, since it occurs in lightly anaesthetized immobile animals (Fenaux, unpublished). This is the first phase of house building.

During the second phase, which only lasts for 600 ms, the whole of the trunk enters the house rudiment and enlarges it, and then the tail curves around and enters it, to bring the animal completely inside the rudiment.

The third phase consists of sequences of tail undulations, which stretch and swell the house by pumping it up. Each sequence of this phase lasts 700 ms and the entire phase lasts for about 25 s (see also section 3.4.1). At the end of the third phase, the house is finished and the animal can start feeding. The volume of the house is about three hundred times the volume of the trunk of the animal (Fenaux and Hirel, 1972).

9.8. Growth

After the tail twist, somatic cell number is already fixed for most organs and the juvenile grows by increasing cell size. In the same culture, specimens of the same age may grow at different rates, and differ in size (Table 9.1); spawning can occur in animals of different sizes. The specimens cultured at 22°C hatched about 3 h after fertilization. They twisted their tails and built their first houses after 8 h. The first spawning occurred on the third day and after the fifth day all animals had spawned. At 14°C, hatching occurred 6 h after fertilization; the twist of the tail and building of the first house took about 18 h. The first spawning was recorded on the tenth day and on the twelfth day all the specimens had spawned.

There is a remarkably high daily specific growth rate during the first 24 h. At 22°C, it is up to 5.9, and even at 14°C, the maximum is no less than 2.68. It decreases rapidly thereafter, and in almost all mature specimens it is only 0.2 at 14°C. There is an allometric relation between the length of the tail and the length of the trunk, and this changes after the tail twist has taken place (Fenaux, 1976).

Table 9.1 Chronology of the major events in the development of *O. dioica* at 14 and 22°C (after Fenaux, 1976).

Days	Hours	Tail length		Major events	
		14°C	22°C	14°C	22°C
	0			Fertilization	Fertilization
	1				
	2				
	3		90 ± 3		Hatching
	4		195 ± 10		
	5		250 ± 7		
	6	90 ± 3	290 ± 8	Hatching	
	7	150 ± 6	345 ± 11		
	8	180 ± 8	370 ± 6		Shift of tail, first house
	9	220 ± 17			
	10	260 ± 9			
	11	310 ± 12			
	18	370 ± 7		Shift of tail, first house	
1	24	410 ± 15	620 ± 80		
2		630 ± 60	1000 ± 140		
3		1040 ± 70	1610 ± 370		First spawning
4		1500 ± 85	2300 ± 115		
5		1800 ± 40			Last spawning
6		2000 ± 140			
7		2100 ± 70			
8		2200 ± 10			
9		2260 ± 30			
10		2310 ± 30		First spawning	
11					
12				Last spawning	

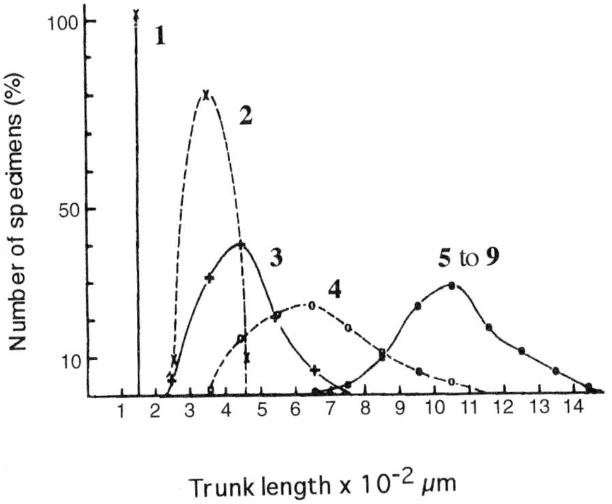

Fig. 9.7. Change in trunk size with time, in specimens of a population from a single parental couple of *O.dioica*. Numbers on the curves indicate the number of days from hatching. The last curve shows size and maturity of all the specimens which spawned between fifth and ninth day. After Fenaux *et al.* (1986).

At higher temperatures, for example at 29°C in Kingston Harbor, Jamaica, Hopcroft and Roff (1995) observed daily specific growth rates ranging from 3.0 to the extraordinary value of 23.0! They concluded that in the 20 l plastic microcosms used, generation time was slightly > 1 day, and definitely < 2 days. In the microcosms, growth rate slowed over the 4–5 day periods studied, so that generation times of slightly > 1 day seem most probable. As these authors point out, growth rates indicating up to 23-fold daily increase in weight exceed those of any other metazoans, even including salps (Heron, 1972; Heron and Benham, 1984).

9.9. Population dynamics

Several parameters that characterize a population arising from a single pair of *Oikopleura dioica* were studied in a culture system (Fenaux *et al.*, 1986). Egg volume varied by a factor of 1.4. The daily size of individuals in a population followed a Gaussian distribution, and the spread of this distribution increased with time (Fig. 9.7). In a population that arose at 18°C from 252 eggs, mortality reduced the number of individuals to 162 after 5 days. During the following 4 days, the appendicularians reached sexual maturity and died after the gametes were released. The first individuals to mature were males, 115 h after fertilization. The average trunk size of mature animals was 942 ± 95 μm for males and 1165 ± 104 μm for females. All appendicularians released their gametes and died 215 h after fertilization. The number of eggs per female showed considerable variation, ranging between 41 and 362 μm, with an average of 170 ± 78 μm.

Several other species have been cultured in the laboratory, some for many generations, e.g. *O.longicauda* (Fenaux and Gorsky, 1983) and *O.cophocerca* (Fenaux, unpublished); others, such as *Fritillaria formica*, *O.fusiformis*, *O.albicans* (Fenaux and Gorsky, unpublished) and *O.labradoriensis* (Galt, unpublished) through one full cycle or parts of the cycle. The processes of development are very similar in the different species of the genus, differing only in a few biologically unimportant details.

Almost a decade ago, Galt and Fenaux (1990a) wrote 'it is time for increased experimental research on physiological and cellular processes and regulation of larvacean development. Larvaceans are abundant, ecologically important animals. Their simple body plan, and transparency and the amenability of some species to laboratory handling makes them suitable subjects for study of a variety of developmental and physiological problems'.

Since then, the group has turned out to have even more claim to be of value as experimental animals: it seems probable that appendicularians occupy a special phylogenetic position as the sister group to the vertebrata (Chapter 16), offering interesting material for genetic studies, particularly in view of the short generation time and the distinct division between trunk and tail.

TEN

The role of Appendicularia in marine food webs

G. Gorsky and R. Fenaux

10.1. Nutritional strategy of appendicularians

Appendicularians are among the commonest members of the zooplankton community, often the second or the third most abundant group in plankton samples. They are found in all oceans, from coastal surface waters to the deep sea (Chapter 15). Since they are fragile organisms, they are often destroyed or severely damaged when collected with plankton nets. In consequence, there have been relatively few studies attempting to assess their environmental impact. Nevertheless, in recent years it has become obvious that their remarkable filtration systems, and rapid generation times, mean that they almost certainly play a significant role in the marine food web.

The unique filtration system in the house that they secrete (Chapters 6 and 8) involves two external filters, before food particles are trapped (as in all tunicates) on the pharyngeal mucous filter. Thus before entry into the oesophagus and digestion by the appendicularian, particulate organic matter undergoes three sieving processes.

First, coarse inlet filters exclude large particles, and then a complex food concentrating filter traps suspended particles and concentrates them. The concentrated mass of food particles is then sucked into the mouth via a buccal tube, and collected on the pharyngeal filter. This and food particles on it, are drawn into the oesophagus, and digested in the gut.

Evidently, the pore size of these three filters will determine the size range of particles on which the appendicularians feeds. Table 6.2 (p. 111) shows what is presently known of the dimensions of the meshes of the inlet and food concentrating filters in different species. The inlet filter meshes vary from some 13×13 μm in the smaller oikopleurids to 50×170 μm in the larger species. The food concentrating filters are very much finer, ranging from 0.15×0.9 μm in small oikopleurids to 0.2×1.0 μm in large *O.vanhoeffeni.* Unfortunately there is only a single measurement of the dimensions of the pharyngeal filter, which in *O.vanhoeffeni,* according to Deibel and Powell, (1987a), is 3.3×0.35 μm: much coarser than the pore size of the food concentrating filter!

Deibel and Lee (1992) and Morris and Deibel (1993) explain this paradox by the fact that the food particles enter the buccal cavity already aggregated, so the pharyngeal filter remains efficient even for particles smaller than its pore size. However, there is considerable doubt that the measurements for this pharyngeal filter are correct, for unlike all other tunicate pharyngeal filters examined, that of *O.vanhoeffeni* appeared irregular. There is certainly need for re-examination of the pharyngeal filters in this and other species.

It seems quite clear from these observations of filter mesh size that appendicularians can capture submicron sized particles. In consequence, appendicularians can effectively obtain energy from the microbial ecosystem and transfer it directly to larger metazoans, thus bypassing the size-related biomass transfer in marine food webs. As will be seen below, recent work by Flood *et al.* (1992) and Bedo *et al.* (1993) suggests an even shorter food chain between colloidal organic matter and higher trophic levels, including commercial fish.

10.2. Appendicularian impact on the microbial loop

Several authors have described the potential of appendicularians for grazing on bacteria sized prey (Flood, 1978; Deibel *et al.*, 1985, 1992; Bedo *et al.*, 1993). King *et al.* (1980) observed that *O.dioica* in an enclosed water column ingested a maximum of 6% of the bacterial standing stock daily, not sufficient to regulate bacterial populations but important enough to transfer a significant amount of bacterial carbon directly to higher trophic levels. Alldredge (1977) showed that of the seven appendicularian species co-occurring in the Gulf of California, five consumed predominantly small flagellates and particles and bacteria (bacteria were found in the faeces). Two large species, *Megalocercus huxleyi* and *Oikopleura intermedia,* filtered mostly diatoms.

Flood *et al.* (1992) showed that appendicularians can filter colloidal organic matter by measuring the clearance rate of melanin particles (from *Sepia* ink). These particles were 0.13 ± 0.023 μm in diameter, i.e. smaller than the mesh size of the food concentrating filter (0.22 ± 0.04 μm). The clearance rate of colloids was only 2–11% of the clearance rate of plastic beads 10–20 μm in diameter. As the mesh size of smaller species is smaller, for example 0.15 ± 0.02 μm for *O.dioica,* an inverse relationship between colloid retention efficiency and the size of appendicularians seems likely. There is certainly a relationship between particle size and retention efficiency, as Deibel and Lee (1992) demonstrated for *O.vanhoeffeni,* using particles 0.6–7 μm in diameter. The retention efficiency for the largest particles was 91 ± 3%, for the 3 μm sized particles 88 ± 6%, and for ~0.6 μm plastic beads 44 ± 8%. If the prediction that the smaller the appendicularian, the higher its efficiency in filtering submicronic particles is confirmed, this will enforce reconsideration of a number of concepts concerning energy transfer in the pelagic food chain. The possibilities are that:

(i) appendicularians short-circuit the microbial food web by directly exploiting picoplankton production;
(ii) they actively contribute to the dissolved organic carbon (DOC) flux, considering that an important part of the DOC occurs in the high molecular weight range (Koike *et al.* 1990);
(iii) small size fritillarians and juveniles of smaller oikopleurids may play a significant role in the re-aggregation processes of dissolved organic matter (DOM).

According to Flood *et al.* (1992) the retention efficiency of > 0.2 μm sized colloidal DOC of *O.vanhoeffeni* may be about 10%. Clogging of the feeding filter and the re-aggregation evidently could increase the retention efficiency of submicron particles. However, Bedo *et al.* (1993) found that the ratio between 0.2 and 0.75 μm beads in the gut and faecal pellets was significantly less than in the surrounding food suspension and in the house, explained by the loss of the smaller beads through the pharyngeal filter (Fig. 10.1).

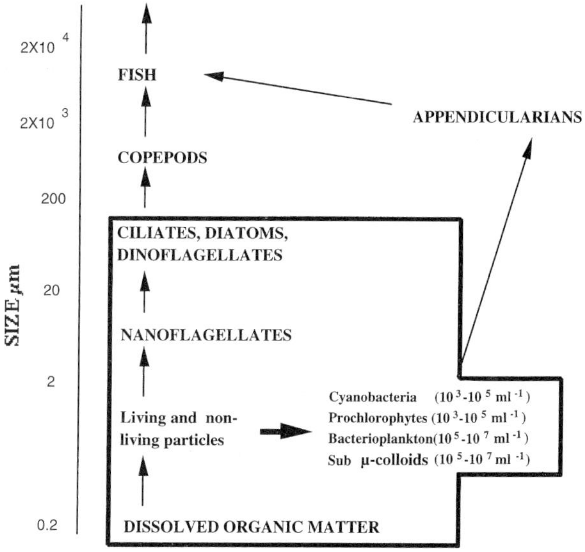

Fig. 10.1. Role of appendicularians in the marine food web. The microbial loop is outlined. Dotted area, types and abundances of submicron size particles in the upper layers (Koike *et al.*, 1993) aggregated by appendicularians. Modified from Fenchel (1988).

In general, for any predator of a given size there is also a minimum size of prey below which the capture is inefficient. Typical length ratios between prey and predator vary from 1:10 to 1:100. Appendicularians retaining submicronic particles exhibit the rare ratio of 1:10 000. The shorter the food chain, the more energy is transferred to higher trophic levels. In this regard the classical food chain is the 'diatom–copepod–fish' chain. In the case of the microbial loop the quantity of energy which ends up at the fish level is inversely proportional to the increase in the number of trophic levels. This is in oligotrophic systems, where the base of the food chain consists of small cells. Thus where the food web is dominated by the the microbial loop, the ecological role of appendicularians must be considerable.

The nutritional adaptations described above have the following implications:

(i) they allow appendicularians to use a food resource not available to competing zooplankton (Table 10.1);
(ii) they bypass the microbial loop and directly transfer dissolved or small sized particulate organic matter (e.g. submicron colloids, prochlorophytes, cyanobacteria, bacterioplankton and nanoflagellates) to higher trophic levels.
(iii) they permit or facilitate the adaptation of appendicularians to niches, such as the aphotic depths, where the main food source is small detritic aggregates.

Table 10.1 Potential competitors of appendicularians for food resources; species and minimum food particle size (modified from Fortier *et al.*, 1993).

Taxon	Food size (μm)	Reference
Salps		
Cyclosalpa polae	2.1	Harbison and McAlister, 1979
Cyclosalpa affinis	3.4	Harbison and McAlister, 1979
Cyclosalpa floridana	1.5	Harbison and McAlister, 1979
Pegea confoederata	3.3	Bone *et al.*, 1991
Pegea confoederata	2.8	Harbison and Gilmer, 1976
Salpa maxima	2.6	Harbison and Gilmer, 1976
Salpa fusiformis	< 5	Silver and Bruland, 1981
Thalia democratica	1.2	Mullin, 1983
Doliolids		
Dolioletta gegenbauri	0.2–5	Crocker *et al.*, 1991
Pteropods		
Cavolinia tridentata	1–5	Gilmer, 1974
Euphausiids		
Euphausia pacifica	5	Parsons *et al.*, 1967
Euphausia superba	3	Schnack, 1985
Copepods		
Acartia clausi	2.5–8	Nival and Nival, 1976
Calanus pacificus	8	Parsons *et al.*, 1967
Neocalanus plumchrus	2.5	Frost *et al.*, 1983
Neocalanus cristatus	3	Frost *et al.*, 1983
Rhincalanus gigas	3	Schnack, 1985
Cladocerans		
Penilia avirostris	2	Turner *et al.*, 1988
Tintinnids		
Stenosemella ventricosa	> 1.3	Rassoulzadegan and Etienne, 1981

10.3. Predation on appendicularians

Appendicularians are eaten by many planktonic carnivores and are particularly important food for many larval fish. A brief list of different predators is given in Table 10.2; a more extensive list will be found in Fenaux *et al.* (1990).

As seen in Table 10.2, both adult and larval fishes feed on appendicularians. Many fish are almost entirely dependent upon them at some stages in their life history; the availability of *O.dioica* can determine successful larval development of several commercially important fish species. Last (1980) showed that *O.dioica* and *Fritillaria borealis* are eaten by pleuronectiform larvae, forming between 40 and 75% of their prey, whilst Ikewaki and Tanaka (1993) found that the diet of larvae of the pleuronectid *Paralichthys olivaceus* consisted of up to 87% *Oikopleura* spp. and 2.4% *Fritillaria* spp.

Appendicularians are also attractive food items for smaller adult fishes. Hobson (1974) observed coral reef fishes actively feeding on appendicularians and Genin *et al.*, (in preparation) observed selective predation on appendicularians on a Red Sea coral reef by *Pseudanthias squamipinnis* (Serranidae). Interestingly, appendicularians were preferred to copepods.

In favourable conditions appendicularians form dense blooms. Other dense concentrations of appendicularians are produced by breeding patches (Alldredge, 1982). Both represent valuable food concentrations for different predators, but as the spawning of mature appendicularians is easily triggered by slight turbulence, predation will not prevent gamete release.

10.3.1. *Use of discarded houses by unicellular organisms*

As larvaceans filter non-selectively, an entire microbial ecosystem is introduced into the house which becomes a microhabitat (Davoll and Silver, 1986; Bedo *et al.*, 1993). According to these authors, the enrichment factor for bacteria, flagellates and ciliates found in the house is from 100 to 1000 times higher then their ambient concentrations. Davoll and Silver (1986) proposed a successional history of discarded appendicularian houses, divided into three phases: (1) an inoculation phase — with the appendicularian actively filtering in the house; (2) the microcosm phase — after abandonment of the house; (3) the remnant phase — after its utilization by the microbial ecosystem. Carbon incorporated into free bacteria in the pelagic environment is recycled within hours, while the microbial community concentrated in

Table 10.2 Major predators of appendicularians (a comprehensive list is given in Fenaux *et al.*, 1990).

Predators of appendicularian	Reference
Invertebrates	
Foraminifera	Anderson, 1993
Chaetognaths	Feigenbaum and Maris, 1984
Hydromedusae	Mills, personal communication
Scyphomedusae	Fancett, 1988
	Larson, 1991
Ctenophora	Hirota, 1974
Copepoda	Ohtsuka and Onbé, 1989
Fish	
Scombridae	Bullen, 1908
	Ozawa *et al.*, 1991
Clupeidae	Hardy, 1924
	Jespersen, 1928
	Savage, 1937
	Radovitch, 1952
Gadidae	Hop *et al.*, 1994
	Nedreaas, 1987
	Raitt and Adams, 1965
Sparidae	Shimamoto and Watanabe, 1994
Argentinidae	Halliday, 1969
Pleuronectidae	Ikewaki and Tanaka, 1993
	Shelbourne, 1962
	Last, 1978
Salmonidae	Manzer, 1969
Bathylagidae	Cailliet and Ebeling, 1990
Myctophidae	Gorelova, 1975

the larvacean house continues to evolve during days or weeks. Gorsky *et al.* (1984) found that the retention half-time of houses discarded by appendicularians feeding on Americium-labelled diatoms at 13°C was 219 h. Morris and Deibel (1993) consider that the appendicularian feeding filter functions as a tangential filter and as such, it is concentrating colloidal material by the exclusion of water. The high concentration factor (up to 1000 times; Flood, 1991b; Morris and Deibel, 1993) enhances the aggregation rates and creates favourable conditions for the development of a microbial ecosystem in the house.

10.3.2. *Metazoan predation on discarded houses*

Discarded houses are consumed by many predators, such as medusae, ctenophores, siphonophores, chaetognaths, adult fish and fish larvae, as well as by benthic organisms. Alldredge (1976a), Hamner and Robison (1992), Ohtsuka and Kubo (1991) Ohtsuka *et al.* (1993) and Steinberg *et al.* (1994) point out that the mesozooplankton selectively grazes different parts of appendicularian houses. Several copepod species were identified on the external part of the larvacean house. Some copepods feed on internal structures of the house and on its associated microbial ecosystem. Particles trapped in the internal filters are selectively consumed by different planktonic organisms, such as the copepods *Oncea mediterranea, Microsetella norvegica* and *Paracalanus aculeatus,* as well as by ostracods and euphausiid larvae (Alldredge, 1972, 1976a). Suspended appendicularian houses constitute solid substrates in the water column, so many copepod species that have a benthic type of feeding strategy can take advantage of this. Appendicularian houses contribute to marine snow, as do macroaggregates of different origin. Lampitt *et al.* (1993) provided experimental evidence that marine snow is a valuable food source for planktonic crustaceans. Ostracods, copepods, and amphipods actively ingested marine snow particles. As Steinberg *et al.* (1994) emphasized, the mid-water detritus contains a unique but neglected invertebrate community. This community can therefore play an active role contributing to the aggregation-dispersion processes in the open ocean.

10.4. Appendicularians in aphotic layers

Data on the distribution patterns of appendicularians mostly concern the euphotic layer. The recent development of new observational methods provide opportunities for investigations of their distribution to greater depths. At present 69 appendicularian species have been described (Fenaux, 1993a), but only a few species dominate the appendicularian biomass in the euphotic zone.

Five species alone make up 88% of the appendicularian biomass collected over a year off Villefranche-sur-Mer. These same five species are also the most frequently observed throughout the year (see section 14.2).

Advances in our knowledge of aphotic communities have been directly related to technological progress. Using manned submersibles off California, Barham (1979) discovered a mid-water appendicularian population at depths between 150 and 375 m with houses up to 100 cm across and Silver and Alldredge (1981) described several layers of high concentrations of marine snow in the meso- and bathypelagic zones. Aggregates sampled from 1000 and 1650 m depths contained parts of appendicularian houses. Fenaux and Youngbluth (1990) and Fenaux (1993b) described two new mesopelagic appendicularian genera collected by the submersible *Johnson Sea Link* off the Bahamas. The first, *Mesochordaeus,* includes a new species *M.bahamasi,* the second, *Mesoikopleura* includes four species, three of them new: *M.enterospira* and *M.youngbluthi* sampled in the Bahamian mesopelagic waters at depths between 710 and 860 m, and *M.gyroceanis* (Fenaux, 1993b) collected by the French *Cyana* submersible in the western Mediterranean. Another new midwater appendicularian, *Oikopleura villafrancae,* was described by Fenaux (1992) from the Ligurian sea (northwestern Mediterranean). Successive submersible cruises (Laval *et al.,* 1989, 1992; Gorsky *et al.,* 1991) and the long term survey made in this zone using the Underwater Video Profiler (Gorsky *et al.,* 1992) showed that this species, with densities varying between 1 and 10 individuals m^{-3}, is endemic to coastal mesopelagic waters between 350 and 650 m depth, feeding on suspended organic matter near the continental slope. Nevertheless, during the submersible cruises epipelagic species were observed at mesopelagic depths.

Karl and Knauer (1984) found off California a peak in the abundance of freshly discarded larvacean houses between 900 and 1400 m. They used sediment traps to investigate the production, decomposition and transport of organic matter from the surface to a depth of 2000 m. Their data indicated the existence of several layers of intense biological activity. These layers were situated in the oxygen minimum zone (between 700 and 900 m) and in the sub-oxygen minimum zone (1000–1400 m deep). They concluded that the deep maxima in RNA and DNA synthesis of microorganisms associated with particles are the result of *in situ* biological activity. The above observations are in accord with those made during the *Gyrocyan* submersible cruise in 1989 in the south western Mediterranean, where a deep layer of appendicularians was discovered between 900 and 1500 m. The *M.gyroceanis* holotype (Fenaux, 1993b) was collected at the depth of 1200 m. Several giant *Bathochordaeus* houses m^{-3} were reported by Barham (1979) from mesopelagic waters in the Pacific Ocean off Mexico and by Hamner and Robison (1993) in Monterey Bay (California). Youngbluth (1984) noted near Bermuda a density of 10 mesopelagic appendicularians m^{-3}. The same concentration was reported by Gorsky *et al.* (1991) in the Mediterranean for *O.villafrancae.*

The discovery of these mesopelagic and bathypelagic appendicularian populations, enumerated above, demonstrates the adaptive capacity of these filter feeders to grow and reproduce in deep, detritus-based ecosystems. Little is known of the metabolic adaptations of deep organisms to their environment. The reduction of metabolic costs might be one strategy, since according to Alldredge (1975) the energy invested in house construction is positively correlated to its solidity and (using similar estimates for the abundant Mediterranean oikopleurids) we found that the firmness of the house decreases with increasing size (Gorsky, unpublished). The size of deep-sea species corresponds to or exceeds that of pelagic species, hence it is possible that house secretion in deep-sea species does not require much energy.

Results obtained on accessible epipelagic appendicularians indicate that they lack mechanisms (such as resting eggs and overwintering stages) which enable populations to survive long periods of low food supply (Gorsky, 1980; King, 1982). The existence of species adapted to life in mesopelagic and deep waters suggests the presence of accumulation layers rich in detritus of ingestible size, satisfying their dietary requirements. However, our restricted knowledge of appendicularian distribution and abundance in deep layers precludes any reliable assessment of their importance in processes governing the deep food chains.

10.5. Production of marine snow by appendicularians

When the secretion of a new house is completed in collapsed form around the trunk, the appendicularian abandons the house it occupies and inflates the new one (Fenaux, 1985). House production is temperature and food dependent. *Oikopleura dioica* produces from 3 houses day^{-1} at 13°C to 16 houses day^{-1} at 23°C (Fenaux, 1985). According to Taguchi (1982), *O.longicauda* produced 5.33 houses day^{-1} at 26°C. In cold Newfoundland waters, Deibel (1988) observed that *O.vanhoeffeni* produced from < 1 to 6 houses day^{-1} in a temperature range from –1.6 to 4.5°C. Filtering rate and efficiency are discussed elsewhere (see Chapter 8); this section considers the fate of material aggregated by appendicularians.

Table 10.3 Proportion of chlorophyll *a* filtered but not ingested and ^{241}Am-labelled algae in discarded *Oikopleura dioica* houses (methods are described in references given).

Algal species	Per cent trapped	Methodology	Reference
Isochrysis galbana	28.6	Chl *a*	Gorsky, 1987
Isochrysis galbana	24.9	Chl *a*	Gorsky, 1987
Thalassiosira pseudonana	37.0	^{241}Am	Gorsky *et al.*, 1984
Thalassiosira pseudonana	25.0	^{241}Am	Gorsky *et al.*, 1984
Thalassiosira pseudonana	32.0	^{241}Am	Gorsky *et al.*, 1984

The discarded house contains ~30% of the filtered matter sequestered in its filters (Table 10.3). In addition, faecal pellets, which are usually ejected outside the house, may adhere to its internal structures and remain inside the house (Taguchi, 1982), increasing its energetic value.

The energy content of a discarded house compared to other components of the marine snow is important, because of the way it is colonized by the bacterial flora. The initial house community is acquired and concentrated from the beginning of the filtering activity in the house. Moreover, according to Bedo *et al.* (1993), the filtering activity increases the concentration of colloid matter in the house, creating favourable conditions for rapid development of the colonizing microbial ecosystem. In the open sea, abandoned houses constitute an important part of the 'marine snow'. In Newfoundland, during the period of appendicularian blooms, the discarded houses of *O.vanhoeffeni* and *O.labradoriensis* form the major part of the 'slub'. This is an accumulation of mucus on fishing nets making them visible to the fish. The catch efficiency of the net is then greatly reduced (see Buggeln, 1980; Mahoney, 1981).

10.5.1. *Role of discarded houses in the marine environment*

Macroscopic aggregates are significant components of the marine snow and consequently important transport agents of organic matter to deep layers and to the sea floor. Relatively few studies have considered the quantitative significance of macroaggregates such as abandoned appendicularian houses. Alldredge (1979) reported that appendicularian houses comprised a mean of 6% of the macroscopic aggregates in the Santa Barbara Channel and 24% of those in the Gulf of California. According to Silver and Alldredge (1981), larvacean feeding filter plays an important role in cementing floculent suspended aggregates. These authors estimate that the carbon concentration of marine snow exceeds that in surrounding water by at least three orders of magnitude and that the marine snow is one of the sources of the heterogeneity of carbon estimates in the water column. The contribution of marine snow to the processes of re-mineralization or to vertical fluxes depends upon the prevailing physical and biological parameters. In the northeast Atlantic during the spring of 1990 Turley and Mackie (1994) found between 2–25 × 10^8 bacteria, 1 to 5 × 10^7 cyanobacteria and 1–33 × 10^6 flagellates ml^{-1} in aggregates > 0.5 mm. The microbial ecosystem activity in the aggregates accelerates their re-mineralization. On the other hand, Karl and Bird (1993) observed that during the spring–summer bloom periods, the bacterial production rates in the Antarctic coastal marine ecosystem contributed only minimally to the total ecosystem production. Their results were confirmed by Pomeroy and Wiebe (1993), who demonstrated at the latitudes 30–80°S a pronounced reduction in bacterial growth rate at the commonly found substrate concentrations, combined with low temperatures. Furthermore, according to Pomeroy and Deibel (1986), bacterial growth remains low during the spring bloom in the cold waters off Newfoundland (between 1 and 2°C). They postulated that consequently more primary production finds its way to large consumers, such as the abundant larvacean *Oikopleura vanhoeffeni* and hence through this pathway to the deep layers. Alldredge *et al.* (1986) measured the contribution of bacteria attached to marine snow. According to their results, the contribution of attached bacteria to total bacterial production averaged only 8 ± 7%, but occasionally reached 26%. Alldredge (1981) calculated the grazing impact of appendicularians whilst simultaneously measuring the biomass and the rate of *in situ* increase of food particles. At densities ranging from hundreds to thousands of individuals m^{-3}, *O.dioica* cleared 1–37% of the given volume. Only at the highest population densities did appendicularians consume more than the daily production of their particular food. The fact that the impact of appendicularians on their food stock is only high periodically does not diminish their importance in the transport and transformation of organic matter in the water column. The turnover and production rates of micron size particles are rapid (< 10^{-2} years). The re-packaging of these small particles into larger ones prevents their rapid re-mineralization and enriches the stock of digestible particles for higher trophic levels. Since appendicularians are present in all oceans and at all depths, often in high concentrations, they contribute permanently to the shift of the particle size spectrum towards larger dimensions.

10.5.2. *Sinking of discarded larvacean houses*

Silver and Alldredge (1981) measured the sinking rates of *O.dioica* houses in the laboratory at 5 and 16°C. The rates measured were 64.6 ± 5.8 and 57 ± 3.5 m day^{-1} respectively. The houses in the 5°C experiments were larger than those at 16°C. The sinking velocities of discarded *O.longicauda* houses were also studied in the laboratory by Taguchi (1982). These experiments were carried out in still water columns and gave a mean sinking velocity of 189 ± 31 m day^{-1}. However, Taguchi observed in the field that on a calm day the discarded houses floated on the surface, containing a gas bubble probably produced by the phytoplankton trapped in the house or by microbial activity. The origin of filtered particles may be an important factor in controlling the specific gravity of the discarded houses. According to Alldredge *et al.* (1987), surface mixing can also slow down the sinking rate of particles. Gorsky *et al.* (1984) measured the sinking rates of discarded houses of *O.dioica* at eight temperatures, from 4 to 26°C. Sinking rates ranged from 26 ± 14 m day^{-1} for small houses at 4°C to 157 ± 20 m day^{-1} for large houses at 26°C and were linearly related to the water density. Alldredge and Gotschalk (1988) measured the size-specific sinking rates of undisturbed aggregates directly *in situ.* They found a mean sinking rate of 74 ± 39 m day^{-1}. Sinking rate was an exponential function of aggregate size and weight. Hamner and Robison (1992), using manned submersibles and a ROV, observed giant larvacean houses in the Monterey Canyon, California. The organisms apparently continuously secreted a mucous sheet up to 2 m long (but see p. 121, Chapter 6). The authors calculated that the occupied houses could be maintained at a specific depth for as long as a month. Particles trapped in or adsorbed on the occupied mucous structure stopped sinking. But once abandoned, the house collapses to a compact mass which sinks rapidly (800 m day^{-1}). These collapsed houses have been observed on the sea bed in Monterey Bay down to 3500 m.

Davoll and Youngbluth (1990) measured the heterotrophic activity on mesopelagic larvacean houses and, using available data from the literature, calculated their contribution to vertical transport. The potential carbon flux from these aggregates amounted to 8% of the total carbon flux. However, for realistic estimates of the role of appendicularian houses in geochemical processes, more information is needed with regard to their spatial and temporal distribution and to house secretion in euphotic and aphotic layers.

10.5.3. Faecal flux

Landry *et al.* (1994) studied the mesozooplankton grazing activity in the Southern California Bight during three winter–spring and three autumn cruises (1985–1988). They found that appendicularians were always a numerically important component in the zooplankton community, with densities between 2×10^4 and 5×10^4 individuals m^{-2} in the integrated upper 70 m of the water column. According to the above authors, the observed oikopleurids were major grazers during all cruises, but during the autumn cruises, the gut pigment content of their population was higher than that of copepods. They concluded that: (i) small non-migratory zooplankton species are generally more important grazers than larger, migratory taxa; (ii) small tunicates are efficient grazers of primary producers and may frequently be more important than crustaceans; (iii) the size structure of the zooplankton varies seasonally, the larger forms being more important in late winter and spring, while the smaller forms are dominant in autumn. Therefore, a seasonal shift in size structure of the mesozooplankton community would affect faecal pellet size and thus the particle flux.

Urban *et al.* (1992) investigated seasonal variations in the contents of faecal pellets of *Oikopleura vanhoeffeni* and confirmed its status as a 'generalist' suspension feeder. The pellets, which are elliptical and compact, are enveloped by the remains of the pharyngeal filter. Appendicularians eject some of the pellets from the house via the outlet sphincter (Fenaux, 1986), but some may remain in the house. According to Taguchi (1982) for *O.longicauda* and Theron and Gorsky (1996) for *O.dioica* and *O.albicans*, appendicularians release faecal pellets at regular intervals. Taguchi (1982) found that *O.longicauda* produced one faecal pellet every 146 s in the laboratory under optimal feeding conditions (at 26°C). Using sediment trap results, Taguchi (1982) estimated that the annual faeces production rate for *O.longicauda* was 243 ± 105 pellets day^{-1} in Kaneohe bay (Hawaii). The annual average volume of sedimented appendicularian faecal pellets in Kaneohe Bay, Hawaii, was 2 ± 1 cm^3 m^{-2} day^{-1}, or 66% of all zooplankton faecal pellets in units of volume.

For *O.dioica* at 18.5 ± 0.2°C one faecal pellet was produced every 161 ± 41 s and approximately 530 faeces individual^{-1} day^{-1}. Sinking rates of faecal pellets (measured in an undisturbed water column) were a function of water density, ranging from 25 m day^{-1} at 4°C to 166 m day^{-1} at 26°C (Gorsky *et al.*, 1984). These authors found the faecal pellet volume of an *O.dioica* of trunk length 650 μm to be $4.6 \pm 1.0 \times 10^4$ μm^3, assuming the shape to be cylindrical. Since the volume of a *T.pseudonana* cell used as food was 61 μm^3, each faecal pellet could contain ~750 ^{241}Americium-labelled algae. In fact 500 cells were counted by the radiotracer methodology, corresponding to a compacting efficiency of 67%.

As Urban *et al.* (1993b) showed in the ocean, however, values for faecal pellet content, density and the packaging index (% open area) in *O.vanhoeffeni* vary seasonally. Winter faeces were denser and more compact than spring faeces filled with diatoms. The sinking velocity of the faeces was closely related to their density, which depended on the composition of the diet. Similar observations were made by Dagg and Walser (1987) on the sinking velocity of copepod faecal pellets produced in low food concentrations. Their sinking velocity was reduced due to their low density. Thus, the sinking rate is related to the quantity and the nature of the ingested food. Consequently, for modelling faecal fluxes, it is important to determine the faecal density.

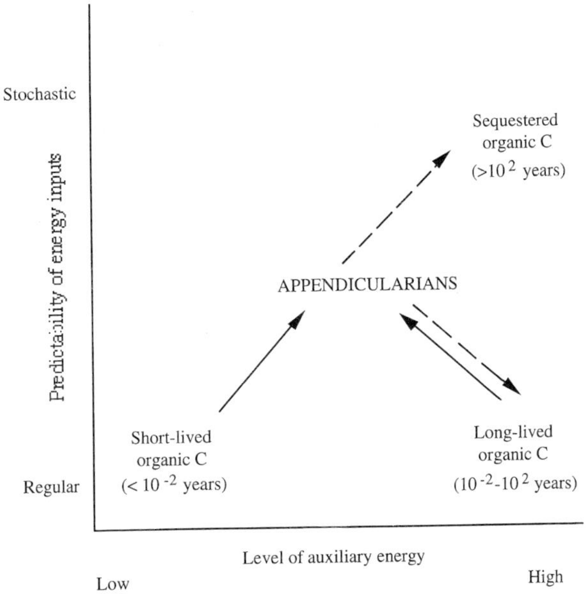

Fig. 10.2. Conceptual model link three biogenic carbon pools with different turnover times in the oceans to the level of auxiliary (mechanical) energy and the predictability of its delivery into the pelagic environment. Appendicularians contribute to channel part of the organic carbon toward longer lived pools. Solid arrow, carbon ingestion; dashed arrow, carbon export. Modified from Legendre *et al.* (1993).

10.6. Role of appendicularians in carbon cycling

Legendre *et al.* (1993) and Fortier *et al.* (1994) defined three pools of biogenic carbon according to their turnover times: short-lived, long-lived, and sequestered. The first (< 5 μm) with a turnover time of $< 10^{-2}$ years, is composed of small eucaryotes and small sized components of the microbial food web. The second pool, with a turnover time of from 10^{-2} to 10^2 years, includes a wide range of organisms, from large cells (> 5 μm) to large animals, including fish. The third pool is characterised by a turnover time of $> 10^2$ years and is mainly represented by buried organic matter or by inorganic sediments of biogenic origin.

Small detrital particles produced in the superficial layer are rapidly recycled. Larger, dense aggregates can reach deeper layers where the residence time of carbon will increase. Thus, organisms with high packaging activity will contribute most to the transfer of carbon from the pool with a short turnover time to a pool with a longer one. This shift of carbon to pools with longer residence times will be influenced by the size ratio between predator and prey. Since appendicularians package particles down to four orders of magnitude smaller than their own size, they contribute substantially to the carbon turnover time (see Fig. 10.2) The sinking rates of faecal pellets and houses (measured mainly on small species) of epipelagic populations are in the same range as those of copepod faecal pellets (Small *et al.*, 1979). Higher temperature increases the sinking rate, but also tends to accelerate decomposition. Moreover, copepods are known for their penchant for coprorhexy and coprophagy (Paffenhöfer and Knowles, 1979). Although copepods have not been reported feeding on larvacean faecal pellets, this possibility certainly exists.

Alldredge (1979) calculated a 91 m day^{-1} maximum sinking rate for aggregates 3 mm in diameter. The calculated flux based on these results is one order of magnitude larger than the flux measured by sediment traps in the same area. She suggested that turbulence and discontinuity layers may slow or halt the descent of some aggregates. Hence the contribution of epipelagic appendicularians to carbon sequestration in deep waters would be expected to be low. This is not consistent with observations that many houses and faeces do reach mesopelagic and deep layers. The explanation of this contradiction may lie in the permanent occurrence of larvaceans under the mixed layer. Marine snow produced by these populations can reach deep waters and their repackaging activity may play a significant role in channelling short lived carbon toward sequestration. In the western Barents Sea outlet, Zeller *et al.* (1994) found that up to 30% of the organic carbon in sediment traps moored at 610, 1840 and 1950 m consisted of appendicularian, ostracod and euphausiid faecal pellets. They observed an increase in flux rates in the deep traps, attributed to the flux of lithogenic elements. The authors emphasise the significant down slope particle transport of energy-rich organic matter supplied by the pelagic producers on the shelf. Gonzales *et al.* (1994) deployed sediment traps in the central Barents Sea at a depth of 158 m. The vertical particle flux was low throughout the year. They identified larvaceans as important producers of faecal pellets in the upper 150 m, even when the overall input of organic matter to the benthos was low. Thus, under appropriate conditions, appendicularian populations may be responsible for faecal chlorophyll sedimentation pulses to intermediate or greater depths.

10.7. Conclusions

The fragility of appendicularians has meant that laboratory and *in situ* studies are less numerous than for other zooplankton groups. However, several special aspects of their interaction with the pelagic food web have been determined.

(i) Their particular filter feeding strategy, which includes high retention efficiencies for submicrometre size particles, allows appendicularians to use a food resource not available to other filter feeders. This may explain the successful adaptation of appendicularians to aphotic depths and to oligotrophic environments.

(ii) Appendicularians inhabiting superficial layers exhibit high metabolic activity and rapid short-term population responses to events of enhanced production. The fastest generation time was found for *Oikopleura dioica* in Kingston Harbor, Jamaica, by Hopcroft and Roff (1995). Remarkably, the minimum generation time at 29°C did not exceed 24 h with a 23-fold increase in biomass over this period. This ranks as one of the fastest generation times amongst metazoans (see also p. 103, Chapter 5).

(iii) The sinking velocity of faecal pellets is obviously closely related to their density, which depends on the food composition and concentration. The contribution of appendicularian faecal production to the vertical fluxes of organic matter can be important in time and space. In Kaneohe Bay, Hawaii, 66% of total zooplankton faecal pellets in units of volume were larvacean faeces (Taguchi, 1982). The size structure of appendicularian populations varies seasonally, the larger forms being more important in cold waters, while the smaller forms are dominant in warm oligotrophic environments. Therefore, a seasonal shift in size structure of appendicularian community would affect the vertical flux of faecal pellets.

(iv) Appendicularians do not store energy in the form of lipid deposits (Deibel *et al.*, 1992; Gorsky and Palazzoli, 1989) and lack any metabolic adaptation to reduced food concentrations (Gorsky, 1980; Fenaux, 1985). As they need a continuous food supply, the presence of permanent appendicularian communities in mesopelagic and deep strata indicate the occurrence of aphotic accumulation layers rich in detritic matter of ingestible size and quality.

(v) Because of the high N and C content of their tissues, the absence of a carapace and their relatively slow escape reactions, appendicularians constitute an important fraction of the diet of pelagic predators, including fish.

(vi) By their aggregation activity, their houses and their faecal production appendicularians contribute significantly to the shift toward large particles in the suspended particle size spectrum and to the vertical fluxes of organic matter. The increasing quantity and nature of the data obtained using new techniques suggest that another pathway of vertical transport can occur in the water column, viz. vertically cascading transport with appendicularians as possible key indicators of layers where processes of re-packaging take place.

Physical processes such as stratification, wind stress, upwelling, fronts, geostrophic circulation, eddies and other hydrological factors affect the structure of the marine food web and the role of appendicularians in it. The influence of physical processes on the spatial and temporal heterogeneity of appendicularian populations is poorly understood, and must underlie the (unknown) mechanisms of 'seeding' appendicularian populations. Small populations must survive in occasional food patches and benefit from the geostrophic circulation to colonise coastal waters during favourable periods.

The discovery of mesopelagic appendicularian populations in the open ocean (including oligotrophic areas) suggests that current estimates concerning strata are incorrect, being significantly biased low. Our techniques are evidently inadequate. The superficial microbial communities may become inhibited during sinking into deep layers and thus reduce the re-mineralization rates of particulate organic matter (POC). This may result in a greater proportion of the POC reaching deep layers. The processes of DNA and protein synthesis of bacterial assemblages may be affected by increasing pressure and decreasing temperature (Turley, 1993).

It is manifest that our understanding of the processes and trophic interactions in the interior of the oceans is tightly coupled to 'state of the art' of marine biotechnology. With fragile organisms like appendicularians the introduction of new *in situ* sampling techniques and appropriate laboratory methods are needed to answer some of the questions raised in this chapter and to prove or disprove some of the suggestions made.

ELEVEN

The abundance, distribution, and ecological impact of doliolids

D. Deibel

11.1. Introduction

The doliolid life cycle is complex (Chapter 1), with obligatory alternation of sexual and polymorphic asexual generations; in different doliolids, there are varying degrees of multiplication at sexual and asexual stages, and in one species there is a special short cycle. In brief, as summarized in Fig. 11.1, sexual gonozooids liberate planktonic eggs (usually cross-fertilized) that develop via a short larval stage into asexual oozooids. From a ventral stolon in the oozooid, buds form which migrate around to a dorsal caudal cadophore. The buds are of two kinds, the trophozooids being devoted to nourishing the colony (by this stage, the old oozooid or old nurse has lost all internal organs except heart and brain), whilst the phorozooids bear buds that develop into the next gonozooid generation. Figures 11.2–11.6 show different stages in the doliolid life cycle. In response to favourable conditions, doliolids can produce vast 'blooms' or 'swarms' of individuals by asexual reproduction by budding from the oozooid stolon.

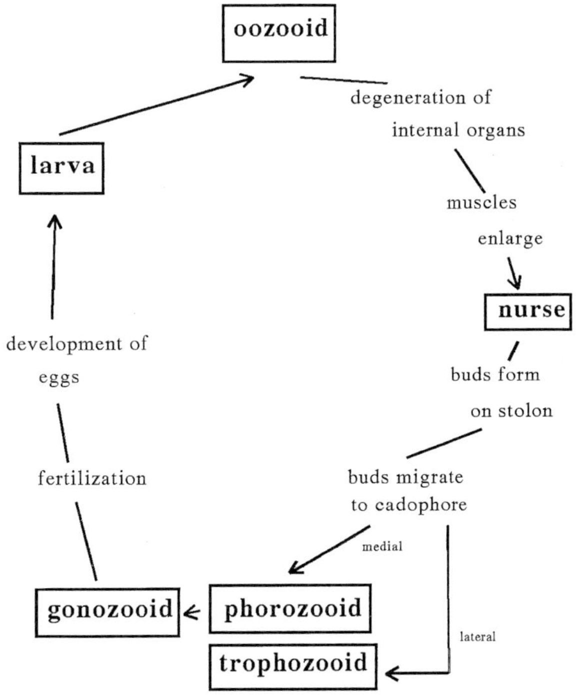

Fig. 11.1. Doliolid life cycle. Adapted from Braconnot (1971).

11.2. Abundance, distribution and biomass

Doliolids have a cosmopolitan distribution, preferring warm, continental shelf waters, rarely being found north or south of 64° (Fig. 11.7; Table 11.1; Garstang, 1933; Berrill, 1950; Hopkins and Torres, 1988). Doliolids are best known from the western Mediterranean Sea and the southeastern continental shelf of the USA (Fig. 11.7). They occur primarily in euphotic waters within the upper mixed layer. Mid-water forms are little known, due perhaps to undersampling (see section

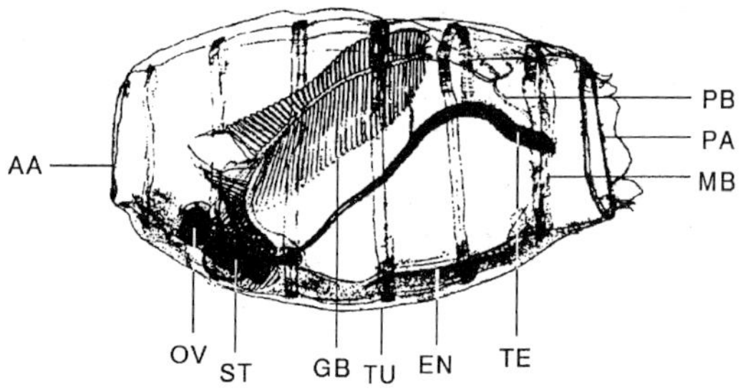

Fig. 11.2. Gonozooid of *Dolioletta gegenbauri* in near-lateral view. Length 10 mm. AA, atrial aperture; EN, endostyle; GB, gill bar; MB, muscle band; OV, ovary; PA, pharyngeal aperture; PB, peripharyngeal band; ST, stomach; TE, testes; TU, tunic. Adapted from Braconnot (1971).

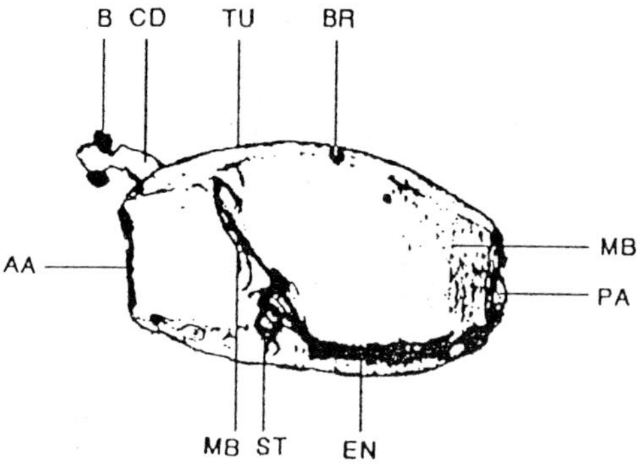

Fig. 11.3. Oozooid of *D.gegenbauri* in lateral view, length 2 mm. B, bud attached to cadophore; BR, brain; CD, cadophore. Other labels as in Fig. 11.2. Adapted from Braconnot (1971).

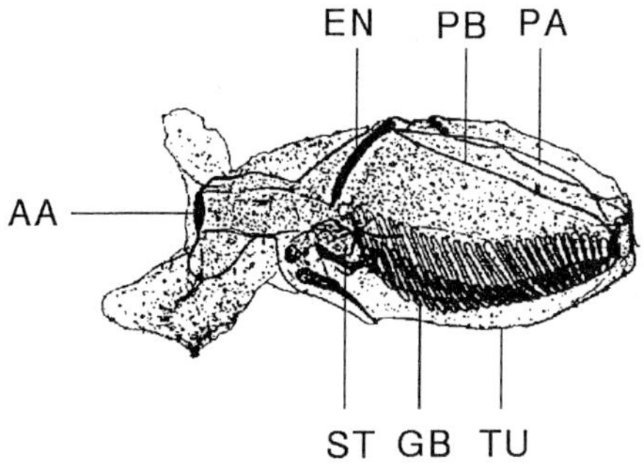

Fig. 11.5. Trophozooid of *D.gegenbauri* in lateral view. About 1 mm long. Labels as in Fig. 11.2. Adapted from Braconnot (1971).

17.3.4). Doliolids may form extremely dense swarms covering hundreds of square kilometres, lasting from days to several weeks (Binet, 1976; Deibel, 1985; Paffenhöfer *et al.*, 1987, 1995). Peak concentrations in these swarms may reach 1,000–4,500 individuals m^{-3}, representing from 25–100 mg C m^{-3} (Binet, 1976; Atkinson *et al.*, 1978; Deibel, 1985; Paffenhöfer and Lee, 1987; Paffenhöfer *et al.*, 1987).

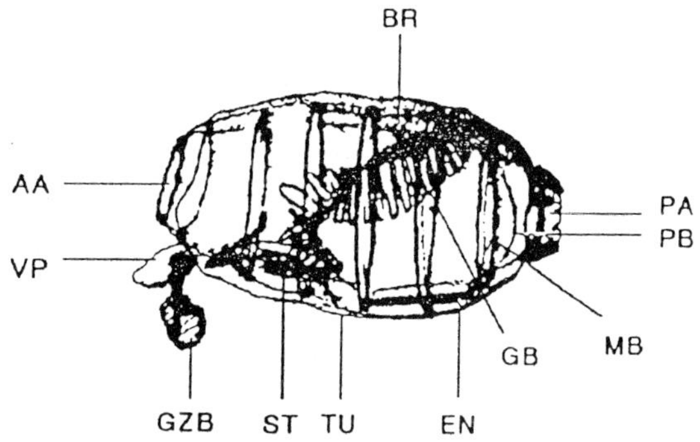

Fig. 11.4. Phorozooid of *Doliolum nationalis* in lateral view. Length 2 mm. GZB, gonozooid bud; VP, ventral peduncle. Other labels as in Figs 11.2 and 11.3. Adapted from Braconnot (1971).

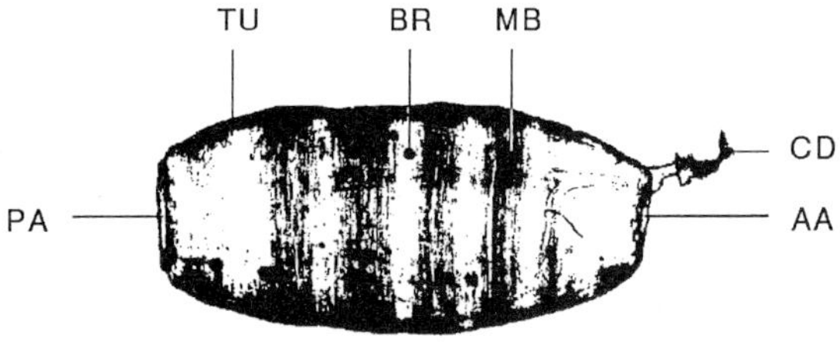

Fig. 11.6. Old nurse of *Doliolum* sp. in lateral view. Length 8 mm. All labels as in Figs 11.2 and 11.3. Adapted from Braconnot (1971).

11.2.1. *Spatial scale*

11.2.1.1. *Horizontal*

Most studies of doliolid distribution have been done at the mesoscale level (i.e. from 10 to several hundred kilometres horizontally; Table 11.1), although there are few continuous records of distribution over these scales. The continuous plankton recorder surveys have revealed > 100 m^{-3} of *Doliolum nationalis* over a 112 km transect of the North Sea between England and Denmark (Lindley *et al.*, 1990; Fig. 11.8). This high abundance was apparently the result of asexual reproduction by a seed population from the north. *Dolioletta gegenbauri* is also common in north western European waters (Hunt, 1968; Roskell, 1986a,b) and is considered to be indicative of oceanic water intrusion, which occurs more frequently in warmer than average years.

Dolioletta gegenbauri seems to be associated with mesoscale intrusions of cool water onto the continental shelf off the coasts of Georgia and California (Table 11.1; Deibel, 1985; Paffenhöfer *et al.*, 1987, 1995; Mackas, 1991). Doliolid concentrations up to ~4700 m^{-3} were found in the mid-shelf region between the 30 and 40 m

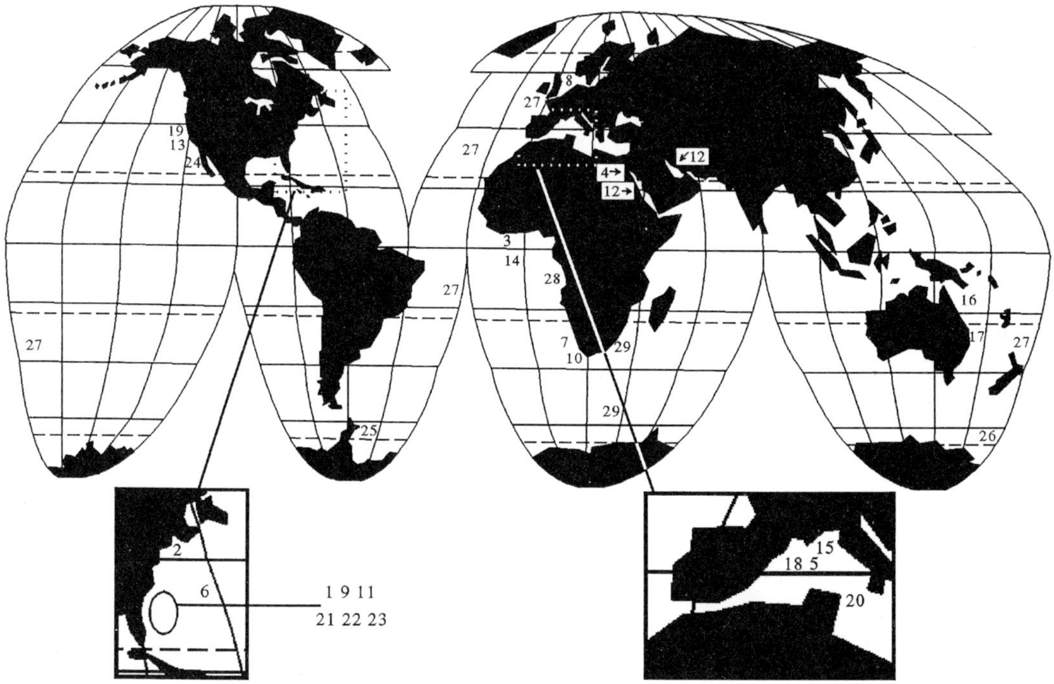

Fig. 11.7. Global distribution of doliolid collections. The numbers refer to those in Table 11.1.

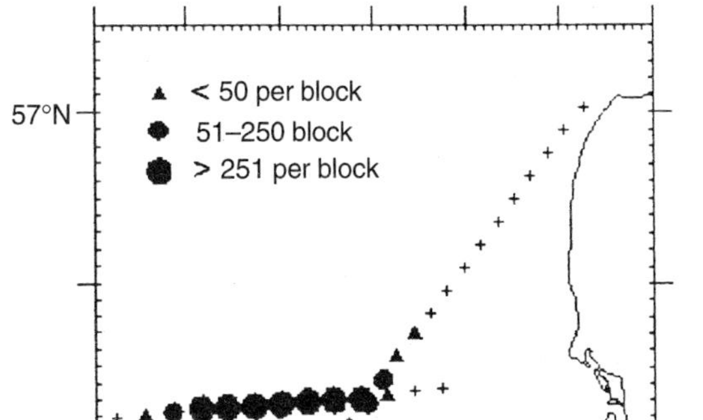

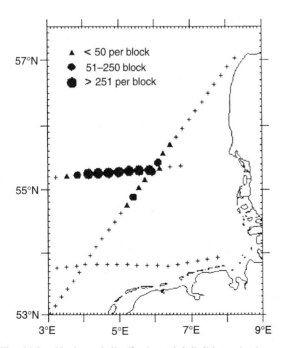

Fig. 11.8. Horizontal distribution of doliolid species in continuous plankton records from west of Denmark in October 1989. No doliolids were found where + is shown. The symbols represent the number of doliolids in each 3 m^3. Adapted with permission from Lindley *et al.* (1990).

isobaths off northern Florida and Georgia, with greatest numbers during summer in near bottom and thermocline waters that originated from intrusions of mid-depth Gulf Stream waters (~16–18°C; Paffenhöfer *et al.*, 1987; Fig. 11.9). These patches were at least tens of kilometres in horizontal extent.

In a subsequent study in the same area during winter, Paffenhöfer *et al.* (1995), using a combination of hydrographic, ADCP and plankton net observations, estimated that doliolids covered 90 km in the cross-shelf direction and 77 km along shore, for a patch area of ~6900 km^2. However, the dense area within this patch, with abundances > 500 m^{-3}, typically covered 20 km in the cross-shelf dimension and 20 m vertically. If this swarm was also 20 km in extent in the along shore dimension (a conservative assumption), then the swarm areal dimension was ≈400 km^2. This winter swarm was in the lee of the Charleston Bump, a quasi-stationary, spin-off eddy of the Gulf Stream, which seems to be an area of persistently high concentrations of doliolids (see also Deibel, 1985).

These observations are similar to ones made by Le Fevre on the northwestern continental shelf off France in 1979 and 1985 (personal communication), where high concentrations of doliolids were found on the coastal side of a strong thermal front between oceanic and shelf waters. Off California, *Dolioletta gegenbauri* was

Table 11.1 Global summary of field studies of the abundance and distribution of doliolids.

Map no.	Location	Species	Stage	Spatial scale	Temporal scale	Environmental correlates	Swarm formation	Peak conc. (m^{-3})	Sampling technique	Reference
1	W. Atlantic Florida shelf		OZ	Metres (vertical)	Minutes	Temperature, salinity, chlorophyll	Yes	≈300	Video on manned submersible	Paffenhöfer *et al.*, 1991
2	W. Atlantic Massachusetts shelf	*Doliolum nationalis*	GZ/PZ	1 cm (vertical) 100 m (horizontal)	Minutes	Temperature, salinity	Yes		Towed video	Davis *et al.*, 1992
3	E. Atlantic Ivory Coast shelf			10 m (vertical) 100 km (horizontal)	Hours/ season	Temperature, salinity currents	Yes	500–1,000		Binet, 1976
4	Gulf of Aqaba	*Doliolina mülleri* *Doliolum denticulatum*	GZ/PZ/ TZ/OZ	10 m (vertical)	Season				WP-2 net	Fenaux and Godeaux, 1970
5	Mediterranean/ Villefranche shelf	*Doliolina mülleri* *Doliolum nationalis* *D.denticulatum* *Doliolides rarum* *Dolioletta gegenbauri*	GZ/PZ/ TZ/OZ	10–100 m (vertical) 10–100 km (horizontal)	Season annual		Yes		Jesperson net, 50 cm diameter	Trégouboff, 1965
6	W. Atlantic/ Bermuda			10–100 m (vertical)	Season	Temperature	No	≈4	Ring net, 1 m diameter, no. 2 and no. 8 nylon	Deevey and Brooks, 1971
7	E. Atlantic/ S. Africa shelf/ Benguela	*Doliolum denticulatum* *D.nationalis* *D.tritonis*	GZ/OZ	10–100 km	Season	Temperature, salinity	No		Discovery nets, N100H & N100B	Van Zyl, 1959
8	North Sea	*Doliolum nationalis*		10–100 km	Season	Temperature	Yes	≈100	Continuous plankton recorder	Lindley *et al.*, 1990
9	W. Atlantic/ Georgia–Florida shelf	*Dolioletta gegenbauri*		10 m (vertical) 10–100 km (horizontal)	Days-week	Temperature currents	Yes	≈4,500	Tucker trawl, 0.16 m^2, 103 μ mesh	Paffenhöfer *et al.*, 1987
10	S. Africa shelf/ Benguela	*Doliolum denticulatum* *D.nationalis* *D.mirabilis* *Dolioletta gegenbauri*	PZ/GZ/ OZ	10–100 km	Monthly/ season	Temperature	Yes		Discovery nets N100H, N100B, and N70V	de Decker, 1973; Lazarus and Dowler, 1979

Table 11.1 *(Continued)*

Map no.	Location	Species	Stage	Spatial scale	Temporal scale	Environmental correlates	Swarm formation	Peak conc. (m^{-3})	Sampling technique	Reference
		D.tritonis								
11	W. Atlantic/ Georgia shelf	*Dolioletta gegenbauri*	PZ/GZ/ TZ/OZ	25 m (vertical) 10–100 km (horizontal)	Days	Temperature salinity	Yes	≈3,200	Tucker Trawl, 0.16 m^2, 110 μm mesh	Deibel, 1985
12	E. Mediterranean/ Gulf of Aqaba, Seuz and Persian Gulf	*Doliolum denticulatum* *D.nationalis* *Doliolina mülleri* *D.intermedium* *Dolioletta gegenbauri*	PZ/GZ/ TZ/OZ	50 m (vertical) 10–100 km (horizontal)	Season	Temperature salinity	No		WP-2 net, 200 μm mesh	Godeaux, 1973; Fenaux and Godeaux, 1970
13	E. Pacific/ California Current	*Dolioletta gegenbauri* *Doliolina mülleri* *D.obscura* *D.rarum* *D.undulatum* *Doliolum denticulatum* *D.rubescens*	PZ/GZ/ TZ/OZ	10–100 km			No			Berner, 1967
14	E. Atlantic/ Senegal coast	*Doliolum* spp. *D. nationalis* *Dolioletta gegenbauri*	GZ/PZ/ OZ	10–100 km	Days	Temperature, salinity	No	≈5	Ring net, 1 m diameter, 500 & 1000 μm mesh	Seguin and Ibanez, 1974
15	Mediterranean/ Villefranche shelf	*Doliolum nationalis* *D.denticulatum*		10–100 km	Season	Temperature	Yes		Ring net, 1 m diameter, 1670 μm mesh	Braconnot, 1963
16	W. Pacific/ Great Barrier Reef	*Doliolum denticulatum* *Doliolina tritonis* *Dolioletta gegenbauri*	GZ/PZ/ OZ	10 m (vertical) 10 km (horizontal)	Season		No			Russell and Colman, 1935

Table 11.1 *(Continued)*

Map no.	Location	Species	Stage	Spatial scale	Temporal scale	Environmental correlates	Swarm formation	Peak conc. (m^{-3})	Sampling technique	Reference
17	W. Pacific/ E. Australian shelf	*Doliolum denticulatum* *Dolioletta gegenbauri*		50 m (vertical) 10–100 km (horizontal)		Temperature, salinity	No		Discovery nets, N70, N100 & N200	Sheard, 1965
18	Mediterranean/ Gulf of Lyon	*Doliolum nationalis*	GZ/PZ	10–100 km	Days/ week	Temperature, salinity	Yes		Discovery nets	Braconnot and Casanova, 1967
19	E. Pacific/ California Current	*Dolioletta gegenbauri*		20 m (vertical) 10–100 km (horizontal)	Days	Temperature, salinity, currents	Yes		MOCNESS, 1 m^2, 100 μm mesh	Mackas *et al.*, 1991;
20	Mediterranean/ Sardinia shelf	*Doliolum nationalis*	PZ	10–100 km			Yes	500		Braconnot, 1967
21	W. Atlantic/ Georgia-Florida shelf	*Dolioletta gegenbauri*	PZ/GZ	20 m (vertical) 10–100 km (horizontal)	Days	Temperature, salinity, currents	Yes	>4000	Tucker trawl, 0.16 m^2, 103 μm mesh	Paffenhöfer and Lee, 1987
22	W. Atlantic/ Georgia-Carolina shelf	*Dolioletta gegenbauri*		20 m (vertical) 10–100 km (horizontal)	Days	Temperature, salinity	Yes	≈800	Tucker trawl, 0.16 m^2, 100 μm mesh + diaphragm pump	Paffenhöffer *et al.*, 1984; Paffenhöffer, 1985
23	W. Atlantic/ Florida shelf	*Dolioletta gegenbauri*	GZ	10–100 km	Days	Temperature, salinity	Yes	≈1500	Ring net, 50 cm diameter, 253 μm mesh	Atkinson *et al.*, 1978
24	E. Pacific/ California Current	*Doliolum denticulatum* *Dolioletta gegenbauri*		50 m (vertical) 10–2,000 km (horizontal)	Season	Temperature, salinity	Yes	≈10	Ring net, 1 m diameter, 650 μm mesh	Berner and Reid, 1961
25	Antarctic	*Doliolina intermedium var. resistible*		50 m (vertical) 10–100 km (horizontal)		Temperature, salinity	No		Tucker trawl, 0.16 m^2, 162 μm mesh	Hopkins and Torres, 1988
26	Antarctic	*Doliolina intermedium var. resistibile*		Single station			No			Garstang, 1933

OZ, oozooid stage; GZ, gonozooid stage; PZ, phorozooid stage; TZ, trophozooid stage. No entry indicates that information was not available in the primary publication.

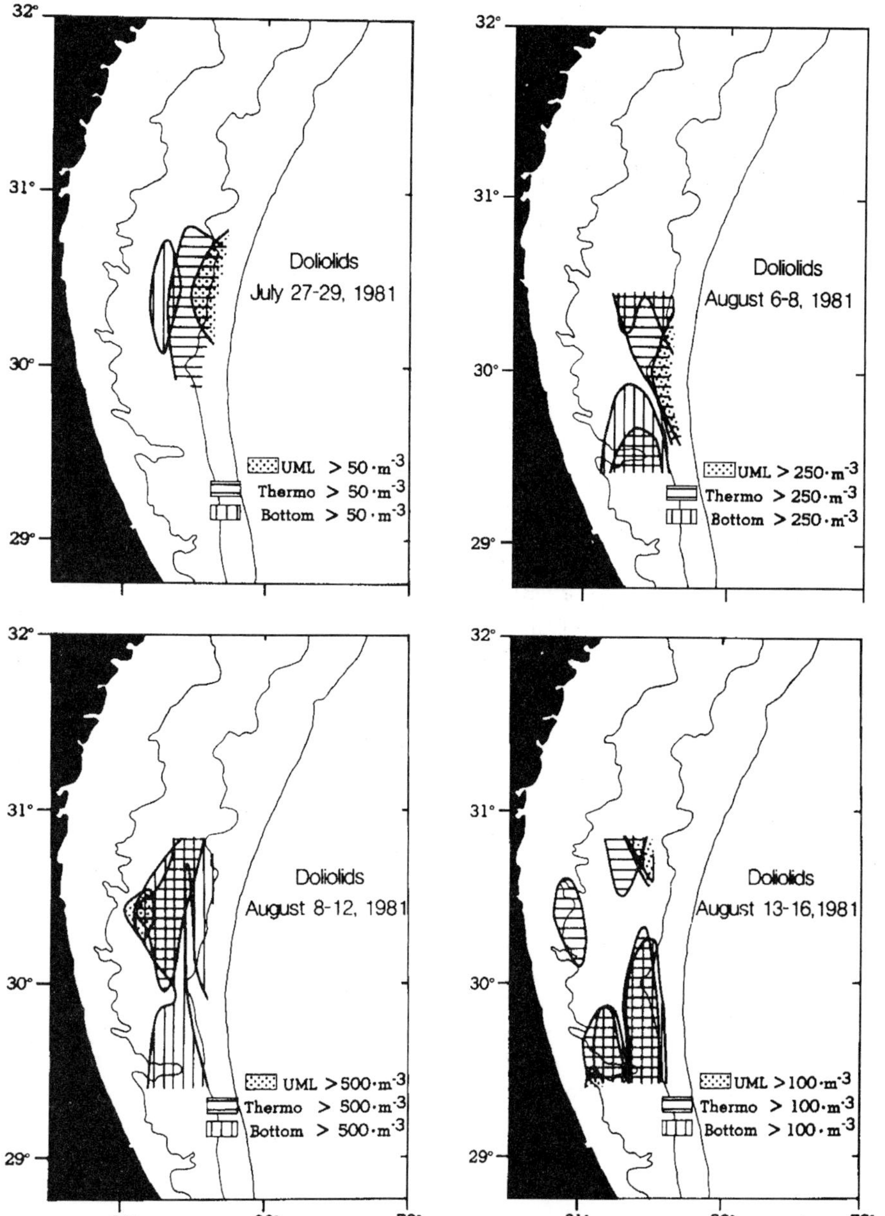

Fig. 11.9. Horizontal and vertical distribution of doliolids off the coast of Georgia and Florida in July–August, 1981. Concentrations are shown for the upper mixed layer (0–20 m), the thermocline (20–30 m) and the bottom layer (30–45 m). Adapted from Paffenhöfer *et al.* (1987).

very abundant at near shore stations in the upper mixed layer, where the water column was 40–60 m deep (Mackas *et al.*, 1991). Doliolids were also common along the offshore axes of mesoscale jets of cold water.

There is also evidence of the association of doliolids with continental shelf fronts and upwelling off the coast of South Africa (Table 11.1; De Decker, 1973). *Doliolum denticulatum* was at times dense in mid-shelf waters over the Agulhas Bank, offshore of the upwelled waters of the Benguela Current, in water temperatures > 14°C.

Video has been used on manned and unmanned submersibles to determine fine scale distribution of doliolids in both horizontal and vertical dimensions (see below for vertical results). This technology has great promise for future ecophysiological studies which will help determine the environmental factors that lead to

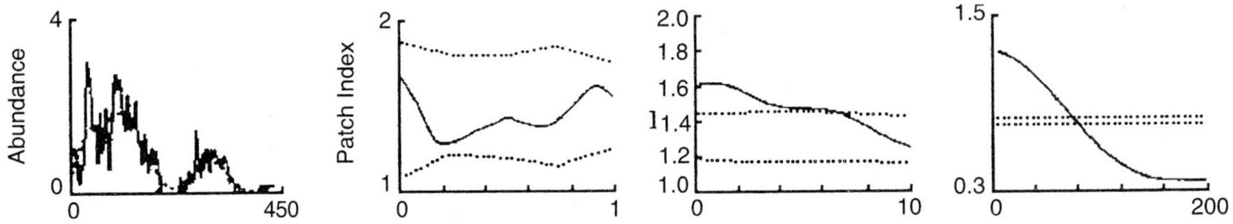

Fig. 11.10. The abundance and patchiness of doliolids south of Massachusetts, determined by a towed video device. The data in the abundance plot has been smoothed with a 10 m window to remove microscale patchiness. Abundance is given in numbers of individuals l^{-1}. The three columns showing the patch index are for the 1 10, and 200 m domains. Deviations of the index outside the 95% confidence limits (dotted lines) indicate significant patchiness. The patch index is given as the mean number of doliolids l^{-1}. Deviations below the lower confidence limits indicate lower abundance than expected. Adapted with permission from Davis *et al.* (1992).

patch formation. It appears that doliolids can form patches on scales as small as 1–5 m in shelf waters south of Woods Hole, Massachussetts, in August (Davis *et al.*, 1992; Table 11.1; Fig. 11.10). Peak abundances were 4000 m^{-3}, very similar to peak abundances determined by Deibel (1985) and Paffenhöfer and Lee (1987) off the coasts of Florida and Georgia using plankton nets made of 110 μm mesh. Larger scale patches were also identified, on the order of 160–180 m (Fig. 11.10). Horizontal patchiness on this scale would not be detected using plankton nets. It is likely that patches of this size result from the interaction of physical processes with asexual reproduction (Davis *et al.*, 1992).

11.2.1.2. *Vertical*

Video recordings made during dives of the manned submersible *Johnson Sealink* on the continental shelf off Florida, have revealed doliolids to be most abundant in middle and outer shelf waters at the base of the upper mixed layer in the thermocline between surface and intruded bottom waters (Paffenhöfer *et al.*, 1991; Table 11.1; Fig. 11.11). This confirms earlier reports of data collected with plankton nets (Deibel, 1985; Paffenhöfer *et al.*, 1987). In some cases, doliolid concentration changed by several hundred m^{-3} in a vertical distance of only 2 m (Fig. 11.11). Doliolids were found in waters with little chlorophyll *a*, indicating that phytoplankton were not abundant. The concentration of doliolids into layers was attributed to active swimming behaviour (Paffenhöfer *et al.*, 1991), although there is no evidence that they are capable of directed swimming.

Doliolids may change their vertical distribution seasonally on the continental shelf off the Ivory Coast, West Africa (Binet, 1976; Table 11.1; Fig. 11.12). In the late autumn (May), 90% of the population was at a depth greater than 20 m. However, in early winter (June), 90% of the population was above 40 m depth and at the beginning of spring (September) the doliolids were spread evenly over the upper 40 m of the water column. There was no indication of vertical migration on a diel

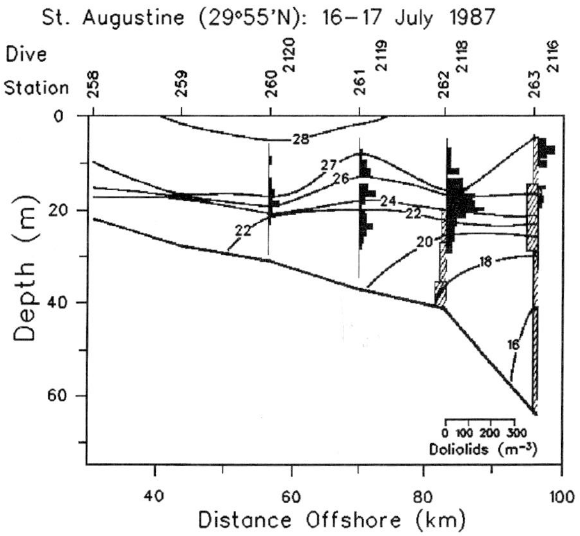

Fig. 11.11. Vertical distribution of doliolids (solid) along a transect off St Augustine, Florida, as determined from a manned submersible. The relative abundance of the colonial diatom *Thalassiosira subtilis* (hatched) and temperature isotherms (°C) are also shown. Adapted with permission from Paffenhöfer *et al.* (1987).

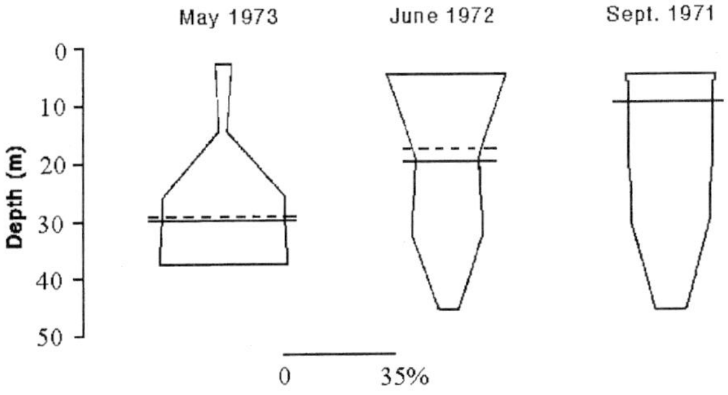

Fig. 11.12. The vertical distribution of doliolids off the West African coast determined by net hauls. The scale bar shows the percentage of the total population at each depth. The dotted line shows the vertical position of the 24°C isotherm and the solid line separates the surface current from the deeper counter current. Adapted with permission from Binet (1976).

time scale. There is similar evidence from off the south eastern US coast, where doliolids are found primarily in near-bottom waters in summer (Paffenhöfer and Lee, 1987; Paffenhöfer *et al.*, 1987) but in middle and upper mixed layer waters in the winter and spring (Deibel, 1985; Paffenhöfer *et al.*, 1995). One hypothetical explanation for this seasonal shift in depth distribution of doliolids is related to seasonal differences in the depth of maximum phytoplankton concentrations, which are in upper waters in the winter and spring but in near-bottom intrusion waters in summer (Paffenhöfer *et al.*, 1995).

There was some indication that the doliolids were located differentially in the Guinea Current and the deeper counter-current, so as to take advantage of phytoplankton production in the shear zone between the two currents, at ~20–40 m depth. Binet (1976) identified asexual reproduction as the mechanism allowing doliolids to take advantage of short-term pulses in phytoplankton production.

There is new evidence that the life history stages of doliolids may possess specific depth preferences. On the middle continental shelf off Georgia and South Carolina, Paffenhöfer *et al.* (1995) found larvae and oozooids beneath the pycnocline in near-bottom waters 40–80 m deep, phorozooids just above the pycnocline and gonozooids near the surface in the upper mixed layer. This vertical distribution pattern was remarkably persistent irrespective of the total doliolid concentration and of the state of water column stratification.

11.2.2. *Temporal scale*

11.2.2.1. *Days–weeks*

Only a few studies have been done on the temporal persistence of swarms of doliolids. *Doliolum denticulatum* forms swarms in the Great Barrier reef lagoon that can last from less than 1 month to several months in austral summer and autumn (i.e. December–June; Russell and Colman, 1935). This estimate is temporally coarse, however, due to monthly sampling.

There is much evidence showing doliolid patches to persist on time scales of weeks to months. Doliolid patches off the coast of Georgia and Florida may persist for ~14 days, with an increase in concentration of 5-fold over the first 10 days and 4-fold over the final 4 days (Fig. 11.9; Paffenhöfer and Lee, 1987; Paffenhöfer *et al.*, 1987). Doliolid swarms in the eastern Mediterranean Sea persist for 1–2 months, with peaks of abundance within this time period lasting < 2 weeks (Binet, 1976). A similar result has been found for doliolids off the coast of Villefranche-sur-Mer, in which swarms occurred in November and December with a period less than the sampling frequency of 2 weeks (Braconnot, 1963).

There has only been a single study documenting doliolid abundance over a diel time scale (Fig. 11.13; Atkinson *et al.*, 1978). At a station off St Augustine, Florida, on 2 successive days, the concentration of doliolids increased from 7 to 10-fold between midnight (minimum) and noon (maximum). The mechanism accounting for this diel variability was not discussed.

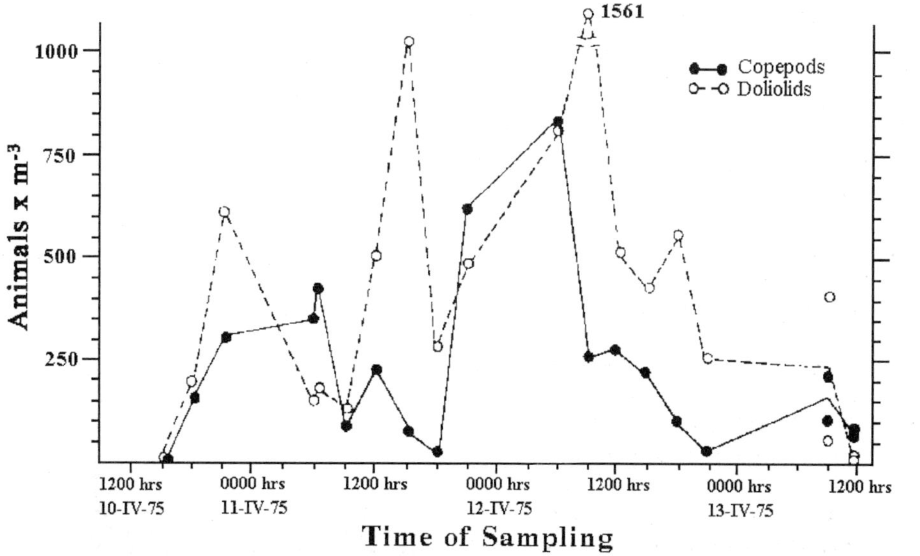

Fig. 11.13. Diel variability in the concentration of copepods and doliolids off the coast of St Augustine, Florida, determined with net hauls. Each increment on the *x*-axis represents 3 h. Adapted with permission from Atkinson *et al.* (1978).

Video and other *in situ* techniques can provide additional detail of patch persistence on the scale of days. Doliolid abundance may change by over an order of magnitude in only 2 days at a fixed station off Florida (Paffenhöfer *et al.*, 1991). Deibel (1985) used subsurface drogues and plankton nets to follow a patch of *Dolioletta gegenbauri* on the middle continental shelf off Georgia for 3 days. Over this time the concentration of doliolids did not change substantially, and patch formation was attributed to asexual reproduction in association with a phytoplankton bloom.

11.2.2.2. *Seasonal*

Doliolids seem to be most abundant in winter and spring, associated with seasonally increasing temperatures and perhaps with spring phytoplankton blooms. Doliolids are most common off South Africa in spring and summer when water temperatures are increasing (Van Zyl, 1959). Only *Doliolum nationalis* appears to have an inverse relationship with mean integral water temperature in the upper mixed layer (Van Zyl, 1959). Also, *D.nationalis* was most common in the North Sea in October, when upper mixed layer temperatures were declining (Lindley *et al.*, 1990). In the western Mediterranean, doliolids were observed in all months, with highest numbers in the upper 200 m in the coldest season, from September through April (Braconnot, 1963; Trégouboff, 1965; Table 11.1). The phorozooid stage was by far the most common, suggesting that asexual reproduction was occurring at an accelerating rate during the autumn. This is similar to the distribution of *Doliolum denticulatum* in the eastern Mediterranean, where it is most abundant from October to March (Godeaux, 1973, 1975).

Doliolum denticulatum has a bimodal seasonal distribution in waters surrounding the Great Barrier reef, with a primary maximum in December and secondary maximum in April–May (Russell and Colman, 1935). These peaks were found to be dominated by small gonozooids and large nurses, indicative of asexual reproduction.

Doliolids are also common in the great subtropical, anticyclonic gyres of the Pacific Ocean, including the Kuroshio, Westwind Drift, California Current and North Equatorial Current in the northern hemisphere and the East Australia Current, Westwind Drift, Peru Current and South Equatorial Current in the southern hemisphere. *Doliolum denticulatum* and *Dolioletta gegenbauri* co-inhabit these waters, with the latter continually moving outward from the gyres into water of appropriate temperature at depths of < 100 m, above the winter thermocline. *Dolioletta gegenbauri* was rarely found at temperatures < 14°C and did not increase in concentration at temperatures > 22°C. Seasonally, it was least abundant and furthest south in March and most abundant and furthest north in September.

11.2.2.3. *Environmental correlates*

The occurrence of doliolids has often been related to water temperature. Doliolids are rare in polar waters. *Doliolina resistibile* has been collected at 55°S and +1°C and *Doliolum denticulatum* at 34–38°S and 19–21°C (Godeaux and Meurice, 1978). *Doliolina resistibile* is the only doliolid found to be living at surface water temperatures < 3°C. There have been no reports of physiological limits to doliolid survival at near-zero temperatures.

The maximum concentrations of doliolids off the Ivory Coast, Africa, occurred in the spring at the time of the annual minimum water temperature (Binet, 1876). During the rest of the year, numbers were highest when coastal upwelling of relatively cool water occurred and when phytoplankton crops were at a maximum. *Doliolum denticulatum* was identified as an endemic slope water species off eastern Australia and occurred, as did *Dolioletta gegenbauri*, in waters > 17.6°C (Sheard, 1965). Sheard considers all of the east Australian doliolids to be indicative of subtropical water advecting into the area. High concentrations of *Doliolum nationalis* have been found near shore in the Gulf of Lyon, western Mediterranean (50–70% of the total macrozooplankton), but very low concentrations occurred on the continental shelf beyond (Braconnot and Casanova, 1967). Highest concentrations of this doliolid were coincident with reduced coastal salinities, with greatest numbers between 37.3 and 38.0‰.

11.2.2.4. *Swarms*

Few studies have focused on the formation and persistence of swarms of doliolids. Using a combination of moored current meters and ship-based measurement of hydrographic and planktonic variables, Paffenhöfer and Lee (1987) studied the development and persistence of patches of doliolids off the coasts of Georgia and Florida. A patch was formed in summer seaward of the 30 m isobath and was advected to the north over several weeks. Concentration within the patch covering 2800 km^2 was > 50 individuals m^{-3}. Within the next 2 weeks, flow changed from northward and offshore to southward and onshore, winds decreased and doliolids increased to cover the entire continental shelf between 29°30′ and 30°50′N (9300 km^2). Highest concentrations were usually found in the thermocline and near bottom layers, where phytoplankton concentrations were highest (i.e. > 1 mm^3 l^{-1}). This represents an increase in the concentration of zooids of from 5 to 10-fold over 2 weeks, resulting primarily from asexual reproduction (i.e. phorozooids and gonozooids dominated the population). These populations were thought to have derived from seed populations in Gulf Stream upwellings downstream from Cape Canaveral, Florida. Maximum concentrations of doliolids reached > 4000 m^{-3}. It is clear that off the southeastern USA, doliolids prefer cool, particle-rich water, typical of coastal upwelling in the subtropics (Paffenhöfer *et al.*, 1984).

The relative concentration of the various stages of the life cycle can be used to infer how swarms are formed. High concentrations of gonozooids and young oozooids would point to sexual reproduction, while high concentrations of old nurses and phorozooids would suggest asexual reproduction. Braconnot (1967) reported 300–500 phorozooids m^{-3} of *Doliolum nationalis* in the Mediterranean Sea southwest of the island of Sardinia, with no gonozooids. Other studies of swarms have also reported high ratios of phorozooids to gonozooids, including 200:1 by Trégouboff (1965), 400:1 by Russell and Hastings (1933) and 1500:1 by Casanova (1966). Deibel (1985) has used the relative concentration of gonozooids to old nurse stages as an index of the intensity of asexual reproduction. At total zooid concentrations of < 100 m^{-3} the ratio was 5–10:1, at total concentrations of 100–500 m^{-3} it was 20–600:1 and at total concentrations > 1000 m^{-3} it was 50–1,000:1. It is obvious from these observations that swarms generally are the result of asexual reproduction.

11.2.2.5. *Biomass*

Only a few investigators have included determination of the biomass of doliolids in addition to numerical abundance. Because tunicates are gelatinous with a high content of salt and chemically bound water (Madin *et al.*, 1981), the recommended measure of biomass is the content of organic carbon or nitrogen as determined by high temperature combustion in an elemental analyzer (Schneider, 1990). The biomass of doliolids off the coast of Georgia, ranged over three orders of magnitude, from 0.06 to 70 μg C l^{-1} (Deibel, 1985; Fig. 11.14). This peak biomass is equivalent to 31% of the average quantity of total particulate organic carbon in these waters as collected on a GF/C glassfibre filter (i.e. 228 μgC l^{-1}). There is no published information on the biomass of other macrozooplankton in Georgia continental shelf waters with which to compare these estimates of doliolid biomass (Paffenhöfer, personal

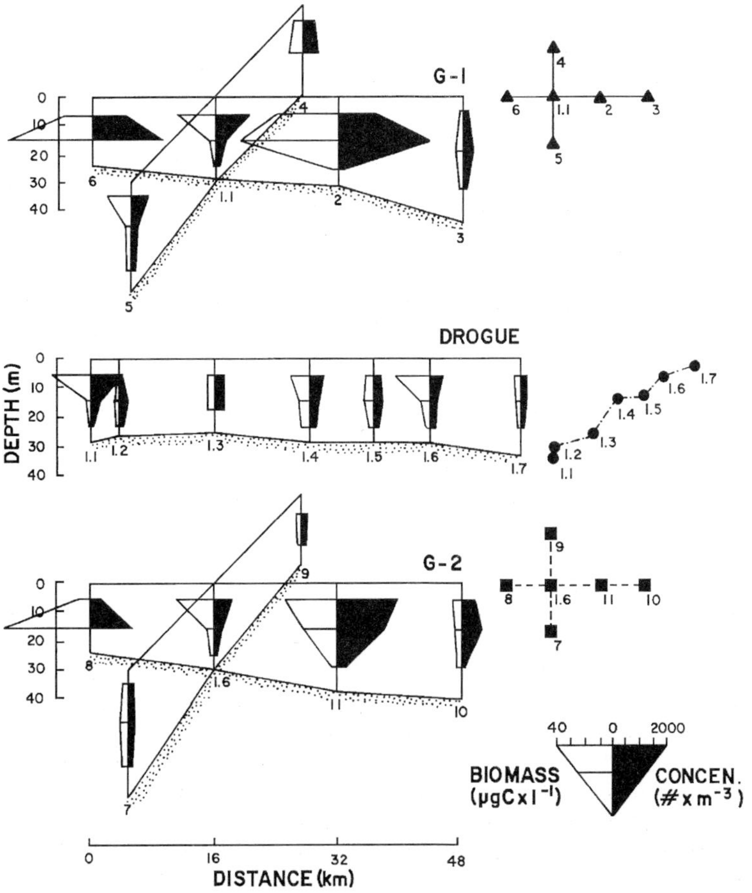

Fig. 11.14. Vertical and horizontal distribution of *D.gegenbauri* in terms of numbers and biomass off the coast of Georgia, as determined with net hauls. The station numbers are shown beneath each plot and plan views of the station pattern are shown on the right. The drogue samples were taken over 18 h, while following a subsurface drogue. Adapted from Deibel (1985).

communication). Mackas *et al.* (1991) determined the biomass of *Dolioletta gegenbauri* in the California Current to range from ≈1–8.8 μg C l^{-1}, lower than the peak biomass off Georgia. This biomass in the California Current is equivalent to 12–75% of the biomass of all other zooplankton combined (Mackas *et al.*, 1991).

Gulf Stream intrusions onto the Georgia–Florida continental shelf supply a large portion of the nitrogen requirements of phytoplankton populations in middle and outer shelf waters (Atkinson *et al.*, 1984). Generally, the phytoplankton blooms that occur in response to intrusions are considered to be lost to shelf biota, being re-entrained into the Gulf Stream by offshore flow (Yoder *et al.*, 1983). The usual reason given for this lack of coupling between primary production and zooplankton grazing is the slow reproduction and growth of shelf copepod populations (Paffenhöfer *et al.*, 1984). But we now know that doliolids can respond quickly to phytoplankton blooms by reproducing asexually (Deibel, 1985). What grazing impact might doliolids have on intrusion phytoplankton populations?

If we express the biomass of doliolids off Georgia in terms of elemental nitrogen, we can estimate the nitrogen flux from Gulf Stream intrusions through phytoplankton blooms to swarms of doliolids. The onshore NO_3-nitrogen flux in an average intrusion lasting 21 days is 3200 metric tons (Atkinson *et al.*, 1984). Making the conservative assumption that doliolid biomass increases by an order of magnitude in response to an intrusion (Paffenhöfer *et al.*, 1987), we have doliolid biomass increasing by 60 μg C l^{-1} over 21 days. Given a C:N ratio for doliolids of 4.4 (Madin *et al.*, 1981); this is equivalent to an increase of nitrogen biomass of 13.6 μg l^{-1}. We will assume that an average doliolid swarm covers 5000 km^2 of the Georgia shelf (Paffenhöfer and Lee, 1987; Paffenhöfer *et al.*, 1987) and is 20 m thick in the vertical (Deibel, 1985), giving a volume for this model swarm of 100 km^3. By multiplication and units conversion, the increase in nitrogen of 13.6 μg l^{-1} over a swarm occupying 100 km^3 is 1500 metric tons. Thus, this somatic biomass is equivalent to 47% of the total NO_3 nitrogen advected onshore by the intrusion over 21 days. At the peak of the swarm, faecal pellet production by doliolids (Caron *et al.*, 1989) can utilise the remaining 53% of the phytoplankton nitrogen within the intrusion in only 3.8 days. This suggests that doliolid faeces are a major flux of particulate nitrogen during Gulf Stream intrusions and that doliolid somata are a primary sink of nitrogen. Thus, we recommend caution when drawing the conclusion that phytoplankton blooms in intrusions are not grazed (Atkinson *et al.*, 1984).

11.2.2.6. *Assessment of current knowledge and future developments*

Unfortunately, most of the studies of the distribution and abundance of doliolids have been done with nets that could not be closed and that did not have a flow meter (Table 11.1). Therefore our knowledge is largely qualitative. In addition, most of these studies have been limited to surveys over a 'cruise' time scale of 7–21 days, so that existing observations of changes in abundance may suffer temporal aliasing. We need more information on doliolid distribution from *in situ* instruments at daily to weekly time scales. This is difficult to do for an organism that is an opportunistic colonist which can form patches in a matter of days. However, it is likely that important ecosystem properties such as the time-averaged vertical and horizontal flux of nitrogen are dominated by 'event' scale processes, such as Gulf Stream intrusions and the resultant aperiodic blooms of phytoplankton and doliolids.

We know the rates of physiological processes of doliolids, such as ingestion, assimilation and faecal pellet production, much more precisely than we know the distribution and abundance of doliolids in space and time. Thus, future research efforts should be focused primarily on methods for the determination of distribution and abundance of doliolids in the field. Likely technologies to assess doliolid biomass at the pertinent space and time scales are instruments such as the Longhurst–Hardy plankton recorder and *in situ* video. Perhaps the ideal instrument package would join the plankton recorder with a video recording device. High technology video instruments, such as the 'Critter Cam' (Strickler, unpublished), have the promise to document gelatinous zooplankton and phytoplankton distribution in vertical and horizontal dimensions over spatial scales of 5 μm to several millimetres and time scales of minutes to days. Incorporated with other measures of water column structure (e.g. CTD with *in situ* fluorometer) and vertical currents (Acoustic Doppler Current Profiler) a great deal could be achieved toward the determination of doliolid abundance and distribution at relevant space and time scales. Prototype video systems are already being tested, and it is anticipated that investigators will be able to identify zooplankton to the species level from video records (Paffenhöfer *et al.*, 1991). Davis *et al.* (1992) suggest that recent advances in video image processing, with real-time digitization, thresholding, convolutions, and edge detection, will allow real-time sorting of plankton into taxonomic categories. Since doliolids are gelatinous and contain primarily water, they have limited acoustic signatures (Flagg and Smith, 1989). Thus, video technology has the most promise for assessing distribution and abundance.

11.2.2.7. *Summary*
Doliolids inhabit primarily continental shelf waters in the winter and spring, often in association with mid-shelf fronts. Most doliolids prefer temperatures between 14 and 24°C, but *Doliolina resistibile* is found in polar oceans at temperatures < 3°C. Doliolids often form swarms covering hundreds of square kilometres lasting from several days to several weeks. They achieve this explosive population growth via asexual reproduction, and peak concentrations may reach 4500 m^{-3} with a biomass of 100 μg carbon l^{-1}. *In situ* video has shown that patch scales may approach 10 m horizontally, and have shown order of magnitude changes in concentration over just 2 m vertically. Video technology has great promise for further study of the space and time scales of doliolid populations.

11.3. Role as particle consumers and producers

11.3.1. *Grazing*

Doliolids pump water at high rates (Deibel, 1982; Crocker *et al.*, 1991) and retention efficiencies are likely high (Crocker *et al.*, 1991; Deibel, unpublished results). Given that doliolids may often occur in dense swarms (see above) their grazing impact may be very large. Doliolids may reproduce asexually in waters with high particle concentrations (> 1 mm^3 l^{-1}; Paffenhöfer *et al.*, 1984, 1995; Deibel, 1985; Paffenhöfer, 1985; Paffenhöfer and Lee, 1987) and during the lifetime of the swarm they may reduce particle concentrations between 2 and 50 μm diameter to very low levels (Fig. 11.15; Deibel, 1985). They may compete directly with juvenile copepods for food as their preferred size of particle is similar to that of copepods. The technology exists to determine particle concentration in the thin layers (< 2 m in the vertical) in which doliolids often occur (Paffenhöfer and Lee, 1991). It is likely that in these layers the water is nearly filtered clean by doliolids (Deibel, 1985).

By multiplying laboratory determined grazing rates and field estimates of abundance, Deibel (1985) has shown that doliolid swarms off the coast of Georgia, USA, should have the capability to clear from < 1 to 118% of their resident water volume each day ($\bar{X} \pm$ SD = 17 ± 28%), with greatest grazing pressure at the peak doliolid concentration of ≈3200 m^{-3}. Using a similar rationale, Paffenhöfer *et al.* (1995) estimated a population clearance rate of 25–75% during winter in middle shelf waters off South Carolina. Off the coast of California, *D.gegenbauri* may clear up to 10% of their resident water volume day^{-1} when present at 800 individuals m^{-3} (Crocker *et al.*, 1991). A model by Jackson (1980) shows that in the North Pacific Central Gyre, at concentrations < 1 m^{-3}, doliolids filter only 0.5% of their resident water volume day^{-1}. Given new measurements of doliolid abundance of > 4500 m^{-3} (Paffenhöfer and Lee, 1987), grazing pressure may equal or exceed 100% day^{-1} for at least short periods of time. This grazing impact is important, because doliolids have the ability to ingest free-living bacteria as well as large diatoms and may constitute a 'shunt' of the microbial loop, directly and efficiently transferring microbial biomass to large faecal pellets and to the large metazoan predators of doliolids (Crocker *et al.*, 1991). For example, at concentrations of 100–850 m^{-3}, there was a statistically significant, inverse relationship between the abundance of doliolids and heterotrophic and autotrophic flagellates (Paffenhöfer *et al.*, 1995). These flagellates were dominated by spherical cells 2–6 μm in diameter, at the lower end of the range retained by doliolids. In this instance, nanoflagellate numbers were depleted to a threshold of 1×10^3 ml^{-1}.

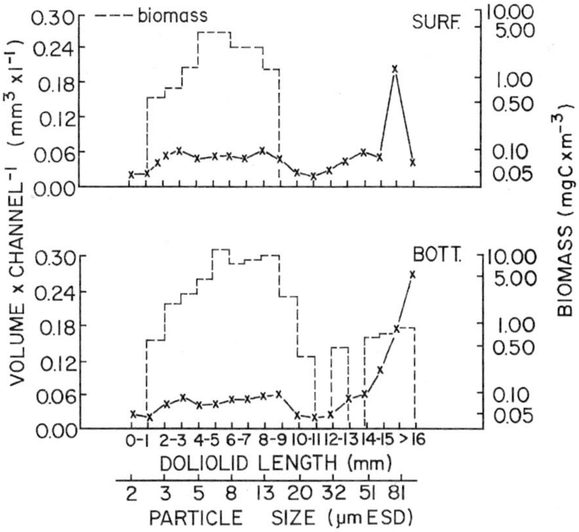

Fig. 11.15. Histogram of the biomass of *D.gegenbauri* in 1 mm size classes and the particle volume versus particle size spectra from the Coulter counter, for the East–West transect of grid 1. The Coulter counter channel > 81 μm ESD is a cumulative channel for all particles > 81 μm counted by the 400 μm orifice tube. The discrete depth samples for particle counts came from the midpoint of the oblique zooplankton tows. Particle volume versus particle size spectra from all remaining discrete depth samples are available from Deibel. Adapted from Deibel (1985).

11.3.1.1. *Assessment of knowledge and future possibilities*
Studies of the grazing rate of doliolids are far too few to generalize concerning trends and underlying mechanisms. Although feeding by the various zooids has been documented (Deibel, 1985; Crocker *et al.*, 1991; Tebeavand Madin, 1994), not even preliminary studies

of the functional response to food concentration, size and quality have been made. New *in situ* techniques for measuring feeding will be needed to complement recent advances with *in situ* video for determining doliolid abundance and distribution. We know little of the size selection of doliolids (Deibel, unpublished results), or of the actual composition of the diet in nature. Since differences in feeding mode exist between doliolids, appendicularians and salps, it is risky to make generalizations between them about details of feeding. Estimates from the California Current and off Georgia indicate that doliolid grazing impact may be substantial, from an average of ~10% day^{-1} to a peak of 100% day^{-1}. However, we are extremely ignorant about this fundamental aspect of doliolid ecology, and much work is needed.

11.3.2. *Particle production: faecal pellets*

1.3.2.1. *Pellet production*

Faecal pellet production by doliolids has never been quantified. Pellets of *Dolioletta gegenbauri* are loosely formed aggregates with a mean size of 0.26 mm^3 and a dry weight of 10–50 μg, containing ~20% carbon by weight (Pomeroy and Deibel, 1980; Bruland and Silver, 1981; Pomeroy *et al.*, 1984). They may be a flat rectangular shape but often are irregularly shaped, resembling upper mixed layer flocculent organic aggregates (Pomeroy and Deibel, 1980; Bruland and Silver, 1981). Under the light microscope they are generally nearly completely transparent. Newly expelled faeces often contain viable phytoplankton and bacteria within a mucous matrix (Pomeroy and Deibel, 1980). In subtropical waters, doliolid faeces are rapidly colonized by bacteria from both water column and gut populations (Pomeroy and Deibel, 1980). Bacterial colonization is followed by rapid growth of bacterivorous protozoans (Pomeroy and Deibel, 1980). The primary density of pellets has never been determined, but they are visibly less dense than faeces of salps and copepods (Pomeroy and Deibel, 1980; Bruland and Silver, 1981). This may result in lower than expected sinking velocities for doliolid faeces in comparison to that of other macrozooplankton.

11.3.2.2. *Sinking velocity*

There have been several studies of the stillwater sinking velocity and content of doliolid faeces. Faeces from *Dolioletta gegenbauri* sink at velocities ranging from 41 to 405 m day^{-1}, from 20 to 50% of the velocity expected for similar sized pellets from euphausiids and salps (Bruland and Silver, 1981; Deibel, 1990; Fig. 11.16). Reynolds numbers are low (< 4), indicating that sinking pellets are influenced primarily by viscous forces (Deibel, 1990). Empirical curve fitting indicates that sinking velocity increases as the first power of faeces

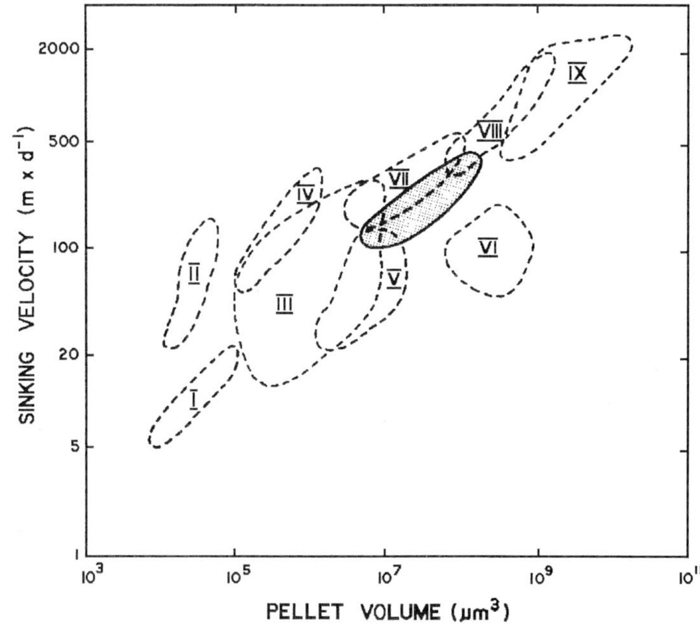

Fig. 11.16. Logarithmic plot of sinking velocity (m day)$^{-1}$ versus faecal pellet volume (μm^3) for a variety of macrozooplankton, adapted from a figure by Bruland and Silver (1981). I, copepod nauplii and copepodids (Paffenhöfer and Knowles, 1979); II, *Oikopleura dioica* (Gorsky *et al.*, 1984); III, small and large copepods (Small *et al.*, 1979); IV, naturally occurring faecal material (Smayda, 1969); V, *Pontella meadii* (Turner, 1977); VI, *Dolioletta gegenbauri* (Bruland and Silver, 1981); VII, euphausiids (Small *et al.*, 1979); VIII, *Corolla spectabilis* (Bruland and Silver, 1981); IX, salps (Bruland and Silver, 1981; Madin, 1982). Shaded area data for *D.gegenbauri* from Deibel (1990). From Deibel (1990).

diameter (Deibel, 1990). These sinking velocities suggest that doliolid faeces leave the upper mixed layer intact within no more than 36 h of their production, supplying energy and materials to midwater or demersal food webs. However, DOM released in the first 8 h may stimulate growth of water column bacteria (Pomeroy *et al.*, 1984). In addition, reingestion (i.e. coprophagy) may intercept faeces before they have left the upper mixed layer, seeding new faeces with the microbial populations that were colonizing the old ones (Pomeroy *et al.*, 1984). Reingestion must occur within 2–3 days for significant amounts of energy to be returned to the doliolid population (Pomeroy *et al.*, 1984). Also, sinking velocities seem to depend on the nature of the particles ingested, and the most flocculent amorphous faeces may be recycled by bacteria within the upper mixed layer (Deibel, 1990).

A recent flow analysis model has indicated that when doliolids are in swarm concentrations there are fundamental implications for the magnitude and nature of the sinking flux (Michaels and Silver, 1988). The flux of faecal material on an annual basis can be quite high, up to 22% of the total nitrogen exports from the upper mixed layer. But nearly all of this flux occurs on only those few days each year when the doliolids are present in swarm concentrations. In addition, when doliolid faecal material dominates daily vertical flux there is a much larger contribution of microplankton and nanoplankton to the total flux than when copepod faeces dominate flux. This reinforces the need to know more about the size and persistence of swarms of doliolids, because isolated swarms taking place aperiodically during the year may contribute an inordinately large component to the annual flux. Michaels and Silver (1988) caution that typical time scales of measurement (i.e. that of the usual oceanographic cruise) may result in large errors in estimating annual flux because they may miss important, but infrequent, events. Recent work in the North Atlantic provides a test of these models, in that small cyanobacteria ($\sim$1 μm diameter) were present in high concentrations in the sediments at 4.5 km depth when a salp swarm predominated in the upper mixed layer (Pfannkuche and Lochte, 1993). The cyanobacteria were present in high concentrations in salp faecal pellets collected in the upper mixed layer. This is one of the first cases to report a tunicate-mediated transfer of a production signal in the upper mixed layer directly to the abyssal benthos at $>$ 4 km depth.

11.4. Role in nutrient regeneration: excretion

Little is known about the magnitude or nature of excretion by doliolids. The only data of which I am aware is that of Biggs (1977). For doliolids containing from 0.1–1.0 mg protein (BSA equivalents), the mean ($\pm$ SD) respiration rate was 24.2 $\pm$ 7.7 μl O_2 individual^{-1} h^{-1} (n = 7) and the mean ammonia excretion rate was 0.072 $\pm$ 0.060 μg-at NH_4^+ individual^{-1} h^{-1} (n = 6) (both rates were calculated from the original data table available from Biggs). The resulting atomic O:N ratio was 45 $\pm$ 26, indicating a metabolism based primarily on the oxidation of lipids or carbohydrates, and suggesting little food limitation (Conover and Mayzaud, 1975; Mayzaud and Conover, 1988). Since lipids are a minor constituent of pelagic tunicates generally (Madin *et al.*, 1981; Deibel *et al.*, 1992), this suggests that dietary carbohydrates may be used for routine catabolism. Furthermore, it is likely that much of the protein within doliolids is complexed with mucopolysaccharides in the tunic (Madin *et al.*, 1981), and thus may not be metabolically labile in the short term.

Doliolid protein-specific respiration and excretion rates were higher than most other gelatinous zooplankton, including hydromedusae, schyphomedusae, ctenophores, heteropods, pteropods and salps (Biggs, 1977). The same trend was found for the O:N ratio (Biggs, 1977; Cetta *et al.*, 1986; Kremer *et al.*, 1986). High ratios may occur as a result of: (1) discontinuous ammonium excretion; (2) excretion of dissolved nitrogen in other forms; (3) the formation of solid waste products, such as uric acid crystals (Heron, 1976); (4) increased respiration (but not excretion) as a result of increased swimming contractions when making contact with the container walls (i.e. an artifact of containment). At the present time we lack the knowledge to distinguish between any of these four possibilities.

High individual excretion rates for doliolids suggests that they may play an important role in the nitrogen cycle of coastal seas. Biggs (1977) estimated that at a concentration of 1 m^{-3}, gelatinous zooplankton in the Sargasso Sea would supply 39–63% of the ammonium requirements of the phytoplankton. If we extend Biggs' excretion rates for doliolids to the scale of a typical swarm off the coast of Georgia (2,000 individuals m^{-3} as a conservative estimate; Paffenhöfer and Lee, 1987), rates would be $\approx$144 μg-at NH_4^+ m^{-3} h^{-1}, or 800% of the daily ammonia nitrogen requirements of typical shelf phytoplankton populations under upwelling conditions (18 μg at N m^{-3} h^{-1}; Yoder *et al.*, 1983). At a typical non-swarm concentration of 20 individuals m^{-3}, doliolids would supply $\approx$10% of the daily ammonia N requirements of shelf phytoplankton populations. Clearly, at the time of Gulf Stream upwelling, excretion by swarm populations of doliolids may act to prolong diatom blooms by providing recycled nitrogen to replace dwindling supplies of nitrate N. Because of their explosive population growth in response to upwelling (Deibel, 1985), doliolids may act to sequester upwelled nitrogen on the continental shelf before physical processes result in dissipation or reintrainment of the intrusion offshore (reintrainment occurs with ca. a monthly period; Atkinson *et al.*, 1984).

11.5. Role as prey: links to subsequent trophic levels

Whilst salps are known to be prey for a variety of invertebrate and vertebrate predators (Chapter 12), much less is known about predation on doliolids. Seapy (1980) has reported that adult heteropods (*Carinaria* spp.) prey heavily on salps and doliolids and Larson *et al.*, (1989) report predation by epipelagic narcomedusae on salps and doliolids, with heaviest predation on doliolids by *Solmaris corona, Cunina duplicata* and *C.proboscidia.*

Although pelagic tunicates are composed primarily of water, on a dry weight basis they contain a high proportion of nitrogen, and salps have tunics composed of protein–polysaccharide complexes (Madin *et al.*, 1981). Thus, it is likely that doliolids are a good source of dietary protein for predators.

11.6. Role as competitors: effects on other zooplankton

There are many reports of an inverse relationship between the abundance of doliolids and that of other macrozooplankton, especially copepods (Van Zyl, 1959; Ogawa and Nakahara, 1979; Deibel, 1985; Paffenhöfer *et al.*, 1995). Paffenhöfer *et al.* (1995) present several hypotheses that may account for this relationship. These include: (1) competition for small food particles; (2) predation on copepod eggs and nauplii; (3) active vertical migration by the copepods. Gut content analyses have shown that *Dolioletta gegenbauri* may contain many copepod eggs (Paffenhöfer *et al.*, 1995). Abundances of *D.gegenbauri* off the coast of Georgia, USA, and hatching time of copepod eggs, suggest that each egg should encounter a doliolid at least once before hatching. This would produce extremely high predation pressure. Larger salps may even prey directly on adult copepods in the Antarctic (Hopkins and Torres, 1989). Increased knowledge of competitive exclusion will be dependent on better information on the duration, extent, and causes of doliolid swarms.

TWELVE

The parasites and predators of Thaliacea

G.R. Harbison

12.1. Introduction

This chapter is intended to provide an overview of organisms that have been found to interact with thaliaceans. Since so many species parasitize or feed on thaliaceans, the bulk of this paper will consist of lists, and I will limit my comments, except to point out areas where my own observations supplement or contradict reports in the literature. At the outset, I would like to stress that the distinctions between predators and parasites are somewhat arbitrary, particularly for the smaller invertebrate metazoans. As Laval (1980) noted for hyperiid amphipods, such distinctions are largely based on the size of the predator: if it is small relative to its prey, the relationship appears parasitic.

It should be remembered that most thaliaceans are relatively large and slow moving, so that their defences against predation, if they exist, are likely to be subtle. Gelatinous zooplankton in general, and Thaliacea in particular, are similar to terrestrial plants. It is relatively easy for organisms to feed on them, but they have tissues that are low in nutritive value per unit weight. Gelatinous zooplankton are difficult to identify in gut content studies, since they are so easily degraded into unrecognizability. Thus, in most such studies, crustaceans, which have indigestible exoskeletons, are easily identified, while the much more digestible gelatinous organisms are overlooked, or classified as 'digested remains'. Because of this bias, it is generally thought that Thaliacea and other gelatinous zooplankton are not suitable sources of food to most marine organisms. I hope that the data presented here, together with recent reviews on medusivorous fishes by Ates (1988) and Arai (1988) will finally lay that misconception to rest.

12.2. Bacteria

Luminous bacteria are obligate parasites of species of *Pyrosoma*. The brief literature on this association has been reviewed by Monniot (1990).

12.3. Protozoa

Only a handful of protozoan parasites have been found in thaliaceans (Table 12.1). Undoubtedly there are many more, but they have been little investigated outside the Mediterranean. Théodorides (1989) and Monniot (1990) have recently reviewed this literature, so only a few comments are appropriate here.

The dinoflagellate, *Oodinium amylaceum*, is far more common than reports in the literature suggest. The specimens that I have collected are morphologically similar to species that are found on other soft-bodied planktonic organisms such as chaetognaths and ctenophores (McLean and Nielsen, 1989; Mills and McLean, 1991). I have also collected *Oodinium* spp. on the pteropod, *Corolla* sp., and on the ctenophore, *Leucothea* sp. Whether these dinoflagellates really belong to several species and are highly specific ectoparasites, or they all belong to the same species and are really non-specific micropredators is unclear. Described species of *Oodinium* fall into two general size categories. Those found on appendicularians and on *Creseis* sp. have maximal dimensions ranging between 130 and 180 μm, whereas those found on salps, chaetognaths, ctenophores and medusae have maximal sizes ranging between 320 and 450 μm. I assume that all of the specimens I have collected on salps belong to the same

Table 12.1. Protistan parasites of Thaliacea

Species	Host	Location	Reference
Phylum Mastigophora			
Oodinidae			
Oodinium amylaceum (Bargoni)	*Thalia democratica*	Mediterranean	Bargoni, 1894
	Salpa maxima	Mediterranean	Chatton, 1920
	Salps	Mediterranean	Trégouboff & Rose, 1957
	Cyclosalpa quadriluminis	Indian Ocean	Personal observation
	Cyclosalpa affinis, Salpa maxima	Equatorial Atlantic	Personal observation
	Pegea bicaudata, Thetys vagina, Salpa fusiformis	Northeast Atlantic	Personal observation
Oodinium poucheti Lemmerman	Salps	Mediterranean	Trégouboff & Rose, 1957
Phylum Ciliophora			
Order Microthoracida			
Conchophryidea			
Conchophrys davidoffi Chatton	*Pyrosoma giganteum*	Mediterranean	Chatton, 1911
Order Apostomatida			
Ephelotidae			
Tunicophrya sessilis (Collin)	*Pyrosoma elegans*	Mediterranean	Collin, 1912
	Pyrosoma elegans	Mediterranean	Trégouboff, 1916
Dendrosomatidae			
Actinobranchium salparum (Entz)	*Thalia democratica*	Mediterranean	Entz, 1884
	Thalia democratica	Mediterranean	Collin, 1912
	Thalia democratica	Mediterranean	Trégouboff, 1916
Actinobranchium salparum pyrosomae (Trégouboff)	*Pyrosoma elegans*	Mediterranean	Trégouboff, 1916
Phylum Sporozoa			
Thalicolidae			
Thalicola ensiformis (Bargoni)	*Thalia democratica*	Mediterranean	Bargoni, 1894
	Thalia democratica	Mediterranean	Ormières, 1964
	Thalia democratica	Mediterranean	Théodoridès & Desportes, 1975
Thalicola flava (Roboz)	*Pegea confoederata*	Mediterranean	Roboz, 1886
	Pegea confoederata	Mediterranean	Ormières, 1964
	Pegea confoederata	Mediterranean	Théodoridès & Desportes, 1975
	Pegea confoederata	Indian Ocean	Godeaux, personal communication
Thalicola filiformis Théodoridès & Desportes	*Cyclosalpa virgula*	Mediterranean	Théodoridès & Desportes, 1975
Thalicola salpae (Frenzel)	*Salpa maxima*	Mediterranean	Frenzel, 1885
	Salpa maxima	Mediterranean	Ormières, 1964
	Salpa fusiformis, S. maxima, Ihlea punctata	Mediterranean	Théodoridès & Desportes, 1975

species, first described by Bargoni (1894). Since McLean and Nielsen (1989) and Mills and McLean (1991) do not discuss the differences between their specimens and *O.amylaceum*, my identifications are only provisional.

The gregarine genus *Thallicola* is restricted to salps, and four species have been described. Previous reports indicate that they are always found inside the guts of salps (Monniot, 1990). However, Prof. Jean Godeaux has kindly sent me scanning electron micrographs of gregarines that closely resemble *T.flava*. These were found associated with the nucleus, but outside the gut. Ormières (1964) also found cysts that he believed were gregarines outside the lumen of the gut. Gregarines belonging to different families have also been collected from the guts of several hyperiid amphipods (*Vibilia spp.*, *Phronima* spp. and *Phronimella elongate*) and copepods (*Sapphirina* sp.) (Ormières, 1964; Théodorides and Desportes, 1968). Since all of these crustaceans feed on salps (see below), I think it quite likely that many of these are identical with gregarines found in salps, but have differing morphologies in their crustacean hosts. Indeed, Ormières (1964) commented on the similarities of certain cysts found in *Sapphirina* sp. with those found in salps.

At least one protozoan is unequivocally a predator on salps. The colonial radiolarian, *Collozoum caudalum* Swanberg and Anderson, preys on *Thalia democratica* in the equatorial Atlantic (Swanberg and Anderson, 1985).

11.4. Cnidaria

Many cnidarians probably consume Thaliacea, but only a handful are known (Table 12.2). This is simply due to the fact that the only way to investigate predation by cnidarians is to observe it directly, either in the field or in the laboratory. Cnidarians usually void their gut contents when stimulated, so animals must be collected individually, and gut contents examined as soon as possible after collection. The paucity of data in Table 12.2 reflects the small number of such studies.

In salp swarms, medusae (*Aequorea* spp. and *Pelagia noctiluca*) are frequently observed consuming large numbers of salps. These medusae lack a satiety response and continually stuff salps into their mouths. This excessive feeding results in the production of large boluses of partially digested salps, and the production of copious quantities of marine snow. I have also frequently seen cerianthid larvae feeding in salp swarms.

Extremely small medusae sometimes resemble parasites. Small *Aequorea* spp., for example, have been collected attached to several different salp species. When on the salps, they are small flattened discs, resembling small *Lampea pancerina*. If they are pried off the salp, a circular scar demonstrates that the medusa was ingesting the tissue of the salp. The behaviour of juvenile *Aequorea*, and the frequency with which adults are seen in salp swarms, strongly suggests that salps constitute a major food source for this medusa.

11.5. Ctenophora

Only four genera of ctenophores are presently known to consume salps (Table 12.3)

One genus, *Lampea*, has a curious parasitic stage which was originallty described as *Gastrodes parasiticum*. Subsequent work (Harbison and Carré, personal observations) has shown that this is really a developmental stage of *Lampea pancerina*. Another species, *Lampea lactea*, is an highly specific parasite and predator on *Salpa cylindrica*. It is evident from their behaviour that *Lampea* species depend on salps as their primary food source.

Beroe species are commonly seen eating salps and doliolids. The systematics of this ctenophore genus are too poorly known at present to identify specimens to species. In spite of the limited number of observations in Table 12.3, I think it likely that practically any ctenophore will ingest thaliaceans if it is able to capture them. Because most lobates and cestids appear to specialize on small prey (Harbison *et al.*, 1978), it is unlikely that thaliaceans constitute a significant food source for these taxa. However, thaliaceans may be important in the nutrition of many cydippids and beroids.

11.6. Mollusca

Several species of heteropods have been reported to eat thaliaceans (Table 12.4). Indeed, Seapy (1980) found that salps and doliolids were the dominant prey of *Carinaria cristata*. Although Hirsch (1915) reported that *Pterotrachea* sp. eats salps, at least two species in the genus, *P.coronata and P.hippocampus*, do not eat salps in the laboratory (Lalli and Gilmer, 1989). Thus, there may be a high degree of feeding specificity among heteropods.

I have been able to find only two reports of thaliaceans in the gut contents of cephalopods (Table 12.4). This is hardly surprising, since the ground up material in their gut contents is extremely difficult to identify. Besides these reports, the octopod, *Argonauta* sp., has been found living in *Pegea socia* in the Gulf of Mexico (Banas *et al.*, 1982), and I have also collected male argonauts in *Pegea socia* and one female in *Pegea confoederata* in the North Atlantic. Since I have seen *Argonauta* sp. eating *Pelagia noctiluca*, it is quite possible that these octopods use salps for food as well as shelter. Likewise, both male and female *Ocythoe tuberculata* have been collected in *Pegea bicaudata, Thetys vagina, Cyclosalpa virgula*

Table 12.2 Cnidarian predators of thaliaceans.

Species	Prey	Location	Reference
Class Hydrozoa			
Order Trachylina			
Geryoniidae			
Geryonia sp.	*Salpa cylindrica* solitary	Coral Sea	Personal observation
Liriope sp.	*Salpa cylindrica*	Northwest Atlantic	Personal observation
Aeginidae			
Aegina citrea Eschscholtz	Salps, ctenophores, hydromedusae	Northwest Atlantic	Larson *et al.*, 1989
Solmarisidae			
Pegantha clara Bigelow	Eat salps in the laboratory	Northwest Atlantic	Larson *et al.*, 1989
Pegantha ?laevis Bigelow	Salps	Northwest Atlantic	Larson *et al.*, 1989
Cuninidae			
Cunina duplicata Maas	Eat salps in the laboratory	Northwest Atlantic	Larson *et al.*, 1989
Cunina globosa Eschscholtz	Eat salps in the laboratory	Northwest Atlantic	Larson *et al.*, 1989
Cunina proboscidea Metschnikoff & Metschnikoff	Eat *Haeckelia rubra* in the lab.	Northeast Pacific	Mills & Goy, 1988
	Salps & doliolids	Northwest Atlantic	Larson *et al.*, 1989
	Brooksia rostrata	Northwest Atlantic	Personal observation
Solmissus albescens (Gegenbaur)	*Cavolinia* sp., ctenophores	Mediterranean	Mills & Goy, 1988
	Salps, ctenophores	Mediterranean	Personal observation
Order Hydroida			
Pandeidae			
Leuckartiara sp.	*Salpa cylindrica* solitary	Coral Sea	Personal observation
Laodiceidae			
Orchistoma sp.	*Salpa cylindrica, Cavolinia longirostris*	Northwest Atlantic	Personal observation
Aequoreidae			
Aequorea sp.	*Thalia democratica*	Mediterranean	Sigl, 1912
	Salpa cylindrica	Bahamas	Hamner *et al.*, 1975
	Salpa cylindrica, Pegea confoederata, Cyclosalpa affinis, C. polae, Cestum veneris, Ocyropsis spp., *Bolinopsis* sp., Diphyids, *Cavolinia* spp.,	North Atlantic	Personal observation

Table 12.2 *(Continued)*

Species	Prey	Location	Reference
Order Siphonophora			
Apolemiidae			
Apolemia uvaria Lesuer	Salps, 11% of gastrozooids with prey	Northeast Pacific	Purcell, 1981
Class Scyphozoa			
Order Semaeostomeae			
Pelagiidae			
Pelagia noctiluca (Forsskål)	Salps	Northeast Atlantic	Delap, 1907, in Larson, 1987
	Doliolids, cnidarians, crustaceans, larvaceans	Mediterranean	Malej, 1982
	Salps, hydromedusae, siphonophores, ctenophores, polychaetes, crustaceans	North Atlantic	Larson, 1987
	Pegea confoederata, Salpa cylindrica, Thalia democratica, doliolids, chaetognaths, foraminifera, fish, *Amphinema rugosum*	Northwest Atlantic	Personal observation
	Salps	Coral Sea	Personal observation
Class Anthozoa			
Cerianthidae			
Cerianthid larvae	*Cyclosalpa polae, Ihlea asymmetrica*, ctenophores	North Atlantic	Personal observation

Table 12.3 Ctenophoran predators of Thaliacea.

Species	Prey	Location	Reference
Phylum Ctenophora			
Class Tentaculata			
Order Cydippida			
Pleurobrachiidae			
Pleurobrachia bachei Agassiz	Doliolids minor	California	Hirota, 1974
Lampeidae			
Lampea pancerina (Chun)	*Salpa fusiformis, S. maxima, Pegea confoederata*	Mediterranean	Korotneff, 1888, 1891
	Salpa fusiformis	Japan	Komai, 1922
	Cyclosalpa floridana	Northwest Atlantic	Harbison *et al.*, 1978
	Salpa fusiformis	North Atlantic	Harbison & Carré, personal observations
Lampea lactea (Mayer)	*Salpa cylindrica*	Vietnam	Dawydoff, 1937
	Salpa cylindrica	Northeast Atlantic	Harbison *et al.*, 1978
	Slapa cylindrica	Northeast Atlantic	Harbison & Carré, personal observations
Order Lobata			
Eurhamphaeidae			
Eurhamphaea vexilligera Gegenbaur	*Thalia democratica*, siphonophores	Northwest Atlantic	Personal observation
Class Nuda			
Order Beroida			
Beroidae			
Beroe spp.	Eat salps in the laboratory	Mediterranean	Greve *et al.*, 1976
	Salps & ctenophores	North Atlantic	Harbison *et al.*, 1978
	Salps, doliolids as well as ctenophores	Northeast Atlantic	Personal observation

Table 12.4 Molluscan predators of thaliaceans.

Species	Prey	Location	Reference
Class Gastropoda			
Order Mesogastropoda			
Carinariidae			
Carinaria Cristata (Linnaeus)	Salps, doliolids dominant prey	Southern California	Seapy, 1980
Cardiapoda placenta (Lesson)	*Salpa cylindrica*	Bahamas	Hamner *et al.*, 1975
Pterotracheidae			
Firoloida desmaresti Lesueur	Salps, scyphozoa, siphonophores		Hirsch, 1915
Pterotrachea sp.	Salps, scyphozoa, siphonophores		Hirsch, 1915
Order Nudibranchia			
Glaucidae			
Glaucus atlanticus Forster	*Salpa fusiformis*	Gulf of Guinea	Le Borgne, 1982
Class Cephalopoda			
Order Teuthoidea			
Loliginidae			
Loligo opalescens	Thaliacea, hydromedusae minor	Northeast Pacific	Brodeur *et al.*, 1987
Ommastrephidae			
Illex illecebrosus	Salpidae minor	Georges Bank	Bowman, personal communication

and *Salpa maxima* in the Mediterranean and off the California coast (Jatta, 1896; Lo Bianco, 1909; Hardwick, 1970).

11.7. Arthropoda

Two major groups of arthropods, copepods and hyperiid amphipods, are commonly found with thaliaceans. The primary group of copepods that appears to be associated with thaliaceans is the Sapphirinidae (Table 12.5). This family has long been known to be parasites and predators of salps. Indeed, Sigl (1913) attempted to determine patterns of association between sapphirinids and salps from plankton collections. I have not reproduced her results in Table 12.5 as I feel that they tend to cloud the presently unclear patterns of specificity. Some species appear to be genuinely non-specific in their thaliacean associations. Good examples of this are *Sapphirina auronitens*, which occurs on colonial radiolarians as well as on salps, and *S.angusta*, which has once been found on a colonial radiolarian, as well as on 10 different salp species. Other species however appear highly specific. For example, of 128 separate records of association (where at least one fourth copepodid or older stage was found), *S.iris* was found on a species of *Pegea* 126 times and once only on *Cyclosalpa pinnata* and *C.polae*. Llikewise, *S.gastrica* appears to associate only with *Pegea* species, although this is presently based on eight records only. Although there are insufficient data at present, *S.ovatolanceolata* may be associated specifically with *Thalia democratica*

I have collected only four species in sufficient numbers to draw conclusions about their host specificity: *S.auronitens, S.angusta, S.iris,* and *S.sali.* Of these, only *S.iris* shows a high degree of specificity.

Heron (1973) described the behaviour of *S.angusta* on *Thalia democratica*, from field and laboratory observations. The juveniles behave as parasites, since they are small in relation to the salps. However adult females are predators, and can consume most of the edible tissue of *Thalia democratica* in a few hours. It would take much longer, of course, for larger salps to be consumed.

Heron (1973) suggested that the remarkable iridescence of the males made them easy to see so that the females could leave the salps to mate. I have often seen vertical stacks of these males while diving. They space themselves out about a metre apart, one above the other, and stacks of up to five males have been observed. They are extremely easy to see, since they are so iridescent. Although it might be assumed that their high visibility would make them easy for predators to see, they are difficult to catch, since a change in orientation quickly makes them invisible. This vertical stacking may be used cooperatively to attract females away from their salp hosts, and the female may go to the male with the brightest display.

In addition to sapphirinids, I have collected isolated specimens *of Farranula gracilis* on *Thalia democratica, Miracia efferata* on *Pegea confoederata, Oncaea venusta* on *Brooksia rostrata*, and *Oncaea mediterranea* on *Traustedtia multitentaculata.* These species are commonly collected on other organisms, larvacean houses, or marine snow, so I assume that their associations with salps are transient and have little significance.

In the southern ocean the large copepod *Rhincalanus gigas* has been found in a feeding association with *Salpa thompsoni* (Perissinoto and Pakhomov, 1997), where the copepods feed on the food string. It is possible that under conditions of very high particle concentrations during the onset of a phytoplankton bloom, the copepods may benefit the salps by unclogging their filtering system, hence that they are then symbionts.

Hyperiid amphipods also commonly associate with thaliaceans. Madin and Harbison (1977) and Laval (1980) have documented and reviewed previous reports of these associations, and thus they do not need to be reviewed in detail here. Some points are worthy of emphasis, however.

Two genera, *Lycaea* and *Vibilia*, are largely restricted to salps and pyrosomes, and it is likely that they have coevolved with their thaliacean hosts. Evidence for tight specificity is clearest in the genus *Lycaea*, where *L.vincenii* is found primarily on *Salpa cylindrica*, and *L.nasula* is found exclusively on *Cyclosalpa affinis* (Madin and Harbison, 1977). Species of *Lycaea* consume the tissues of their hosts, so these patterns of specificity may be established by allelochemicals produced by the salps. Other species in the genus appear to have broader specificities, but this may be due to a poor understanding of their taxonomy (Madin and Harbison, 1977). One species, *Lycaea serrala*, has never been collected on thaliaceans, and thus may live differently from other members of the genus.

Species of *Vibilia* have a characteristic behaviour on salps, in which they sit near the opening of the oesophagus and eat the food string (Madin and Harbison, 1977). If the salp is starved, species of *Vibilia* will consume the tissues of their host. In spite of this stereotyped behaviour, patterns of specificity are presently unclear, perhaps because the taxonomic problems with this genus are even greater than with *Lycaea.* Members of the genus *Cyllopus*, which are closely related to *Vibilia*, have not been proved unequivocably to associate with salps. However, Weigmann-Haass (1983) considers such an association probable, based on morphology, and Lancraft *et al.* (1989) found a close correspondence between the vertical distribution of *Cyllopus lucasii* and its presumed host, *Salpa thompsoni.*

Several other species of hyperiid amphipods are often found on salps. In these cases, the importance of the

Table 12.5 Associations of Sapphirinidae with Thaliacea.

Species	Host	Location	Reference
Sapphirinidae			
Sapphirina angusta Dana	*Cyclosalpa pinnata, Pegea bicaudata, Salpa maxima*	Mediterranean	Giesbrecht, 1892
	Salps	Mediterranean	Steuer, 1908
	Thalia democratica	Southeast Australia	Heron, 1973
	Pegea socia	Gulf of Mexico	Banas *et al.*, 1982
	Pegea confoederata, Cyclosalpa pinnata, C. polae, C. affinis, Salpa maxima, S. aspera, S. cylindrica, Brooksia rostrata, Ihlea punctata, Thalia democratica, radiolarians	North Atlantic	Personal observation
Sapphirina auronitens Claus	*Pyrosoma* sp., *Thalia democratica*	Mediterranean	Giesbrecht, 1892
	Pegea sp., *P. confoederata, Cyclosalpa affinis, Salpa maxima, S. cylindrica, S. fusiformis*, radiolarians, *Rhizosolenia* mats	North Atlantic	Personal observation
Sapphirina bicuspidata Giesbrecht	Doliolid, radiolarian	North Atlantic	Personal observation
Sapphirina gastrica Giesbrecht	*Pegea confoederata, P. socia*	North Atlantic	Personal observation
Sapphirina iris Dana	*Cyclosalpa pinnata, Pegea bicaudata, Salpa maxima*	Mediterranean	Giesbrecht, 1892
	Salps	North Atlantic	Wolfenden, 1911
	Pegea bicaudata	Japan	Furuhashi, 1966
	Pegea sp., *P. confoederata, P. bicaudata, P. socia Cyclosalpa pinnata, C. polae*	North Atlantic	Personal observation
Sapphirina lactens Giesbrecht	Nurses of *Dolchinia mirabilis*	Mediterranean	Giesbrecht, 1892
	Doliolid	North Atlantic	Personal observation
Sapphirina ovatolanceolata Dana	*Thalia democratica*	Mediterranean	Giesbrecht, 1892
	Thalia democratica	Mediterranean	Graeffe, 1902
	Salps	Mediterranean	Stever, 1908
	Thalia democratica	North Atlantic	Personal observation
Sapphirina sali Farran	*Pegea bicaudata, Cyclosalpa pinnata, C. polae, C. affinis, Salpa maxima, S. cylindrica, Ihlea punctata*	North Atlantic	Personal observation
Sapphirina scarlata Giesbrecht	*Thalia democratica*	North Atlantic	Personal observation

associations is less clear than for species of *Lycaea* and *Vibilia*. Species *of Brachyscelus* are frequently found on salps as adults, but juveniles develop on medusae (Harbison *et al.*, 1977; personal observations). Likewise, adult *Oxycephalus* appear to be common predators on salps, but the juveniles develop on ctenophores. Juvenile *Parathemisto* are also found on salps, but they are also found on medusae, and the adults appear to be free-living. I have also collected specimens of *Hyperiella dilatata* on *Salpa thompsoni* near Cape Evans, Antarctica, but these amphipods also occur on medusae and pteropods (Larson and Harbison, 1990; McClintock and Janssen, 1990).

Perhaps the best known association of hyperiid amphipods with thaliaceans is the curious behaviour of species of *Phronima and Phronimella*. While these species are predators of salps and other gelatinous zooplankton (Laval, 1980; personal observations), they also fashion barrels (used as shelters and nurseries) from the tests of salps, pyrosomes and siphonophores (Laval, 1978).

Two other hyperiid amphipods have been occasionally found on thaliaceans: *Anchylomera blossevillei* on pyrosomes (Risso, 1816), and *Pseudolycaea pachypoda* on salps and pyrosomes (Chevreux and Fage, 1925; Laval, 1980). However, I have frequently observed the former species swimming freely, and the latter has also been found on the medusa, *Lyriope tetraphylla* (Harbison *et al.*, 1977). It is probable that these species do not depend primarily on thaliaceans for food or shelter.

11.7. Echinoderms

One of the more remarkable reports in the literature is that of Duggins (1981), who found that the sea urchin, *Strongylocentrotus franciscanus*, switched from eating macroalgae to *Salpa fusiformis*, so that salps constituted over 66% of the diet for four weeks in Torch Bay, Alaska. The clear preference of this sea urchin for salps resembles the switching of the herbivorous rockfish, *Sebastes mystinus*, to gelatinous zooplankton during periods of upwelling (Hallacher and Roberts, 1985).

11.8. Vertebrates

Kashkina (1986) reviewed some of the literature on fish predation on salps, and thus did not consider pyrosomes and doliolids. I have considerably extended her literature search, and have increased her list about 4-fold (Table 12.6). The obvious conclusion that one can draw from an inspection of the table is that practically any planktivorous fish of sufficient size probably eats thaliaceans. However, more noteworthy is the fact that over 70 species in eleven different orders consume large quantities of thaliaceans. Thaliaceans are merely a subset of the gelatinous zooplankton, and I have excluded those reports where thaliaceans were not found in the gut contents. Many of the fishes which eat thaliaceans also eat other gelatinous zooplankton, such as medusae, siphonophores, ctenophores and heteropods. I am in the course of preparing a much larger paper on the adaptations of gelatinivorous fishes, which will deal with the functional morphology of the various groups, and so will not discuss this aspect of their biology here. It seems clear that gelatinivory is a fundamental life style of many marine fishes, and that it has evolved independently in many groups.

In areas where thaliaceans are particularly abundant many fishes are reported to feed heavily on them. Two such areas are in the South Pacific around New Zealand and Australia, and the Southern Ocean. Twenty one species consume large numbers of thaliaceans around New Zealand and Southern Australia and Tasmania, and eight species feed heavily on salps in the Southern Ocean. Nevertheless, inspection of Table 12.6 reveals that species that feed heavily on thaliaceans are found in all parts of the world.

Since species of the stromateoid, *Tetragonurus*, depend on thaliaceans for food and shelter (Janssen and Harbison, 1981), the records of some other stromateoids are extremely suggestive. For example, I have found *Amarsipus carlsbergi* twice, associated with the same species of salp. It is tempting to speculate that it must live in much the same way as do species of *Tetragonurus*, because it closely resembles them in external morphology. In contrast to most other stromateoids, which are very deep bodied, amarsipids and tetragonurids are relatively long and slender, which may be an adaptation to living in the body cavities of thaliaceans (Janssen and Harbison, 1981).

Most members of the Centrolophidae that are known to eat gelatinous zooplankton feed heavily on salps and pyrosomes. Particularly in the New Zealand area, centrolophids appear to specialize on pelagic tunicates.

The argentinoids rival the stromateoids in the diversity of species that feed on thaliaceans and other gelatinous zooplankton. It is likely that this group has evolved specialized adaptations for the consumption of gelatinous zooplankton.

Other groups which appear to specialize on thaliaceans include the Oreosomatidae and perhaps also the Pentacerotidae. Luvarids and molids are also specialists on gelatinous zooplankton.

Many scorpaenids feed heavily on thaliaceans, and at least one species, *Sebastes mystinus*, appears to prefer gelatinous zooplankton to its normal diet of kelp. Likewise, many lutjanids feed on salps and pyrosomes, although only one species appears to feed heavily on them. Quite possibly their feeding on thaliaceans is

Table 12.6 Fishes that eat Thaliaceans.

Species	Prey	Location	Reference
Class Elasmobranchii			
Order Squaliformes			
Squalidae			
Centrophorus moluccensis Bleeker	*Pyrosoma* sp. minor	Indian Ocean	Bass *et al.*, 1976
Etmopterus baxteri Garrick	Thaliacea minor	New Zealand	Clark *et al.*, 1989
Squalus acanthias Linnaeus	*Salpa* sp., *Pleurobrachia* sp., *Chrysaora* sp. all minor	North Sea	Rae, 1967
	Salps, siphonophores, ctenophores minor	Northwest Atlantic	Bowman *et al.*, 1984
Class Holocephali			
Order Chimaeriformes			
Callorhynchidae			
Callorhynchus milii (Bory de St. Vincent)	Salps, Scyphozoa	New Zealand	Graham, 1939
Chimaeridae			
HYDROLAGUS OGILBYI (Waite)	Pyrosomes, decapods major prey	New Zealand	Shuntov, 1979
Hydrolagus lemures Whitley	Gelatinous plankton in diet	New Zealand	Shuntov, 1979
Class Actinopterygii			
Order Notacanthiformes			
Notacanthidae			
Notacanthus sexspinis Richardson	Mainly copepods, but *Pyrosoma* sp. minor	Southwest Africa	Macpherson, 1983
Order Anguilliformes			
Congridae			
Bathyuroconger vicinus (Vaillant)	Benthic feeder, but *Pyrosoma* sp. 18% weight in largest size class	Southwest Africa	Macpherson, 1983
Order Clupeiformes			
Clupeidae			
Alosa pseudoharengus (Wilson)	Thaliacea 1%, Cnidaria 1.5% total volume	Norhwest Atlantic	Bowman, personal communication
Clupea harengus Linnaeus	*Ihlea asymmetrica, Dolioletta gegenbauri*	North Atlantic	Fraser, 1949
	Ihlea punctata	North Atlantic	Fraser, 1962a*
	Salpa fusiformis willingly eaten	Norwegian Sea	Pavshtiks & Rudakova, 1962*
Ethmalosa fimbriata (Bowdich)	*Lensia* sp. + *Obelia* sp. = 1.4% of diet, *Dolioletta* minor	Coast of Ghana	Blay & Eyeson, 1982
Order Salmoniformes			
Suborder Argentinoidei			
Argentinidae			
ARGENTINA ELONGATA Hutton	Salpidae, Pyrosomatidae principal food	New Zealand	Shuntov, 1979*
	Salps 97% Index of Relative Importance (IRI)	S.E. New Zealand	Clark, 1985
ARGENTINA SILUS Ascanius	Salps main prey	North Atlantic	Hickling, 1927 in Wood & Raitt, 1968
	Salpa fusiformis	North Atlantic	Wood & Raitt, 1968
	Salps, *Clione*	North Atlantic	Wheeler, 1969
	Salps/ctenophores 24%, unidentifiable 38% total number	Northeast Atlantic	Mauchline & Gordon, 1983b*
	Siphonophores 1%, Ctenophora 2%, Thaliacea 1%, animal remains 11% total volume	Northwest Atlantic	Bowman, personal communication
Argentina sphyraena Linnaeus	*Pyrosoma elegans* frequent	Mediterranean	Valiani, 1940
	Primarily a bottom feeder, also eats pyrosomes	North Atlantic	Wheeler, 1969

Table 12.6 *(Continued)*

Species	Prey	Location	Reference
Bathylagidae			
BATHYLAGUS ANTARCTICUS Günther	Coelenterates 10% of diet, *Salpa thompsoni* minor	Weddell Sea	Hopkins & Torres, 1989
	Siphonophores, salps frequent	Weddell Sea	Lancraft *et al.*, 1991
Bathylagus euryops Goode & Bean	Salps/ctenophores 4.1%, unidentifiable 56% total number	Northeast Atlantic	Mauchline & Gordon, 1983b
LEUROGLOSSUS STILBIUS (Gilbert)	Salps major prey	California coast	Cailliet, 1972*
	Salps 30%, *Oikopleura* 20% volume, siphonophores & radiolarians minor	California	Cailliet & Ebeling, 1990
Alepocephalidae			
ALEPOCEPHALUS AUSTRALIS Barnard	Thaliacea (*Salpa thompsoni, Ihlea magalhanica*) 58%, hyperiid amphipods (*Vibilia stebbingi, Cyllopus magellanicus*) 32%, Scyphozoa 5% IRI	New Zealand	Clark *et al.*, 1989
ALEPOCEPHALUS BAIRDII Goode & Bean	Pyrosomes 3.3% frequency occurrence in adults, 88% in juveniles, coelenterates 79% in adults, 13% in juveniles	E. Central Atlantic	Golovan' & Pakhorukov, 1975
	Periphylla periphylla 37%, hyperiids 37% frequency	North Atlantic	Du Buit, 1978
	Salps or ctenophores 25% by vol., salpidae, medusae definitely present	Northeast Atlantic	Mauchline & Gordon, 1983*
	Food mostly coelenterates, also tunicates	Northeast Atlantic	Markle & Quéro, 1984
Alepocephalus productus Gill	*Pyrosoma*, cephalopoda		Verighina & Golovan', 1978
ALLEPOCEPHALUS ROSTRATUS Risso	Pyrosomes 32%, ctenophores, *Atolla* 38% frequency	E. Equatorial Atlantic	Golovan' & Pakhorukov, 1980
	Unidentifiable soft tissues	Northeast Atlantic	Mauchline & Gordon, 1983b
	Pyrosoma sp. 27%, scyphozoa 6% weight in some size classes	Southwest Africa	Macpherson, 1983
CONOCARA FIOLENTI (Sazonov & Ivanov)	Salps 98% wt.	Bahamas	Crabtree & Sulak, 1986
CONOCARA MACROPTERUM (Vaillant)	Salps 34% wt.	Bahamas	Crabtree & Sulak, 1986
Rouleina attrita (Vaillant)	*Pyrosoma*, Cephalopoda, shrimps		Verighina & Golovan', 1978
Talismania bifurcata Parr	*Pyrosoma*, jellyfishes		Verighina & Golovan', 1978
XENODERMICHTHYS COPEI (Gill)	Salps & lots of unidentifiable jelly	Northeast Atlantic	Mauchline & Gordon, 1983
Suborder Salmonoidei			
Salmonidae			
ONCORHYNCHUS KETA (Walbaum)	Sea squirts 30% of diet in September	Bering Sea	Karpenko & Piskunova, 1984
	Salps 56% gut contents	Gulf of Alaska	Pearcy *et al.*, 1984
Oncorhynchus kisutch (Walbaum)	Many unidentified hyperiids	Northeast Pacific	Peterson *et al.*, 1982
	Cnidaria up to 7% total weight, Thaliacea minor	Northeast Pacific	Brodeur *et al.*, 1987
	Salps, *Vellela*, ctenophores, siphonophores, thecosomes	Northeast Pacific	Brodeur & Pearcy, 1990
Oncorhynchus tshawytscha (Walbaum)	Salps, ctenophores, siphonophores, *Limacina*	Northeast Pacific	Brodeur & Pearcy, 1990

Table 12.6 *(Continued)*

Species	Prey	Location	Reference
Order Stomiiformes			
Suborder Gonostomatoidei			
Sternoptychidae			
Argyropelecus aculeatus Valenciennes	*Salpa fusiformis*, Doliolida, Thaliacea, *Vibilia*, siphonophores minor	Gulf of Mexico	Hopkins & Baird, 1985b
Argyropelecus affinis Garman	*Doliolum* sp. 0.7% total no.	Southern California	Hopkins & Baird, 1977*
Argyropelecus hemigymnus Cocco	*Vibilia*, much unidentfiable material	Northeast Atlantic	Roe *et al.*, 1984
Argyropelecus sladeni Regen	Siphonophores 0.3%, *Vibilia* 1% total number	Southern California	Hopkins & Baird, 1977
Maurolicus muelleri (Gmelin)	Salpidae <0.1% biomass	Tasmania	Young & Blaber, 1986
Sternoptyx diaphana Hermann	Thaliacea, *Vibilia* minor	Gulf of Mexico	Hopkins & Baird, 1985b
Suborder Photichthyoidei			
Phosichthyidae			
Ichthyococcus ovatus (Cocco)	Tunicates only gut contents	Indian Ocean	Bradbury *et al.*, 1971
Order Aulopiformes			
Alepisauridae			
ALEPISAURUS FEROX Lowe	Salps, doliolids, pyrosomes minor, heteropods frequent	North Atlantic	Haedrich, 1964
	Salps, ctenophores, medusae, heteropods, squids, pelagic amphipods	California	Fitch & Lavenberg, 1968
	Salps, *Pyrosoma spinosum*, heteropods, pteropods common, doliolids, medusae also found	Sugura Bay, Japan	Kubota & Uyeno, 1970
Alepisaurus sp.	Salps & pyrosomes 16%, medusae 6% freq. of occur.	Northwest Atlantic	Matthews *et al.*, 1977
Order Myctophiformes			
Myctophidae			
Ceratoscopelus townsendi (Eigenmann & Eigenmann)	Salps, pyrosomes minor	Indian Ocean	Leg & Rivaton, 1969
CERATOSCOPELUS WARMINGII (Lütken)	Salps 4%, doliolids 0.2%, siphonophores 11% total no.	Gulf of Mexico	Hopkins & Baird, 1977*
	Salps (mainly *Thalia democratica*) eaten outside warm-core eddy, also siphonophores	Tasman Sea	Brandt, 1981
	Salps up to 12% total number, also siphonophores	Equatorial Atlantic	Kinzer & Schulz, 1985
	Salps 5%, siphonophores 10%, medusae 2% frequency, salps 15% total weight in one area	Equatorial Atlantic	Duka, 1986
Diaphus danae Tåning	Salpidae 0.1% biomass	Tasmania	Young & Blaber, 1986
Diaphus dumerilii (Bleeker)	Salps 1%, siphonophores 2% total number	Gulf of Mexico	Hopkins & Baird, 1977*
Diaphus garmani Gilbert	Salps minor	Equatorial Pacific	Nakamura, 1970*
Diaphus rafinesquei (Cocco)	Salps, pyrosomes 4% wt.	E. Indian Ocean	Leg & Rivaton, 1969
Diaphus taaningi Norman	*Doliolum* sp. 1% total weight	Cariaco Trench	Baird *et al.*, 1975
	Doliolum 0.1%, *Oikopleura* 71% total number	Cariaco Trench	Hopkins & Baird, 1977

Table 12.6 *(Continued)*

Species	Prey	Location	Reference
Diaphus theta Eigenmann & Eigenmann	Salps minor	Northeast Pacific	Tyler & Pearcy, 1975
Gymnoscopelus braueri (Lönnberg)	Salps minor	Weddell Sea	Lancraft *et al.*, 1991
Lampanyctodes hectoris (Günther)	Salpidae 0.4%, Siphonophora 0.6% biomass	Tasmania	Young & Blaber, 1986
Lampanyctus alatus Goode & Bean	Thaliaceans & coelenterates minor	Gulf of Mexico	Hopkins & Baird, 1985a
Lepidophanes guentheri (Goode & Bean)	Slaps 0.2%, siphonophores 1% total number	Gulf of Mexico	Hopkins & Baird, 1977*
Myctophum affine (Lütken)	Salps 4% total volume	Equatorial Atlantic	Greze, 1983
MYCTOPHUM ASPERUM Richardson	Salps 34%, siphonophores 5% total volume	Equatorial Atlantic	Greze, 1983
SCOPELOPSIS MULTIPUNCTATUS	Salps, pyrosomes, medusae minor	E. Indian Ocean	Leg & Rivaton, 1969
Brauer	Salps (mainly *Thalia democratica*) minor prey outside warm-core eddy, also siphonophores eaten	Tasman Sea	Brandt, 1981
Stenobrachius leucopsaurus (Eigenmann & Eigenmann)	Salps & siphonophores very minor	California	Cailliet & Ebeling, 1990
Symbolophorus boops (Richardson)	Salps	Southern Ocean	Rowedder, 1979
Order Gadiformes			
Suborder Gadoidei			
Gadidae			
Gadus morhua Linnaeus	Salps	Northeast Atlantic	Fraser, 1960, 1961, 1962a*
Melanogrammmus aeglefinus	*Iasis zonoria* minor	Northwest Atlantic	McKenzie & Homans, 1937*
(Linnaeus)	*Iasis zonaria* minor	Northwest Atlantic	Homans & Needler, 1943
	Thaliacea minor	Northwest Atlantic	Bowman & Michaels, 1984
MICROMESISTIUS AUSTRALIS Norman	Salps important	New Zealand	Shpak, 1976*
	Salps	New Zealand	Inada & Nakamura, 1975
	No salps, but Hyperiidae 25%	South Atlantic	Shust, 1978
	Salps, pyrosomes common food	New Zealand	Shuntov, 1979*
	Iasis zonaria, Salpa thompsoni <1% IRI, but hyperiid amphipods 40% IRI	S.E. New Zealand	Clark, 1985a
Micromesistius poutassou (Risso)	*Vibilia armata*	Mediterranean	Tournier, 1968
Pollachius virens (Linnaeus)	Urochordata 0.5% total volume	Northwest Atlantic	Bowman, personal communication
Phycidae			
Urophycis tenuis (Mitchill)	Thaliacea minor	Northwest Atlantic	Bowman & Michaels, 1984
Merlucciidae			
Macruronus novaezelandii (Hector)	Salps, pyrosomes common food	New Zealand	Shuntov, 1979*
	Iasis zonaria, Salpa thompsoni 5% IRI, but hyperiid amphipods 19% IRI	S.E. New Zealand	Clark, 1985
	Pyrosoma atlanticum 0.3% frequency of occurrence	Tasmania	Blaber & Bulman, 1987
Merluccius productus (Ayres)	Thaliacea minor	Northeast Pacific	Brodeur *et al.*, 1987

Table 12.6 *(Continued)*

Species	Prey	Location	Reference
Suborder Macrouroidei			
Macrouridae			
Caelorinchus aspercephalus Waite	Salpidae 9% IRI	N.E. New Zealand	Clark, 1985
Caelorinchus fasciatus (Günther)	Salps, pyrosomes in diet	New Zealand	Shuntov, 1979*
	Thaliacea minor, hyperiid amphipods 86% IRI	New Zealand	Clark *et al.*, 1989
Caelorinchus mirus (McCulloch)	Salps, pyrosomes in diet	New Zealand	Shuntov, 1979*
Coryphaenoides armatus (Hector)	Medusae minor	Northeast Pacific	Pearcy & Ambler, 1974
CORYPHAENOIDES RUPESTRIS Gunnerus	Salps 30% weight	North Atlantic	Podrazhanskaya, 1969*
	Salps/ctenophores minor	Northeast Atlantic	Mauchline & Gordon, 1984a
Lepidorhynchus denticulatus (Richardson)	Salps, pyrosomes in diet	New Zealand	Shuntov, 1979*
	Salpidae minor, but hyperiid amphipods 27% IRI	S.E. New Zealand	Clark, 1985
MACROURUS CARINATUS Günther	Thaliacea (*Salpa thompsoni, Iasis zonaria, Thetys vagina, Ihlea magalhanica*) 52%, hyperiid amphipods 38% IRI		
Macrourus holotrachys Günther	Thaliacea minor	South Atlantic	Dudochkin, 1988
Order Ophidiiformes			
Suborder Ophidioidei			
Ophidiidae			
Selachophidium guentheri Gilchrist	*Pyrosoma* sp. 9% weight	Southwest Africa	Macpherson, 1983
Order Beloniformes			
Exocoetidae			
Cheilopogon agoo (Temminck & Schlegel)	Hyperiid amphipods, pteropods, radiolarians	Japan	Suyehiro, 1942
Exocoetus monocirrhus Richardson	Salpidae minor	Tropical Pacific	Gorelova, 1980
Exocoetus volitans Linnaeus	Salps minor	Tropical Atlantic	Evans & Sharma, 1963
Flying fishes	Eat salps		Fraser, 1962b
	salps important	Caribbean	Hall, in Foxton, 1966
Hirundichthys affinis (Günther)	*Salpa cylindrica* minor	West Indies	Lewis *et al.*, 1962
Scomberesocidae			
Cololabis saira (Brevoort)	Thaliacea in 14% of stomachs	Japan	Yasuda, 1960
	Vibilia armata up to 4%, unidentified material up to 80% weight	Northeast Pacific	Brodeur *et al.*, 1987
Scomberesox saurus (Walbaum)	Heteropods, amphipods, radiolarians, *Sapphirina*	Mediterranean	Lo Bianco, 1909
Order Beryciformes			
Suborder Berycoidei			
Trachichthyidae			
Hoplostethus sp.	Salps, pyrosomes in diet	New Zealand	Shuntov, 1979*

Table 12.6 *(Continued)*

Species	Prey	Location	Reference
Suborder Stephanoberycoidei			
Melamphaidae			
Poromitra crassiceps (Günther)	Salps, unidentified soft tissue, amphipods, copepods	Northeast Atlantic	Mauchline & Gordon, 1984b
SCOPELOGADUS BEANII (Günther)	Siphophores in 7% of stomachs, 89% unidentified tissues, possibly salps	Northeast Atlantic	Mauchline & Gordon, 1984b
	Gelatinous zooplankton (mainly salps) >20%, hyperiid amphipods >35% volume	Northwest Atlantic	Gartner & Musick, 1989
Order Cetomimiformes			
Cetomimidae			
Gyrinomimus sp.	Crustaceans, but salps & fish found in stomachs, perhaps due to feeding in net		Paxton, 1989
Order Zeiformes			
Oreosomatidae			
ALLOCYTTUS NIGER James, Inada & Nakamura	*Salpa thompsoni 37%, hyperiid amphipods (Vibilia stebbingi, Cyllopus magellanicus, Themisto gaudichaudii)* 40% IRI	New Zealand	Clark *et al.*, 1989
ALLOCYTTUS VERRUCOSUS (Gilchrist)	Salps, pyrosomes primary food, coelenterates common	New Zealand	Shuntov, 1979*
NEOCYTTUS RHOMBOIDALIS (Gilchrist)	Tunicata 45%, Coelenterata 7% frequency	S.E. New Zealand	Shuntov, 1979*
	Salps 21%, *Pyrosoma atlanticum* 15% total calories	Tasmania	Blaber & Bulman, 1987
PSEUDOCYTTUS MACULATUS Gilchrist	Thaliacea (mainly *Salpa thompsoni*) 88%, hyperiid amphipods 9% IRI, Scyphozoa minor	New Zealand	Clark *et al.*, 1989
Order Scorpaeniformes			
Suborder Scorpaenoidei			
Scorpaenidae			
HELICOLENUS DACTYLOPTERUS (Delaroche)	Pyrosomes & crabs	West Africa	Cadenat, 1954
	Pyrosomes in 3 of 58 fish with food in guts	Morocco	Collegnon & Aloncle, 1960
	Benthic feeder, but *Pyrosoma* in 1 fish	Northwest Africa	Merrett & Marshall, 1981
	Pyrosoma up to 6% weight	Southwest Africa	Macpherson, 1983
	Salpidae 46% total volume	Georges Bank	Bowman, personal communication
Helicolenus lengerichi Norman	Mesopelagic fishes, pyrosomes, shrimps, squids	25°S, 84°W	Golovan' & Pakhorukov, 1987
	Pyrosoma usually minor, but up to 20% diet rarely	Southeast Pacific	Golovan' *et al.*, 1991
HELICOLENUS PAPILLOSUS (Bloch & Schneider)	Salps, pyrosomes principal food	New Zealand	Shuntov, 1979*
HELICOLENUS PERCOIDES Richardson	Salp <0.1%, *Pyrosoma atlanticum* 19% caloric content	Tasmania	Blaber & Bulman, 1987
Sebastes alutus (Gilbert)	*Vibilia*	Northeast Pacific	Brodeur & Percy, 1984
Sebastes diploproa (Gilbert)	*Vibilia propinqua* common	Northeast Pacific	Brodeur & Pearcy, 1984
SEBASTES ENTOMELAS (Jordan & Gilbert)	Salps dominant prey	North California	Adams, 1987
Sebastes flavidus (Ayres)	Siphonophores, Ctenophores, medusae, *Vibilia*	Northeast Pacific	Brodeur & Pearcy, 1984

Table 12.6 *(Continued)*

Species	Prey	Location	Reference
SEBASTES MYSTINUS (Jordan & Gilbert)	Salps, medusae major prey	California	Gotshall *et al.*, 1965
	Salps, doliolids, larvaceans 52%, Siphonophores, medusae 0.7% volume	California	Love & Ebeling, 1978
	Pelagic tunicates 79% during upwelling, algae 88% total weight during non-upwelling	California	Hallacher & Roberts, 1985
	Maximum diet volumes: Thaliacea 35%, *Velella velella* 5%, siphonophores 18%, hydromedusae 0.3%, Scyphozoa 19%, ctenophores 6%, heteropods 2%, pteropods 3%, hyperiid amphipods 2%, plants 52%	California	Hobson & Chess, 1988
SEBASTES SERRANOIDES (Eigenmann & Eigenmann)	Salps, doliolids, larvaceans 5% volume	California	Love & Ebeling, 1978
	Pelagic tunicates 21% weight	California	Hallacher & Roberts, 1985
Suborder Anoplopomatoidei			
Anoplopomatidae			
ANOPLOPOMA FIMBRIA (Pallas)	Cnidaria up to 19%, Thaliacea up to 10% total weight, Ctenophora minor	Northeast Pacific	Brodeur *et al.*, 1987
	Salpidae & Cnidaria minor	California	Cailliet *et al.*, 1988
Suborder Cottoidei			
Cyclopteridae			
PARALIPARIS CALLIDUS Cohen	Salps, medusae or ctenophores major prey	Northwest Atlantic	Wenner, 1979
PARALIPARIS COPEI Goode & Bean	Salps, medusae or ctenophores major prey	Northwest Atlantic	Wenner, 1979
Order Perciformes			
Suborder Percoidei			
Polyprionidae			
Polyprion oxygeneios (Bloch & Schneider)	Pyrosomes important	S. New Zealand	Bary, 1960
Serranidae			
CAPRODON LONGIMANUS (Günther)	Salps important	Tasman Sea	Dudarev, 1979*
	Salps up to 60% total volume, always major prey	Tasman Sea	Markina, 1984*
	Feed almost exclusively on doliolids	New Zealand	Kingsford & MacDiarmid, 1988
CAESIOPERCA LEPIDOPTERA (Bloch & Schneider)	Feed almost exclusively on doliolids	New Zealand	Kingsford & MacDiarmid, 1988
Paralabrax clathratus (Girard)	Salps, doliolids, larvaceans 8%, Siphonophores, medusae 0.4% volume	California	Love & Ebeling, 1978
Paranthias furcifer (Valenciennes)	Salps 12.9% total volume	West Indies	Randall, 1967
Epigonidae			
Epigonus denticulatus (Dieuzeide)	Salps 6% frequency	South Pacific	Pavlova, 1979*
Epigonus lenimen (Whitley)	*Pyrosoma atlanticum* 0.2% caloric content	Tasmania	Blaber & Bulman, 1987
Epigonus telescopus (Risso)	Pyrosomes 7.2% total weight	Mediterranean	Macpherson, 1981
Apogonidae			
Phaeoptyx conklini (Silvester)	Hyperiids, 18.8%, larvaceans, 7.3%	West Indies	Randall, 1967

Table 12.6 *(Continued)*

Species	Prey	Location	Reference
Malacanthidae			
Lopholatilus chamaeleontceps Goode & Bean	*Iasis zonaria* numerous, but mainly eats crabs	Northwest Atlantic	Linton, 1901
	Salps, but primarily benthic feeder	Northwest Atlantic	Bigelow & Schroeder, 1953
	Pyrosomes	Northwest Atlantic	personal observations
Carangidae			
Elegatis bipinnulata (Quoy & Gaimard)	Seen eating colonial salps	Pacific, Costa Rica	Hunter & Mitchell, 1967*
PARASTROMATEUS NIGER (Bloch)	Thaliacea 43%, undigested matter 26%, prawns 15%, amphipods 4% of diet	Indian Ocean	Sivaprakasam, 1963
	Salps 43% of diet	Bombay	Lodh *et al.*, 1988
Usacaranx lutescens (Richardson)	Salps	New Zealand	Graham, 1939
Bramidae			
Brama sp.	*Pegea confoederata*	North Atlantic	personal observations
Brama sp.	Salps, amphipods, euphausiids, squids, fishes	S.E. Pacific	Pavlov, 1991
Emmelichthyidae			
EMMELICHTHYS NITIDUS CYANESCENS (Guichenot)	Pyrosomes 24.2%	25° 42′S, 85° 26′W	Golovan' & Pakhorukov, 1987
EMMELICHTHYS STRUHSAKERI Heemstra & Randall	Salps up to 31% of diet	Coral Sea	Markina & Boldyrev, 1980*
Lutjanidae			
Aprion virescens Valenciennes	Salps	Western Indian Ocean	Talbot, 1960
Apsilus dentatus Guichenot	Tunicates	Caribbean	Allen, 1985
Etelis carbunculus Cuvier	Pelagic urochordates	Indo-Pacific	Allen, 1985
Lutjanus bohar (Forsskal)	Pyrosomes, doliolids, *Cavolinia* minor	Western Indian Ocean	Talbot, 1960
	Urochordates	Indo-Pacific	Allen, 1985
Lutjanus campechanus (Poey)	*Pyrosoma* sp. called 'tapioca' by fishermen, common in stomachs	Caribbean	Anon, 1970 in Kami, 1973
	Planktonic urochordates	Atlantic, Caribbean	Allen, 1985
Lutjanus rivulatus (Cuvier)	In one sample, many fish filled with salps	Western Indian Ocean	Talbot, 1960
Lutjanus sanguineus (Cuvier & Valenciennes)	Salps, doliolids, pteropods, medusae	Western Indian Ocean	Talbot, 1960
Lutjanus vivanus (Cuvier)	Tunicates, pelagic urochordates	Western Atlantic	Allen, 1985
Ocyurus chrysurus (Bloch)	Salps 3%, siphonophores 7%, ctenophores 3%, heteropods 1%, pteropods 6% total volume	West Indies	Randall, 1967
Pristipomoides auricilla (Jordan, Evermann & Tanaka)	*Pyrosoma* sp. in 3 of 5 stomachs, salps in 1	Guam	Kami, 1973
	Pelagic tunicates, salps	Indo-Pacific	Allen, 1985
PRISTIPOMOIDES FILAMENTOSUS (Valenciennes)	*Pyrosoma* sp. in 5 of 5 stomachs, major prey, salps in 3 stomachs	Guam	Kami, 1973
Pristipomoides flavipinnis Shinohara	*Pyrosomea* sp. in 1 of 6 stomachs	Guam	Kami, 1973
	Pelagic tunicates	Indo-Pacific	Allen, 1985

Table 12.6 *(Continued)*

Species	Prey	Location	Reference
Pristipomoides multidens (Day)	Urochordates	Indo-Pacific	Allen, 1985
Pristipomoides sieboldii (Bleeker)	*Pyrosoma* sp.	Guam	Kami, 1973
	Pelagic urochordates	Indo-Pacific	Allen, 1985
Pristipomoides zonatus (Valenciennes)	Planktonic urochordates	Indo-Pacific	Allen, 1985
Rhomboplites aurorubens (Cuvier)	Salps, coelenterates minor	Northwest Atlantic	Grimes, 1979
Caesionidae			
CAESIO CUNING (Bloch)	Salps, doliolids, pteropods, heteropods	Great Barrier Reef	Hamner *et al.*, 1988
CAESIO ERYTHROGASTER Cuvier & Valenciennes	Salps, tunicates primary diet	Great Barrier Reef	Williams & Hatcher, 1983
Sparidae			
CHRYSOPHRYS AURATUS (Bloch & Schneider)	Salps predominated in one set of fish, but otherwise minor, scyphozoa & hydromedusae also minor	New Zealand	Godfriaux, 1969
	Salps 2% of diet	New Zealand	Godfriaux, 1974a
	Guts often filled with *Thalia democratica*	New Zealand	Cassie, 1956
DENTEX MACROPHTHALMUS (Bloch)	*Pyrosoma* sp. 50% weight in smallest size class	Southwest Africa	Macpherson, 1983
Dentex marocanus Valenciennes	Pyrosomes, crustaceans, molluscs	West Africa	Cadenat, 1954
DENTEX sp.	Pyrosomes, medusae, crustaceans	West Africa	Cadenat, 1954
Stenotomus chrysops (Linnaeus)	Salps minor	Northwest Atlantic	Bowman & Michaels, 1984
Scorpididae			
Scorpis lineolatus Kner	Salps	South Australia	Coleman, 1980
Ephippididae			
CHAETODIPTERUS FABER (Broussonet)	Salps 12.6%, sponges 33% total volume	West Indies	Randall, 1967
Platax teira (Forsskal)	Salps, sea jellies, algae, zooplankton	Northeast Australia	Coleman, 1981
Chaetodontidae			
Chaetodon corallicola Snyder	Salps 0.2% diet volume	Hawaii	Hobson, 1974
Chaetodon miliaris Quoy & Gaimard	Salps 35%, hyperiids 0.4% diet volume	Hawaii	Hobson, 1974
Chaetodon unimaculatus Bloch	*Thalia democratica*	Hawaii	Yount, 1958*
Pentacerotidae			
PENTACEROS JAPONICUS Döderlein	Salps & pyrosomes 33% vol.	25° 42′ S 85° 26′ W	Golovan' & Pakhorukov, 1987
PSEUDOPENTACEROS RICHARDSONI (Smith)	Salps, pyrosomes	Hawaii	Pontekorvo, 1974*
	Salps, pyrosomes	Hawaii	Borets, 1975*
	Salps important	Hawaii	Fedosova, 1976*
Pomacentridae			
Chromis dispilus (Griffin)	Doliolids 2%, gelatinous zooplankton 1% of diet	New Zealand	Kingsford & MacDiarmid, 1988
Chromis multilineatus (Guichenot)	Tunicates 3%, Siphonophores 2%, Pteropods 2% total vol.	West Indies	Randall, 1967
Pomacentrus fuscus Cuvier	Tunicates 0.2%, scyphozoans 1% total volume	West Indies	Randall, 1967
Cheilodactylidae			
Nemadactylus macropterus (Bloch & Schneider)	Salps 1.5%, coelenterates 0.1% of diet	New Zealand	Godfriaux, 1974c
	Salps 0.6% total volume	New Zelanad	Godfriaux, 1974b

Table 12.6 *(Continued)*

Species	Prey	Location	Reference
Suborder Labroidei			
Labridae			
Clepticus parrai (Bloch & Schneider)	Salps 4.7% vol., also siphonophora	West Indies	Randall, 1967
Suborder Notothenioidei			
Nototheniidae			
Dissostichus eleginoides Smitt	Cnidaria 1%, salps 0.3% occurrence	Crozet & Kerguelen	Duhamel & Hureau, 1985
GOBIONOTOTHEN GIBBERIFRONS	Salps 0.5%, pyrosomes 0.6%, ctenophores 20% of fish	South Georgia	Permitin & Tarverdiyeva, 1972*
(Lönnberg)	Salps secondary food	South Orkney	Permitin & Tarverdiyeva, 1978
	Salps 35% freq. occur	South Shetlands	Tarverdiyeva & Pinskaya, 1980
	Tunicata, Coelenterata minor	Antarctic Peninsula	Daniels, 1982
	Salps minor	South Shetlands	Casaux *et al.*, 1990
LEPIDONOTOTHEN KEMPI (Norman)	Salps 8%, ctenophores 13%, cnidaria 3% of fish	South Georgia	Permitin & Tarverdiyeva, 1972
	Salps primary food, cnidaria tertiary food	South Orkney	Permitin & Tarverdiyeva, 1978*
	Salps, ctenophores	South Georgia	Shust & Pinskaya, 1978
	Salps 18%, cnidaria 4% frequency of occurrence	Antarctic Peninsula	Tarverdiyeva & Pinskaya, 1980
	Tunicata 1% volume	Antarctic Peninsula	Daniels, 1982
LEPIDONOTOTHEN LARSENI (Lönnberg)	Salps secondary food	South Orkney	Permitin & Tarvediyeva, 1978*
	Salps 64% frequency of occurrence	South Shetland	Tarverdiyeva & Pinskaya, 1980
	Tunicata, Ctenophora, Coelenterata minor	Antarctic Peninsula	Daniels, 1982
	Salps important	Lena Bank	Shandikov, 1986
Lepidonotothen nudifrons (Lönnberg)	Salps 2%, ctenophores 1% of fish	South Georgia	Permitin & Tarverdiyeva, 1972
	Tunicata minor	Antarctic Peninsula	Daniels, 1982
LEPIDONOTOTHEN SQUAMIFRONS	Salps, ctenophores important	Kerguelen	Duhamel, 1981
(Günther)	Ctenophora, Cnidaria, salps minor	Crozet Island	Duhamel & Pletikosic, 1983
	Ctenophora 7%, Cnidaria 3%, salps 34% occurrence	Crozet	Duhamel & Hureau, 1985
	Ctenophora 14%, Cnidaria 24%, salps 40% occurrence	Kerguelen	Duhamel & Hureau, 1985
Notothenia coriiceps Richardson	Tunicata, Coelenterata, Ctenophora minor	Antarctic Peninsula	Daniels, 1982
	Salps 0.7%, hyperiid amphipods 65% weight	Southern Ocean	Williams, 1983
	Salps minor	South Shetlands	Casaux *et al.*, 1990
NOTOTHENIA MICROLEPIDOTA Hutton	Salps, pyrosomes 27% frequency	S.E. New Zealand	Shuntov, 1979*
	Salps 38%, hyperiid amphipods 22% IRI	New Zealand	Clark, 1985
NOTOTHENIA ROSSII Richardson	Salps minor, ctenophores major	South Georgia	Tarverdiyeva, 1972*
	Salps minor	South Georgia	Permitin & Tarverdiyeva, 1972*
	Ctenophora 27%, hyperiid amphipods 55% freq. occur.	South Georgia	Hoshiai, 1979
	Salps 21% frequency of occurrence	South Shetlands	Tarverdiyeva & Pinskaya, 1980
	Cnidaria 14%, Ctenaria 7%, Salps 4% occurrence	Kerguelen	Duhamel, 1981
	Ctenophora 13%, Cnidaria 42%, salps 18% occurrence	Kerguelen	Duhamel & Hureau, 1985
Paranotothenia magellanica (Foster)	Thaliacea, Hydrozoa minor	Tierra del Fuego	Moreno & Jara, 1984
Patagonotothen brevicauda (Lönnberg)	Thaliacea moderate	Tierra del Fuego	Moreno & Jara, 1984

Table 12.6 *(Continued)*

Species	Prey	Location	Reference
Patogonotothen longipes (Steindachner)	Thaliacea moderate	Tierra del Fuego	Moreno & Jara, 1984
Patagonotothen tessellata (Richardson)	Thaliacea, Hydrozoa moderate	Tierra del Fuego	Moreno & Jara, 1984
Trematomus bernacchii Boulanger	Tunicata 3% total volume	Antarctic Peninsula	Daniels, 1982
TREMATOMUS EULEPIDOTUS Regan	Salps tertiary food	South Orkney	Permitin & Tarverdieva, 1978*
	Salps 29% frequency of occurrence	South Shetlands	Tarverdiyeva & Pinskaya, 1980
	Salps minor	Weddell Sea	Kock *et al.*, 1984
	Salpa thompsoni 4%, doliolids 8%	Antarctic	Kozlov & Naumov, 1988
	Hyperiid amphipods, pteropods (*Clione* & *Cavolinia*?) major prey, salps secondary, ctenophores minor	Southern Ocean	Roshchin, 1991
Trematomus scotti (Boulenger)	Tunicata, coelenterata minor	Antarctic Peninsula	Daniels, 1982
Bathydraconidae			
RACOVITZIA GLACIALIS Dollo	Salps major prey	S. Weddell Sea	Hubold & Ekau, 1990
Suborder Trachinoidei			
Ammodytidae			
Ammodytes dubius Reinhardt	Unidentified Urohordata 4%, Cnidaria 0.1% total volume	Northwest Atlantic	Bowman, personal communication
Pinguipedidae			
PARAPERCIS COLIAS (Bloch & Schneider)	*Salpa fusiformis*	New Zealand	Thompson, 1948*
	Pyrosoma sp. in all but a few of several hundred fish	New Zealand	Bary 1960
Suborder Gobioidei			
Gobiidae			
Sufflogobius bibarbatus (von Bonde)	Molluscs 39%, Annelids 30%, Foraminifera 13%, *Pyrosoma* sp. 7% weight	Southwest Africa	Macpherson, 1983
Suborder Acanthuroidei			
Luvaridae			
LUVARIS IMPERIALIS Rafinesque	Pyrosomes, medusae, ctenophores & fish	Northeast Pacific	Fitch & Lavenberg, 1968
	One specimen full of salps	Madeira	Wheeler, 1969
	Salps, medusae		Wheeler, 1975
	Salps, medusae, ctenophores		Wheeler, 1978
Acanthuridae			
ACANTHURUS THOMPSONI (Fowler)	Salps 19%, siphonophores 10% diet volume	Kona, Hawaii	Hobson, 1974
Suborder Scombroidei			
Trichiuridae			
Aphanopus carbo Lowe	Pyrosomes	Madeira	Currie, in Fraser, 1962b
	Pyrosomes frequent	Madeira	Maul, in Foxton, 1966

Table 12.6 *(Continued)*

Species	Prey	Location	Reference
Euthynnus pelamis (Linnaeus)	Salps, siphonophores minor	Central Pacific	King & Iverson, 1962
	Tunicates minor	Eastern Pacific	Alverson, 1963
	Salps minor	Central Pacific	Nakamura, 1965
	Salpidae 2%, siphonophores <0.1% volume	West Africa	Dragovitch & Potthoff, 1972
MACKEREL	*Iasis zonaria* major prey	Japan	Tokioka & Bhavarayana, 1979
SCOMBER JAPONICUS Houttuyn	Salps	Northwest Atlantic	Bigelow & Schroeder, 1953
	Doliolum in 21% of guts	Northwest Pacific	Kun, 1954*
	Salps & pyrosomes in 26% of fish, also *Diphyes*	Japan	Takano, 1954
	Salpa fusiformis in 15 of 20 stomachs – major prey	Sea of Japan	Nishimura, 1958*
	Thaliacea up to 21%, Cnidaria up to 7%, Ctenophora up to 2% total weight	Northeast Pacific	Brodeur *et al.*, 1987
Scomber scombrus Linnaeus	Urochordata 0.1% total volume	Northwest Atlantic	Bowman, personal communication
Thunnus alalunga (Bonnaterre)	*Iasis zonaria, Pyrosoma atlanticum, Pelagia noctiluca Velella velella, Chelophyes appendiculata*, pteropods heteropods, all minor components of diet	Northeast Atlantic	Legendre, 1940
	Salps 4%, siphonophores 1%	Central Pacific	King & Iverson, 1962
	Salps minor	West coast of U.S.	Pinkas *et al.*, 1971
	Tunicates 5% occurrence		Le Gall, 1974
Thunnus albacares (Bonaterre)	Salps, pyrosomes minor	Central Pacific	Reintjes & King, 1953
	Salps, pyrosomes minor	Central Pacific	King & Iverson, 1962
	Tunicates minor	Eastern Pacific	Alverson, 1963
	Salpidae 0.4% volume	West Africa	Dragovitch & Potthoff, 1972
Thunnus maccoyii (Castelnau)	Salps minor	South Australia	Serventy, 1956
Thunnus obesus (Lowe)	Salpidae 4.2%, *Pyrosoma* 1.2%	Central Pacific	King & Iverson, 1962
Thunnus thynnus (Linnaeus)	Pyrosomes, pteropods	Japan	Kishinouye, 1923
	Pyrosoma atlanticum very common	Bimini	Krumholz, 1959
	Salpidae 0.6% total volume	Bimini	Dragovitch, 1970
Suborder Stromateoidei			
Amarsipidae			
Amarsipus carlsbergi Haedrich	Jellyfish-like tissues, chaetognaths	Indo-Pacific	Haedrich, 1969
	Cyclosalpa affinis	Seychelles	Janssen & Harbison, 1981
	Cyclosalpa affinis	Coral Sea	Personal observation
Centrolophidae			
Centrolophus niger (Gmelin)	Young associate with medusae & salps	South Africa	Smith & Heemstra, 1986
HYPEROGLYPHE ANTARCTICA (Carmichael)	*Pyrosoma atlanticum* major prey	Tasmania	Cowper, 1960
	Pyrosoma atlanticum major prey	S.E. Australia	Winstanley, 1978
Hyperoglyphe perciformis (Mitchill)	Salps, algae	Northwest Atlantic	Linton, 1901
	Salps, Ctenophora	Northwest Atlantic	Bigelow & Welsh, 1925
	Salps, Ctenophora	Northwest Atlantic	Bigelow & Schroeder, 1953

Table 12.6 *(Continued)*

Species	Prey	Location	Reference
ICICHTHYS LOCKINGTONI Jordan & Gilbert	*Salpa* sp. 36%, unidentified 34% weight	Northeast Pacific	Brodeur *et al.*, 1987
SERIOLELLA BRAMA (Günther)	Salps	New Zealand	Graham, 1939
	Iasis zonaria 38%, *Pyrosoma atlanticum* 24%, coelenterates 2%	S.E. New Zealand	Gavrilov & Markina, 1979*
	Salps 21%, pyrosomes 47%, coelenterates 21% frequency	New Zealand	Shuntov, 1979
SERIOLELLA CAERULEA Guichenot	*Iasis zonaria* 58%, *Pyrosoma atlanticum* 40%, coelenterates 0.5%	New Zealand	Gavrilov & Markina, 1979*
	Salps, pyrosomes principal food, coelenterates common	New Zealand	Shuntov, 1979*
SERIOLELLA PUNCTATA (Forster)	*Iasis zonaria* 59%, *Pyrosoma atlanticum* 39%, coelenterates 0.3%	New Zealand	Gavrilov & Markina, 1979*
	Salps 12%, pyrosomes 70% frequency	S.E. New Zealand	Shuntov, 1979*
SCHEDOPHILUS OVALIS (Cuvier & Valenciennes)	Pyrosomes always found in guts	Madiera	Maul, 1964
Nomeidae			
Cubiceps capensis (Smith)	*Cyclosalpa polae*	North Atlantic	Janssen & Harbison, 1981
Cubiceps pauciradiatus Günther	*Salpa cylindrica*	Coral Sea	Personal observation
Psenes cyanophrys Cuvier & Valenciennes	*Pegea confoederata*	Seychelles	Jansen & Harbison, 1981
	Pegea confoederata	N. Atlantic, Coral Sea	Personal observation
Tetragonuridae			
TETRAGONURUS ATLANTICUS Lowe	Salps	North Atlantic	Janssen & Harbison, 1981
TETRAGONURUS CUVIERI Risso	Salps	Mediterranean	Emery 1882
	Pyrosomes, cnidaria, ctenophores	California	Fitch 1949, 1952
	Salps	North Atlantic	Janssen & Harbison, 1981
	Many immatures (150 mm) living inside *Pyrosoma*	Tasmania	Last *et al.*, 1983
TETRAGONURUS PACIFICUS Abe	Salps	North Atlantic	Janssen & Harbison, 1981
Tetragonurus sp.	Collected in cloaca of *Pyrosoma* sp.	Pacific	Parin, 1970, fig. 13b
Stromateidae			
PAMPAS ARGENTEUS (Euphrasen)	Salps, hydromedusae main prey	Arabian Sea	Rege & Bal, 1963
PEPRILUS TRIACANTHUS (Peck)	Thaliacea 41% wt.	Northwest Atlantic	Maurer & Bowman, 1975
	Thaliacea 12% wt.	Northwest Atlantic	Bowman & Michaels, 1984
	Urochordata 29%, *Clione* 17%, Cnidaria 1%, Ctenophora 0.1%, Unidentified 46%	Northwest Atlantic	Bowman, personal communication
Stromateus fiatola Linnaeus	Eats *Cotylorhiza*, other medusae & salps in aquarium	Mediterranean	Lo Bianco, 1909
Order Pleuronectiformes			
Scophthalmidae			
Scophthalmus aquosus (Mitchill)	Salpidae <0.1% total volume	Northwest Atlantic	Bowman, personal communication
Order Tetraodontiformes			

Table 12.6 *(Continued)*

Species	Prey	Location	Reference
Order Tetraodontiformes			
Suborder Balistoidei			
Balistidae			
BALISTES FUSCUS Bloch & Schneider	Salps primary food in fall & winter	Gulf of Australia	Markina, 1973*
Canthidermis maculata (Bloch)	Seen eating colonial salps	Off Costa Rica	Hunter & Mitchell, 1967*
Canthidermis sufflamen (Mitchill)	One specimen with salp or medusa, siphs 8.5%, hyperiids, 3.7%, *Cavolinia* spp. 21% total volume	West Indies	Randall, 1967
Melichthys niger (Bloch)	Salps 2%, siphonophores 3%, pteropods 6% total volume	West Indies	Randall, 1967
Monacanthidae			
Aluterus monocerus (Linnaeus)	Seen eating colonial salps	Off Costa Rica	Hunter & Mitchell, 1967*
Aluterus scriptus (Osbeck)	Seen eating colonial salps	Off Costa Rica	Hunter & Mitchell, 1967*
File fish	Eat salp stomachs	Northwest Atlantic	Janssen & Harbison, 1981
NELUSETTA AYRAUDI (Quoy & Gaimard)	Salps 19%, fish 23%, gastropods 14%, crustaceans 13%	South Australia	Lindholm, 1984
Parika scaber (Forster)	Eat salps & medusae in midwater, but mainly feed on sponges, ascidians & algae	N.E. New Zealand	Russell, 1983
Molidae			
MOLA MOLA Linnaeus)	Salps, ctenophores, medusae, amphipods	Northwest Atlantic	Linton, 1901
	Jellyfish, ctenophores or salps	Northwest Atlantic	Bigelow & Welsh, 1925
	Salps, jellyfish, amphipods	Northwest Atlantic	Nichols & Breder, 1927
	Salps	Northeast Atlantic	Ehrenbaum, 1936*
	Salps, medusae, ctenophores	North Atlantic	Fraser-Brunner, 1951
	Salps, medusae, ctenophores	Northwest Atlantic	Bigelow & Schroeder, 1953
	Medusae, salps, ctenophores	North Atlantic	Wheeler, 1969
	Salps, jellyfishes, ctenophores	North Atlantic	Wheeler, 1978

References in Kashkina (1986) indicated with an asterisk. Species that prey heavily on Thaliacea are capitalized.

merely opportunistic, and only reflects the relative abundance of thaliaceans in the plankton. This is also likely the case with nototheniids, since *Salpa thompsoni* is such a dominant organism in the plankton of the Southern Ocean (Foxton, 1966). The feeding of nototheniids on salps must be regarded as largely opportunistic, since during some seasons. A thalicean diet may pose problems for fish. Mianzan *et al.* (1997) have ascribed mass mortality of *Scomber japonicus* to the fish eating salps that had been feeding (without apparent ill effects to themselves) on the toxic dinoflagellate *Alexandrium tamarense.*

Many species of marine turtles also prey heavily on thaliaceans (Table 12.7). Most eat a wide variety of gelatinous zooplankton, and also gelatinous benthic organisms, such as ascidians and sponges. I include records of feeding on gelatinous organisms other than thaliaceans, since I think that all turtles that feed on gelatinous material probably also eat thaliaceans. Although the green turtle, *Chelonia mydas*, is not known to feed on pelagic tunicates, it does eat medusae and ascidians. Since the green turtle is thought to be more herbivorous than any other marine turtle, I think it likely that salps could be a major component of the diet, particularly during the 'lost year' spent at sea. Medusae with zooxanthellae, salps and pyrosomes all contain high concentrations of vegetable matter, and could provide an essentially herbivorous diet for juvenile green turtles in the open ocean. Microscopic examination of gut contents of young turtles could probably be used to test the validity of this speculation. Hartog (1980) used this technique to study the feeding of the leatherback turtle, *Dermochelys coriacea*, on medusae and siphonophores. In a similar way, examination of the gut contents of juvenile green turtles for the presence of phytoplankton could be used to establish the presence of the remains of gelatinous oceanic filter feeders.

I have found very little information on feeding by marine birds on thaliaceans (Table 12.8), and at present it appears that only albatrosses and seagulls eat them. This is probably because many investigators cannot recognize the remains of gelatinous zooplankton in bird stomachs. Harrison (1984) found in a single study that 11 species of birds, including petrels, puffins, auklets and gulls, consumed medusae. A similar study, directed specifically toward detecting the presence of thaliaceans would probably be equally profitable, particularly in regions where salps and pyrosomes are usually abundant, such as the Southern Ocean and the New Zealand area. Probably any seabird that eats medusae will also eat thaliaceans. Tomo (1971) reports that five species of petrels eat medusae in the Southern Ocean, and Watson (1966) lists four species of frigate birds that eat jellyfish in the tropical Atlantic. The report of Trégouboff and Rose (1957) that seagulls eat the stomachs of salps suggests that the remains of salps could be difficult to identify. Microscopic examination of stomach contents to determine the presence of phytoplankton could be the only way to detect feeding on thaliaceans.

I have been able to find only two reports of predation by mammals on thaliaceans (Table 12.8). This is rather surprising, particularly in the case of whales, since pelagic tunicates are so important in the Southern Ocean, and it seems unlikely that they would pass up such an abundant food source. There are two possible explanations for the absence of pelagic tunicates in the diets of whales: whales avoid feeding in areas where salps are abundant, or pelagic tunicates have simply been overlooked in whale gut content studies. Two reports for the sei whale, *Balaenoptera borealis*, indicate that the latter is more likely to be the case. In one case, *Vibilia armata*, which is known to be associated with salps, was reported as a minor component of the gut contents (Best, 1967). More compelling, however, is the report where the stomachs of three sei whales were found filled with the stromateoid fish, *Tetragonurus cuvieri*, along with the remains of tunicates (Budylenko, 1978). This fish is regarded as rare by ichthyologists, and depends on salps and pyrosomes for food and shelter (Janssen and Harbison, 1981). *Tetragonurus* is obviously not a rare fish to sei whales, and in this case appears to be associated with tunicates as well. Seals have been reported to eat jellyfish (Salvini-Plawen, 1972), and they may also eat thaliaceans, but I have found no reports that document this. Brattström (1972) reported that 'wild mink revelled in stranded salps'. This can be interpreted either that they ate them or that they rolled in them.

11.9. Conclusions

The tables in this paper show that a large number of organisms exploit thaliaceans. These tables probably contain only a small fraction of the interactions that actually occur. I must admit that I was rather surprised to find so many reports in the literature, and I was forced by the editorial deadline in 1992 to conclude my literature search. I am quite confident that I have overlooked many important references, and apologize for these oversights. I did not review the literature on benthic organisms, and expect that many examples of feeding on thaliaceans might be found there. For example, since cerianthid larvae consume salps voraciously, I expect that other anthozoans, such as anemones and corals, feed on them.

In order properly to understand the trophic position of thaliaceans, it will be necessary to obtain quantitative data on feeding rates and gut residence time (Arai, 1988). Such quantification is likely to be impossible for

Table 12.7 Marine turtles that prey on gelatinous zooplankton.

Species	Prey	Location	Reference
Dermochelyidae			
DERMOCHELYS CORIACEA (Vandelli)	Medusae, *Hyperia galba*, plants	Concarneau	Vaillant, 1886
	Scyphomedusae, algae	Ceylon	Deraniyagala, 1930
	Cyanea capillata & *Hyperia medusarum*	Nova Scotia	Bleakney, 1965
	Stomach full of salps	Madeira	Brongersma, 1968
	Scyphomedusa, salps, pyrosomes		Brongersma, 1969
	Cyanea, Hyperia, pteropod shell		Brongersma, 1972
	Medusae, *Libinia spinosa*, a crab parasitic on medusae		Frazier *et al.*, 1985
	Siphophores, *Apolemia*, Scyphozoa	Northeast Atlantic	Hartog, 1980
	Scyphozoa	Northeast Atlantic	Hartog & Nierop, 1984
	Medusae & tunicates		Ernst & Barbour, 1989
	Scyphomedusae, tunicates		Márquez, 1990
Cheloniidae			
CARETTA CARETTA (Linnaeus)	Medusae abundant in stomachs	North Atlantic	Vaillant, 1886
	Jellyfish & salps		Murray & Hjort, 1912
	Molluscs, *Margaritifera vulgaris*, scyphomedusae	Ceylon	Deraniyagala, 1930
	Salpa maxima	North Atlantic	Harant & Vernières, 1934
	Salps		Fraser, 1949
	2 of 3 specimens with salps	Madeira	Brongersma, 1968
	Medusae, salps		Brongersma, 1972
	Physalia, Strombus gigas, Loggerhead sponge	Bermuda	Babcock, 1937
	Salps, *Pyrosoma atlanticum*, gastropods, important, also pelagic coelenterates	Northeast Atlantic	Nierop & Hartog, 1984
	Jellyfish, tunicates, live in open sea		Ernst & Barbour, 1989
	Jellyfish, pteropods, crabs, clams		Márquez, 1990
Chelonia agassizii Bocourt	*Pyrosoma*, jellyfishes, sponges	West coast of Mexico	Márquez, 1990
Chelonia mydas (Linnaeus)	*Botryllus* sp. 75% volume	West Africa	Cadenat, 1957
	Jellyfish, pteropods	North Atlantic	Brongersma, 1967
	Eat mainly plants, but also ascidians, molluscs, sponges, & echinoderms	Brazil	Fereira, 1968
	Jellyfish, sponges, ascidians, but mainly plants		Hirth, 1971
	Pterotrachea sp.		Brongersma, 1972
	Jellyfish, but mainly plants		Ernst & Barbour, 1989
Eretmochelys imbricata (Linnaeus)	*Velella, Salpa*, pteropods	New Zealand	McCann, 1966
	Physalia, ascidians	Bermuda	Babcock, 1937
	Siphonophores, medusae, *Velella*, anemones	Mediterranean	Hartog, 1980
	Coelenterates, sponges, crabs, sea grass, ascidians		Witzell, 1983
	Portuguese-man-of war, pelagic in *Sargassum*		Ernst & Barbour, 1989
	Eat *Chironex fleckeri* in lab	N.E. Australia	Cropp, 1985
Lepidochelys kempi (Garman)	Medusae		Brongersma, 1972
LEPIDOCHELYS OLIVACEA (Eschscholtz)	Jellyfish		Ernst & Barbour, 1989
	Salps major prey	Southern Mexico	Márquez, 1990

Species that are reported to feed heavily on gelatinous zooplankton are capitalized.

Table 12.8 Birds and mammals that feed on thaliaceans.

Species	Prey	Location	Reference
Class Aves			
Diomedea cauta Gould	Salps	Tropical Atlantic	Watson, 1966
Diomedea chrysostoma Forster	*Salpa thompsoni*	South Georgia	Foxton, 1966
Diomedea melanophris Temminck	*Salpa thompsoni*	South Georgia	Foxton, 1966
Sea gulls	Eat stomachs of salps	Mediterranean	Trégouboff & Rose, 1957
Class Mammalia			
Balaenoptea borealis Lesson	*Velella, Vibilia armata* minor	South Africa	Best, 1967
	Stomachs of three filled with *Tetragonurus cuvieri,* also tunicates, squids, *Calanus*	Southeast Atlantic	Golubovsky *et al.*, 1972, in Budylenko, 1978
Mustela vison Schreber	*Salpa fusiformis* – 'wild mink revelled in stranded salps'	Norway	Brattström, 1972

most of the parasites and predators I have listed. To use fishes as an example, most cannot be maintained in the laboratory, so we will still have to rely on gut content analyses. With existing methods, the numbers of gelatinous prey ingested cannot be determined. Therefore, one cannot at present state unequivocally that any group of fishes depends primarily on thaliaceans as a food source, and it is likely that this situation will remain, unless some unpredictable technological advance is made.

However, one can still infer trophic importance indirectly. I think that the fact that several unrelated groups of organisms have evolved specializations for the utilization of thaliaceans as food or shelter provides the most compelling indirect evidence that we now have. Five such cases are now well known. These are the documented, often highly specific associations of species of the copepod, *Sapphirina*, the hyperiid amphipods, *Lycaea* and *Vibilia*, the ctenophore, *Lampea*, and the fish, *Tetragonurus*, with salps.

A great deal of research is still needed before these associations are fully understand, but I think that it is clear that salps play a central role in the life histories of all five groups. I am confident that further research will show that thaliaceans are equally important to many other groups listed in the tables in this paper, and perhaps to other groups that are not listed.

Thaliaceans, because of their relatively large size, undoubtedly have a significant role in structuring the open ocean planktonic community. They provide an essentially benthic habitat for small crustaceans, by providing them with large surfiees in the plankton. They also provide shelter for fishes and octopods, and serve as sources of food for a wide variety of taxa. Although they have seldom been regarded as suitable food because of the high water content of their tissues, it is clear that numerous groups have evolved to exploit this important food resource in the open ocean.

11.10. Acknowledgements

I thank V.L. McAlister for identifying the copepods, and R.E. Bowman for providing unpublished data on fish gut contents. I also thank all of the people who helped in my literature search. These include, but certainly are not limited to: J.E. Craddock, B.H. Robison, R.M.L. Ates, B. Collette, J. Tyler, G. Cailliet and J. Gartner. Research supported in part by National Science Foundation Grant No. OCE-9303417.

THIRTEEN

Bioluminescence in the Appendicularia

Charles P. Galt and Per R. Flood

13.1. Introduction and history

Bioluminescence is a conspicuous feature of many marine organisms, and among the pelagic Tunicata it is well documented in the Pyrosomida (section 1.2.2) and in certain oikopleurid Appendicularia, whose mucoid feeding houses emit light. Giglioli (1870), who reported multicoloured luminescence from the tail of unidentified species of appendicularians, probably actually observed iridescence from the caudal muscle bands, as there is no evidence of caudal luminescence or multiple wavelength emissions from a single animal. We owe to Hans Lohmann (1899a, 1933) the first reliable accounts of light production by appendicularian tunicates. He described brilliant greenish flashes emanating from the trunks of free-swimming *Oikopleura albicans* as they made contact with surfaces or were otherwise agitated. He further reported that during expansion of the house a brilliant light emanated continuously from the animal's trunk, (he speculated that such luminescence would serve to deter predators during this vulnerable operation).

The expanded house (as Lohmann observed) also produces light from unknown but consistently organized sources in response to agitation or prodding. Finally, Lohmann noted that even discarded houses emitted flashes upon stimulation. Lohmann suggested that light was produced by 'fluorescent' secretions from the oral glands and from parts of the oikoplast epithelium. Since other species, including members of the Fritillariidae, produced similar secretions, Lohmann (1933) supposed that all appendicularians were luminescent. Harvey (1952) concluded from Lohmann's

Table 13.1 Oikopleuridae that possess oral glands and subchordal cells.

Luminescence demonstrated	Luminescence not demonstrated
Oikopleura albicans	*O.caudaornata O.cophocerca*
O.dioica	*O.villafrancae indentacauda*
O.labradoriensis	*O.gaussica*
O.villafrancae	*O.inflata*
O.rufescens	*O.parva*
O.vanhoeffeni	*Folia gracilis*
Stegosoma magnum	*F.mediterranea*

Luminescence emanating from house rudiment inclusion bodies has been conclusively demonstrated in the species in the left column. Inclusion bodies are known for all but *O. caudaornata, O.inflata* and possibly *O.villafrancae indentacauda.* It is almost certain that all three morphological features and bioluminescence are demonstrable in all species in this table.

descriptions that appendicularians surely possessed 'self-luminosity' and remarked that the possible role of luminous bacteria had not been investigated. Except for scattered observations of presumed appendicularian luminescence from shipboard (Tarasov, 1956; Staples, 1966), appendicularian luminescence was apparently ignored for the next 25 years. We here expand upon Lohmann's original observations as we review current knowledge of the physical, chemical, and biological bases of light emission in appendicularians, speculate on its adaptive role, and suggest future directions of study of this unusual phenomenon.

13.2. Occurrence of bioluminescence among appendicularian species

As Galt *et al.* (1985) pointed out, houses of any appendicularian species may become colonized with potentially luminous microorganisms (Davoll and Silver, 1986; Davoll and Youngbluth, 1990), such as bacteria, dinoflagellates, and protozoa. Thus even non-luminescent species may produce houses that become secondarily luminescent (Widder *et al.*, 1989), much as other kinds of organic aggregates or marine snow may emit light (Orzech and Nealson, 1984). Intrinsic luminescence is found only in certain oikopleurid appendicularian species that share a combination of easily recognized morphological traits: oral glands, subchordal cells, and *Hautungskorper* ('inclusion bodies'; see section 13.3. below) (Fig. 13.1). In these species, rudimentary and expanded houses emit brief flashes of blue-green light when the houses are stimulated, normally by agitation. Whereas oral glands and subchordal cells are usually recognizable in fixed specimens collected with plankton nets, the *Hautungskorper*, or inclusion bodies, are often unidentifiable. However, when identifiable, they are useful, species-specific, taxonomic aids to identification (Lohmann and Buckmann, 1926; Lohmann, 1896a, 1933; Buckmann and Kapp, 1975).

Luminescence (always emanating from house rudiment inclusion bodies) has been conclusively demonstrated from the seven species in Table 13.1, and possibly from *O.villafrancae indentacauda* (Flood and Galt, in preparation). Inclusion bodies are therefore known to exist in at least 12 of the 15 species characterized by oral glands and subchordal cells. There is little reason to doubt the presence of inclusion bodies in the remaining species. Moreover, it is now clear that oral glands, along with luminescent inclusion bodies, play a functional role in light emission. We concur with the conclusion of Galt *et al.* (1985) that intrinsic bioluminescence is probably found only in those appendicularian species that possess this triad of taxonomic characteristics. Because oral glands are the most easily recognizable trait of the three, their presence is a quick indicator of intrinsic bioluminescence in appendicularian species. Three of the species listed as probably luminescent by Galt *et al.* (1985), viz. *O.drygalskii, O.valdiviae,* and *O.weddelli,* are now considered synonymous with *O.gaussica* (Fenaux, 1993).

According to the latest taxonomic review of Appendicularia (Fenaux, 1993; see also Chapter 18), oral glands and subchordal cells are present in 15 of the 32 species of Oikopleuridae (Table 13.1).

13.3. Topography of luminescence in rudimentary and expanded houses

Intrinsic luminescence originates from small, granular point sources in the acellular, mucous houses of appendicularians. These luminescent sources are subunits of the so-called *Hautungskorper* ('shedding bodies'; Lohmann and Bückmann, 1926; Lohmann 1896a,b), 'house etchings' (Bückmann and Kapp, 1975), or 'inclusions' (Galt and Sykes, 1983; Galt *et al.*, 1985) that have been described in several appendicularian species. Hautungskorper form species-specific patterns of deposits in the house rudiment and were so named from a presumed role in expansion of the mucous layers of the house rudiment to a functional house (Lohmann and Bückmann, 1926; Lohmann, 1933; Körner, 1952). However, since numerous Oikopleuridae lack *Hautungs-*

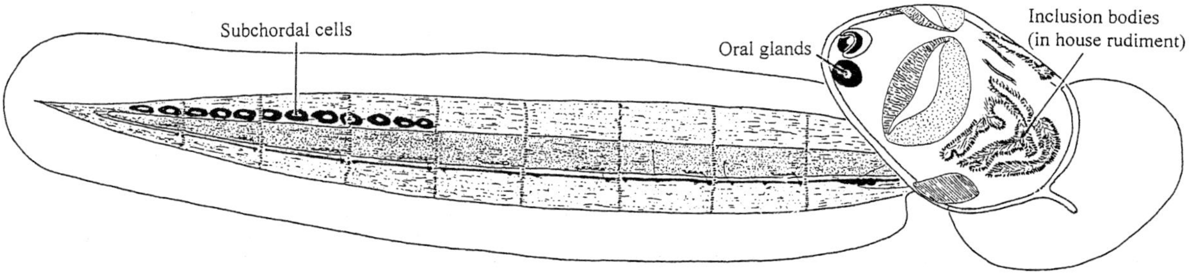

Fig. 13.1. Oikopleurid appendicularian illustrating the location of inclusion bodies, oral glands and subchordal cells.

korper and yet produce fully functional, oikopleurid-type houses, we see no necessary relation of these structures to the deployment of the house. Rather, based on their functional association with bioluminescence (Galt and Sykes, 1983; Galt *et al.*, 1985), we here adopt the term (luminescent) inclusion bodies. Bioluminescence emanates from inclusion bodies that are incorporated in the mucoid house rudiment as it is secreted by the appendicularian and not from the tissues of the animal itself (cf. Galt, 1978). Almost without exception the entire pattern of inclusion bodies is bilaterally symmetrical and is located on the dorsolateral and posterior part of the house rudiment, behind Fol's oikoplast and dorsal to Eisen's oikoplast, two recognizable regions of the epithelium that secretes the house (see Chapters 2 & 6). In *Oikopleura dioica* there is also a row of some 6–10 inclusion bodies on the anteroventrolateral part of the rudiment, beneath the oral glands. Despite this gross similarity in distribution of inclusion bodies, the detailed pattern is highly species specific in all species examined (Fig. 13.2). Nevertheless, within a species the patterns reveal slight bilateral, individual, and ontogenetic variations (e.g. *O.labradoriensis*; Fig 5,3,4 13.3). Variations also exist between successive house rudiments from the same animal. As the animal grows, the pattern becomes increasingly complex. In *O.labradoriensis* the pattern starts out as a simple, short band of parallel, rod-shaped inclusion bodies in the youngest animals and becomes increasingly long and complex so that in house rudiments of large animals, it contains as many as 1,000 inclusion bodies on each side of the rudiment (Fig. 13.4). In *O.dioica* the pattern is simpler, starting as a single, dorso-lateral, ovoid inclusion body on each side of the first house of post-metamorphic *O.dioica.* In the adult, the pattern comprises five rows, each with 5–20 pillow-shaped inclusion bodies. Variation in the pattern between successive rudiments of the same animal has been observed also in this species (Flood, unpublished). The inclusion bodies are located immediately below the external limiting membrane of the house rudiment. This protective layer is shed at the start of the house expansion process and the inclusion bodies are soon exposed directly to the environment surrounding the house. During and after the house expansion process

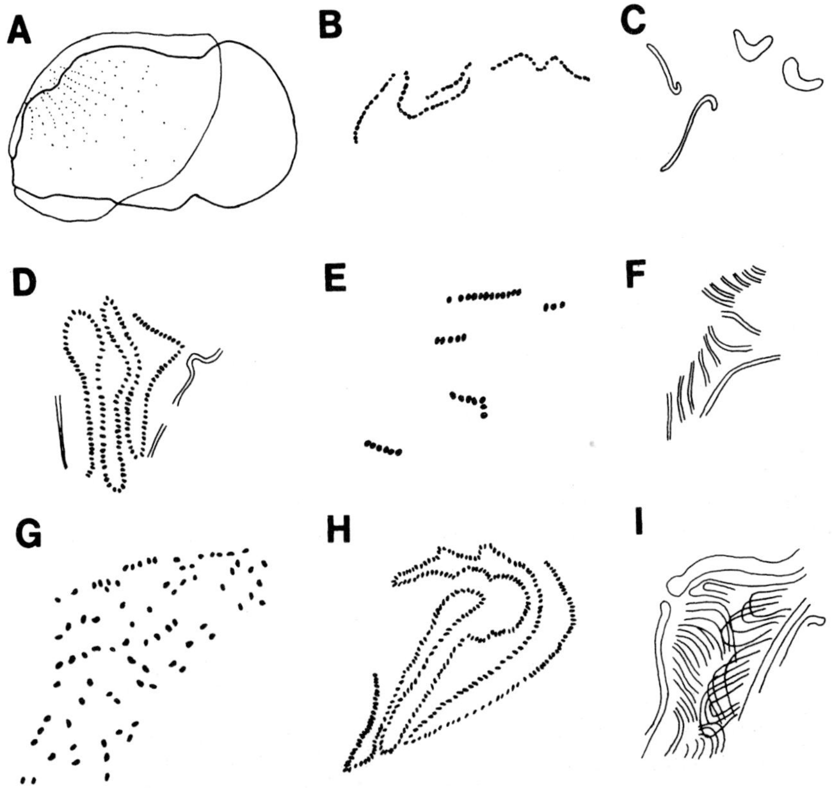

Fig. 13.2. Pattern of inclusion bodies in house rudiments of oikopleurid appendicularians. (a) *Oikopleura rufescens*, (b) *Stegosoma magnum*, (c) *Oikopleura cophocerca*, (d) *O.gaussica*, (e) *O.dioica*, (f) *O.parra*, (g) *O.vanhoeffeni*, (h) *O.labradoriensis* and (i) *O.albicans.* In (a) inclusion bodies are shown in outline of house rudiment overlying animal's trunk. Other patterns are not drawn to scale but each lies over approximately the anterior two-thirds of the animal's trunk. (a) and (b) redrawn from Galt *et al.* (1985). (c)–(i) from Lohmann and Bückmann (1926).

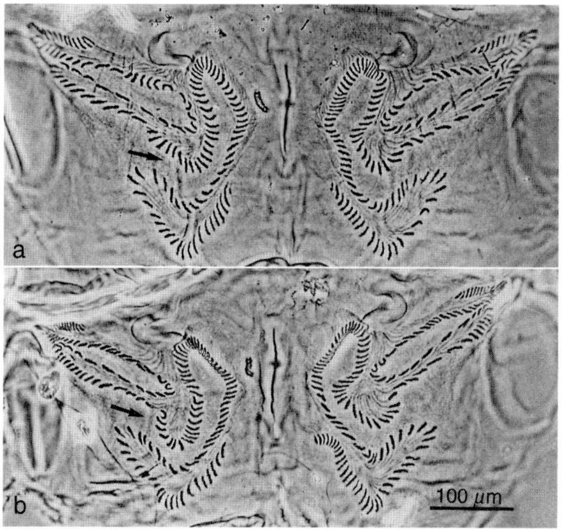

Fig. 13.3. *Oikopleura labradoriensis.* Phase-contrast micrographs of inclusion body patterns on two successive house rudiments (a) and (b) dissected from the same animal of 780 µm trunk length and flattened on a slide. The left and right inclusion body patterns are shown here on either side of the dorsal midline. Comparison of (a), from the outer and older rudiment, with (b), from the inner, newer rudiment, shows that successive rudiments — produced within a few hours of each other — have near to, but not perfectly identical, inclusion body patterns (arrows) and numbers (130 and 139 bodies in (a) 142 and 137 bodies in (b).

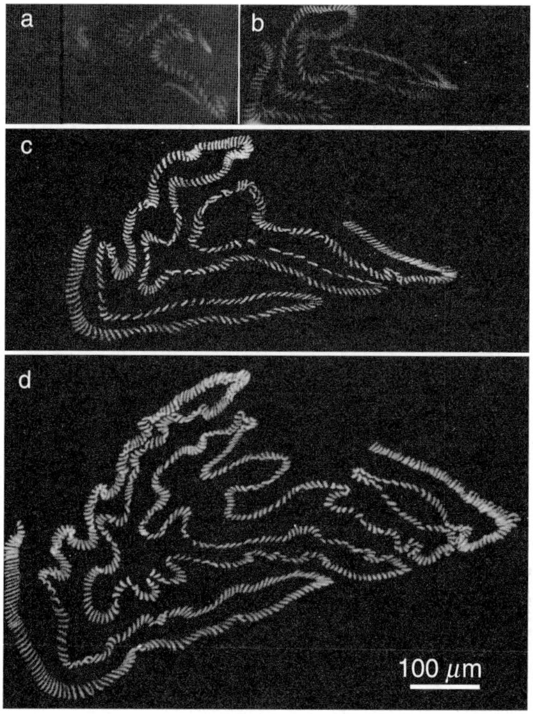

Fig. 13.4. *Oikopleura labradioriensis.* Autofluorescence images of inclusion body patterns on right sides of house rudiments dissected from animals of trunk length 430 µm (a), 780 µm (b), 1300 µm (c) and 2040 µm (d). Note increasing complexity and number of (a)–(d) inclusion bodies in pattern with increasing animal size. Scale bar in (d) applies to (a)–(c) as well.

the inclusion bodies become difficult to study, as their spacing increases and numerous foreign particles settle on and adhere to the house. The only reliable ways to identify the inclusion bodies then are to view yellow autofluorescence from their contained granules; to record their light emission upon artificial stimulation or to examine clean houses expanded in filtered seawater under strong dark field or differential interference contrast illumination. Thus, in expanded houses, a similar, but derived pattern of luminescent sources is dispersed over the surface of the house (Fig. 13.5). In the case of *O.albicans,* luminescent, filamentous streamers trail behind the expanded house (Galt and Fenaux, unpublished).

13.4. Structural organization of luminescent inclusion bodies

In some species (e.g. *O.dioica* and *O.villafrancae*) the inclusion bodies are laid down as rows of pillow-shaped structures closely applied to the external house surface (Fig. 13.6a). In other species (like *O.vanhoeffeni*) they occur as loosely organized rows of bean- or kidney-shaped structures in the rudiment (Fig. 13.6b) and as long and sickle-shaped structures in the expanded

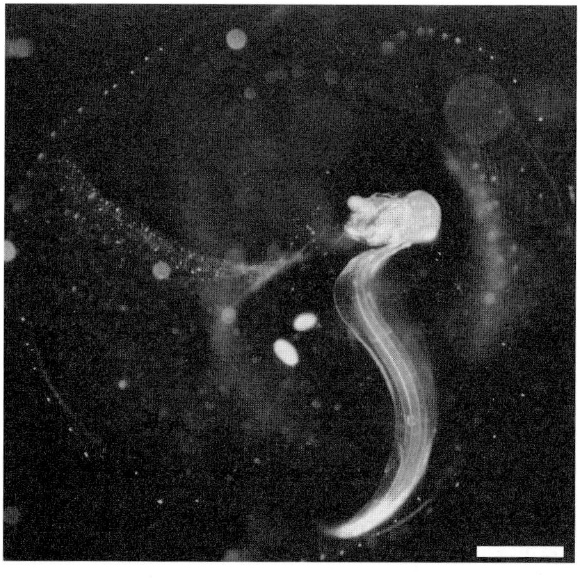

Fig. 13.5. Dark field macrograph of live *Oikopleura dioica* in its house. Note distribution of inclusion bodies (regularly spaced white spots) along circumference of house. Scale bar = 1 mm.

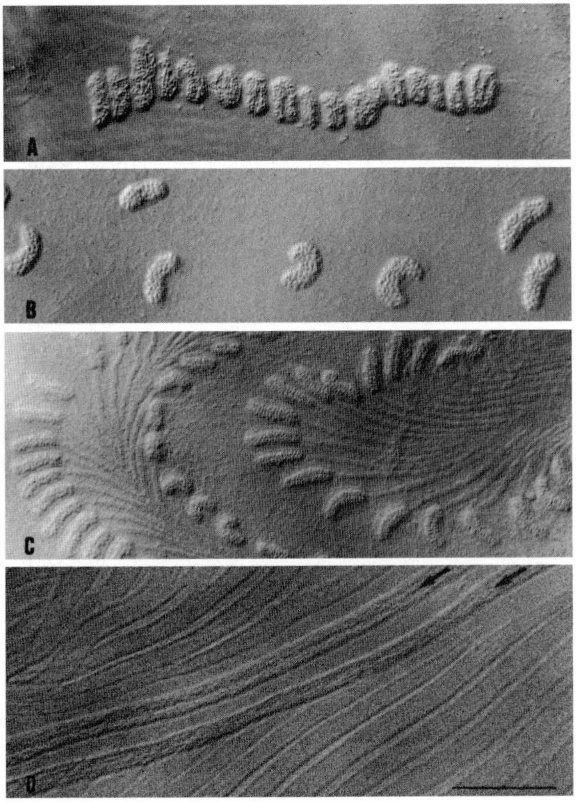

Fig. 13.6. Differential interference contrast light micrographs of: (a) Row of pillow-shaped inclusion bodies in *O.dioica.* (b) Irregularly dispersed bean-shaped inclusion bodies in *O.vanhoeffeni.* (c) Curved bands of parallel-stacked rod-shaped inclusion bodies, with attached threads in *O.labradoriensis.* (d) Linear pattern of band-shaped inclusion bodies in *O.albicans.* A granular content may only be discerned in some of these inclusion bodies (*arrows*). Scale bar = 50 μm applies to (a)–(d).

house. These sickle-shaped bodies appear to be attached to the external surface of the house at both ends, so that the midregion protrudes into the surrounding water (Flood, unpublished). The highest level of organization of inclusion bodies is probably found in *O.labradoriensis,* where each body is rod-shaped, approximately 5 μm wide and 20 μm long, and laid down parallel to the next in a single, complexly folded series on each side of the house rudiment (Fig. 13.6c). Each body is attached at one end to the external surface of the expanded house by a flexible hinge. From the other, distal end of the body a long (> 100 μm), slender thread projects into the surrounding water. However, at least in partially expanded houses, the distal ends of these threads return to the external house surface to be anchored at some distance from where the rods are attached (Flood, unpublished). In *O.albicans* (Lohmann, 1899; Galt and Fenaux, unpublished) the inclusion bodies take the form of six major, long, slender paired lines on the dorso-lateral surface of the house rudiments (Fig. 13.6d). In addition, there are, in the expanded house, fluorescent mucoid secretions that appear to originate from the oral glands and migrate during house expansion to a pair of very long, thin filaments that ultimately are freed from the house, save for an attachment at one end, and trail behind it as thin streamers. These trailing filaments fluoresce yellow-green under blue excitation and emit light when stimulated (agitation, distilled water, or Triton-X100). The trailing filaments are up to twice as long as the house itself (e.g. 30 mm trailing from a 15 mm house). In the above species, each inclusion body generally consists of a demarcated area of extracellular ground substance in which numerous (20–100) spherical, dense, refractile granules, 1–3 μm in diameter, are dispersed. In the unfixed state, and after formaldehyde fixation of short duration, the granules emit a yellow-green fluorescence (> 515 nm) when excited with blue light, (Galt and Sykes, 1983; Galt *et al.*, 1985; Flood, unpublished; Galt unpublished) and, in some cases (Flood, unpublished), a blue fluorescence with ultraviolet excitation.

Apart from the work of Sykes (1982), suggesting a relationship of the refractile granules to mitochondria, there are no accounts of the ultrastructural bases of appendicularian luminescence. The following description is based on unpublished results obtained by one of us (PRF). Whereas the shape of individual inclusion bodies often differs between species, the internal organization of inclusion bodies reveals great similarity among the three species (*O.dioica, O.labradoriensis* and *O.vanhoeffeni*) examined so far by transmission electron microscopy (TEM) (Fig 13.7). The boundary of an inclusion body is marked only by a prominently denser packing of extracellular filaments and granules in its ground substance as compared to that of the surrounding house rudiment. No trace of a cell membrane or equivalent structure has been detected at this boundary. Embedded in this finely granular, extracellular matrix are found numerous membrane-bound, electron dense bodies. These evidently correspond to the refractile granules observed by light microscopy. However, since these bodies have a complex internal structure, the term 'granule' is inappropriate at the TEM level. Further, recent results indicate that the bioluminescence is directly associated with these bodies (Flood, unpublished; Galt, unpublished) rather than to the entire inclusion body as such (Galt and Sykes, 1983). Accordingly, we apply the descriptive term 'lumisome' to these luminescent microsources, though we presently imply no structural or chemical similarity to lumisomes, scintillons (Fogel *et al.*, 1972), or photosomes (Nicolas *et al.*, 1990) described for other phyla. The lumisomes vary slightly in size and appearance in the three species examined so far (Fig. 13.7), but all are bounded by a

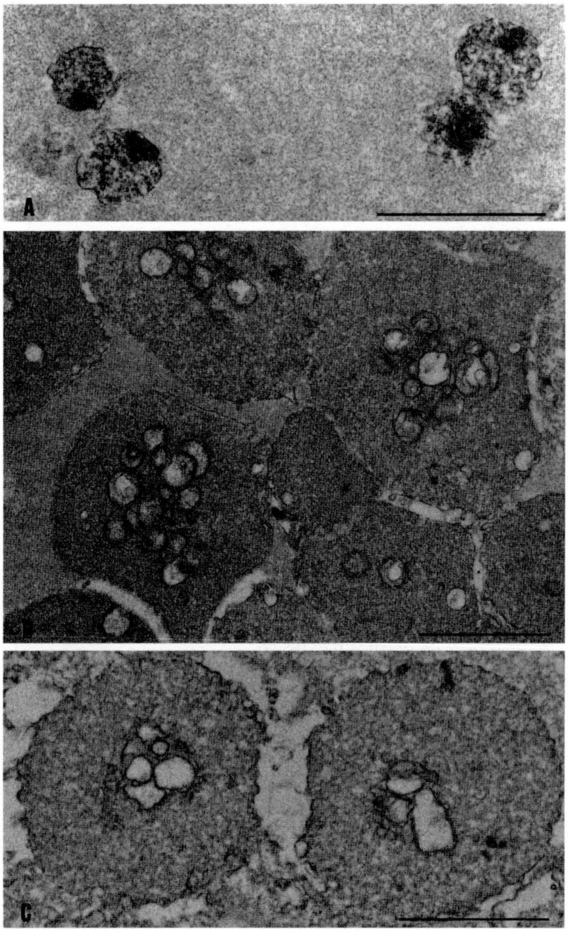

Fig. 13.7. Transmission electron micrographs of ultrathin sections through the granules (lumisomes) contained within inclusion bodies in: (a) *O.dioica*, (b) *O.vanhoeffeni* and (c) *O.labradoriensis*. Scale bars = 1 μm.

cell membrane, are filled by a densely packed, finely granular matrix, and contain minute. vesicles, each bounded by a double membrane (Sykes, 1982). The boundary membrane of the lumisome is difficult to fix intact but appears to be a typical lipid bilayer membrane with no trace of associated cell wall material. The minute internal vesicles of the lumisome reveal some resemblance to miniature mitochondria. However, this diagnosis must await verification by other techniques, since their minute size usually precludes typical cristae internal to their double boundary membrane. In *O.vanhoeffeni* (Fig. 13.7b) the lumisomes reach their largest size (up to 3 μm in diameter) and highest degree of packing. They occupy up to 90% of the volume of each inclusion body. Each lumisome contains some 20–100 internal, double walled vesicles, gathered in a single group near the centre of the lumisome. The space between these vesicles and the boundary membrane of the lumisome is occupied by a granular matrix whose degree of packing is uniform within each lumisome, but varies markedly between neighbouring lumisomes within the same inclusion body. In *O.labradoriensis* (Fig. 13.7c) the lumisomes are smaller, ~2 μm in diameter, contain fewer internal vesicles, but are otherwise similar in appearance and packing density to those of *O.vanhoeffeni*. Single filaments sometimes penetrate the matrix area of the lumisomes and lead toward their contained vesicles. It is unknown if these filaments connect to the long threads that are associated with the inclusion bodies of this species. The thread-like extensions themselves consist of a finely fibrillar extracellular material ensheathing one or a few tubular structures. These extracellular tubules are wider in diameter than ordinary cytoplasmic microtubules. In *O.dioica* (Fig. 13.7a) the lumisomes are small and irregular in outline occupying less than 50% of the inclusion body volume. The lumisome matrix granules appear to be slightly larger than in the other two species. Minute vesicles are located at the periphery of the lumisome.

13.5. Origin and maturation of inclusion bodies and their lumisomes

Our observations (Flood and Galt, unpublished) that inclusion bodies are not always luminescent or fluorescent in very recently formed house rudiments implies some kind of maturation or activation of luminescent sites (inclusion bodies or lumisomes). Furthermore, the virtually invariable co-occurrence of oral glands and subchordal cells in species that possess luminescent inclusion bodies (*Hautungskorper*) has led to studies of the involvement of these other organs in appendicularian luminescence. Lohmann (1933) mentioned the oral glands as possible sources of bioluminescent material. However, he suggested no relationship between luminescence and inclusion bodies. The oral gland was later suggested by Fredriksson and Olsson (1981) to be itself luminescent and to produce a secretion of luminescent cell fragments that make up the inclusion bodies of the house rudiments (see section 6.3.3). However, we have been unable to confirm the luminescence of the oral gland and its secretion, and Fredriksson and Olsson (1981) provided no explanation of how the secreted cell fragments could end up as species-specific patterns of inclusion bodies in regions of the house rudiment remote from the oral gland. The relation of this gland to bioluminescent inclusion bodies has therefore remained obscure. However, recent experiments on *O.labradoriensis* by one of us (PRF) show convincingly that the oral gland secretion contributes material to the inclusion bodies. Microsurgical destruction of the oral gland on one side of several specimens consistently prevented the formation of typical, conspicuous inclusion bodies on the same side of the house rudiments later synthesized. However, careful examination with phase contrast optics revealed a faint template of the inclusion bodies in their normal position on the same side of the house rudiment. These 'ghost' inclusion bodies contained no trace of the numerous refractile lumisomes that were present in the normal inclusion bodies of the untreated side (Fig. 13.8a; Flood, unpublished). Time lapse video microscopy of the same species (Andersen and Flood, unpublished) has shown that the oral gland produced a long, slender, pseudopodial secretion that penetrated the ventral oikoplastic epithelium, and coursed postero-laterally to extend to the dorso-lateral surface of the nascent house rudiment. This cytoplasmic process appeared to follow a preformed channel below the external surface of the nascent house rudiment and to spread out over the entire dorso-lateral area of the house rudiment where 'ghost' inclusion bodies were already present. Within this wide portion of the pseudopod, cytoplasmic streaming and ruffling membranes were prominent. Soon after the cytoplasmic process reached a group of 'ghost' inclusion bodies, minute granules present in the cytoplasmic process settled within the 'ghosts' (Fig. 13.8b). These granules were at first non-fluorescent but soon grew larger and more refractile and ended up as typical lumisomes of mature morphology and autofluorescence, as found in most finished rudiments (Fig. 13.8c). Long before this growth and differentiation of refractile lumisomes was completed, the pseudopod pinched off at the oral gland and began to degenerate. Cytoplasmic remnants, usually containing submicron sized granules with a weak autofluorescence of the same wavelength characteristics as the matured large and refractile lumisomes, could frequently be traced as an extension of the inclusion body pattern in the direction of the oral gland. With TEM, such trailing granules appeared as largely empty membrane-bound vesicles (Flood, unpublished). In spite of their autofluorescence, such mini-granules have never been observed to emit light. Thus it is clear that functional inclusion bodies, which comprise the species-specific geometric pattern on house rudiments and later on the expanded houses, are formed through the concerted actions of the house-secreting oikoplastic epithelium and the oral glands. The former lays down the template pattern of 'ghost' inclusion bodies and the latter provides a secretion that seeds these inclusion bodies with rapidly growing and maturing lumisomes. At present we have no evidence to connect the subchordal cells functionally with bioluminescence. Delsman (1911, 1912) found the oral glands and subchordal cells to

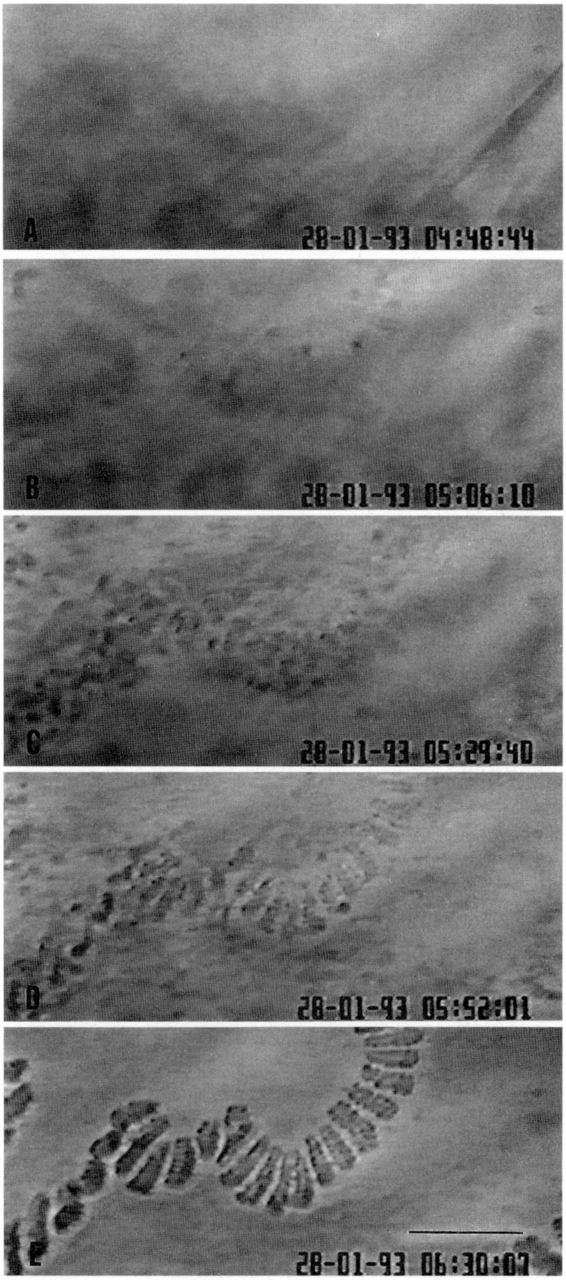

Fig. 13.8. Selected frames from a time lapse video sequence revealing the differentiation of typical inclusion bodies in a nascent house rudiment of an immobilized *O.labradoriensis* as seen by phase contrast light microscopy. (A) Just before the arrival of the cytoplasmic process from the oral gland. 'Ghost inclusion bodies' are hardly visible on the videoprint. (B) The cytoplasmic process from the oral glands has spread across most of the field of view. (C) and (D) successive stages in the maturation of the inclusion bodies. (E) Fully matured inclusion bodies. Scale bar = 50 μm applies to (A)–(E).

have a common embryonic origin in cells migrating from the tail endoderm. Fredriksson and Olsson (1981) speculated from this developmental link that the subchordal cells were luminescent and later (1991) suggested a functional relationship between subchordal cells and oral glands based on similarities in cytology, surface activity, and affinity for *in vivo* staining by neutral red. However, they also reported structural differences between the two cell types that could imply functional differences, and they no longer suggested luminescence as a function of subchordal cells. In numerous visual and photometric observations of light from six species of appendicularians, we have never recorded light from the tail.

13.6. Structural and physiological changes upon light emission

Careful microscopic examination and image intensification of inclusion bodies in house rudiments of *O.dioica* and *O.labradoriensis* before, during, and after mechanical (tapping of microscope stage) and chemical (0.1–1.0% Triton-X100 in seawater) stimulation revealed changes during light emission (Flood, unpublished). Light flashes lasting about 300 ms (see below) originated from the lumisomes contained in the inclusion bodies, but the light-emitting area reached a diameter approximately three times the size of the lumisome during most of the flash duration. Apparently, the peak intensity of light was emitted from within the lumisome, and the intensity was significantly lower in the halo or diffusion zone around the lumisome. Most significantly, the lumisome lost most of its contrast in the microscope, or disappeared altogether, at the moment light was emitted (Fig. 13.9a–c) (Flood, unpublished). When viewed with the TEM, such discharged lumisomes appeared disintegrated. Their boundary membrane was gone, and they could be identified only as a cluster of small and irregularly sized vesicles dispersed in a coarsely granular matrix of lower electron density than in the ground substance of the surrounding inclusion body. From these observations we conclude that the light discharge is caused by the disintegration of the lumisome boundary membrane and diffusion of ions or substances across this broken barrier. With image intensification, single lumisomes never emitted light more than once. These results corroborate our evidence that rudiments and houses that have been stimulated to luminesce to exhaustion by shearing forces (mechanical stirring) show no ability to recover, even after undisturbed incubation for > 12 h at normal temperatures (Galt, Smith and Grober, unpublished). This implies that the appendicularian bioluminescence system is a transitory system in two respects: the animal incorporates its bioluminescence system in houses that are abandoned after 4–6 h, and each lumisome seems capable of generating only one flash of light, which leads to its breakdown.

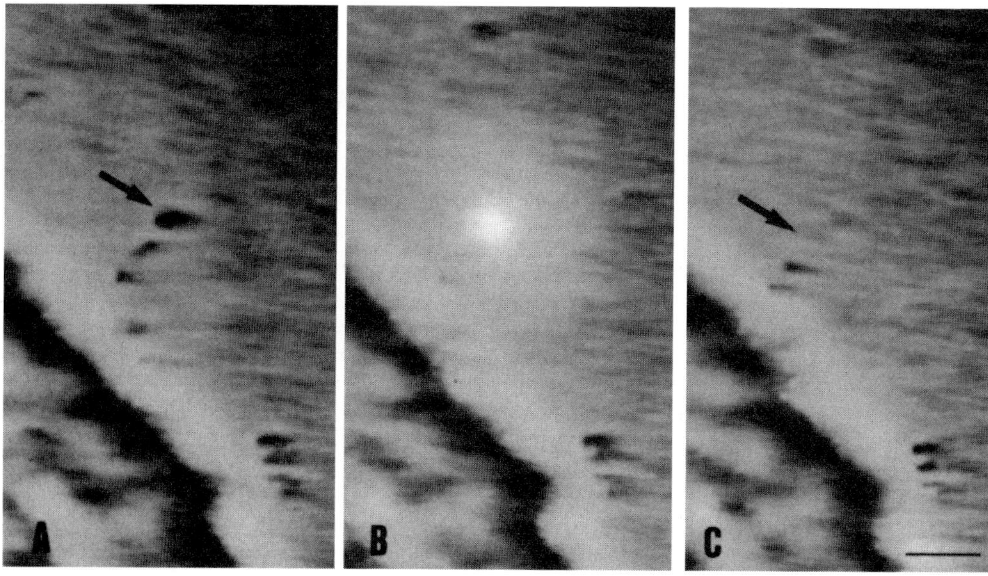

Fig. 13.9. Selected frames from an image intensified video sequence of a house rudiment attached to the trunk of an immobilized *O.dioica*, before (A), during (B) and after (C) light emission caused by mechanical stimulation. Note the disappearance of one inclusion body (**arrows**) after light emission. A small part of the animal's trunk is seen in the lower left corner. Scale bar = 50 μm.

13.7. Emission characteristics: inhibition, kinetics, intensity, spectral emission

Luminescence is not inhibited by light, at least in *O.dioica, O.labradoriensis,* and *O.albicans.* Houses from animals collected at midday yield luminescence similar in kinetics, intensity, and total stimulatable light (TSL) to houses from animals collected at night (Galt, Smith and Grober, in preparation). Exposing the animals or their houses to light in the laboratory or on the microscope during processing seems not to influence the luminescent emission. The kinetics of light emission from individual lumisomes after mechanical or Triton-X100 stimulation of microdissected house rudiments have been studied by image intensified video microscopy of house rudiments of *O.dioica* and *O.labradoriensis* by one of us (PRF). Using the NTSC video format (60 half-frames/s) and a video clock reading to 0.01 s, an overall time resolution between 20 and 30 ms was obtained. The light discharge from a lumisome was a spike with a duration of ~300 ms, a rise time of 50–80 m and a slower decay phase. Frequently, a very dim afterglow persisted after the 300 ms, but soon became indistinguishable from stray light from discharging neighbouring lumisomes. The size of the light-emitting disc began as equivalent to the diameter of the lumisome in the first video frame, but expanded to about three times this diameter in the third frame and remained at this size for the next 2 or 3 frames, before it faded slowly away more or less evenly over the entire cross-section (Fig. 13.10a,b). These details could be discerned without causing full saturation of the central pixels of the illuminated disk during the peak of the flash and are, accordingly, interpreted as real events rather than caused by instrument limitations (blooming, etc.). Single lumisomes were never seen to emit light more than once, and the term 'discharge' of light therefore seems appropriate. Luminescent discharges from comparable specimens of house rudiments have also been recorded by one of us (PRF) in a luminometer calibrated to read 7×10^9 quanta s^{-1} with a full scale sensitivity down to 1 nA. Light flashes (Fig. 13.11a) often reached an amplitude of 0.01 μA in $<$ 100 ms and returned completely to the baseline in $<$ 400 ms. The integrated quantum content of one flash was about 10^7 quanta. Although we lack proof that these recorded flashes represent the lumisome discharges viewed with image intensification, their similar kinetics support this view. We have several records from carefully removed rudiments of 1000–2000 such flashes per rudiment before exhaustion of their bioluminescence. In response to brief squirts of seawater, fresh rudiments gave an immediate strong emission of light of $<$ 1 s duration followed by a weaker afterglow for up to 10 s (Fig. 13.11b). This peak and shoulder response is interpreted as the summed effect of more or less synchronously discharged multiple lumisomes. Twenty repeated syringe needle squirts of seawater at the rudiments within the luminometer cuvette somehow activated the rudiments to discharge the remaining lumisomes at increasing frequencies over several minutes even without further stimulation. These semi-spontaneous or delayed flashes varied considerably in amplitude even when derived from the same house rudiment. On the other hand, most flashes followed the same time course with a rapid rising phase, a slower declining phase, and a total duration of ~300 ms. However, at high gain an elevated baseline activity of light emission was frequently observed for the subsequent 3–5 s. Prolonged increases in baseline emission of light over several minutes were often seen in rudiments stressed by repeated seawater squirts or in rudiments stressed by chemical means. This prolonged baseline increase was usually associated with an increased frequency of flashing (Fig. 13.11c). When intact houses are stimulated by the turbulence of small squirts of water or a stirring motor in a light-tight photomultiplier chamber for measurement of total stimulable luminescence (TSL), light is emitted as trains of partially summated bursts over several seconds (Fig. 13.11d). Individual flashes ranged from about 40 to 300 ms, with rise times and half decay times of about 10–60 ms (Galt, 1978; Galt and Sykes, 1983; Galt *et al.*, 1985). Amplitudes of such summated flashes reached 10^{12} quanta s^{-1} in houses of *O.labradoriensis* (Galt and Grober, 1985). The emission spectrum of appendicularian bioluminescence, which is blue-green to the human eye, has been determined for only one species, *O.dioica* (Widder *et al.*, 1983). In a sample consisting of several animals, peak emission was at 483 nm, with a relatively broad bandwidth of 95 nm (full width at half-maximum amplitude). This emission is consistent with that of the majority of pelagic organisms (Herring, 1983; Widder *et al.*, 1983) and is within the range of the absorption maxima of visual pigments of pelagic animals (Young, 1981). Light from individual inclusion bodies (each containing some 20–100 lumisomes) from *O.labradoriensis* and *O.dioica* has also been recorded directly on diapositive colour film and revealed slight differences in hue between the two species (with light from *O.dioica* appearing more green than light from *O.labradoriensis*), but not between inclusion bodies of the same species (Flood, unpublished).

13.8. Modes of stimulation

There is little evidence that appendicularian houses emit light in the absence of external stimuli. Photo-

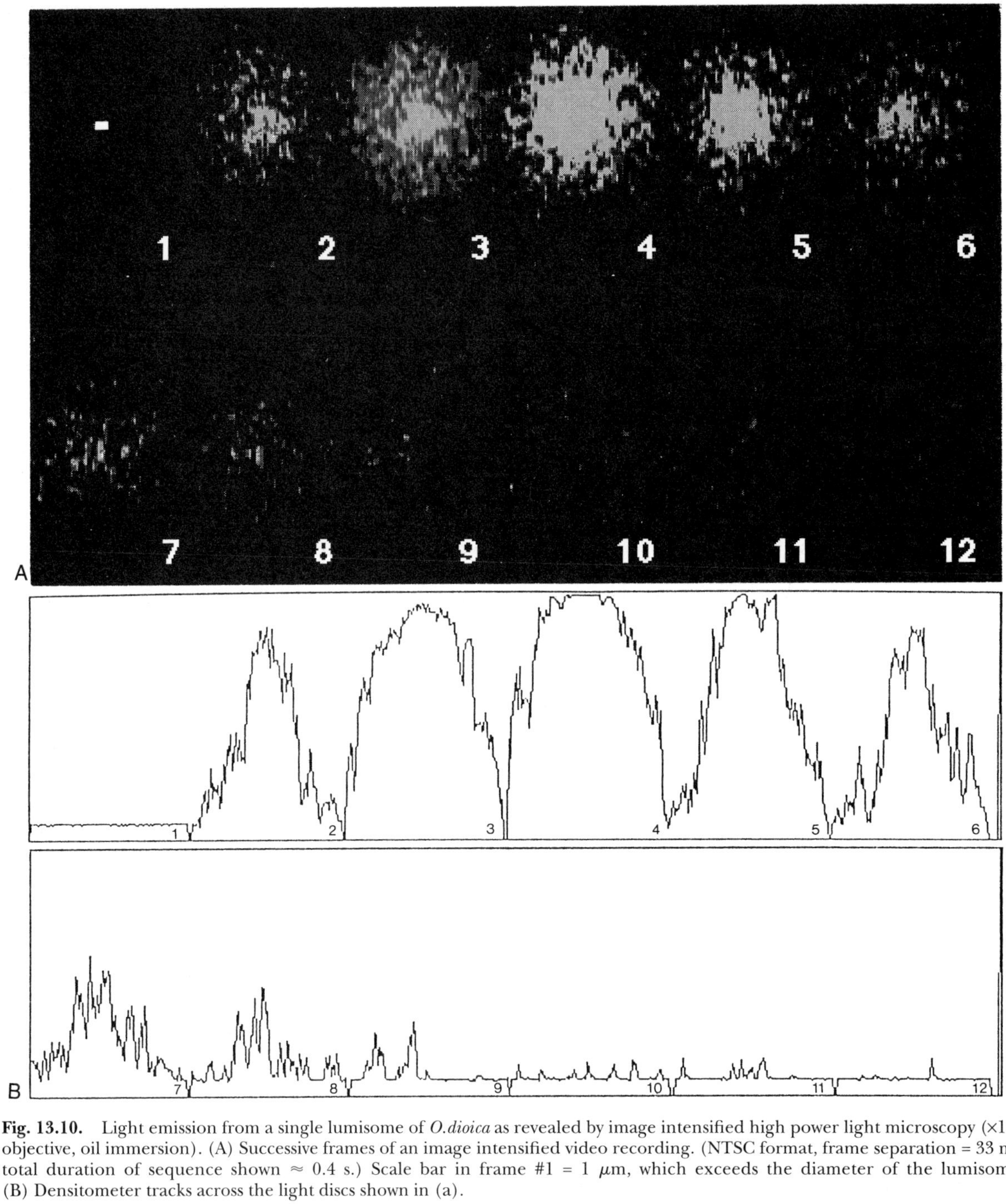

Fig. 13.10. Light emission from a single lumisome of *O.dioica* as revealed by image intensified high power light microscopy (×100 objective, oil immersion). (A) Successive frames of an image intensified video recording. (NTSC format, frame separation = 33 ms, total duration of sequence shown ≈ 0.4 s.) Scale bar in frame #1 = 1 μm, which exceeds the diameter of the lumisome. (B) Densitometer tracks across the light discs shown in (a).

multiplier recordings of house rudiments picked from the trunks of *O.dioica* and *O.labradoriensis* reveal occasional weak flashes of light. However, the flashes are fewer the less disturbed the rudiment is, and most emit no flashes, even when they later prove to emit much light in response to mechanical stimulation. In numerous observations of several species, we have never observed or recorded spontaneous luminescence during house expansion (Galt and Smith, unpublished, for *O.dioica*, and *O.labradoriensis*; Galt and Fenaux, unpublished, for *O.albicans*). The luminescent system is sensitive to very slight turbulence or disturbance, so that apparently spontaneous flashes from swimming animals or their houses may result from contact with vessel walls

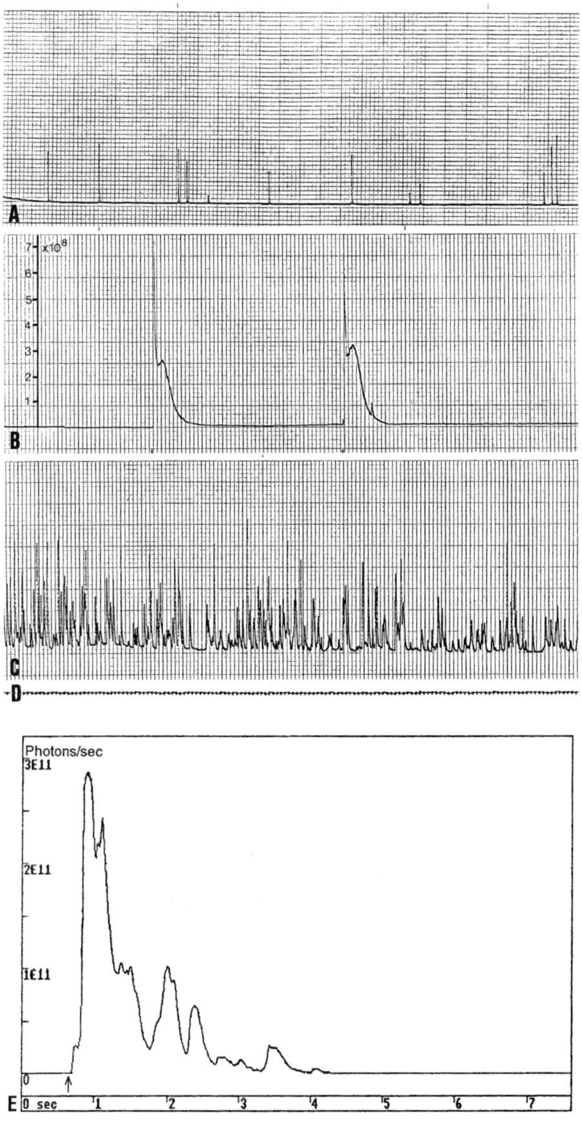

Fig. 13.11. Luminometer trace recordings of light emission from *O.labradoriensis*. (A) Reveals (semi) spontaneous light flashes of approximately 0.3 s duration in a house rudiment previously slightly disturbed mechanically. (B) Shows summated light discharges with prominent afterglow during mechanical stimulation of a house rudiment by squirting 10 μml seawater into the (1.5 ml) cuvette. (C) Shows increased baseline activity and several flash-like discharges in a more severely disturbed house rudiment. (D) Time marks of 1 mm/s applies to all tracks above. Vertical scale in (B), in approximate number of photons/s, also applies to (A) and (C). (E) Shows trains of partially summated bursts of light in response to continuous stirring of a clean house of *O.dioica* in a 20 ml vial. Integration of such responses over time gives Total Stimulable Luminescence (TSL).

or slight currents in the viewing chamber or *in situ*, from the animal's own swimming motions. However, given the rapid motions and turbulence generated by the animal during house expansion, the lack of luminescence in expanding houses is difficult to reconcile with the extreme sensitivity of rudimentary and fully developed houses to prodding or weak currents. Nonetheless, we propose that the spontaneous and continuous luminescence described by Lohmann (1899a, 1933) resulted from contact with the walls of the observation vessel or the surface during the increasingly vigorous swimming that accompanies expansion of the house. The stimulus that normally elicits luminescence from appendicularian houses under field conditions is probably agitation or turbulence in the surrounding water. However, in the laboratory flashes can be triggered from houses or house rudiments in a variety of ways. Luminescence can be elicited from the mucous house rudiments adhering to the animal's trunk, or from rudiments dissected from the trunk. Simply tapping the slide on which the rudiment lies in a drop of seawater, touching a coverslip placed on the rudiment, or adding drops of distilled water or other agents to the rudiment in seawater all cause luminescent responses. Prodding or agitating the occupied or discarded expanded houses also elicits a luminescent response from similar point sources dispersed over the surface of the house (Flood and Galt, in preparation). These dispersed point sources are distributed in species-specific linear patterns over the entire external surface of expanded houses (Galt and Flood, 1984; Flood, in preparation). Effective modes of *in vitro* excitation include mechanical (prodding or other agitation of the house), physicochemical (distilled water and detergents such as Triton-X100), and ultrasonic (laboratory sonicators and cleaners) stimuli. Preliminary attempts to elicit luminescence with electrical stimulation of house rudiments have been unsuccessful (Galt, unpublished).

13.9. Biochemical basis of appendicularian luminescence

Little is yet known on this subject. A classical luciferin–luciferase reaction is probably involved, as extracts of heat-inactivated house rudiments of *O.labradoriensis* usually caused new light emission when added to house rudiments previously stimulated to exhaustion of their light emission by Triton-X100 or mechanical stirring. Likewise, methanol extracts of fresh *O.labradoriensis* house rudiments consistently caused light emission when diluted by water and added to similarly exhausted house rudiments, but not when added to heat inactivated or methanol-treated house rudiments (Flood, unpublished). The bioluminescence system of *O.labradoriensis*, and probably all other luminescent appendicularians, seems to belong to the family of coelenterazine-driven systems, since purified coelenterazine caused new light emission from previously exhausted house rudiments and since methanol extracts of fresh house rudiments emitted light when added in diluted form to purified luciferase from a sea pen (*Ptilosarcus* sp.) (Flood and Klabusay, unpublished). The luciferase of appendicularians is probably concentrated inside the lumisomes, as these were stained heavily by Coomassie blue and thus contained much more protein than the surrounding extracellular matrix of the inclusion bodies. The lumisomes also had a weak tendency to oxidize diaminobenzidine to form the precipitate characteristically found in peroxidase reactions (Flood, unpublished). This is in agreement with the notion that 'all luciferases share the feature of being oxygenases' (Hastings, 1983a, p. 309). More recent information may suggest that the oxidation is a function of substrate structure. Accordingly, rather than all luciferases being oxygenases, it may be said that: 'the emission of light from a living organism is now clearly established to be the result of an oxidative chemical reaction' (McCapra, 1990, p. 265). The coelenterazine or coelenterazine-like luciferin of appendicularians is probably also located within the lumisomes, maybe even complexed to the luciferase. The prominent yellow-green autofluorescence of the intact lumisomes, and the lack of such fluorescence from fully spent lumisomes perhaps supports such an hypothesis. The excitation mechanism leading to light emission from appendicularian lumisomes is still obscure. Depolarization by 0.5 M KCl caused no light emission, and KCN or NaN_3 blocking of metabolism overnight had no effect on light emission. Likewise, prolonged incubation in non-ionic media or in calcium- and magnesium-free media (with EDTA) did not block light emission. Still, elevated extracellular Ca^{2+} concentration and Ca^{2+} ionophores caused increased baseline activity of light emission and increased frequency of flashing from most house rudiments when tested in a sensitive luminometer (Flood, unpublished). We incline to the assumption that yet-unidentified chemical changes, caused by the breakdown of the lumisome boundary membrane in response to mechanical stimulation, somehow initiate the interaction between luciferin and luciferase and the emission of light. Probably the enzyme and its substrate will have to be isolated in true solution before further details may be elucidated.

13.10. Endogenous origin of luminescence in appendicularians

One may speculate whether the appendicularian bioluminescence system is wholly intrinsic to these metazoans or whether it depends on 'borrowed' chemicals or organisms, as is often found in other systems (Nealson and Hastings, 1979; Warner and Case, 1980; Frank *et al.*, 1984; McFall-Ngai and Ruby, 1991). This question is particularly pertinent since the closest bioluminescent relative of appendicularians, the zooids of the pelagic tunicate *Pyrosoma* sp., apparently base their bioluminescence on modified bacteria (see section 4.2.1), capable of emitting flashes of light, under the control of the host organism (Mackie and Bone, 1978; Leisman *et al.*, 1980; Nealson and Hastings, 1980). The lumisomes of appendicularians, as described in the previous sections, are comparable to bacteria in size and they originate from the oral gland, which is located close to the oral cavity where all food particles must pass. They also undergo a rapid and pronounced maturation process to acquire their bioluminescent property long after they have left the oral gland. However, in spite of several attempts, it has been impossible to find direct evidence for a bacterial origin of appendicularian lumisomes. A test for bacterial luciferase was negative (Galt, 1978).

Appendicularians grown in sterile filtered seawater for several days (Galt, 1978; Galt and Sykes, 1983; Galt *et al.*, 1985; Galt and Smith, 1990), or overnight in the presence of antibiotics or metabolic inhibitors like KCN and NaN_3 (Flood, unpublished), display normal bioluminescence. DNA-specific fluorochromes, like DAPI and Hoechst 33258, indicate that DNA, if present at all, must be an order of magnitude less abundant than in commensal bacteria attached to the house rudiments (Flood and Galt, unpublished). Previous published ultrastructural observations on the oral glands and their secretion (Fredrikson and Olsson, 1981) and tests with a coelenterazine-related luciferin and luciferase (Flood and Klabusay, unpublished) also lend no support to a bacterial hypothesis. We are, however, still puzzled by the complex involvement of the oral glands in the deployment of a cellular process to distant parts of the house rudiment and by the involvement of this process in the origin and maturation of highly specialized lumisomes in the otherwise completely acellular house rudiment. Detailed ultrastructural studies of these events during all stages of the 'secretion' and maturation process, are therefore in progress (Andersen and Flood, unpublished).

13.11. Geographical distribution of bioluminescent appendicularians and their contribution to environmental luminescence

The appendicularians with inclusion bodies and subchordal cells known or presumed to be luminescent (Table 13.1) include species from a variety of pelagic habitats (Galt and Tisdale, 1983; Fenaux, 1993; Youngbluth *et al.*, in preparation). These range from epipelagic coastal (*O.dioica*), oceanic (nearly all the remaining luminescent species), warm and warm–temperate (*O.rufescens, O.mediterranea, Stegosoma magnum*, and others), cold polar and subpolar (*O.labradoriensis, O.vanhoeffeni* and *O.gaussica*), to mesopelagic waters (*O.caudaornata, O.inflata, O.villafrancae* and *O.villafrancae indentacauda*). The dominant species in coastal (*O.dioica*), north-polar (*O.labradoriensis* and *O.vanhoeffeni*), south-polar (*O.gaussica*) and subpolar surface waters are luminescent (Galt and Tisdale, 1983). In tropical and subtropical oceanic waters, luminescent species typically rank third or fourth behind the non-luminescent and often strongly dominant *O.longicauda* and *O.fusiformis*. Finally, of perhaps 8–10 recently discovered mesopelagic species (Fenaux 1993), at least three are probably luminescent. Thus bioluminescent appendicularians are widely distributed, often abundant members of epipelagic (particularly coastal and high latitude) and mesopelagic communities and must therefore be considered in studies of bioluminescence in the water column. The luminescence capacity or bioluminescence potential of an organism may be measured as the total light output (photons) resulting from the mechanical stimulation of that organism to exhaustion. This variable, designated 'total stimulatable luminescence' (TSL), is useful in determining the potential contribution of a species to near-surface luminescence. For appendicularian houses, which are incapable of recharging the luminescent system, TSL, when correlated with certain biological parameters, also gives information about the luminescent system itself. Galt and Grober (1985) and Galt and Smith (1990) used a calibrated photometer to collect extensive TSL data for both *O.dioica* and *O.labradoriensis* houses as a function of species, animal and house size, age of the house after its production, and other variables (Fig. 13.12). TSL ranged from about 10^{10} to nearly 10^{13} photons house^{-1}. Mean TSL was significantly greater for *O.labradoriensis* than for *O.dioica*, although if size was factored out in an analysis of covariance, the interspecific differences were less significant (*O.dioica* is a small species compared to *O.labradoriensis*). Mean TSL was significantly greater for clean houses produced in the laboratory in particle-free seawater than for field collected houses. There was no significant difference in TSL between the first and second successive clean houses produced by the same animal. The effect of animal size on TSL of its house was especially strong for *O.labradoriensis*, where TSL increases significantly with size of animal (and therefore with size of house, which increases with animal size) and with the number of inclusion bodies in the house. We did not measure TSL from individual inclusion bodies, but we estimate from whole house TSL data combined with inclusion body counts in similar sized houses, that each inclusion body emits about 10^9 photons when stimulated maximally. This compares well with the 10^7 photons recorded

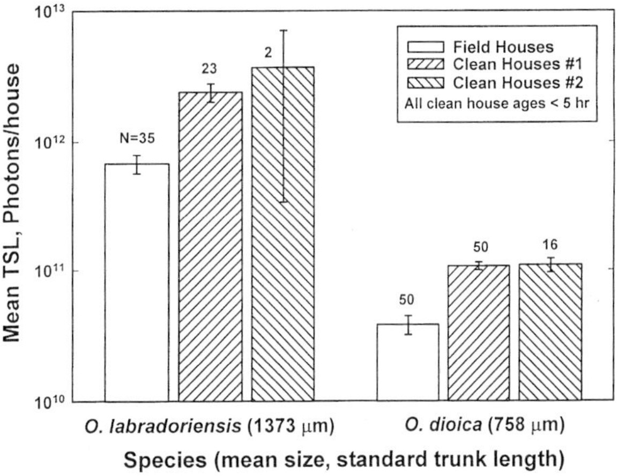

Fig. 13.12. Total Stimulable Luminescence (TSL, means with standard error bars) of field-collected houses and successive (#1, 2) laboratory-produced (clean) houses of *O.labradoriensis* and *O.dioica* of indicated mean sizes. (Standard Trunk Length, STL, from mouth to posterior extent of stomach). Clean houses yielded significantly more light than field-collected houses, and houses of *O.labradoriensis* produced significantly more light than those of *O.dioica*, though this effect may be due to size differences between the two species.

from presumed single lumisome discharges (Flood, unpublished), since each inclusion body may contain some 100 lumisomes. TSL decreased with the age of a house after it was expanded. There was little decline in TSL during the first 4–6 h, during which time the house would normally be occupied. After 6 h, however, for *O.labradoriensis*, TSL decreased markedly with house age, down about 100-fold in 2 days and indistinguishable from instrument noise in 3–4 days. Galt and Smith (1990) estimated luminescent appendicularian population contribution to water column bioluminescence potential as follows. The mean TSL for *O.dioica* and *O.labradoriensis* is about 10^{11} and 10^{12} photons house^{-1} respectively (Galt and Smith, 1990; Galt, Smith and Grober in preparation). Abundances of these two species are in the range of about 10^2–10^4 and 10–10^2 m^{-3}, respectively. We assume that each animal produces about 5 houses day^{-1} (Fenaux, 1985, for *O.dioica*; Galt, unpublished, for *O.labradoriensis*), and that there are about nine discarded houses for each occupied house in the water column (Alldredge, 1976a). Then the rough bioluminescence potential for appendicularian luminescence, based on these two species alone, is of the order of 10^{14}–10^{16} photons m^{-3}. Assuming that the mean appendicularian densities apply through a 10 m water column, the bioluminescence potential is some 10^{15}–10^{17} photons m^{-2}, similar to the total near-surface luminescence potential of about 10^{15} photons m^{-2} reported by Batchelder and Swift (1989) and Batchelder *et al.* (1992), and considerably higher than recently published values for colonial Radiolaria

(3×10^{12} photons m^{-2}), and up to 10–100 times greater than estimates of total near-surface bioluminescence potential from all plankton combined (about 2×10^{14} photons m^{-2} in the western Sargasso Sea; Latz *et al.*, 1987). We conclude that appendicularians can make significant contributions to water column luminescence, and that the luminescence from occupied and discarded houses must be taken into account and not just the numbers of animals themselves.

13.12. Adaptive value of luminescence in appendicularians

It is likely that bioluminescence confers adaptive advantage to the appendicularian species that possess it, although at present there is no direct evidence of any differential fitness attributable to the presence or lack of luminescence. Reviews (Buck, 1978; Morin, 1983; Young, 1983) suggest several functional roles for bioluminescence, but most are unlikely to apply to appendicularians, simply because of aspects of their biology and the properties of their light emission. Appendicularian houses emit light only when agitated, and the flash duration is short; appendicularians possess no known photoreceptive capacity; and appendicularians filter phytoplankton. Therefore, functions that depend on emitter-initiated, sustained emission of light or that require vision, such as intraspecific communication for reproduction or aggregation, or illuminating or luring prey, cannot apply. Neither is concealment by counter-illumination tenable. Luminescence may be an aposematic signal that warns a predator of unpalatable prey (Porter and Porter, 1979; Grober, 1988). In recent studies (Galt, 1995), juveniles of several species of reef fish rejected houses of certain non-luminescent species, providing some evidence for the non-palatibility of houses. However, both luminescent and non-luminescent species are common prey of numerous kinds of predators (see Table 10.2).

McAllister (1967) suggested that the numerous rows of photophores in midshipman fish may mimic luminescent ctenophores. Similarly, the rows of luminescent inclusion bodies on the houses of oikopleurids may mimic a large, gelatinous tentaculate carnivore rather than a relatively small herbivore enclosed in a nutrient-rich house. The ease with which luminescent inclusion bodies slough off the surfaces of appendicularian houses in the laboratory (Galt and Smith, unpublished) implies that loss of inclusion bodies is fairly common in gentle currents or in the turbulence associated with an approaching fish predator. In another form of visual deception, the sudden release of what may appear to be numerous flashing dinoflagellates may mask the presence of what is in fact a much larger potential prey organism. The functions of appendicularian luminescence that seem most tenable are predator deterrence resulting from contact with or turbulent agitation of the house. Mensinger and Case (1992) provided evidence supporting the 'burglar alarm' hypothesis, in which luminescence of prey (dinoflagellates) under attack by a predator (a crustacean) increases the vulnerability of the first predator to its own secondary predator (a fish). Such a role is plausible for appendicularians, but has not been tested. The luminescent appendicularian house may function as a sort of sacrificial lure or decoy, a 'deceptive diversion' (Morin, 1983) that distracts a predator while the appendicularian leaves the house and swims to the safety of darkness. Indeed, even the houses of non-luminescent species reduce predation on the oikopleurid in this manner (Lohmann, 1933; Galt, 1995). Finally, luminescence in appendicularian houses may repel predators that attack the house (Morin's, 1983, 'contact flash predator deterrent') by overloading their visual system ('blinding' eyes that are adapted to low light levels), otherwise startling or distracting the predator, or altering its behaviour in such a way as to interfere with the attack (Buskey and Swift, 1983, 1985; Buskey *et al.*, 1983). Such deterrence of predation has been documented for copepods feeding on luminescent dinoflagellates (Esaias and Curl, 1972; White 1979). The presence of species-specific geometric patterns of luminescence on appendicularian houses (Galt *et al.*, 1985; Flood and Galt, in preparation) may suggest specificity between predator and prey. Interestingly, the non-luminescent species *O.longicauda* exhibits a behavioural escape response in reaction to being approached by a probe, pipette, or predator in which the entire occupied house tumbles away from the point of stimulus (Alldredge, 1976b; Galt, 1995). This reaction has not been reported in other species and particularly not in the luminescent species we have studied, perhaps indicating that non-luminescent species have evolved other, behavioural mechanisms of reducing predation in the absence of luminescence. Lohmann (1899a, 1933) suggested that a dramatic display of luminescence from *O.albicans* during expansion of its house deterred predators at a time when the animal is particularly vulnerable. We have not observed luminescence during numerous examinations of house expansion by *O.dioica*, *O.labradoriensis*, and *O.albicans* (Flood, Galt and Flood, unpublished; Galt and Smith, unpublished; Galt and Fenaux, unpublished). In contemplating a predator deterrent role of appendicularian luminescence, one may consider the known predators of appendicularians and their sensory capabilities. Luminescence benefits the emittor only at sufficiently low light levels. This restriction requires that the candidate predator detects prey at low light levels, either non-visually or with sensitive photoreceptors. Nevertheless, vision must be adequate for the predator to respond to a luminescent stimulus.

13.13. Some considerations for future research

Much effort has still to be put into working out the ontogenetic development of the bioluminescent system to understand the involvement of the oral gland in appendicularian luminescence. The co-occurrence of subchordal cells in all bioluminescent appendicularian species also calls for further investigation. Another problem of morphogenetic and ecological significance is the species specificity of the geometrical patterns of luminescent inclusion bodies. Is there any relation to the specific adaptive value of light emission among the several species? A detailed examination of the chemical aspects and molecular excitation mechanisms of appendicularian luminescence will be equally important, especially since it seems possible that a unique system, different from all other known bioluminescence systems, is involved. As will have been evident from section 13.12 above, the adaptive role of appendicularian bioluminescence remains entirely mysterious; it is not difficult to suggest functions, but hard indeed to prove that they apply to appendicularian bioluminescence. Larval and adult fish are probably the major predators on both luminous and non-luminous appendicularians, (see p. 110) and feed largely visually. As yet, the interactions between fish larvae and luminescent appendicularians in their houses (at different light levels) remain to be studied, and this approach might prove rewarding.

13.14. Acknowledgements

We thank the director and staff of Friday Harbor Laboratories, University of Washington, for excellent working facilities and support during several visits. We also thank our collaborators J. Blinks, M. Latz, and E. Ridgeway for advice and stimulating discussions leading to this review. Research by CPG was supported in part by grant N00014-85-K-0591 from the US Office of Naval Research.

FOURTEEN

The cladistic biogeography of salps and pyrosomas

R.W.M. Van Soest

14.1. Introduction

The biogeography of salps and pyrosomas has so far been treated only by the classical approach of tabulating individual distributions, combining them into a limited number of inferred areas of endemism, and comparing these with abiotic parameters such as temperature and ocean current patterns (Van Soest, 1975, 1979; Van der Spoel and Heyman, 1983). This ecological approach emphasizes salp and pyrosoma recent ecology, ignoring the historical causes underlying the distribution patterns. However, it cannot answer such questions as: why are there no Arctic salp species?; where and when did the (sub)Antarctic species arise?; why are so many taxa restricted to the Indo-West Pacific?

Such questions might perhaps be answered by inferring possible scenarios for the observed distribution patterns from the geological history of the oceans (e.g. Van Soest, 1975). However, although this approach may work well in certain land and freshwater areas and much less well in benthic marine areas, in the pelagic environment it is virtually useless since there are great uncertainties about palaeoecology and palaeocurrent patterns earlier than the Pleistocene.

A promising rival approach is cladistic biogeography (e.g. Humphries and Parenti, 1986; Nelson, 1986). This involves constructing phylogenies of monophyletic groups, placing the distributions of the individual taxa on the cladograms (area cladograms), so acquiring the distributional history of the groups and then analysing the resultant patterns. The cladograms provide information on the history of speciation and on past distributions, which can then be compared with geological data. By seeking a universal pattern in the available area cladograms (Wiley, 1988), which will reflect the history of the areas of endemism, the geological data is no longer needed. In the case of pelagic organisms, this yields scenarios of diversification in the open ocean realm.

A first attempt at cladistic biogeography for pelagic organisms was made recently by Van der Spoel *et al.* (1990) using Euphausiids. It was inspired by the attempt to discover whether central gyral distributions may be considered the historically stable areas or 'hydroplates' (Van der Spoel *et al.*, 1990) in the pelagic environment, whose history would give clues to past distributions and diversification. Unfortunately, this study had some basic flaws (cladograms were presented as solved, but in fact contained many unsolved parts), which affected the conclusions to such an extent that the leading questions could not be answered. However, the study had the considerable merit of showing pelagic biogeographers that historical approaches are entirely possible.

Pelagic tunicates, and salps in particular, are promising subjects for cladistic biogeographical studies. Taxa and distributions are well known, diversity is low, and some character systems with potential for phylogenetic inference have been made operational. Several of the larger taxa (*Salpa, Thalia, Cyclosalpa,* Pyrosomatidae) potentially contain information on the history of the areas in which they occur. For pyrosomas a cladogram is already available (Van Soest, 1981), while for the salp genus *Thalia* an attempt at numerical character analysis has been published by Dauby and Godeaux (1987). This chapter explores the possibilities for historical explanations of observed pelagic tunicate distribution patterns, by constructing cladograms and combining these into a general area cladogram *sensu* Wiley (1988).

14.2. Methods and sources

Making a cladogram of a group involves the recognition of synapomorphic characters, the recognition of one or more outgroups with which to polarize in-group characters, and a numerical cladistic analysis using the principle of parsimony for conflicting character distributions (see for example. Wiley, 1981, for a general introduction to the subject). Below, this will be done for *Salpa, Thalia* and *Cyclosalpa.* The Pyrosomatidae cladogram is taken from Van Soest (1981).

Constructing taxon area cladograms involves determining the areas of endemism of the taxa. While this may be a simple task with certain freshwater or terrestrial taxa, it is much more difficult in the marine habitat. The major problem is the overlapping or widespread distribution, which is the rule rather than the exception with marine organisms, and holoplanktonic organisms are the ultimate examples of this. These overlapping distributions make it difficult to recognize discrete areas of endemism whence taxa may have spread over larger areas. Van der Spoel and Heyman (1983) presented a comprehensive view of the distribution patterns in plankton, which showed clearly that within group variation as well as amongst group variation in distributions is large. Still, distributions are restricted, even if large and overlapping, so the notion of areas of endemism is probably valid also for planktonic organisms. Two rival approaches to identify areas of endemism will be compared and discussed.

A simple approach to determining areas of endemism would be to take all the records of each taxon (see also discussion in Henderson, 1991). However, that would result in a large number of unique, slightly different, areas of distribution for each of the taxa involved, and analysis of their relationships would be seriously affected by the (lack of) comprehensiveness of the distribution data. To avoid this problem areas of endemism are inferred from recurring roughly similar areas in various pelagic groups (in this case all thaliacean species). Thus, the present areas of endemism usually unite distributions of several species (Platnick, 1991). However, in order not to compromise true areas of endemism by adding distributions too divergent from each other, a number of areas of endemism used in this study are based upon a single species. Recognized areas of endemism will in many cases be partly or wholly overlapping with one another. Because these partially overlapping areas may very well have originated at different times, they are taken as the basis for scoring taxa in their area of endemism. Single records are considered to be representative for an area of similar abiotic environment (e.g. the single record of *Pyrosoma godeauxi* is considered to be a sign of occurrence in the 'Western Indian Ocean' area). Likewise, gaps in certain distributions which are obviously due to poor sampling intensity are ignored. For example, a species occurring in the Western Indian Ocean and the Western Pacific Ocean but apparently lacking in the Eastern Indian Ocean is scored in the same area of endemism as a species that is continuously distributed over those areas.

The second approach is to adopt the areas given in 'Biogeographical subdivision of the world oceans based on salp distribution' (Van Soest, 1975, fig. 12). Areas are identified as potential areas of endemism on the basis of one or more taxa restricted to them. In such a way 11 non-overlapping areas have been identified. However, most salp and pyrosoma species have distributions spreading over two or more of these areas and thus a serious methodological problem in constructing the matrix of taxa and distributions is apparent, involving Assumptions 1 and 2 of Nelson and Platnick (1981; cf. Humphries and Parenti, 1986). Assumption 0 (Zandee and Roos, 1987; Wiley, 1988) will be adopted in this analysis.

To avoid confusion, in Fig. 14.1 all areas of endemism in both approaches are given as one or more of the numbered areas in Van der Spoel and Heyman's (1983) 'Areal division of the oceans' (fig. 194). All Thaliacean distributions, including those of the taxa studied below, are listed in Table 14.1 using the 22 areas of Van der Spoel and Heyman (1983). Sources: salps, Van Soest, 1975; pyrosomas, Van Soest, 1981; doliolids, Borgert, 1894; Neumann, 1913; Godeaux, 1973a,b, 1987, 1988a,b; Esnal, 1978; Esnal and Simone, 1982. Infraspecific taxa (e.g. the subspecies *Thalia democratica indopacifica* and *Cyclosalpa quadriluminis parallela*) are omitted, but *Salpa gerlachei* and *Pegea socia* are accepted provisionally as discrete species (see Chapter 17).

Finding a general area cladogram using Wiley's (1988) biogeographic parsimony analysis (BPA) involves combining the matrices of areas and taxa of the taxon area cladograms of *Cyclosalpa, Salpa, Thalia* and Pyrosomatidae into a single large matrix and feeding this to the numerical cladistic computer program PAUP 3.0 (Swofford, 1988). The program will determine the most parsimonious relationships of the areas of endemism; in the case of several equally parsimonious solutions, a consensus cladogram is produced. Scenarios of diversification and dispersal of individual taxa can be inferred 'objectively' from the general area cladogram. These scenarios may then be compared with geological information.

14.3. Areas of endemism of Thaliacea

14.3.1. *Approach 1*

From the sources listed above, the following areas of endemism were recognized in the first approach (the numbers refer to those in Fig. 14.1); areas pertaining solely to salps of other than the genera studied as well as only to doliolids are omitted.

A. Widespread in all three oceans from subarctic/subantarctic to tropical latitudes: *Salpa fusifomis, Thalia democratica, Pyrosoma atlanticum, Doliolina krohni* (4–22).

B. Widespread in all three oceans from 40°N to 40°S, following approximately the 10°C isotherm in the north and the 15°C isotherm in the south: *Salpa fusiformis, Pegea confoederata, Thalia orientalis, Ihlea punctata, Doliolum denticulatum, Doliolum nationalis, Dolioletta gegenbauri* (4–17,21,22).

C. Widespread in all three oceans from subtropical to tropical latitudes, 35°N to 35°S, following the 15°C isotherm in the north (including the Mediterranean) and the 20°C isotherm in the south: *Cyclosalpa polae, Helicosalpa virgula, Iasis zonaria* (4–10,12–16, 19–22).

D. Widespread in all three oceans from boreal to tropical latitudes, 40°N to 40°S, following approximately the 15°C isotherm, but excluding the Mediterranean and the Red Sea: *Salpa aspera, Ritteriella retracta, Tethys vagina* (4–20).

E. Widespread in all three oceans from subtropical to tropical latitudes, 30°N to 30°S, following the 20°C isotherm: *Salpa younti, Ritteriella amboinensis, Traustedtia multitentaculata, Weelia cylindrica, Brooksia rostrata, Brooksia berneri* (4–17, 19–21).

F. Widespread distribution in all three oceans from subtropical to tropical latitudes in the northern hemisphere, but restricted to tropical latitudes in the southern hemisphere: *Cyclosalpa affinis* (4–8,12–16,18, 20,22).

G. Widespread distribution in all three oceans from 40°N to 40°S, but absent in the northern Pacific, and the Mediterranean: *Pyrostremma agassizi* (4–6,9,12–16, 18,21).

H. Western Indian Ocean tropical distribution: *Pyrosoma godeauxi, Dolioletta chuni, Dolioletta mirabilis, Dolioletta valdiviae, Doliolina indica, Doliolina sigmoides* (10,12,13).

I. Indo-West Pacific tropical distribution: *Salpa tuberculata, Ritteriella picteti, Cyclosalpa ihlei, Thalia rhomboides. Helicosalpa komaii, Helicosalpa younti* (10, 12–15,19).

J. Indo-Pacific tropical distribution: *Cyclosalpa sewelli, Metcalfina hexagona, Pyrosomella verticillata, Pyrosomella operculata* (9–16,18).

K. Antarctic–Subantarctic distribution: *Salpa thompsoni, Ihlea magalhanica, Doliolina resistibile* (1,2,8–11,17,18).

L. High Antarctic distribution: *Salpa gerlachei, Ihlea racovitzai* (2).

M. East Pacific tropical distribution: *Cyclosalpa strongylenteron, Doliolina obscura, Doliolina separata, Doliolina undulata* (16).

N. Atlantic bi-antitropical distribution (lacking in the tropical Atlantic and in all other areas): *Cyclosalpa pinnata, Pegea socia* (4,5,7).

O. Southern Ocean antiboreal distribution: *Thalia longicauda* (8–11,18).

P. West Pacific tropical distribution: *Thalia rhinoceros, Thalia sibogae* (14–16).

Q. Widespread distribution in all three oceans from subtropical to tropical latitudes, excluding the tropical Atlantic (bi-antitropical): *Pegea bicaudata, Pegea socia, Pyrostremma spinosum.* (4,5,8–15,18,19,22).

R. Antiboreal Atlantic–tropical Indian Ocean distribution: *Pyrosoma ovatum* (8,9,12).

S. Tropical and northern Indo-Pacific: *Cyclosalpa quadriluminis* (12–16,19,20).

T. Tropical Atlantic–Indian Ocean distribution: *Cyclosalpa danae, Cyclosalpa floridana, Thalia cicar, Pyrosoma aherniosum* (4,6–8; 10, 12–16, 18–19, 21).

U. Tropical Atlantic–Indian Ocean and northern Pacific distribution: *Cyclosalpa bakeri* (4,7,12,15,16,19,20).

V. Tropical Atlantic–subtropical Indo-West Pacific distribution: *Salpa foxtoni* (4,10–15,18,19).

14.3.2. *Approach 2*

Areas of endemism based on salps (from Van Soest, 1975, fig. 12, except his areas I [Arctic, no thaliaceans known from it] and XI [Northern Temperate Pacific, merged with area X because of insufficient thaliacean distribution data]). Characteristic species for the areas are noted and the Van der Spoel and Heyman notation is given in parentheses. Many of the areas can be further characterized by clinal forms of widespread salp species (cf. Van Soest, 1975; De Visser, 1986; De Visser and Van Soest, 1987).

I. Boreal Atlantic: *Pegea socia* (4,5,?22).
II. Gulf Stream–Canary Current area: *Doliopsoides meteori* (6).
III. Tropical Atlantic: *S.fusiformis* with low muscle fibre number (7).
IV. Southern Ocean subtropical: *Thalia longicauda* (8–11,18).
V. Antiboreal: *Ihlea magalhanica* (1)
VI. High Antarctic: *Ihlea racovitzai* (2)
VII. Indian Ocean: *Pyrosoma godeauxi* (12–14)
VIII. West Pacific: *Thalia rhinoceros* (15)
IX. East Pacific: *Cyclosalpa strongylenteron* (16)
X. Northern Pacific: *Thalia democratica* with high muscle fibre number (19,20).

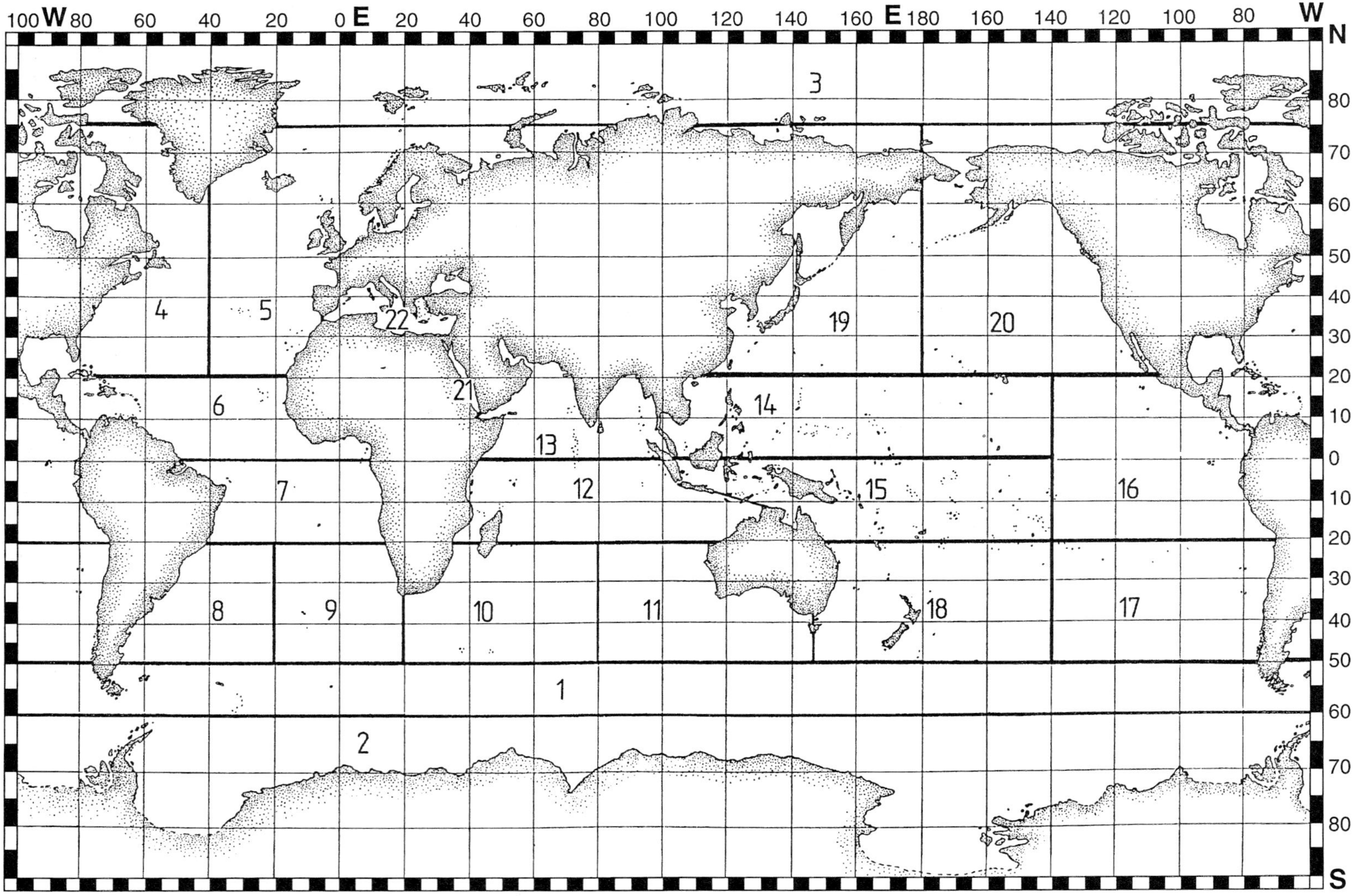

Fig. 14.1. Areal division (numbers 1–22) of the oceans (after Van der Spoel and Heyrnan, 1983), used for the tabulation of thaliacean distributions. In the text thaliacean areas of endemism are presented as coded area numbers referring to this map.

Table 14.1 Distributions of Thaliacea over the 22 areas of Fig. 14.1.

Species	Area																					
	1	2	3	4	5	6	7	8	9	10	11	12	13	14	15	16	17	18	19	20	21	22
SALPIDAE																						
Salpa fusiformis	–	–	–	x	x	x	x	x	x	x	x	x	x	x	x	x	x	x	x	x	x	x
*Salpha thompsoni**	x	x	–	–	–	–	–	x	–	–	–	–	–	–	–	–	x	x	–	–	–	–
*Salpa gerlachei**	–	x	–	–	–	–	–	–	–	–	–	–	–	–	–	–	–	–	–	–	–	–
Salpa aspera	–	–	–	x	x	x	x	x	x	x	x	x	x	x	x	x	x	–	x	x	–	–
Salpa younti	–	–	–	x	–	x	x	–	x	x	x	x	x	x	x	x	–	x	x	x	–	–
Salpa maxima	–	–	–	x	x	x	x	x	x	x	x	x	x	x	x	x	x	–	–	–	–	x
Salpa tuberculata	–	–	–	–	–	–	–	–	–	–	–	x	–	–	x	–	–	–	–	–	x	–
Weelia cylindrica	–	–	–	x	x	x	x	x	x	x	x	x	x	x	x	x	x	–	x	x	x	–
Ritteriella amboinensis	–	–	–	–	x	–	–	–	–	x	–	x	x	x	x	x	–	x	x	x	x	–
Ritteriella picteti	–	–	–	–	–	–	–	–	–	x	–	x	x	x	–	–	–	–	–	–	–	–
Ritteriella retracta	–	–	–	x	x	x	x	x	x	x	–	x	x	x	x	x	x	x	x	x	–	–
Metcalfina hexagona	–	–	–	–	–	–	–	–	x	x	x	–	x	x	x	x	–	x	–	–	–	–
Ihlea punctata	–	–	–	x	x	x	x	x	x	x	x	x	x	x	x	x	–	–	x	x	–	x
Ihlea magalhanica	x	–	–	–	–	–	–	x	x	x	x	–	–	–	–	–	x	x	–	–	–	–
Ihlea racovitzai	–	x	–	–	–	–	–	–	–	–	–	–	–	–	–	–	–	–	–	–	–	–
Thetys vagina	–	–	–	x	x	x	x	x	x	x	–	x	x	x	x	x	–	x	x	x	–	x
Iasis zonaria	x	–	–	x	x	x	x	x	x	x	x	x	x	x	x	x	x	x	x	x	x	x
Pegea confoederata	–	–	–	x	x	x	x	x	x	x	x	x	x	x	x	x	x	x	x	x	x	x
Pegea socia	–	–	–	x	x	–	–	–	–	–	–	–	–	–	–	–	–	–	–	–	–	–
Pegea bicaudata	–	–	–	x	x	–	–	x	x	x	–	x	x	x	–	–	–	–	x	–	–	x
Traustedtia multitentaculata	–	–	–	x	x	x	x	–	–	x	x	x	x	x	x	x	–	–	x	x	–	–
Thalia democratica	–	–	–	x	x	x	x	x	x	x	x	x	x	x	x	x	x	x	x	x	x	x
Thalia rhinoceros	–	–	–	–	–	–	–	–	–	–	–	–	–	x	x	x	–	–	–	–	–	–
Thalia rhomboides	–	–	–	–	–	–	–	–	–	x	–	x	–	x	x	x	–	–	–	–	x	–
Thalia sibogae	–	–	–	–	–	–	–	–	–	–	–	–	–	–	x	–	–	–	–	–	–	–
Thalia orientalis	–	–	–	x	x	x	x	x	–	x	–	–	x	–	x	x	x	x	x	x	x	x
Thalia cicar	–	–	–	x	–	x	x	x	–	–	–	x	–	x	–	–	–	–	x	–	x	–
Thalia longicauda	–	–	–	–	–	–	–	x	x	x	x	–	–	–	–	–	–	–	–	–	–	–
Brooksia rostrata	–	–	x	x	x	–	x	–	–	–	x	x	x	x	x	–	x	–	–	–	x	x
Brooksia berneri	–	–	–	x	–	–	–	–	–	–	–	–	x	x	x	–	–	–	–	–	–	–
Cyclosalpa pinnata	–	–	–	x	x	–	x	–	–	–	–	–	–	–	–	–	–	–	–	–	–	x
Cyclosalpa sewelli	–	–	–	–	–	–	–	–	–	x	–	x	x	x	x	x	x	–	–	–	–	–
Cyclosalpa polae	–	–	–	x	x	x	x	x	x	x	–	x	x	x	x	x	–	–	x	–	–	x
Cyclosalpa quadriluminis	–	–	–	–	–	–	–	–	–	–	–	x	x	x	x	x	–	–	x	x	–	–
Cyclosalpa affinis	–	–	–	x	x	x	x	x	–	–	–	x	x	x	x	x	–	x	–	x	–	x
Cyclosalpa danae	–	–	–	–	–	x	–	–	–	–	–	–	x	–	x	–	–	–	–	–	–	–
Cyclosalpa floridana	–	–	–	x	–	x	x	–	x	–	x	x	x	x	x	x	–	–	x	–	–	–
Cyclosalpa strongylenteron	–	–	–	–	–	–	–	–	–	–	–	–	–	–	–	x	–	–	–	x	–	–
Cyclosalpa ihlei	–	–	–	–	–	–	–	–	–	x	–	–	–	x	–	–	–	–	–	–	–	–

Table 14.1 *(Continued)*

Species	Area																				
	1	2	3	4	5	6	7	8	9	10	11	12	13	14	15	16	17	18	19	20	21	22
Cyclosalpa bakeri	–	–	–	x	–	–	x	–	–	–	–	x	–	–	x	x	–	–	x	x	–	–
Cyclosalpa foxtoni	–	–	–	x	–	–	–	–	–	x	x	x	x	x	x	–	–	x	x	–	–	–
Helicosalpa virgula	–	–	–	x	x	x	x	–	–	x	–	x	x	x	x	–	–	–	x	x	–	x
Helicosalpa younti	–	–	–	–	–	–	–	–	–	–	–	x	–	–	x	–	–	–	–	–	–	–
Helicosalpa komaii	–	–	–	–	–	–	–	–	–	–	–	–	*x*	–	*x*	–	–	–	*x*	–	–	–
PYROSOMATIDAE																						
Pyrosoma atlanticum	–	–	–	x	x	x	x	x	x	x	x	x	x	x	x	x	x	x	x	x	–	x
Pyrosoma aherniosum	–	–	–	–	–	x	x	–	x	x	–	x	x	–	x	–	–	x	x	–	–	–
Pyrosoma ovatum	–	–	–	–	–	–	–	–	x	x	–	–	x	–	–	–	–	–	–	–	–	–
Pyrosoma godeauxi	–	–	–	–	–	–	–	–	–	x	–	–	–	–	–	–	–	–	–	–	–	–
Pyrosomella verticillata	–	–	–	–	–	–	–	–	–	x	x	–	x	x	x	x	x	–	–	–	–	–
Pyrosomella operculata	–	–	–	–	–	–	–	–	–	–	–	x	–	x	–	x	–	–	–	–	–	–
Pyrostemma agassizi	–	–	–	x	x	x	–	–	x	–	–	x	x	x	x	x	x	–	x	–	–	–
Pyrostremma spinosum	–	–	–	x	x	–	–	–	x	x	x	x	x	x	x	–	–	x	–	–	–	–
DOLIOLIDAE																						
Doliolum denticulatum	–	–	–	x	x	x	x	x	x	–	x	–	x	x	x	–	–	x	–	–	x	x
Doliolum nationalis	–	–	–	x	x	x	–	x	–	–	–	x	x	x	x	–	–	–	–	–	x	x
Dolioletta chuni	–	–	–	–	–	–	–	–	–	x	–	x	–	–	–	–	–	–	–	–	–	–
Dolioletta gegenbauri	–	–	–	x	x	x	x	–	–	x	–	x	x	x	x	–	–	x	–	–	x	x
Dolioletta mirabilis	–	–	–	–	–	–	–	–	–	–	–	x	x	–	–	–	–	–	–	–	–	–
Dolioletta tritonis	–	–	–	–	–	–	–	–	–	–	–	x	x	x	x	–	–	–	–	–	x	–
Dolioletta valdiviae	–	–	–	–	–	–	–	–	–	x	–	–	x	–	–	–	–	–	–	–	–	–
Doliolina indica	–	–	–	–	–	–	–	–	–	–	–	x	–	–	–	–	–	–	–	–	x	–
Doliolina intermedia	–	–	–	–	x	x	–	–	–	x	–	x	x	–	–	x	–	–	–	–	–	x
Doliolina krohni	–	x	–	–	–	–	x	–	x	x	–	x	–	–	–	–	–	–	–	–	x	x
Doliolina mülleri	–	–	–	x	x	x	–	–	x	x	-	x	x	–	–	x	–	–	–	–	–	x
Doliolina obscura	–	–	–	–	–	–	–	–	–	–	–	–	–	–	–	x	–	–	–	–	–	–
Doliolina resistibile	–	x	–	–	–	–	–	–	–	–	–	–	–	–	–	–	–	–	–	–	–	–
Doliolina separata	–	–	–	–	–	–	–	–	–	–	–	–	–	–	–	x	–	–	–	–	–	–
Doliolina sigmoides	–	–	–	–	–	–	–	–	–	x	–	–	x	–	–	–	–	–	–	–	–	–
Doliolina undulata	–	–	–	–	–	–	–	–	–	–	–	–	–	–	–	x	–	–	–	–	–	–
Doliopsis rubescens	–	–	–	x	–	–	–	–	–	–	–	–	–	–	–	–	–	–	–	–	–	x
Doliopsoides horizoni	–	–	–	–	–	–	–	–	–	–	–	–	–	–	–	x	–	–	–	–	–	–
Doliopsoides meteori	–	–	–	–	–	x	–	–	–	–	–	–	–	–	–	–	–	–	–	–	–	–
Dolioloides rarum	–	–	–	–	–	–	x	–	–	–	–	x	–	–	–	–	–	–	–	–	–	x

*Note: *Salpa gerlachei* and *S. thompsoni* may be regarded as clinal varieties of *S. fusiformis* (see Chapter 17)

14.4. Phylogenetic analyses of *Salpa*, *Cyclosalpa*, and *Thalia*

14.4.1. *Phylogenetic relationships within the Salpidae*

To date, no formal character analysis of the whole family has been made, and it is beyond the scope of this chapter to do so here. However, to make a successful character analysis of the genera *Salpa, Cyclosalpa,* and *Thalia* the nearest outgroups of these genera are needed for polarity decisions. Metcalf (1918) and Ihle (1958) suggested some phylogenetic groupings which, for want of better schemes, might serve to find these outgroups. They correctly pointed out that the Salpidae may be divided into small groups of genera on such characters as the symmetry of the musculature in aggregate individuals (blastozooids), the number of body muscles, especially in solitary individuals, and the extent to which the body muscles encircle the body. Suggested groupings based on the number of body muscles are: *Salpa, Weelia, Ritteriella, Thetys, Ihlea,* and *Metcalfina* (six in aggregates (blastozooids), usually many in the solitary (oozooid) stage; *Thalia, Traustedtia, Pegea,* and *Iasis* (five or four); *Cyclosalpa, Helicosalpa* and *Brooksia* (usually seven). Although the ancestral number of body muscles is unknown, other characters unite all or some of the members of these groups, so we may assume them to be monophyletic. *Salpa, Weelia,* and *Ritteriella* share very similar 'fusiform' aggregates, which is a clear synapomorphy for the three. *Weelia* normally has the same number of body muscles as *Salpa* (nine), but in *Weelia* this is apparently variable (eight or nine). It is likely that *Weelia* is the sister group of *Salpa* (in fact some authors, including Godeaux in Chapter 17, do not recognize a separate genus *Weelia*), with *Ritteriella* as the sister group of both.

Thalia solitary individuals share with *Traustedtia* the possession of long palps protruding well beyond the test contours. Other features such as the relatively thin muscle fibres and the very low total number (30–100) of muscle fibres make it quite likely that both are sister groups. *Cyclosalpa* and *Helicosalpa* share the possession of long luminous organs in the solitary individuals and in both the alimentary canal is elongated dorsally to the end of the gill bar. It is obvious that *Helicosalpa* is the sister group of *Cyclosalpa*

14.4.2. *Character analysis of the genus* Salpa

14.4.2.1. *Solitary individuals (oozooids)*

1. The number of body muscles in all species of *Salpa* is always fixed at nine (state 1b), whereas in the outgroups (*Weelia* and *Ritteriella*) this varies (state 1a).
2. The oral musculature of *Salpa* has an extra muscle (termed M2 by Winkler, 1975) (state 2b), which is lacking in *Weelia* (state 2a).
3. The fusion of the first four body muscles in the mid-dorsal line is total in the outgroup (*Weelia*) (state 3a); *S.fusiformis, S.thompsoni, S.gerlachei,* and *S.aspera* have MIV free (state 3b); *S.younti* has only MI and II fused (state 3c); *S.maxima* and *S.tuberculata* have all four muscles free (state 3d).
4. The last two body muscles (in *Salpa* MVIII and MIX) are parallel in the outgroup *Weelia* and in *S.maxima* and *S.tuberculata* (state 4a); in *S.aspera* and *S.younti* they are clearly incurved to each other in the mid-dorsal line (state 4b); in *S.fusiformis, S.thompsoni,* and *S.gerlachei* they are fused in the mid dorsal line (state 4c).
5. The oral musculature of *Weelia* shows the muscles called M6 and M7 by Winkler (1975) attached to M1 (state 5a); in *S.maxima S.aspera,* and *S.younti* these are attached to M5 (state 5b); in *S.thompsoni, S.gerlachei,* and *S.fusiformis* M5, M6, and M7 are free (state 5c) (*S.tuberculata* was not investigated).
6. The test or tunic of *Weelia, S.maxima,* and *S.fusiformis* is entirely smooth (state 6a); in *S.thompsoni, S.gerlachei, S.aspera,* and *S.younti* the tunic has a characteristic pattern of longitudinal ridges (state 6b); in *S.tuberculata* there are four echinated cushions around the atrial opening (state 6c).
7. The shape of the dorsal tubercle is a simple C in *Weelia, S.fusiformis, S.thompsoni* and *S.gerlachei* (state 7a); in *S.aspera* it is a more elaborate L-shape (state 7b); in *S.younti, S.maxima,* and *S.tuberculata* the shape is more convoluted and resembles a C (state 7c).

14.4.2.2. *Aggregate individuals (blastozooids)*

8. In all *Salpa* species there are six body muscles (state 8b), whereas *Weelia* has only five; possibly the ancestral number is six as *Ritteriella* also has six, but it has MI and MII fused to a large extent, unlike the condition in *Salpa.*
9. The fusion of body muscles in the mid dorsal line in *Weelia* is total (state 9b); in *S.aspera, S.younti, S.fusiformis, S.thompsoni,* and *S.gerlachei* MI–IV are fused, but this is also the case in *Ritteriella,* so it is possibly the ancestral condition (state 9a); in *S.maxima* and *S.tuberculata* the body muscles verge toward each other but do not fuse (state 9c).
10. MV and MVI are free in *Ritteriella, S.maxima, S.tuberculata, S.aspera,* and *S.younti* (state 10a); they are fused in the middorsal line in *S.fusifomis, S.thompsoni,* and *S.gerlachei* (state 10b).
11. The tunic is smooth in *Weelia, S.maxima,* and *S.fusiformis* (state 11a); it has echinated ridges in *S.aspera, S.younti, S.thompsoni* and *S.gerlachei* (state 11b); in *S.tuberculata* there are two echinated cushions at the dorsal side (state 11c).
12. The tunic shows a distinct bulge at the location of the nucleus in *S.maxima* and *S.tuberculata* (state 12b),

Table 14.2 Taxon-character matrix of the genus *Salpa*.

Species	Character no.											
	1	2	3	4	5	6	7	8	9	10	11	12
Weelia	a	a	a	a	a	a	a	a	a	a	b	a
S.maxima	b	b	d	a	b	a	c	a	c	a	a	b
S.tuberculata	b	b	d	?	a	c	c	b	c	a	c	b
S.aspera	b	b	b	b	b	b	b	b	a	a	b	a
S.younti	b	b	c	b	b	b	c	b	a	a	b	a
S.fusiformis	b	b	b	c	c	a	a	b	a	b	a	a
S.thompsoni	b	b	b	c	c	b	a	b	a	b	b	a
S.gerlachei	b	b	b	c	c	b	a	b	a	b	b	a

whereas this is not found in *Weelia, S.fusiformis, S.thompsoni, S.gerlachei, S.aspera,* and *S.younti.*

The character states above were assembled in a matrix (Table 14.2) and examined with PAUP 3.0. This analysis yielded a single most parsimonious solution of 22 steps, presented in Fig. 14.2 as the best estimate of the phylogenetic relationships within the genus *Salpa.*

14.4.3. *Character analysis of the genus* Thalia

14.4.3.1. *Solitary individuals (oozoids)*

1. There are five body muscles in *Traustedtia* (state 1a) and six (or five + intermediate muscle) in *Thalia* (state 1b).
2. The last two body muscles are parallel in *Traustedtia* and *T.longicauda* (state 2a); they are touching in *T.rhomboides* and *T.sibogae* (state 2b) and fused in *T.democratica, T.rhinoceros, T.orientalis,* and *T.cicar* (state 2c).
3. Muscle fibre number in *Traustedtia, T.longicauda, T.democratica, T. rhinoceros, T.rhomboides* and *T.sibogae* is clearly over 35 and may go up to 106 (state 3a); in *T.cicar* and *T.orientalis* the number is below 40 and may be lower than 30; accordingly the muscles are relatively thin and fused over a wide stretch dorsally (state 3b).
4. Atrial palps are absent in *Traustedtia* and *T.longicauda* (state 4a); they are undivided in *T.democratica* and *T.rhinoceros* (state 4b), they are bifid in *T.rhomboides, T.sibogae, T.orientalis,* and *T.cicar* (state 4c).
5. Posterior projections are smooth in *Traustedtia* (state 5a), and have small denticulations in all *Thalia* species; in *T.longicauda* they are vestigial (state 5b), whereas they are clearly visible in all other species (state 5c)
6. Posterior projections are limp in *Traustedtia* and *T.longicauda* (state 6a), whereas they are stiff in the other species (state 6b).

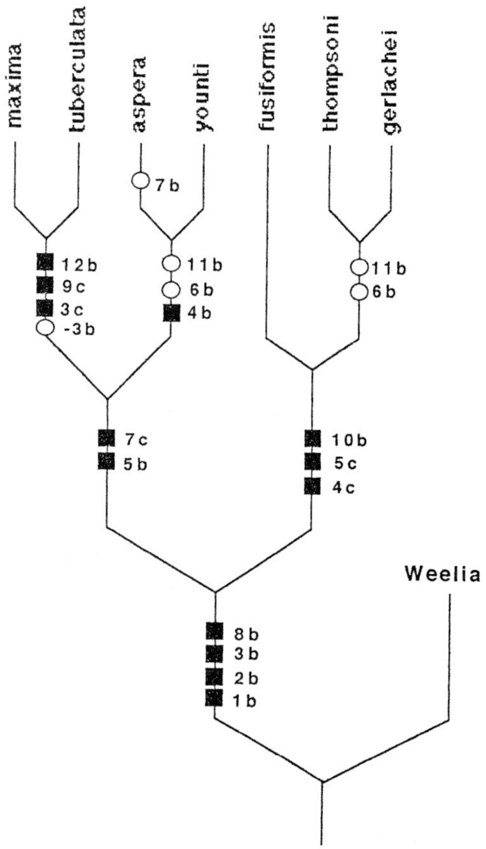

Fig. 14.2. Cladogram of the genus *Salpa*. Numbers indicate characters/states discussed in the text.

7. Lateral projections are absent in *Traustedtia, T.longicauda, T.sibogae* and *T.orientalis* (state 7a), and present in *T.democratica, T.rhinoceros, T.rhomboides,* and *T.cicar* (state 7b).
8. Medio-ventral projections are absent in *Traustedtia* and *T.longicauda* (state 8a), and present in the other species (state 8b).

14.4.3.2. *Aggregate individuals (blastozooids)*

9. In *Traustedtia* only four muscle bands are found (state 9a), whereas in *Thalia* there are five (state 9b).
10. The muscle fibre number of *Traustedtia* and *T.longicauda* is variable and always over 16 (state 10a), whereas the other species have a fixed number of 16 (state 10b).
11. The tunic is smooth in *Traustedtia, T.longicauda, T.democratica, T.rhinoceros, T.orientalis,* and *T.cicar* (state 11a); in *T.rhomboides* and *T.sibogae* the tunica bears spined ridges (state 11b).
12. In *Traustedtia, T.longicauda, T.democratica,* and *T.rhinoceros* the nucleus bears a prominent 'projection' (which is probably homologous to similar structures in, for example, *Cyclosalpa* which are blood-forming organs) (state 12a); such a structure is absent in *T.rhomboides, T.sibogae, T.orientalis,* and *T.cicar* (state 12b).
13. Posteriorly, the tunic is more or less elongated into a triangular shape in *Traustedtia, Thalia democratica, T.rhinoceros, T.rhomboides,* and *T.sibogae* (state 13a); in *T.longicauda, T.orientalis,* and *T.cicar* the tunic is rounded (state 13b).

Character analysis of the matrix of Table 14.3 using PAUP 3.1) yielded a single most parsimonious tree of 19 steps, presented in Fig. 14.3. This differs from the results obtained by Dauby and Godeaux (1987), who performed a distance analysis. Such an analysis does not use the outgroup for character polarization and also the characters used and the distinguished character states were different. For that reason a comparison of the trees is not very useful. There is agreement over the position of *T.longicauda, T.democratica,* and *T.rhinoceros,* while the position of *T.rhomboides* is quite different in the two. The close morphological relationship with *T.sibogae* and the shared bifid atrial palps of the group comprising *T.rhomboides, T.sibogae, T.orientalis,* and *T.cicar* is taken as evidence that the present analysis is the more likely representation of the phylogenetic relationships.

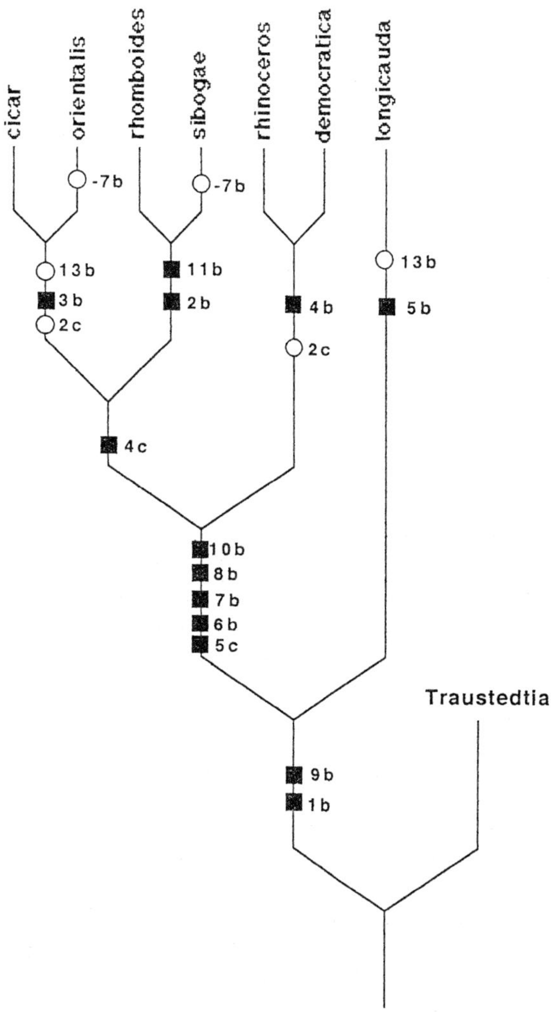

Fig. 14.3. Cladogram of the genus *Thalia.* Numbers indicate characters/states discussed in the text.

Table 14.3 Taxon-character matrix of the genus *Thalia.*

Species	Character no.												
	1	2	3	4	5	6	7	8	9	10	11	12	13
Traustedtia	a	a	a	a	a	a	a	a	a	a	a	a	a
Thalia.longicauda	b	a	a	a	b	a	a	a	b	a	a	a	b
T.democratica	b	c	a	b	c	b	b	b	b	b	a	a	a
T.rhomboides	b	b	a	c	c	b	b	b	b	b	b	b	a
T.sibogae	b	b	a	c	c	b	a	b	b	b	b	b	a
T.orientalis	b	c	b	c	c	b	a	b	b	b	a	b	b
T.cicar	b	c	b	c	c	b	b	b	b	b	a	b	b

14.4.4. *Character analysis of the genus* Cyclosalpa

14.4.4.1. *Solitary individuals (oozoids)*

1. The ventral longitudinal muscles present in all *Cyclosalpa* species (state 1b), are absent in the outgroup *Helicosalpa* (state 1a).
2. Muscle bands I–VII are interrupted in the middorsal line in *Helicosalpa* and most *Cyclosalpa* (state 2a), with the exception of *C.danae* and *C.affinis*, which have MIII-MVII continuous (state 2b).
3. In *Helicosalpa, C.polae, C.quadriluminis, C.bakeri, C.strongylenteron,* and *C.foxtoni* MVI extends forward along the middorsal line (state 3a); in *C.pinnata, C.sewelli, C.affinis, C.danae, C.floridana,* and *C.ihlei* this extension is not developed (state 3b).
4. In those species which have a forwardly extended MVI there are some in which the left and right MVI stay separate (in *Helicosalpa, C.bakeri,* and *C. foxtoni*) (state 4a), and some in which these are fused in the mid-dorsal line (*C.polae* and *C.quadriluminis*) (state 4b).
5. The lateral luminous organs are continuous in *Helicosalpa* and *C.floridana* (state 5a); they are interrupted by the muscle bands in *C.pinnata, C.sewelli, C.polae, C.quadriluminis, C.bakeri, C.foxtoni,* and *C.ihlei* (state 5b); in *C.affinis* and *C.danae* they are absent (state 5c) (the peculiar rounded organs in *C.danae* are not considered as homologous to the longitudinal luminous organs of the other species).
6. The shape of the dorsal tubercle is highly convoluted in *Helicosalpa, C.foxtoni, C.polae, C.quadriluminis, C.affinis, C.danae,* and *C.ihlei* (state 6a); in *C.sewelli, C.floridana, C.bakeri,* and *C.foxtoni* the shape is simple (state 6b).

14.4.4.2. *Aggregate individuals (blastozooids)*

7. In *Helicosalpa* aggregate individuals are attached to each other with an oval disc (state 7a), while all *Cyclosalpa* species are linked through a peduncle (state 7b).
8. In *Helicosalpa* aggregate individuals are arranged in chains as in other salp genera (state 8a), whereas in all *Cyclosalpa* species they are arranged in whorls (state 8b).
9. Muscle bands I and II are fused in *Helicosalpa* and most *Cyclosalpa* species (state 9a), with the exception of *C.affinis* and *C.danae* where they are free (state 9b).
10. Muscle bands III and IV are fused in *Helicosalpa* and most *Cyclosalpa* species (state 10a), with the exception of *C.affinis, C.danae,* and *C.quadriluminis* which have them free (state 10b).
11. Luminous organs are absent in *Helicosalpa, C.affinis, C.danae* (the rounded organs found in this species are not considered homologous to the longitudinal luminous organs of the other species), *C.floridana, C.bakeri, C.foxtoni, C.strongylenteron,* and *C.ihlei* (state 11a), while they are present in *C.pinnata, C.sewelli, C.polae,* and *C.quadriluminis,* (state 11b).
12. In *C.bakeri, C.foxtoni,* and *C.strongylenteron* (possibly also in *C.ihlei*) an extension of MIV known as the visceral muscle is present (state 12b), which is not found in the other species nor in *Helicosalpa* (state 12a).
13. In *Helicosalpa, C.bakeri, C.foxtoni,* a posterior extension of the tunica is present containing the testis and/or a haemopoetic blood-forming organ (state 13a); in *C.bakeri* and *C.foxtoni* this is also found, but here there are two posterior extensions (state 13b); in *C.pinnata, C.sewelli, C.polae, C.quadriluminis, C.affinis,* and *C.danae* this extension is absent (state 13c).
14. In *Helicosalpa, C.bakeri, C.foxtoni,* a clearly distinguishable haemopoetic organ is present. In *Helicosalpa* it is found near the caecum (state 14a), in the other two species it lies in one of the two posterior extensions (state 14b). No such organ is known in the other species (state 14c), although in *C.strongylenteron* and *C.ihlei* it is not certain whether it is present or not.
15. A recognizable gut caecal extension is found in *Helicosalpa, C.bakeri, C.foxtoni,* and *C.strongylenteron* (possibly also in *C.ihlei* (state 15a); in the other species this is absent (state 15b).
16. The gut is coiled in *Helicosalpa, C.affinis, C.floridana, C.bakeri, C.foxtoni, C.strongylenteron,* and *C.ihlei* (state 16a), whilst it is straight in *C.pinnata, C.sewelli, C.polae,* and *C.quadriluminis* (state 16.b).
17. The dorsal tubercle is highly convoluted in *Helicosalpa, C.pinnata, C.danae, C.polae, C.affinis,* and *C.ihlei* (state 17a); in *C.floridana, C.bakeri, C.foxtoni, C.sewelli, C.polae,* and *C.quadriluminis* its shape is simple (state 17b).
18. As in other salp genera, *Helicosalpa, C.bakeri, C.foxtoni, C.strongylenteron,* and *C.ihlei* have asymmetrical individuals (state 18a); in the other species they are completely bilaterally symmetrical (state 18b).

The character analysis of Table 14.4 using PAUP 3.0 yielded a single most parsimonious tree of 29 steps (Fig. 14.4)

14.5. Phylogenetic relationships of Pyrosomatidae

These were determined by Van Soest (1981), where the character analysis is given. Three genera and eight species are recognized; their relationships are shown in Fig. 14.5 (from Van Soest, 1981, fig. 21).

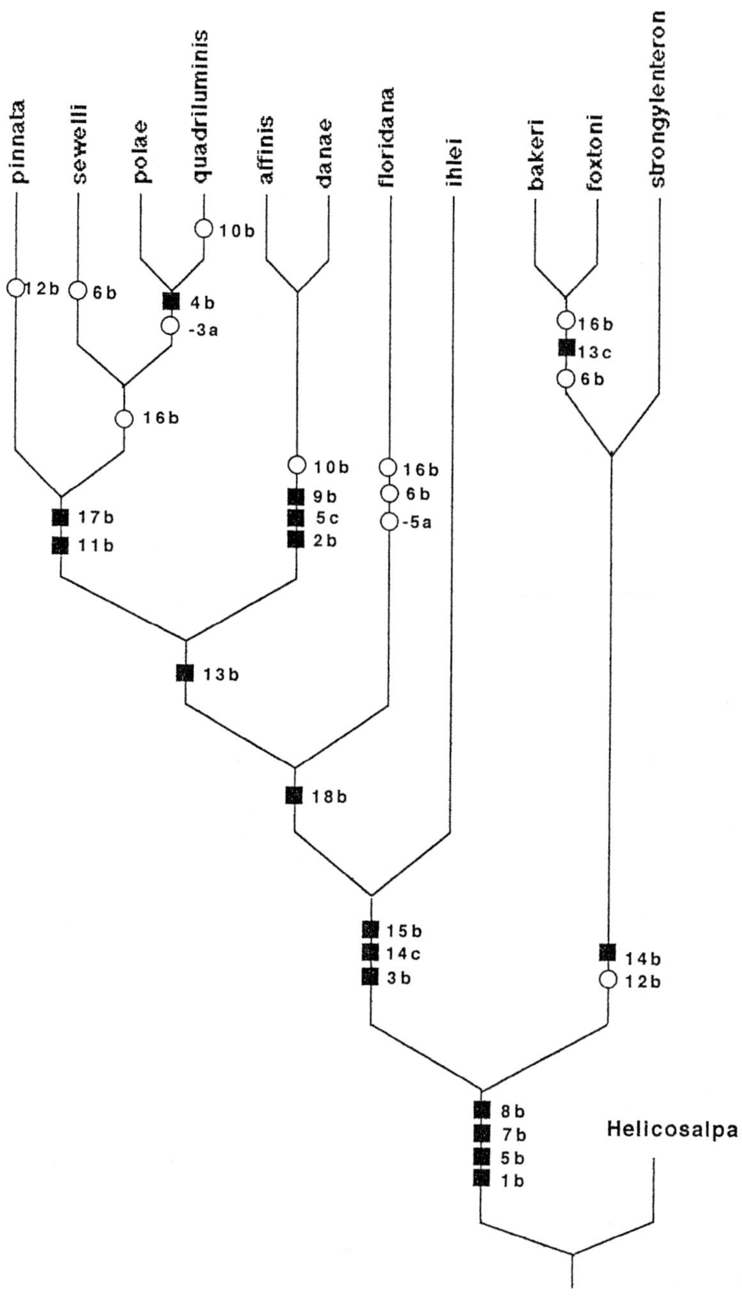

Fig. 14.4. Cladogram of the genus *Cyclosalpa*. Numbers indicate characters/states discussed in the text.

14.6. Area cladistic analysis

The presence or absence of the taxa studied and their presumed ancestors in areas of endemism A–V and I–X are given in Tables 14.5 and 14.6. Ancestral distributions are derived from the distribution of their offspring. These area–taxa matrices were run in PAUP 3.0

The resulting general area cladogram of the first approach (Fig. 14.6) consists of a series of repeated area relationships each of roughly the following composition: widespread distribution over all three oceans (groups G + Q + A + T, B, C + F, U + V, D + E) are more closely related to a restricted distribution in the Indian Ocean and/or West Pacific (groups R + H, I + P, S + J, M) than to Southern Ocean distributions (groups O, K + L). East and West Pacific distributions are fairly closely related. The enigmatic antitropical Atlantic distribution (N) appears similar to the Southern Ocean distribution, or

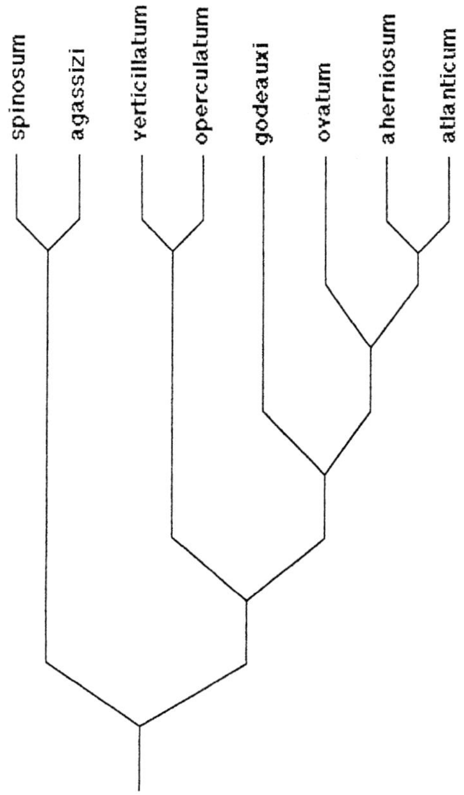

Fig. 14.5. Cladogram of the family Pyrosomatidae. After Soest (1981).

in other words, tropical/subtropical distributions are more closely related to each other than to cool water distributions. Such a pattern accords with both ecological and historical data.

The hydrography of the oceans, in particular temperature distributions, suggests that widespread distributions may result from the dispersal of temperature-tolerant species from one ocean to another, whilst restricted tropical Indo-Pacific distributions represent stenothermic species unable to disperse through the cooler Southern Ocean waters, and Southem Ocean distributions represent stenothermic cold water species unable to penetrate the warmer parts of the oceans and thus unable to reach the Arctic–Boreal seas. The historical explanation is that after the break-up of Pangea into a northern continental mass Laurasia and a southern land mass Gondwana, the plankton diverged into a northern, warm water fauna (with a Tethyan distribution) and a southern cool water fauna (with a Gondwanan distribution). Possibly, the birth of a circum-Antarctic current system south of Australia and South Africa enhanced isolation of the Southern Ocean fauna. Further break-up of Laurasia and the opening up of the Atlantic, as well as the northwardly directed movement of Africa and Australia caused East–West speciation within the warm water fauna, while the southern cool water fauna became impoverished by declining temperatures in the second half of the Tertiary. This historical scenario is supported by the area cladogram of Fig. 14.6, which predicts that tropical–subtropical taxa diverged more recently than the divergence of warm versus cool taxa. This picture also offers an explanation of why there are no Arctic pelagic Tunicata. At the time these faunas diverged into warm and cool water faunas the Arctic Ocean was not yet formed.

The cladogram of the second approach (Fig. 14.7), which conforms in methodology more precisely to Wiley's (1988) BPA approach, is unfortunately a rather uninformative diagram. There are no groups of areas, thus making interpretation difficult. In general, more peripheral areas [Antarctic, (anti-)Boreal, North Pacific, etc.] are at the base of the tree, suggesting also that divergence of peripheral areas was earlier in time than that of central areas. However, since no area groupings are found, that conclusion cannot be drawn. Ecologically, this general area cladogram is also difficult to interpret, as it is unclear why Southern Ocean distributions should be more closely related to tropical distributions than (for example) to antiboreal distributions.

14.7. Conclusions

In conclusion, it seems inescapable that the first approach yields results that are ecologically and historically acceptable: Southern Ocean species are suggested to have diverged earlier in time than recent warm water species. The general area cladogram strongly suggests that the majority of thaliacean distributions arose during the Early and Middle Tertiary break-up of Laurasia and Gondwana and the later rearrangement of the continents. This conclusion is accordance with the views put forward by Pierrot-Bults and Van der Spoel (1979, fig. 2)

14.8. Scenarios of speciation

With the conclusions reached above we can develop scenarios of divergence and distribution of salps and pyrosomas in a more objective way than done by Soest (1975). The general pattern given above compared with the individual cladograms of the genera make the following scenarios the most likely.

14.8.1. Salpa *spp.*

The *S.fusiformis, S.thompsoni, S.gerlachei* divergence can be confidently situated in the time that the Southern Ocean became isolated in the period between the Cretaceous and the Middle Tertiary. Since this divergence is younger than that between *S.maxima, S.tubercu-*

Table 14.4 Taxon-character matrix of the genus *Cyclosalpa*.

Species	Character no.																	
	1	2	3	4	5	6	7	8	9	10	11	12	13	14	15	16	17	18
Helicosalpa	a	a	a	a	a	a	a	a	a	a	a	a	a	a	a	a	a	a
C.pinnata	b	a	b	?	b	a	b	b	a	a	b	b	b	c	b	a	b	b
C.sewelli	b	a	b	?	b	b	b	b	a	a	b	a	b	c	b	b	b	b
C.polae	b	a	a	b	b	a	b	b	a	a	b	a	b	c	b	b	b	b
C.quadriluminis	b	a	a	b	b	a	b	b	a	b	b	a	b	c	b	b	b	b
C.affinis	b	b	b	?	c	a	b	b	b	b	a	a	b	c	b	a	a	b
C.danae	b	b	b	?	c	a	b	b	b	b	a	a	b	c	b	a	a	b
C.floridana	b	a	b	?	a	b	b	b	a	a	a	a	a	c	b	b	a	b
C.bakeri	b	a	a	a	b	b	b	b	a	a	a	b	c	b	a	b	a	a
C.foxtoni	b	a	a	a	b	b	b	b	a	a	a	b	c	b	a	b	a	a
C.strongylenteron	b	a	a	a	?	a	b	b	a	a	a	b	a	?	a	a	a	a
C.ihlei	b	a	b	?	b	a	b	b	a	a	a	?	a	?	?	a	a	a

Table 14.5 Area-taxon matrix of the first approach.

Areas*	Taxa†																														
	1	2	3	4	5	6	7	8	9	10	11	12	13	14	15	16	17	18	19	20	21	22	23	24	25	26	27	28	29	30	31	32
A	0	0	0	0	1	0	0	0	0	1	0	0	0	0	0	0	0	0	0	0	0	0	0	0	0	0	0	0	0	0	0	0
B	0	0	0	0	0	0	0	1	0	0	1	1	0	0	0	0	0	0	0	0	0	0	0	0	0	0	0	0	0	0	0	0
C	0	0	0	0	0	0	0	0	0	0	0	0	0	0	0	1	0	0	0	0	0	0	0	0	0	1	0	1	1	1	1	0
D	0	0	1	0	0	0	0	0	1	0	1	1	0	0	0	0	0	0	0	0	0	0	0	0	0	0	0	0	0	0	0	0
E	0	0	0	1	0	0	0	0	1	0	0	0	0	0	0	0	0	0	0	0	0	0	0	0	0	0	0	0	0	0	0	0
F	0	0	0	0	0	0	0	0	0	0	0	0	0	0	0	0	0	1	0	0	0	0	0	0	0	0	1	0	1	1	1	0
G	0	0	0	0	0	0	0	0	0	0	0	0	0	0	0	0	0	0	0	0	0	0	0	0	0	0	0	0	0	0	0	0
H	0	0	0	0	0	0	0	0	0	0	0	0	0	0	0	0	0	0	0	0	0	0	0	0	0	0	0	0	0	0	0	0
I	0	1	0	0	0	0	0	1	0	0	1	1	0	0	0	0	0	0	0	0	1	0	0	0	0	0	0	0	0	0	1	0
J	0	0	0	0	0	0	0	0	0	0	0	0	0	0	1	0	0	0	0	0	0	0	0	0	1	0	0	1	1	1	1	0
K	0	0	0	0	0	1	0	0	0	1	0	1	1	0	0	0	0	0	0	0	0	0	0	0	0	0	0	0	0	0	0	0
L	0	0	0	0	0	0	1	0	0	1	0	1	1	0	0	0	0	0	0	0	0	0	0	0	0	0	0	0	0	0	0	0
M	0	0	0	0	0	0	0	0	0	0	0	0	0	0	0	0	0	0	0	0	0	0	0	1	0	0	0	0	0	0	0	0
N	0	0	0	0	0	0	0	0	0	0	0	0	0	1	0	0	0	0	0	0	0	0	0	0	1	0	0	1	1	1	1	0
O	0	0	0	0	0	0	0	0	0	0	0	0	0	0	0	0	0	0	0	0	0	0	0	0	0	0	0	0	0	0	0	0
P	0	0	0	0	0	0	0	0	0	0	0	0	0	0	0	0	0	0	0	0	0	0	0	0	0	0	0	0	0	0	0	0
Q	0	0	0	0	0	0	0	0	0	0	0	0	0	0	0	0	0	0	0	0	0	0	0	0	0	0	0	0	0	0	0	0
R	0	0	0	0	0	0	0	0	0	0	0	0	0	0	0	0	0	0	0	0	0	0	0	0	0	0	0	0	0	0	0	0
S	0	0	0	0	0	0	0	0	0	0	0	0	0	0	0	0	1	0	0	0	0	0	0	0	0	1	0	1	1	1	1	0
T	0	0	0	0	0	0	0	0	0	0	0	0	0	0	0	0	0	0	1	1	0	0	0	0	0	0	1	0	1	1	1	0
U	0	0	0	0	0	0	0	0	0	0	0	0	0	0	0	0	0	0	0	0	0	1	0	0	0	0	0	0	0	0	0	1
V	0	0	0	0	0	0	0	0	0	0	0	0	0	0	0	0	0	0	0	0	0	0	1	0	0	0	0	0	0	0	0	1

*Areas A–V are listed in the text with their coded area numbers.
†Taxa 1–62 are derived from the cladograms of *Salpa* (Fig. 14.2) (numbers 1–13), *Thalia* (Fig. 14.3) (numbers 35–47), *Cyclosalpa* (Fig. 14.4) (numbers 14–34) and Pyrosomatidae (Fig. 14.5) (numbers 48–62).

Table 14.5 *(Continued)*

Areas*	Taxa†																													
	33	34	35	36	37	38	39	40	41	42	43	44	45	46	47	48	49	50	51	52	53	54	55	56	57	58	59	60	61	62
A	0	0	0	1	0	0	0	0	0	0	0	1	0	1	1	0	0	0	0	0	0	0	1	1	1	1	0	1	0	1
B	0	0	0	0	0	0	0	0	1	1	0	0	1	1	1	0	0	0	0	0	0	0	0	0	0	0	0	0	0	0
C	0	1	0	0	0	0	0	0	0	0	0	0	0	0	0	0	0	0	0	0	0	0	0	0	0	0	0	0	0	0
D	0	0	0	0	0	0	0	0	0	0	0	0	0	0	0	0	0	0	0	0	0	0	0	0	0	0	0	0	0	0
E	0	0	0	0	0	0	0	0	0	0	0	0	0	0	0	0	0	0	0	0	0	0	0	0	0	0	0	0	0	0
F	0	1	0	0	0	0	0	0	0	0	0	0	0	0	0	0	0	0	0	0	0	0	0	0	0	0	0	0	0	0
G	0	0	0	0	0	0	0	0	0	0	0	0	0	0	0	0	1	0	0	0	0	0	0	0	0	0	0	0	1	1
H	0	0	0	0	0	0	0	0	0	0	0	0	0	0	0	0	0	0	0	1	0	0	0	0	0	1	0	1	0	1
I	0	1	0	0	0	1	0	0	0	0	1	0	1	1	1	0	0	0	0	0	0	0	0	0	0	0	0	0	0	0
J	0	1	0	0	0	0	0	0	0	0	0	0	0	0	0	0	0	1	1	0	0	0	0	0	0	0	1	1	0	1
K	0	0	0	0	0	0	0	0	0	0	0	0	0	0	0	0	0	0	0	0	0	0	0	0	0	0	0	0	0	0
L	0	0	0	0	0	0	0	0	0	0	0	0	0	0	0	0	0	0	0	0	0	0	0	0	0	0	0	0	0	0
M	1	1	0	0	0	0	0	0	0	0	0	0	0	0	0	0	0	0	0	0	0	0	0	0	0	0	0	0	0	0
N	1	1	0	0	0	0	0	0	0	0	0	0	0	0	0	0	0	0	0	0	0	0	0	0	0	0	0	0	0	0
O	0	0	1	0	0	0	0	0	0	0	0	0	0	0	1	0	0	0	0	0	0	0	0	0	0	0	0	0	0	0
P	0	0	0	0	1	0	1	0	0	0	1	1	1	1	0	1	0	0	0	0	0	0	0	0	0	0	0	0	1	1
Q	0	0	0	0	0	0	0	0	0	0	0	0	0	0	1	0	0	0	0	0	0	0	0	0	0	0	0	0	0	0
R	0	0	0	0	0	0	0	0	0	0	0	0	0	0	0	0	0	0	0	0	1	0	0	0	1	1	0	1	0	1
S	0	1	0	0	0	0	0	0	0	0	0	0	0	0	0	0	0	0	0	0	0	0	0	0	0	0	0	0	0	0
T	0	1	0	0	0	0	0	1	0	1	0	0	1	1	1	0	0	0	0	0	0	1	0	1	1	1	0	1	0	1
U	1	1	0	0	0	0	0	0	0	0	0	0	0	0	0	0	0	0	0	0	0	0	0	0	0	0	0	0	0	0
V	1	1	0	0	0	0	0	0	0	0	0	0	0	0	0	0	0	0	0	0	0	0	0	0	0	0	0	0	0	0

Table 14.6 Area-taxon matrix of the second approach.

Areas*	Taxa†																													
	1	2	3	4	5	6	7	8	9	10	11	12	13	14	15	16	17	18	19	20	21	22	23	24	25	26	27	28	29	30	31
I Boreal Atlantic	1	0	0	0	1	0	0	1	0	1	1	1	0	1	0	1	0	1	0	0	0	0	0	0	1	1	1	1	1	1	1
II Canary Current area	1	0	1	0	1	0	0	1	1	1	1	1	0	0	0	1	0	1	0	0	0	0	0	0	0	1	1	1	1	1	1
III Tropical Atlantic	1	0	1	1	1	0	0	1	1	1	1	1	0	0	0	1	0	1	1	1	0	1	1	0	0	1	1	1	1	1	1
IV Southern Ocean	1	0	1	1	1	0	0	1	1	1	1	1	0	1	0	0	0	0	0	0	0	0	0	0	1	0	0	1	1	1	1
V Antiboreal	0	0	0	0	1	1	0	0	0	1	0	1	1	0	0	0	0	0	0	0	0	0	0	0	0	0	0	0	0	0	0
VI Antarctic	0	0	0	0	0	1	1	0	0	1	0	1	1	0	0	0	0	0	0	0	0	0	0	0	0	0	0	0	0	0	0
VII Indian Ocean	1	1	1	1	1	0	0	1	1	1	1	1	0	0	1	1	1	1	1	1	1	1	1	0	1	1	1	1	1	1	1
VIII West Pacific	1	1	1	1	1	0	0	1	1	1	1	1	0	0	1	1	1	1	1	1	1	1	1	0	1	1	1	1	1	1	1
IX East Pacific	1	0	1	1	1	0	0	1	1	1	1	1	0	0	1	1	1	1	0	0	0	0	1	0	1	1	1	1	1	1	1
X Northern Pacific	0	0	1	1	1	0	0	0	1	1	1	1	0	0	0	0	1	1	0	0	0	1	0	0	0	1	1	1	1	1	1

Areas*	Taxa†																													
	32	33	34	35	36	37	38	39	40	41	42	43	44	45	46	47	48	49	50	51	52	53	54	55	56	57	58	59	60	61	62
I Boreal Atlantic	0	0	1	0	1	0	0	0	0	1	1	0	1	1	1	1	0	0	0	0	0	0	0	1	1	1	1	0	1	0	1
II Canary Current area	0	0	1	0	1	0	0	0	0	1	1	0	1	1	1	1	1	1	0	0	0	0	0	1	1	1	1	0	1	1	1
III Tropical Atlantic	1	1	1	0	1	0	0	0	1	1	1	0	1	1	1	1	0	1	0	0	0	0	1	1	1	1	1	0	1	1	1
IV Southern Ocean	1	1	1	1	1	0	0	0	0	1	1	0	1	1	1	1	1	1	0	0	0	1	0	1	1	1	1	0	1	1	1
V Antiboreal	0	0	0	1	1	0	0	0	0	0	0	0	1	0	1	1	1	0	0	0	0	1	0	1	1	1	1	0	1	1	1
VI Antarctic	0	0	0	0	0	0	0	0	0	0	0	1	0	0	0	0	0	0	0	0	0	0	0	0	0	0	0	0	0	0	0
VII Indian Ocean	1	1	1	0	1	0	1	0	1	1	1	0	1	1	1	1	1	1	1	1	1	1	1	1	1	1	1	1	1	1	1
VIII West Pacific	1	1	1	0	1	1	1	1	1	1	1	0	1	1	1	1	1	1	1	1	0	0	1	1	1	1	1	1	1	1	1
IX East Pacific	1	1	1	0	1	0	0	0	0	1	1	0	1	1	1	1	0	1	1	1	0	0	0	1	1	1	1	1	1	1	1
X Northern Pacific	1	1	1	0	1	0	0	0	0	0	0	0	1	0	1	1	0	0	0	0	0	0	0	1	1	1	1	0	0	0	1

*Areas 1–X are listed in the text with their coded area numbers.
†Taxa 1–62 are derived from the cladograms of *Salpa* (Fig. 14.2) (numbers 1–13), *Thalia* (Fig. 14.3) (numbers 35–47), *Cyclosalpa* (Fig. 14.4) (numbers 14–34) and Pyrosomatidae (Fig. 14.5) (numbers 48–62).

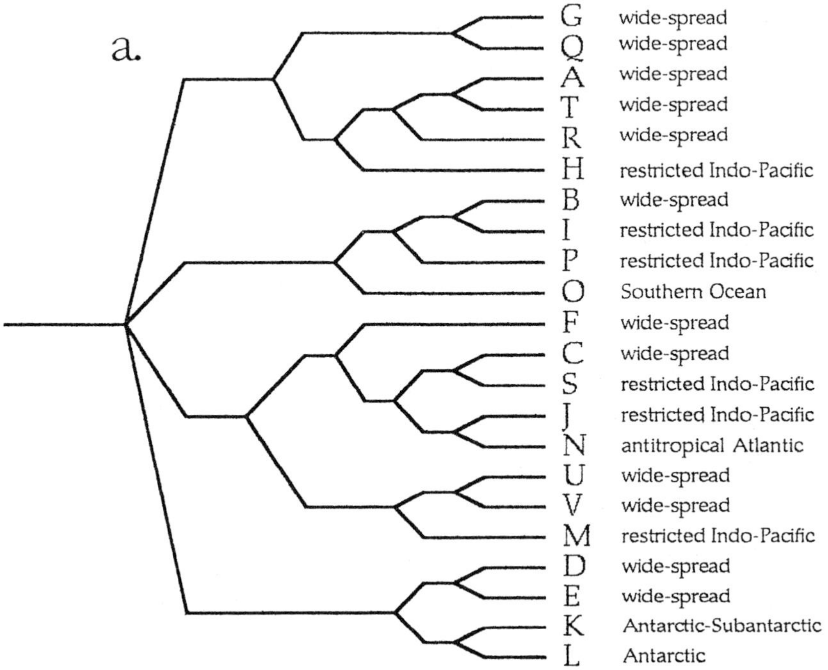

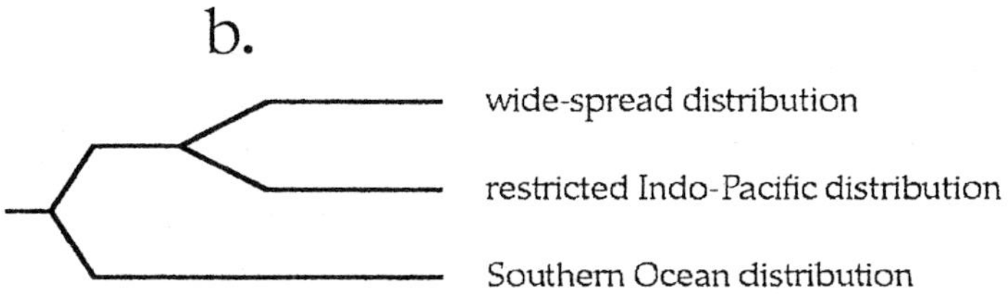

Fig. 14.6. (a) Area cladogram of the first approach involving areas of endemism A–V, discussed in the text. (b) Area cladogram of the first approach simplified and reduced by omitting redundant area relationships.

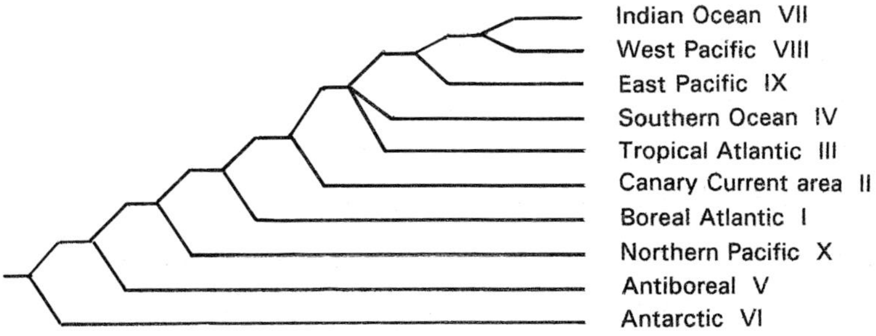

Fig. 14.7. Area cladogram of the second approach involving areas of endemism I–IX.

lata, S.aspera, S.younti on the one hand and *S.fusiformis, S.thompsoni, S. gerlachei* on the other, we must assume that the two groups are of Mesozoic age. The similarity between *S.maxima* and *S.tuberculata* combined with the restricted distribution of *S.tuberculata* suggests a fairly young divergence, but further speculations cannot be founded upon facts. *Salpa tuberculata* may turn out to be a neritic species. The divergence of *S.aspera* and *S.younti* may be supposed to be older than that of *S.maxima* and *S.tuberculata*, because of the overlapping distribution. Possibly, the Pleistocene scenario of Van Soest (1975) applies, but this is not well established. *Salpa thompsoni* and *S.gerlachei* are close, allopatric and may not even be different species (Casareto and Nemoto, 1987). Muscle fibre variation in number probably constitutes speciation in progress since the last Ice Age ('Sisyphean speciation; De Visser, 1986), which may or may not lead eventually to discrete species. It is possible that differences in muscle fibre number reflect more extensive differences (cf. *Pegea socia* and *P.confoederata*; Madin and Harbison, 1978), but this can only be confirmed by further (genetic) evidence.

14.8.2. Thalia *spp.*

Thalia longicauda and the *T.democratica* group probably diverged in the period between the Cretaceous and the Middle Tertiary. Further speciation events must be of more recent origin: *T.democratica* and *T.rhinoceros* may represent a Pleistocene speciation event involving the emergence of Wallacea (Fleminger, 1986). Further differentiation within *T.democratica* (*T.democratica indopacifica* Van Soest, 1975) must then be of subrecent origin. Divergence of the Indo-Pacific species *T.rhomboides* and *T.sibogae* may have proceeded along the same line, although there are analogies with *Salpa maxima* and *S.tuberculata*, because *T.sibogae* is very probably a neritic species (see Tokioka, 1979, for a discussion of neritic speciation). The divergence of *T.cicar* and *T.orientalis* is not readily reconstructed.

14.8.3. Cyclosalpa *spp.*

Here scenarios are not easily developed because of the extensive overlap of most distributions, even those of closely related species. The group *C.bakeri, C.foxtoni, C.strongylenteron* possibly constitutes a case of inter-ocean east–west speciation, because it is strongly represented in the East Pacific. Possibly also caused by this is the infraspecific Pacific east–west pair *Cyclosalpa quadriluminis quadriluminis* and *C.q.parallela.* Further east–west speciation is shown by *C.pinnata* (Atlantic) and *C.sewelli* (Indo-Pacific). No Southern Ocean species occur within the group, nor in the outgroup. If all salp genera were of comparable age (but there is no firm evidence that they are), then extinction of *Cyclosalpa* spp. in the Southern Ocean might be suggested.

14.8.4. *Other genera of Salpidae*

Although no formal character analyses are available for the genera with three species, *Ritteriella, Pegea, Ihlea* and *Helicosalpa,* their distributions show similarities with the genera considered in 13.7.1–3, and similar scenarios are assumed to apply. *Ihlea punctata* and the pair *I.racovitzai, I.magalhanica* probably started to diverge at the Laurasian–Gondwanan divergence, whilst *I.racovitzai* and *I.magalhanica* possibly demonstrate the past occurrence of a vicariance event separating Eastern and Western Antarctic oceans (e.g. the breaking off and the northward movement of Australia). *Pegea socia* and *P.confoederata* show allopatric distribution in the North Atlantic (Madin and Harbison, 1978), but it is at present unclear what the exact distributions of the two species are because they can only be certainly distinguished when alive. *Brooksia berneri* and *B.rostrata* have wide overlapping distributions. When the monospecific genera *Weelia, Metcalfina, Thetys, Iasis,* and *Traustedtia* arose remains unclear.

14.8.5. *Pyrosomatidae*

There are no Southern Ocean pyrosomas, and in view of the morphological divergence within the group, this is best explained in terms of the extinction of Southern Ocean representatives in the cooling period of the Lower Tertiary. A further indication that pyrosomas are stenothermic warm water animals is seen in the restriction of one of the three genera, *Pyrosomella* (with two species) and one species of *Pyrosoma* to the Indo-West Pacific. Van Soest (1981) suggested the extinction of Atlantic relatives of these taxa during the Miocene–Pliocene cooling of the Atlantic. Diversity in the group is low and speculation about divergence times of the present taxa is hampered by overlapping distributions of the remaining five species.

14.8.6. *Doliolidae*

It is clear that the systematics and biogeography of the Doliolidae lag behind that of the other pelagic tunicates, and that a monographic treatment of the group is much needed. Hence it is premature to try to compare their distribution patterns with those of salps and pyrosomas.

14.9. Discussion

Clearly only a very general result has been obtained from this area cladistic study of pelagic tunicate faunas. Widespread tropical and subtropical distributions in all three oceans are more closely related to each other (and thus are more recently diverged) than each of them with the Southern Ocean distribution, but further area relationships are not clear. Recurrent widespread distributions in a large proportion of the species and genera studied prevent a more detailed outcome. Since overlapping distributions are the rule rather than the exception (cf. Van Soest 1979; over 50% of all plankton species have a 'moderately wide distribution' over the tropical and subtropical parts of all three oceans), it is likely that future area cladistic analyses using other holoplanktonic groups will suffer from the same problems. Perhaps certain groups of crustaceans will yield more details (cf. Van der Spoel *et al.*, 1990), as they seem to have more restricted distributions. Still, the results show a general accordance with the ruling paradigm of allopatric speciation (Pierrot-Bults and Van der Spoel, 1979), and there is no evidence for sympatric speciation (McGowan, 1973), nor for 'planktopatric' speciation (Shih, 1986), or 'punctuated' speciation (Herman, 1986).

Recent literature on cladistic biogeography (e.g. Myers and Giller, 1988; Ladiges *et al.*, 1991) tends to consider widespread overlapping areas of distribution as problematical data needing additional treatment using Assumptions 1 and 2 (Nelson and Platnick, 1981). Others assume that widespread distributions signify close area relationships and proceed according to Assumption 0 (Zandee and Roos, 1987; Wiley, 1988). Marine distributions often show large areas of overlap, in plankton distributions for over 50% of the species; moreover, plankton distributions are often wholly overlapping. Using Assumptions 1 and 2 would involve the 'creation' of a disproportionately large number of imaginary area relationships, which seems to prevent reaching a justifiable result. A similar problem prevents the use of the Component method (Page, 1990). The present attempt at BPA (Wiley, 1988) using Assumption 0 failed to produce an informative result and it seems clear that adding more taxon cladograms of other plankton groups will not improve the result. Apparently, the overlapping areas of endemism constitute a basic problem for Assumptions 0, 1 and 2.

For that reason, the areas of endemism were considered *a priori* as discrete areas here (cf. the first approach above), despite the often extensive overlap. In this way, the problem of widespread distributions was avoided and the results of this BPA approach are now quite informative. In the field of marine cladistic biogeography, where extensively overlapping distributions are frequent, it is suggested that this 'No Assumptions' approach should be adopted.

FIFTEEN

Appendicularian distribution and zoogeography

R. Fenaux, Q. Bone, and D. Deibel

15.1. Introduction

Although a good deal is known of the broad outlines of appendicularian distribution in different oceans, only at a few favoured sites has sampling been sufficiently detailed, and carried on for long enough, to gain a reasonably 'complete' idea of their horizontal and vertical distribution over time. Sampling presents some difficulties, and much of the information from particular areas on the presence or absence of species and their abundance must be regarded as provisional only. Most of our knowledge of global appendicularian distribution has come from oceanographic expeditions during the 19th and 20th centuries, which means that each site (station) was usually sampled only once. The few studies that have achieved repeated sampling from the same station over time have shown marked seasonal qualitative variations in the catch, hence the absence of a particular species in a single sample should not be taken as absolute proof that that species is absent from the area.

However this is not the only sampling difficulty. Due to size differences between species, and increase in size with age within species, the mesh size of the plankton nets used will influence the nature of the samples obtained, both qualitatively and quantitatively. A very clear example of the magnitude of this sampling bias was provided by Fenaux (1986b), who took simultaneous vertical samples in the Mediterranean with two nets of mesh size 200 and 53 μm. Many fewer specimens were caught by the net with the larger mesh, indeed, to agree with the catch of the 53 μm net, that of the 200 μm net had to be multiplied by a factor of 2.9 for Oikopleuridae and 4.6 for the smaller Fritillariidae. Another difficulty in sampling has been raised by recent interesting submersible and diver observations, which indicate that appendicularians may be extremely abundant in very thin layers only a few metres thick, and these layers may be patchy (Deibel *et al.*, unpublished). All of these sampling problems, in particular the temporal and spatial inhomogeneity of appendicularian populations, means that most of the data we have on appendicularian distribution is incomplete.

There is only a brief account of global appendicularian distribution in this chapter, and instead, a more detailed account is given of a few areas where sampling has been most effectively carried out, and our knowledge of appendicularian distribution is most complete. A bibliography (up to 1989) of papers dealing entirely or in part with appendicularian distribution was given by Fenaux *et al.* (1990, pp. 75–104).

15.2. Abundance and species succession

Some areas have been regularly sampled over several years, whilst others have been sampled more intensively over a shorter period (usually over 1 year). It is from a combination of studies of this kind that recent progress has been made in understanding the distribution and successional nature of appendicularian species.

These studies have been mainly in temperate regions such as the Adriatic (Skaramuca, 1980, 1983; Baranovic *et al.*, 1992), southern Bay of Biscay (Acuña and Anadón, 1992; Acuña, 1994), and the western Mediterranean (Fenaux, 1961, unpublished; see below). There have also been important studies in regions where there is regular interchange between warm and cold waters, as in southwest Hokkaido Japan (Shiga, 1985), and, finally, studies in the Arctic (Pavshtiks, 1972; Mumm, 1993)).

The best-known temperate area is probably the Bay of Villefranche (western Mediterranean, Ligurian Sea), which is rich in species, and (thanks to the presence of appendicularian specialists) has been studied in considerable detail over recent years, and in lesser detail, for over 50 years. To study the annual species succession in some detail, appendicularian populations at Villefranche were followed from the beginning of April 1978 until the end of March, 1979 (Fenaux, unpublished; see below). Daily vertical net tows from 75 m to the surface were taken whenever practicable using a 50 μm mesh Phyto net (similar to the WP2 net except for mesh size; Fenaux and Palazzoli, 1979). Sampling was at site B, at the entrance to the bay of Villefranche, where the depth is 80 m (Fenaux, 1963). The collections were grouped into 52 weekly samples (each consisting of from two to five separate samples). In all, 210 samples containing 371, 118 specimens were collected. (see Table 15.1 and Fig. 15.4).

Thirty eight species are presently known from the Mediterranean; 25 of these were found among the samples. Of those not collected during the survey period, eight are rare in the area: *Mesoikopleura haranti, Appendicularia tregouboffi, Tectillaria fertilis, Folia mediterranea, Oikopleura intermedia, Fritillaria charybdae, F.urticans* and *Kowalevskia oceanica.* These have only been found as isolated individuals or in very small numbers during 28 years of intermittent collection. Two large species (*Stegosoma magnum* and *Megalocercus abyssorum*) are not generally abundant, and as adults are poorly sampled by the Phyto net.

A striking feature to emerge from the study was that of the 25 species in the samples, only five species made up nearly 90% of the total number of specimens. These were: *Fritillaria borealis sargassi, Oikopleura longicauda, O.dioica, O.fusiformis,* and *F.pellucida typica.* What is more, these five species occurred in over 80% of the samples, and each made up more than 10% of the total in the different samples. If *F.formica digitata, F.borealis intermedia, F.haplostoma,* and *F.megachile* were added to the five most abundant species, then these nine species together made up more than 97% of the individuals in the samples. Their abundance naturally varied seasonally (Fig. 15.1); the greatest number of individuals (805 m^{-3} was collected in August, and the least (9 m^{-3}) in April. The seasonal means (in individuals m^{-3}) were: spring 86 ± 33, summer 250 ± 113, autumn 110 ± 109,

Table 15.1. Quantitative results for each species, in decreasing order (based on 210 samples taken over a year). The numbers is front of each correspond to those in Fig. 15.4, which displays the same data graphically.

	Species	Total number	Abundance (%)	Frequency (%)	n m^3
1	*F.borealis sarg.*	104,763	28.23	95.19	39.91
2	*O.longicauda*	70,440	18.98	90.38	26.83
3	*O.dioica*	61,353	16.53	96.15	23.37
4	*O.fusiformis*	52,120	14.04	82.69	19.85
5	*F.pellucida typica*	38,495	10.36	92.79	14.66
6	*F.formica digitata*	12,316	3.33	30.70	4.69
7	*F.borealis intermedia*	8,593	2.32	38.94	3.27
8	*F.haplostoma*	6,960	1.88	42.79	2.65
9	*F.megachile*	4,123	1.11	42.31	1.57
10	*A.sicula*	3,192	0.86	43.27	1.21
11	*O.graciloides*	2,398	0.65	37.50	0.91
12	*O.rufescens*	1,249	0.34	12.50	0.48
13	*F.formica tuberculata*	1,221	0.33	32.69	0.47
14	*O.parva*	1,031	0.28	16.83	0.39
15	*O.cophocerca*	912	0.25	24.04	0.35
16	*F.tenella*	756	0.20	25.96	0.29
17	*O.albicans*	520	0.14	13.46	0.20
18	*F.aequatorialis*	448	0.12	11.54	0.17
	K.tenuis	96	0.03	4.81	0.04
	F.venusta	48	0.01	2.96	0.018
	F.messanensis	40	0.01	0.95	0.015
	F.fagei	16	0.004	0.48	0.006
	F.mediterranea	16	0.004	0.95	0.006
	F.fraudax	8	0.002	0.48	0.003
	F.gracilis	4	0.001	0.48	0.001
Total		371.118			

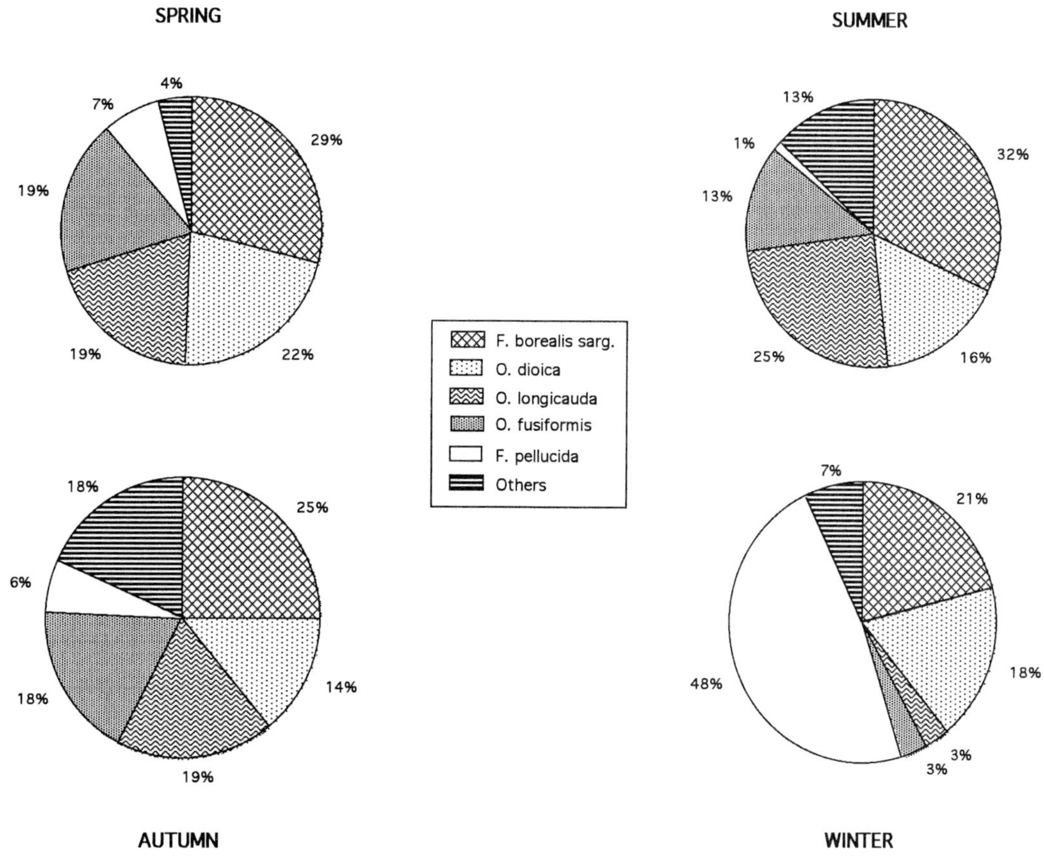

Fig. 15.1. Seasonal variations over a year of the five most abundant appendicularian species at Villefranche-sur-Mer collected by fortnightly vertical samples from 75 m to the surface.

and winter 100 ± 85. After smoothing the fluctuations of the annual curves, on average there was a period of relative abundance from the beginning of July until the end of October. In contrast, in the Adriatic, Baranovic *et al.* (1992) found the maximum abundance (over the period 1960–1982) in March.

These five most abundant species at Villefranche are also those most abundant in the Adriatic (Skaramuca, 1983), where they made up 91% of the total appendicularians, and in the southern Bay of Biscay (Acuña and Anadón, 1992). Similarly, in Volcano Bay, Southwest Hokkaido, Shiga (1985) found the same five species to be most abundant (together with the cold water *O.labradoriensis*). Different authors have employed different types of net, with mesh sizes ranging from 50 to 330 μm, hence direct comparison between data sets of the actual abundance, (particularly of *smaller* forms) is rather difficult, to say the least.

Perhaps the most notable feature, however, is that the species succession in different areas is the same. Fenaux (1961a) showed that the dominance of the different species in Villefranche-sur-Mer followed a defined sequence starting with *Fritillaria borealis* in December to January, followed by *F.pellucida typica* until the middle of March, then by *Oikopleura dioica* until the end of April, *O.fusiformis* in May and *O.longicauda* until the end of November (Fig. 15.2). This sequence continued over the 4 year study and the timing varied very little.

More recently, a very similar seasonal sequence was observed by Acuña and Anadón (1992) in the south of the Bay of Biscay, off the northern Spanish coast. At 13 stations, appendicularians were sampled using vertical WP2 (200 μm mesh) net tows throughout the euphotic layer at near monthly intervals from January to December. Temperature, salinity, and chlorophyll concentrations were also measured at a range of depths.

The seasonal succession from January to December was: *F.borealis*, *O.dioica*, *F.pellucida*, *O.fusiformis*, and *O.longicauda*. Principal component and cluster analyses showed a community gradient structure coupled to temperature at the depth of the chlorophyll maximum (rather than to surface temperature).

Acuña and Anadón put forward as a working hypothesis that the *overall* abundance of appendicularians is

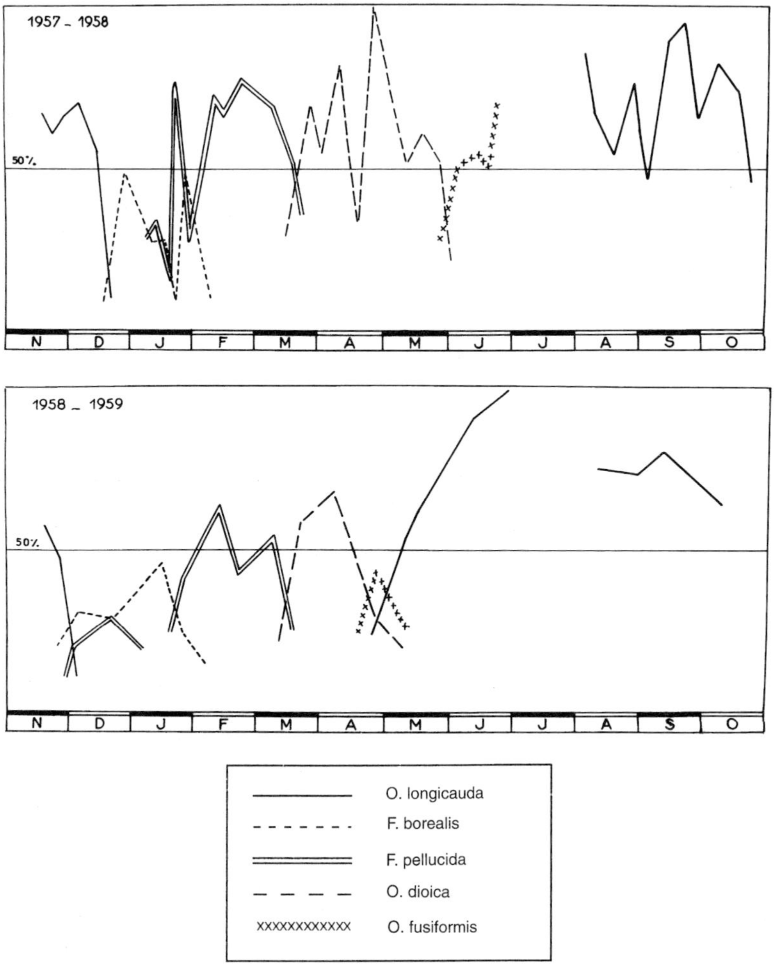

Fig. 15.2. Sequential relative dominance of the most numerous species in the bay of Villefranche-sur-Mer. Modified from Fenaux (1961).

dependent on primary production, whilst the *relative* abundance of the different species depends on the temperature at which primary production occurs. In view of the small degree of metabolic regulation in appendicularians (Gorsky *et al.*, 1987), Acuña and Anadón inferred that they are very sensitive to water temperature; the physiological adaptations to temperature being species specific. This hypothesis of species-specific physiological temperature adaptation fitted well with the position of the different species in the series, and with the data from Acuña's (1994) work on vertical distribution (see next section), but it is unclear whether it is generally applicable. First, it is evident that different stocks of the same species are adapted to different temperatures. For example, *O.dioica* lives at 29°C in Kingston harbour, Jamaica (Hopcroft and Roff, 1995), but the Mediterranean stock of the species (most abundant at water temperatures around 19°C) can only survive at this temperature after slow acclimation (Gorsky, personal communication).

Secondly, observations in other areas suggest that although some species are stenothermal, others seem to be eurythermal, tolerating a wide temperature range. For example, Shiga (1985) pointed out that in the populations he studied in Volcano Bay, although *O.labradoriensis* and *Appendicularia sicula* only appeared over a narrow range of temperature and salinity, *F.borealis* f *typica* and *O.dioica* seemed to have a much wider range. *Fritillaria borealis* f *typica* remained most abundant above the thermocline over a wide temperature range, in September at 15°C, and in March at 2°C. The population of *O.dioica* studied by Uye and Ichino (1995) in Fukuyama harbour (a eutrophic inlet of the Inland Sea of Japan), occurred between 8.9 and 28.2°C.

The remarkable data set (as yet unpublished) collected by Fenaux at Villefranche, mentioned at the beginning of this section, deserves further analysis. The success of the recent application of a Markov regression model for ordinal ecological series (Ménard *et al.*, 1993) to the analysis of the fluctuations in salp abundance at Villefranche (Ménard *et al.*, 1994), suggests that a similar approach to the appendicularian data would be of great interest.

15.3. Vertical distribution

The vertical distribution of appendicularians was first described by Lohmann (1896a). The extensive material of the 'Plankton Expedition' enabled him to point out that the upper 200 m contained most of the specimens, and that few were found at greater depths. However, these and the following observations were based on non-closing nets hauled vertically to the surface from different depths: the species present in the deepest hauls and absent from shallower hauls were attributed to the deeper layers. During the Sud Polar Expedition, off the 'Kaiser Wilhelm' coast (Antarctic), Lohmann and Bückmann (1926) found the richest layer between 100 and 200 m. In the same area, but in the open sea, this 'rich' layer was between 200 and 300 m. These first 'general' observations have more recently been much extended by studies in particular areas.

15.3.1. *Epipelagic zone (surface to 500 m)*

The most detailed studies of vertical distribution in temperate waters have been made by Fenaux (1963, 1968b–d). Two different sites were studied, with bottom depths of 80 and 600 m. At the shallower site, three segments of the water column (75–50, 50–25 and 25–0 m) were sampled. At the deeper site, the same three segments were sampled, and in addition, 500–300, and 300–150 m. Vertical tows with a closing net were made over 2 years, twice monthly at the first site, and monthly at the second site.

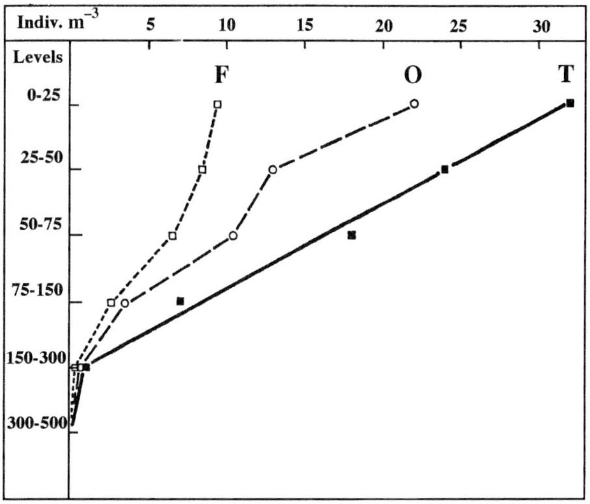

Fig. 15.3. The average number of appendicularians m^{-3} at different depths during 12 successive months at Villefranche-sur-Mer. Modified from Fenaux (1968). O, Oikopleuridae; F, Fritillariidae; T, total appendicularians.

Figures 15.3–15.5 show that the upper mixed layer contains numerous individuals of relatively few species, whereas middepth waters contain few individuals belonging to many species. The greater part of all the specimens collected came from the upper layers, in fact, over 70% of all the specimens collected between 500 m and the surface were found at < 80 m depth. Thirty one percent of all the specimens (16 species) were collected between

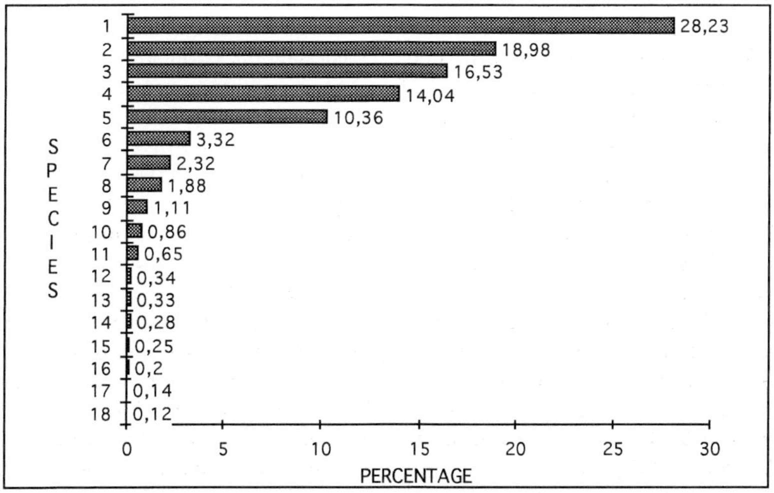

Fig. 15.4. Diagram of the abundance of the first 18 most abundant species based on the numbers in Table 15.1.

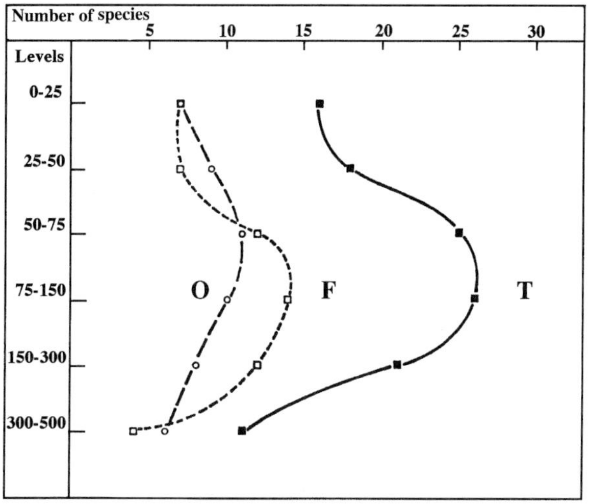

Fig. 15.5. Bathymetric variations in the number of species collected during 12 successive months at Villefranche-sur-Mer. Modified from Fenaux (1968). O, Oikopleuridae; F, Fritillariidae; T, total appendicularians.

the surface and 25 m, whilst in contrast, between 300 and 150 m there were 21 different species, but they made up only 6% of the total number of specimens. Appendicularians show little diel vertical migration (Palma, 1986), and in this study, nycthemeral movements were only of small amplitude (about 10 m).

The distribution of the three appendicularian families in the water column was different. Fritillariidae were distributed in thin strata in the surface layers and so only appeared homogeneously distributed when the samples included a layer at least 25 m thick. They were usually more numerous between 75 and 50 m. Oikopleuridae were present in all the samples, and during the months May–August they were almost the only appendicularians in the water column from 75 m to the surface. Lastly, Kowalevskiidae were found mainly from May to October. They were relatively abundant in May and June, particularly between 50 and 25 m. In contrast to the other families they were more numerous in the open sea than in coastal waters.

Gorsky (personal communication) has observed at Villefranche a thin (1–2 m thick) extended patch with many *Kowalewskia*, this pattern of discontinous distribution in thin layers is plainly very difficult to sample effectively, or indeed, to discover at all, except by underwater observations. Since divers off Newfoundland have observed similar thin patches of *Oikopleura vanhoeffeni* (Deibel, unpublished), such a thin patchy distribution may characterize other families also.

Acuña (1994) examined the summer vertical distribution of appendicularians in the southern Bay of Biscay, from samples collected by vertical WP2 net hauls during August 1990, when chlorophyll *a*, temperature, salinity, and particle sizes were also measured. Species correlation matrices and principal component analyses showed a vertical gradient of species abundance. From the surface to the deep layer this was: *Oikopleura longicauda, O.fusiformis, Fritillaria pellucida,* and *O.rufescens. Oikopleura dioica* was insufficiently numerous for significant conclusions to be drawn.

Acuña concluded that the hypothesis that species-specific temperature optima determined the *temporal* (seasonal) species succession could also be applied to the *spatial* (vertical) scale, but was careful to note that the correlation of appendicularian abundance with temperature does not necessarily imply causation. In fact, as seen below, it is usual to consider some species as stenothermal, but others are classed as eurythermal and do not appear to have narrow temperature preferences. An interesting feature of Shiga's (1985) observations on vertical distribution was (as Acuña, 1994, pointed out) that they hinted at seasonal downwards migration into layers of suitable temperature, as the thermocline moved deeper in the water column. In Logy Bay, Newfoundland, Deibel has found a similar seasonal change in the depth distribution of *O.vanhoeffeni* (Deibel *et al.*, in preparation).

15.3.2. *Mesopelagic zone (500–1000 m)*

During the last 30 years, observations and sampling with armoured diving suits and bathyscaphes have revealed the true distribution of the plankton in the mesopelagic and bathypelagic (below 1000 m) levels.

However, it is only in the past decade that a new mesopelagic ecosystem has been revealed by sampling using manned submersibles capable of gently catching fragile specimens *in situ,* (Youngbluth, 1984; Youngbluth *et al.*, 1990). A large percentage of the appendicularians in this ecosystem are new endemic mesopelagic forms, some belonging to new genera, such as *Mesochordaeus bahamasi* (Fenaux and Youngbluth, 1990) and *Mesoikopleura enterospira, M.gyroceanis* and *M.youngbluthi* (Fenaux, 1993). Others were placed in the genus *Oikopleura: O.villafrancae,* with the variety *indentacaudata. Inopinata inflata* and *Mesopelagica caudaornata* (Fenaux and Youngbluth, 1991) were subsequently both moved to *Oikopleura* by Fenaux (1993b). With the increasing number of mesoplanktonic samples collected with similar techniques, the number of new species and even new genera will certainly increase in the future. As filter feeders, mesopelagic species seem likely to play an important role in the recycling of sinking particles originating from the euphotic layer. These particles include living and detrital phytoplankton cells, microzooplankton and zooplankton faecal pellets, and large aggregates of marine snow. In addition, the recent discovery of the possibility that appendicularians can use colloidal particles as food (Flood *et al.*, 1992) provides another possible source of sustenance for mesopelagic species. Since appendicularians are fed upon by many endemic predators, such as ctenophores, siphonophores, and adult and larval fish, colloid utilization by appendicularia may provide an important 'short-circuit' of the food web, directly transmitting energy from very small particles to much larger organisms of the mesopelagic zone.

15.3.3. *Bathypelagic zone*

If the mesopelagic appendicularian fauna is poorly known, the bathypelagic fauna is hardly known at all. Gorsky (personal communication) has observed (from the submersible *Cyana*) a dense appendicularian population off the Algerian coast between 800 and 1500 m. Appendicularians have been reported at 2000 m from Russian submersibles (Youngbluth, personal communication). There are obvious technical problems in studying the bathypelagic plankton, and it seems probable that submersible observation and collection will remain the best source of information, despite its cost and difficulty. As Hamner (1992) pointed out, the discarded filter nets of the large *Bathychordaeus* sp. may be abundant lying on the bottom (see Chapter 6).

15.4. Geographical distribution

Appendicularians are present in all oceans and seas world-wide, apart from the Dead, Aral, Azov and Caspian Seas (Table 15.3). As noted above, little is presently known of the mesopelagic forms and next to nothing of bathypelagic appendicularians.

Although most species can be regarded as warm water forms, most often eurythermic, appendicularians occur in both the Antarctic and Arctic oceans. These few cold water species may be either stenothermic or eurythermic. In the Antarctic, there is a single *Oikopleura, O.gaussica (syn. O.valdiviae, O.drygalskii, O.weddeli)*, a single *Fritillaria, F.antarctica*, and two *Pelagopleura, P.magna and P.australis.* Lohmann and Hentschel (1939) cited eight other new *Oikopleura* species and a new *Fritillaria*, but as they unfortunately gave no descriptions, these can only be considered as *nomina nuda*.

In the Arctic Ocean, only *F.polaris* seems to be stenothermic. *O.labradoriensis* and *O.vanhoeffeni* are eurythermic cold water forms originating from Arctic waters, but may be found far to the south during winter or within cold currents, such as the Labrador current. *Fritillaria borealis typica*, is also eurythermic, and is the only bipolar species.

However, Lohmann (1931) pointed out that *O.gaussica* from the Antarctic is very close to *O.vanhoeffeni* and *O.labradoriensis* of Arctic seas (both closely related), and believed that all three of these oikopleurid species could be regarded as but local variations of a single bipolar species. All other epipelagic appendicularian species are warm water forms and belong to the 15 existing epipelagic genera, which include the three genera *Oikopleura, Fritillaria,* and *Pelagopleura* that have cold water representatives. In practice, however, most appendicularian populations consist mainly of species in the two genera, *Oikopleura* and *Fritillaria.* For example, numerous samples from Atlantic warm waters are made up of *Oikopleura* (75%) and *Fritillaria* (25%), whilst all the other genera represent < 2% of the total (Lohmann, 1933, 1934). The mesopelagic *Oikopleura* species and those of other genera may be stenothermic, but no doubt stocks may differ. In the Mediterranean for instance, deep waters are always warmer than 12°C, whereas elsewhere they are much colder.

15.4.1. *Antarctic Ocean*

The Antarctic Ocean is here considered as the waters between the Antarctic continent and the Antarctic convergence. With this definition, which excludes subantarctic waters, 23 species of appendicularia have been collected in the area. Descriptions, maps, and references to these species are given by O'Sullivan (1983). In addition to the cold water species discussed above in the introduction, some 'warm water' species, such as *O.fusiformis, O.longicauda,* and *O.parva,* also occur south of the Antarctic convergence, as well as *Sinisteroffia scrippsi* and *Stegosoma magnum.* Numerous *Fritillaria* species have also been found: *F.aberrans, F.drygalskii, F.formica* (which may be a new subspecies; Tokioka, 1964), *F.haplostoma + F.haplostoma glandularis, F.megachile, F.pellucida typica, F.tenella,* and *F.venusta.* A single *Kowalewskia* sp. in poor condition was found by Tokioka, who referred it tentatively to *K.tenuis.* Lohmann (1928) termed such warm water forms 'tropical visitors' and thought they might live in the deeper layers which have a slightly higher temperature, whilst the endemic species lived in the upper 100 m. Tokioka (1961) was uncertain that there could be such a vertically separated distribution, because the ambient physico-chemical characteristics are not very different between 200–400 m and the surface.

He suggested instead that either the endemic species are regional ecological forms of some warm water species, or that the warm water species found in the Antarctic are themselves regional varieties, as was shown for *F.haplostoma* and *F.formica.*

15.4.2. *Arctic Ocean*

Few appendicularian species has been collected in the Arctic compared with the Antarctic. The first were described by Lohmann (1895, 1896a,b) from the speci-

Table 15.3. World distribution of appendicularian taxa.

Taxa	Atlantic Ocean	Pacific Ocean	Indian Ocean Red Sea	Mediterranean Sea	Arctic Ocean	Antarctic Ocean
Bathochordeaus charon	X	X	X			
B.stygius	X	X				
Mesochordaeus bahamasi	X					
Althoffia tumida	X	X	X			
Chunopleura microgaster			X			
Folia gracilis	X	X		X		X
F.mediterranea		X		X		
Megalocercus abyssorum	X	X	X	X		
M.huxleyi		X	X			
Mesoikopleura enterospira	X					
M.gyroceanis				X		
M.haranti				X		
M.youngbluthi	X					
Oikopleura albicans	X	X	X	X		
O.caudaornata	X					
O.cophocerca	X	X	X	X		
O.dioica	X	X	X	X		
O.fusiformis	X	X	X	X		X
O.fusiformis cornutogastra	X	X	X			
O.gaussica	X					
O.gracilis	X	X	X	X		
O.inflata	X					
O.intermedia	X	X	X	X		
O.labradoriensis	X	X			X	
O.longicauda	X	X	X	X		X
O.parva	X	X	X	X	X	X
O.rufescens	X	X	X	X		
O.vanhoeffeni	X	X			X	
O.villafrancae				X		
O.villafrancae indentacauda				X		
Pelagopleura australis						X
P.gracilis	X	X				
P.magna						X
P.oppressa	X					
P.verticalis	X	X	X			
Sinisteroffia scrippsi		X				X
Stegosoma magnum	X	X	X	X		X
Appendicularia sicula	X	X	X	X		
A.tregouboffi				X		
Fritillaria aberrans	X	X	X			X
F.abjornseni	X	X	X			X
F.aequatorialis	X		X	X		
F.antarctica						X
F.arafoera		X	X			
F.borealis typica	X	X			X	X
F.borealis intermedia	X	X	X	X		
F.borealis sargassi	X	X	X	X		
F.charybdae		X		X		
F.drygalskii	X			X		X
F.fagei			X	X		
F.formica digitata	X	X	X	X		X
F.formica tuberculata	X		X	X		
F.fraudax	X	X	X	X		
F.gracilis	X	X	X	X		
F.haplostoma	X	X	X	X		X

Table 15.3. *(Continued)*

Taxa	Atlantic Ocean	Pacific Ocean	Indian Ocean Red Sea	Mediterranean Sea	Arctic Ocean	Antarctic Ocean
F.helenae	X					
F.megachile	X	X	X	X		X
F.messanensis			X	X		
F.pacifica		X	X			
F.pellucida omani			X			
F.pellucida typica	X	X	X	X		X
F.polaris					X	
F.tenella	X	X	X	X		X
F.urticans				X		
F.venusta	X	X	X	X		X
Tectillaria fertilis	X	X	X	X		
T.taeniogonia		X	X			
Kowalewskia oceanica	X		X	X		
K.tenuis	X	X		X		X
Number of Oikopleuridae	26	23	17	17	3	9
Number of Fritillariidae	19	19	22	21	2	10–11
Number of Kowalevskiidae	2	1	1	2	0	1?
Total number of taxa	47	43	40	39	5	19–21

mens caught during the the *Plankton* and *Grönland* expeditions. Even south to 60°N, the species he described are still the only ones known from Arctic waters, with the exception of *Fritillaria polaris* discovered in 1934 by Berstein in the Kara Sea. There are three cold water species: *Oikopleura labradoriensis, O.vanhoeffeni,* and *F.borealis typica.*

It is important to recognize that appendicularian (and, indeed, all plankton) distribution is not just a simple function of latitude. For example, the cold water form *O.vanhoeffeni* occurs off Newfoundland at 47°N, and the warm water form *O.parva* was recorded by Lohmann (1905) off Spitzbergen at 80°N (presumably as a result of the influence of the North Atlantic Drift in the waters around Spitzbergen). Similarly, to the south of Greenland a few *O.parva* were collected with *O.vanhoeffeni* and *O.labradoriensis* (Wesenberg-Lund, 1926). Obviously, large scale current systems are going to complicate latitudinal distribution patterns. The Labrador and North Pacific Currents carry *O.labradoriensis* and *O.vanhoeffeni* (probably the first appendicularian ever described, by Chamisso and Eysenhardt, 1821) south to 50°N along the east and west American coasts (Galt, 1970; Buchanan and Browne, 1981; Mahoney, 1981; Deibel, 1988). There are few appendicularian species on the north-east coast of Canada and off Newfoundland: *O.vanhoeffeni, O.labradoriensis, O.dioica,* and *F.borealis typica* (Davis, 1982; Frost *et al.*, 1983; Taggart and Frank, 1987), but their numbers can be enormous. For example, in continental shelf waters off Labrador, the biomass of *F.borealis* and *O.vanhoeffeni* can be 6–10% of the total zooplankton biomass (Buchanan and Browne, 1981). This is somewhat higher than values estimated by Mumm (1993) for the Nansen basin of the Arctic Ocean, northeast of Spitzbergen, of 1%. Both *F.borealis* and *O.vanhoeffeni* have been found in the Arctic ocean, the Canadian archipelago and Baffin Bay, and have been considered as indicator species for Arctic waters (Kramp, 1942; Grainger, 1965). For example, Udvardy (1954) found *O.vanhoeffeni* in the strait of Belle Isle and in the northern Gulf of St Lawrence, areas heavily influenced by the Labrador Current. Likewise, Longhurst *et al.* (1984) found *O.vanhoeffeni* to be most abundant in two areas leading from the high Arctic into Baffin Bay, i.e. Jones Sound and Kane Basin. *Oikopleara vanhoeffeni* was most abundant in Arctic surface water in Jones Sound and in intermediate depth Arctic Ocean water in the Kane Basin (Longhurst *et al.*, 1984).

There are naturally very considerable seasonal variations, in the Arctic as elsewhere. For example, off the west coast of Greenland, during March, samples from 200 m to the surface each contained between 12,000 and 52,000 *F.borealis typica,* whilst numerous similar samples from May to December had less than 200 specimens per sample (Lohmann and Bückmann, 1926). Close to the North Pole, Pavshtiks (1972), gave an analysis of 150 samples taken under ice north of 85°N over several years (1954–1956) and showed that the percentage of *Oikopleura* spp. and *F.borealis* increased, in August (Summer) up to 20% of the total zooplankton observed. In the north east water polynya at 80°N oikopleurids are most abundant in late summer, but the seasonal cycle of fritillariids is not yet known (Deibel, unpublished).

15.4.3. *Atlantic Ocean*

Forty three species of appendicularians are presently known in the Atlantic Ocean. The north eastern region of the northern zone is species poor, both quantitatively and qualitatively, and most are found in the equatorial zone, in the Gulf of Guinea, north of Dakar and along the north east coast of Brazil to Rio de Janeiro. This southern distribution was well studied by Lohmann and Hentschel (1939). The warm waters are characterized by the following species (in decreasing order): *Oikopleura longicauda, O.fusiformis, O.rufescens,* and *O.cophocerca.* Towards the north, the order changes because *O.longicauda,* which is more strictly a warm water species than *O.fusiformis,* gives precedence to the latter and its distribution in the south seems to be limited by the 15°C isotherm (Lohmann and Hentschel, 1939). Nevertheless *O.longicauda* was sampled near the 60°N parallel off the south east of Iceland, below 200 m at a temperature of 12.5°C (Wesenberg-Lund, 1926). In coastal waters or in those currents strongly influenced by coastal waters, such as the Guinea Current, *O.dioica* predominates over all other species of appendicularians.

During 1950, 70% of the appendicularians collected off the west coast of Morocco were *O.dioica* (Furnestin, 1957). The most abundant fritillariids in these samples were *F.formica, F.pellucida,* and *F.borealis sargassi. Fritillaria haplostoma,* which is generally scarce in open waters, was abundant in some coastal areas, such as around the Fernando de Noronha islands north of Recife (Brazil) and off Guinea, (equatorial Africa) around the Fernando Po islands, (Lohmann and Hentschel, 1939).

O.dioica and *F.borealis typica* are found in the Baltic Sea, (characterized by low salinity water), whilst in the Kattegat three additional species occur: *O.vanhoeffeni, F.venusta* and *A.sicula* (Bückmann, 1926). These five species have been found in the North Sea off Bergen, with *O.parva* and *F.gracilis*? (Runnström, 1931).

15.4.4. *Pacific Ocean*

A detailed review of appendicularian distribution in the different areas has been given by Tokioka (1960), and thus a few examples only will be considered. In the Northern Pacific region, *Oikopleura longicauda* is still usually the most abundant species, nevertheless, its distribution is different from that in the Atlantic: it is predominant among the warm water species even in temperate or subarctic mixed waters. Tokioka considered this as one of the most noteworthy characters of the appendicularian fauna of the North Pacific. After *O.longicauda,* the most important oikopleurids in the open ocean are *O.rufescens, Megalocercus huxleyi, Stegosoma magnum,* and *O.cophocerca.* Among the Fritillariidae, *F.borealis sargassi* and *F.pellucida* are the most abundant, followed by *F.formica digitata, F.fraudax,* and *F.tenella.* As in the Atlantic, *F.haplostoma* is widely distributed, but is generally not abundant.

The western Subarctic Gyre, a very poor area, separates the northern North Pacific from the Bering Sea. Shiga (1982) examined vertical samples (usually taken from 150 m to the surface) from both regions during eight summers between 1957 and 1972. The most abundant species was *O.labradoriensis,* with maximum densities of 100 m^{-3} in the Bering Sea and 328 m^{-3} in the North Pacific. *Oikopleura vanhoeffeni* and *F.borealis typica* were also found. The same author (Shiga, 1985) again used vertical samples to study the annual appendicularian cycle between 1974 and 1980 in Volcano Bay, south west Hokkaido, Japan, where there is a periodic annual interchange between the cold (Oyashio) and warm (Tsugaru) water. By far the most abundant species was *F.borealis typica,* maximum densities in the Bay being over 10^3 individuals m^{-3} in March. Other species (*O.labradoriensis, O.dioica, O.longicauda,* and *O.fusiformis*) reached maxima of around 100–250 individuals m^{-3} apart from *O.parva, F.pellucida,* and *Appendicularia sicula,* which were always less than 3 individuals m^{-3}.

The southern Pacific is less well known. South of the equator, at 4°S and 152°E (off the Bismarck archipelago), Lohmann (1931) found 23 species of appendicularians in the samples collected by Dahl in 1896–1897. Eight of these represent almost 90% of the total number: *O.longicauda, F.fusiformis, O.rufescens, O.dioica, F.borealis sargassi, F.pellucida, F.formica,* and *F.haplostoma.* Along the west coast of Australia, Thompson (1948) collected 22 species, of which eight represented 95% of the total. The list is similar to that of Lohmann, but *Megalocercus huxleyi* and *Stegosoma magnum* take the place of *F.formica* and *F.haplostoma.* Tokioka (1960) noted the curious absence of *Appendicularia sicula* from this list. Along the coast of Peru, *O.longicauda* represents 94% of the total number of appendicularians, followed by *O.intermedia* at only 2% (Fenaux, 1968b).

Fenaux and Dallot (1980), studying samples taken from Point Conception (34°N) to the south of Baja California (22°N), found 27 taxa; 10 of them representing 93% of all the appendicularians collected: *O.longicauda* 50%, *F.venusta* 11%, *F.tenella* 11%, and smaller numbers of *F.pellucida* f *typica, O.fusiformis, O.rufescens, Stegosoma magnum, O.dioica, O.cophocerca,* and *Tectillaria fertilis.* Concordant correlations using several numerical methods were seen between groups of species and different water masses. The association of the species *O.longicauda, O.fusiformis, and O.cophocerca* had been previously noted in the Pacific off Australia (Sheard, 1965). Together with *F.tenella, F.venusta,* and *F.pellucida typica,* they were placed in a 'general' group. Fenaux and Dallot proposed that this 'general' group of mixed species was characteristic of the hydrological transition zone of the waters of the central Pacific. The southern (meridional) group of associated species consisting of *O.cornutogastra, O.rufescens, O.parva, Stegosoma magnum, F.borealis sargassi,* and *Tectillaria fertilis* corresponded to equatorial Pacific waters. Fenaux and Dallot's analyses gave no support to Tokioka's (1960) suggestion that the relative abundance of *O.fusiformis* and *O.longicauda* was a useful index for distinguishing different water masses.

15.4.5. *Mediterranean Sea*

Thirty eight appendicularian species are presently known from the Mediterranean. Some species seem to be Mediterranean endemics, e.g. *Fritillaria urticans, Mesoikopleura haranti, M. gyroceanis,* and *Appendicularia tregouboffi*. The genera *Bathochordaeus, Pelagopleura,* and *Althoffia,* and the two cold water species *Oikopleura labradoriensis and O.vanhoeffeni,* have never been found. *Oikopleura fusiformis cornutogastra, Fritillaria helenae, F.drygalskii,* and *F.abjornseni* are also absent. The rarity of *O.rufescens* is notable.

In the different Mediterranean regions, as elsewhere, a few species dominate the appendicularian fauna. Six species represent at least 90% of the total number in the western Mediterranean: *O.longicauda, F.borealis sargassi + F.borealis intermedia, F.pellucida, O.dioica* and *O.cophocerca.* At present 32 species are known from Villefranche-sur-Mer (Fenaux, 1963), 27 from Messina (Lohmann, 1899b) and 17 from Algiers (Bernard, 1958). Similarly, in the Adriatic, of the 27 species known (Skaramuca, 1980, 1983), five species (*O.longicauda, O.fusiformis, O.dioica F.pellucida,* and *F.borealis*) represent more than 91% of the total. A strong west–east gradient in appendicularian population density has been shown, correlated with the river Po inflow in the northwest part of the Adriatic (Fenaux, 1972), and Baranovics *et al.*, (1992) have discussed the species distribution in relation to the hydrology of different regions of the Adriatic. Recently, Zagami *et al.* (1996) have reported striking daily differences in zooplanktonic biomass in the Straits of Messina, where there are complex current systems. Appendicularians (species not given) varied over 24 hr (3 hourly samples) in January from 1.95 individuals m^{-3} to 177.4 individuals m^{-3}, largely following the changes in direction of current flow.

Twenty nine are now known from the eastern (oriental) Mediterranean (Fenaux, 1974), which was previously inadequately sampled (Fenaux, 1967). Only *O.dioica* occurs in the Black Sea; the report of *O.cophocerca* (Zernov, 1913) was a mistaken determination.

15.4.6. *Indian Ocean, Persian Gulf and Red Sea*

There are 40 species in this area, though one, *Chunopleura microgaster* (Lohmann, 1914), is doubtful. According to Tokioka (1960), the reality of this endemic monospecific genus, based on a very small number of badly preserved specimens, is suspect. Investigations on appendicularian distribution in the region may be divided historically into four main periods. After early studies off Zanzibar and the Seychelles, at the end of the last century, and from the west coast of Australia in 1909, the results of the two main German expeditions of 1898/99 and 1901/1903 were published, with some delay (Lohmann, 1914; Lohmann and Büchmann, 1926; Lohmann, 1931). 30 appendicularian species were described; however, only parts of the area were sampled. There followed a period up to 1952 during which little was published; finally, the period of the International Indian Ocean Expedition (IIOE), which began in 1964, added 5 species as well as much knowledge of appendicularian distribution and abundance. There are two notable features of the appendicularian fauna in the area. First, in the occidental region there is a more or less regular decrease in species number from the south (21 species) to north (two species only). Secondly, as Fenaux (1973) pointed out, there was a very significant latitudinal increase in the number of individuals caught in the central part of the Indian Ocean during the cruise of the RV *Anton Bruun*, as seen in Table 15.2. From south of 42°S to 20°N a steady increase was found, except in the equatorial counter current between the equator and 10°S.

Table 15.2 The numbers of appendicularians collected in the central region of the Indian Ocean during the fifth cruise of the RV *Anton Bruun* (1964).

Area	Number of appendicularians caught
South of 42°S	1
40°S–30°S	7
30°S–20°S	16
20°S–10°S	132
10°S to the Equator	106
Equator to 10°N	485
10°N–20°N	611

From Fenaux (1973).

15.5. Seasonal variation

Seasonal variations in abundance have been examined in several different regions, not only in terms of the numbers of appendicularians, but also in terms of total zooplankton biomass. For example, Jansa (1985) found in vertical hauls off Mallorca (western Mediterranean) that appendicularians varied between 1 and 20% of *total* zooplankton biomass, whilst Uye and Ichino (1995) found the biomass of *O.dioica* in an inlet of the Sea of Japan to vary between 0.21 mgC m^{-2} (in August) and a maximum of 11.4 mg C m^{-2} at the end of June.

15.5.1. *Arctic and Antarctic*

In colder waters, seasonal temperature and light changes presumably underlie changes in appendicularian populations, by changing the productivity of the pico- and nanoplankton on which they feed (see p. 163). In the high Arctic there are only small variations in surface temperature, from –1.7 to 0.9°C in summer and –1.8 to –1.9°C in winter, but off Newfoundland, winter surface temperatures are around –1.7°C whilst summer surface temperatures are around 17°C. According to Lohmann (1933, 1934), in cold waters where seasonal surface temperatures vary little, light changes (inducing nanoplankton blooms) are perhaps the most important factor determining variations in appendicularian populations. Lohmann's view has received support from Fukuchi *et al.* (1985), who showed that under the Antarctic sea ice, appendicularians increase markedly in number during August (Winter).

15.5.2. *North Pacific*

An analysis of monthly plankton samples taken over several years in and off Volcano Bay, southwest Hokkaido (Shiga, 1985), showed that the periodic exchange between the spring entry of cold Oyashio water and the autumn entry of warm Tsugaru water largely control the seasonal and vertical distribution of appendicularians. In spring, *F.borealis typica* and *O.labradoriensis* were abundant, whereas *O.longicauda, O.fusiformis,* and *O.dioica* were dominant in the autumn. the relationship between water masses and appendicularians was well explained by a series of temperature/salinity curves (Fig. 15.6).

15.5.3. *Tropics*

In tropical areas, continuous observations over at least 1 year are rare, but the available data show that temperature variations (smaller than in temperate areas) induce smaller variations in appendicularian abundance than in cold and temperate regions (Fenaux, 1970, 1980). For example, in Kingston Harbor, Jamaica, annual temperature is almost constant at 29 ± 1°C (Hopcroft and Roff, 1995), and at Lime Cay, just south of the harbour, it only varies between 27 and 29°C (Clarke and Roff, 1990). In an interesting study at Lime Cay, Clark and Roff took vertical samples of different planktonic taxa over a

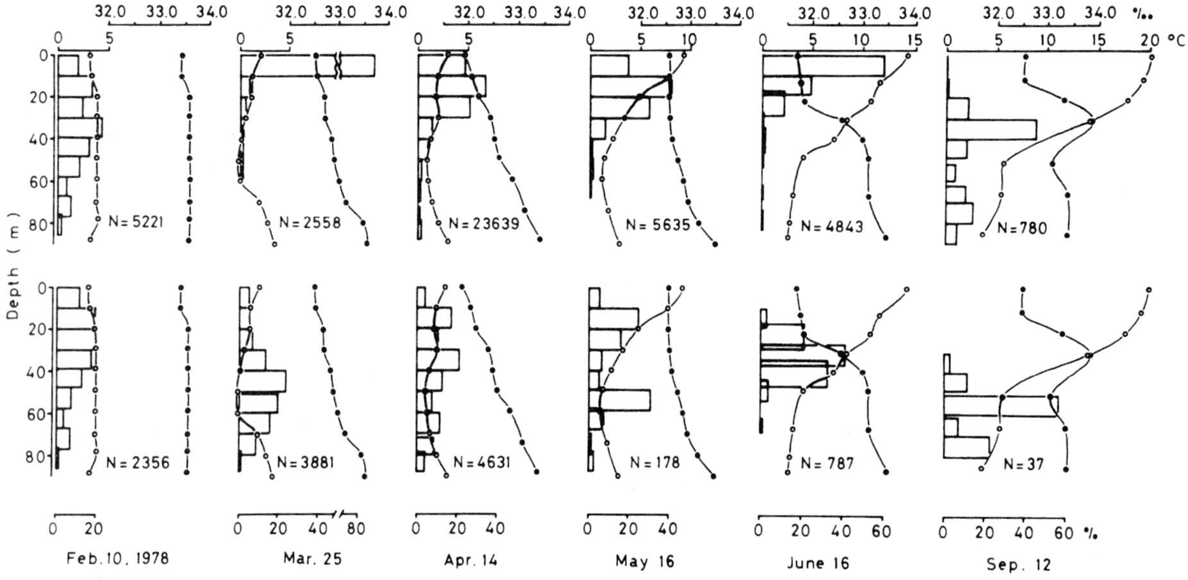

Fig. 15.6. The vertical distributions (%) of *Fritillaria borealis f. typica* (upper series, and *Oikopleura labradoriensis* (lower) between February and September 1978 at a station in Volcano Bay, Hokkaido, Japan, together with salinity (•–•) and temperature (○–○) profiles. N, total number of specimens in the whole water column. From Shiga (1985).

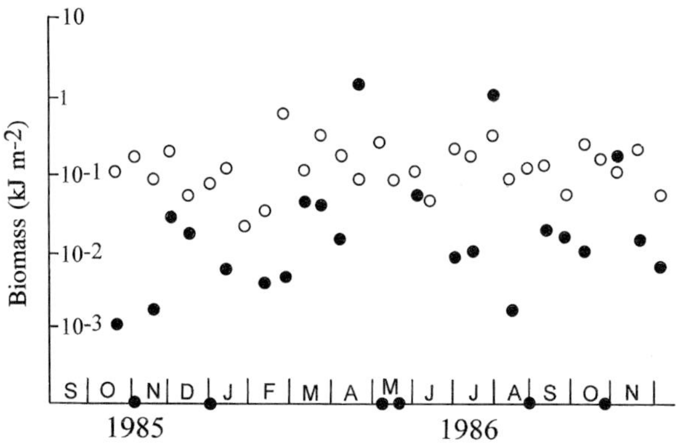

Fig. 15.7. Biomass of oikopleurid (○) and fritillariid (•) larvaceans at Lime Cay, south of Kingston, Jamaica, from November 1985 to December 1986. After Clarke and Roff (1990).

1 year period, using 600 and 200 μm mesh nets, and observed that whilst oikopleurid larvaceans (undetermined but probably mainly *O.dioica*) varied much in abundance (from < 1500 individuals m^{-2} to a maximum of 25 780 individuals m^{-2}, no particular pattern was evident over the year (Fig. 15.7). This contrasted markedly with fritillariids, where extreme fluctuations were found, between the two maxima of 88 150 individuals m^{-2} and 59 950 individuals m^{-2} in April and August (Fig. 15.7). In view of the mesh size of the nets used, these numbers for fritillariids were probably biased low. Hopcroft and Roff (1995) found that the coefficient of variance (0.39) of the annual picoplankton (< 2 μm) through the year suggested that it was relatively stable, hence these fluctuations were not related to food availability

In the gulf of Akaba (Red Sea), the distribution of several abundant appendicularian species is directly or indirectly related to temperature (Fenaux, 1973).

15.6. Other aspects of distribution

15.6.1. *Blooms*

Several blooms of appendicularians have been reported; the first observed seems to have been by Quoy and Gaimard (1836), whilst the most famous example was given by Seki (1973). Seki described a 'red tide' of *Oikopleura dioica* in Sannich Inlet (Vancouver Island, Canada). This developed over several days in July 1968, and spread several metres wide for several kilometres. The population density in the first metres under the surface reached 25 600 individuals m^{-3}. This bloom occurred during a heavy bloom of phytoplankton. Fenaux (unpublished) observed a similar accumulation at the same site at the end of May 1973. This remarkable population density has, however, recently been exceeded in another temperate inlet (of the Sea of Japan), where Uye and Ichino (1995) found in July 1987 53 200 individuals m^{-3}!

It is hardly surprising that appendicularians can form blooms, when their astonishingly short generation times are considered. For *O.dioica* Esnal *et al.* (1985) estimated a generation time of around 1 day (!) for the population living in Brazilian coastal waters at 27–28°C. Hopcroft and Roff (1995) estimated a similar time for the population in Kingston Harbor, Jamaica, at 29°C and Uye and Ichino (1995) a minimum time of 1.2 days (at 28.2°C). These generation times are temperature dependent: at high water temperatures they naturally much exceed those for *O.dioica* at lower temperatures. For example, Uye and Ichino (1995) calculated generation times of 15.8 days at 10°C, 7.8 days at 15°C and 1.9 days at 25°C. It is certainly most striking to find metazoans with such short generation times, and hence able to exploit favourable conditions very rapidly.

Small scale horizontal vortices with swarms of *O.longicauda* have been reported off the California coast by Owen (1966). Alldredge (1982) observed mature specimens of the same species off southern California during spring. They formed very dense windrows (up to 3565 individuals l^{-1}), swimming without houses. This stage of mature specimens without houses has been constantly observed in culture (Fenaux and Gorsky, 1983). Similar observations were reported, on the same species and on *Stegosoma magnum*, by Galt and Fenaux (1990). These concentrations of animals without houses may suggest that some appendicularians have a special reproductive behaviour, forming breeding aggregations that facilitate cross-fertilization, but this suggestion awaits further study.

15.6.2. *Appendicularians as indicators of water masses and currents*

Appendicularian species, or species groups including appendicularians, have been reported as indicators of different water masses and currents by numerous authors, including: Atlantic (Lohmann, 1896a); Japan Sea (Aida, 1907); Southern California (Essenberg, 1926; Fenaux and Dallot, 1980); Newfoundland (Udvardy, 1954); Gulf of Bengal (Ganapati and Bhavanarayana, 1958); New Zealand (Bary, 1960); Brazilian waters (Forneris 1965); North Atlantic (Bückmann, 1970). As an example, Forneris (1965) distinguished coastal waters (salinity < 35‰, temperature > 19°C), shelf waters (salinity < 35–36‰, temperature 22°C), and tropical waters (salinity 36‰, temperature > 20°C). The predominant species in coastal waters were *O.longicauda, O.fusiformis,* and *O.dioica*, the first two (together with *F.haptostoma*) being found also in tropical waters. Shelf waters contained predominantly *O.longicauda* and *O.fusiformis*, whilst *F.pellucida* replaced *F.haptostoma.*

Lohmann (1895) noted the influence of the river Amazon far from its mouth. At 100 nautical miles off the land all the fritillarians, except *F.formica*, disappeared, 'although temperature and salinity did not change very much'. *Fritillaria formica* disappeared at about 75 miles to the east of the mouth and only three *Oikopleura* persisted: *O.longicauda* (= *O.velifera*), *O.fusiformis*, and *O.dioica*. Inside the mouth of the Amazon, where the salinity varies from 22.3 to 11.4‰ between ebb and flow, *O.dioica* was collected in abundance. This species was even found about 50 miles inside the mouth of the river where the salinity was lower than 4.3‰. Costello and Azam (1983) showed that during large tides, appendicularians (*O.dioica* being the dominant species) enter the estuary of North Inlet (South Carolina) at a density higher than 20 000 individuals m^{-3}. Recently, Shiga (1993b) showed that in the Northern Bering Sea during the summers of 1983 and 1986, *Oikopleura vanhoeffeni* was exclusively distributed within the Bering shelf water. Finally, in deep waters, mesopelagic appendicularians may be indicators of layers rich in organic matter (Gorsky *et al.*, 1990, 1991).

15.7. Conclusion

This chapter should have made clear, if the reader had not already inferred it, that even in the best studied areas, agreement is far from having been reached about the proximate causes of abundance of appendicularian species, or of their succession in time. Whilst it is probably true that both temperature and salinity are normally important variables, there certainly seem to be some species (*O.dioica* for example) that are both eurythermic and euryhaline. Much of the difficulty in disentangling the (interacting) factors determining appendicularian distribution may result from the inadequacy of plankton nets for sampling animals that are distributed discontinuously on a relatively small scale. Several groups are seeking to remedy this by using towed video plankton recorders (VPR) to sample gelatinous zooplankton on the small scales required. Already VPRs have been used off Georges Bank (Newfoundland) to discover extremely high concentrations of oikopleurids in very thin horizontal layers (Norrbin, personal communication). Dr G. Gorsky's group at Villefranche is developing a novel video system to examine the vertical distribution of appendicularians (Gorsky, personal communication). At present, the stumbling block in VPR use is the tedious manual analysis of many hours of video tape, but it is reasonable to hope that successful automated image analysis systems will be developed before long. Finally, the short generation times of *O.dioica* (and other species?) at high water temperature, and the (relative) ease of culturing them, raises the interesting possibility of genetic studies.

SIXTEEN

Molecular phylogeny of tunicates. A preliminary study using 28S ribosomal RNA partial sequences: implications in terms of evolution and ecology

R. Christen and J.-C. Braconnot

16.1. Introduction

Phylogenetic information can now be derived from the comparison of nucleic acid sequences that are strictly homologous in different taxa. Among available gene sequences, ribosomal RNAs (rRNA) have provided some of the most valuable phylogenetic information. Three rRNA molecules have been generally used, namely the 5S rRNA, 16–18S rRNA and 26–28S rRNA. Phylogenies derived from the 5S molecule are now thought to be less reliable because of an inappropriate rate of evolution (Halanych, 1991). In contrast, both the 16–18S rRNA and the 23–28S rRNA are now widely used (Sogin *et al.*, 1986; Baroin *et al.*, 1988; Field *et al.*, 1988; Perasso *et al.*, 1989; Christen *et al.*, 1991a; Lafay *et al.*, 1992). The $5'^{-1}$ end of the 28S rRNA molecule (i.e. about 400 nucleotides) is an interesting region for the rapid investigation of phylogenetic problems, since it comprises domains of widely different rates of evolution that allow molecular phylogenies to be derived for species that radiated less than 50 million years ago (for example some of the echinoid radiations; see Smith *et al.*, 1992), as well as radiations that took place more than 600 million years ago, like the origin of Metazoa or the radiations of Porifera (Christen *et al.*, 1991b; Lafay *et al.*, 1992).

This chapter reports the result of sequencing the 5′-end of the 28S rRNA molecule for a number of different tunicates, including thaliaceans, ascidians, and appendicularians. These sequences have been aligned by comparisons with homologous sequences belonging to different classes of echinoderms, acraniates (amphioxus), and vertebrates. Several phylogenetic methods (maximal parsimony, maximum likelihood, and neighbour joining) have been used to analyse the internal relationships between tunicates and to study the relationships of tunicates to echinoderms, acraniates and vertebrates. This work was completed before the recent paper by Wada and Satoh (1994) including an appendicularian and a salp. Gratifyingly, the analyses of the 28S rRNA and 18S rDNA are in good accord.

16.2. Ascidian, thaliacean, and appendicularian relationships

In order to emphasise the special position of appendicularians shown by our analysis, the term 'tunicates' will henceforth be used to include ascidians and thaliaceans only, and will exclude appendicularians. Figure 16.1 is a combined scheme that describes the phylogenetic relationships within the restricted set of tunicates, appendicularians, and amphioxus (see the figure legend for details). Because outgroups are missing from this analysis (but see Fig. 16.2), the monophyletic origin and the respective positions of tunicates and appendicularians cannot be assessed from these data. However, the removal of distantly related species is a prerequisite for correctly analysing relationships between closely related species (Smith *et al.*, 1992). All phylogenetic methods identified not only broad classical taxonomic divisions, but also the subdivisions into subclasses and orders as represented on the right of Fig. 16.1 (see also Table 16.1). However, because representatives of several taxa are missing (compare Fig. 16.1 and Table 16.1), and because the groupings of ascidians and tunicates respectively as monophyletic units were not extremely robust, these results can only be regarded as suggestive rather than definitive. The monophyly of ascidians was recognized by neighbour joining and loosely by maximum likelihood ($P < 0.05$), while in maximum parsimony a single most parsimonious tree was obtained

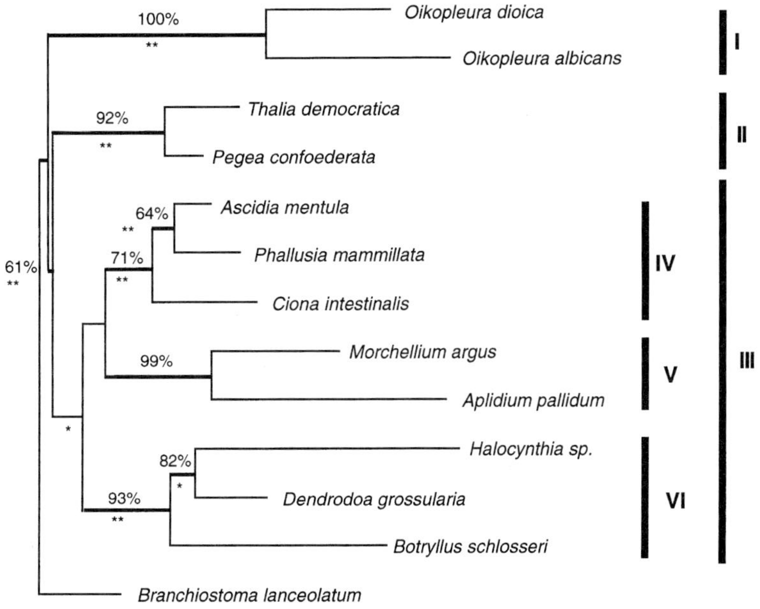

I : Appendicularia
II : Thaliacea
III: Ascidia
IV: Phlebobranchia
V: Aplousobranchia
VI: Stolidobranchia

Fig. 16.1. Phylogenetic relationships between tunicates. The unrooted tree shown in this figure was obtained by the neighbour joining method, using the parts of the rRNA sequences shown in Fig. 16.2. Maximum likelihood and maximum parsimony were also used to ascertain the robustness of the various groups. Branches shown in heavy lining were also observed in the most parsimonious tree. For maximum likelihood, ** indicates a branch significantly different from 0 to $P < 0.01$ and * at $P < 0.5$. A bootstrap analysis undertaken with maximum parsimony (PAUP, heuristic option) gave percentages indicated above each branch when greater than 50%. Classical taxonomic divisions (see Table 16.1) are identified on the right of the figure.

Table 16.1 Classical systematics of Tunicata.

Class	Subclass	Order	Family	Species
Thaliacea				
	Salpida			*Pegea confoederata, Thalia democratica*
	Doliolida			
	Pyrosomida			
Ascidia				
	Enterogona			
		Aplousobranchia		*Aplidium pallidum, Morchellium argus*
		Phlebobranchia		*Ascidia mentula, Ciona intestinalis*
				Phallusia mammilata
	Pleurogona			
		Stolidobranchia		*Botryllus schlosseri, Dendrodoa grossularia, Halocynthia* sp.
Appendicularia				
			Oikopleuridae	*Oikopleura dioica*
				Oikopleura albicans
			Kowalevskiidae	
			Fritillariidae	

Species are shown whose sequences have been used in the phylogenetic analysis.

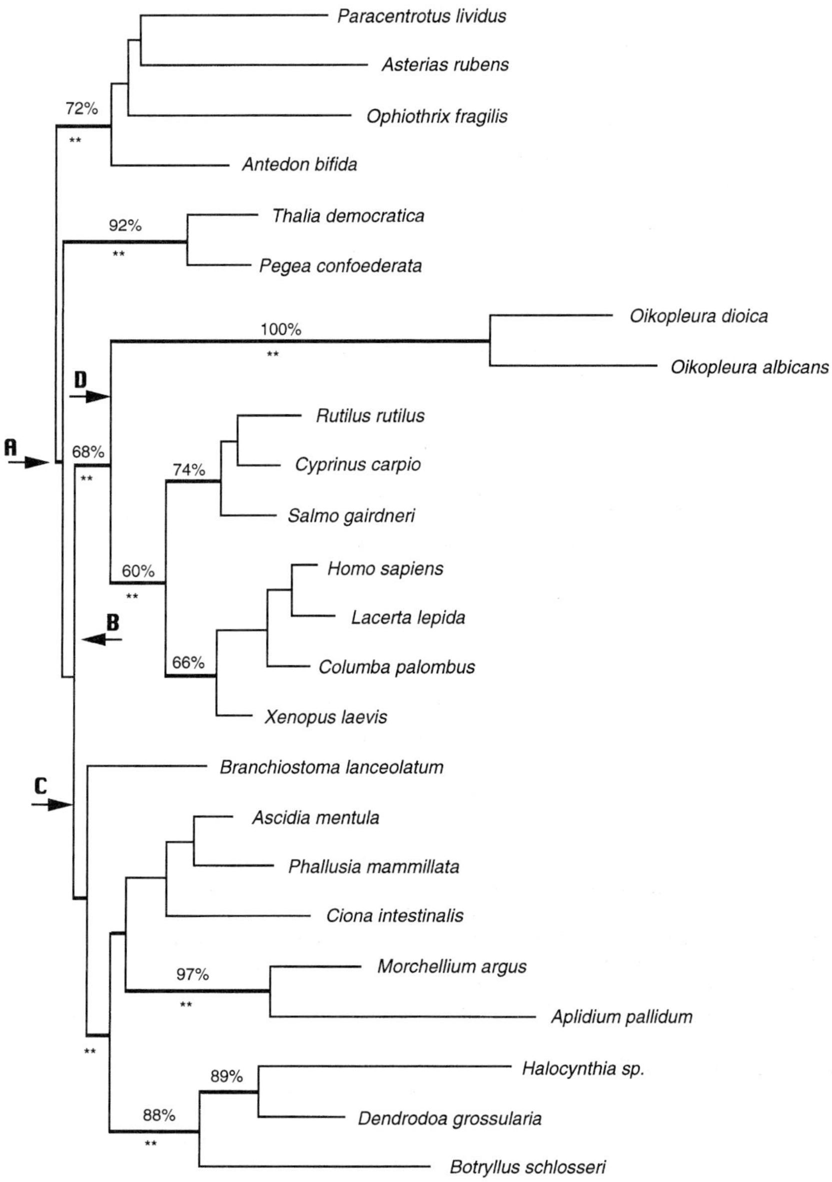

Fig 16.2. Same conventions for Fig. 16.1. Maximum likelihood and maximum parsimony were also used to ascertain the robustness of the various groups. Branches shown in heavy lining were also observed in the most parsimonious tree. For maximum likelihood. ** indicates a branch significantly different from O at $P < 0.01$ and * at $P < 0.5$. A bootstrap analysis undertaken with maximum parsimony (PAUP, heuristic option) gave percentages indicated above each branch when greater than 50%. Classical taxonomic divisions (see table 16.1) are identified on the right of the figure.

(length 184, consistency index 0.766) that identified the Phlebobranchs as the sister group to all tunicates. At one more step, seven trees were obtained that differed mostly in the order of branching of the thaliaceans, phlebobranchs, aplousobranchs, and stolidobranchs. This result suggests that the relationships between the different orders of ascidians cannot be ascertained clearly in this analysis. Thaliaceans were identified as a very robust monophyletic unit, but unfortunately the two sequences analysed belong to the same salp subclass. It is likely that the two other classes of thaliaceans would have shown deep radiations (as observed for the various orders of ascidians). In order to root the tree shown in Fig. 16.1, it was necessary to use outgroup species. The requirement for unambiguous outgroups often leads to the use of distantly related species; following such an approach, one would have chosen in this study species such as (for example) insects, molluscs, or nematodes.

However, our previous work (Smith *et al.*, 1992) showed that outgroups which are too distant bring serious problems of rooting. Distant outgroup sequences indeed comprise many positions for which several successive mutations have occurred, leading to more noise than signal. It is thus necessary: (i) to choose outgroups as closely related as possible to the species studied; (ii) to exclude outgroup species that have an especially fast evolutionary clock. In our case, echinoderms were obvious candidates as outgroups, because they are classically described as more closely related to chordates and protochordates (deuterostomians) than to any other invertebrate (protostomians). Among echinoderms, holothurians have a fast molecular clock and were excluded; we have thus retained one representative for each remaining class: *Paracentrotus lividus* (echinoid), *Asterias rubens* (asteroid), *Ophiothrix fragilis* (ophiuroid), *Antedon bifida* (crinoid).

From the analysis shown in Fig. 16.2, several results may be emphasized.

(i) Appendicularians and vertebrates form a very robust monophyletic unit, the two *Oikopleura* spp. forming a sister group to all vertebrates. This monophyletic unit was identified altogether by neighbour joining, parsimony, and maximum likelihood, and showed a high percentage of retention in the bootstrap analysis.

(ii) All ascidians formed a robust monophyletic unit, identified by all methods. Only the bootstrap analysis was unable to confirm the cohesion of this taxon; this last result and the uncertainty observed in Fig. 16.1 suggest that the molecular information that identifies ascidians as a monophyletic unit may be reduced to a few positions; a more detailed analysis would perhaps show that ascidians share a few derived characters that are not present in thaliaceans and vertebrates, for which reason the inclusion of the vertebrate sequences reinforced the signal for obtaining a monophyletic group of ascidians.

Amphioxus, thaliaceans, ascidians, and the group vertebrates-appendicularians were seen as radiations that separated from each other in a relatively short period of time. It is thus particularly difficult to place the common root for all these groups, and to determine their exact order of radiation. For example, Fig. 16.2 shows different possible rooting alternatives that cannot presently be excluded:

(i) root A is the root as represented in this figure;
(ii) root B would lead to a monophyletic group of tunicates and amphioxus;
(iii) root C would place the thaliaceans as the sister group to vertebrates and appendicularians.

Only root D is clearly rejected by all the methods that have been used.

16.3. Validity of the data

Our data indicate a late radiation of appendicularians as the *sister group* of the vertebrates. The eight ascidian species chosen were well separated along classical taxonomic lines. The difficulty observed in resolving the deep relationships between Ascidia, Thaliacea, and chordates (including Appendicularia in this last taxon) probably results from two different factors:

(i) a lack of genetic information (due to the absence of the other subclasses of thaliaceans and/or to a low number of informative sites);

(ii) the use of inappropriate methods to place the root.

This last has been previously discussed (Smith *et al.*, 1992) using 28S rRNA data, and the echinoid classes as a model system, the conclusion being that rooting is often difficult when branches among the ingroups have different rates of evolution (as is the case here; see Figs 16.1 and 16.2). Longer sequences might provide more information, but it is not always clear that the signal-to-noise ratio is improved by increasing the length of sequence analysed. The quality of the sequences (i.e. no saturation of mutation, and knowledge of the ancestral state allowing polarization of the characters) are probably more important. Finally, obtaining sequences from more animals is definitively the crucial key to obtaining a good phylogeny. This problem is especially true for thaliaceans because we lack data from two out of three of the subclasses.

16.4. Comparison with 18S rDNA analyses

Figure 16.4. shows the results obtained by Wada and Satoh (1994) from analyses of 18 rDNA sequences. As in our analyses, appendicularians are separated from ascidians and thaliaceans, but only one salp sequence was available for the thaliaceans (*Thalia democratica*) and their use of distant outgroups (platyhelminths and molluscs) is a possible reason for the lack of resolution observed in their phylogenetic tree despite the use of longer sequences.

16.5. Conclusions

16.5.1. *Morphological evolution*

Aside from gaining more information for positioning the root of the tree shown in Fig. 16.2, obtaining more salp and doliolid sequences would be of importance in terms of understanding their evolution.

For example, older classifications sometimes separated salps into hemimyaria and desmomyaria,

```
               ------------------------------------------------------------------------------------------------------
O.dioica  NNNGACCUCA GAUUAGUCGA GAGCACUCGC UGAAUUUAAG CAUAUUAUUA AGCGAAGGAA AAGAAACUAA CAAGUAUUCC CCCAGUAAUG GCGAAUGAAC
O.albican NNNGACCUCA GAUUAGACGA GACGACUCGC UGAAUUUAAG CAUAUUAUUA AGCGAAGGAA AAGAAAAUAA CAAUGAUUCC UUUAGUAAUG GCGAAUGAAC
B.lanceot NNNGACCUCA GCUCAGGCGA GACGACCCGC UGAAUUUAAG CAUAUUACUA AGCGGAGGAA AAGAAACUAA CAAGGAUUCC CUCAGUAACG GCGAGUGAAG
T.democra NNUGACCUGA GAUCGG*CGG GACGACCCGC UGAACUUAAG CAUAUCACUA AGCGGAGGAA AAGAAACUAA CAAGGAUUCC CCUAGUAACG GCGAGUGAAG
A.mentula NNCGGCCUGA GGUCGGACGA GGCUACCCGN UGAAUUUAAG CNUAUUACUA AGCGGAGGAA AAGAAACUAA CGAGGAUUCC CUCAGUAACG GCGAGUGAAG
C.intesti NNNNNCCUGA GGUCGGACGA GGCUACCCGC GGAAUUUAAG CAUAUUACCA AGCGGAGGAA AAGAAACUAA CGAGGAUUCC CUCAGUAACG GCGAGUGAAG
P.mammill NNCGGCCUGA GGUCGGNCGG UGUUACUCGG UGAAUUUAAG CAUAUUACUA AGCGGAGGAA AAGAAACUAA CGAGGAUUCC CUCAGUAACG GCGAGUGAAG
H.sp.     NNCGACCUGA GUUCAGGCGA GAGCNCCCGN UGAAUUUNAG CUUAUUAUUA AGCGGAGGAA GAGAAACCGA ACCGGAUUCC UUGAGUAACG GCGAGUGAAU
D.grossul NNNGACCUGA GUUCAGGCGA GAGCACCCGC UGAAUUUAAG CAUAUUAUUA AGCGGAGGAA AAGAAACCAA CCGGGAUUCC CGGAGUAACG GCGAGUGAAU
B.schloss NNGCACCUGA GUUCAGGCGA GGCUACCCGC UGAAUUUAAG CAUAUUAUUA ANCGGUGGAA AAGAAACCAA CNGGGAUUCU CCGAGUAACG GCGAGUGAAG
M.argus   NNUGACCUGA GGUCGGACGA GAGCNCCCGC CGAAUUUAAG CAUAUUACUA AGCGGAGGAA AAGAAACUAA CGAGGAUUCC CUCAGUAACG GCGAGUGAAG
A.pallidu NNUGACCUGA GGUCGGACGA GAGCACCCGC CGAAUUUAAG CAUAUUACUA AGCGGAGGAA AAGAAACUAA CNAGGAUUCC CUCAGUAACG GCGAGUGAAG
P.confede NGGGACCUGA GGUCGG*CGG GACGACCCGC UGAACUUAAG CAUAUCACUA AGCGGAGGAA AAGAAACUAA CGAGGAUUCC CUCAGUAACG GCGAGUGAAG

          -----    --------------------                    ----------------------                            ---
O.dioica  AGGGA****A AAGCUCAAAG CCGAAUCGUU UUGUUUUUGG AGCAGAGCG* CUAUGUGGCA UAUAGCUCGA GCAGUGGGUG AGUC*****G UGCGAGUUUA
O.albican AAGGA****A AAGCUCAGUG CUAAACCGCU UUCUUUUUGG AGGAAAGUG* AAAUGUGGCA UAAAUUUUUC GCAAUGGGGG AGUG*****A UGCGAAUUUA
B.lanceot CGGGA****A AAGCCCAGCN NNAAAUCCCG CAUCCCACG* GGGCGCGGG* AACUGUGGCG UUCGGGAG*G UGCUCUCCGU CCG*UCG**A CCGUUGCCCA
T.democra CGGGA****U AAGCCCAGCG CUAAAUCCGA GCCUUUUUGG AGGGCGAGG* AAAUGUAGCG UAUGGAAUUC MGCAUGCGGC GAACCA***A CGACUGCCCA
A.mentula UGGGA****A GUGCCCAUCG CCGAAUCCGC GUGCCUCCUC GGCGCGCGG* AAUUGUGGCG UAAAGAAUUC GCUUCGGCGA GCG*UCG**G UCGCCGCCCG
C.intesti UGGGA****A GAGCCCAUCG CCGAAUCCGC GCGUUCUUGG AGCGUGCGG* AAUUSUGGCG UACGGAAUU* SUCUGCGCGC GCG*UC**** CUGCUGCCCA
P.mammill CGGGA****U GUGCCCAUCG CCGAAUCCGC GUGCUCUCGA GGCGCUCG** AAUUGUGGCG UAAAGAAUUC GCUUCNGUGC GCG*UCG**U UCGCUGCCCG
H.sp.     CGG*A****U CCGUCCAACG UCGAAUCCGG CGCMUCUCGG AGCGCGCCG* AGCUGUGACG UACGGAA*GU CUCUGUGCGG CCG*UCG**G CGUGCGCCGG
D.grossul CGGGA****A GAGUCCAGCG UCGAAUCUN* *GCUCUUGGA GCGCGCCG** AGUUGUGACG UACGGAA*GU CCCUGUGCGA CUG*UCG**G CGAGUGUCGG
B.schloss CGAGA****U CAGUCCAGCG UCGAACCUGC GCGUUCUCGA AGCGCGCCG* GAAUGUGACG UACGGAAGAC CCCUGUGCGG UCGCUCG**G CGGGCGCCGU
M.argus   CGGGACCGUC UAGCCCAKCG CCGAAUCAGC GC*UUUCGGA *GCGGCCCG* ACCUGUGGCG UACAGACCUG GGCRCGCGGG CGAGCNN**N GGCCGGCCCG
A.pallidu CGGGAU*ACU AGGCCCAGCN CCGAAUCGGC CCGCUNCGGA GCGGCUCNNN AAAUNUNGCN NAAAN*CCCG GGCNAGCGGG YYCUCGUCGC GGCCGACCCG
P.confede CGGGA****C AAGCCCAGCG CUGAAUCCAA AGCCUUUUGA *GUGUGCGG* AAAUGUAGCG UACGGAA**G CGUCCUGUAC ACC*UUG**G UCUCUGCCCA

          ----------------------------                                                                 -------------
O.dioica  AGUACAUUUG AAUGUGUCUC UAC*CCAUAG *CAGGUGU*G AGUCCCAUCU *GACGAG*** **CGCGGCGA GCCAAAGCGA GAGCUUAGAG UCGAGUUGUU
O.albican AGUACAUUUG AAUGUGUCUU UAC*CCGUAG *CAGGUGU*G AGGCCCAUAU UGAGGAG*** ***CAUUGCA ACCAAAGCNN ANANNUAGAG UCNNNUUNUU
B.lanceot GGUCCUUCUG AUCGAGGCCU CUC*CCAGAG *CGGGUGU*C AGGCCCAUGG CGGCGAC*** *GGCGGCGGU UCGNCUGCGC CUCCUCGGAG UCGGGUUGUU
T.democra CGUUCUUCUG AUCGAGGCCU UUUCCCAGAG *CGGGUGU*U AGGCCCUUGG CGGCGGU*** *CGUAAUUCG UCCUAGCGGU UUUCCCGGAG UCGGGUUGUU
A.mentula CGUCCUUCUG AUCGAGGCCU CGUCCCAGAG *CGGGUGU*C AGGCCCAUGG AKGGBGK*** *GNCKKACCG CGCUACGCGU CUUCCUAGAG UCGGGUUGUU
C.intesti CGUUCUUCUG AUCGAGGCUU CGCUCCCUAG *CGGGKGU*C AGGCCCUUGG CGGCGGC*** GGGUAGCGCG CUCUCGGCGU CUUCCCGGAG UCGGGUUGUU
P.mammill CGUCCUUCUG AUUGAGGCCU CGUCCCAGAG *CGGGUGU*C AGGCCCAUGG CGGUGGC*** ***GAGGCGC GGCUGUACGU CUUCCCAGAG UCGGGUUGUU
H.sp.     CGUCCUUCUG AUCGAGGCCU CGUCCCGAAG CCGGGWGUCC AGGCCCCACA AGGGNNCU** *CGCCGCGCU CGCUUGCAGU CCUCUCGGAG UCGGGUUGUU
D.grossul UGUCCUUCUG AUCGAGGCCU CGUCCCAUGG *CGGGUGU*C AGGCCCAUAA GGGCGCU*** *UGCCGCGGU CGCUUCGGGU CUUCCCGGAG UCGGGUUGUU
B.schloss AGUCCUUCUG AUCGAGGCCU CAUCCCAGUG *CGGGUGU*C AGGCCCACGA GGGCGUCC** *GNNCCGUGC CGCUUCGGGU CUUCCCGGAG UCGGGUUGUU
M.argus   AGUCCCCCUG AUCGAGGCCG UCGCCCGCGG *AGGGUSU*C AGGCCCGUGG CGGUCGGACG GCCCGCGCCC ACGCCCGAGC GGUCUCGGAG UCGGGUUGUU
A.pallidu AGUCUCUCUG AUCGGAGCCG GUGCCCCUGG *AGGGUGU*C AGGCCCCUUG GGUCGGN*** NCGCGCGGUC ACGCCCGGGC G*UCUCGGAG UCGGGUUGCU
P.confede CGUUCUUCUG AUUGAGACCU UAUCCCAUAG *CGGGUGU*U AGGCCCUUAG CGGUGGAGG* CCAUUGGUGU ACUCGGGCGU CUUCCCGGAG UCGGGUUGUU

          ---------------------------------------------------------------------------
O.dioica  UGGGAAUGCA GCUCUAAGCG GGUGGUAAAC CCCAUCUAAA GCUAAAUACU GAUAUGAGAC CGAUAGCNAA CAAGU
O.albican UGGGAAUGCA GCUCUAAGUG GGUGGUAAAC CCCAUCUAGG GCUAAAUACU GAUGUGGGAC CGAUAGCNAA CAAGU
B.lanceot UGGGAAUGCA GCCCAAAGCG GGUGGUAAAC UCCUCCUAAG GCUAAAUACG GACACGAGAC CGAUAGNNNN NNNNN
T.democra UGGAAAUGCA GCCCUAAGCG GGUGGUAAAC UCCAUCUAAG GCUAAAUACU GGCUCGAGAC CGAUAGCGAA CAANN
A.mentula UGGGAAUGCA GCCCAAAGCG GGUGGUAAAC UCCAUCUAAG GGUAAAUACU GUCGCGAGAC CGAUAGNNAA CAANN
C.intesti UGGGAAUGCA GCCCAAAGCG GGUGGUAAAC UCCAUCUAAG GSUAAAUACN GUCGCGAGAC CGAUAGCGAA CAANN
P.mammill UGGGAAUGCA GCCCAAAGCG GGUGGUAAAC UCCAUCUAAG GCUAAAUACU GUCGCGAGAC CGAUAGCGAA CAANN
H.sp.     UGGGAAUGCA GCCCGAAGUG GGUGGUAAAC UCCACCUACG ACUAAAUACC GUCGCGAGAC CGAUAGCGAA NNNNN
D.grossul UGGGAAUGCA GCCCAAAGCG GGUGGUAAAC UCCACCUAAG ACUAAAUACG GUCGCGAGAC CGAUAGCGGA CAANN
B.schloss UGAGAAUGCA GCCUAAAGCG GGUGGUAAAC UCCAUCUAAG ACUAAAUACG GUCACNAGAC GCAYAGCCAA CNNNN
M.argus   UGGGAAUGCA GCCCGAAGCG GGUGGUAGAC UCCAUCUAAG GCUGAAUACG GUCGCGAGAC CGACAGCGAA CAANN
A.pallidu UGGGAAUGCA GCACGAAUGC GGUGGCACGC UCCACCCGUG GCUUAACACG GUCGCGAGAC CGACAGCGAA NNNNN
P.confede UGGAAAUGCA GCCCAAAGCG GGUGGUAAAC UCCAUCUAAG GCUAAAUACU GGCUCGAGAC CGAUAGCNAA CAANN
```

Fig. 16.3. Sequences of the 5′-end of the 28S RNA have been aligned by comparison with a broader database including species from various metazoan phyla. Domains used for constructing Fig. 16.1 are indicated above. Stars are deletions necessary for optimal alignment.

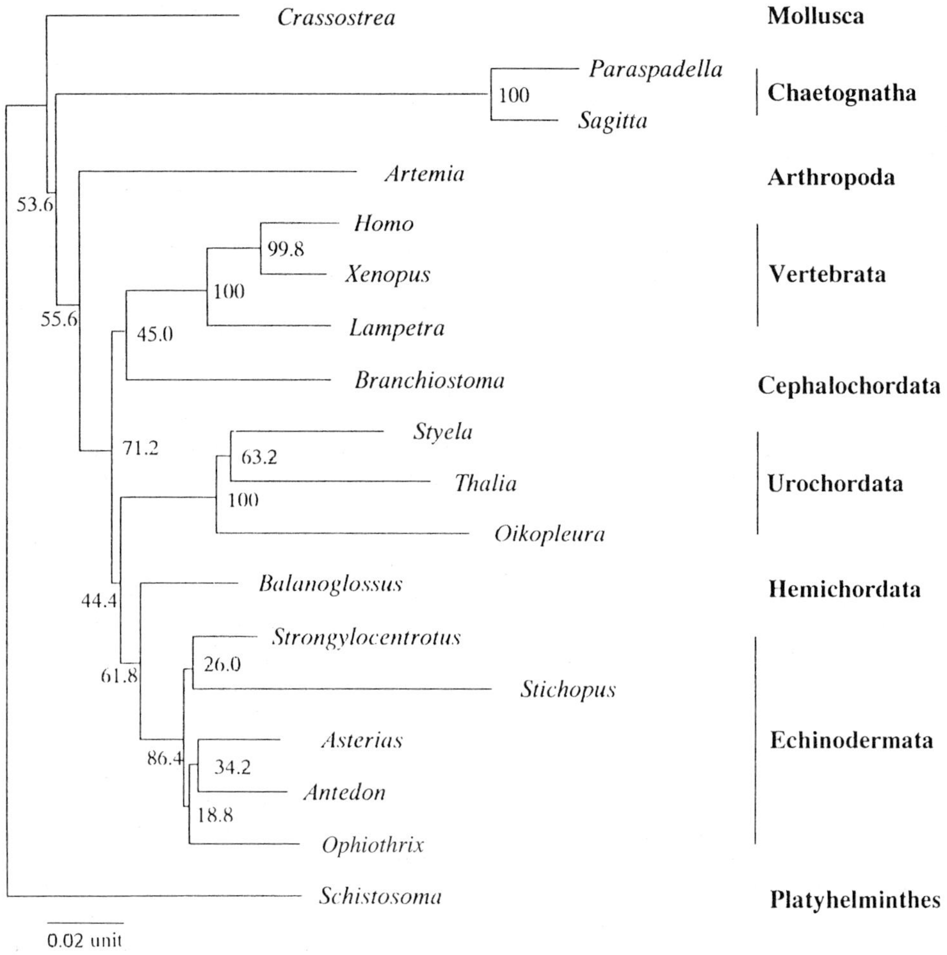

Fig. 16.4. Phylogenetic relationships among chordates and protochordates, inferred from comparisons of 18S rDNA sequences aligned by maximum likelihood. The tree was constructed by neighbour joining as in Fig. 16.1. Numbers indicate the degree of support for internal branches of the tree assessed by bootstrapping; they show the percentage of times a node was supported in 500 bootstrap pseudoreplications. Scale bar, evolutionary distance of 0.02 nucleotide substitutions per position in the sequence. From Wada and Satoh (1994).

according to the arrangement of their musculature (Table 16.2), but it is presently uncertain which may be the less advanced (section 16.1). Again, the relationships between the different thaliacean sub-classes, e.g. doliolids and salps, remain unclear, though several lines of morphological evidence suggest independent derivation from sessile ascidians (see section 16.1).

16.5.2. *Evolutionary ecological considerations*

Metazoa comprise the most complex forms of life. In contrast to their ancestors (most protozoan species), metazoa are usually obligatorily sexual, their normal reproductive cycle consisting of the fertilization of eggs by sperm, and not involving fission or budding. Since

Table 16.2 Summary of morphological traits of Thalicea.

Character	Thaliacea							
	Doliolids				Salps			Pyrosomas
	Doliolum	*Doliolina*	*Dolioletta*	*Ihlea*	*Salpa*	*Thalia*	*Pegea*	*Pyrosoma*
larval notochord	Functional	Functional	Reduced					
Muscles	Circular	Circular	Circular	Circular	Interrupted	Interrupted	Reduced	Reduced

this character is apparently shared by all metazoan phyla today, it can be concluded that it is a primitive character. In evolutionary terms, the possibility of reproducing extremely rapidly and efficiently through division or active budding from a single individual represents an extraordinarily important short term advantage (Williams and Mitton, 1973; Smith, 1978). The loss of this extremely advantageous character in most metazoa was presumably outweighed by the gain of another selective advantage; it has thus been proposed that obligatory sexual reproduction was a necessary character for the appearance and complex differentiation of the tissues present in metazoa (Christen *et al.*, 1991b). In terms of ecological efficiency, there is a considerable selective pressure for the reintroduction of asexual reproduction, one response to this pressure being the appearance of parthenogenesis, where eggs can develop without activation by sperm. Parthenogenesis has appeared not only in insects, such as aphids and phasmids, but also in cichlid fish. However, in metazoa, parthenogenesis leads to the loss of genetic recombination of two different genomes, and genetic innovations from different individuals can no longer be easily united in a single organism. Hence, it is generally thought that parthenogenesis, although ecologically efficient, is a long term evolutionary 'dead end' that may lead to the future extinction of a species if external conditions alter.

A second way around the loss of 'efficiency' resulting from exclusively asexual reproduction, is the tunicate solution of alternation of generations between a sexual generation and a rapidly multiplying asexual generation formed by budding from a stolon. It is this metagenesis that enables salps to colonize their surroundings rapidly, so forming immense blooms or swarms in favourable conditions (see Chapter 7).

Doliolids also show metagenesis, but most species follow sexual reproduction by less striking budding than salps. One species (*Doliolum nationalis*) seems to have been able to completely uncouple the budding process from gamete fusion, by developing a 'short cycle' with seemingly indefinitely prolonged budding of phorozooid individuals without intervention of a sexual generation (Braconnot, 1967). This special short cycle of *D.nationalis* is described in section 1.3.5. Ecological efficiency, measured by the ability to invade and rapidly occupy an ecological niche, is extremely high for this particular species, which through this process at some periods of the year becomes predominant in the macroplankton. It would be extremely interesting to know if this coupling of rapid asexual reproduction (here a budding process) with *intermittent* sexual reproduction is a recent development or if it appeared a long time ago; an ancient appearance would contradict the theory suggested above, and would require an investigation of the reasons for which high ecological efficiency did not result in a large diversification. It is suggested that the appearance of budding as a competitive alternative to sexuality for reproduction could lead to a global dedifferentiation (loss of complexity) in the organism in which the phenomenon occurs when the ecological niche is stable.

It is also interesting that sexual activity (phorozoids develop gonads) occurs in particular places: the mouths of the river Po (Braconnot and Katavic, 1979) and of the Rhone, as well as in the bay of Villefranche in autumn, after heavy rains (Braconnot and Casanova, 1967). Following fertilization, a chordate larvae is obtained and a cycle identical to that observed in other species is obtained (Braconnot, 1977). The causes for the appearance of this sexual activity remain to be understood, a possible explanation being that gene recombination (through sexual reproduction) is triggered by stress (lower salinity or chemical pollution). This species would have thus reached a 'quasi-perfect' mode of reproduction: active budding when the conditions are favourable, gene recombination when diversification is required.

Finally, because all tunicates share a common morphology and ecological niche as marine filter feeders, knowledge of a precise phylogeny, and the determination of which radiations led to greater diversification, could help in understanding both the functioning of marine ecosystems and the mechanisms of adaptation/selection within them: which small changes have been significant improvements that led either to the creation of a new ecological niche, or to the displacement of other species from a pre-existing niche.

16.6. Acknowledgments

This work was supported by the CNRS and a grant from the Conseil Générale to R.C.

SEVENTEEN

The relationships and systematics of the Thaliacea, with keys for identification

J. Godeaux

17.1. Thaliacean relationships

17.1.1. *Introduction*

Almost exactly a century ago Brooks (1893) remarked that it should not be difficult to place the relationships of the Thaliacea upon a sound and permanent basis, as the forms which are to be considered are so very few and our knowledge of them so complete. Whilst it is true that there are few different thaliacean types, and that knowledge of their morphology (and other aspects) has increased since Brooks wrote, their relationships are still unclear. For the present, until further molecular data becomes available, speculations about thaliacean relationships are based on comparative embryology and morphology, (including sperm morphology).

Three general questions naturally arise when thaliacean relationships are considered: (i) have the Thaliacea evolved from a free living or a sessile ancestor?; (ii) is this ancestor common to the three thaliacean orders?; (iii) what are the relationships between the three orders?

Comparative embryology and anatomy favour the hypothesis of a return to the pelagic way of life from *different* sessile ancestors, i.e. the thaliaceans are *polyphyletic,* and do not form a natural group (see Godeaux 1957–1958). This was the view of Uljanin (1884), who further supposed that the ascidians from which the thaliaceans were derived, themselves arose from appendicularian-like forms. The brief survey of the anatomical and embryological data which follows outlines the basis for this conclusion.

17.1.2. *General*

Thaliaceans have all the typical tunicate characters: outer tunic, ventral endostyle, neural gland, and a cardiopericardium with periodic reversal of beat. In adult morphology, they seem as a whole closest to the aplousobranch ascidians: simple gill, neural gland below the ganglion and a reduced abdomen behind the pharyngeal cavity. The different thaliacean groups, however, show their ascidian affinities to a greater or lesser degree, so that they are more obvious in some than in others.

The oviparous doliolids still have a tadpole stage, although simplified, and there is an obvious metamorphosis. Since all doliolid stages are planktonic a motile distributory larva is no longer required and the doliolid larva is evidently vestigial.

The pyrosomas have an oozooid similar to an ascidian, but have lost the tadpole larval stage, and metamorphosis is lacking by tachygenesis. In salps, the early ascidian-like oozooid, although present, is obscured by the calymmocytes which envelop the different embryonic anlagen (see below).

17.1.3. *Ascidians*

The starting point of an ascidian tunicate is an embryo reduced to the preblastoporal region, with a single pair of coelomic cavities, secondarily fusing on the midventral line to form the future cardiopericardium. The larval body consists of a globular trunk or cephalenteron (Brien, 1948) and a transient locomotor tail. The free-swimming life of the tadpole usually lasts for a few hours only, and often the metamorphosis (and even blastogenesis) begins early (even prior to settlement), especially in incubating ascidians (e.g. the Aplousobranchiata). The larval stage has even been completely lost in some Molgulidae

In the course of the metamorphosis a pair of primary protostigmata appears on both sides of the pharynx (de Selys-Longchamps, 1901), sometimes already divided into rows of stigmata (Julin, 1904). These two protostigmata are quickly divided into four secondary protostigmata by languets growing from the dorsal margin (e.g. *Clavelina lepadiformis*). De Selys-Longchamps (1901) suggested that the presence of the two protostigmata in the ascidians resulted from a precocious division of a primitive protostigma, which is in fact occasionally observed in *Ciona* and *Ascidiella* (Willey, 1893; de Selys-Longchamps, 1901). The stage with a single protostigma is preserved in the thaliaceans (Julin, 1904), possibly indicating that they arose early from the basal tunicate stock.

17.1.4. *Doliolids*

In doliolids a kind of simplified embryo persists, devoid of neural and endoblastic anlagen in the tail, with either striated muscle fibres and some 40 notochordal cells (as in the ascidian tadpole) in *Doliolina mülleri* and *Doliolum denticulatum* and just clusters of undifferentiated chordal and mesoblastic cells in the anuran tadpole of *Dolioletta gegenbauri*. The neural cavity of the cephalenteron has no sense organs, and it is very doubtful that the statocyst, which only occurs in the oozooid (developing from an ectodermal cupula on the left side), has any homology with the statocyst of the ascidian tadpole.

A pair of peribranchial cavities of ectodermal origin appears in the dorso-posterior zone of the cephalenteron. The peribranchial exhalant apertures fuse into an odd and median atrial siphon; the backward movement of this siphon pushes the tail ventrally in front of muscle VII. The muscle rings arise from lateral mesoblastic plates, and most of them (I–IV in front and VII–IX behind) are simply siphonal muscles adapted to locomotion by jet propulsion.

The different parts of the digestive tract are progressively formed from the endoblastic mass. In the primitive *Doliolina mülleri*, the intestine is U-shaped, as it is in all sessile animals (Bryozoa, endoprocts, *Phoronis*, *Cephalodiscus*, Crinoidea and, of course, ascidians).

The pair of protostigmata runs dorso-ventrally as in the ascidians (e.g. *Clavelina lepadiformis*). The wall bearing the protostigma divided into four stigmata has pivoted by 90°, with the branchial slits facing the buccal siphon and the pharyngeal cavity. In this way the transit of water through the body and therefore swimming by jet propulsion is facilitated. In the blastozooids, the stigmata are more numerous but have the same arrangement as in the oozooid. The doliolid feeding mechanism, in which the dorsal volute of the peripharyngeal bands rotates the filter into a string entering the oesophagus (p. 82), is closely comparable to that in very young *Clavelina lepadiformis*, as figured by Werner and Werner (1954).

The stolon has a complex structure, being composed not only of the epicardia (as in aplousobranch ascidians) but also of diverse mesoblastic, cardioblastic, and cloacal outgrowths, whose development is still poorly understood (Neumann, 1906).

17.1.5. *Pyrosomas*

Pyrosomas have often been considered as primitive, even being regarded by some authors as the pelagic stem group of the Thaliacea. In fact, numerous characters are in favour of a sessile origin. A tadpole larva is completely absent. In relation to ovoviviparity and tachygenesis, the adult organs appear at once even when the embryo is greatly simplified as in the *Pyrosoma ambulata* subfamily. In the *P.fixata* subfamily, development leads to an ascidian-like animal sitting on an enormous yolk drop, but nothing in its organization suggests a swimming stage. It is noteworthy that the two siphons are very close to each other, being separated only by the neural complex, as in most ascidians. There is no trace of a possible backward migration of the atrial siphon as observed elsewhere, e.g. in doliolids.

Most of the tissues remain embryonic with the exception of the cardiopericardium producing the blood flow through the stolon lacunae, and of the ciliary canal of the neural complex (whose role still remains unknown). It is curious that the canal is present even in the reduced cyathozooid of *Pyrosoma atlanticum*, where the cilia beat until the engulfment of the embryo into the tetrazooid colony.

The sole protostigma, a simple hole in *P.atlanticum*, is stretched antero-posteriorely and divided into a row of vertical stigmata crossed by several longitudinal bars or sinuses in *Pyrostremma vitjasi* and *P.spinosum*. The pharyngeal cavity is ample, with a ventral endostyle (more or less incurved), and several dorsal languets. The gill basket is pierced by numerous dorso-ventral slits of stigmata, looking very similar to that of a phlebobranch such as *Ciona*, but of different origin (corresponding to a single antero-posterior protostigma; Julin, 1912a). Such a gill has the superficial appearance of a phlebobranch ascidian gill. The particular orientation of the protostigma cannot be explained by the backward movement of the atrial siphon as suggested by Brien (1948, p. 863), rather it is a primary character. The differences between the simple gill of the lower aplousobranch ascidians and that of the pyrosomas suggest that the two are derived from different, albeit closely related, ancestral forms. Indeed, although pyrosomas cannot be directly related to any living ascidian, pyrosomas most clearly resemble ascidians in structure.

As blastogenesis is a simple strobilization, the stolon components retain their embryonic potentiality. No ascidian presents such a condition, even in the primitive family Polyclinidae, where all the internal organs of the blastozooid originate from the internal vesicle resulting from the fusion of the two epicardia.

In the blastozooid, the hinder position of the atrial siphon is secondary, accompanied by the ventral movement of the stolon (Godeaux, 1957, 1958). The digestive loop is bent as in the oozooid (and sessile animals). The tip of the stolon is surrounded by the elaeoblast of enigmatic function, but surely unlike the tail of a hypothetical tadpole, as larval organs are always missing in the blastozooid, whose development is direct (Brien 1928).

Pyrosomas display a similar condition to that seen in the Hydrozoa: an essentially sessile oozooid (resembling an ascidian) and a pelagic and somewhat different blastozooid. The colony derives from a remote sessile, budding and colonial ancestor whose oozooid has progressively lost its importance in the life cycle owing to the acquisition of ovoviviparity.

17.1.6. *Salps*

In salps both the oozooid and the blastozooid are pelagic and have the same importance in the life cycle. The different stages of embryogenesis are masked by the calymmocytes which envelop the clusters of blastomeres. There is no tadpole stage: the different organ anlagen organize themselves into the follicular mass and progressively converge and build the embryo (Brien, 1928, 1948). The young oozooid is linked to the blastozooid by means of the so-called placenta. The atrial siphon appears dorsally (*Cyclosalpa pinnata*, Brooks, 1893; *Thalia democratica*, Salensky, 1883; Brien, 1928, 1948) and secondarily migrates towards the hind part of the animal. Furthermore, the interoscular space is narrow at the beginning of development and the atrial siphon lies near to the the neural complex. Later, these two organs separate further than the distance between the nervous system and the buccal siphon. According to Julin (1904), the unique undivided branchial slit corresponds to a longitudinal protostigma (cf. *Pyrosoma atlanticum*) having rotated by 90°. The intestinal loop, behind the pharynx, is usually twisted in a more or less dense nucleus.

Salps are active swimmers with powerful muscles, originating from lateral blastemas, and in contrast to doliolids, the locomotor muscles do not represent modified sphincter musculature.

The structure of the stolon is complex: an ectodermal muff contains the axial endostylar outgrowth (the future digestive system) and several mesodermal strands, respectively the anlagen of the atrial cavity, of the muscles, of the nervous system, of the blood cells, of the gonads and of the cardiopericardium. In cross-section, the stolon is quite similar to that of pyrosomas. As in pyrosomas, the nervous system derives from the mesoblast and, possibly, the cardiopericardium is an outgrowth of the maternal organ (Brien, personal communication). In both cases, blood cells, muscles and gonads are of mesoblastic origin. Thus the main difference between the budding of salps and pyrosomas lies in the origin of the peribranchial–atrial cavities.

As in the pyrosoma, the budding is of the endostylar type, unknown in living ascidians. The stolon undergoes successive waves of divisions giving off a series of whorls or chains of blastozooids. Strobilization in this case is not complete owing to the role of the mesoblast. Pyrosomas and salps appear closely related but represent two different evolutionary trends, which have followed more or less parallel courses.

(i) Pyrosomas have developed ovoviviparity with, as consequences, the disappearance of the tadpole stage and the progressive reduction of the oozooid which remains sheltered but free in the maternal body (a condition somewhat similar to that in the higher plants with a reduced gametophyte and a full-grown sporophyte). (ii) Salps have retained both stages of their cycle developing true viviparity, thanks to a placenta, an original solution to the problem of the development of the oozooid.

17.1.7. *Other evidence for thaliacean polyphyly*

The conclusions of the recent studies by Holland and Franzen on the fine structure of tunicate spermatozoa are in accord with this view of the polyphyly of Thaliacea. According to Holland (1989), doliolids arose from an early stock of clavelinids, the stem group of the solitary ascidians. *Clavelina lepadiformis*, considered to be a less evolved ascidian, shows primitive features in its sperm morphology (Franzen, 1992). Moreover, *Clavelina lepadiformis* is a budding but not colonial ascidian; its blastozooids are not embedded in a common tunic but instead are easily separated (as are the blastozooids carried by the doliolid nurse).

Salps are suggested to have arisen from a later stock of colonial ascidians like the didemnids (Holland, 1988). Similarly, pyrosomas arose from colonial ascidians. As suggested by the structure of their spermatozoa, pyrosomas have possibly not given rise to any other tunicate. Nevertheless it is not yet possible to decide whether salps arose from pyrosomas, or salps and pyrosomas arose independently from colonial ascidians (Holland, 1990).

17.1.8. *Conclusion*

Doliolids derive from a sessile ascidian close to the aplousobranch ascidians, pyrosomas and salps from ascidian-like sessile ancestors, probably related if not actually identical, whose traces can be detected by scrutinizing the development of the oozooid. Thus the pelagic life of the Thaliacea is a secondary phenomenon, and furthermore the Thaliacea is an artificial class and is the result of convergence between different evolutionary lines.

On the other hand, appendicularian sperm morphology displays primitive features supporting the hypothesis that these animals represent a very early branch of the tunicate stem (Holland *et al.*, 1988). Molecular studies of thaliacean affinities on the lines of those begun for ascidians by Wada *et al.* (1992) and for the pelagic groups by Wada and Satoh (1994) should resolve thaliacean relationships within the next few years, although there may be difficulty in obtaining suitable material.

17.2. Systematics of Thaliacea

The class Thaliacea consists of three orders, Doliolida, Pyrosomatida, and Salpida.

All are marine, holoplanktonic, microphagous animals. They all show alternation of generations (metagenesis): the oozooid producing buds, the blastozooid hermaphroditic (and in some cases also able to bud). Pharyngeal and cloacal cavities are ample, the abdomen is very reduced, the post-abdomen is always wanting. Doliolida and Salpida swim actively with prominent locomotor muscle bands. Development resembles that of an aplousobranchiate ascidian, although somewhat condensed. The tadpole stage is only seen in Doliolids.

The different thaliaceans look very similar, but as emphasized earlier, this may be the result of convergent evolution. Pyrosomas and Salps are probably related, but Doliolids seem to have arisen from another primitive stem. The class Thaliacea is still retained here for convenience, but the reader should recognize that the arrangement is artificial and may be changed subsequently.

17.3. Systematic account with keys for Identification

17.3.1. *Order Pyrosomatida*

The order Pyrosomatida consists of the single family Pyrosomatidae with less than 10 species. The genus *Pyrosoma* was created by Péron (1804), who described *Pyrosoma giganteum*, a species common in the Atlantic Ocean, which he considered as a single cylindrical animal. The name reflects the fact that pyrosomas are amongst the most luminescent of all marine organisms. Other specimens, caught in the Mediterranean near Nice, were studied by Lesueur (1815), who recognized their colonial nature, and especially by Savigny (1816), who proved their relationships with the salps and placed them in the family Luciae, with three species, now merged in the single *Pyrosoma atlanticum* Péron, 1804. This original separation into three species was due to the considerable changes in form as the colony grows, *P.elegans* being the younger form with regular whorls of blastozooids, *P.giganteum* the oldest form with a length of over 40 cm. Colonies are sexually mature around 20 cm. Another species, *P.spinosum*, was later identified by Herdman (1888) from the Challenger collections. A large number of papers devoted to the embryology, blastogenesis and systematics of these animals appeared during the last century (see Godeaux, 1957). Neumann (1913a) distinguished two groups: (i) the *Pyrosoma ambulata* group, the buds of which are carried through the tunic matrix by phorocyte; (ii) the *Pyrosoma fixata* group, where phorocyte cells are wanting.

Although their relationship is obvious, the two groups of pyrosomas correspond to two different genera: *Pyrosoma* (e.g. *P.atlanticum* Péron, 1904) and *Pyrostremma*, Garstang (1928) with *Pyrostremma spinosum* (Herdman, 1888).

Neumann's two groups were raised by Van Soest (1979a, 1981) to subfamily status: the *P.ambulata* group to the Pyrosomatinae (two genera: *Pyrosoma* and *Pyrosomella*) and the *P.fixata* group to the Pyrostremmatinae (a single genus: *Pyrostremma*). The main differences between the two subfamilies consist in the more or less condensed development of the embryos and in some details of the blastozooid structure. The oozooid is a short-lived transient and blastogenetic stage; the blastozooids are hermaphrodite and blastogenetic and are responsible for propagation of the species and the growth of the colony. In contrast to the other pelagic tunicates, pyrosomas build permanent hollow colonies; the blastozooids resulting from asexual propagation are essentially independent but remain embedded side by side in a common tunic, as in the aplousobranchiate ascidians. The tunic varies in consistency according to the species, either firm and cartilaginous as in *Pyrosoma atlanticum* or soft as in *Pyrostremma spinosum*, the zooids of which are easily set free.

As already stressed, blastozooids of the two subfamilies are very similar, but nevertheless exhibit different somatic characters of systematic importance.

In the Pyrostremmatinae triangular spines are visible below the buccal aperture. The cloacal diaphragm of the colony is wanting. In *Pyrostremma spinosum* (Herdman, 1888), the colony bears a very long posterior whip-like tail; in *P.agassizi* (Ritter and Byxbee, 1905) the common cloacal aperture is provided with four fragile short quadrangular processes.

In the Pyrosomatinae, the outer wall of the colony is either smooth (*Pyrosomella verticillata* (Neumann, 1909) and *P.operculata* (Neumann, 1908), or provided with more or less protruding projections containing the buccal cavity (*Pyrosoma atlanticum* Péron, 1804; *P.aherniosum* Seeliger, 1895; *P.ovatum* Neumann, 1909; *P.godeauxi* van Soest, 1981).

In the Pyrostremmatinae, the stigmata are oblique with regard to a curved or even bent endostyle. The buccal aperture displays a crown of tentacles. The intestinal loop is brought down horizontally, with the anus facing the cloacal siphon, above which an appendix is present. The zooids are relatively muscular: dorsal muscles radiating below the ganglion, ventral muscles arising from below the buccal siphon, cloacal vertical muscles at the level of the pharynx and connected with the adjacent zooids.

In the Pyrosomatinae, the stigmata are vertical and the endostyle is straight or only slightly curved. There is a single ventral tentacle by the oral aperture. The intestinal loop is vertical with the anus to the left. The musculature is weak, the cloacal muscles are behind the gut and the dorsal musculature is reduced to a few fibres. There is no cloacal appendix. A pair of haemopoietic organs lies in the dorsal blood lacuna; these organs lie near the gut in the Pyrostremmatinae.

The order Pyrosomatida consists of a single family, Pyrosomatidae, and two subfamilies: Pyrostremmatinae with the single genus *Pyrostremma* (*Propyrosoma*) and two species, Pyrosomatinae with two genera, *Pyrosomella* and *Pyrosoma*, and six species.

Pyrosoma systematics are based on the revision by van Soest (1979a, 1981) after Neumann (1913a,b) and Metcalf and Hopkins (1919). The following characters are involved: shape, consistency, aspects of the outer surface of the colony, position and number of stigmata, form of the endostyle, size of the buccal and atrial siphons, number of branchial bars and dorsal languets, position of the intestinal loop, presence or absence of a diaphragm at the common cloacal aperture, protandry or protogyny, location of the developing embryo.

17.3.2. *Dichotomous key for identification of pyrosomas*

1. Colonies entirely smooth, small sized (<15 cm), regular whorls of zooids, the youngest ones near the common atrium 2
   Colonies with the entire outer surface bearing protruding test spines or projections, zooids not in regular whorls 3
2. Cloacal siphon less than a third of the branchial basket, length of the sexually mature colony up to 5 cm, up to 33 stigmata, 17 bars and 4–5 dorsal languets, protogyny, embryo in the right peribranchial cavity (Figs 17.1 and 17.5c) ....*Pyrosomella verticillata*
   Cloacal siphon extremely long, longer than the branchial basket, provided with an operculum,

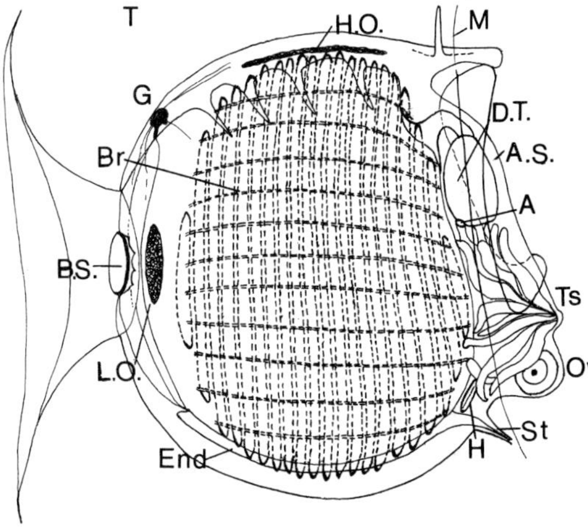

Fig. 17.1. Zooid of *Pyrosomella verticillata*. A, anus; A.S., atrial siphon; Br, branchial basket with vertical stigmata and longitudinal bars; B.S., buccal siphon; D.T., digestive tract; End, endostyle; G, brain; H, heart; H.O., haemopoietic organ; L.O., luminous organ; M, muscle; Ov, ovary; St, stolon; T, common tunic; Ts, testis. Modified after Neumann (1913a).

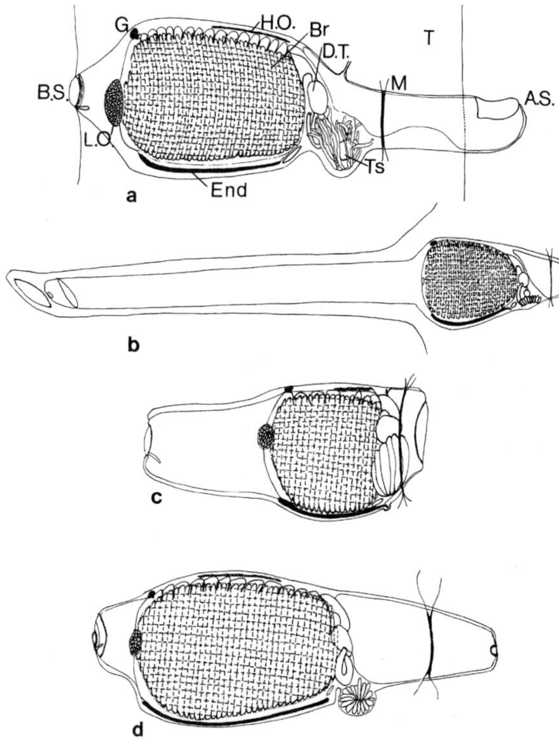

Fig. 17.2. Zooids of *Pyrosoma ambulata* gp. (a) *Pyrosomella operculata*. A.S., atrial siphon; Br, branchial basket with vertical stigmata and longitudinal bars; B.S., buccal siphon; DT, digestive tract; End, endostyle; G, brain H.O., haemopoietic organ; L.O., luminous organ; M, muscle; T, common tunic; Ts, testis. Modified after Neumann (1913a) and Van Soest (1981). (b) *Pyrosoma ovatum*. (c) *Pyrosoma aherniosum*. (d) *Pyrosoma atlanticum*. Modified after Van Soest (1981).

length of the sexually mature colony up to 13 cm, up to 45 stigmata, 20 bars and 16 dorsal languets, protogyny (Fig. 17.2a) ....*Pyrosomella operculata*

3. Test denticulation below the buccal aperture low and triangular, gill slits oblique, endostyle incurved or angular, cloacal muscles on the branchial basket 4
   Oral siphon contained in more or less long test projections, gill slits vertical, endostyle straight or slightly curved, cloacal muscles on the cloacal siphon..5
4. Test thick, opaque, soft, zooids longer than high, angular endostyle, up to 53 stigmata, 34 bars and 22 dorsal languets, colony reaching a very large size (>12 m). (Figs 17.4 and 17.5b)..*Pyrostremma spinosum*
   Test thin, transparent, zooids as high as long, incurved endostyle, up to 50 stigmata, 30 bars and 5–6 dorsal languets, four short quadrangular processes at the opening of the colony (Fig. 17.3) ....*Pyrostremma agassizi*
   (*Note*: according to van Soest *P. vitjasi* is a synonym of *P. agassizi.*)
5. Colony ovate to semi-globular..................................6
   Colony cylindrical, finger-shaped 7
6. Colony of tough consistency, opaque, provided with a well-developed diaphragm, length up to 5 cm, individuals with an extremely long oral siphon, up to 40 stigmata, 18 bars and 9–12 dorsal languets, protandry (Fig. 17.2b)......................*Pyrosoma ovatum*
   Colony soft, transparent, weakly developed diaphragm, a few limp smooth test projections, length up to 12 cm, up to 30 stigmata and 18 bars, slight protandry..............................*Pyrosoma godeauxi*
7. Colony small with few whorls of zooids, sexually mature at 2 cm, protogyny, oral siphon as broad as the branchial basket, up to 40 stigmata, 18 bars and 7 dorsal languets, embryo developing in the right peribranchial cavity (Fig. 17.2c and 17.5d) ....*Pyrosoma aherniosum*
   Colony larger (up to 60 cm), sexually mature at 12 cm, oral siphon much narrower than the branchial basket, numerous more or less long test processes, up to 50 stigmata, 26 bars and 6–12 dorsal languets, embryo developing in the cloacal siphon (Figs 17.2d and 17.5a) ....*Pyrosoma atlanticum*

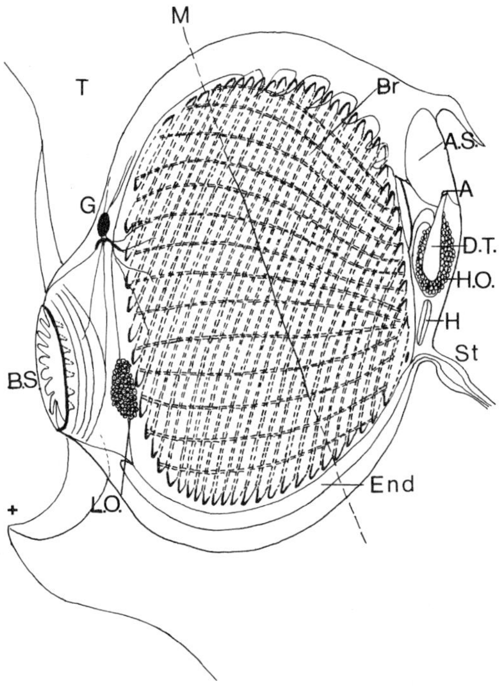

Fig. 17.3. Zooid of *Pyrostremma agassizi*. A, anus; A.T., atrial siphon; Br, branchial basket with slightly oblique stigmata and longitudinal bars; B.S., buccal siphon; D.T., digestive tract; End, endostyle; H, heart; H.O., haemopoietic organ, L.O., luminous organ; St, stolon; T, common tunic; +, triangular test denticulation. M, muscle. Modified after Neumann (1913a).

17.3.3. *Order Doliolida*

The order Doliolida contains three distinct families, of very different size: the Doliolidae, with four genera and some twenty species, the Doliopsoididae, with one genus and two species, and the Doliopsidae (Anchinidae), with one genus and one species.

Doliolida are marine holoplanktonic, microphagous animals mainly living in the epipelagic and upper

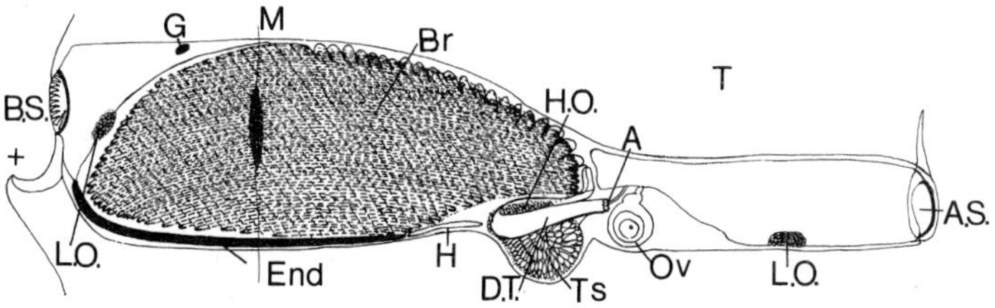

Fig. 17.4. Full grown zooid of *Pyrostremma spinosum*. A, anus; A.S., atrial siphon; Br, branchial basket; B.S., buccal siphon; D.T., digestive tract; End, endostyle; G, brain; H, heart; H.O., haematopoietic organ; L.O., luminous organs; M, muscle; Ov, ovary; T, common tunic; Ts, testis; +, triangular test denticulation (most of the muscles and the nerves are omitted). Modified after Neumann (1913a).

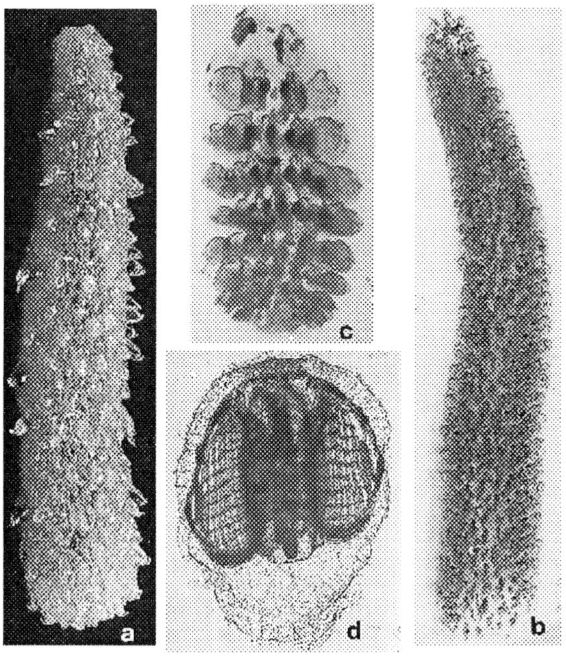

Fig. 17.5. Photographs of *Pyrosoma*. (a) colony of *Pyrosoma atlanticum* (length 9.5 cm). (b) Young colony of *Pyrostremma spinosum* (length 11.5 cm). (c) Colony of *Pyrosomella verticillata* (length 2.1 cm). (d) Tetrazooid colony of *Pyrosoma aherniosum*

mesopelagic layers. A few species are found in deeper waters and are still not yet accurately described. Their life cycle is very complex and remains incompletely known for most species. The Mediterranean species of Doliolidae have been best studied and have no less than six different and successive morphological stages (Trégouboff and Rose, 1957). In this family, the egg gives rise to a caudate larva recalling the ascidian tadpole, although rather simplified. The doliolid tadpole is actually an embryo enclosed in a follicular envelope and is less differentiated. This embryo gradually metamorphoses into a barrel-shaped oozooid, provided with nine annular muscles. After degeneration of most of its viscera (endostyle, branchial septum, intestinal loop, etc.), the oozooid becomes an actively swimming old nurse. Its ventral stolon successively gives off three kinds of buds migrating to the dorsal process: the spoon-like and very simplified trophozooids, the barrel-shaped phorozooids (with eight muscle rings) which carry the gonozooid probuds on their ventral stalks, attaching them to the nurse. From the eggs and sperm of the mature gonozooids, released into the sea, the next generation of larvae are formed.

The systematics of the family involves different characters: the shape of the intestinal loop, the number and the size of the muscular ring, the length of the endostyle, the shape and the extension of the branchial septum and, for the gonozooids only, the size, shape and position of the testis, always lying on the left side of the animal.

The genera and species of doliolids are identified as phorozooid and gonozooid forms. Only four oozooids have been described up to now, corresponding to the four genera defined by Garstang (1933), namely *Dolioloides, Doliolina, Doliolum* and *Dolioletta*. This discrepancy can be explained by the existence of cryptic species at the level of the oozooids (Godeaux, 1961), a hypothesis strengthened by Braconnot (1964, 1970a,b, 1971a, 1974a), who succeeded in rearing several oozooids from the egg.

17.3.4. *Dichotomous key for identification of doliolids*

Eight or nine muscles in parallel, transverse rings..family Doliolidae

Eight muscles in separate rings, except MV and MVI which are open dorsally and folded in an arc......................................family Doliopsoididae

Muscles reduced to the sphincters and to a pair of lateral S-shaped bands.........family Doliopsidae (Anchinidae)

Family Doliolidae

A. Larvae* (Fig. 17.6)

1. Follicular envelope fusiform, elongated larva with an active tail 2
   Follicular envelope spherical, dumpy larva with a poorly differentiated tail (Fig. 17.6c) genus *Dolioletta*
2. An empty ampulla between trunk and tail (Fig. 17.6a)..genus. *Doliolina*
   Ampulla missing (Fig. 17.6b) ....genus *Doliolum*

B. Fully developed animals*

1. Barrel-shaped individuals with viscera.....................2
   Barrel-shaped individuals generally devoid of viscera apart from heart, buds on a dorso-posterior process. Nurses,..6
   Spoon-like individual with reduced muscles and a ventral fixation stalk, buccal aperture very wide..Trophozooids, 8
2. Nine muscle rings, four pairs of gill slits, dorso-posterior process in relation with M VII ....Oozooids, 3
   Eight muscle rings, always more than four pairs of gill slits, temporary ventral process in front of M VII, gonads may be present or absent...................Phorozooids and gonozooids, 9

C. Oozooids* (Fig. 17.7)

3. Intestinal tube stretched sagittally 4
   Upright U- or S-shaped intestinal tube (Fig. 17.7a) ....genus *Doliolina*
4. Oesophagus concavity upwards (Fig. 17.7b) genus *Dolioloides*

[*Note*: a drawing by Uljanin (1884) suggests that the larva of the genus *Dolioloides* (17.6d) is also provided with a caudal ampulla.]

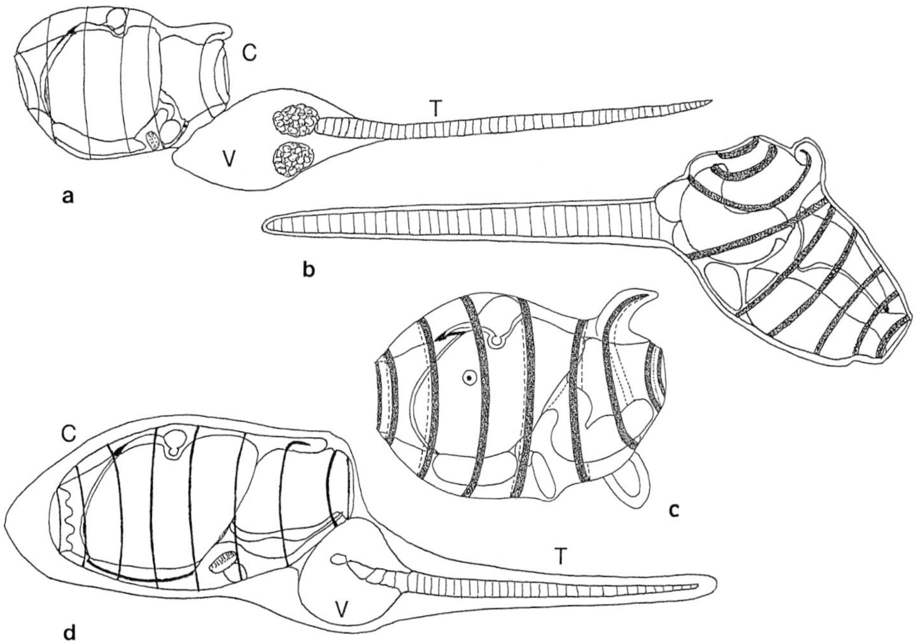

Fig. 17.6. Larvae. (a) *Doliolina mülleri.* C, cephalenteron; T, tail; V, caudal vesicle. (b) *Doliolum denticulatum.* (c) *Dolioletta gegenbauri.* (d) *Dolioloides rarum.*

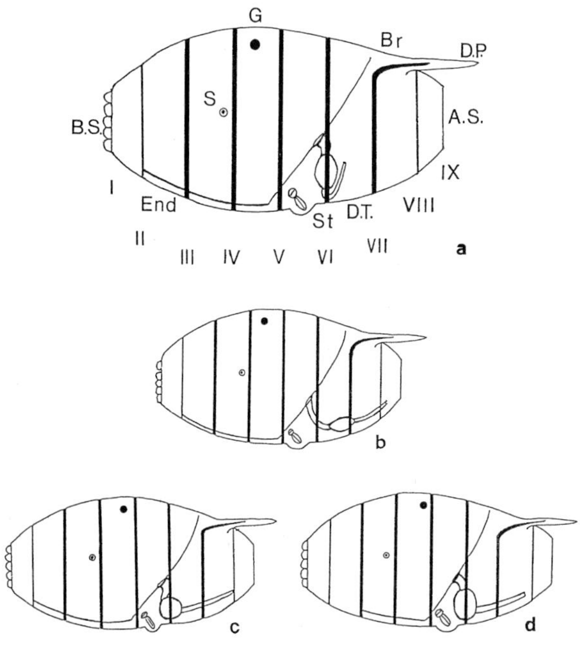

Fig. 17.7. Oozooids. (a) *Doliolina* sp. (b) *Dolioloides* sp. (c) *Doliolum* sp. (d) *Dolioletta* sp. A.S., atrial siphon; Br, branchial septum; B.S., buccal siphon; D.P., dorsal process; D.T., digestive tract; End, endostyle; G, brain; S, statocyst; St, stolon; I–IX, body muscles.

Oesophagus concavity downwards 5

5. Endostyle from M II to M V (Fig. 17.7c) ....genus *Doliolum*
   Endostyle from M III to M V (Fig. 17.7d) ....genus *Dolioletta*

D. Nurses (Fig. 17.8)*

6. Muscle bands twice as wide as the interspaces, viscera intact............................genus *Dolioloides*
   Muscles wide, although separated 7
   M II to VIII fused in a continuous sheet, length up to 2 mm (Fig. 17.8b) ....genus *Doliolum*
7. M III wider or equal to M IV, statocyst always present, two pairs of anterior nerves, robust appearance, reaching a large size (up to 40 mm) (Fig. 17.8c) ....genus *Dolioletta*
   M IV wider than M III or all muscles very narrow (1/4 of the interspace), statocyst (most often) missing, a single pair of anterior nerves, frail appearance, length less than 10 mm (Fig. 17.8a) ....genus *Doliolina*

[*Note*: the nurse of *Dolioloides rarum* is not known with certainty (Garstang, 1933). It seems to retain its viscera intact. Nurses with slender muscles and

**Note*: the species of the different genera of oozooids cannot yet be distinguished (cryptic species).

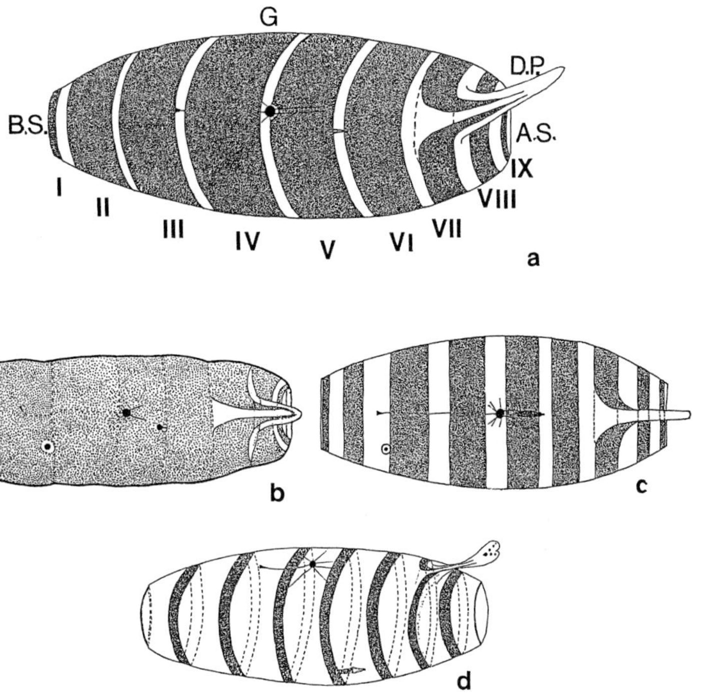

Fig. 17.8. Nurses. (a) *Doliolina type mülleri.* (b) *Doliolum denticulatum.* (c) *Dolioletta gegenbauri.* (d) *Doliolina type resistibile.* A.S., atrial siphon; B.S., buccal siphon; D.P., dorsal process; G, nervous ganglion; I–IX, body muscles. Note: not to same scale.

broad interspaces (Fig. 17.8d) belong to the group *Doliolina indicum* (western Indian Ocean, Red Sea), *D.intermedium* (north Atlantic) and *D.resistibile* (Southern Ocean). Those with wide muscles and narrow interspaces belong to the group *Doliolina mülleri–D.krohni.* It is not yet possible to separate the nurses belonging to the genus *Doliolum (D.denticulatum* and *D.nationalis*) and those belonging to the genus *Dolioletta (D.gegenbauri, D.tritonis,* etc.) (Godeaux 1961).]

E. Trophozooids (Fig. 17.9)

8. Gill slits less numerous (±10) anal aperture below the oesophageal funnel (Fig. 17.9a) ..genus *Doliolina*
   Gill slits numerous (10–40) anal aperture at the level of the oesophageal funnel (Fig. 17.9b) ....genera *Dolioletta* and *Doliolum*
   (*Note*: the trophozooids of genera *Dolioletta* and *Doliolum* are still indistinguishable.)

F. Phorozooids and gonozooids (Figs 17.10–17.12)
Both forms of all species are almost identical, except that the phorozooids have no gonads but bear the buds of the gonozooids on their ventral stalk.

9. Stretched intestinal loop ....genus *Dolioloides,* 10
   U- or S-shaped upright intestinal loop..genus *Doliolina,* 11

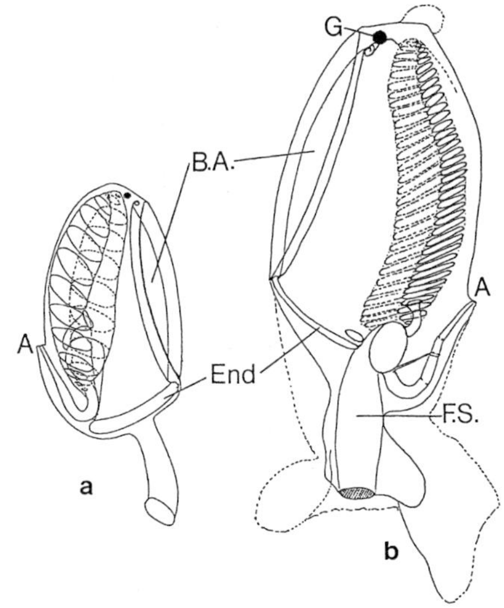

Fig. 17.9. Trophozooids. (a) *Doliolina mülleri.* (b) *Dolioletta gegenbauri* (after Braconnot 1970b). A, anus; B.A., buccal aperture; End, endostyle; F.S., fixation stalk; G, brain.

   Dextral arched intestine ....genus *Doliolum,* 18
   Close coiled intestinal loop..........genus *Dolioletta,* 19

10. Branchial septum curving behind M V, five gill slits, endostyle M II 1/4–M IV 3/4 (testis

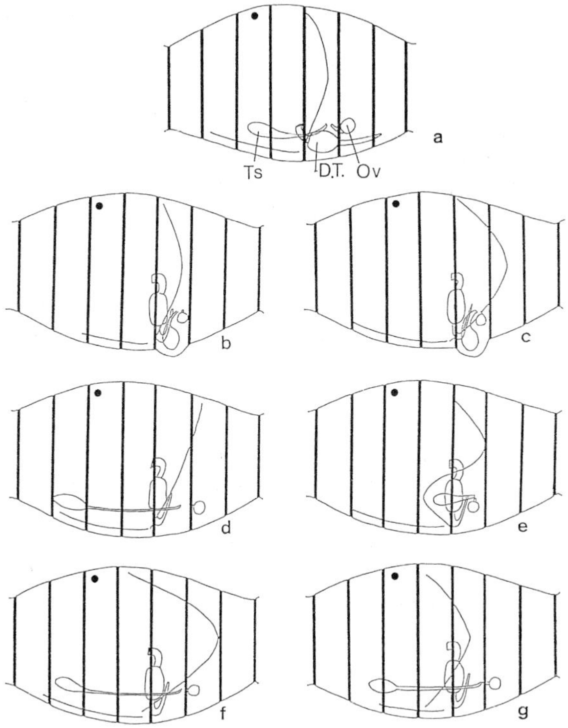

Fig. 17.10. Blastozooids (phorozooids and gonozooids). (a) *Dolioloides rarum.* (b) *Doliolina mülleri.* (c) *Doliolina krohni.* (d) *Doliolina indicum.* (e) *Doliolina sigmoides.* (f) *Doliolina intermedium.* (g) *Doliolina resistibile.* D.T., digestive tract; Ov, ovary; Ts, testis. After Garstang (1933).

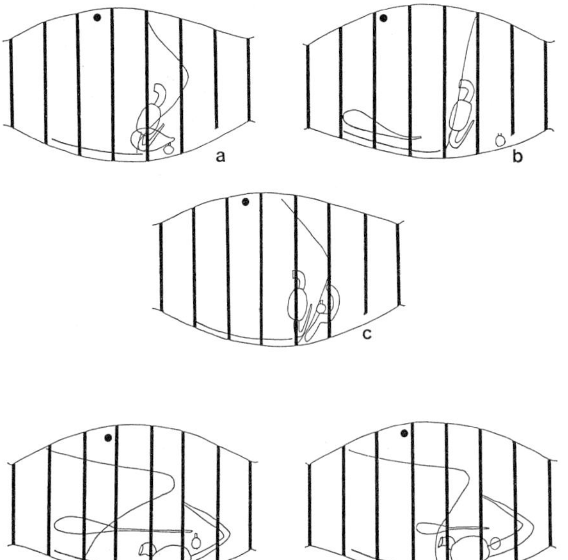

Fig. 17.11. Blastozooids (phorozooids and gonozooids). (a) *Doliolina obscura.* (b) *Doliolina separata.* (c) *Doliolina undulata.* After Tokioka and Berner (1958). (d) *Doliolum denticulatum.* (e) *Doliolum nationalis.* After Garstang (1933).

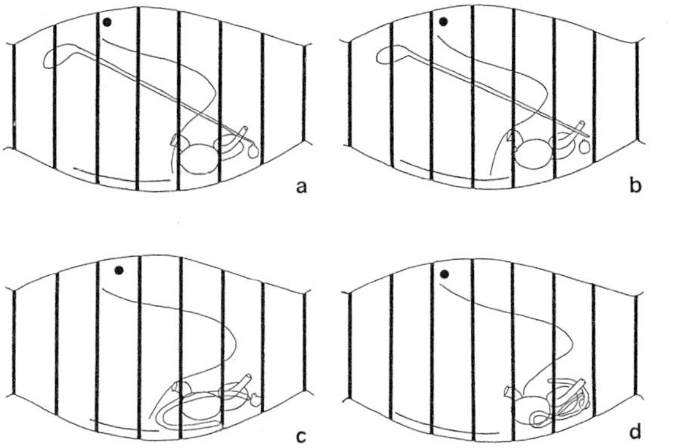

Fig. 17.12. Blastozooids (phorozooids and gonozooids). (a) *Dolioletta gegenbauri.* (b) *Dolioletta tritonis.* (c) *Dolioletta mirabile* (*chuni* ?). (d) *Dolioletta valdiviae.* After Garstang (1933).

horizontal, swollen in front of MIV) (Fig. 17.10a) .... *Dolioloides rarum*

11. M VII forming a complete hoop. *Doliolina perfecta*, 12
    M VII interrupted ventrally ..... *Doliolina imperfecta*, 17
12. Branchial septum less curved between M V and M VI 13
    Branchial septum distinctly bent between M IV and M V 15
13. Branchial septum oblique and nearly straight from M VI above to M V below 14
    Branchial septum slightly curved from M V above to M V below, 10–14 gill slits, endostyle M II 3/4–M IV 2/3 (testis pyriform, vertical, protruding ventrally below the intestine) (Fig. 17.10b) ..... *Doliolina mülleri*
14. Five gill slits, black spots covering the intestine, endostyle M II 1/3–M IV 2/3 (testis horizontal, swollen between M II–M III) (Fig. 17.10d) .... *Doliolina indicum*
    Twelve to 45 gill slits, endostyle M II 3/4–M IV 2/3 (testis pyriform, vertical, protruding ventrally below the intestine) (Fig. 17.10c) .... *Doliolina krohni*
15. Branchial septum S-shaped, attached above and below at M V, but incurved in front of the digestive canal, endostyle M II–M IV 2/3 (testis horizontal, short, swollen in front of M V) (Fig. 17.10e) .... *Doliolina sigmoides*
    Branchial septum extending from M IV above to M V below, the top of the arch between M V and M VI 16
16. Twelve to 45 gill slits, endostyle M II–M IV 2/3 (testis horizontal, tubular, extending beyond M II) (Fig. 17.10f) .... *Doliolina intermedium*
    Thirty to 40 gill slits, endostyle M II 1/2–M IV 2/3, (testis horizontal, tubular, reaching M II) (Fig. 17.10g) .... *Doliolina resistibile*
17. Branchial septum S-shaped between M IV and M V, 40 gill slits, endostyle M II 1/4–M V, reddish pigment cells concealing intestinal loop (and

gonads) (testis very short, swollen in front of M VI) (Fig. 17.11a) .... *Doliolina obscura*
Branchial septum oblique from M VI above to M V below, 10 gill slits, endostyle long M II–M V (testis horizontal, long, tubular, swollen between M II and M III) (Fig. 17.11b) .... *Doliolina separata*
Branchial septum from M IV dorsally to M VI posteriorly and to M V ventrally, 30–40 gill slits, endostyle long M II–M V (testis sausage-shaped, oblique between M V and M VI) (Fig. 17.11c) .... *Doliolina undulata*
[*Note: Doliolina obscura, D.separata* and *D.undulata* were named *Doliolina imperfecta* by reason of the ventral gap of M VII, as opposed to the other species of the genus having a complete hoop (*Doliolina perfecta*) (see Tokioka and Berner 1958a,b).]

18. Branchial septum strongly arched from M II dorsally to M V 2/3 posteriorly and M III ventrally, endostyle short M II–M IV, (testis horizontal, long, swollen in front of M III, sometimes beyond M II (Fig. 17.11d) .... *Doliolum denticulatum*
Branchial septum, strongly arched, from M II dorsally, to M V 2/3 posteriorly and M V to MV 2/3 posteriorly and MV ventrally endostyle short M II–M IV, (testis horizontal, short, swollen behind M IV) (Fig. 17.11e) .... *Doliolum nationalis*
(*Note:* phorozooids of *D.nationalis* from the Sea of Japan have a conspicuous red spot overlying the brain.)
19. Branchial septum strongly arched, not passing M VI posteriorly, up to 70 gill slits 20
Branchial septum strongly arched passing beyond M VI posteriorly, up to 75 gill slits 21
20. Branchial septum from M III dorsally to M V ventrally, endostyle M II 1/2–M IV 1/2 (testis tubular, oblique, sometimes passing beyond M II) (Fig. 17.12a) .... *Doliolетta gegenbauri*
Branchial septum from M III dorsally to M IV ½ ventrally, endostyle M II ½–M IV ½ (testis tubular, oblique, sometimes passing M II) (Fig. 17.12b) .... *Doliolетta tritonis*
21. Branchial septum M III–M IV, endostyle short M II 4/5–M V, M VIII ventrally open, (testis tubular, coiled between M IV and M VI) (Fig. 17.12c) .... *Doliolетta mirabilis (chuni)*
Branchial septum M III–M V, endostyle long M II 4/5–M IV 4/5, (testis tubular, coiled between M V and M VI) (Fig. 17.12d) .... *Doliolетta valdiviae*

Family DOLIOPSOIDIDAE

The family Doliopsoididae seems to be monogeneric with very few species described so far: *Doliopsoides meteori* Krüger, 1939 and *D.horizoni* Tokioka and Berner, 1958a,b. Another possible species, *D.bahamensis* is described briefly below (Godeaux, unpublished).

The genus was erected by Krüger (1939) for some 13 gonozooids collected in the south Atlantic Ocean (30–40°S), mostly below 400 m depth, at 10 stations between Africa and America. The few gonozooids and the single phorozooid of the second species were caught in the eastern Pacific Ocean (8–10°S) and described by Tokioka and Berner (1958b). An isolated gonozooid was also recorded from the subtropical convergence area (38°S) by Godeaux and Meurice (1978); it is somewhat different from Krüger's specimens. More recently, three gonozooids were caught by R. Harbison in deep water of the north western Atlantic Ocean, off the Bahamas, and named *Doliopsoides bahamensis* by Godeaux (see Fig. 17.13a).

The biological cycle of the Doliopsoididae is poorly known. Metagenesis seems to exist, as phorozooids and gonozooids have been identified by Tokioka and Berner (1958b). Development probably takes place in deep water, where most of the specimens have been collected.

Doliopsoides bahamensis resembles a doliolid blastozooid: barrel-shaped body, siphons with lappets at both

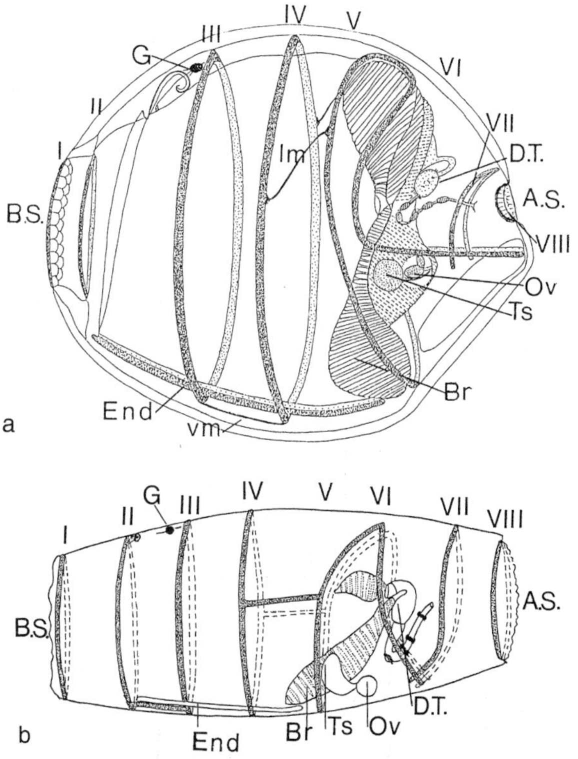

Fig. 17.13. (a) *Doliopsoides bahamensis* blastozooid (Harbison and Godeaux, unpublished). A.S., atrial siphon; Br, branchial septum; B.S., buccal siphon; D.T., digestive tract (with faecal pellets); End, endostyle; G, brain; Ov, ovary; Ts, testis; I–VIII, body muscles. (b) *Doliopsoides meteori* gonozooid (after Krüger, 1939). Lettering as for (a), apart from D.T., where no faecal pellets are visible in the digestive tract.

ends of the main axis, eight narrow muscles separated by broad gaps, buccal and cloacal cavities separated by a branchial septum (some 50 gill slits) divided into two parts (20 gill slits above, 30 below), endostyle extending from M II to M IV 3/4 in front of the cardiopericardium. The digestive tract is U-shaped (a primitive character?) in the sagittal plane of the body. The oesophagus is bent and has a long helix not seen in the other species. As in *D.horizoni*, the stomach bears several finger-like protuberances (blind caeca), possibly also present in *D.meteori*, but not reported by Krüger, whose material was in poor condition. The pyloric gland is clearly visible, but according to the specimen it is either simple or ramifying (not shown in Fig. 17.13a). The species is hermaphroditic, the gonads are below the gut, with the testis anterior to the ovary. A single oocyte is visible in the ovary and the species may be protogynous, since in the smaller specimen the oocyte (500 μm) is larger than the entire testis, whilst in the larger, the testis is bi-lobate (830 × 500 μm) and seems ripe (Fig. 17.13a).

The main differences between *Doliopsoides* sp. and any doliolid blastozooid concern the nervous system and the pattern of the musculature. The brain lies in front of M III and not behind. A large dorsal spiral volute of the peripharyngeal bands lies in front of the brain, in the second intermuscular zone. Only the muscles M I–M IV and M VIII form a complete ring, the muscles M V and VI are open dorsally and their extremities fuse in a pair of characteristic dorsal arches. Moreover, M VI and M VII are open ventrally and cross each other; in some specimens, the free ends of M VI are united by a fibrous strand. The single phorozooid known of *D.horizoni* has a short ventral protuberance into which both ends of M VII project. Furthermore, a sigmoid lateral muscle (lm) links M IV and M V on both sides and a thin medioventral muscle (vm), running below the endostyle is stretched either between M II and M III in *D.meteori* (Fig. 17.13b) or between M III and M IV in all the other specimens (possibly the sole morphological difference between the two species?).

Family Doliopsidae (Anchinidae)

This family also seems to be monogeneric. *Anchinia savigniani*, described by Rathke (1835) from papers left by Eschscholtz, is probably identical to *Anchinia rubra* Vogt, 1852, to *Doliopsis rubescens* Vogt, 1854 and to *Doliopsis savigniana* Krüger, 1939. That species, abundant in the past in winter and spring at Villefranche-sur-Mer (Kowalevsky and Barrois, 1883; Barrois, 1885), is now extremely rare (Braconnot, 1970a). A few specimens have also been identified in collections from the Atlantic Ocean (Krüger, 1939; Godeaux, unpublished).

Most of the specimens investigated were in the form of stalked buds at different stages of development, spread out on solid whitish 'laces' floating in the water. Three successive generations of similar blastozooids were early recognized, a condition recalling somewhat that known in the Doliolidae: asexual 'trophozooids', then 'phorozooids' with abortive gonads and finally 'gonozooids' provided with fertile gonads, all arising from the longitudinal stolon borne by the 'lace'. They have no stolon.

Neither the larva nor the budding oozooid are known and a large part of the biological cycle of the Doliopsidae is still a mystery.

Doliopsis exhibits the general morphological pattern of a doliolid. All the characteristic organs are present; nevertheless there are a few differences (Fig. 17.14). The body is more globular, being two or three times deeper than long. The musculature is reduced to the buccal and atrial sphincters and to a pair of short S-shaped lateral bands. The dorsal process above the atrial siphon and the ventral attachment stalk are transient organs, disappearing during the growth of the individual.

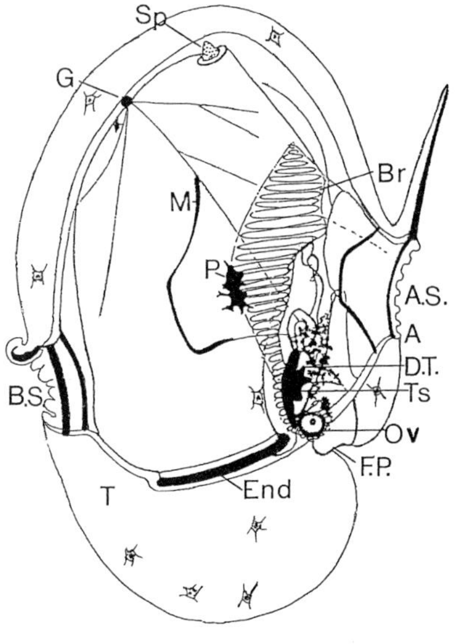

Fig. 17.14. *Doliopsis rubescens* gonozooid. After Kowalevsky and Barrois (1883). A, anus; A.S., atrial siphon; Br, branchial septum; B.S., buccal siphon; D.T., digestive tract surrounded by testis; End, endostyle; F.S., fixation stalk; G, brain; M, muscle; Ov, ovary; P, pigment spot; Sp, spiral organ; T, tunic; Ts, testis.

17.4.1. *Order Salpida*

Salps are a small group of marine, holoplanktonic, microphagous animals. Some 40 species have been described up to now. They are widespread, especially in tropical and warm temperate oceanic waters. Only a few species are known from the Antarctic.

In most metagenetic animals, such as for example the colonial and social ascidians, it is usually not possible to distinguish the two generations. In salps, however, the alternating solitary oozooids and aggregate blastozooids were often described as different species since they were so dissimilar. Thus Forsskål (1775) named both *Salpa maxima* (oozooid) and *S.africana* (blastozooid) and *Thalia democratica* (oozooid) and *T.mucronata* (blastozooid). Today, a single name is retained, following the principle of priority; the name is not always that of the oozooid.

Nevertheless, to determine salps, it is easier to consider the oozooid (identified by its stolon) and the blastozooid (bearing one or several eggs) separately. The blastozooid form is usually simpler than the corresponding oozooid. The body musculature is less developed, with a maximum of six muscles. Moreover, both forms have evolved separately, and convergences are obvious between blastozooids belonging to different genera (e.g. *Salpa* and *Ritteriella*).

The characters used for systematics are various, as follows.

(i) Number and arrangement of body muscles (fusing, touching, approaching, interrupted, etc.), number of muscle fibres (even if the body form is symmetrical, the number is generally different on each side; it is therefore preferable to consider the total number of fibres, both sides together); moreover, the number of fibres in one species (e.g. *Salpa fusiformis*) can vary clinally with regard to the temperature at the sampling site (van Soest, 1972).

(ii) Shape of the alimentary canal, either elongated or bent into a more or less compact loop to form a nucleus.

(iii) Outline of the aperture of the dorsal tubercle.

(iv) Body processes provided either with muscles or with a haemocoelian cavity.

(v) Shape and form of the stolon (blastozooids arranged either in whorls or along a chain).

(vi) Number of eggs or embryos (1–4).

(vi) Thickness, protuberances, ridges, crests, denticulations of the test, if present.

(vii) Number of accessory eyes above the brain (unfortunately the reddish pigment is quickly destroyed by formalin).

[*Note*: details of the musculature can be made visible by staining the specimen either with chlorantine red (1/100 or less) or bengal rose; with these dyes, the organs are coloured, but the test is not. They are thus more suitable than toluidine blue, which stains the test metachromatically.]

The classification given below follows that of Ihle and Ihle-Landenberg (1933), who upgraded to the rank of genera the different subgenera defined by Metcalf (1918).

The order Salpida consists of the sole family Salpidae with two subfamilies, Cyclosalpinae and Salpinae (Yount, 1954).

The Cyclosalpinae, with the genera *Cyclosalpa* and *Helicosalpa*, form a rather homogeneous group. The oozooid always has seven body muscles, interrupted ventrally in most species and sometimes also dorsally. Lateral spindle-shaped light organs are generally present. The main part of the alimentary canal is a straightened tube overlying the branchial bar and with its opening close to the ganglion. The ventral stolon bears buds arranged in successive whorls (*Cyclosalpa*) or on a linear chain (*Helicosalpa*).

The blastozooid has four body muscles (rarely three by fusion of the first two), all continuous dorsally. Light organs are exceptional. Ventrally a peduncle (remnant of the stalk attaching the young individual to the axis of the stolon) may either contain (*Cyclosalpa*) or not (*Helicosalpa*) the terminations of the intermediate muscles and of some of the body muscles. The shape of the alimentary canal is variable, but it is never bent into a nucleus. There is a single egg on the right between muscles III and IV. Testis variable in shape and position.

The Salpinae, although conforming to this general scheme, are more diverse; the group is less homogeneous, with some ten clearly distinct genera, often monotypic.

The oozooid, more or less rectangular in profile, has from four to more than 20 body muscles, interrupted ventrally, sometimes dorsally, often approaching or fusing characteristically. No light organs. The alimentary canal is free from the branchial bar and in most cases bent into a more or less dense loop (nucleus). The stolon is either ventral and straight or curled around the digestive loop. The buds and blastozooids are always arranged in a double symmetrical file along the axis of the linear stolon (chain); the mirror symmetry (enantiomorphism) of the blastozooids is more or less pronounced.

The blastozooid is often fusiform owing to the presence of a pointed test process at both ends. It has from four to six body muscles, often fused in dorsal groups. No light organs. Usually the alimentary canal is a more or less dense loop. In most cases, the ovary contains a single egg (there are three or four eggs in a few species), lying posteriorly on the right in front of the last muscle. Testis variable, often close to the gut.

17.4.2. *Dichotomous key to salp species*

Animals bearing a stolon emerging ventrally or posteriorly, gonads always absent, oozooid or solitary form (*proles solitaria*, P.S.) 1

Animals bearing one or several eggs or embryos on the posterior right side, stolon always absent, blastozooid, aggregate form or *proles gregata* (P.G.) 23

A. Oozooids or solitary forms (Figs 17.15–17.18)

1. Alimentary canal straight, overlying the branchial bar, anus anterior, seven annular body muscles, light organs usually present 2
Alimentary canal horizontal and straightened backwards, anus posterior, light organs always missing 9
2. Alimentary canal bent into a more or less compact loop always behind the branchial bar, light organs always absent 10
Muscles parallel, all interrupted mid-ventrally.........3
Muscles I–V fused into a ventral mass, M VI interrupted and M VII continuous ventrally, inconspicuous light organs between M II and VI (Fig. 17.15b) ....*Cyclosalpa floridana*
3. Muscles I–VII also interrupted dorsally, independent, light organs well developed............................4
Some muscles continuous or fused dorsally............8
4. Muscles I–VII running parallel 5
Muscles I–VII approaching ventrally, M VI extending dorsally forward and separately between M V to M I. 6
5. Five light organs between M I and M VI (Fig. 17.15a) ....*Cyclosalpa pinnata*
Four light organs between M II and M VI *Cyclosalpa sewelli*
6. Muscles I linked dorsally with M VI by *two* longitudinal muscles, M I and M V forming ventrally two pairs of approaching but independent longitudinal muscles, two atrial palps, one of the

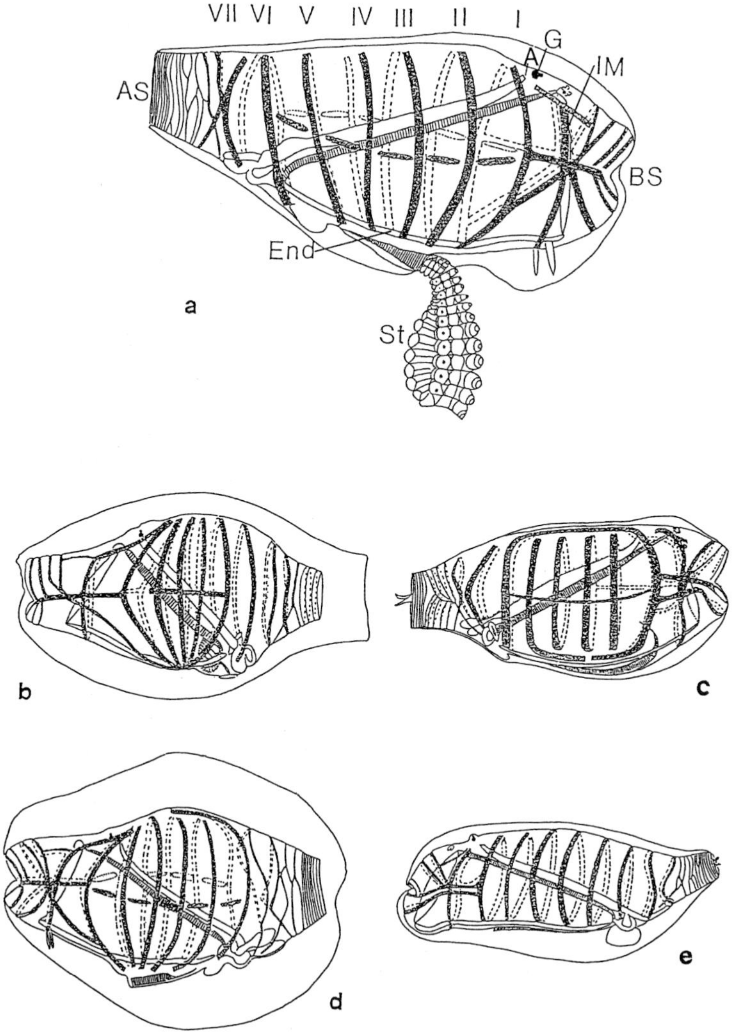

Fig. 17.15. Cyclosalpinae, oozooids. After Metcalf (1918). (a) *Cyclosalpa pinnata.* A, anus; AS, atrial siphon; BS, buccal siphon; End, endostyle; G, nervous ganglion; St, stolon; I–VII, body muscles; IM, intermediary muscle. (b) *Cyclosalpa floridana.* (c) *Helicosalpa (Cyclosalpa) virgula.* (d) *Cyclosalpa bakeri.* (e) *Cyclosalpa affinis.*

largest cyclosalpas (up to 14 cm overall) (Fig. 17.15c) .... *Cyclosalpa (Helicosalpa) virgula*
Muscles I and VI not linked 7

7. Light organs well developed between M I and M VI (Fig. 17.15d) .... *Cyclosalpa bakeri*
Light organs much reduced and situated *on* body muscles II–V .... *Cyclosalpa foxtoni*

8. Muscles III–VII continuous dorsally, light organs missing, two minute atrial palps (Fig. 17.15e) .... *Cyclosalpa affinis*
Muscles VI dorsally fused into a *single* longitudinal muscle extending towards muscles III, five light organs .... *Cyclosalpa polae*

9. More than 10 body muscles, often asymmetrical, open ventrally, fused or exchanging fibres dorsally (Fig. 17.16a) .... *Ritteriella retracta*
More than 20 body muscles, same disposition .... *Ritteriella picteti*

10. Long anterior proboscis, below the lower lip, bearing four longitudinal muscles. Seven continuous body muscles in two dorsal groups (M I–III and M IV–VII); M III and IV fused laterally (Fig. 17.16c) .... *Brooksia rostrata*
No anterior proboscis 11

11. No more than five body muscles (intermediate muscles not included) 12

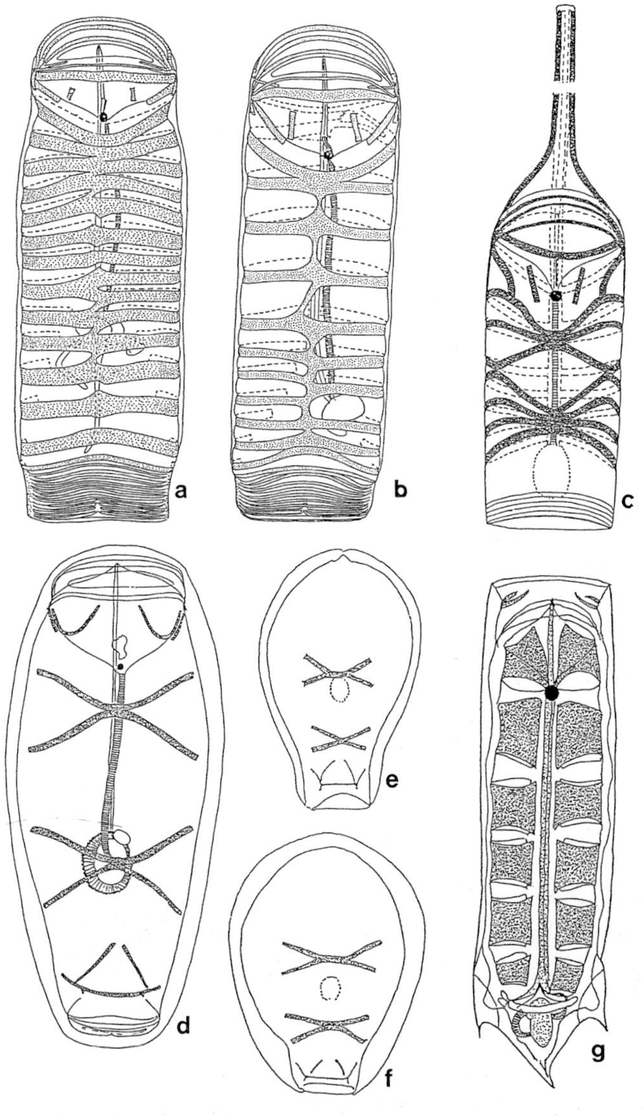

Fig. 17.16. Salpinae, oozooids of various genera. (a) *Ritteriella retracta.* (b) *Ritteriella amboinensis.* After Meurice (1970, 1974). (c) *Brooksia rostrata.* (d) *Pegea confoederata.* (e) *Pegea bicaudata.* (f) *Pegea socia.* After Madin and Harbison (1978). (g) *Iasis zonaria.*

More than five body muscles (intermediate muscles not included) 18

12. Four short body muscles in two groups, firmly fusing dorsally, stolon making a single turn around the nucleus, animal reaching a large size (up to 12 cm overall) (Fig. 17.16d) .... *Pegea confoederata*
Four short body muscles in two groups, M I and II fusing dorsally, M III and IV approaching or touching, but not fusing, test thick, stolon making a single turn around the nucleus, body length up to 7 cm (Fig. 17.16e) .... *Pegea bicaudata*
Four short body muscles in two groups, M I and II fusing briefly dorsally, M III and IV just touching, body plump, mature stolon making two complete turns around the nucleus, length up to 14 cm (Fig. 17.16f) .... *Pegea socia*
Five body muscles 13

13. Five body muscles parallel and continuous dorsally and ventrally (Fig. 17.17a) .... *Thalia longicauda*
Five body muscles interrupted dorsally and ventrally, broad, approaching but not touching, length up to 6 cm (Fig. 17.16g) .... *Iasis zonaria*
Some muscles fused or touching 14

14. Body muscles in complete rings. M I–III and M IV–V fused dorsally 15
Muscles widely interrupted ventrally, M I–III fused for a great length, M IV and V separate and bifurcated (Fig. 17.17f) .... *Traustedtia multitentaculata*

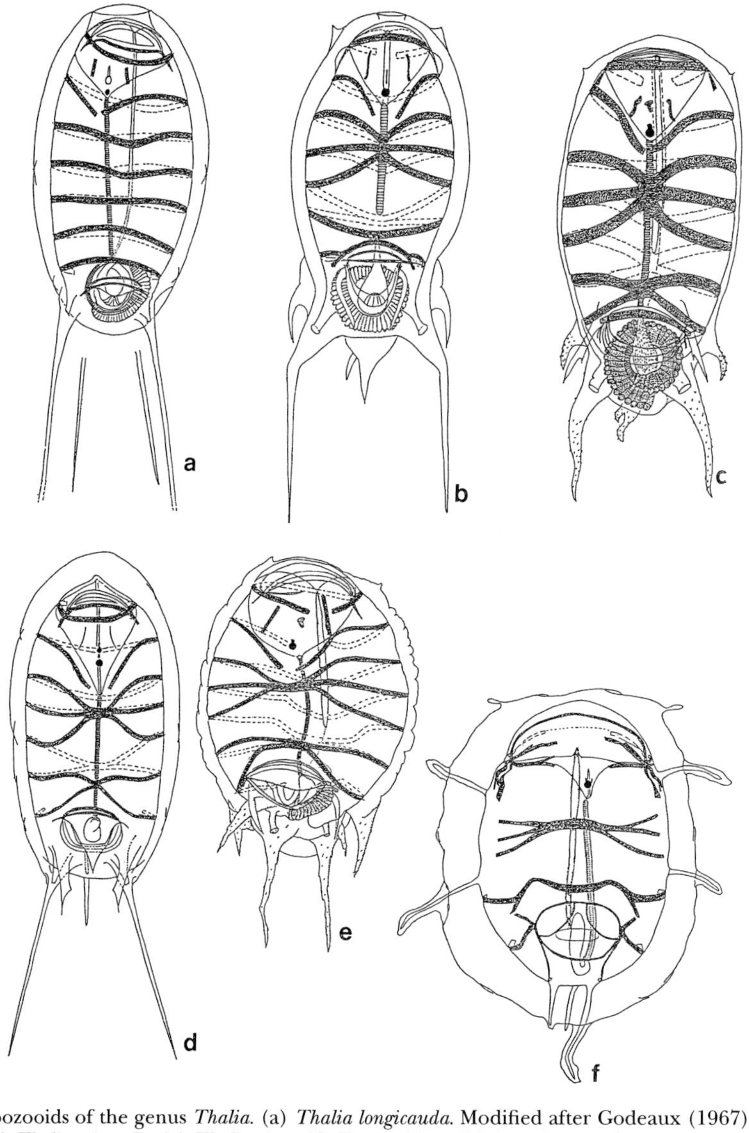

Fig. 17.17. Salpinae, oozooids of the genus *Thalia.* (a) *Thalia longicauda.* Modified after Godeaux (1967). (b) *Thalia democratica.* (c) *Thalia rhomboides.* (d) *Thalia orientalis.* (e) *Thalia cicar.* (f) *Traustedtia multitentaculata.*

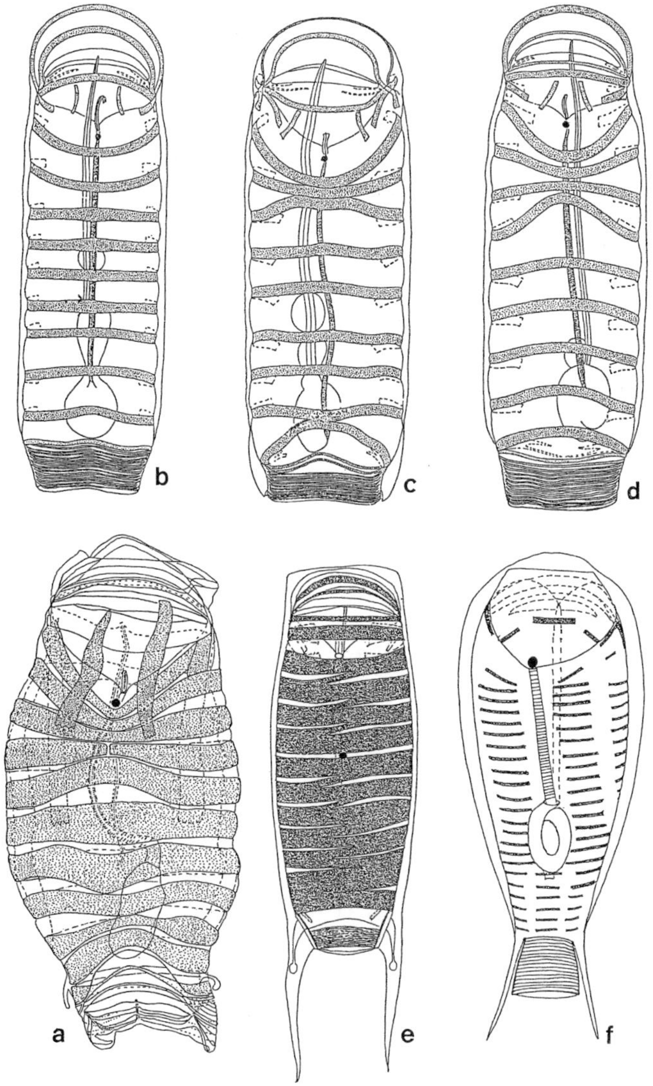

Fig. 17.18. Salpinae, oozooids of various genera. (a) *Ihlea punctata (asymmetrica)*. (b) *Salpa maxima.* (c) *Salpa fusiformis.* (d) *Salpa cylindrica.* After Meurice (1974). (e) *Metcalfina hexagona.* (f) *Thetys vagina.*

15. Cloacal palps simple (Fig. 17.17b) ..*Thalia democratica*
 Cloacal palps bifurcate 16
16. Fibres of the body muscles very numerous (>150) (Fig. 17.17c). ....*Thalia rhomboides*
 Fibres of the body muscles less numerous (<70) 17
17. Two posterior projections very long, lateral test projections missing (Fig. 17.17d) ....*Thalia orientalis*
 Two posterior projections very short, two lateral test projections present (Fig. 17.17e) ....*Thalia cicar*
18. Seven broad body muscles in complete rings, touching dorsally, laterally and ventrally ..*Ihlea magalhanica*
 Eight broad body muscles (intermediate muscle not included) forming complete rings excepted for M VIII open ventrally (Fig. 17.18.a)..........................*Ihlea punctata (asymmetrica)*
 At least nine body muscles 19
19. Nine body muscles ventrally interrupted 20
 More than nine body muscles................................22
20. Nine body muscles parallel to each other (Fig. 17.18b)..*Salpa maxima*
21. Some body muscles fused or approaching dorsally; the others parallel.
 Muscles I and II close or fused, V III and IX approaching..*Salpa younti*
 Muscles I–III and VIII–IX fused or closely approaching dorsally (Fig. 17.18c)*..................*Salpa fusiformis*

Note: *Salpa gerlachei* and *S.thompsoni* are here considered as clinal forms of *S.fusiformis* (see Foxton, 1961, 1962)

Muscles I–III fused or closely approaching .... *Salpa aspera*
Muscles I–IV fused or closely approaching (Fig. 17.18d) .... *Salpa cylindrica*

22. Nine to 12 broad, asymmetrical muscles, widely interrupted ventrally, continuous dorsally, exchanging fibres (Fig. 17.18e) .... *Metcalfina hexagona*
Ten to 13 thin muscles, often asymmetrical, touching or exchanging fibres dorsally, the three or four first ones as complete rings, the following ones ventrally open, digestive loop vertical (Fig. 17.16b) .... *Ritteriella amboinensis*
At least 16 muscles widely interrupted ventrally, interrupted dorsally, the first and the last ones in two pieces, animal reaching a large size (up to 24 cm overall) (Fig. 17.18f) .... *Tethys vagina*

B. Blastozooids or aggregate forms (Figs 17.19–17.22)

23. Alimentary canal is either a straight tube or a widely open loop, four body muscles at most, a more or less prominent, anterior and ventral peduncle .... Cyclosalpinae
24. Alimentary canal in a more or less compact loop, usually more than four body muscles, no ventral peduncle .... Salpinae, 31
Alimentary canal straight below the endostyle, anus anterior close to the peduncle 25
Alimentary canal loosely coiled, light organs absent 28
25. A pair of light organs between M II and III 26
Two pairs of light organs between M II–III and M III–IV 27
26. Muscles I and II strongly fused, M III and IV approaching mid-dorsally; from 52 to 75 muscles fibres overall, dorsal tubercle slightly convoluted (Fig. 17.19a) .... *Cyclosalpa pinnata*
Muscles I–II and III–IV strongly fused dorsally, from 14–19 muscle fibres overall, dorsal tubercle arched .... *Cyclosalpa sewelli*
Muscles I–II and III–IV strongly fused dorsally, from 20–29 muscle fibres overall, dorsal tubercule in form of ε .... *Cyclosalpa polae*
27. Muscles I–II and III–IV fused dorsally, from 29–38 muscle fibres overall .... *Cyclosalpa quadriluminis*
Muscles I–II fused, M III and IV approaching or touching dorsally, from 35–58 muscles fibres overall .... *Cyclosalpa quadriluminis parallela*
28. Body muscles symmetrical 29
Body muscles strongly asymmetrical, dextral, and sinistral individuals 30
29. Body muscles parallel dorsally, interrupted ventrally, M I and M IV united to the next siphonal sphincter, intestinal loop protruding posteriorly (Fig. 17.19b) .... *Cyclosalpa affinis*
Only three body muscles (I and II completely fused?). M III and IV fused dorsally and approaching M I (+ M II ?), M I (+ M II ?) touching M III ventrally and joining the intermediate muscle, M III and IV touching ventrally (Fig. 17.19c) .... *Cyclosalpa floridana*
30. *Dextral* individual: M I and II strongly fused dorsally, free on the right side and fused again before entering the peduncle. On the left side, M II asymmetrical and joining M III, M III and M IV fused dorsally (5d). *Sinistral* individual: mirror image* .... *Cyclosalpa bakeri*
Dextral individual: M II–IV fused mid-dorsally, M I and II fused on the giving a ventral branch and a lateral branch lining the atrial siphon. *Sinistral* individual: mirror image† (Fig. 17.19e) .... *Helicosalpa virgula*
31. Four body muscles 32
Five body muscles 33
Six body muscles 39
32. Body muscles very short, in two groups (M I–II and M III–IV) firmly fused dorsally, body plump, animal reaching a large size (up to 11 cm overall) (Fig. 17.20a) .... *Pegea confoederata*
Body muscle very short, in two groups (M I–II and M III–IV), touching but not fusing, two (sometimes one) more or less long and equal posterior processes with mantle extensions, test very thick, body length up to 98 cm (Fig. 17.20.b) .... *Pegea bicaudata*
Body muscle very short, in two groups, barely touching dorsally, test fairly thick and firm, length up to 13 cm (Fig. 17.20c) .... *Pegea socia*
Body muscles in two groups strongly fused dorsally, widely interrupted ventrally, M IV branching laterally, nucleus protruding between two long posterior processes (Fig. 17.20d) .... *Traustedtia multitentaculata*
Muscle I with more than 12 fibres, more than 120 fibres overall, same general appearance as above .... *Ritteriella picteti*
Dextral individual: body asymmetrical, four muscles on the right, three on the left, joining ventrally and dorsally in a single mass (+ intermediate muscle). *Sinistral* individual: mirror image (Fig. 17.20e) .... *Brooksia rostrata*

**Note*: the aggregate form of *C.foxtoni* not very different, that of *C.ihlei* still unknown (see Soest, 1974a).

†*Note*: the aggregate form of *Helicosalpa younti* is still unknown. That of *H.komaii* is similar to *H.virgula* but its testis is spherical (see Soest, 1974a).

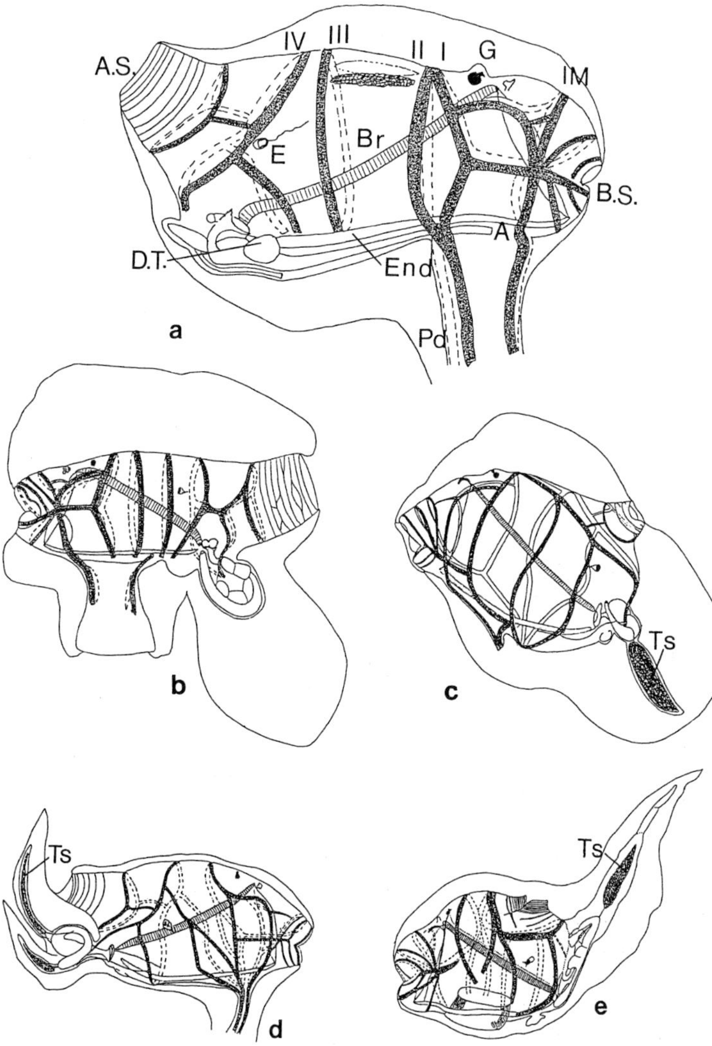

Fig. 17.19. Cyclosalpinae, blastozooids. After Metcalf (1918). (a) *Cyclosalpa pinnata.* A, anus; A.S. atrial siphon; Br, branchial bar; B.S., buccal siphon; D.T., digestive tract; E, embryo; End, endostyle; G, brain; Pd, peduncle; Ts, testis; I–IV, body muscles; IM, intermediary muscle. (b) *Cyclosalpa affinis.* (c) *Cyclosalpa floridana.* (d) *Cyclosalpa bakeri.* (e) *Helicosalpa* (*Cyclosalpa*) *virgula.*

33. Five body muscles, parallel or approaching dorsally 34
    Five body muscles continuous, fusing or touching dorsally 35
34. Body muscles broad, parallel, M I open dorsally, all widely interrupted ventrally, three or four eggs or embryos, test firmly tied to ectoderm, an asymmetrical tunical protuberance behind, length up to 4.5 cm (Fig. 17.20f) .... *Iasis zonaria*
    Body muscles narrow and very short, interrupted dorsally, M I–III approaching, M IV–V parallel, animal reaching a large size (up to 20 cm) (Fig. 17.20g) .... *Thetys vagina*
35. Two groups of fused muscles (I–III and IV–V) 36
    Body muscles fused differently and displaying a strong asymmetry 38
36. Two groups of body muscles (I–III and IV–V) separate dorsally 37
    Two groups of body muscles fused or closely approaching dorsally (Fig. 17.23e) ....*Salpa cylindrica*
37. From 27 to 38 muscle fibres .... *Thalia longicauda*

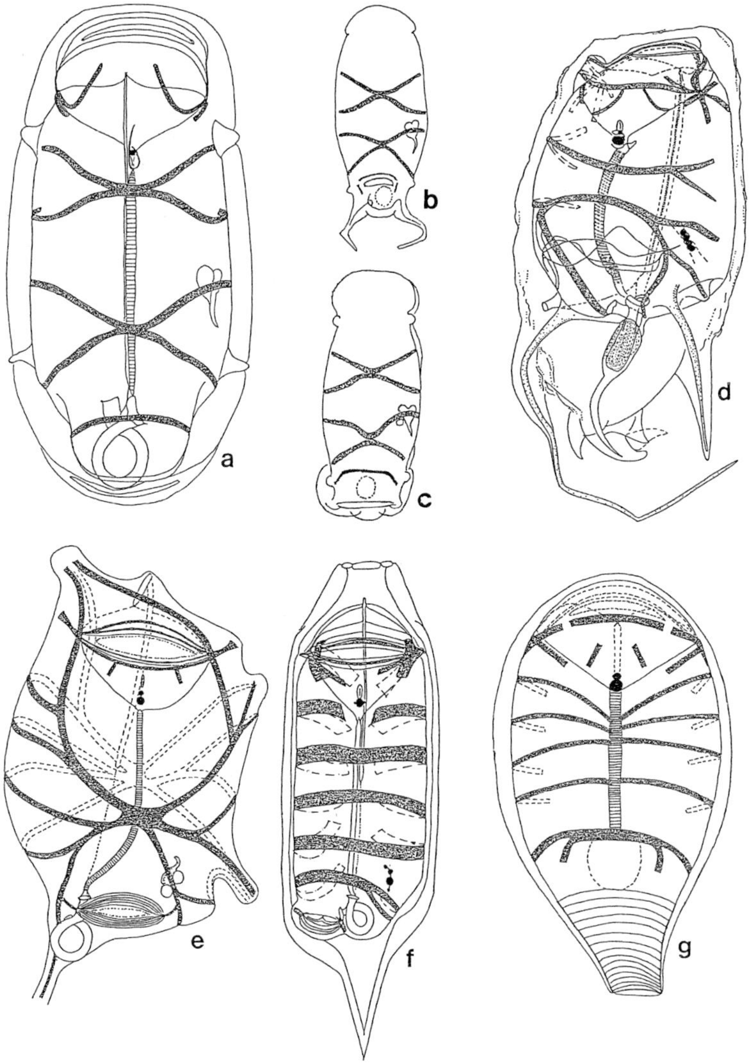

Fig. 17.20. Salpinae, blastozooids of various genera. (a) *Pegea confoederata.* (b) *Pegea bicaudata.* (c) *Pegea socia.* After Madin and Harbison (1978). (d) *Traustedtia multitentaculata.* (e) *Brooksia rostrata.* (f) *Iasis zonaria.* (g) *Thetys vagina.*

About 16 muscle fibres overall in I–V (Fig. 17.21b) .... *Thalia democratica*
(Fig. 17.21c) .... *Thalia rhomboides*
(Fig. 17.21d) .... *Thalia cicar*
(Fig. 17.21e) .... *Thalia orientalis*

38. Body muscles open ventrally, M I and II (and sometimes M III) fused dorsally, M III–IV and V parallel, M V bifurcate laterally (Fig. 17.22a) .... *Ihlea punctata (asymmetrica)*
Body muscles broad, asymmetrical, M III and IV fused on one side, M V bifurcate with one branch lining the atrial siphon .... *Ihlea magalhanica*
39. Muscles I and II fusing or approaching dorsally, all muscles interrupted ventrally, test firm, length up to 6 cm (Fig. 17.22b) .... *Metcalfina hexagona*
Three dorsal groups of fused muscles M I and II, M III and IV, M V and VI, the two first also fused or closely touching 40
40. Muscles I and II more strongly fused than III and IV, the two groups closely approaching or touching dorsally, the third group wide apart from them 41
Muscles I–II and muscles III–IV equally fused dorsally, both groups fused or touching dorsally, the third group wide apart from them 42
41. Muscle I with 5–7 fibres, some 40 fibres overall, M IV and V widely separate laterally (Fig. 17.23a) .... *Ritteriella amboinensis*
Muscle I with 12 fibres, from 60–119 fibres overall, M IV and V approaching but not fused laterally (Fig. 17.23b) .... *Ritteriella retracta*

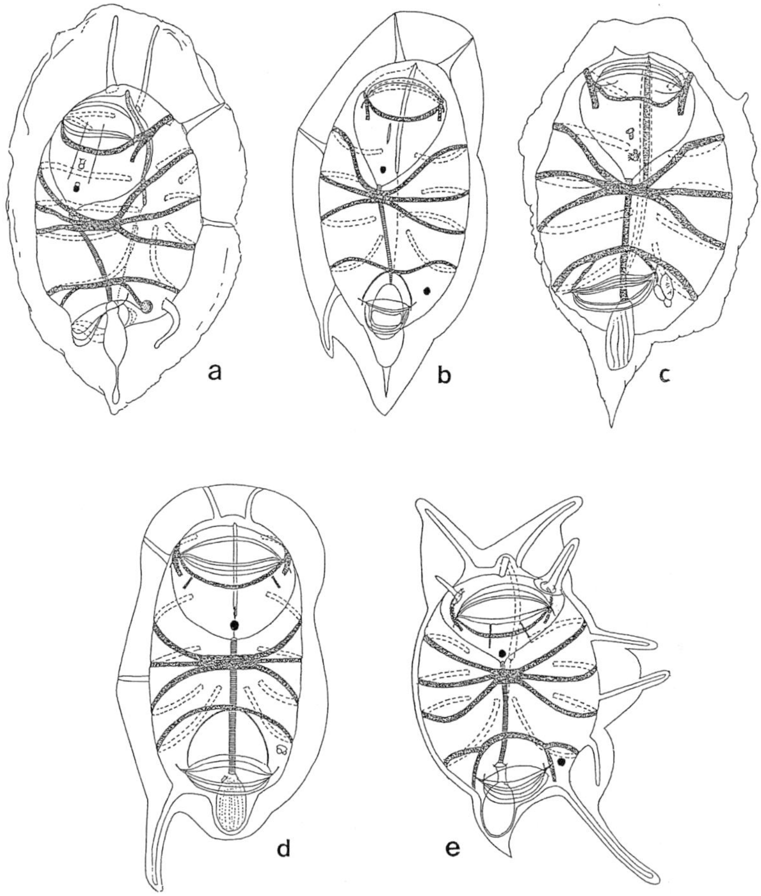

Fig. 17.21. Salpinae, blastozooids of the genus *Thalia*. (a) *Thalia longicauda*. After Godeaux (1967). (b) *Thalia democratica*. (c) *Thalia rhomboides*. (d) *Thalia cicar*. (e) *Thalia orientalis*.

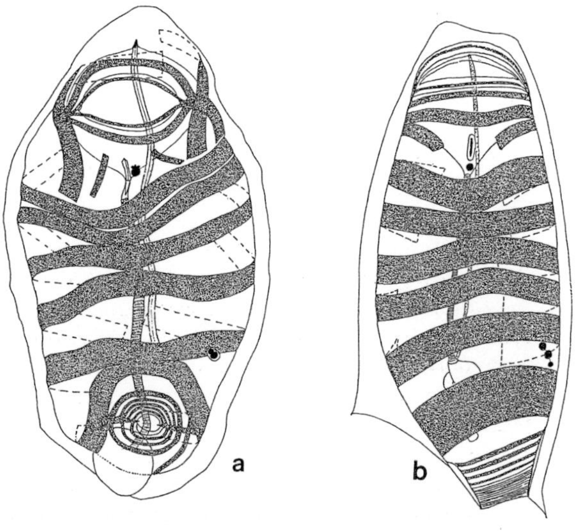

Fig. 17.22. Salpinae, blastozooids. (a) *Ihlea punctata* (*asymmetrica*). (b) *Metcalfina hexagona*.

Muscle I with more than 12 fibres, more than 120 fibres overall, same general appearance as above .... *Ritteriella picteti*

42. Muscles I–IV fused mid-dorsally, M IV and M V joining or approaching on both sides 43
Groups I–II and III–IV closely approaching mid-dorsally, M IV and V not approaching laterally .... *Salpa maxima*

43. Muscles IV and V joining on both sides .... *Salpa fusiformis*
Muscles IV and V simply approaching on both sides .... *Salpa aspera*

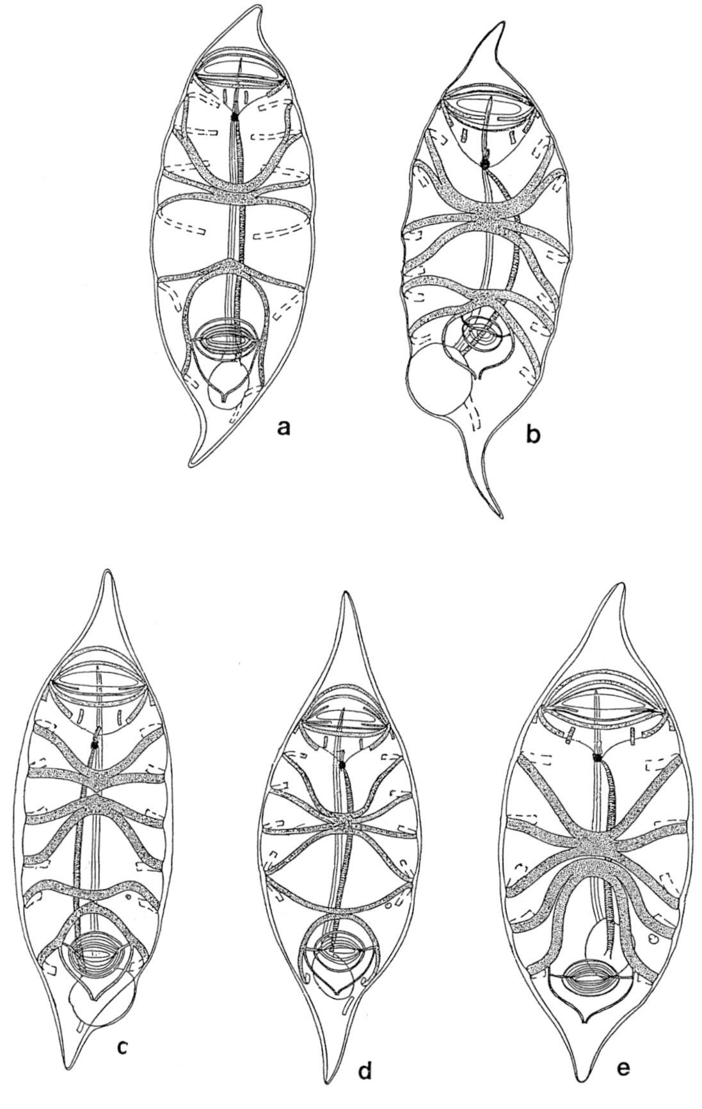

Fig. 17.23. Salpinae, blastozooids of the genera *Ritteriella* and *Salpa*. (a) *Ritteriella amboinensis*. (b) *Ritteriella retracta*. (c) *Salpa maxima*. (d) *Salpa fusiformis*. (e) *Salpa cylindrica*. After Meurice (1974).

EIGHTEEN

The classification of Appendicularia

R. Fenaux

18.1. Foreword

The Classification of Appendicularia (Tunicata). History and Current State (Fenaux, 1993) may be consulted for a more extended and detailed account of the classification, and its historical development, than is possible in this chapter.

18.2. Principles and problems

The keys for determination given here follow the principles explained by Lohmann and Bückmann (1926) and revised by Lohmann in 1933–1934, with additions and modifications. For the Oikopleuridae they are primarily based on the structure of the digestive tract, but for the Fritillariidae on very diverse criteria. For these reasons, and particularly for the Fritillariidae, numerous characters must be used to determine the species.

Because they are particularly fragile, appendicularians are frequently determined with difficulty. Problems with collection and preservation techniques are responsible for much of this difficulty (Fenaux, 1976a), but even with living or perfectly preserved specimens, there may be some doubt of the exact determination, for several main reasons.

(1) Some taxa have been determined from a few damaged or badly preserved specimens and the original description of different parts of the animal is incorrect.

(2) Various key characters vary with age, for example, the ciliated spiracular rings in the genus *Pelagopleura* are thought to be circular in young specimens and oval in older specimens.

(3) There may be variations in the shape of certain organs, or the animals themselves, creating an almost continuous spectrum of characters from one species to another or indeed within the same species. Tokioka (1956b) created a group 'the *Fritillaria haplostoma* complex', including different species. Again, about 10 varieties of *F. borealis* have been described by several authors, but the variations in the shape of the ovary and testis and of features of the tail seem to be continuous and I retain therefore, following Tokioka (1960), only three forms related to different environmental conditions.

(4) In all cases (apart from the few species which have been cultured in the laboratory), very young immature specimens are difficult (and sometimes impossible) to determine

In addition to these difficulties, the following should be noted.

(1) Essenberg (1926) described 30 new species which were all later placed in synonymy by Lohmann or Tokioka.

(2) Several new Antarctic and Subantarctic species (*Oikopleura falklandica, O. frigida, O. magellanica, O. meteori, O. mirabilis, O. oblonga, O. rigata, O. simplex* and *Fritillaria nana*) are mentioned in Lohmann and Hentschel (1939, p. 219) but have never have been described or figured, thus they have to be regarded as *nomina nuda*. The material was destroyed during World War II.

(3) Two genera cannot be accepted without reservations. *Chunopleura* Lohmann, 1914 is only known from a few specimens, all collected in very poor condition. *Sinisteroffia* (Tokioka, 1957) may simply be an artifact due to asymmetric penetration of the fixative.

(4) Two new appendicularians from mesopelagic waters were placed by Fenaux and Youngbluth in 1991 in the new genera *Inopinata*, with the species *inflata*, and *Mesopelagica*, with the species *caudaornata*. This present revision of the classification of the Oikopleuridae

induces me to consider the specimens as only new species of the genus *Oikopleura*.

(5) *Bathochordaeus* specimens, collected by manned submersibles and perfectly preserved, showed the validity of the two species *B.charon* and *B.stygius* (Fenaux, to be published).

(6) This classification includes a new division of the subfamily Oikopleurinae into two supergenera (Fenaux, 1993a).

18.3. Determination keys to generic level (see below)

18.3.1. *Definition of Appendicularia, Appendiculariae, Copelata or Perennichordata*

Free swimming pelagic tunicates, composed of a trunk and a tail; the tail with a persistent notochord during the adult stage. Part of the trunk is covered by a secretory epithelium which secretes a gelatinous house enclosing the animal and containing highly specialized devices for filter feeding.

18.3.2. *Family Oikopleuridae Lohmann, 1915*

Trunk ovoid. Endostyle straight. A pair of branchial passages with ciliated rings connects the pharyngeal cavity with the exterior, each ending in a spiracular aperture. Spiracles situated in the rectum region. Stomach wall composed of very numerous small cells with a row of a few large cells. Fol's oikoplasts are present on both parts of the anterodorsal oikoplastic layer.

Subfamily BATHOCHORDAEINAE Lohmann, 1915 (New diagnosis *in* Fenaux and Youngbluth, 1990).

Trunk oval, dorso-ventrally compressed. Mouth lying antero-dorsally. Endostyle short and squat. Buccal glands absent. Large elongate branchial passage, the ciliary rings practically level with the external epithelium. The peripharyngeal bands do not join dorso-medially in the pharynx, but end separately on either side of the oesophageal opening. Fol's fibroblasts are both preceded and followed by a crescent of giant cells (colloplasts). The anteroventral oikoplastic region is reduced to a small portion ending just after the endostyle. The tail widens progressively from its base to its distal extremity, which ends in a large slightly rounded thin plane.

Trunk flattened, distinctly wider than high, trilobate anteriorly, the anterior region larger than the posterior. Mouth provided with an anterior bulge and a small posterior lip. Pharyngeal cavity straight. The tubular oesophagus narrows before passing to the antero-dorsal region of the stomach, slightly to the left of the median axis. Stomach and intestine large and continu-

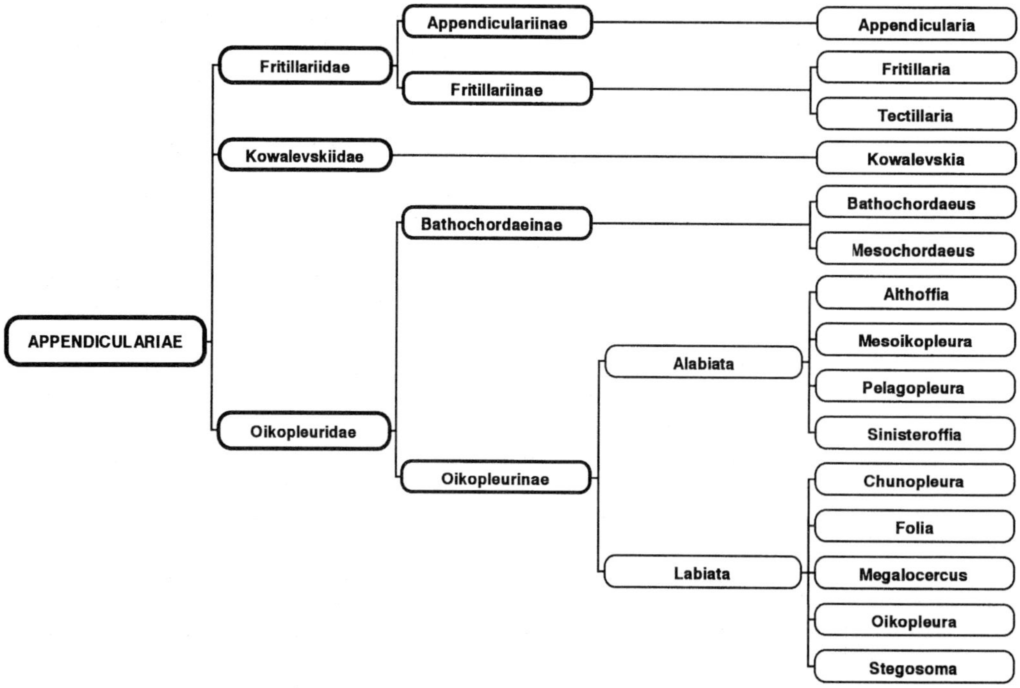

Fig. 18.1. Dichotomous appendicularian classification.

ous, making a horizontal crescent filling almost the entire width of the trunk. The rectum curves to the left towards the middle of the pharyngeal floor in a gentle slope. The anus lies behind the spiracles........................Genus *Bathochordaeus* Chun, 1900

Trunk slightly wider than high. Mouth surrounded by four lips, the most anterior the longest. Large pharyngeal cavity. Spiracles very elongated, extending about half the length of the trunk. Oesophagus short, funnel-like, with a large opening. Stomach ovoid, vertical, slightly turned to the left. Intestine short, on the right of the stomach, continuing in a rectum returning towards the middle of the pharyngeal floor in an almost vertical aspect. Anus between the hinder end of the spiracles. In the right inner third quarter of the tail there are numerous, small subchordal cells..Genus *Mesochordaeus* Fenaux and Youngbluth, 1990.

Subfamily OIKOPLEURINAE Lohmann, 1896
Mouth situated forward, with or without lower lip. Endostyle elongate. Buccal glands present or absent. Peripharyngeal bands join dorsomedially in the pharynx. Branchial passages with complete circular or oval ciliated rings (except genus *Megalocercus*). Fol's fibroblasts are only preceded by a row of giant cells (colloblasts). Antero-dorsal oikoplastic layer extends to the anus.

A. Mouth with lower lip
Supergenus Labiata Fenaux, 1993
(a) Digestive tract compactly coiled into a nucleus. Stomach, which covers the terminal intestine, is laterally widened in a right and a left lobe. The dorsal oesophagus ends in the left lobe of the stomach..........................Genus *Oikopleura* Mertens, 1830
(b) Digestive tract incompletely developed, owing to the absence of one or both of the lateral lobes.

(1) Left lobe of the stomach well developed but not the right. Ciliary rings of the branchial passages complete. Buccal glands and subchordal cells present....................Genus *Folia* Lohmann, 1892
(2) Left lobe of the stomach much reduced. The middle part of the stomach lying on the median part of the intestine. The cilia of the branchial passages are restricted to a small section in the anterior region. No buccal glands or subchordal cells......Genus *Megalocercus* Chun, 1887

(c) Digestive tract forms a loop broadly open to the front. The loop stands vertically. Oesophagus and anus are on the longitudinal axis of the trunk. Behind the oesophageal aperture in the loop, a stomachal pouch corresponding to the left lobe (containing the row of giant cells) emerges by a narrow duct.

(1) Pouch large. Buccal glands and subchordal cells present.............................Genus *Stegosoma* Chun, 1887
(2) Pouch slender and elongate. Buccal glands and subchordal cells absent.. ....Genus *Chunopleura* Lohmann, 1914

B. Mouth without lip, surrounded by a wreath of sensory cilia.
Stomach without pouch-shaped process. The middle part of the loop is developed into a stomach. No buccal glands. Tail with amphichordal cells which differ in arrangement not only with the species but also in specimens of the same species.
Supergenus Alabiata Fenaux, 1993
(a) Distance between mouth aperture and anterior extremity of the endostyle smaller than distance between anus and posterior extremity of the endostyle. Ciliated rings of the branchial passages oval or elongated.

(1) Stomach located in the sagittal axis of the trunk, with cardia at the top and pylorus at the bottom....................Genus *Pelagopleura* [Lohmann, in] Lohmann and Bückmann, 1926
(2) Stomach located beside the left wall of the trunk..........................Genus *Sinisteroffia* Tokioka, 1957
(3) Stomach located horizontally, with cardia on the right and the pylorus on the left, both in the anterior part.............................Genus *Althoffia* Lohmann, 1892

(b) Distance between mouth aperture and anterior extremity of the endostyle larger than distance between anus and posterior extremity of the endostyle. Stomach more or less oblique in comparison with sagittal plan of body. Ciliated rings of the branchial passages circular..............Genus *Mesoikopleura* Fenaux, 1993

18.3.3. *Family Fritillariidae Lohmann, 1915*

Trunk dorso-ventrally compressed or spindle-shaped. Endostyle curved upwards. The ciliated spiracular rings connect the pharyngeal cavity with the exterior directly without a tubular branchial passage. Spiracles situated distinctly in the anterior part of the pharyngeal cavity. Stomach wall composed of a few large cells. No Fol's fibroblasts are visible in the dorsal oikoplastic layer. Ventral oikoplastic layer reduced to a small anterior area.

Subfamily APPENDICULARIINAE Seeliger, 1895
Mouth without lip, but surrounded by cilia. Endostyle short, slightly curved, widened at the anterior extremity. Dorsal oikoplastic epithelium extended back to cover the digestive nucleus. The tail becomes progressively narrower at the proximal part and is notched at the distal end.

One genus, characteristics as for the subfamily ....Genus *Appendicularia* Fol, 1874

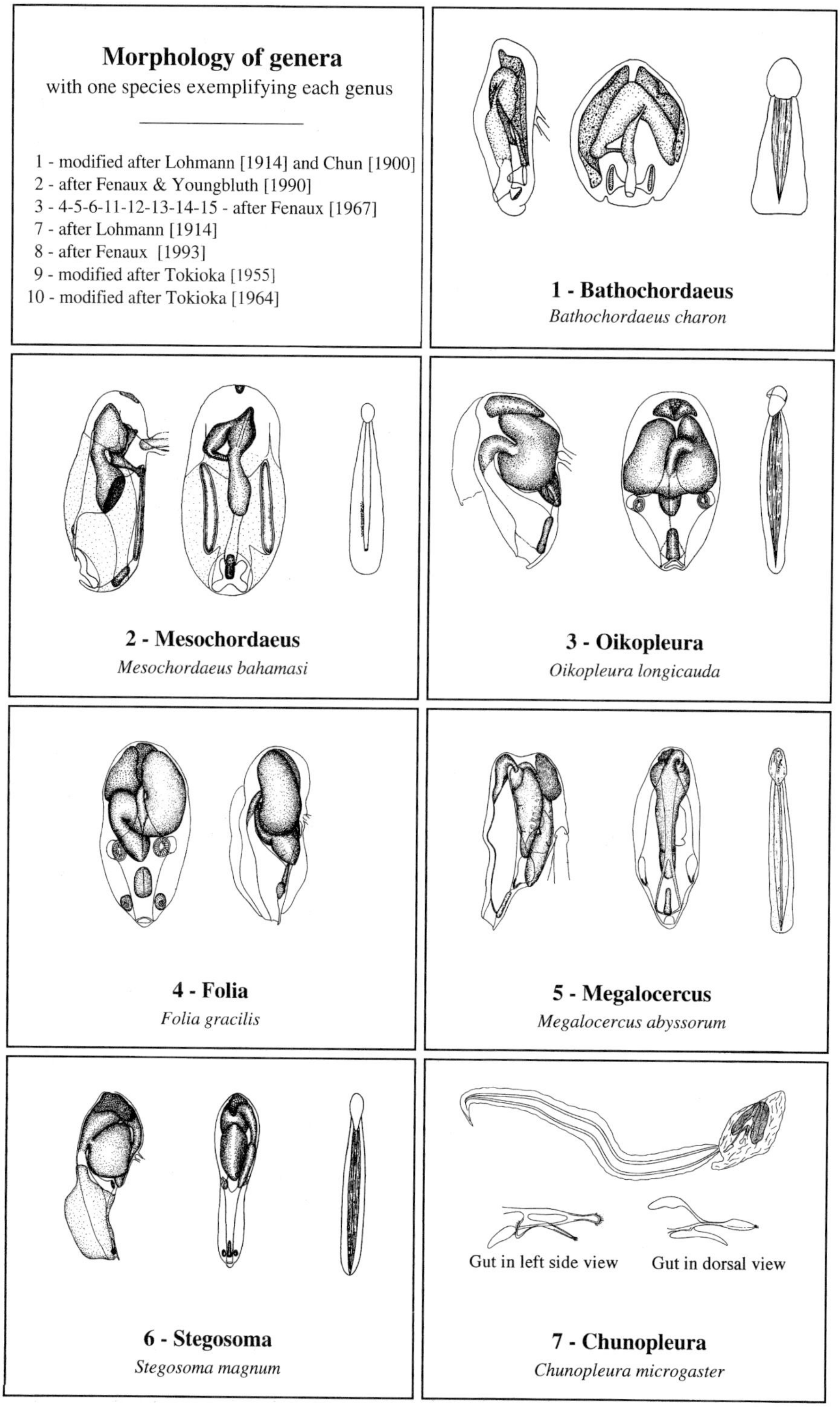

Fig. 18.2. The morphology of the different appendicularian genera, with a single species exemplifying each genus. (1) Modified after Lohmann (1914) and Chun (1900); (2) after Fenaux and Youngbluth (1990); (3–6 and 11–15) after Fenaux (1967); (7) after Lohmann (1914); (8) after Fenaux (1993); (9) modified after Tokioka (1955); (10) modified after Tokioka (1964).

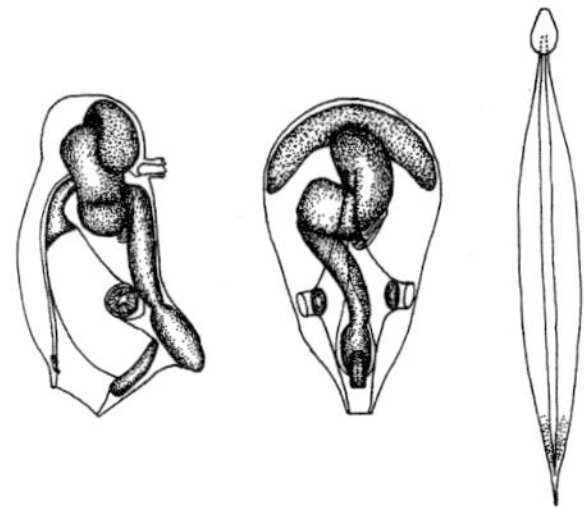

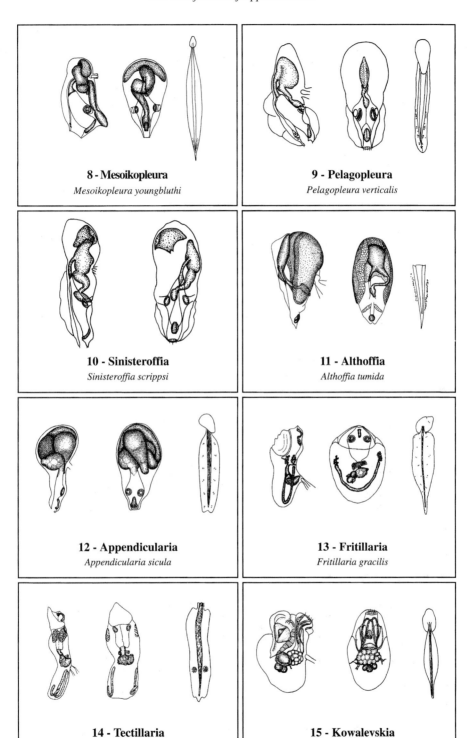

8 - Mesoikopleura
Mesoikopleura youngbluthi

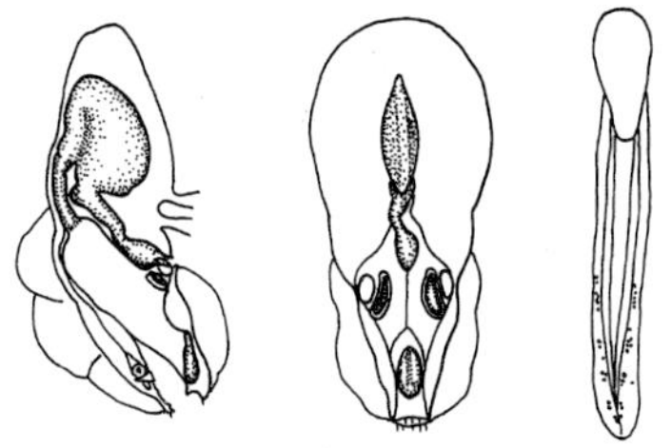

9 - Pelagopleura
Pelagopleura verticalis

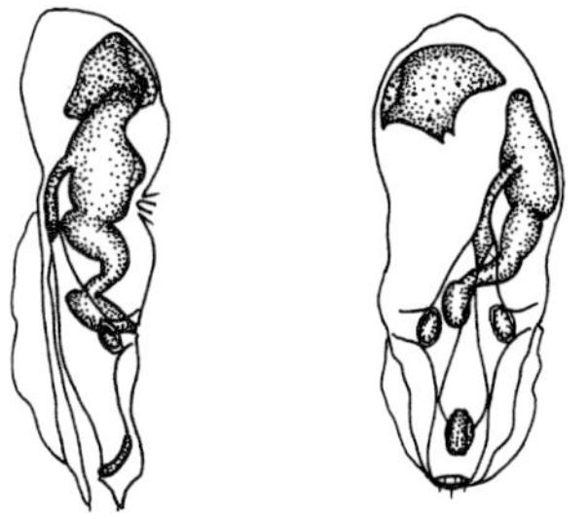

10 - Sinisteroffia
Sinisteroffia scrippsi

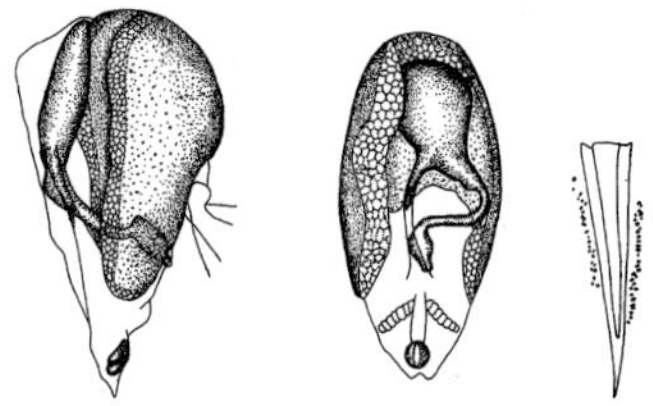

11 - Althoffia
Althoffia tumida

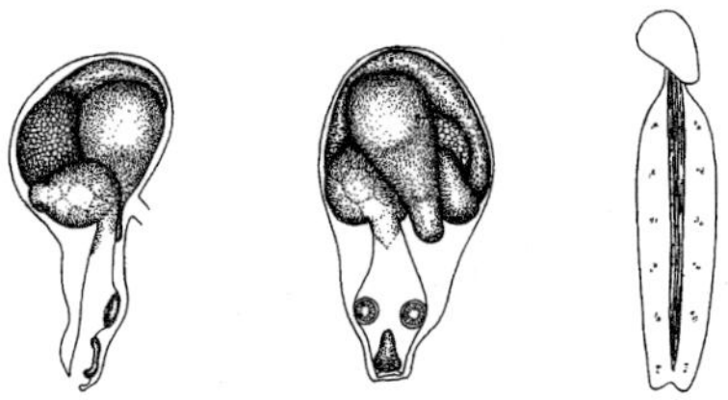

12 - Appendicularia
Appendicularia sicula

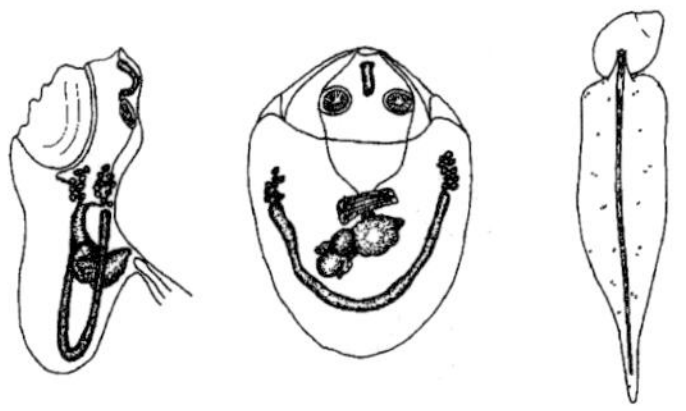

13 - Fritillaria
Fritillaria gracilis

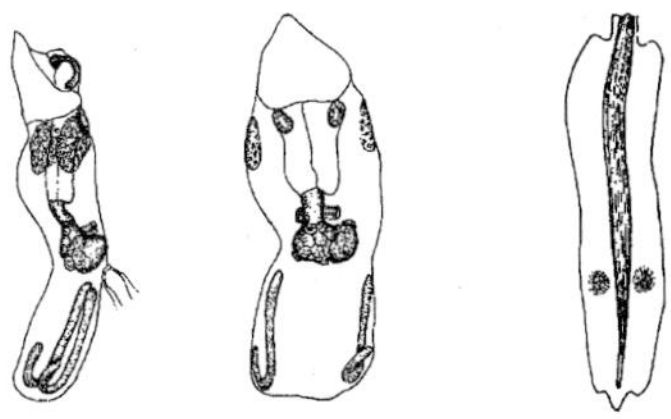

14 - Tectillaria
Tectillaria fertilis

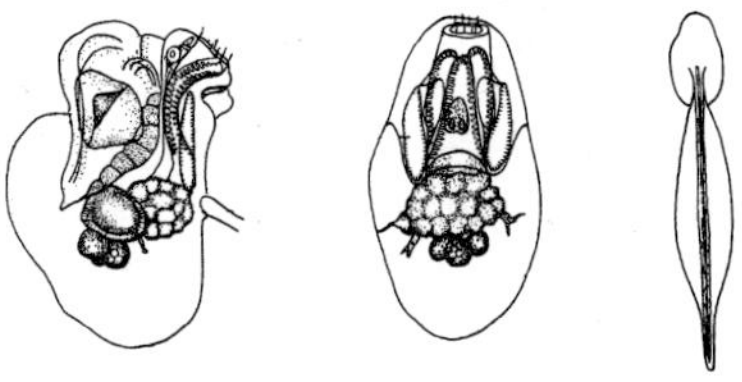

15 - Kowalevskia
Kowalevskia oceanica

Subfamily Fritillariinae Seeliger, 1895
Snout-shaped mouth, often with lobate lips. Anterior part of the trunk wide and flat, with a wide oikoplastic cavity. The tail narrows abruptly at the proximal part.

A. Two ovaries located in the pharyngeal cavity, in front of the digestive nucleus. The two testes are completely or partially situated behind the gut. Ciliated rings of the spiracles small and circular. Dorsal oikoplastic epithelium forms a broad shield, roofing over the mouth aperture. Pair of subterminal notches near the acuminate distal end of the fin. One round group of small condensed amphichordal cells at the last third on each side ....Genus *Tectillaria* [Lohmann, in] Lohmann and Bückmann, 1926
B. The gonads are all located behind the gut. Ciliated rings of the spiracles of various forms. Dorsal oikoplastic epithelium covering the anterior and middle areas of the pharyngeal cavity, the posterior border curving upwards....................................Genus *Fritillaria* Fol, 1872

18.3.4. *Family Kowalevskiidae (Koweleskidae Lahille, 1888; emend. Lahille, 1890)*

Trunk short, shaped like the bowl of a pipe. No endostyle, no heart. The pharyngeal cavity divided longitudinally into three compartments by two rows of ciliated combs. The ciliated rings and spiracular openings wide and oval. Stomach wall composed of a few large cells. Oikoplastic layer characterized by a spectacularly large spherical button-shaped cell dorsally. Tail resembles leaf of a willow (*Salix*).
One genus, characteristics as for the family...............
..............Genus *Kowalevskia* Fol, 1872; emend. Fol, 1874

18.4. Keys from genera to species

Note: Capital letter(s) in bold face, printed after the name, show the frequency of capture in the regions concerned (see Tables 15.1–15.2): **CC**, very common; **C**, common; **R**, rare; **RR**, very rare (one or a few specimens known only). **M**, mesopelagic. This naturally only affords a very general indication of abundance, since the specimens caught vary with season and with sampling methods, etc.

18.4.1. *Family Oikopleuridae Lohmann, 1915*

Subfamily Bathochordaeinae Lohmann, 1915 (new diagnosis in Fenaux and Youngbluth, 1990)
Genus *Bathochordaeus* Chun, 1900
A. The pharyngeal cavity is tubular, diameter similar to that of the oesophagus. The branchial passages start from this cavity by a narrow aperture and enlarge as a funnel-shaped duct to reach a wide oval ciliary ring.............................*B.charon* Chun, 1900 [**R M**]
B. The pharyngeal cavity widens just after the endostyle up to the oesophagus, by a large communication with the branchial passage. The spiracles are almost as wide as the elongate oval ciliary ring ....*B.stygius* Garstang, 1937 [**R**]
Genus *Mesochordaeus* Fenaux and Youngbluth, 1990
One species, characteristics as for the genus.................
...........*M.bahamasi* Fenaux and Youngbluth, 1990 [**RR M**]

Subfamily Oikopleurinae Lohmann, 1896
Supergenus Labiata Fenaux, 1993
Genus *Oikopleura* Mertens, 1830
I. Buccal glands and subchordal cells absent....................
..................................Subgenus *Coecaria* Lohmann, 1933
A. Small caecum directed backwards.
Ripe gonads largely extended backwards, giving the trunk a characteristic shape ....
....*O.gracilis* Lohmann, 1896 [**C**]
B. Finger-shaped caecum directed upwards.
(a) Close to the cardial region. Trunk with a velum strongly developed and extended over the pharyngeal cavity, often torn or destroyed ....
....................................*O.longicauda* (Vogt, 1854) [**CC**]
(b) Separated from the cardial region by a wide rounded bight. Mouth opens upwards.........................
..................................*O.intermedia* Lohmann, 1896 [**C**]
C. Caecum developed upwards and backwards.
(a) In side view, outlines curved backwards.................
....*O.fusiformis* Fol, 1872 [**CC**]
(b) In side view, outlines rectilinear. Mouth opens upwards..........*O.fusiformis cornutogastra* Aida, 1907 [**C**]
II. Buccal glands and subchordal cells present................
.................................Subgenus *Vexillaria* Lohmann, 1933
A. Numerous subchordal cells.
(a) Small, lying irregularly in an expanded area; on immature specimens the cardia behind the left lobe of the stomach ......*O.vanhoeffeni* Lohmann, 1896 [**CC**]
(b) Vesicular, arranged more or less in two dorso-ventral rows.
(1) Cells separated in young, and compressed in mature animals, much expanded along the tail. Left lobe of the stomach with a small caecum directed backwards ....
...............................*O.albicans* (Leuckart, 1854) [**C**]
(2) Vesicular along the distal third of the tail. Small caecum only dorsally...
.......................*O.labradoriensis* Lohmann, 1892 [**CC**]
(3) Vesicular, disposed in alternate rows along the distal half of the tail. Digestive part of the trunk much inflated protruding forwards as far as the posterior part of the endostyle................................
.....*O.inflata* (Fenaux and Youngbluth, 1991) [**R M**]

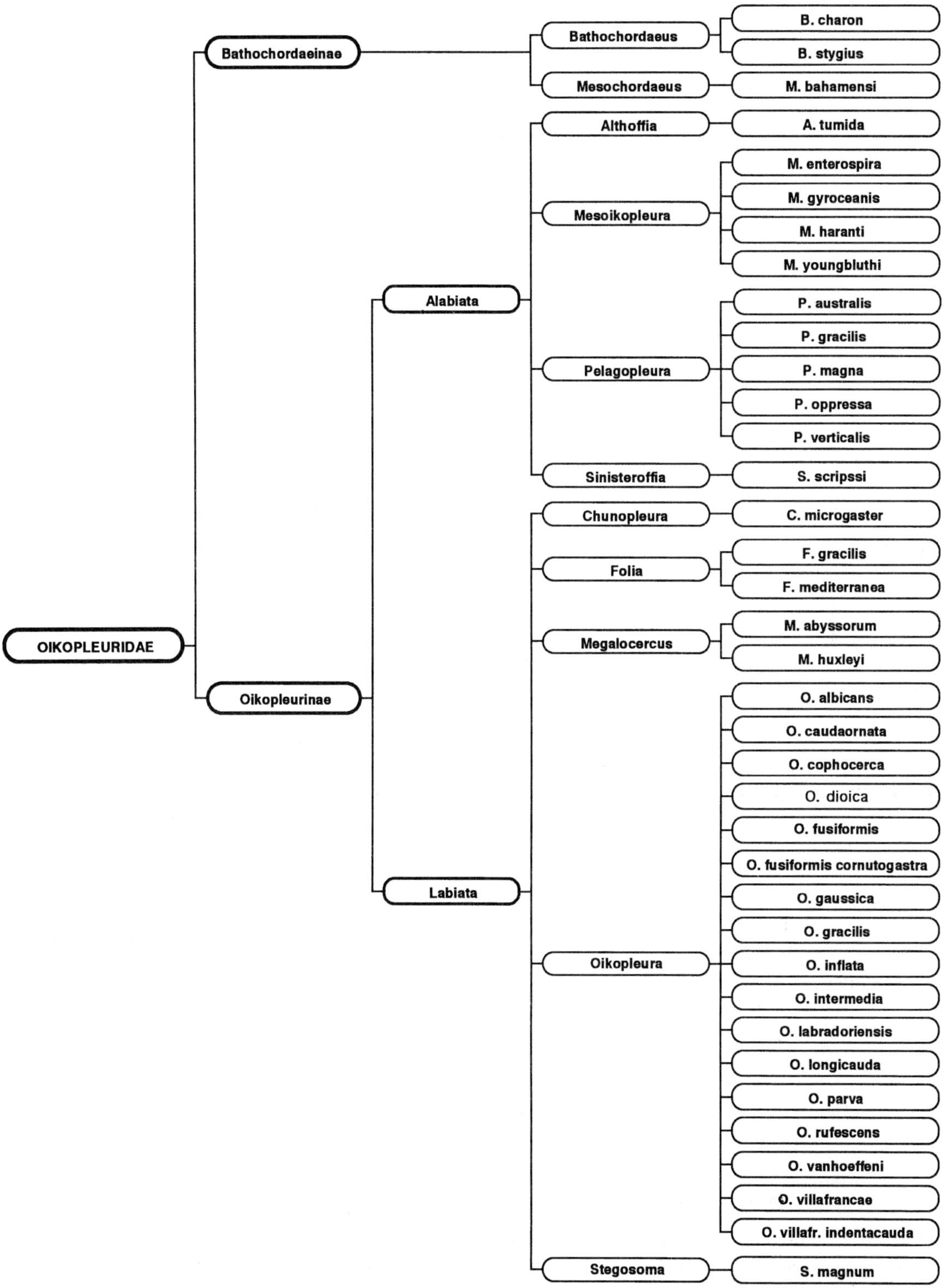
OIKOPLEURIDAE
Bathochordaeinae
Bathochordaeus
B. charon
B. stygius
Mesochordaeus
M. bahamensi
Oikopleurinae
Alabiata
Althoffia
A. tumida
Mesoikopleura
M. enterospira
M. gyroceanis
M. haranti
M. youngbluthi
Pelagopleura
P. australis
P. gracilis
P. magna
P. oppressa
P. verticalis
Sinisteroffia
S. scripssi
Labiata
Chunopleura
C. microgaster
Folia
F. gracilis
F. mediterranea
Megalocercus
M. abyssorum
M. huxleyi
Oikopleura
O. albicans
O. caudaornata
O. cophocerca
O. dioica
O. fusiformis
O. fusiformis cornutogastra
O. gaussica
O. gracilis
O. inflata
O. intermedia
O. labradoriensis
O. longicauda
O. parva
O. rufescens
O. vanhoeffeni
O. villafrancae
O. villafr. indentacauda
Stegosoma
S. magnum

(c) Arranged in a single row.

(1) Large vesicular subchordal cells along the distal half of the tail. Lanceolate extension of fin on each side of the tail at the point of maximum width .... *O.caudaornata* ....(Fenaux and Youngbluth, 1991) **[R M]**

(2) About 10 amoeboid-shaped cells in side view in the last quarter of the tail. Lobes of the stomach strongly compressed together ....*O.villafrancae* Fenaux, 1992 **[C M]**

(3) Same as below with right surface of the distal end of the tail indented and pleats just below the beginning of the wider part...................................... ......*O.villafrancae indentacauda* Fenaux, 1992 **[R M]**

B. Subchordal cells few, arranged in a single row.

(a) Without postcardial caecum on the left stomach lobe.

(1) One spindle-shaped cell. Rectum nearly vertical, anus under the anterior wall of the stomach..... ....*O.rufescens* Fol, 1872 **[C]**

(2) Two spindle-shaped cells. Posterior wall of the left lobe of the stomach nearly vertical. Unisexual.. ....*O.dioica* Fol, 1872 **[CC]**

(3) Six to 14 subchordal cells oval or capsular. Left lobe of the stomach rounded in outline. Tokioka (1964) demonstrated rather clearly that the four species described from the Antarctic (*O.gaussica* Lohmann, 1905; *O.valdiviae* Lohmann, 1905; *O.drygalskii* Lohmann, 1926 and *O.weddelli* Lohmann, 1928) are specimens of *O.gaussica* in different conditions of preservation or maturity ....*O.gaussica* Lohmann, 1905 **[R]**

(b) With small post-cardial caecum on the left stomach lobe.

(1) Four spindle-shaped cells................................... ....*O.parva* Lohmann, 1896 **[R]**

(2) Between five and eight small globular cells, disposed in two distinct groups..................................... ....*O.cophocerca* (Gegenbaur, 1855) **[CC]**

Genus *Folia* Lohmann, 1892

A. Stomach vertical, oesophagus enters dorsally. Small buccal glands. More than 10 small vesicular subchordal cells ....*F.gracilis* Lohmann, 1892 **[R]**

B. Stomach horizontal, oesophagus enters on the right. Large buccal glands. Five spindle shaped subchordal cells....................*F.mediterranea* (Lohmann, 1899b) **[RR]**

Genus *Megalocercus* Chun, 1887

A. Left lobe tubular, reaching forwards, covering the greater part of the intestine. Posterior end of the endostyle not close to the anus.. ....*M.abyssorum* Chun, 1887 **[R]**

B. Left lobe wide, quadrangular, only covering a small part of the intestine. Posterior end of the endostyle close to the anus ....*M.huxleyi* ([Ritter, in] Ritter & Byxbee, 1905) **[C]**

Genus *Stegosoma* Chun, 1887

Only one species, characteristics as for the genus........... ....*S.magnum* (Langerhans, 1880) **[R]**

Genus *Chunopleura* Lohmann, 1914

Only one species, characteristics as for the genus........... ....*C.microgaster* Lohmann, 1914 **[RR]**

Supergenus Alabiata Fenaux, 1993

Genus *Pelagopleura* [Lohmann, in] Lohmann and Bückmann, 1926

A. Trunk vertically curved. Endostyle short. Branchial passages with rounded ciliated rings (immature specimens). Stomach with a caecum posteroventrally curved. Distal extremity of the tail notably acuminate................. ....*P.australis* (Bückmann, 1924) **[R M]**

B. Trunk spindle-shaped, elongated by expansion of the gonads backwards in mature animals. Endostyle short. Branchial passages with oval ciliated rings. Stomach triangular in side view, expanded backwards. Triangular ovaries on each side of the dorsal part of the testis......... ....*P.gracilis* (Lohmann, 1914) **[R M]**

C. Endostyle elongate, thin. Branchial passages with elongate ciliary rings. Stomach roughly oval and elongate. Intestine thin, very long. Distal extremity of the tail acuminate ....*P.magna* Tokioka, 1964 **[R M]** *non P.magna* ....[Lohmann, in] Lohmann and Bückmann, 1926.

D. Trunk dorso-ventrally compressed. Endostyle elongate. Branchial passages with small oval ciliated rings. Stomach spherical. Distal extremity of the tail sharply acuminate..................*P.oppressa* (Lohmann, 1914) **[R M]**

E. Endostyle short and wide. Branchial passages with oval ciliated rings. Stomach roughly rounded in outline. Ovary and testes are both paired. Tail with a smoothly acuminate distal extremity... ....*P.verticalis* (Lohmann, 1914) **[R M]**

Genus *Sinisteroffia* Tokioka, 1957

Only one species, characteristics as for the genus........... ....*S.scrippsi* Tokioka, 1957 **[RR]**

Genus *Althoffia* Lohmann, 1892

Only one species, characteristics as for the genus........... ....*A.tumida* Lohmann, 1892 **[RR]**

Genus *Mesoikopleura* Fenaux, 1993

A. Endostyle elongate. Stomach notably oblique. Intestine spiral ....*M.enterospira* Fenaux, 1993 **[RR M]**

B. Stomach communicates widely with the proximal intestine which has a bump at the upper part and a knob below. Tail with visible shoulders and (4+3) fusiform amphichordal cells.. ....*M.gyroceanis* Fenaux, 1993 **[RR M]**

C. Stomach irregularly bag-shaped, almost in sagittal axis. Intestine almost horizontal. Gonads horseshoe-shaped....................*M.haranti* (Vernières, 1934) **[RR M]**

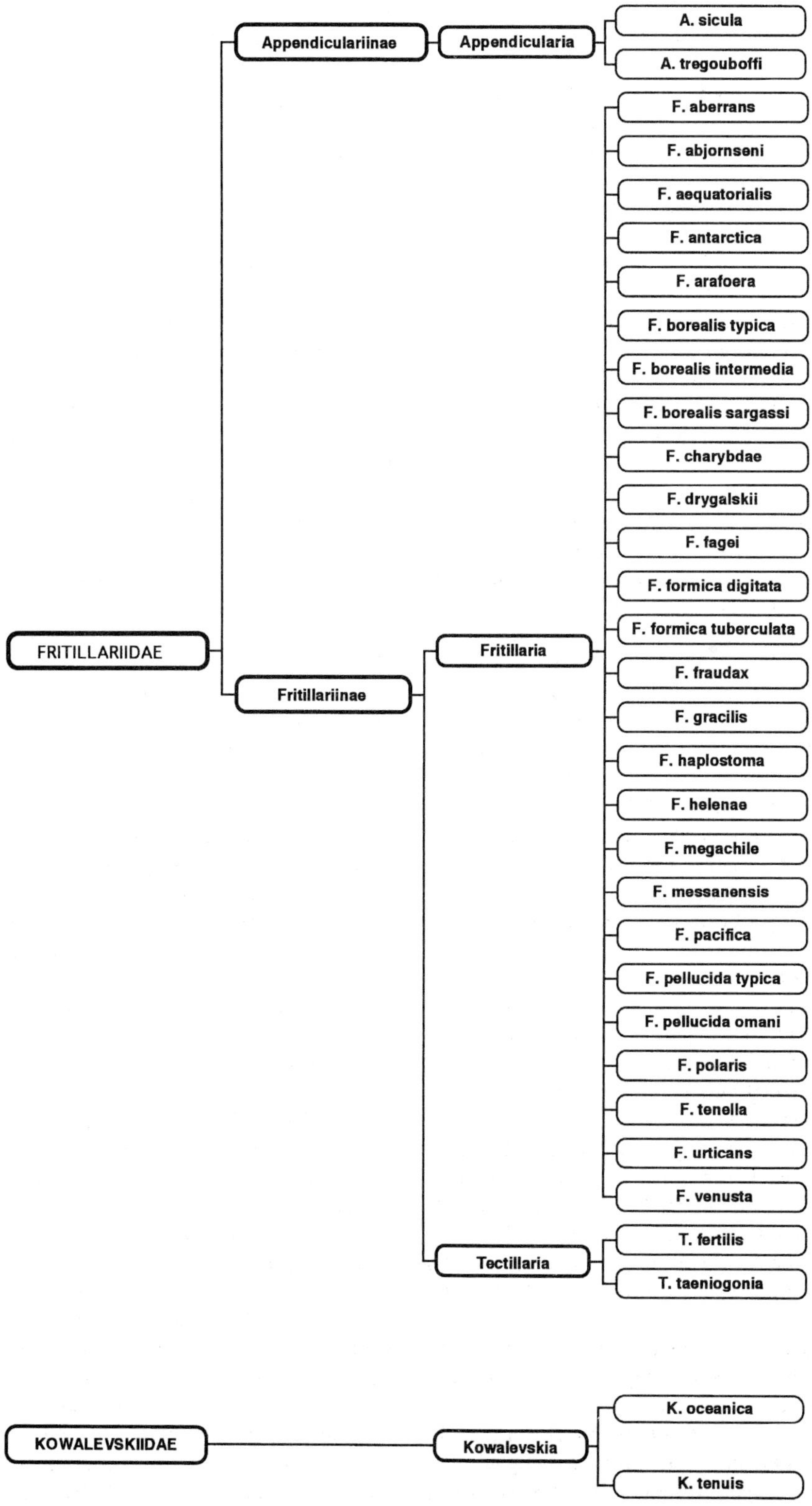
Appendiculariinae
Appendicularia
A. sicula
A. tregouboffi
F. aberrans
F. abjornseni
F. aequatorialis
F. antarctica
F. arafoera
F. borealis typica
F. borealis intermedia
F. borealis sargassi
F. charybdae
F. drygalskii
F. fagei
F. formica digitata
F. formica tuberculata
FRITILLARIIDAE
Fritillaria
F. fraudax
Fritillariinae
F. gracilis
F. haplostoma
F. helenae
F. megachile
F. messanensis
F. pacifica
F. pellucida typica
F. pellucida omani
F. polaris
F. tenella
F. urticans
F. venusta
Tectillaria
T. fertilis
T. taeniogonia
K. oceanica
KOWALEVSKIIDAE
Kowalevskia
K. tenuis

D. Endostyle short. Intestine almost horizontal. Rectum extend partly out of the trunk. Tail very long, more than nine times the length of the trunk. In mature specimens the gonads may be as wide as the length of the trunk...... ....*M.youngbluthi* Fenaux, 1993 [**R M**]

18.4.2. *Family Fritillariidae Lohmann, 1915*

Subfamily APPENDICULARIINAE Seeliger, 1895
Genus *Appendicularia* Fol, 1874; *non Appendicularia* Chamisso, 1821 *nomen nudum*
A. Vast rectum, pear-shaped, occupying a big part of the digestive tract. No anus visible. Gonads located above the digestive tract. Tail moderately wide.......................... ....*A.sicula* Fol, 1874 [**C**]
B. Small rectum with anus visible. Gonads located behind the digestive tract. Tail with very narrow musculature.......................*A.tregouboffi* Fenaux, 1960 [**R**]

Subfamily FRITILLARIINAE Seeliger, 1895
Genus *Tectillaria* [Lohmann, in] Lohmann and Bückmann, 1926
A. Axis of the digestive tract transverse. Testes thick, cylindrical and curved upward posteriorly. Ovaries at the level of the spiracles.........*T.fertilis* (Lohmann, 1896) [**R**]
B. Axis of the digestive tract almost longitudinal. Testes thin, elongated in a loop from the posterior part of the trunk to the ovaries. Ovaries between the digestive nucleus and the spiracles.. ....*T.taeniogona* (Tokioka, 1957) [**RR**]
Genus *Fritillaria* Fol, 1872; *non Fritillaria* Quoy and Gaimard, 1833 *nomen nudum*
A. Ciliated rings of the branchial passages circular.
(a) Distal extremity of the tail acuminate.
(1) Trunk elongated and flat. Ciliated ring of the spiracles almost in contact on the ventral median line.........................*F.abjornseni* Lohmann, 1909 [**R**]
(2) Trunk elongated, bent upwards in the digestive nucleus region. Ciliated spiracular rings not in contact on the ventral median line.............*F.formica*
(2.1) Posterior edge of the dorsal oikoplastic layer located much before the stomach. Median lobe of the upper lip elongated and provided anteriorly with two bundles of cilia. Testis oval in mature specimens.. ......*F.formica tuberculata* [Lohmann, in] Lohmann ....and Bückmann, 1926 [**C**]
(2.2) Posterior edge of the dorsal oikoplastic layer extending to the stomach. Small median lobe of the upper lip short and provided anteriorly with two finger-shaped bulges. Testis triangular with rounded angles in mature specimens....... ....*F.formica digitata* [Lohmann, in] Lohmann and ....Bückmann, 1926 [**C**]
(3) Trunk spindle-shaped, elongated and slightly bent. Digestive tract longitudinal. Large upper lip. Ovary spherical followed by an elongate cylindrical testis ...*F.haplostoma* Fol, 1872; emend. Fol, 1874 [**CC**]
(4) Trunk quite wide, a little bent. The posterior edge of the dorsal oikoplastic epithelium reaches the anterior wall of the stomach. Ovary spherical, testis elongated. Proximal end of the tail has sloping shoulders instead of angular shoulders as in other fritillarians..........*F.arafoera* Tokioka, 1956 [**R**]
(5) Trunk oval, compressed dorso-ventrally. One testis c-shaped opening forwards, parallel to the posterior and lateral walls of the trunk; one ovary mass ahead of each extremity. Distal part of the tail lanceolate, very broad musculature.......................... ....*F.gracilis* Lohmann, 1896 [**R**]
(6) Trunk a little elongate, rounded at the posterior extremity. Gonads curiously described with a ring-shaped testis preceding a spherical ovary. Tail similar to the tail of *F.gracilis*, but the acuminate distal part is very short. Species only found in the Kara Sea in 1930 and 1931, never subsequently..............*F.polaris* Bernstein, 1934 [**RR**]
(b) Distal extremity of the tail notched or truncated.
(1) Without amphichordal glands. The two postero-lateral regions of the trunk with one cuticular extension, generally persistent only in young specimens.
(1.1) Elongated, compressed dorso-ventrally, more or less laterally narrowed at the level of the stomach and enlarged in the posterior part. Mouth with prominent upper lip. One ovary and one testis..*F.borealis*
(1.1a) Trunk little enlarged posteriorly. Ovary spherical, ahead of the oval elongate testis. Musculature of the tail very narrow, terminally acuminate ...*F.borealis typica* Lohmann, 1896 [**CC**]
(1.1b) Trunk roughly 8-shaped with posterior part well enlarged. Testis Y shaped. Spherical ovary located above the left branch. Musculature of the tail wide, especially in the first proximal half, ending abruptly.. ....*F.borealis sargassi* Lohmann, 1896 [**CC**]
(1.1c) Trunk similar to *sargassi*, and gonads to *typica*. Musculature of the tail with an intermediary width between the two other forms and blunt terminally... ....*F.borealis intermedia* Lohmann, 1905 [**CC**]
(1.2) The shape of the trunk as well as of the gonads is close to *F.borealis typica*, but the upper lip is more extended and there is a more or less

*Esnal *et al.* (1996) suggest that *F.borealis sargassi* be regarded as a separate species to *borealis* rather than as a form of *borealis.*

triangular glandular cell, behind each spiracle. The musculature of the tail is wide ....*F. messanensis* [Lohmann, in] Bückmann 1924 **[R]**

(2) With amphichordal glands. The posterior extremity of the trunk with two cuticular extensions, generally persistent to maturity. Glandular cells behind the endostyle and between the spiracles.

(2.1) Trunk spindle-shaped elongate, never bent. Digestive tract longitudinal. The cuticular extensions are small, located medially and often lacking in mature animals. Two pairs of amphichordal cells, contiguous and flat ....*F. megachile* Fol, 1872 **[C]**

(2.2) Trunk compressed dorso-ventrally. Testis asymmetrical, expanded forwards to the right. Ovary spherical on the left. The cuticular expansions are conical and well separated. Tail with wide musculature and two pairs of non contiguous amphichordal glandular cells with visible exit canals...*F. pellucida*

(2.2a) Trunk almost rectangular. Digestive nucleus small. Testis r-shaped, massive. Cuticular extensions rather long ....*F. pellucida typica* Busch, 1851 **[CC]**

(2.2b) Trunk oval. Digestive nucleus small. Testis Y-shaped. Cuticular expansions short ....*F. pellucida omani* Fenaux, 1967 **[C]**

(2.3) Trunk elongated, compressed dorso-ventrally, almost rectangular. Gonads symmetrical, anterior spherical ovary followed by the elongated mass of the testis. Two triangular cuticular expansions. Tail with narrow musculature and two urn-like multicellular amphichordal glands open distally..*F. tenella* Lohmann, 1896 **[C]**

(2.4) Trunk wide, dorso-ventrally compressed, almost rectangular, laterally narrowed at the level of the digestive nucleus. Gonads globular in young specimens; in mature, testis trapezoidal with horn-shaped lateral lobes. Ovary developed along the testis. In the main indentation of the tail, there is a second very small indentation. A single long flat amphichordal cell on each side, sometimes difficult to see ....*F. venusta* Lohmann, 1896 **[C]**

B. Ciliated rings of the branchial passages oval, more or less elongate. Distal extremity of the tail acuminate.

(a) Tail with amphichordal cells.

Trunk laterally narrowed at the level of the stomach. Endostyle very curved. Ciliated ring of the spiracles large and elliptic. Testis forms a laminar mass curled up dorsally along the lateral walls. The anterior margin is concave. The two string-like ovaries are located along the anterior margin and become united in mature animals. Wide tail, distal extremity with two lateral notches. A group of three flattened amphichordal cells of different size stand in an oblique line on each side of the distal half of the tail ....*F. pacifica* Tokioka, 1958 **[R]**

(b) Tail without amphichordal cells.

(1) Ciliated rings of the branchial passages slot-shaped.

(1.1) One testis, one ovary.

(1.1a) Trunk elongate, with the anterior part widest. The distal end is acuminate. Mouth with projecting upper lip. Ovary more or less spherical in front of a long cylindrical testis. Tail characterized by the musculature which consists of 10–12 muscular bands separated from one another in adult specimens.....*F. aberrans* Lohmann, 1896 **[R]**

(1.1b) Shape of the trunk close to *F. aberrans*. Mouth with wide lips. Heart very peculiar with ventral side made up of a row of four big cells. Very small cuticular appendage on each side of the rounded distal part. Testis ovoid preceded by a short tube-shaped ovary. Tail with medium width musculature ....*F. fagei* Fenaux, 1961 **[R]**

(1.2) Two testes, two ovaries.

(1.2a) Trunk elongate, dorso-ventrally compressed, without lateral constriction. Rounded at the posterior extremity. Transversal digestive nucleus with, to the left, a glandular appendix almost as large as the stomach. Two plate-shaped testes, each surrounded by a moniliform ovary. Tail wide near the root, very acuminate at the distal extremity, narrow musculature ....*F. charybdae* [Lohmann, in] Lohmann and ....Bückmann, 1926 **[R]**

(1.2b) Trunk elongate, dorso-ventrally compressed, without lateral constriction, acuminate at the posterior extremity. Each plate-shaped testis vertical, preceded by the mass of an ovary. Surface of the trunk and the tail covered by small button-shaped cells, with the center in relief. On living animals the buttons can be made to extrude a filament (by addition of fresh water), reminiscent of the action of cnidarian nematocysts...............................*F. urticans* Fol, 1872 **[RR]**

(2) Ciliated rings of the branchial passages oval and elongated.

(2.1) Digestive nucleus transverse, endostyle strongly curved, trunk compressed dorso-ventrally.

(2.1a) Two globular processes in the pyloric region. Voluminous rectum almost vertical, so the anus is behind the stomach. Ribbon-shaped testis slightly curved forwards. Moniliform ovary surrounding the testis. Musculature of the tail narrow ....*F. fraudax* Lohmann, 1896 **[R]**

(2.1b) Characteristics close to *F.fraudax*, but digestive nucleus is larger and the musculature of the tail is wider. Possibly a cold water form of *F.fraudax*..............*F.antarctica* Lohmann, 1905 [**R**]
(2.2) Digestive nucleus oblique or longitudinal. Endostyle slightly curved. Trunk elongate.
(2.2a) Rather narrow, a little wider in the pharyngeal region; slightly bent. Spiracles oval. Testis oval and elongate, surrounded on the 'equator' by a moniliform ovary..
.........................*F.aequatorialis* Lohmann, 1896 [**R**]
(2.2b) Wide oval spiracles. One testis horseshoe-shaped, opening backward and partitioned. Two ovaries cord-like, wound round the testis.............
............................*F.helenae* Bückmann, 1924 [**RR**]
(2.2c) Wide oval spiracles. One flat rectangular testis, two sausage-like ovaries, one on each side of the testis ....
..*F.drygalskii* [Lohmann, in] Bückmann, 1923 [**R**]

18.4.3. *Family Kowalewskiidae (Kowalewskidae Lahille, 1888; emend. Lahille, 1890)*

Genus *Kowalevskia* Fol, 1872; emend. Fol, 1874
A. No hood. Tail very long, seven to 10 times the trunk length. A single spherical ovary and a single oval testis...*K.tenuis* Fol, 1872 [**C**]
B. A hood. Tail short, four to five times the trunk length. Two testes and one ovary ....
....*K.oceanica* Lohmann, 1899 [**R**]

References

Acuña, J.-L. 1994. Summer vertical distribution of appendicularians in the central Cantabrian Sea (Bay of Biscay). *Journal of the Marine Biological Association of the United Kingdom*, **74**, 585–601.

Acuña, J.-L. and Anadón, R. 1992. Appendicularian assemblages in a shelf area and their relationship with temperature. *Journal of Plankton Research*, **14**, 9, 1233–1250.

Acuña, J.-L. and Deibel, D. 1994. Feeding experiments with living oikopleurid tunicates in the northeast Greenland polynya. *EOS, Transactions of the American geophysical Union*, **75**, 127 (abstract).

Acuña, J.-L., Deibel, D. and Sooley, S. 1994. A simple device to transfer large and delicate planktonic organisms. *Limnology and Oceanography*, **39**, 2001–2003.

Acuña, J.-L., Deibel, D. and Morris, C.C. 1996. Particle capture mechanism of the pelagic tunicate *Oikopleura vanhoeffeni Limnology and Oceangraphy*, **41**, 1800–1814.

Acuña, J.-L., Bedo, A W., Harris, R.P. and Anadón, R. 1995. The seasonal succession of appendicularians (Tunicata: Appendicularia) off Plymouth. *Journal of the Marine Biological Association of the United Kingdom*, **75**, 755–758.

Adams, P.B. 1987. Diet of widow rockfish *Sebastes entomelas* in northern California. In *Widow Rockfish, Proceedings of a Workshop* (W.H. Lenarz and D.R. Gunderson, eds) Tiburon, California, December 11–12, 1980. NOAA Technical Reports, National Marine Fishery Service, no. 48, pp. 37–41.

Aida, T. 1907. Appendicularia of Japanese waters. *Journal of the College of Science, Imperial University, Tokyo*, **23**(5), 25 pp.

Alldredge, A.L. 1972. Abandoned larvacean houses: a unique food source in the pelagic environment. *Science*, **177**, 885–887.

Alldredge, A.L. 1975. Quantitative natural history and ecology of appendicularians and discarded appendicularian houses. PhD Thesis, University of California, Davis, 149 pp. (University microfilms, Ann Arbor, MI, no. 76–7823).

Alldredge, A.L. 1976a. Discarded appendicularian houses as sources of food, surface habitats, and particulate organic matter in planktonic environments. *Limnology and Oceanography*, **21**(1), 14–23.

Alldredge, A.L. 1976b. Field behaviour and adaptive strategies of appendicularians (Chordata: Tunicata). *Marine Biology*, **38**, 29–39.

Alldredge, A.L. 1976c. Appendicularians. *Scientific American*, **235**, 94–100.

Alldredge, A.L. 1977. House morphology and mechanisms of feeding in the Oikopleuridae (Tunicata, Appendicularia). *Journal of Zoology, London*, **181**, 175–188.

Alldredge, A.L. 1979. The chemical composition of macroscopic aggregates in two neritic seas. *Limnology and Oceanography*, **24**, 855–866

Alldredge, A.L. 1981. The impact of appendicularian grazing on natural food concentrations *in situ*. *Limnology and Oceanography*, **26**(2), 247–257.

Alldredge, A.L. 1982. Aggregation of spawning appendicularians in surface windrows. *Bulletin of Marine Science*, **32**(1), 250–254.

Alldredge, A.L. 1984. The quantitative significance of gelatinous zooplankton as pelagic consumers. In *Flows of Energy and Materials in Marine Ecosystems. Theory and Practice.* (M.J.R. Fasham, ed.), pp. 407–433. Plenum Press, New York.

Alldredge, A.L. and Gotschalk, C.C. 1988. The contribution of marine snow of different origins to biological processes in the water column. *EOS, Transactions of the American geophysical union*, **69**, 44 (abstract).

Alldredge, A.L. and Madin, L.P. 1982. Pelagic tunicates. Unique herbivores in the marine plankton. *Bioscience*, **32**, 655–663.

Alldredge, A.L. and Silver, M.W. 1988. Characteristics, dynamics and significance of marine snow. *Progress in Oceanography*, **20**, 41–82.

Alldredge, A.L. and Youngbluth, M.J. 1985. The significance of macroscopic aggregates (marine snow) as sites of heterotrophic bacterial production in the mesopelagic zone of the subtropical Atlantic. *Deep-Sea Research*, **32**, 1445–1456.

Alldredge, A.L., Cole, J.J. and Caron, D.A. 1986. Production of heterotrophic bacteria inhabiting macroscopic organic aggregates (marine snow) from surface waters. *Limnology and Oceanography*, **31**, 68–78

Alldredge, A.L., Gotschalk, C.C. and MacIntyre, S.S. 1987. Evidence for sustained residence of macrocrustacean fecal pellets in surface waters off Southern California. *Deep Sea Research*, **34**, 1641–1652.

Allen, G.R. 1985. FAO species catalogue. Snappers of the world. An annotated and illustrated catalogue of lutjanid species known to date. *FAO Fisheries Synopses*, **6** (125), 208 pp.

Allman, G.J. 1859 On the peculiar appendage of Appendicularia, styled 'haus' by Mertens. *Quarterly Journal of Microscopical Science*, **7**, 86–89.

Alverson, F.G. 1963. The food of yellowfin and skipjack tunas in the eastern Pacific Ocean. *Bulletin of the Inter-American Tropical Tuna Commission*, **7**, 295–396.

Andersen, V. 1985. Filtration and ingestion rates of *Salpa fusiformis* Cuvier (Tunicata: Thaliacea): effects of size, individual weight and algal concentration. *Journal of Experimental Marine Biology and Ecology*, **87**, 13–29.

Andersen, V. and Nival, P. 1986a. A model of the population dynamics of salps in coastal waters of the Ligurian Sea. *Journal of Plankton Research*, **8**, 1091–1110.

Andersen, V. 1986b. Effect of temperature on the filtration rate and percentage of assimilation of *Salpa fusiformis* Cuvier (Tunicata: Thaliacea). *Hydrobiologia*, **137**, 135–140.

Andersen, V. and Nival, P. 1986. Ammonia excretion rate of *Salpa fusiformis* Cuvier (Tunicata: Thaliacea): effects of individual weight and temperature. *Journal of experimental Marine Biology and Ecology*, **99**, 121–132.

Andersen, V. and Nival, P. 1988. A pelagic ecosystem model simulating production and sedimentation of biogenic particles: role of salps and copepods. *Marine Ecology Progress Series*, **44**, 37–50.

Andersen, V. and Sardou, J. 1994. *Pyrosoma atlanticum* (Tunicata, Thaliacea): diel migration and vertical distribution as a function of colony size. *Journal of Plankton Research*, **16**, 337–349.

Andersen, V., Sardou, J. and Nival, P. 1992. The diel migrations and vertical distributions of zooplankton and micronekton in the Northwestern Mediterranean Sea. 2. Siphonophores, hydromedusae and pyrosomids. *Journal of Plankton Research*, **14**, 1155–1169.

Anderson, O.R. 1993. The trophic role of planktonic Foraminifera and Radiolaria. *Marine Microbial Food Webs*, **7**, 31–51.

Anderson, P.A.V. 1979. Epithelial conduction in salps. I. Properties of the outer skin pulse system of the stolon. *Journal of Experimental Biology*, **80**, 231–239.

Anderson, P.A.V. and Bone, Q. 1980. The communication between individuals in salp chains. II. Physiology. *Proceedings of the Royal Society of London B*, **210**, 569–574.

Anderson, P.A.V., Bone, Q., Mackie, G.O., and Singla, C.L. 1979. Epithelial conduction in salps. II. The role of nervous and non-nervous conduction system interactions in the control of locomotion. *Journal of Experimental Biology*, **80**, 241–250

Angel, M.V. 1984. Detrital organic fluxes through pelagic ecosystems. In *Flows of Energy and Materials in Marine Ecosystems. Theory and Practice* (M.J.R. Fasham, ed.), pp. 475–516. Plenum Publishing, New York.

Angel, M.V. 1989a. Does mesopelagic biology affect the vertical flux? In *Productivity of the Ocean: Present and Past* (W.H. Berger, V.S. Smetacek and G. Wefer, (eds), pp. 155–173. John Wiley & Sons, Chichester, UK.

Angel, M.V. 1989b. Vertical profiles of pelagic communities in the vicinity of the Azores Front and their implications to deep ocean ecology. *Progress in Oceanography*, **22**, 1–46.

Arai, M.N. 1988. Interactions of fish and pelagic coelenterates. *Canadian Journal of Zoology*, **66**, 1913–1927.

Arkett, S.A. 1987. Ciliary arrest controlled by identified central neurons in a urochordate (Ascidiacea). *Journal of Comparative Physiology*, **161A**, 837–847.

Ates, R.M.L. 1988. Medusivorous fishes, a review. *Zoölogisches Mededeelingen*, **62** (3), 29–42.

Atkinson, L.P., Paffenhöfer, G.-A. and Dunstan, W.M. 1978. The chemical and biological effect of a Gulf Stream intrusion off St. Augustine, Florida. *Bulletin of Marine Science*, **28**, 667–679

Atkinson, L.P., O'Malley, P.G., Yoder, J.A. and Paffenhöfer, G.-A. 1984. The effect of summertime shelf break upwelling on nutrient flux in southeastern United States continental shelf waters. *Journal of Marine Research*, **42**, 969–993.

Babcock, H.L. 1937. The sea turtles of the Bermuda Islands, with a survey of the present state of the turtle fishing industry. *Proceedings of the Zoological Society of London, Series A*, **107**, 595–601.

Baird, R.C., Hopkins T.L. and Wilson, D.F. 1975. Diet and feeding chronology of *Diaphus taaningi* (Myctophidae) in the Cariaco Trench. *Copeia*, 1975, 356–365.

Baker, A.N. 1971. *Pyrosoma spinosum* Herdman, a giant tunicate new to New Zealand waters. *Records of the Dominion Museum, Wellington*, **7**, 107–117.

Banas, P.T., Smith, D.E. and Biggs, D.C. 1982. An association between a pelagic octopod *Argonauta* sp. and aggregate salps *Pegea socia*. *US National Marine Fisheries Service Fishery Bulletin*, **80** (3), 648–650.

Baranovic, A., Vucetic, T. and Pucher-Petkovic, T. 1992. Long-term fluctuations of zooplankton in the middle Adriatic Sea (1960–1982). *Acta Adriatica*, **33**, 85–120.

Bargoni, E. 1894. Di un foraminifero parassitico nelle Salpe (*Salpicola amylacea* n.g.,n.sp.) e considerazioni sul corpusculi amilacei dei protozoi superiori. *Richerche Laboratorio Anatomia Normale Roma*, **4** (1/2), 43–62.

Barham, E.G. 1979. Giant larvacean houses: observations from deep submersibles. *Science*, **205**, 1129–1131.

Baroin, A., Perasso, R., Qu, L.H., Brugerolle, G., Bachellerie, J.P. and Adoutte, A. 1988. Partial phylogeny of the unicellular eukaryotes based on rapid sequencing of a portion of 28S ribosomal RNA. *Proceedings of the National Academy of Sciences USA*, **85**, 3474–3478.

Barrington, E.J.W. 1965. *The Biology of Hemichordata and Protochordata*, 167 pp. Oliver and Boyd, Edinburgh, UK.

Barrois, J. 1885 Recherches sur le cycle génétique et le bourgeonnement de l'Anchinie. *Journal de l'Anatomie et de la Physiologie*, **21**, 193–267.

Bary, B.M. 1960. Notes on ecology, distribution and systematics of pelagic tunicates from New Zealand. *Pacific Science*, **14**, 101–121.

Bass, A.J., D'Aubrey, J.D. and Kistnasamy, N. 1976. Sharks of the east coast of southern Africa. 6. The families, Oxynotidae, Squalidae, Dalatiidae and Echinorhinidae. *Investigation Report, Oceanographic Research Institute, Durban*, 45 pp.

Bassot, J.-M. and Bilbaut, A. 1977. Bioluminescence des élytres d'*Acholoe*. Luminescence et fluorescence des photosomes. *Biologie Cellulaire*, **28**, 163–168.

Batchelder, H.P. and Swift, E. 1989. Estimated near-surface mesoplanktonic bioluminescence in the western North Atlantic during July 1986. *Limnology and Oceanography*, **34**, 113–128.

Batchelder, H.P., Swift, E. and Van Keuren, J.R. 1992. Diel patterns of planktonic bioluminescence in the northern Sargasso Sea. *Marine Biology*, **113**, 329–339.

Bathmann, U.V. 1988. Mass occurrence of *Salpa fusiformis* in the spring of 1984 off Ireland: implications for sedimentation processes. *Marine Biology*, **97**, 127–135.

Baussant, T. 1993. Analyse spatio-temporelle du microplancton en Mer Ligure (Méditerranée Nord-Occidentale), Thèse de Doctorat, Université Pierre et Marie Curie, Paris France.

Bedo, A.W., Acuña, J.-L., Robins, D. and Harris, R.P. 1993. Grazing in the micronic and sub-micronic particle size range: the case of *Oikopleura dioica* (Appendicularia). *Bulletin of Marine Science*, **53**, 2–14.

Beers, J.R. 1966. Studies on the chemical composition of the major zooplankton groups in the Sargasso Sea off Bermuda. *Limnology and Oceanography*, **11**, 520–528.

Bernard, M. 1958. Systématique et distribution saisonnière des Tuniciers pélagiques d'Alger. *Rapports et Procès-verbaux des réunions — Commission Internationale pour l'Exploration Scientifique de la Mer Méditerranée*, **14**, 211–231.

Berner, L.D. 1967. Distribution atlas of Thaliacea in the California Current Region. *Californian Cooperative Oceanic Fisheries Investigations Atlas*, **8**, 1–322

Berner, L.D. and Reid, J.L., Jr 1961. On the response to changing temperature of the temperature-limited plankter *Doliolum denticulatum* Quoy and Gaimard 1835. *Limnology and Oceanography*, **6**, 205–215.

Berner, L. 1954. On the previously undescribed aggregate form of the pelagic Tunicate, *Ritteriella picteti* (Apstein, 1904). *Pacific Science*, **8**, 121–124.

Bernstein, T. 1934. Zooplankton Karskogo morya po materialam ekspeditsii arkticheskogo Instituta na 'Sedove' 1930 goda i 'Lomonosove' 1931 goda. (Zooplankton des nördlichen Teiles des Karischen Meeres). *Trudy Arkticheskogo Instituta*, **9**, 3–58 (in Russian, German abstract).

Berrill, N.J. 1950a. *The Tunicata, with an Account of the British Species*, 354 pp. The Ray Society, London.

Berrill, N.J. 1950b. Budding in *Pyrosoma*. *Journal of Morphology*, **87** (3), 537–552.

Berrill, N.J. 1950c. Budding and development in *Salpa*. *Journal of Morphology*, **87**, 553–606.

Best, P.B. 1967. Distribution and feeding habits of baleen whales off the Cape Province. *Investigation Report Division of Sea Fisheries of South Africa*, **57**, 1–44.

Bidigare, R.R. and Biggs, D.C 1980 The role of sulphate exclusion in buoyancy maintainance by siphonophores and other oceanic gelatinous zooplankton. *Comparative Biochemistry and Physiology*, **66A**, 467–471.

Bigelow, H.B. and Schroeder, W.C. 1953. Fishes of the Gulf of Maine. *Fishery Bulletin*, **53** (74).

Bigelow, H.B. and W.W. Welsh. 1925. Fishes of the Gulf of Maine. *Bulletin of the United States Bureau of Fisheries*, **40** (1), 1–567.

Biggs, D.C. 1977. Respiration and ammonium excretion by open ocean gelatinous zooplankton. *Limnology and Oceanography*, **22**, 108–117.

Binet, D. 1976. Contribution à l'ecologie de quelques taxons du zooplancton de Côte d'Ivoire II. Dolioles, salpes, appendiculaires. *Documents scientifiques du centre de récherches Océanographiques d'Abidjan*, **7**, 45–61

Blaber, S.J.M. and Bulman, C.M. 1987. Diets of fishes of the upper continental slope of eastern Tasmania, content, calorific values, dietary overlap and trophic relationships. *Marine Biology*, **95**, 345–356.

Blay, J. and Eyeson, K.N. 1982. Feeding activity and food habits of the shad, *Ethmalosa fimbriata* (Bowdich), in the coastal waters of Cape Coast, Ghana. *Journal of Fisheries Biology*, **21**, 403–410.

Bleakney, J.S. 1965. Reports of marine turtles from New England and eastern Canada. *Canadian Field Naturalist*, **79**, 120–128.

Bogoraze, D. and Tuzet, O. 1969a. Ultrastructure des muscles cardiaques et des muscles de la queue chez *Oikopleura longicauda* Vogt. *Comptes Rendus Hebdomadaires des Séances de l'Académie des Sciences*, **268D**, 1417–1419.

Bogoraze, D. and Tuzet, O. 1969b. Ultrastructure du muscle de la queue de l'appendiculaire *Oikopleura longicauda* Vogt. Les limites cellulaires; les disques intercalaires. *Cahiers de Biologie Marine*, **10**, 365–374.

Bogoraze, D. and Tuzet, O. 1974. Contributions à l'étude du tube digestif de l'Appendiculaire *Oikopleura longicauda* Vogt. *Annales des Sciences Naturelles (Zoologie et Biologie Animale)*, **16** (12), 55–95.

Bolles Lee, A. 1891. On a little known sense organ in Salpa. *Quarterly Journal of Microscopical Science*, **32**, 89–96.

Bollner, T., Holmberg, K. and Olsson, R. 1986. A rostral sensory mechanism in *Oikopleura dioica* (Appendicularia). *Acta Zoologica (Stockholm)*, **67**, 235–241.

Bollner, T., Storm-Mathisen, J. and Ottersen, O.P. 1991. GABA-like immunoreactivity in the nervous system of *Oikopleura dioica* (Appendicularia). *Biological Bulletin, Woods Hole*, **180**, 119–124.

Bone, Q. 1959. Observations upon the nervous systems of pelagic tunicates. *Quarterly Journal of Microscopical Science*, **100**, 167–181.

Bone, Q. 1982. The role of the outer conducting epithelium in the behaviour of salp oozooids. *Journal of the Marine Biological Association of the United Kingdom*, **62**, 125–132.

Bone, Q. 1985. Epithelial action potentials in *Oikopleura* (Tunicata:Larvacea). *Journal of Comparative Physiology*, **156A**, 117–123.

Bone, Q. 1987. Tunicates. In *Nervous systems in Invertebrates* (M.A. Ali, ed.), NATO ASI series, Vol. 141, pp. 527–557. Plenum Press, New York.

Bone, Q. 1989a. Evolutionary patterns of axial muscle systems in some invertebrates and fish. *American Zoologist*, **29** (1), 5–18.

Bone, Q. 1989b. On the muscle fibres and locomotor activity of doliolids. *Journal of the Marine Biological Association of the United Kingdom*, **69**, 597–607.

Bone, Q. and Mackie, G.O. 1975. Skin impulses and locomotion in *Oikopleura* (Tunicata: Larvacea). *Biological Bulletin, Woods Hole*, **149**, 267–286.

Bone, Q. and Mackie, G.O. 1977. Ciliary arrest potentials, locomotion and skin impulses in *Doliolum* (Tunicata: Thaliacea). *Rivista di Biologia Normale e Patologica*, **3**, 181–191.

Bone, Q. and Mackie, G.O. 1982. Urochordata. In *Electrical Conduction and Behaviour in 'Simple' Invertebrates* (G.A.B. Shelton, ed.), pp. 473–535. Clarendon Press, Oxford.

Bone, Q. and Ryan, K.P. 1973. The structure and innervation of the locomotor muscles of salps (Tunicata: Thaliacea). *Journal of the Marine Biological Association of the United Kingdom*, **53**, 873–883.

Bone, Q. and Ryan, K.P. 1974. On the structure and innervation of the muscle bands of *Doliolum* (Tunicata: Cyclomyaria). *Proceedings of the Royal Society of London B*, **187**, 315–327.

Bone, Q. and Ryan, K.P. 1979. The Langerhans receptor of *Oikopleura* (Tunicata: Larvacea). *Journal of the Marine Biological Association of the United Kingdom*, **59**, 69–75.

Bone, Q. and Trueman, E.R. 1983. Jet propulsion in salps (Tunicata: Thaliacea). *Journal of Zoology, London*, **201**, 481–506.

Bone, Q. and Trueman, E.R. 1984. Jet propulsion in *Doliolum* (Tunicata: Thaliacea). *Journal of Experimental Marine Biology and Ecology*, **76**, 105–118.

Bone, Q., Anderson, P.A.V. and Pulsford, A. 1980. The communication between individuals in salp chains. I. Morphology of the system. *Proceedings of the Royal Society of London B*, **210**, 549–558.

Bone, Q., Braconnot, J.-C. and Carré C. 1997a. On the heart and circulation in *Doliolum* (Tunicata, Thaliacea). *Scientia Marina*, **61**, 189–194.

Bone, Q., Braconnot, J.-C. and Carré C. 1997b. On the filter-feeding of *Doliolum* (Tunicata, Thaliacea). *Journal of Experimental Marine Biology and Ecology*, **214**, 179–193.

Bone, Q., Fenaux, R. and Mackie, G.O. 1977a. On the external surface in Appendicularia. *Annales de l'Institut d'Océanographie*, **53**, 237–244.

Bone, Q., Flood, P.R., Mackie, G.O. and Singla, C.L. 1977b. On the organisation of the sarcotubular systems in the caudal muscle cells of larvaceans (Tunicata). *Acta Zoologica (Stockholm)*, **58**, 187–196.

Bone, Q., Gorsky, G. and Pulsford, A.L. 1979. On the structure and behaviour of *Fritillaria* (Tunicata: Larvacea). *Journal of the Marine Biological Association of the United Kingdom*, **59**, 399–411.

Bone, Q., Pulsford, A. and Amoroso, E.C. 1985. The placenta of the salp (Tunicata: Thaliacea). *Placenta*, **6**, 53–64.

Bone, Q., Braconnot, J.-C. and Ryan, K.P. 1991. On the pharyngeal feeding filter of the salp *Pegea confoederata* (Tunicata: Thaliacea). *Acta Zoologica (Stockholm)*, **72**, 55–60.

Bone, Q., Inoue, I. and Tsutsui, I. 1997. *Doliolum* muscle fibres: contraction and relaxation in the absence of a sarcoplasmic reticulum. *Journal of Muscle Research and Cell Motility*, **18**, 375–380.

Borets, L.A. 1975. Progress in the study of the biology of boarfish *Pentaceros richardsoni* (Smith) [in Russian]. *Studies in Fish Biology and Commercial Oceanography*, **6**, 82–90.

Borgert, A. 1894. Die Thaliacea der Plankton-Expedition. C. Vertheilung der Doliolen. *Ergebnisse der Plankton-Expedition der Humboldt-Stiftung*, **2**, 1–66.

Bowlby, M.R., Widder, E.A. and Case, J.F. 1990. Patterns of stimulated bioluminescence in two pyrosomes (Tunicata: Pyrosomatidae). *Biological Bulletin, Woods Hole*, **179**, 340–350.

Bowman, R.E. and Michaels, W.L. 1984. Food of seventeen species of northwest Atlantic fish. US National Marine Fisheries Service Northeast Fisheries Center, NOAA Technical Memorandum NMFSF/NEC-28. 183pp.

Bowman, R.E., Eppi, R. and Grosslein, M. 1984. Diet and consumption of spiny dogfish in the Northwest Atlantic.

International Council for the Exploration of the Sea. C.M. 1984/G, Vol. 27. 16 pp.

Braconnot, J.-C. 1963. Étude du cycle annuel des Salpes et Dolioles en rade de Villefranche-sur-Mer. *Journal du Conseil International pour l'Exploration de la Mer*, **28**, 21–36.

Braconnot, J.C. 1964. Sur le développement de la larve de *Doliolum denticulatum* Quoy et Gaymard. *Comptes Rendus Hebdomodaires des Séances de l'Académie des Sciences Paris*, **259**, 4361–4363.

Braconnot, J.-C. 1967. Sur la possibilité d'un cycle court de développement chez le Tunicier pélagique *Doliolum nationalis* Borgert 1893. *Comptes Rendus Hebdomodaires des Séances de l'Académie des Sciences Paris*, **264**, 1434–1437.

Braconnot, J.-C. 1968. Sur le développement de la larve du Tunicier pélagique Doliolide: *Doliolum* (*Dolioletta*) gegenbauri Ulj. 1884. *Comptes rendus Hebdomodaires des Séances de l'Academie des Sciences Paris*, **267**, 629–630.

Braconnot, J.-C. 1970a. Contribution à l'étude biologique et écologique des Tuniciers pélagiques Salpides et Doliolides, Thèse de Doctorat d'Etat, Paris, 2 Vols, 111 pp.

Braconnot, J.-C. 1970b. Contribution a l'étude des stades successifs dans le cycle des tuniciers pélagiques Doliolides. I. les stades larvaire, oozoide, nourrice et gastrozoide. *Archives de Zoologie Expérimentale et Générale*, **111**, 629–668.

Braconnot, J.-C. 1971a. Contribution à l'étude des stades successifs dans le cycle des Tuniciers pélagiques Doliolides II. Les stades phorozoïde et gonozoïde des Doliolides. *Archives de Zoologie Expérimentale et Générale*. **112**, 5–31.

Braconnot, J.-C. 1971b. Contribution à l'étude biologique et écologique des Tuniciers pélagiques Salpides et Doliolides. I. Hydrologie et écologie des Salpides. *Vie et Milieu, Serie B*, **22**, 257–286.

Braconnot, J.-C. 1971c. Contribution à l'étude biologique et écologique des Tuniciers pelagiques Salpides et Doliolides. II. Hydrologie et écologie des Doliolides. *Vie et Milieu*, **22** (3), 437–467.

Braconnot, J.-C. 1973. Contribution à l'étude des stades successifs dans le cycle des Tuniciers pélagiques Salpides en Méditerranée. *Bulletin de l'Institut Océanographique de Monaco*, **71** (1424), 1–27.

Braconnot, J.-C. 1974a. Sur la réalité du cycle sexué chez le Tunicier pélagique: *Doliolum nationalis* Borgert, 1893, avec la première description de sa larve. *Comptes Rendus Hebdomadaires des Séances de l'Académie des Sciences Paris*, **278D**, 1759–1760.

Braconnot, J.-C. 1974b. Le tunicier pélagique *Pyrosoma atlanticum* Péron 1809 en mer Ligure (Méditerranée orientale). *Rapports de la Commission Internationale pour l'Exploration Scientifique de la Mer Méditerranée*, **22** (9), 97–99.

Braconnot, J.-C. 1977. Sur le cycle sexué chez le tunicier pelagique *Doliolum nationalis* avec la première description du stade oozoide. *Comptes Rendus Hebdomadaires des Séances de l'Académie des Sciences Paris*, **278D**, 284, 835–837.

Braconnot, J.-C. and Casanova, J.-P. 1967. Sur le tunicier pélagique *Doliolum nationalis* Borgert 1893 en Méditerranée Occidentale (Campagne du '*Président-Théodore-Tissier*', September–October 1958). *Revue des Travaux de l'Institut des Pêches Maritimes*, **31**, 393–401.

Braconnot, J.-C. and Jegu, M. 1981. Le cycle de *Thalia democratica* (Salpidae), croissance et durée de chaque génération. *Rapports et Procès-verbaux des Réunions — Commission Internationale pour l'Exploration Scientifique de la Mer Méditerranée*, **27**, 197–198.

Braconnot, J.-C and Katavic, J. 1979. L'existence du cycle sexué dans certaines populations du tunicier pélagique *Doliolum nationalis* en Mediterranée occidentale et en Adriatique. *Rapports et Procès-verbaux des Réunions. Commission Internationale pour l'Exploration Scientifique de la Mer Méditerranée*, **25–26**, 163–164.

Braconnot, J.-C., Choe, S.-M. and Nival, P. 1988. La croissance et le développement de *Salpa fusiformis* Cuvier (Tunicata Thaliacea). *Annales de l'Institut Océanographique Paris*, **64**, 101–114.

Bradbury, M.G., Abbott, D.P., Bovbjerg, R.V., Mariscal, R.N., Fielding, W.C., Barber, R.T., Pearse, V.B., Proctor, S.J., Ogden, J.C., Wourms, J.P., Taylor, L.R., Jr, Christofferson, J.G., Christofferson, J.P., McPhearson, R.M., Wynne, M.J. and Stromborg, P.M., Jr 1971. Studies on the fauna associated with the deep scattering layers in the equatorial Indian Ocean, conducted on R/V *Te Vega* during October and November 1964. In *Proceedings of an International Symposium on Biological Sound Scattering in the Ocean* (G.B. Farquhar, ed.). Maury Center for Ocean Science, Washington, DC, No. 5, pp. **409–452**.

Brandt, S.B. 1981. Effects of a warm-core eddy on fish distributions in the Tasman Sea off East Australia. *Marine Ecology Progress Series*, **6** (1), 19–33.

Brattström, H. 1972. On *Salpa fusiformis* Cuvier (Thaliacea) in Norwegian coastal and offshore waters. *Sarsia*, **48**, 71–90.

Brien, P. 1928. Contribution à l'étude de l'embryogénèse et de la blastogénèse des Salpes. *Recueil des Travaux de l'Institut de Zoologie Torley-Rousseau*, **2**, 5–116.

Brien, P. 1948. Embranchement des Tuniciers. In *Traité de Zoologie* (P.P. Grassé, ed.), Vol. 11, pp. 553–895. Masson et Cie, Paris.

Brodeur, R.D. and Pearcy, W.G. 1984. Food habits and dietary overlap of some shelf rockfishes (genus *Sebastes*) from the northeastern Pacific Ocean. *Fishery Bulletin*, **82** (2), 269–293.

Brodeur, R.D. and Pearcy, W.G. 1990. Trophic relations of juvenile Pacific salmon off the Oregon and Washington coast. *Fishery Bulletin*, **88**, 617–636.

Brodeur, R.D., Lorz H.V. and Pearcy, W.G. 1987. Food habits and dietary variability of pelagic nekton off Oregon and Washington, 1979–1984. *NOAA Technical Report, National Marine Fisheries Service*, no. 57, p. 32.

Brongersma, L.D. 1967. *British Turtles, Guide for the Identification of Stranded Turtles on British Coasts*. British Museum (Nat. Hist.), London.

Brongersma, L.D. 1968. Notes upon some turtles from the Canary Islands and from Madeira. *Proceedings of the Köninklichen Nederlandischen Akademie van te Wetenschappen Amsterdam* C, **712**, 128–136.

Brongersma, L.D. 1969. Miscellaneous notes on turtles II A–B. *Proceedings of the Köninklichen Nederlandischen Akademie van te Wetenschappen Amsterdam* C, **721**, 76–102.

Brongersma, L.D. 1972. European Atlantic turtles. *Zoologische Verhandlungen*, **121**, 1–318.

Brooks, W.K. 1876. On the development of *Salpa*. *Bulletin of the Museum of Comparative Zoology*, **3**, 291–348.

Brooks, W.K. 1893. The genus Salpa: a monograph. *Memoirs from the Biological Laboratory of the Johns Hopkins University*, **2**, 1–396.

Browne, R. 1756. *The Civil and Natural History of Jamaica*. White, London.

Bruland, K.W. and Silver, M.W. 1981. Sinking rates of fecal pellets from gelatinous zooplankton (salps, pteropods, doliolids). *Marine Biology*, **63**, 295–300.

Buchanan, R.A. and Browne, S.A.M. 1981. Zooplankton of the Labrador coast and shelf during summer 1979. *Consultant's Report for Petro-Canada Explorations Ltd*, 78 pp. LGL Ltd, St. Johns, Newfoundland.

Buck, J. 1978. Functions and evolutions of bioluminescence. In *Bioluminescence in Action* (P.J. Herring, ed.), pp. 419–460. Academic Press, London.

Bückmann, A. 1923. *Beitrag zur Kenntnis der Appendicularien auf Grund der Ausbeute der Deutschen Südpolarexpedition.* 10 pp. A. Hoffmann, Leipzig. (Auszug aus der Doktorarbeit).

Bückmann, A. 1924. Bemerkungen über Appendicularien aus der Ausbeute der Deutschen Südpolarexpedition. *Zoologischer Anzeiger,* **59**, 7–8, 205–212.

Bückmann, A. 1926. Copelata. *Tierwelt der Nord- und Ostsee,* **12**, 1–20.

Bückmann, A. 1970. Die Verbreitung der Kaltwasser- und der Warmwasserfauna der Appendicularien im nördlichen Nordatlantischen Ozean im Spätwinter und Spätsommer 1958. *Marine biology,* **5**, 35–56.

Bückmann, A. and H. Kapp. 1975. Taxonomic characteristics used for the distinction of species of Appendicularia. *Mitteilungen aus dem Hamburgischen Zoologischen Museum und Institut,* **72**, 201–228.

Budelmann, B.U. 1977. Structure and function of the angular acceleration receptor systems in the statocysts of cephalopods. *Symposium of the Zoological Society of London,* **38**, 309–324.

Budylenko, G.A. 1978. On sei whale feeding in the Southern Ocean. *Report of the International Whaling Commission,* **28**, 379–385.

Buggeln, R.G. 1980. The correlation of 'slub' on fishing nets with the zooplankton species, *Oikopleura vanhoffeni,* in coastal Newfoundland. *Final Report to the Development Branch, Newfoundland and Labrador Department of Fisheries,* October 1980, 9 pp.

Bullen, G.E. 1908. Plankton studies in relation to the western mackerel fishery. *Journal of the Marine Biological Association of the United Kingdom* (N.S.), **8**, 3, 269–302.

Burighel, P., Lane, N.J., Martinucci, G.B., Fenaux, R. and Dallai, R. 1989. Junctional diversity in two regions of the epidermis of *Oikopleura dioica* (Tunicata, Larvacea). *Cell and Tissue Research,* **257**, 3, 529–535.

Busch, W. 1851. *Beobachtungen über Anatomie und Entwickelung einiger wirbellosen Seethiere,* Vol. VIII, 143 pp. A. Hirschwald, Berlin.

Buskey, E.J. and Swift, E. 1983b. Behavioral responses of the coastal copepod *Acartia hudsonica* (Pinley) to simulated dinoflagellate bioluminescence. *Journal of Experimental Marine Biology and Ecology,* **72**, 43–58.

Buskey, E.J. and Swift, E. 1985. Behavioral responses of oceanic zooplankton to simulated bioluminescence. *Biological Bulletin, Woods Hole,* **168**, 263–275.

Buskey, E., Mills, L. and Swift, E. 1983. The effects of dinoflagellate bioluminescence on the swimming behavior of a marine copepod. *Limnology and Oceanography,* **28**, 575–579.

Cadenat, J.A. 1954. Notes d'ichthyologie ouest-Africaine. 7. Biologie. Régime alimentaire. *Bulletin de l'Institut Fondamental d'Afrique Noire, Serie A,* **16**, 564–583.

Cadenat, J. 1957. Observations de cétacés, siréniens, chéloniens et sauriens en 1955–1956. *Bulletin de l'Institut Fondamental d'Afrique Noire, Serie A,* **19**, 1358–1375.

Cailliet, G.M. 1972. The study of feeding habits of two marine fishes in relation to plankton ecology. *Transactions of the American Microscopical Society,* **91**, 88–89.

Cailliet, G.M. and Ebeling, A.W. 1990. The vertical distribution and feeding habits of two common midwater fishes (*Leuroglossus stilbius* and *Stenobrachius leucopsaurus*) off Santa Barbara. *Californian Cooperative Oceanic Fisheries Investigations Reports,* **31**, 106–123.

Cailliet, G.M., Osada, E.K. and Moser, M. 1988. Ecological studies of sablefish in Monterey Bay. *California Fish and Game,* **74**, 132–153.

Carlisle, D.B. 1950. Alcune osservazioni sulla meccanica dell'alimentazione della *Salpa. Pubblicazione della Stazione Zoologica di Napoli,* **22**, 146–154.

Caron, D.A., Madin L.P. and Cole, J.J. 1989. Composition and degradation of salp fecal pellets: implications for vertical flux in oceanic environments. *Journal of Marine Research,* **47**, 829–850.

Casanova, J.P. 1966. Pêches planctoniques superficielles et profondes en Méditerranée occidentale: VII Thaliacés. *Revue Travaux Institut Pêches Maritimes,* **30**, 385–390.

Casareto, B.E. and Nemoto, T. 1986. Salps of the Southern Ocean (Australian sector) during the 1983–84 summer, with special reference to the species *Salpa thompsoni* Foxton 1961. *Proceedings of the 7th Symposium on Polar Biology, National Institute for Polar Biology,* no. **40**, pp. 221–239.

Casareto, B.E. and Nemoto, T. 1987. Latitudinal variation of the number of muscle fibres in *Salpa Thompsoni* (Tunicata, Thaliacea) in the Southern Ocean, implications for the validity of the species *Salpa gerlachi. Proceedings of the 9th Symposium on Polar Biology, National Institute for Polar Biology,* pp. 90–104.

Casaux, R.J., Mazotta, A.S. and Bairera-Oro, E.R. 1990. Seasonal aspects of the biology and diet of nearshore nototheniid fish at Potter Cove, South Shetland Islands, Antarctica. *Polar Biology,* **11**, 63–72.

Cassie, R.M. 1956. Spawning of the snapper, *Chrysophrys auratus* Forster in Hauraki Gulf. *Transactions of the Royal Society of New Zealand,* **84**, 309–328.

Cetta, C.M., Madin, L.P. and Kremer, P. 1986. Respiration and excretion by oceanic salps. *Marine Biology,* **91**, 529–537.

Chamisso, A. de and Eysenhardt, C.G. 1821. De animalibus quibusdam de classe vermium linneana, in circumnavigatione terrae, auspicante comite N. Romanzoff, duce Ottone de Kotzebue, annis 1815–1818, peracta, observatis. *Nova Acta Physico-medica Academiae Caesareae Leopoldino-Carolinae Naturae Curiosorum,* **10** (2), 343–374.

Chatton, E. 1911. II. Ciliés parasites des Cestes et des Pyrosomes. *Archives de Zoologie Experimentale Génerale, Serie 5,* **8**, viii–xx.

Chatton, E. 1920. Les Péridiniens parasites. Morphologie, reproduction, éthologie. *Archives de Zoologie Experimentale Génerale,* **59**, 1–475.

Chevreux, E. and Fage, L. 1925. *Amphipodes,* Faune de France no. 9. Librairie de la Faculté des Sciences, Paris, France.

Cho, B.C. and Stancyk, F. 1988. Heterotrophic bacterioplankton production measurement by the tritiated thymidine incorporation method. *Archiv für Hydrobiologie,* **31**, 153–162.

Christen, R., Ratto, A., Baroin, A., Perasso, R., Grell, K.G. and Adoutte, A. 1991a. An analysis of metazoan origins, using comparisons of partial sequences of the 28S ribosomal RNA, reveals an early emergence of triploblasts. *The EMBO Journal,* **10**, 499–503.

Christen, R., Ratto, A., Baroin, A., Perasso, R., Grell, K.G. and Adoutte, A. 1991b. Origins of metazoans. A phylogeny deduced from sequences of the 28S ribosomal RNA. In *The Early Evolution of Metazoa and the Significance of Problematic Taxa* (A. Simonetta and S. Conway Morris, eds), pp. 1–13. Cambridge University Press, Cambridge.

Chun, C. 1887. Die pelagische Thierwelt in grösseren Meerestiefen und ihre Beziehungen zu der Oberflächenfauna. *Bibliotheca Zoologica,* **1**, 1–72.

Chun, C. 1900. *Aus den Tiefen des Weltmeeres,* Vol. VII, pp. 1–549. G. Fischer, Jena.

Clark, M.R. 1985. The food and feeding of seven fish species from the Campbell Plateau New Zealand. *New Zealand Journal of Marine Freshwater Research,* **19**, 339–364.

Clark, M.R., King, K.J. and McMillan, P.J. 1989. The food and feeding relationships of black oreo, *Allocyttus niger,* smooth oreo, *Pseudocyttus maculatus* and eight other fish species from the continental slope of the south-west Chatham Rise. *New Zealand Journal of Fisheries Biology*, **35**, 465–484.

Clarke, A., Holmes, L.J. and Gore, D.J., 1992. Proximate and elemental composition of gelatinous zooplankton from the Southern Ocean. *Journal of Experimental Marine Biology and Ecology*, **155**, 55–68.

Clarke, C. and Roff, J.C. 1990. Abundance and biomass of herbivorous zooplankton off Kingston, Jamaica, with estimates of their annual production. *Estuarine and Coastal Shelf Science*, **31**, 423–437.

Coale, K.H. and Bruland, K.W. 1985. ^{234}Th:^{238}U disequilibria within the California Current. *Limnology and Oceanography*, **30**, 22–33.

Coleman, N. 1980. *Australian Sea Fishes South of 30°S.*, 302 pp. Doubleday, Sydney.

Coleman, N. 1981. *Australian Sea Fishes North of 30°S.*, 297 pp. Doubleday, Sydney.

Collegnon, J. and Aloncle, H. 1960. Le régime alimentaire de quelques poissons benthiques des côtes marocaines· *Bulletin de l'Institut de Pêches Maritimes, Maroc*, **5**, 17–29.

Collin, B. 1912. Etude monographique sur les Acinétiens — morphologie, physiologie, systematique. *Archives de Zoologie Experimentale et Génerale*, **51**, 1–457.

Colombera, D. and Fenaux, R. 1973. Chromosome form and number in the Larvacea. *Bolletino di Zoologia*, **40**, 347–353.

Conover, R.J. 1966. Assimilation of organic matter by zooplankton. *Limnology and Oceanography*, **11**, 338–345.

Conover. R.J. and Mayzaud P. 1975. Respiration and nitrogen excretion of neritic zooplankton in relation to potential food supply. In *Proceedings of the 10th European Symposium of Marine Biology, Ostend*, Vol. 2, pp. 151–163.

Costello, J. and Azam, S.E. 1983. Tidal influence upon appendicularian abundance in North Inlet estuary, South Carolina. *Journal of Plankton Research*, **5**, 263–277.

Cowper, T.R. 1960. Occurrence of *Pyrosoma* on the continental slope. *Nature, London*, **187**, 878–879.

Crabtree, R.E. and Sulak, KJ. 1986. A contribution to the life history and distribution of Atlantic species of the deep-sea fish genus *Conocara* (Alepocephalidae). *Deep-Sea Research*, **33A** (9), 1183–1202.

Crocker, K.M., Alldredge, A.L. and Steinberg, D.K. 1991. Feeding rates of the doliolid, *Dolioletta gegenbauri*, on diatoms and bacteria. *Journal of Plankton Research*, **13**, 77–82.

Cropp, B. 1985. Australia's invisible stinger, a jellyfish feared more than the shark. *Oceans*, **18**, 3–7.

Curl, H., Jr 1962. Analyses of carbon in marine plankton organisms. *Journal of Marine Research*, **20**, 181–188.

Cuvier, G. 1804. Mémoire sur les Thalides et sur les Biphores. *Annales de la Musée d'Histoire Naturelle de Paris*, **4**, 360–382.

Dagg, M.J. and Walser, W.E. 1987. Ingestion, gut passage, and egestion by the copepod *Neocalanus plumchrus* in the laboratory and in the subarctic Pacific Ocean. *Limnology and Oceanography*, **32**, 178–188.

Dam, H.G. and Peterson, W.T. 1988. The effect of temperature on the gut clearance rate constant of planktonic copepods. *Journal of Experimental Marine Biology and Ecology*, **123**, 1–14.

Damas, D. 1904. Contribution à l'étude des Tuniciers. *Archives de Biologie*, **20**, 745–833.

Daniels, R.A. 1982. Feeding ecology of some fishes of the Antarctic peninsula. *Fishery Bulletin*, **80**, 575–588.

Daponte, M.C. and Esnal, G.B. 1994. Differences in embryological development in 2 closely related species *Ihlea racovitzai* and *Ihlea magalhanica* (Tunicata, Thaliacea). *Polar Biology*, **14**, 455–458.

Daponte, M.C., Capitanio, L.F., Machinandiarena, L. and Esnal, G.B. 1993. Planktonic tunicates (Chordata, Tunicata) of the RTMA 'Evrika' in the southwestern Atlantic Ocean (1988). *Iheringia, Série Zoologie, Porto Alegre*, **74**, 71–80.

Dauby, P. and Godeaux, J. 1987. Application de l'analyse multidimensionelle aux espèces du genre *Thalia* (Tunicata, Thaliacea). *Annales de la Société Royale Zoologique de Belgique*, **107**, (2), 175–180.

Davis, C.C. 1982. A preliminary quantitative study of the zooplankton from Conception Bay, insular Newfoundland, Canada. *International Revue Gesammten Hydrobiologie*, **67**, 713–747.

Davis, C.S., Gallager, S.M. and Solow, A.R. 1992. Microaggregations of oceanic plankton observed by towed video microscopy. *Science*, **257**, 230–232.

Davoll, P.J. 1982. Protists and bacteria from abandoned larvacean houses. *EOS, Transactions of the American Geophysical Union*, **63**(45), 944.

Davoll, P.J. 1984. *Microcosms in the pelagic zone: a study of the microbial community on larvacean house aggregates from Monterey Bay, PhD thesis, University of California, Santa Cruz*, (University microfilms international, Ann Arbor, MI, no. 84–16240).

Davoll, P.J. and Silver, M.W. 1986. Marine snow aggregates: life history sequence and microbial community of abandoned larvacean houses from Monterey Bay, California. *Marine Ecology Progress Series*, **33**, 111–120.

Davoll, P.J. and Youngbluth, M.J. 1990. Heterotrophic activity on appendicularian (Tunicata: Appendicularia) houses in mesopelagic regions and their potential contribution to particle flux. *Deep-Sea Research*. **37**, 285–294.

Dawydoff, C. 1937. Les *Gastrodes* des eaux indochinoises et quelques observations sur leur cycle évolutif. *Comptes Rendus Hebdomadaires des Séances de l'Académie des Sciences Paris*, **204**, 1088–1090.

de Decker, A. 1973. Agulhas Bank plankton. In *Ecological Studies. Analysis and Synthesis* (B. Zeitzschel, ed.), pp. 188–219. Springer-Verlag, Berlin.

Deevey, G.B. and Brooks, A.L. 1971. The annual cycle in quantity and composition of the zooplankton of the Sargasso Sea off Bermuda. II. The surface to 2,000 m. *Limnology and Oceanography*, **6**, 927–943.

Deibel, D. 1982a. Laboratory determined mortality, fecundity and growth rates of *Thalia democratica* Forskal and *Dolioletta gegenbauri* Uljanin (Tunicata, Thaliacea). *Journal of Plankton Research*, **4**, 143–153.

Deibel, D. 1982b. Laboratory-measured grazing and ingestion rates of the salp, *Thalia democratica* Forskal, and the doliolid, *Dolioletta gegenbauri* Uljanin (Tunicata, Thaliacea). *Journal of Plankton Research*, **4**, 189–201.

Deibel, D. 1985a. Clearance rates of the salp *Thalia democratica* fed naturally occurring particles. *Marine Biology*, **86**, 47–54.

Deibel, D. 1985b. Blooms of the pelagic tunicate, *Dolioletta gegenbauri*: are they associated with Gulf Stream frontal eddies. *Journal of Marine Research*, **43**, 211–236.

Deibel, D. 1986. Feeding mechanism and house of the appendicularian *Oikopleura vanhoeffeni*. *Marine Biology*, **93**, 429–436.

Deibel, D. 1988. Filter feeding by *Oikopleura vanhoeffeni*: grazing impact on suspended particles in cold ocean waters. *Marine Biology*, **99**, 177–186.

Deibel, D. 1990. Still-water sinking velocity of fecal material from the pelagic tunicate *Dolioletta gegenbauri*. *Marine Ecology Progress Series*, **62**, 55–60.

Deibel, D. and Lee, S.H. 1992. Retention efficiency of submicrometer particles by the pharyngeal filter of the pelagic tunicate *Oikopleura vanhoeffeni*. *Marine Ecology Progress Series*, **81**, 25–30.

Deibel, D. and Paffenhöfer, G.-A. 1988. Cinematographic analysis of the feeding mechanism of the pelagic tunicate *Doliolum nationalis*. *Bulletin of Marine Science*, **43**, 404–412.

Deibel, D. and Powell, C.V.L. 1987a. Ultrastructure of the pharyngeal filter of the appendicularian *Oikopleura vanhoeffeni*: implications for particle size selection and fluid mechanics. *Marine Ecology Progress Series*, **35**, 243–250.

Deibel, D. and Powell, C.V.L. 1987b. Comparison of the ultrastructure of the food-concentrating filter of two appendicularians. *Marine Ecology Progress Series*, **39**, 81–85.

Deibel, D. and Turner, J.T. 1985. Zooplankton feeding ecology: contents of fecal pellets of the appendicularian *Oikopleura vanhoeffeni*. *Marine Ecology Progress Series*, **27**, 67–78

Deibel, D., Dickson M.L. and Powell, C.V.L., 1985. Ultrastructure of the mucous feeding filter of the house of the appendicularian *Oikopleura vanhoeffeni*. *Marine Ecology Progress Series*, **27**, 79–86.

Deibel, D., Cavaletto J.F., Riehl M. and Gardner, W.S., 1992. Lipid and lipid class content of the pelagic tunicate *Oikopleura vanhoeffeni*. *Marine Ecology Progress Series*, **88**, 297–302.

Delage, Y. and Hérouard, E. 1898. Les Procordés. *Traité de Zoologie Concrète*, Vol. 7, 379 pp. Schleicher Frères, Paris.

Delsman, H.C. 1911. Beiträge zur Entwicklungsgeschichte von *Oikopleura dioica*. *Verhandelingen uit het Rijksinstituut voor het Onderzoek der Zee*, **3** (2), 1–24.

Delsman, H.C. 1912. Weitere Beobachtungen über die Entwicklung von *Oikopleura dioica*. *Tijdschrift der Nederlandsche Dierkundige vereeniging Leiden*, **12**, 197–205.

Denton, E.J. and Shaw, T.I. 1961. The buoyancy of gelatinous marine animals. *Journal of Physiology, London*, **161**, 14–15P.

Deraniyagala, P.E.P. 1930. The Testudinata of Ceylon. *Ceylon Journal of Science B, (Spolia Zeylanica)*, **16**, 43–88.

Dhandapani, P. 1981. The role of thaliacea in the marine plankton ecosystem. *In Proceeding of the Symposium on Ecological Animal Population, Zoological Survey of India*, Part 1, pp. 51–71.

Doak, W. 1971 The Poor Knight Islands (New Zealand). *Fathom*, **1** (7), 28–31.

Downs, J.N. 1989. Implications of the phaeopigment, carbon, and nitrogen content of sinking particulates for the origin of export production, pp. 1–196. PhD dissertation, University of Washington, Seattle.

Drach, P. 1948. La notion de Procordé et les embranchements des Cordés. In *Traité de Zoologie* (P.P. Grassé, ed.), Vol. 11, pp. 545–551. Masson et Cie, Paris.

Dragovitch, A. 1970. The food of bluefin tuna (*Thunnus thynnus*) in the western North Atlantic. *Transactions of the American Fisheries Society*, **99**, 726–732.

Dragovitch, A. and Potthoff, T. 1972. Comparative study of the food of skipjack and yellowfin tuna off the coast of West Africa. *Fishery Bulletin*, **70**, 1087–1110.

Drits, A.V., Arashkevich, E.G. and Semenova, T.N. 1992. *Pyrosoma atlanticum* (Tunicata, Thaliacea): grazing impact on phytoplankton standing stock and role in organic carbon flux. *Journal of Plankton Research*, **14**, 799–809.

Du Buit, M.H. 1978. Alimentation de quelques poissons teléostéens de profondeur dans la zone du seuil de Wyville Thomson. *Oceanologica Acta*, **1**, 129–134.

Dudarev, V.A. 1979. Conditions of existence and biological characteristics of *Caprodon* of the Tasman Sea. *Studies on Fish Biology and Commercial Oceanography*, no. 10, 99–105 (in Russian).

Dudochkin, A.S. 1988. The food of the grenadier, *Macrourus holotrachys* in the southwestern Atlantic. *Journal of Ichthyology*, **28**, 72–76.

Duggins, D.O. 1981. Sea urchins and kelp: the effects of short term changes in urchin diet. *Limnology and Oceanography*, **26**, 391–394.

Duhamel, G. 1981. Characteristiques biologiques des principales espèces du plateau continental des Iles Kerguelen, *Cybium*, **5**, 19–32.

Duhamel, G. and Hureau, J.C. 1985. The role of zooplankton in the diets of certain sub-Antarctic marine fish. In *Antarctic Nutrient Cycles and Food Webs* (W.R. Siegfried, P.R. Condy and R.M. Laws, eds), pp. 421–429. Springer-Verlag, Berlin.

Duhamel, G. and Pletikosic, M. 1983. Données biologiques sur les Notothenidae des Iles Crozet. *Cybium*, **7**, 43–57.

Duka, L.A. 1986. Feeding of *Ceratoscopelus warmingii* (Myctophidae) in the tropical Atlantic. *Journal of Ichthyology*, **26**, 89–95.

Dunbar, R.B. and Berger, W.H., 1981. Fecal pellet flux to modern bottom sediment of Santa Barbara Basin (California) based on sediment trapping. *Geological Society of America Bulletin*, **92**, 212–218.

Ebara, A. 1954. The periodic reversal of the heart-beat in *Salpa fusiformis*. *Science Reports of the Tokyo Bunrika Daigaku*, **7** (110), 199–209.

Ehrenbaum, E. 1936. Naturgeschichte und wirtschaftliche Bedeutung des Seefische Nordeuropas. In *Handbuch der Seefischerei Nordeuropas*. H. Lübbert and E. Ehrenbaum, eds), Vol. 2, p. 337. Schweizerbart'sche Verlag, Stuttgart.

Eisen, G. 1874. *Vexillaria speciosa* n. sp. Ett bidrag till Appendikulariornas anatomi. *Kungliga Svenska Vetenskapens Akademiens Handlingar*, **12**, 9, 15.

Emery, C. 1882. Contribuzioni all'Ittiologia. IV. Sulle condizioni di vita di giovani individui de Tetragonurus cuvieri. *Mittheilungen aus der Zoologischen Station zu Neapel*, **3**, 281–283.

Entz, G. 1884. Ueber Infusorien des Golfes von Neapel. *Mittheilungen aus der Zoologischen Station zu Neapel*, **5**, 289–444.

Ernst, C.H. and Barbour, R.W. 1989. *Turtles of the World*. Smithsonian Institution Press, Washington.

Esaias, W.E. and Curl, H.C. 1972 Effect of dinoflagellate bioluminescence on copepod ingestion rates. *Limnology and Oceanography*, **17**, 901–906.

Esnal, G.B. 1978. Caracteristicas generales de la distribucion de Tunicados pelagicos del Atlantico Sudoccidental, con algunas observaciones morfologicas. *Physis*, **38A**, 91–102.

Esnal, G.B. and Castro, R.J. 1985. *Oikopleura albicans* (Leuckart, 1853), un estudio biométrico. (Tunicata: Appendicularia). *Neotropica, La Plata*, **31**, 111–117.

Esnal, G.B. and Simone, L.C. 1982. Doliolidos (Tunicata, Thaliacea) del Golfo de Mexico y Mar Caribe. *Physis*, **40A**(99), 51–57.

Esnal, G.B., Capitanio, F.L. and Simone, L.C. 1996. Concerning intraspecific taxa in *Fritillaria borealis* Lohmann (Tunicata, Appendicularia). *Bulletin of Marine Science*, **59**, 461–468.

Essenberg, C.E. 1926. Copelata from the San Diego region. *University of California Publications in Zoology*, **28**, 22, 399–521.

Evans, F. and Shaarma, R.E. 1963. The Exocoetidae of the Petula Transatlantic Expedition. *Journal of the Linnean Society (Zoology)*, **45**, 53–59.

Fancett M.S. 1988. Diet and prey selectivity of Scyphomedusae from Port Phillip Bay, Australia. *Marine Biology*, **98**, 4, 503–509.

Fedele, M. 1920. Nuovo organo di senso nei Salpidae. *Monitore Italiano Zoologico*, **31**, 10–17.

Fedele, M. 1921. Sulla nutrizione degli animali pelagici Ricerche sui Doliolidae. *Memorie Rendiconti Comitate Talassografico Italiano Venezia*, **78**, 1–26.

Fedele, M. 1923a. Le attività dinamiche ed i rapporti nervosi nella vita dei Doliolidae. *Pubblicazione della Stazione Zoologica di Napoli*, **4**, 129–240.

Fedele, M. 1923b. Simmetria ed unità dinamica nelle catene di *Salpa*. *Bollettino della Società di Naturalisti di Napoli*, **36**, 20–32.

Fedele, M. 1925. Contributo alla conoscenza dei rapporti neuro-muscolari. Le espansioni motrici intercalari nei Thaliacea. *Bollettino della Società di Naturalisti di Napoli*, **37**, 250–257.

Fedele, M. 1932. Muscoli ed attività nei Thaliacea. *Bollettino della Società di Naturalisti di Napoli*, **44**, 237–250.

Fedele, M. 1933a. Richerche sulla natura dei ritmi muscolari negli invertebrati. *Archives de Science Biologique*, **20**, 107–143.

Fedele, M. 1933b. Sul complesso della funzioni che intervengono nel meccanismo ingestivo dei Salpidae. *Rendiconti del Accademia dei Lincei*, **27**, 241–245.

Fedele, M. 1933c. Sulla nutrizione degli animali pelagici III. Richerche sui Salpidae. *Bollettino della Società di Naturalisti di Napoli*, **45**, 49–118.

Fedele, M. 1933d. Sul ritmo muscolare somatico delle Salpe. *Bolletino della Societa Italiano di Biologia Sperimentale*, **8**, 475–478.

Fedele, M. 1938a. Le assimitrie neuro-sensoriali e i limiti e significato della 'enantiomorfa' nelle salpe aggregate. *Archivio Zoologico Torino*, **24**, 443–526.

Fedele, M. 1938b. Per una migliore conoscenza comparativa della fibre muscolare striata. La muscolatura somatica dei *Thaliacea* studiata dal punto di vista strutturale ed etologico. *Archivio Zoologico Italiano*, **25**, 385–436.

Fedosova, R.A. 1976. Some data on the nutrition of the boarfish *Pentaceros richardsonii* (Smith) at the banks of the Hawaiian Range. *Studies on Fish Biology and Commercial Oceanography*, **7**, 29–36 (in Russian).

Feigenbaum, D. 1991. Food and feeding behavior. In *The Biology of Chaetognaths* (Q. Bone, H. Kapp, and A.C. Pierrot-Bults, eds), pp. 45–54. Clarendon Press, Oxford, UK.

Fenaux, R. 1960. Un appendiculaire nouveau: *Appendicularia tregouboffi* n. sp., récolté dans le plancton de Villefranche-sur-Mer. *Bulletin de la Société Zoologique de France*, **85**, 120–122.

Fenaux, R. 1961a. Existence d'un ordre cyclique d'abondance relative maximale chez les Appendiculaires de surface (Tuniciers pélagiques). *Comptes Rendus Hebdomadaires des Séances de l'Académie des Sciences*, **253**, 2271–2273.

Fenaux, R. 1961b. *Fritillaria fagei* n. sp. un Appendiculaire nouveau découvert dans le plancton de Villefranche-sur-Mer. *Rapports et Procés-verbaux des Réunions — Commission Internationale pour l'Exploration Scientifique de la Mer Méditerranée*, **16**, 147–148.

Fenaux, R. 1961c. Rôle du pylore chez *Fritillaria pellucida* Busch (Appendiculaire). *Comptes Rendus Hebdomadaires des Séances de l'Académie des Sciences*, **252**, 2936–2938.

Fenaux, R. 1963. Ecologie et biologie des Appendiculaires méditerranéans (Villefranche-sur-Mer). *Vie et Milieu*, supplément 16. 1–142.

Fenaux, R. 1965. Contribution à la connaissance d'un Appendiculaire peu commun, *Megalocercus abyssorum* Chun, 1888. *Vie et Milieu*, **15** (4), 979–991.

Fenaux, R. 1966. Synonymie et distribution geographique des appendiculaires. *Bulletin de l'Institut Océanographique Monaco*, **66**, 1–23.

Fenaux, R. 1967a. Une variété de *Fritillaria pellucida* (Busch), 1851, récoltée dans la mer d'Oman. *Cahiers O.R.S.T.O.M., Série Océanographie*, **4** (2), 147–151.

Fenaux, R. 1967b. Les Appendiculaires des mers d'Europe et du bassin méditerranéen. *Faune de l'Europe et du Bassin Méditerranéen*, **2**, 1–116.

Fenaux, R. 1968a. Le méchanisme alimentaire chez les Tuniciers. *Annales de Biologie*, **7–8**, 345–367.

Fenaux, R. 1968b. Appendiculaires. (Campagne de la *Calypso* au large des côtes atlantiques de l'Amérique du Sud (1961–1962) (première partie)). *Annales de l'Institut Océanographique*, Paris (N.S.), **45** (2), 33–46.

Fenaux, R. 1968c. Distribution verticale de la fréquence chez quelques Appendiculaires. *Rapports et Procès-verbaux des Réunions — Commission Internationale pour l'Exploration Scientifique de la Mer Méditerranée*, **19**, 513–515.

Fenaux, R. 1968d. Quelques aspects de la distribution verticale chez les Appendiculaires en Méditerranée. *Cahiers de Biologie Marine*, **9**, 23–29.

Fenaux, R. 1968e. Algunas Apendicularias de la costa Peruana Sur quelques Appendiculaires des côtes péruviennes. *Boletim del Instituto del Mar del Peru*, **9**, 536–552.

Fenaux, R. 1969. Les Appendiculaires du golfe du Bengale. Expédition internationale de l'océan Indien (Croisières du '*Kistna*', juin–août 1964). *Marine Biology*, **2**, 252–263.

Fenaux, R. 1970. Les Appendiculaires de Madagascar (région de Nosy-Bé). Variations saisonnières. *Cahiers O.R.S.T.O.M., Série Océanographie*, **7** (4), 29–37.

Fenaux, R. 1971a. La couche oïkoplastique de l'Appendiculaire *Oikopleura albicans* (Leuckart) (Tunicata). *Zeitschrift für Morphologie der Tiere*, **69** (2), 184–200.

Fenaux, R. 1971b. Sur les Appendiculaires de la Méditerranée orientale. *Bulletin du Muséum National d'Histoire Naturelle* (2), **42** (6), 1208–1211.

Fenaux, R. 1972. Variations saisonnières des Appendiculaires de la région nord Adriatique. *Marine Biology*, **16**, 310–319.

Fenaux, R. 1973. Appendicularia from the Indian Ocean, the Red Sea and the Persian Gulf. In *The Biology of the Indian Ocean* (B. Zeitzschel, ed.), pp. 409–414. Springer-Verlag, Berlin.

Fenaux, R. 1976a. The appendicularians. In *Monographs on Oceanographic Methodology* (H.F. Steadman, ed.), pp. 309–312. UNESCO Press Paris.

Fenaux, R. 1976b. Cycle vital d'un Appendiculaire: *Oikopleura dioica* Fol, 1872. Description et chronologie. *Annales de l'Institut Océanographique, Paris* (N.S.) **52**, 89–107.

Fenaux, R. 1980. Premières données sur l'écologie des Appendiculaires du golfe d'Elat. *Israel Journal of Zoology*, **28**, 177–192.

Fenaux, R. 1985. Rhythm of secretion of oikopleurid's houses (International symposium on marine plankton, Shimizu, Japan, 22 July–1 August 1984). *Bulletin of Marine Science*, **37**, 498–503.

Fenaux, R. 1986a. The house of *Oikopleura dioica* (Tunicata, Appendicularia): structure and functions. *Zoomorphology*, **106**, 224–231.

Fenaux, R. 1986b. Influence de la maille du filet sur l'estimation des populations d'Appendiculaires *in situ*. *Rapports et Procès-verbaux des Réunions — Commission Internationale pour l'Exploration Scientifique de la Mer Méditerranée*, **30** (2), 203.

Fenaux, R. 1989. Les mécanismes de l'alimentation chez les Appendiculaires. *Océanis*, **15**, 33–39.

Fenaux, R. 1992. A new mesopelagic appendicularian: *Oikopleura villafrancae* sp. nov. *Journal of the Marine Biological Association of the United Kingdom*, **72**, 911–914.

Fenaux, R. 1993a. The classification of Appendicularia (Tunicata): history and current state. *Mémoires de l'Institut Océanographique, Monaco*, **17**, 1–123.

Fenaux, R. 1993b. A new genus of midwater appendicularian: *Mesoikopleura* with four species. *Journal of the Marine Biological Association of the United Kingdom*, **73**, 635–646.

Fenaux, R. and Dallot, S. 1980. Répartition des Appendiculaires au large des côtes de Californie. *Journal of Plankton Research*, **2**, 145–167.

Fenaux, R. and Gorsky, G. 1981. La fécondité de l'Appendiculaire *Oikopleura dioica* Fol, 1872. *Rapports et Procés-verbaux des Réunions — Commission Internationale pour l'Exploration Scientifique de la Mer Méditerranée*, **27**, 195–196.

Fenaux, R. and Godeaux, J. 1970. Répartition verticale des Tuniciers pélagiques au large d'Eilat (Golfe d'Aqaba). *Bulletin de la Societé Royale des Sciences de Liège*, **39**, 200–209.

Fenaux, R. and Gorsky, G. 1983. Cycle vital et croissance de l'Appendiculaire *Oikopleura longicauda* (Vogt), 1854. *Annales de l'Institut Océanographique, Paris* (N.S.), **59**, 107–116.

Fenaux, R. and Hirel, B. 1972. Cinétique du Déploiement de la logette chez l'Appendiculaire *Oikopleura dioica* Fol, 1872. *Comptes Rendus Hebdomadaires des Scéances de l'Académie des Sciences*, **275D**, 449–452.

Fenaux, R. and Malara, G. 1990. Taux de filtration de l'appendiculaire *Oikopleura dioica* Fol 1872. *Rapports et Procès-verbaux des Réunions — Commission Internationale pour l'Exploration Scientifique de la Mer Méditerranée*, **32**(1), 20.

Fenaux, R. and Palazzoli, I. 1979. Estimation *in situ* d'une population d'*Oikopleura longicauda* (Appendicularia) à l'aide de deux filets de maille différente. *Marine Biology*, **55**, 197–200.

Fenaux, R. and Palazzoli, I. 1983. Importance de la biomasse des Appendiculaires pour évaluer leur rôle dans le réseau trophique du milieu pélagique. *Rapports et Procès-verbaux des Réunions — Commission Internationale pour l'Exploration Scientifique de la Mer Méditerranée*, **28**, 191–193.

Fenaux, R. and Youngbluth, M.J. 1990. A new mesopelagic appendicularian, *Mesochordaeus bahamasi* gen. nov., sp. nov. *Journal of the Marine Biological Association of the United Kingdom*, **70**, 755–760.

Fenaux, R. and Youngbluth, M.J. 1991. Two new mesopelagic appendicularians: *Inopinata inflata* gen. nov., sp. nov., *Mesopelagica caudaornata* gen. nov., sp. nov. *Journal of the Marine Biological Association of the United Kingdom*, **71**, 613–621.

Fenaux, R., Bedo, A. and Gorsky, G. 1986. Premières données sur la dynamique d'une population d'*Oikopleura dioica* Fol, 1872 (Appendiculaire) en élevage. *Canadian Journal of Zoology*, **64**, 1745–1749.

Fenaux, R., Galt, C.P. and Carpine-Lancre, J. 1990. Bibliographie des Appendiculaires (1821–1989). *Mémoires de l'Institut Océanographique, Monaco*, **15**, 1–129.

Fernex, F.E., Braconnot, J-C., Dallot, S. and Boisson, M. 1996. Is ammonification rate in marine sediment related to plankton composition and abundance? A time series study in Villefranche Bay (NW Mediterranean). *Estuarine, Coastal and Shelf Science*, **43**, 359–371.

Fenchel, T. 1984. Suspended marine bacteria as a food source. In *Flows of Energy and Materials in Marine Ecosystems. Theory and Practice* (M.J.R. Fasham, ed.), pp. 301–315. Plenum Press, New York.

Fenchel, T. 1988. Marine plankton food chains. *Annual Review of Ecology and Systematics*, **19**, 19–38.

Fereira, M.M. 1968. Sobre a alimentacão da aruana, *Chelonia mydas* Linnaeus, ao longo da costa estado do Ceara. *Arquivos Estacao Biologica Marinha Universidade Federal do Ceara*, **8**, 83–86.

Fiala-Medioni, A. and Pequignat, E. 1975. Mise en évidence d'activités enzymatiques dans la branchie des filtreurs benthiques (Ascidies). *Comptes Rendus Hebdomadaires de l'Académie des Sciences de Paris*, **281**, 1123–1126.

Fiala-Medioni, A. and Pequignat, E. 1980. Direct absorption of aminoacids and glucose by the branchial sac and the digestive tract of benthic filter feeders (Ascidians). *Journal of Zoology*, **192**, 403–419.

Field, K.G., Olsen, G.J., Lane, D.J., Giovannoni, S.J., Ghiselin, M.T., Raff, E.C., Pace, N.R. and Raff, R.A. 1988. Molecular phylogeny of the animal kingdom. *Science*, **239**, 748–753.

Fisher, N.S., Nolan, C.V. and Gorsky, G. 1991. The retention of cadmium and zinc in appendicularian houses. *Oceanologica Acta*, **14**, 427–430.

Fisher, N.S., Bjerregaard, P. and Fowler, S.W. 1983. Biokinetics of Americium in marine plankton. *Rapports et Procès-verbaux des Réunions — Commission Internationale pour l'Exploration Scientifique de la Mer Méditerranée*, **28**, 7, 247–249.

Fitch, J.E. 1949. Some unusual occurrences of fish on the Pacific Coast. *California Fish and Game*, **35**, 41–59.

Fitch, J.E. 1952. Toxicity and taxonomic notes on the squaretail *Tetragonurus cuvieri*. *California Fish and Game*, **38**, 251–252.

Fitch, J.E. and Lavenberg, R.J. 1968. *Deep-water Fishes of California*, 155 pp. University of California Press, Berkely, CA.

Flagg, C.N. and Smith, S.L. 1989. On the use of the acoustic Doppler current profiler to measure zooplankton abundance. *Deep-Sea Research*, **36**, 455–474.

Fleminger, A, 1986. The Pleistocene equatorial barrier between the Indian and Pacific Oceans and a likely cause for Wallace's line. In *Pelagic Biogeography* (A.C. Pierrot-Bults, S. van der Spoel, B.J. Zahuranec and R.K. Johnson, eds), *UNESCO Technical Papers in Marine Science*, Vol. 7, pp. 84–97. UNESCO, Paris.

Flood, P.R. 1973. Ultrastructural and cytochemical studies on the muscle innervation in Appendicularia. *Journal de Microscopie*, **18**, 317–326.

Flood, P.R. 1978. Filter characteristics of appendicularian food catching nets. *Experientia*, **34** (2), 173–175.

Flood, P.R. 1981. On the ultrastructure of mucus. *Biomedical Research*, **2**, 49–53.

Flood, P.R. 1983. The gelatinous house of *Oikopleura dioica* (Appendicularia, Tunicata); its architecture and water filtration mechanism. *Program/Abstracts — Western Society of Naturalists*, **64**, 18.

Flood, P.R. 1991a. A simple technique for preservation and staining of the delicate houses of appendicularian tunicates. *Marine Biology*, **108**, 105–110.

Flood, P.R. 1991b. Architecture of, and water circulation and flow rate in, the house of the planktonic tunicate *Oikopleura labradoriensis*. *Marine Biology*, **111**, 95–111.

Flood, P.R. 1994. Appendicularian houses — Architectural wonders of the sea. In *Evolution of Natural Structures* (Proceedings of the 3rd International Symposium Sonderforschungsbereich 230), pp. 151–156. Universität Stuttgart und Universität Tübingen.

Flood, P.R. and Afzelius, B.A. 1978. The spermatozoon of *Oikopleura dioica* Fol (Larvacea, Tunicata). *Cell and Tissue Research*, **191**, 27–37.

Flood, P.R. and Fiala-Medioni, A. 1981. Ultrastructure and histochemistry of the food trapping mucous film in benthic filter-feeders (Ascidians). *Acta Zoologica (Stockholm)*, **62**, 53–65.

Flood, P.R., Bjugn, R. and Frölich, K.W. 1983. Structural aspects of mucus feeding filters and their relation to mucoid substances in general. *Journal of Ultrastructure Research*, **85**, 104.

Flood, P.R., Deibel, D. and Morris, C.C. 1990. Visualization of the transparent gelatinous house of the pelagic tunicate *Oikopleura vanhoeffeni* using *Sepia* ink. *Biological Bulletin, Woods Hole*, **178**, 118–125.

Flood, P.R. Deibel, D. and Morris, C.C. 1992. Filtration of colloidal melanin from sea water by planktonic tunicates. *Nature, London*, **355**, 630–632.

Flood, P.R., Galt, C.P. and Youngbluth, M.J. 1996. An improved technique for field preservation and laboratory analysis of delicate mucous structures from gelatinous zooplankton. In preparation.

Foerder, C.A. and Shapiro, B.M. 1977. Release of ovoperoxidase from sea urchin eggs hardens the fertilization membrane with tyrosin crosslinks. *Proceedings of the National Academy of Sciences USA*, **74**, 4214–4218.

Fogel, M., Schmitter, R.E. and Hastings, J.W. 1972. On the physical identity of scintillons, bioluminescent particles in *Gonyaulax polyedra*. *Journal of Cell Science*, **11**, 305–317.

Fol, H. 1872. Etudes sur les appendiculaires du Détroit de Messine. *Mémoires de la Société de Physique et d'Histoire naturelle de Genève*, **21**, 445–499.

Fol, H. 1874. Note sur un nouveau genre d'Appendiculaires. *Archives de Zoologie Expérimentale et Générale*, **3**, XLIX–LIII.

Fol, H. 1876. Ueber die Schleimdrüse oder den Endostyl der Tunicaten. *Morphologisches Jahrbuch*, **1**, 222–242.

Forneris, L. 1965. Appendicularian species groups and southern Brazil water masses. *Boletim do Instituto Oceanografico, Sao Paulo*, **14**, 53–113.

Forsskål, P. 1775. *Descriptiones animalium*, quae in itinere orientali observavit. Post mortem auctoris edidit C. Niebuhr 2 vols. Möller Hauniae.

Fortier, L. Le Fèvre, J. and Legendre, L. 1994. Export of biogenic carbon to fish and to the deep ocean: the role of large planktonic microphages. *Journal of Plankton Research*, **16**, 809–839.

Fowler, S.W. and Small, L.F. 1972. Sinking rates of euphausiid fecal pellets. *Limnology and Oceanography*, **17**, 293–296.

Fowler, S.W. and Knauer, G.A. 1986. Role of large particles in the transport of elements and organic compounds through the oceanic water column. *Progress in Oceanography*, **16**, 147–194.

Foxton, P. 1961. *Salpa fusiformis* Cuvier and related species. *Discovery Reports*, **32**, 1–32.

Foxton, P. 1966. The distribution and life-history of *Salpa thompsoni* Foxton with observations on a related species, *Salpa gerlachei* Foxton. *Discovery Reports*, **34**, 1–116.

Frank, T., Widder, E.A., Latz, M.I., and Case, J.F. 1984. Dietary maintenance of bioluminescence in a deep-sea mysid. *Journal of Experimental Biology*, **109**, 385–389.

Franqueville, C. 1971. Macroplancton profond (invertébrés) de la Méditerranée nord-occidentale. *Tethys*, **3**, 11–56.

Franzen, Å. 1992. Spermatozoan ultrastructure and spermatogenesis in aplousobranch ascidians, with some phylogenetic considerations. *Marine Biology*, **113**, 77–88.

Fraser, J.H. 1949. The distribution of Thaliacea in Scottish waters 1920–1939. *Scottish Home Department Fisheries Division Scientific Investigations*, **1**, 1–44.

Fraser, J.H. 1960. Plankton investigations from Aberdeen in 1958. *Année Biologique Copenhagen*, **15**, 57–58.

Fraser, J.H. 1961. The oceanic and bathypelagic plankton of the North East Atlantic and its possible significance to fisheries. *Department of Agriculture and Fisheries Scotland Marine Research*, **4**, 1–48.

Fraser, J.H. 1962a. Investigations from Aberdeen in 1960: zooplankton. *Anné Biologique Copenhagen*, **17**, 81.

Fraser, J.H. 1962b. The role of ctenophores and salps in zooplankton and standing crop. *Rapports et Procès-verbaux des Réunions — Commission Internationale pour l'Exploration Scientifique de la Mer Méditeranée*, **153**, 121–123.

Fraser-Brunner, A. 1951. The ocean sunfishes (family Molidae). *Bulletin of the British Museum of Natural History, Zoology*, **1** (6), 87–21.

Frazier, J.M., Meneghel, D. and Achaval, F. 1985. A classification of the feeding habits of *Dermochelys coriaceae. Journal of Herpetology*, **19**, 159–160.

Fredriksson, G. and Olsson, R. 1981. The oral gland cells of *Oikopleura dioica* (Tunicata Appendicularia). *Acta Zoologica (Stockholm)*, **62**, 195–200.

Fredriksson, G. and Olsson, R. 1991. The subchordal cells of *Oikopleura dioica* and *Oikopleura albicans* (Appendicularia, Chordata). *Acta Zoologica (Stockholm)*, **72**, 251–256.

Fredriksson, G., Öfverholm, T. and Ericson, L.E. 1985. Ultrastructural demonstration of iodine binding and peroxidase activity in the endostyle of *Oikopleura dioica* (Appendicularia). *General and Comparative Endocrinology*, **58**, 319–327.

Fredriksson G., Fenaux, R. and Ericson, L.E. 1989. Distribution of peroxidase and iodination activity in the endostyles of *Oikopleura albicans* and *Oikopleura longicauda* (Appendicularia, Chordata). *Cell and Tissue Research*, **255**, 505–510.

Frenzel, J. 1885. Ueber einige in Seethieren lebende Gregarinen. *Archiv für Mikroskopische Anatomie*, **24**, 545–588.

Frost, B.W., Landry, M.R. and Hassett, R.P. 1983. Feeding behaviour of the large calanoid copepods *Neocalanus cristatus* and *N. plumchrus* from the subantarctic Pacific Ocean. *Deep Sea Research*, **30**, 1–13.

Frost, N., Lindsay, S.T. and Thompson, H. 1933. Hydrographic and biological investigations. *Report of the Newfoundland Fishery Research Committee*, **2**, 58–74.

Fukuchi, M., Tanimura, A. and Ohtsuka, H. 1985. Zooplankton community conditions under sea ice near Syowa station, Antarctica. *Bulletin of Marine Science*, **37**, 518–528.

Furnestin, M.L. 1957. Chaetognathes et zooplancton du secteur atlantique marocain. *Revue des Travaux de l'Institut des Pêches Maritimes*, **21** (1–2), 1–356.

Furuhashi, K. 1966. Droplets from the plankton net. XXIII. Records of *Sapphirina salpae* Giesbrecht from the North Pacific, with notes on its copepodite stages. *Publications of the Seto Marine Biological Laboratory*, **14**, 123–127.

Gadomski, D.M. and Boehlert, G.W. 1984. Feeding ecology of pelagic larvae of English sole *Parophrys vetulus* and butter sole *Isopsetta isolepis* off the Oregon coast. *Marine Ecology Progress Series*, **20**, 1–12.

Galt, C.P. 1970. Population composition and annual cycle of larvacean tunicates in Elliott Bay, Puget Sound, 38 pp. MSc thesis, University of Washington, Seattle.

Galt, C.P. 1972. Development of *Oikopleura dioica* (Urochordata: Larvacea): ontogeny of behavior and of organ systems related to construction and use of the house, PhD thesis, WAIX+ 84 pp. University of Washington, Seattle.

Galt, C.P. 1978. Bioluminescence, dual mechanism in a planktonic tunicate produces brilliant surface display. *Science*, **200**, 70–72.

Galt, C.P. 1995. Larvacean houses deter predation by fish and chaetognaths. In *Sensory Ecology and Physiology of Zooplankton Symposium, Honolulu, Hawaii, 8–12 January 1995* (abstract).

Galt, C.P. and Fenaux, R. 1990b. Urochordata Larvacea. In *Reproductive Biology of Invertebrates* (K.G. Adiyodi and R.G. Adiyodi, eds), pp. 471–500. Oxford & IBH Publishing, New Delhi.

Galt, C.P. and Flood, P.R. 1984. Distribution of bioluminescent inclusions in expanded larvacean houses. *American Zoologist*, **24**, 772 (abstract).

Galt, C.P. and Grober, M.S. 1985. Total stimulable luminescence of *Oikopleura* houses (Urochordata, Larvacea). *Bulletin of Marine Science*, **37**, 765 (abstract).

Galt, C.P. and Mackie, G.O. 1971. Electrical correlates of ciliary reversal in *Oikopleura. Journal of Experimental Biology*, **55**, 205–212.

Galt, C.P. and Smith, D.C. 1990. Bioluminescence potential of appendicularian houses (Tunicata: Larvacea): Interspecific effects of size and age of house on total stimulable luminescence. *EOS, Transactions of the American Geophysics Union*, **71**, 175 (abstract).

Galt, C.P. and Sykes, P.F. 1983. Sites of bioluminescence in the appendicularians *Oikopleura dioica* and *O. labradoriensis. Marine Biology*, **77**, 155–159.

Galt, C.P. and Tisdale, L.A. 1983. Global distribution of luminescent larvacean tunicates. *Program/Abstract — Western Society of Naturalists*, **64**, 19.

Galt, C.P., Grober, M.S. and Sykes, P.F. 1985. Taxonomic correlates of bioluminescence among appendicularians (Urochordata: Larvacea). *Biological Bulletin, Woods Hole*, **168**, 125–134.

Ganapati, P.N. and Bhavanarayana, P.V. 1958. Pelagic tunicates as indicators of water movements off Waltair coast. *Current Science*, **27**, 2, 57–58.

Garstang, W. 1928. The morphology of the Tunicata and its bearing on the phylogeny of the Chordata. *Quarterly Journal of Microscopical Science*, **72**, 51–187.

Garstang, W. 1933. Report on the Tunicata. Part. 1. Doliolida. In *British Antarctic ('Terra Nova') Expedition 1910. Natural History Report, Zoology*, Vol. 4 (6), 195–252.

Garstang, W. 1937. On the anatomy and relations of the appendicularian *Bathochordaeus*, based on a new species from Bermuda (*B.stygius*, sp.n.). *The Journal of the Linnean Society, London, Zoology*, **40**, 283–303.

Garstang, W. and Platt, M.I. 1928. On the asymmetry and closure of the endostyle in *Cyclosalpa pinnata*. *Proceedings of the Leeds Philosophical Society*, **1**, 325–334.

Gartner, J.V. and Musick, J.A. 1989. Feeding habits of the deep-sea fish, *Scopelogadus beanii* (Pisces, Melamphaidae), in the western North Atlantic. *Deep-Sea Research*, **36**, 1457–1469.

Gavrilov, G.M. and Markina, N.P. 1979. The feeding ecology of fishes of the genus *Seriolella* (fam. Nomeidae) on the New Zealand Plateau. *Journal of Ichthyology*, **19**, 128–135.

Gegenbaur, C., 1855. Bemerkungen über die Organisation der Appendicularien. *Zeitschrift für Wissenschaftliche Zoologie*, **6**, 406–427.

Genin, A., Gal, G., Linel, D. and Prado-Por, A.S. 1997. Zooplanktivory in coral reefs: predation rates and effects on prey distribution and size. (in preparation)

Georges, D., Holmberg, K. and Olsson, R. 1988. The ventral midbrain cells in *Oikopleura dioica* (Appendicularia). *Acta Embryologica Morphologica Experimentalis*, **9**, 39–47.

Gerber, R.P. and Marshall, N. 1974. Ingestion of detritus by the lagoon pelagic community at Eniwetok Atoll. *Limnology and Oceanography*, **19**, 815–824.

Gianguzza, M. and Dolcemascolo, G. 1984. Formation of the test in the swimming larva of *Ciona intestinalis*: an ultrastructural study. *Journal of Submicroscopic Cytology*, **16**, 289–297.

Gibbons, M.J. 1997. Vertical distribution and feeding of Thalia democratica on the Agulhas bank during March 1994. *Journal of the Marine Biological Association of the United Kingdom*, **77**, 493–506.

Giesbrecht, W. 1892. Systematik and Faunistik der pelagischen Copepoden. *Fauna and Flora des Golfes von Neapel*, **19**,

Giglioli, E.H. 1870. La fosforescenza del mare. Note pelagiche ed osservazioni fatte durante un viaggio di circumnavigazione 1865–68, colla descrizione di due nuove noctiluche. *Atti della real Accademia di Scienza di Torino, Classi dei Scienze Fisica e Matematica*, **5**, 485–505.

Gilmer, R.W. 1974. Some aspects of feeding in the thecosomatous pteropod molluscs. *Journal of Experimental Marine Biology and Ecology*, **15**, 127–144.

Godeaux, J. 1954. Observations sur la glande pylorique des thaliacés. *Annales de la Société Royale Zoologique de Belgique*, **85**, 103–118.

Godeaux, J. 1955. Stades larvaires du *Doliolum*. *Bulletin de l'Académie Royale de Belgique, Serie 5*, **41**, 769–787.

Godeaux, J. 1957–1958 Contribution á la connaissance des Thaliacés (Pyrosome et Doliolum). *Annales de la Société Royale Zoologique de Belgique*, **88**, 1–285.

Godeaux, J. 1961. L'oozoïde de *Doliolum nationalis* Borg. *Bulletin de la Société Royale des Sciences de Liège*, **30**, 5–10.

Godeaux, J. 1962. Tuniciers pélagiques. *Bulletin de l'Institut Royale des Sciences Naturelles de Belgique*, **3**, 1–32.

Godeaux, J. 1965. Observations sur la tunique des tuniciers pélagiques. *Rapports et Procès-verbaux verbaux des Réunions — Commission Internationale pour l'Exploration Scientifique de la Mer Méditerranée*, **18**, 457–460.

Godeaux, J. 1967. Une salpe peu connue, *Thalia longicauda* (Quoy and Gaimard, 1824). *Annales de la Société Royale Zoologique de Belgique*, **97**, 91–102.

Godeaux, J. 1971. L'ultrastructure de l'endostyle des Doliolides (Tuniciers, Cyclomyaires). *Comptes Rendues de l'Académie des Sciences Paris*, **272**, 593–595.

Godeaux, J. 1973a. Tuniciers pelagiques de l'Océan Indien. *Journal of the Marine Biological Association of India*, **14**, 263–292.

Godeaux, J. 1973b. A contribution to the knowledge of the Thaliacean faunas of the eastern Mediterranean and the Red Sea. *Israel Journal of Zoology*, **22**, 39–50.

Godeaux, J. 1975. Les Thaliacés et les milieux hypersalins de la Méditerranée orientale et de la mer Rouge septentrionale. *Rapports de la Commission Internationale de la Mer Méditerranée*, **23**, 113–115.

Godeaux J. 1978. Un exemple de variation clinale en Mer Méditerranée. In *Recherche et Technique au Service de l'Environnement*, pp. 339–345. Cebedoced, Liège.

Godeaux, J. 1981. Etude au microscope électronique de l'endostyle des Doliolidae (Tuniciers cyclomyaires). *Annales de la Société Royale de Zoologie de Belgique*, **111**, 151–162.

Godeaux, J. 1987. Distribution of Thaliacea on a transect from the Gulf of Aden to the central Red Sea during the winter monsoon (March 1979). *Oceanologica Acta*, **10**, 197–204.

Godeaux, 1988a. Thaliacés récoltés en mer d'Arabie, dans le golfe Persique, et dans le golfe d'Aden par le Navire Océanographique *'Commandant Robert Giraud'*. *Bulletin des Seances de l'Académie Royale des Sciences d'Outre-Mer*, **34**, 301–324.

Godeaux, J. 1988b. Thaliacés Mediterranéens, une synthèse. *Bulletin de la Société Royale des Sciences de Liège*, **57**, 359–377.

Godeaux, J. 1989. Functions of the endostyle in the tunicates. *Bulletin of Marine Science*, **45**, 228–242.

Godeaux, J. 1990. Urochordata — Thaliacea. In *Reproductive Biology of Invertebrates* (K.G. Adiyodi and R.G. Adiyodi, eds), Vol. IVB, pp. 453–469. Oxford and I.B.H. Publishing, New Delhi.

Godeaux, J. and Breeur, C. 1976 Remarques sur les populations des salpes de la mer Méditerranée. *Rapports de la Commision Internationale de la Mer Méditerranée*, **23**, 73–74.

Godeaux, J. and Meurice, J.C. 1978. Thaliacés récoltés par la troisième expédition antarctique belge (1966–1967) dans les océans Antarctique et Indien. *Bulletin de la Société Royale des Sciences de Liège*, **47**, 363–385.

Godfriaux, B.L. 1969. Food of the predatory demersal fish in Hauraki gulf I. Food and feeding habits of snapper. *New Zealand Journal of Marine and Freshwater Research*, **3**, 473–504.

Godfriaux, B.L. 1974a. Food of tarakihi in western Bay of Plenty and Tasman Bay, New Zealand. *New Zealand Journal of Marine and Freshwater Research*, **8**, 111–153.

Godfriaux, B.L. 1974b. Food of snapper in western Bay of Plenty, New Zealand. *New Zealand Journal of Marine and Freshwater Research*, **8**, 473–504.

Godfriaux, B.L. 1974c Feeding relationships between tarakihi and snapper in western Bay of Plenty, New Zealand. *New Zealand Journal of Marine and Freshwater Research*, **8**, 589–609.

Goffinet, G. and Godeaux, J. 1992. Ultrastructure de l'endostyle des Salpes. *Rapport de la commission Internationale de la Mer Méditerranée*, **33**, 250.

Golovan', G.A. and Pakhorukov, N.P. 1975. Some data on the morphology and ecology of *Alepocephalus bairdii* Good and Bean (Alepocephalidae) central-eastern Atlantic. *Journal of Ichthyology*, **15**, 44–50.

Golovan', G.A. and Pakhorukov, N.P. 1980. New data on the ecology and morphology of *Alepocephalus rostratus* (Alepocephalidae). *Journal of Ichthyology*, **20**, 77–83.

Golovan', A.A. and Pakhorukov, N.P. 1987. Distribution and behavior of fishes on the Naska and Salay Gomez submarine ranges. *Journal of Ichthyology*, **27**, 71–78.

Golovan', A.A., Pakhorukov, N.P. and Tkhorikov, V.B. 1991. Morphoecological characterization of rosefish, *Helicolenus lengerichi* (Scorpaenidae) from the Nasca Ridge. *Journal of Ichthyology*, **31**, 117–128.

Gonzalez, H., Noethig, E.M., Bathmann, U. and Dunman, M. 1994. Sedimentation in the central Barents Sea (Arctic) during July 1991–June 1992. *EOS, Transactions of the American Geophysical Union*, **75**, 127 (abstract).

Gorbatenko, K.M. and Ilinskii, E.N. 1992. Feeding behaviour of the most common fishes in the Bering Sea. *Journal of Ichthyology*, **32**, 52–60.

Gore, R. 1990. Between Monterey tides. *National Geographic*, **177** (2), 2–43.

Gorelova, T.A. 1975. Zooplankton from the stomachs of juvenile lantern fish of the family Myctophidae. *Oceanology*, **14**, 575–580.

Gorelova, T.A. 1980. The feeding of young flyingfishes of the family Exocoetidae and of the smallwing flyingfish, *Oxyporhamphus micropterus*, of the family Hemirhamphidae. *Journal of Ichthyology*, **20**, 60–71.

Gorman, A.L.F., McReynolds, J.S. and Barnes, S.N. 1971. Photoreceptors in primitive chordates fine structure, hyperpolarizing potentials and evolution. *Science, NY*, **172**, 1052–1054.

Gorsky, G. 1980. Optimisation des cultures d'Appendiculaires. Approche du metabolisme de *O. dioica*, 122 pp. Thèse 3e cycle, Université Pierre et Marie Curie, Paris VI, Paris.

Gorsky, G. 1987. Aspects de l'écophysiologie de l'Appendiculaire *Oikopleura dioica* Fol, 1872 *(Chordata: Tunicata)*, Thèse sciences, 2 vol., 102 pp. Université Pierre et Marie Curie Paris VI, Paris.

Gorsky, G. and Palazzoli, I. 1989. Aspects de la biologie de l'Appendiculaire *Oikopleura dioica* Fol. 1872 (Chordata: Tunicata). (Dynamique du plancton gélatineux, Nice-Acropolis, 27–28 octobre 1988). *Océanis*, **15** (1), 39–49.

Gorsky, G., Palazzoli, I. and Fenaux, R. 1984. Premières données sur la respiration des Appendiculaires (Tuniciers pélagiques). *Comptes Rendus de l'Académie des Sciences* (3), **298**, 531–534.

Gorsky, G., Fisher, N.S. and Fowler, S.W. 1984. Biogenic debris from the pelagic tunicate *Oikopleura dioica* and its role in the vertical transport of a transuranium element. *Estuarine and Coastal Shelf Science*, **18**, 13–23.

Gorsky, G., Palazzoli, I. and Fenaux, R. 1987. Influence of temperature changes on oxygen uptake and ammonia and phosphate excretion, in relation to body size and weight, in *Oikopleura dioica* (Appendicularia). *Marine Biology*, **94** (2), 191–201.

Gorsky, G., Fenaux, R. and Palazzoli, I. 1989. Appendiculaires et leur rôle dans le transport vertical de la matière organique. *Journal du Recherche Océanographique*, **14**, 118–119.

Gorsky, G., Laval, P., Youngbluth, M.J. and Palazzoli, I. 1990. Appendiculaires mésopélagiques, indicateurs potentiels des couches riches en matiére organique. *Rapports et Procès-verbaux des Réunions — Commission Internationale pour l'Exploration Scientifique de la Mer Méditerranée*, **32** (1), 200.

Gorsky, G., Lins da Silva, N., Dallot, S., Laval, P., Braconnot, J.-C. and Prieur, L. 1991. Midwater tunicates: are they related to the permanent front of the Ligurian Sea (NW Mediterranean)? *Marine Ecology Progress Series*, **74**, 195–204.

Gorsky, G., Dallot, S., Sardou, J., Fenaux, R., Carré, C. and Palazzoli, I. 1988. C and N composition of some northwestern Mediterranean zooplankton and micronecton species. *Journal of Experimental Marine Biology and Ecology*, **124**, 133–144.

Gorsky, G., Aldorf, C., Kage, M., Picheral, M., Garcia, Y. and Favole, J. 1992. Vertical distribution of suspended aggregates determined by a new underwater video profiler. *Annales de l'Institut Océanographique, Paris (N.S.)*, **69** (1–2), 275–280.

Gorsky, G., Dallot, S., Sardou, J., Fenaux, R., Carré, C. and Palazzoli, I. 1988. C and N composition of some northwestern Mediterranean zooplankton and micronekton species. *Journal of Experimental Marine Biology and Ecology*, **124** (2), 133–144.

Gotshall, D.W., Smith, J.G. and Holbert, A. 1965. Food of the blue rockfish *Sebastodes mystinus*. *California Fish and Game*, **51**, 147–162.

Gowan, J.A. 1963. Geographic variation in *Limacina helicina* in the North Pacific. In, *Speciation in the Sea* (J.P. Harding and N. Tebble, eds), pp. 1–199. The Systematics Association, London.

Gowing, M.M. 1994. Large virus-like particles from vacuoles of phaeodarian radiolarians, zooplankton guts, and particulate material. *EOS, Transactions of the American Geophysical Union*, **75**, 86 (abstract).

Goy, J. 1977. Migrations verticales du zooplancton. *Résultats de Campagnes à la Mer, CNEXO*, **13**, 71–73.

Graeffe, E. 1902. Übersicht der Fauna des Golfes von Trieste. V. Crustacea. *Arbeiten aus der Zoologisches Institut Wien*, **13** (1), 33–80.

Graham, D.H. 1939. Food of the fishes of Otago Harbour and adjacent sea. *Transactions and Proceedings of the Royal Society of New Zealand*, **68**, 421–436.

Grainger, E.H. 1965. Zooplankton from the Arctic Ocean and adjacent Canadian waters. *Journal of the Fisheries Research Board of Canada*, **22**, 543–564.

Gran, H.H. 1929. Investigations of the plankton outside the Romsdalsfjord 1926–1927. *Rapports et Procès-verbaux des Réunions — Commission Internationale pour l'Exploration de la Mer*, **56**, 1–112.

Greve, W., Stockner, J. and Fulton, J. 1976. Towards a theory of speciation in *Beroe*. In *Coelenterate Ecology and Behavior* (G.O. Mackie, ed.), pp. 251–258. Plenum Press, New York and London.

Greze, E.V. 1983. Feeding of epipelagic Myctophidae (Pisces) of the tropical Atlantic. *Ekologiya Morya*, **14**, 18–23.

Griffin, D.J.C. and Yaldwyn, J.C. 1970. Giant colonies of pelagic tunicates (*Pyrosoma spinosum*) from SE Australia and New Zealand. *Nature, London*, **226**, 464–465.

Grimes, C.B. 1979. Diet and feeding ecology of the vermilion snapper, *Rhomboplites aurorubens* (Cuvier) from North Carolina and South Carolina waters. *Bulletin of Marine Science*, **29** (1), 53–61.

Grober, M.S. 1988. Brittle star bioluminescence functions as an aposematic signal to deter crustacean predators. *Animal Behaviour*, **36**, 493–501.

Grobben, C. 1882. Doliolum und sein Generationswechsel nebst Bemerkungen über den Generationswechsel der Acalephen, Cestoden und Trematoden. *Arbeiten aus dem Zoologischen Institute der Universität Wien*, **4** (2), 201–298.

Haedrich, R.L. 1964. Food habits and young stages of North Atlantic *Alepisaurus* (Pisces, Iniomi). *Breviora*, no. 201, 1–15.

Haedrich, R.L. 1969. A new family of aberrant stromateoid fishes from the equatorial IndoPacific. *Dana Reports*, no. 76. 1–14.

Halanych, K.M. 1991. 5S ribosomal RNA sequences inappropriate for phylogenetic reconstruction. *Molecular Biology and Evolution*, **8**, 249–253.

Hallacher, L.E. and Roberts, D.A. 1985. Differential utilization of space and food by the inshore rockfishes (Scorpaenidae:

Sebastes) of Carmel Bay, California. *Environmental Biology of Fish*, **12**, 91–110.

Halliday, R.G. 1969. Reproduction and feeding of *Argentina sphyraena* (Isospondyli) in the Clyde Sea area. *Journal of the Marine Biological Association of the United Kingdom*, **49**, 785–803.

Hamner, W.M. and Robison, B.H. 1992. *In situ* observations of giant appendicularians in Monterey Bay. *Deep-Sea Research*, **39**, 1299–1313.

Hamner, W.M., Madin, L.P., Alldredge, A.L., Gilmer, R.W. and Hamner, P.P. 1975. Underwater observations of gelatinous zooplankton, sampling problems, feeding biology and behavior. *Limnology and Oceanography*, **20**, 907–917.

Hamner, W.M., Jones, M.S., Carleton, J.H., Hauri, I.R. and Williams, D.McB. 1988. Zooplankton, planktivorous fish and water currents on a windward reef face: Great Barrier Reef, Australia. *Bulletin of Marine Science*, **42**, 459–479.

Harant, H. and Vernières, P. 1934. Tuniciers pelagiques. *Resultats des Campagnes Scientifiques du Prince de Monaco*, **88**, 1–45.

Harbison, G.R. and Campenot, R.B. 1979. Effects of temperature on the swimming of salps (Tunicata, Thaliacea): implications for vertical migration. *Limnology and Oceanography*, **24**, 1081–1091.

Harbison, G.R. and Gilmer, R.W. 1976. The feeding rates of the pelagic tunicate *Pegea confoederata* and two other salps. *Limnology and Oceanography*, **21**, 517–528.

Harbison, G.R. and McAlister, V.L. 1979. The filter-feeding rates and particle retention efficiencies of three species of *Cyclosalpa* (Tunicata, Thaliacea). *Limnology and Oceanography*, **24**, 875–892.

Harbison, G.R., Biggs, D.C. and Madin, L.P. 1977. Associations of Amphipoda Hyperiidea with gelatinous zooplankton. II. Associations with Cnidaria, Ctenophora and Radiolaria. *Deep-Sea Research*, **24**, 465–488.

Harbison, G.R., Madin, L.P. and Swanberg, N.R. 1978. On the natural history and distribution of oceanic ctenophores. *Deep-Sea Research*, **25**, 233–256.

Harbison, G.R., McAlister, V.L. and Gilmer, R.W. 1986. The response of the salp, *Pegea confoederata*, to high levels of particulate material: starvation in the midst of plenty. *Limnology and Oceanography*, **31**, 371–382.

Hardwick, J.E. 1970. A note on the behavior of the octopod *Ocythoe tuberculata*. *California Fish and Game*, **56**, 68–70.

Hardy, A.C. 1924. The herring in relation to its animate environment. 1. The food and feeding habits of the herring with special reference to the East coast of England. *Fishery investigations*, **7**, 1–53.

Hardy, A.C. 1936. Observations on the uneven distribution of oceanic plankton. *Discovery Reports*, **11**, 511–538.

Hardy, A.C. 1956. *The Open Sea*, Vol. 1, *The World of Plankton*, 335 pp. Collins, London.

Hardy, A.C. and Gunther, E.R. 1935. The plankton of the South Georgia whaling grounds and adjacent waters, 1926–1927. *Discovery Reports*, **11**, 1–456.

Harrison, N.M. 1984. Predation on jellyfish and their associates by seabirds. *Limnology and Oceanography*, **29**, 1335–1337.

Hartog, J.C. den. 1980. Notes on the food of sea turtles, *Eretmochelys imbricata* (L.) and *Dermochelys coriacea* (Linnaeus). *Netherlands Journal of Zoology*, **30**, 595–610.

Hartog, J.C. den and van Nierop, M.M. 1984. A study on the gut contents of six leathery turtles *Dermochelys coriacea* (Linnaeus) (Reptilia, Testudines, Dermochelyidae) from British waters and from the Netherlands. *Zoologische Verhandlungen (Leiden)*, **209**, 1–36.

Harvey, E.N. 1952. *Bioluminescence*, 649 pp. Academic Press, New York.

Hastings, J.W. 1983a. Biological diversity, chemical mechanisms, and the evolutionary origins of bioluminescent systems. *Journal of Molecular Evolution*, **19**, 309–321.

Hastings, J.W. 1983a. Chemistry and control of luminescence in marine organisms. *Bulletin of Marine Science*, **33**, 818–828.

Henderson, I.M. 1991. Biogeography without area? In *Australian Biogeography* (P.Y. Ladiges, C.J. Humphries and L.W. Martinelli, eds), *Australian Systematic Botany*, Vol. 4 (1), pp. 59–72.

Herdman, W.A. 1888. Report upon the Tunicata collected during the voyage of H.M.S. Challenger during the years 1873–76. Part 3. *Zoology of the Challenger Expedition*, **27**, 1–166.

Herman, Y. 1986. Modes, tempos and causes of speciation in planktonic Foraminifera. In *Pelagic Biogeography* (A.C. Pierrot-Bults, S. van der Spoel, B.J. Zahuranec and R.K. Johnson, eds), UNESCO *Technical Papers in Marine Science*, Vol. 49, pp. 141–148. UNESCO, Paris.

Heron, A.C. 1972a. Population ecology of a colonizing species: the pelagic tunicate *Thalia democratica* I. Individual growth rate and generation time. *Oecologia*, **10**, 269–293.

Heron, A.C. 1972b. Population ecology of a colonizing species: the pelagic tunicate *Thalia democratica* II. Population growth rate. *Oecologia*, **10**, 294–312.

Heron, A.C. 1973. A specialized predator–prey relationship between the copepod *Sapphirina angusta* and the pelagic tunicate *Thalia democratica*. *Journal of the Marine Biological Association of the United Kingdom*, **53**, 429–435.

Heron, A.C. 1975a. A new type of heart mechanism in the invertebrates. *Journal of the Marine Biological Association of the United Kingdom*, **53**, 425–428.

Heron, A.C. 1975b. Advantages of heart reversal in pelagic tunicates. *Journal of the Marine Biological Association of the United Kingdom*, **55**, 959–963.

Heron, A.C. 1976. A new type of excretory mechanism in the tunicates. *Marine Biology*, **36**, 191–197.

Heron, A.C. and E.E. Benham. 1984. Individual growth rates of salps in three populations. *Journal of Plankton Research*, **6**, 811–828.

Heron, A.C. and Benham, E.E. 1985. Life history parameters as indicators of growth rate in three salp populations. *Journal of Plankton Research*, **7**, 365–379.

Heron, A.C., McWilliam, P.S. and Dal Pont, G. 1988. Length–weight relation in the salp *Thalia democratica* and potential of salps as a source of food. *Marine Ecology Progress Series*, **42**, 125–132.

Herring, P.J. 1983. The spectral characteristics of luminous marine organisms. *Proceedings of the Royal Society of London B*, **220**, 183–218.

Hirota, J. 1974. Quantitative natural history of *Pleurobrachia bachei* in La Jolla Bight. *Fishery Bulletin*, **72**, 295–335.

Hirsch, G.C. 1915. Die Ernährungsbiologie fleischfressender Gastropoden (*Murex*, *Natica*, *Pterotrachea*, *Pleurobranchea*, *Tritonum*). I. Teil. Makroskopischer Bau, Nahrungsaufnahme, Verdauung, Sekretion. *Zoologische Jahrbuch* (*Abteilungen Zoologie und Physiologie das Tiere*), **35**, 357–504.

Hirth, H.F. 1971. Synopsis of biological data on the green turtle, *Chelonia mydas* (Linnaeus) 1758. FAO *Fisheries Synopsis*, no. 85.

Hobson, E.S. 1974. Feeding relationships of teleostean fishes on coral reefs in Kona, Hawaii. *Fishery Bulletin*, **72**, 915–1031.

Hobson, E.S. and Chess, J.R. 1988. Trophic relations of the blue rockfish *Sebastes mystinus*, in a coastal upwelling system off northem California. *Fishery Bulletin*, **86**, 715–743.

Hoekstra, D. and Janssen, J. 1985. Non-visual feeding behaviour of the mottled sculpin, *Cottus bairdi*, in Lake Michigan. *Environmental Biology of Fish*, **12**, 111–117.

Holland, L.Z. 1988. Spermatogenesis in the salps *Thalia democratica* and *Cyclosalpa affinis* (Tunicata: Thaliacea): an electron microscopic study. *Journal of Morphology*, **198**, 189–204.

Holland, L.Z. 1989. Fine structure of spermatids and sperm of *Dolioletta gegenbauri* and *Doliolum nationalis* (Tunicata: Thaliacea): implications for tunicate phylogeny. *Marine Biology*, **101**, 83–95.

Holland, L.Z. 1990. Spermatogenesis in *Pyrosoma atlanticum* (Tunicata: Thaliacea: Pyrosomatida): implications for tunicate phylogeny. *Marine Biology*, **105**, 451–470.

Holland, L.Z. and Miller, R.L. 1994. Mechanism of internal fertilization in *Pegea socia* (Tunicata: Thaliacea), a salp with a solid oviduct. *Journal of Morphology*, **219**, 257–267.

Holland, L.Z., Gorsky, G. and Fenaux, R. 1988. Fertilization in *Oikopleura dioica* (Tunicata, Appendicularia): acrosome reaction, cortical reaction and sperm–egg fusion. *Zoomorphology*, **108**, 229–243.

Holley, M.C. 1986. Cell shape, spatial patterns of cilia, and mucus-net construction in the ascidian endostyle. *Tissue and Cell*, **18**, 667–684.

Holmberg, K. 1982. The ciliated brain duct of *Oikopleura dioica* (Tunicata, Appendicularia). *Acta Zoologica (Stockholm)*, **63** (2), 101–109.

Holmberg, K. 1984. A transmission electron microscope investigation of the sensory vesicle in the brain of *Oikopleura dioica* (Appendicularia). *Zoomorphologie*, **104**, 298–303.

Holmberg, K. 1986. The neural connection between the Langerhans receptor cells and the central nervous system in *Oikopleura dioica* (Appendicularia). *Zoomorphologie*, **106**, 31–34.

Holmberg, K. and Olsson, R. 1984. The origin of Reissner's fibre in an appendicularian, *Oikopleura dioica. Videnskabelige Meddelelser fra Dansk Naturhistorisk Forening*, **145**, 43–52.

Homans, R.E.S. and Needler, A.W.H. 1943. Food of the haddock (*Melanogrammus aeglifinus* Linnaeus). *Proceedings of the Nova Scotian Institute of Science*, **21**, 15–49.

Honjo, S. and Roman, M.R. 1978. Marine copepod fecal pellets: production, preservation and sedimentation. *Journal of Marine Research*, **36**, 45–47.

Hop, H., Gjoesaeter, J., Danielssen, D.S., Howell, B.R. and Moksness, E. 1994. Dietary composition of sympatric juvenile cod, *Gadus morrhua* and juvenile whiting, *Merlangius merlangus* in a fjord of southern Norway. *Aquaculture and Fishery Management*, **25**, 49–64.

Hopcroft, R.R. and Roff, J.C. 1994. Growth rates of tropical marine mesoplankton determined by the creation of artificial cohorts. *EOS, Transactions of the American Geophysics Union*, **75** (3), 187 (abstract).

Hopcroft, R.R. and Roff, J.C. 1995. Zooplankton growth rates: extraordinary production by the larvacean *Oikopleura dioica* in tropical waters. *Journal of Plankton Research*, **17**, 205–220.

Hopkins, T.L. 1985. Food web of an Antarctic midwater ecosystem. *Marine Biology*, **89**, 197–212.

Hopkins, T.L. and Baird, R.G. 1977. Aspects of feeding ecology of oceanic midwater fishes. In *Oceanic Sound Scattering Prediction* (N.R. Anderson and B.J. Zahuranek, eds), pp. 325–360. Plenum Press, New York.

Hopkins, T.L. and Baird, R.C. 1985a. Aspects of the trophic ecology of the mesopelagic fish *Lampanyctus alatus* (Family Myctophidae) in the eastern Gulf of Mexico. *Biological Oceanography*, **3**, 285–313.

Hopkins, T.L. and Baird, R.C. 1985b. Feeding ecology of four hatchetfishes (Sternoptychidae) in the easterm Gulf of Mexico. *Bulletin of Marine Science*, **36** (2), 260–277.

Hopkins, T.L. and Torres, J.J. 1989. Midwater food web in the vicinity of a marginal ice zone in the westerm Weddell Sea. *Deep-Sea Research*, **36A**, 543–560.

Hopkins, T.L. and Torres, J.J. 1988. The zooplankton community in the vicinity of the ice edge, western Weddell Sea, March 1986. *Polar Biology*, **9**, 79–87.

Hopkins, T.L. and Baird, R.G. 1977. Aspects of feeding ecology of oceanic midwater fishes. In *Oceanic Sound Scattering Prediction* (N.R. Anderson and B.J. Zahuranek, eds), pp. 325–360. Plenum Press, New York.

Hoshiai, T. 1979. Feeding behaviour of *Notothenia rossii marmorata* Fischer at South Georgia Station. *Antarctic Records*, **66**, 25–36.

Hubold, G. and Ekau, W. 1990. Feeding patterns of post-larval and juvenile notothenioids in the Southern Weddell Sea (Antarctica). *Polar Biology*, **10**, 255–260.

Humphries, C.J. and Parenti, L.R. 1986. *Cladistic Biogeography*, pp. i–xii and 1–98. Clarendon Press, Oxford.

Hunt, H.G. 1968. Continuous plankton records: contribution towards a plankton atlas of the North Atlantic and the North Sea. Part XI. The seasonal and annual distributions of the Thaliacea. *Bulletin of Marine Ecology*, **6**, 225–245.

Hunter, J.P. and Mitchell, C.T. 1967. Association of fish with flotsam in the offshore water of Central America. *Fishery Bulletin*, **66**, 13–29.

Huntley, M.E., Sykes P.F. and Marin, V. 1989. Biometry and trophodynamics of *Salpa thompsoni* Foxton (Tunicata: Thaliacea) near the Antarctic peninsula in Austral summer, 1983–1984. *Polar Biology*, **10**, 59–70.

Huxley, T.H. 1851. Observations upon the anatomy and physiology of *Salpa* and *Pyrosoma. Philosophical Transactions of the Royal Society of London*, **141** (2), 567–594.

Ihle, J.E.W. 1906. Bijdragen tot de kennis van de morphologie en systematiek der Appendicularien. *Academisch proefschrift*, Universiteit van Amsterdam, 98 p. E.J. Brill, Leiden.

Ihle, J.E.W. 1911. Ueber die Nomenklatur der Salpen. *Zoologischer Anzeiger*, **38**, 585–589.

Ihle, J.E.W. 1912. Desmomyaria. *Das Tierreich*, **32**, 1–67.

Ihle, J.E.W. 1935. Desmomyaria. In *Handbuch der Zoologie* (W. Kükenthal and T. Krumbach, eds), Vol. 5 pp. 401–532. W. de Gruyter Co., Berlin, Germany.

Ihle, J.E.W. 1938. Salpidae. In *Klassen und Ordnungen des Tierreichs* (H.G. Bronn, ed), Vol. 3 (supplement 2, Tunikaten), pp. 69–401. Geist und Portig K.G., Leipzig.

Ihle, J.W. and Ihle-Landenberg, M.E. 1933. Anatomische Untersuchungen über Salpen IV. Allgemeine über den Darmkanal der Salpen. *Zoologischer Anzeiger*, **104**, 194–200.

Ikeda, T. and Bruce, B. 1986. Metabolic activity and elemental composition of krill and other zooplankton from Prydz Bay, Antarctica, during early summer (November–December). *Marine Biology*, **92**, 545–555.

Ikeda, T. and Mitchell, A.W. 1982. Oxygen uptake, ammonia excretion and phosphate excretion by krill and other Antarctic zooplankton in relation to their body size and chemical composition. *Marine Biology*, **71**, 283–298.

Ikewaki, Y. and Tanaka, M. 1993 Feeding habits of the Japanese flounder (*Paralichthys olivaceus*) in the western part of Wakasa Bay, the Japan Sea. *Nippon Suisan Gakkaishi*, **59**, 951–956.

Inada, T. and Nakamura, I. 1975. A comparative study of two populations of the gadoid fish *Micromesistius australis* from the New Zealand and Patagonian–Falkland regions. *Bulletin Far Seas Fishery Research Laboratory*, **13**, 1–26.

Iseki, K. 1981. Particulate organic matter transport to the deep sea by salp fecal pellets. *Marine Ecology Progress Series*, **5**, 55–60.

Ivanova-Kazas, O.M. 1956. On the embryonic development of Pyrosomids (Pyrosomida, Tunicata). *Zoologitsheskij Journal*, **35** (8), 1193–1202 (in Russian, English summary).

Ivanova-Kazas, O.M. 1958–1959. *Pyrosoma vitjasi*, une nouvelle espèce de Pyrosome. *Annales de la Société Royale Zoologique de Belgique*, **8**, 273–279.

Ivanova-Kazas, O.M. 1960. Embryological characterization of Pyrosomata fixata. *Doklady Akademi Nauk SSSR*, **136**, 494–496 (in Russian).

Ivanova-Kasas, O.M. 1962. Sur les formes primitives du développement chez les Pyrosomida. *Cahiers de Biologie Marine*, **3**, 191–208.

Ivanova-Kasas, O.M. 1978. Lower chordates. In *Comparative Embryology of Invertebrate Animals*, Vol. 4, 166 pp. Akademi Nauk SSSR, Moscow, Russia (in Russian).

Jackson, G.A. 1980. Phytoplankton growth and zooplankton grazing in oligotrophic oceans. *Nature, London*, **284**, 439–441.

Jansa, J. 1977. Estudio preliminar del contenido en pigmentos fotosinteticos en el tubo digestivo de apendicularias y salpas. *Boletín del Instituto Español de Oceanografía*, **1**, 7–29.

Jansa, J. 1985. Apendicularias, salpas y plancton en general en la zona W y S de Mallorca. *Boletín Instituto Español Oceanografá*, **2**, 132–154.

Janssen, J. and Harbison, G.R. 1981. Fish in salps: the association of squaretails (*Tetragonurus* spp.) with pelagic tunicates. *Journal of the Marine Biological Association of the United Kingdom*, **61**, 917–927.

Jatta, G. 1896. Cefalopodi viventi nel Golfo di Napoli. *Fauna und Flora des Golfen von Neapel*, **23**, 1264.

Jespersen, P. 1928. Investigations on the food of the herring in Danish waters. *Meddelelser fra Kommisionen for Havundersøgelser*, **2**, 1–150.

Johnson, B.D. and Wangersky, P.J. 1985. Sea water filtration: particle flow and impaction considerations. *Limnology and Oceanography*, **30**, 966–971.

Jørgensen, C.B. 1966a. *The Biology of Suspension Feeding*, pp. 375. Pergamon, London.

Jørgensen, C.B. 1966b. Feeding. In *Marine Biology* (W.T. Edmondson, ed.) Vol. 3, pp. 69–133. New York Academy of Sciences, New York.

Jørgensen, C.B. 1983. Fluid mechanical aspects of suspension feeding. *Marine Ecology Progress Series*, **11**, 89–103.

Jørgensen, C.B. 1984. Effect of grazing: metazoan suspension feeders. In *Heterotrophic Activity in the Sea* (J.E. Hobbie and P.J. leB. Williams, eds), pp. 445–464. Plenum Press, New York.

Josephson, R.K. and Schwab, W.E. 1979. Electrical properties of an excitable epithelium. *Journal of General Physiology*, **74**, 213–236.

Julin, C. 1904. Recherches sur la phylogenèse des Tuniciers. Développement de l'appareil branchial. *Zeitschrift für Wissenschaftliche Zoologie*, **76**, 544–611.

Julin, C. 1912a. Recherches sur le développement embryonnaire de *Pyrosoma giganteum* Les. I. Aperçu général de l'embryogenèse. Les cellules du testa et le développement des organes lumineux. *Zoologische Jahrbücher*, supplement XV (2), 775–863.

Julin, C. 1912b. Les caractères histologiques spécifiques des 'cellules lumineuses' de *Pyrosoma giganteum* et de *Cyclosalpa pinnata*. *Comptes Rendus des Séances de l'Académie des Sciences Paris*, **155**, 525–527.

Kami, H.T. 1973. The Pristipomoides (Pisces, Lutjanidae) of Guam with notes on their biology. *Micronesica*, **9**, 97–118.

Karl, D.M. and Bird, D.F. 1993. Bacterial algal interactions in Antarctic coastal ecosystems. In *Trends in Microbial Ecology* (R. Guerrero and C. Pedros-Alio, eds), pp. 37–40. Spanish Society for Microbiology, Barcelona.

Karl, D.M. and Knauer, G.A. 1984. Vertical distribution, transport, and exchange of carbon in the northeast Pacific Ocean: evidence for multiple zones of biological activity. *Deep-Sea Research*, **31A**, 221–243.

Karpenko, V.I. and Piskunova, L.V. 1984. Importance of macroplankton in the diet of young salmons of the genus *Oncorhynchus* (Salmonidae), and their trophic relationships in the southwestem Bering Sea. *Journal of Ichthyology*, **24** (5), 98–106.

Kashkina, A.A. 1978. Areas of concentration and abundance of salpas in the Atlantic Ocean. *Biologiya Morya*, **3**, 11–16.

Kashkina, A.A. 1986. Feeding of fishes on salps (Tunicata, Thaliacea). *Journal of Ichthyology*, **26**, 57–64.

Kawaguchi, S. 1995. Distribution of salps near the South Shetland Islands during austral summer, 1990–1991 with special reference to krill distribution. *Polar Biology*, **15**, 31–39.

Keats, D.W., Steele D.H. and South, G.R. 1987. Food of winter flounder *Pseudopleuronectes americanus* in a sea urchin dominated community in eastern Newfoundland. *Marine Ecology*, **60**, 13–22.

Kepkay, P.E. 1994. Particle aggregation and the biological reactivity of colloids. *Marine Ecology Progress Series*, **109**, 293–304.

Kiernan, J.A. 1981. *Histological and Histochemical Methods. Theory and Practice*, 344 pp. Pergamon Press, Oxford.

King, K.R. 1981. The quantitative natural history of *Oikopleura dioica* (Urochordata: Larvacea) in the laboratory and in enclosed water columns, PhD thesis, University of Washington, Seattle, WA, 152 pp.

King, K.R. 1982. The population biology of the larvacean *Oikopleura dioica* in enclosed water columns. In *Marine Mesocosms: Biological and Chemical Research in Experimental Ecosystems* (G.D. Grice, ed.), pp. 341–351. Springer-Verlag, Berlin.

King, J.E. and Iverson, R.T.B. 1962. Midwater trawling for forage organisms in the Central Pacific 1951–1956. *Fishery Bulletin*, **62**, 271–321.

King, K.R., Hollibaugh, J.T. and Azam, F. 1980. Predator–prey interactions between the larvacean *Oikopleura dioica* and bacterioplankton in enclosed water columns. *Marine Biology*, **56**, 49–57.

Kingsford, M.J. and MacDiarmid, A.B. 1988. Interrelations between planktivorous reef fish and zooplankton in temperate waters. *Marine Ecology Progress Series*, **48**, 103–117.

Kinzer, J. and Schulz, K. 1985. Vertical distribution and feeding patterns of midwater fish in the central equatorial Atlantic. *Marine Biology*, **85**, 313–322

Kishinouye, K. 1923. Contributions to the comparative study of the so-called scombroid fishes. *Journal of the College of Agriculture, Tokyo*, **8**, 293–457.

Klaatsch, H. 1895. Über Kernveränderungen in Ektoderm der Appendikularien bei der Gehäusebildung. *Morphologisches Jahrbuch*, **23**, 142–144.

Klein, K. 1932. Die Nervenendigungen in der Statocyste von *Sepia*. *Zeitschrift für Zellforschung und Mikroskopisches Anatomie*, **14**, 481–516.

Knoechel, R. and Steel-Flynn, D. 1989. Clearance rates of *Oikopleura* in cold coastal Newfoundland waters: a predictive model and its trophodynamic implications. *Marine Ecology Progress Series*, **53**, 257–266.

Kock, K.H., Schneppenheim, R. and Siegel, V. 1984. A contribution to the fish fauna of the Weddell Sea. *Archiv Fischereiwissenschaft*, **34**, 103–120.

Koike, I., Shigemitsu, H., Kazuki, T. and Kogure, K. 1990. Role of submicrometre particles in the ocean. *Nature, London*, **345**, 242–243.

Koike, I., Hara, S., Terauchi, K., Shibata, A. and Kogure, K. 1993. Marine viruses, their role in upper ocean dissolved organic matter (DOM) dynamics. In *Trends in Microbial Ecology* (R. Guerrero and C. Pedros-Alio, eds), pp. 311–314. Spanish Society for Microbiology, Barcelona.

Komai, T. 1922. *Studies on Two Aberrant Ctenophores, Coeloplana and Gastrodes*, 102 pp. Published by Author, Tokyo, Japan.

Komai, T. 1932. On some salps occuring in the vicinity of Seto, with remarks on the enantiomorphism found in some aggregated forms. *Memoirs of the College of Science, Kyoto University, Series B*, **8**, 65–80.

Körner, W.F. 1952. Untersuchungen über die Gehäusebildung bei Appendicularien (*Oikopleura dioica* Fol). *Zeitschrift für Morphologie und Okologie der Tiere*, **41**, 1–53.

Korotneff, A. 1888. *Cunoctantha* and *Gastrodes*. *Zeitschrift für Wissenschaftlichen Zoologie*, **47**(4), 650–657.

Korotneff, A. 1891. Zoologische Paradoxen. *Zeitschrift für wissenschaftlichen Zoologie*, **51**, 613–628.

Korotneff, A. 1904. Über den Polymorphismus von *Dolchinia*. *Biologisches Zentralblatt*, **24**, 61–65.

Kowalewsky, A. 1867. Entwicklungsgeschichte der einfachen Ascidien. *Mémoires de l'Académie Impériale des Sciences de St.-Pétersbourg*, **10**, 1–19.

Kowalewsky, A. 1875. Ueber die Entwicklungsgeschichte der *Pyrosoma*. *Archiv für Mikroskopische Anatomie*, **11**, 597–635.

Kowalewsky, A. and Barrois, J. 1883. Matériaux pour servir à l'histoire de l'Anchinie. *Journal de l'Anatomie et de la Physiologie*, **19**, 1–23.

Kozlov, A.N. and Naumov, A.G. 1988. Feeding of young of three notothenoid (Notothenoidei) species in the Indian Ocean sector of the Antarctic. *Journal of Ichthyology*, **28**, 70–73.

Kramp, P.L. 1942. Pelagic Tunicata. The Godthaab expedition 1928. *Meddelelser om Grønland*, **80**, 1–10.

Kremer, P. 1977. Respiration and excretion by the etenophore *Mnemiopsis leidyi*. *Marine Biology*, **44**, 43–50.

Kremer, P. and Madin, L.P. 1992. Particle retention efficiency of salps. *Journal of Plankton Research*, **14**, 1009–1015.

Kümmel, G. 1956. Die Feinstruktur des Gehäuses der Appendikularien (*Oikopleura dioica* Fol) und des Mantels der Ascidienlarven (*Botryllus schlosseri* Pall.). *Zoologisches Beiträge* (N.F.), **2**, 431–439.

Kremer, P., Canino M.F. and Gilmer. R.W. 1986. Metabolism of epipelagic tropical ctenophores. *Marine Biology*, **90**, 403–412.

Krishnaswami, S., Baskaran, M., Fowler, S.W. and Heyraud, M. 1985. Comparative role of salps and other zooplankton in the cycling and transport of selected elements and natural radionuclides in Mediterranean waters. *Biogeochemistry*, **1**, 353–360.

Krüger, H. 1939. Die Thaliaceen der 'Meteor' Expedition. *Wissenschaftliche Ergebnisse der Deutschen Atlantischen Expedition Meteor 1925–1927, Biologische Untersuchungen*, **13** (2), 111–152.

Krumholz, L.A. 1959. Stomach contents and organ weights of some bluefin tuna, *Thunnus thynnus* (Linnaeus) near Bimini, Bahamas. *Zoologica*, **44**, 127–131.

Kubota, T. and T. Uyeno. 1970. Food habits of lancetfish *Alepisaurus ferox* (Order Myctophiformes) in Suruga Bay, Japan. *Japanese Journal of Ichthyology*, **17**, 22–28.

Kun, M.S. 1954. Nutritional peculiarities of fingerlings and adult mackerels *Izvestiya Tikhookeanskogo Nauchno-issledovatel 'skogo Instituta Rybnogo Khozyaistva i Okeanografii.*, **42**, 95–108 (in Russian).

Ladiges, P.Y., Humphries, C.J. and Martinelli, L.W. (eds) 1991. *Austral Biogeography, Australian Systematic Botany*, Vol. 4, pp. 59–72.

Lafay, B., Boury-Esnault, N., Vacelet, J. and Christen, R. 1992. An analysis of partial 28S ribosomal RNA sequences suggests early radiations of sponges. *Biosystems*, **28**, 139–1151.

Lahille, F. 1888. Etude systématique des Tuniciers. *Compte Rendu de la 16*[e] Session, Toulouse, 1887, Association Française pour l'Avancement des Sciences, **2**, 667–677.

Lahille, F., 1890. *Recherches sur les Tuniciers des Côtes de France*, 330 pp. Imprimerie Lagarde et Sebille, Toulouse, France.

Lalli, C.M. and Gilmer, R.W. 1989. *Pelagic Snails. The biology of Holoplanktonic Gastropod Mollusks*, 259 pp. Stanford University Press, Stanford .

Lamarck, J.B. 1816. *Histoire naturelle des animaux sans vertèbres*. Vol. 3. *Tuniciers*, pp. 80–130. Vernière, Paris.

Lambert, C.C. and Lambert, G. 1978. Tunicate eggs utilize ammonium ions for flotation. *Science, NY*, **200**, 64–65.

Lampitt, R.S., Noji, T. and Bodungen, B. von 1990. What happens to zooplankton fecal pellets? Implication for material flux. *Marine Biology*, **104**, 737–739.

Lampitt, R.S., Wishner, K.F., Turley, C.M. and Angel, M.V. 1993. Marine snow studies in the Northeast Atlantic Ocean: distribution, composition and role as food source for migrating plankton. *Marine Biology*, **116**, 689–702.

Lancraft, T.M., Torres, J.J. and Hopkins, T.L. 1989. Micronekton and macrozooplankton in the open waters near Antarctic ice edge zones (AMERIEZ, 1983 and 1988). *Polar Biology*, **9**, 225–233.

Lancraft, T.M., Hopkins, T.L., Torres, J.J. and Donnelly, J. 1991. Oceanic micronektonic/macrozooplanktonic community structure and feeding in ice covered Antarctic waters during the winter (AMERIEZ, 1988). *Polar Biology*, **11**, 157–167.

Landry, M.R., Peterson, W.K. and Fagerness, V.I. 1994. Mesozooplankton grazing in the Southern California Bight. I. Population abundances and gut pigment contents. *Marine Ecology Progress Series*, **115**, 55–71.

Langerhans, P. 1880. Über Madeiras Appendicularien. *Zeitschrift für Wissenschaftliche Zoologie*, **34**, (1), 144–146.

Langerhans, P. 1878. Zur Anatomie der Appendicularien. *Monatsberichte der Königlich preussischen Akademie der Wissenschaften zu Berlin*, **1877**, 561–566.

Larson, R.J. 1987. A note on the feeding, growth and reproduction of the epipelagic scyphomedusa *Pelagia noctiluca* (Forskal). *Biological Oceanography*, **4**(4), 447–454.

Larson, R.J. 1991. Diet prey selection and daily ration of *Stomolomphus meleagris*, a filter feeding scyphomedusa from the NE Gulf of Mexico. *Estuarine and Coastal Shelf Science*, **32**, 511–525.

Larson, R.J. and Harbison, G.R. 1990. Medusae from McMurdo Sound, Ross Sea including the descriptions of two new species, *Leuckartiara brownei and Benthocodon hyalinus*. *Polar Biology*, **11**, 19–25.

Larson, R.J., Mills, C.E. and Harbison, G.R. 1989. *In situ* foraging and feeding behaviour of narcomedusae (Cnidaria, Hydrozoa). *Journal of the Marine Biological Association of the United Kingdom*, **69**, 785–794.

Last, P.R., Scott, E.O.G. and Talbot, F.H. 1983. *Fishes of Tasmania*, pp. 1–563. Tasmanian Fisheries Development Authority, Hobart.

Last J.M. 1978. The food of four species of Pleuronectiform larvae in the eastern English Channel and southern North Sea. *Marine Biology*, **45**(4), 359–368.

Last J.M. 1979. The food of larval turbot *Scophthalmus maximus* L. from the west central North Sea. *Journal du Conseil*, **38**, 308–313.

Last, J.M. 1980. The food of twenty species of fish larvae in the west-central North Sea. *Fisheries Research Technical Report*, **60**, 1–44.

Latz, M.I., Frank, T.M., Case, J.F., Swift, E. and Bidigare, R.R. 1987. Bioluminescence of colonial Radiolaria in the western Sargasso Sea. *Journal of Experimental Marine Biology and Ecology*, **109**, 25–38.

Laval, P. 1978. The barrel of the pelagic amphipod *Phronima sedentaria* (Forsk.) (Crustacea, Hyperiidea). *Journal of Experimental Marine Biology and Ecology*, **33**, 187–211.

Laval, P. 1980. Hyperiid amphipods as crustacean parasitoids associated with gelatinous zooplankton. *Oceanography and Marine Biology Annual Review*, **18**, 11–56.

Laval, P. 1995. Hierarchical object-oriented design of a concurrent, individual-bas ed model of a pelagic Tunicate bloom. *Ecological modelling*, **82**, 265–276.

Laval, P. 1996. The representation of space in an object-oriented computational pelagic ecosystem. *Ecological Modelling*, **88**, 113–124.

Laval P., Braconnot, J.C., Carré, C., Goy, J., Morand, P. and Mills, C.E. 1989. Small-scale distribution of macroplankton and micronekton in the Ligurian Sea (Mediterranean Sea) as observed from the manned submersible *'Cyana'*. *Journal of Plankton Research*, **79**, 235–241.

Laval P., Braconnot, J.C. and Linds da Silva, N. 1992. Deep planktonic filter-feeders found in the aphotic zone with the *'Cyana'* submersible in the Ligurian Sea (NW Méditerranean). *Marine Ecology Progress Series*, **79**, 235–241.

Lazarus, B.I. and D. Dowler. 1979. Pelagic Tunicata off the west and south-west coasts of South Africa, 1964–1965. *Fishery Bulletin South Africa*, **12**, 93–119.

Le Borgne, R. 1978. Evaluation de la production secondaire planctonique en milieu océanique par la méthode des rapports C/N/P. *Oceanologica Acta*, **1**, 107–118.

Le Borgne, R. 1982. Zooplankton production in the eastern tropical Atlantic Ocean, Net growth efficiency and P:B in terms of carbon, nitrogen, and phosphorus. *Limnology and Oceanography*, **27**, 681–698.

Le Borgne, R. 1983. Note sur les proliférations de Thaliacés dans le Golfe de Guinée. *Océanographie Tropicale*, **18**, 49–54.

Le Borgne, R. and P. Moll. 1986. Growth rates of the salp *Thalia democratica* in Tikehau atoll (Tuamoto is.). *Océanographie Tropicale*, **21**, 23–29.

Le Gall, J.-Y. 1974. Exposé synoptique des données biologiques sur le germon *Thunnus alalunga* (Bonaterre, 1788). *FAO Fisheries Synopsis*, no. 109.

Legand, M. and J. Rivaton. 1969. Cycles biologiques des poissons mésopélagiques de l'est de l'Océan Indien. Troisième note. Actions predatrice des poissons micronectoniques. *Cahiers O.R.S.T.O.M., Séries Océanographie*, **7** (3), 29–35.

Legendre, L. and Le Fèvre, J. 1992. Interactions between hydrodynamics and pelagic ecosystems: relevance to resource exploitation and climate change. *South African Journal of Marine Science*, **12**, 477–486.

Legendre, L. and Le Fèvre, J. 1995. Microbial food webs and the export of biogenic carbon in the oceans. *Aquatic Microbial Ecology*, **9**, 69–77.

Legendre, L., Le Fèvre, J. and Fortier, L. 1993. Role of microphagous macrozoooplankton in channeling biogenic carbon towards fish and/or the deep ocean. In *Trends in Microbial Ecology* (R. Guerrero and C. Pedros-Alio, eds), pp. 431–434. Spanish Society for Microbiology, Barcelona.

Legendre, R. 1940. La faune pélagique de l'Atlantique au large du Golfe de Gascogne, recueillie dans des estomacs de germons. Troisième Partie: Invertébrés (Céphalopodes exclus). Parasites du germon. *Annales de l'Institut Océanographique*, **20**, 127–310.

Leisman, G., Cohn, D.H. and Nealson, K.H. 1980. Bacterial origin of luminescence in marine animals. *Science*, **208**, 1271–1273.

Lesueur, M. 1815. Mémoire sur l'organisation des Pyrosomes et sur la place qu'ils doivent occuper dans une classification naturelle. *Bulletin Scientifique, Société philomatique, Paris*, **4**, 70–74.

Leuckart R. 1854. Zur Entwickelungsgeschichte der Ascidien. Beschreibung einer schwaermenden Ascidienlarve (Appendicularia). *Zoologische Untersuchungen, Giessen*, **2** (2), 79–93.

Lewis, J.B., Brundritt, J.K. and Fish, A.G. 1962. The biology of the flying fish *Hirundichthys affinis* (Gunther). *Bulletin of Marine Science Gulf Caribbean*, **12**(1), 73–

Lindholm, R. 1984. Observations on the chinaman leather-jacket *Nelusetta ayraudi* (Quoy & Gaimard) in the Great Australian Bight. *Australian Journal of Marine and Freshwater Research*, **35**, 597–599.

Lindley, J.A., Roskell, J., Warner, A.J., Halliday, N.C., Hunt, H.G., John, A.W.G. and Jonas, T.D. 1990. Doliolids in the German Bight in 1989: evidence for exceptional inflow into the North Sea. *Journal of the Marine Biological Association of the United Kingdom*, **70**, 679–682.

Linton, E. 1901. Parasites of fishes of the Woods Hole region. *Bulletin of the United States Fish Commission*, **19**, 405–492.

Lo Bianco, S. 1909. Notize biologiche riguardanti specialmente il periodo di maturita sessuale degli animali del Golfo di Napoli. *Mittheilungen aus der Zoologischen Station zu Neapel*, **19**, 513–761.

Lodh, N.M., Gajbhiye, S.N. and Nair, V.R. 1988. Unusual congregation of salps off Veraval and Bombay, west coast of India. *Indian Journal of Marine Science*, **17**, 128–130.

Lohmann H., 1892. Vorbericht über die Appendikularien der Plankton-Expedition. *Ergebnisse der im Atlantischen Ozean Plankton-Expedition der Humboldt-Stiftung*, 1889, **1A**, 139–149.

Lohmann H., 1895. Ueber die Verbreitung der Appendicularien im Atlantischen Oceane. *Verhandlungen der Gesellschaft Deutscher Naturforscher und Ärzte*, **67**, 113–120.

Lohmann H., 1896a. Die Appendicularien der Plankton-Expedition. *Ergebnisse der Plankton-Expedition der Humboldt-Stiftung*, **2**, 1–148.

Lohmann H., 1896b. Die Appendikularien der Expedition (Zoologische Ergebnisse der Grönland Expedition). *Bibliotheca Zoologica*, **20** (2), 25–44.

Lohmann H., 1899a. Das Gehäuse der Appendicularien, sein Bau, seine Funktion und seine Entstehung. *Schriften des Naturwissenschaftlichen Vereins für Schleswig-Holstein*, **11** (2), 347–407.

Lohmann, H. 1899b. Untersuchungen über den Auftrieb der Strasse von Messina mit besonderer Berücksichtigung der Appendicularien und Challengerien. *Sitzungsberichte der Königlich preussischen Akademie der Wissenschaften zu Berlin, Klasse für Mathematik und Allgemeine Naturwissenschaften*, **20**, 587, 384–400.

Lohmann, H. 1899c. Das Gehäuse der Appendicularien nach seiner Bildungsweise, seinem Bau und seiner Funktion. *Zoologische Anzeiger*, **22**, 206–214.

Lohmann, H. 1905. Die Appendicularien des arktischen und antarktischen Gebiets, ihre Beziehungen zueinander und zu den Arten des Gebiets der warmen Ströme. *Zoologische Jahrbücher*, supplement **8**, 353–382.

Lohmann, H. 1909a. Die Gehäuse und Gallertblasen der Appendicularien und ihre Bedeutung für die Erforschung des lebens in Meer. *Verhandlung Deutsche Zoologiche Gesellschaft*, **19**, 200–239.

Lohmann, H. 1909b. Copelata und Thaliacea. *Fauna Südwest-Australiens*, **2** (10), 143–149.

Lohmann, H. 1914. Die Appendicularien der Valdivia-Expedition. *Verhandlungen der Deutschen Zoologischen Gesellschaft*, **24**, 157–192.

Lohmann, H. 1915. Tunicata. In *Handwörterbuch der Naturwissenschaften*, Vol. 10, pp. 57–90, G. Fischer, Jena.

Lohmann, H. 1928. Beiträge zur Planktonbevölkerung der Weddellsee nach den Ergebnissen der Deutschen Antarktischen Expedition 1911–1912. II. Die Appendicularien-Bevölkerung der Weddellsee. *Internationale Revue der Gesamten Hydrobiologie und Hydrographie*, **20**, 13–72.

Lohmann, H. 1931. Die Appendicularien der Deutschen Tiefsee-Expedition. *Wissenschaftliche Ergebnisse der Deutschen Tiefsee-Expedition auf dem Dampfer 'Valdivia' 1898–1899*, **21**, 1–158.

Lohmann, H. 1933. Erste Klasse der Tunicaten: Appendiculariae. In *Handbuch der Zoologie* (W. Kükenthal and

T. Krumbach, eds), Vol. 5 (2,1), 15–164. W. de Gruyter, Berlin und Leipzig.

Lohmann H. 1934. Erste Klasse der Tunicaten: Appendiculariae. In *Handbuch der Zoologie* (W. Kükenthal and T. Krumbach, eds), Vol 5 (2,3), pp. 193–202. W. de Gruyter, Berlin.

Lohmann H. and Bückmann, A. 1926. Die Appendicularien der deutschen Südpolar Expedition 1901–1903. *Deutsche Südpolar-Expedition 1901–1903,* **18** (Zoologie 10), 63–231.

Lohmann H. and Hentschel E. 1939. Die Appendicularien im Südatlantischen Ozean. *Wissenschaftliche Ergebnisse der Deutschen Atlantischen Expedition auf dem Forschungs- und Vermessungsschiff 'Meteor' 1925–1927,* **13** (3), 153–243.

Longhurst, A., Sameoto, D. and Herman, A. 1984. Vertical distribution of Arctic zooplankton in summer: eastern Canadian Archipelago. *Journal of Plankton Research,* **6**, 137–168

Lopez, M.D., Huntley, M.E. and Sykes, P.F. 1988. Pigment destruction by *Calanus pacificus.* Impact on the estimation of water column fluxes. *Journal of Plankton Research,* **10**, 715–734.

Love, M.S. and Ebeling, A.W. 1978. Food and habits of three switch feeding fishes in the kelp forests off Santa Barbara. *California Fishery Bulletin,* **76**, 257–271.

Mackas, D.L., Washburn, L. and Smith, S.L. 1991. Zooplankton community pattern associated with a California Current cold filament. *Journal of Geophysical Research,* **96**, 14781–14797.

Mackie, G.O. 1995. On the visceral system of *Ciona. Journal of the Marine Biological Association of the United Kingdom,* **76**, 141–151.

Mackie, G.O. and Bone, Q. 1976. Skin impulses and locomotion in an ascidian tadpole. *Journal of the Marine Biological Association of the United Kingdom,* **56**, 751–768.

Mackie, G.O. and Bone, Q. 1977. Locomotion and propagated skin impulses in salps (Tunicata: Thaliacea). *Biological Bulletin, Woods Hole,* **153**, 180–197.

Mackie, G.O. and Bone, Q. 1978. Luminescence and associated effector activity in *Pyrosoma* (Tunicata: Pyrosomida). *Proceedings of the Royal Society of London B,* **202**, 483–495.

Mackie, G.O., Paul, D.H., Singla, C.L., Sleigh, M.A. and Williams, D.E. 1974. Branchial innervation and ciliary control in the ascidian *Corella. Proceedings of the Royal Society of London B,* **187**, 1–353.

Macpherson, E. 1981. Resource partitioning in a Mediterranean demersal fish community. *Marine Ecology Progress Series,* **4**, 183–193.

Macpherson, E. 1983. Ecologia trófica de peces en las costas de Namibia. I. Habitos alimentarios. Resultados expediciónes Cientificas. Suplemento Investigación Pesquera, **11**, 81–137.

Madin, L.P. 1974a. Field studies on the biology of salps (Tunicata, Thaliacea), PhD thesis, 208 pp. University of California, Davis, CA.

Madin, L.P. 1974b. Field observations on the feeding behavior of salps (Tunicata: Thaliacea). *Marine Biology,* **25**, 143–147.

Madin, L.P. 1982. Production, composition and sedimentation of salp fecal pellets in oceanic waters. *Marine Biology,* **67**, 39–45.

Madin, L.P. 1988. Feeding behavior of tentaculate predators: *in situ* observations and a conceptual model. *Bulletin of Marine Science,* **43**, 413–429.

Madin, L.P. 1990. Aspects of jet propulsion in salps. *Canadian Journal of Zoology,* **68**, 765–777.

Madin, L.P. 1996. Sensory ecology of salps (Tunicata, Thaliacea): more questions than answers. *Marine and Freshwater Behaviour and Physiology,* **25**, 175–195.

Madin, L.P. and Cetta, C.M. 1984. The use of gut fluorescence to estimate grazing by oceanic salps. *Journal of Plankton Research,* **6**, 475–492.

Madin, L.P. and Harbison, G.R. 1977. The associations of Amphipoda Hyperiidea with gelatinous zooplankton. I. Associations with Salpidae. *Deep-Sea Research,* **24**, 449–463.

Madin, L.P. and Harbison, G.R. 1978. Salps of the genus *Pegea* Savigny, 1816 (Tunicata, Thaliacea). *Bulletin of Marine Science,* **28**, 335–344.

Madin, L.P. and Kremer, P. 1995. Determination of the filter feeding rates of salps (Tunicata, Thaliacea). ICES *Journal of Marine Science,* **52**, 583–595.

Madin, L.P. and Purcell, J.E. 1992. Feeding, metabolism, and growth of *Cyclosalpa bakeri* in the subarctic Pacific. *Limnology and Oceanography,* **37**, 1236–1251.

Madin, L.P., Cetta, C.M. and McAlister, V.L. 1981. Elemental and biochemical composition of salps (Tunicata: Thaliacea). *Marine Biology,* **63**, 217–226.

Madin, L.P., Kremer, P. and Hacker, S. 1996. Distribution and vertical migration of salps (Tunicata: Thaliacea) near Bermuda. *Journal of Plankton Research,* **18**, 747–755.

Mahoney, E.M. 1981. Observations on *Oikopleura* (Tunicata, Appendicularia) and the contribution of discarded larvacean houses to the slub problem in selected Newfoundland inshore locations. BSc Honours thesis, 56 pp. Memorial University of Newfoundland, St Johns.

Malej, A. 1982. Unusual occurrence of *Pelagia noctiluca* in the Adriatic. I. Some notes on the biology of *Pelagia noctiluca* in the Gulf of Trieste. *Acta Adriatica,* **23**, 97–102 (in Serbo-Croat).

Mancuso, V. 1973. Changes in the fine structure associated with test formation in the ectoderm cells of *Ciona intestinalis* embryo. *Acta Embryologica Experimentalis,* **3**, 247–257.

Mancuso, V. 1974. Formation of the ultrastructural components of the *Ciona intestinalis* tadpole test by animal embryo. *Experientia,* **30**, 1078

Manzer, J.I. 1969. Stomach contents of juvenile Pacific salmon in Chatham Sound and adjacent waters. *Journal of the Fisheries Research Board of Canada,* **26**, 2219–2223.

Markina, N.P. 1973. Seasonal changes in the distribution of plankton of the Great Australian Gulf (on the basis of results of 1968–1969). *Studies of Fish Biology and Commercial Oceanography,* **4**, 50–59 (in Russian).

Markina, N.P. 1984. Nutritional ecology of *Caprodon longimanus* in the Tasman Sea. *Biologiya Morya,* **4**, 32–38 (in Russian).

Markina, N.P. and Boldyrev, V.Z. 1980. Nutrition of redeye on the seamounts of the southwestern Pacific Ocean. *Biologiya Morya,* **4**, 40–45 (in Russian).

Markle, D.F. and Quero, J.C. 1984. Alepocephalidae (including Bathylaconidae, Bathyprionidae). In *Fishes of the North-Eastern Atlantic and the Mediterranean* (P.J.P. Whitehead, M.-L. Bauchot, J.C. Hureau, J. Nielsen and E. Tortonese, eds) Vol. 1, pp. 228–253. UNESCO, Paris.

Márquez, M.R. 1990. *FAO Species Catalogue,* Vol. 11, *Sea Turtles of the World. FAO Fisheries Synopsis* no. 125. FAO, Rome.

Martini, E. 1909a. Studien über die Konstanz histologischer Elemente. 1. *Oikopleura longicauda. Zeitschrift für Wissenschaftliche Zoologie,* **92**, 563–626.

Martini, E. 1909b. Studien über die Konstanz histologischer Elemente. II. *Fritillaria pellucida. Zeitschrift für Wissenschaftliche Zoologie,* **94**, 81–170.

Martini, E. 1909c. Über die Segmentierung des Appendicularienschwanzes. *Verhandlungen der Deutschen Zoologischen Gesellschaft,* **19**, 300–307.

Martini, E 1910. Weitere Bemerkungen über die sogenannte metamere Segmentierung des Appendicularienschwanzes. *Zoologischer Anzeiger,* **35**, 644–652.

Matsueda, H., Handa, N., Inoue, I. and Takano, H. 1986. Ecological significance of salp fecal pellets collected by sediment traps in the eastern North Pacific. *Marine Biology*, **91**, 421–431.

Matthews, F.D., Damkaer, D.M., Knapp, L.W. and Collette, B.B. 1977. Food of western North Atlantic tunas (*Thunnus*) and lancetfishes (*Alepisaurus*). *NOAA Technical Report National Marine Fisheries Service*, **706**, 1–19.

Mauchline, J. and Gordon, J.D.M. 1983. Diets of clupeoid, stomatioid and salmonid fish of the Rockall Trough, Northeastern Atlantic Ocean. *Marine Biology*, **77**, 67–78.

Mauchline, J. and Gordon, J.D.M. 1984a. Diets and bathymetric distributions of the macrourid fish of the Rockall Trough, northeastern Atlantic Ocean. *Marine Biology*, **81**, 107–121.

Mauchline, J. and Gordon, J.D.M. 1984b. Occurrence and feeding of berycomorphid and percomorphid teleost fish in the Rockall Trough. *Journal du Conseil International pour l'Exploration de la Mer*, **41**, 239–247.

Maul, G.E. 1964. Observations on young live *Mupus maculatus* (Gunther) and *Mupus ovalis* (Valenciennes). *Copeia*, **1964** (1), 93–97.

Maurer, R.O., Jr and Bowman, R.E. 1975. *Food Habits of Marine Fishes of the Northwest Atlantic — Data Report. United States National Marine Fisheries Service*, Northeast Fisheries Center, Woods Hole Laboratory Reference Document no. 75–3, 90 pp.

Mayzaud, P. and Conover, R.J. 1988. O:N atomic ratio as a tool to describe zooplankton metabolism. *Marine Ecology Progress Series*, **45**, 289–302.

Mayzaud, P. and Dallot, S. 1973. Respiration et excretion azote du zooplankton. I. Evaluation des niveaux metaboliques de quelques espèces de Méditerranée occidentale. *Marine Biology*, **19**, 307–314.

McAllister, D.E. 1967. The significance of ventral bioluminescence in fishes. *Journal of the Fisheries Research Board of Canada*, **24**, 537–554.

McCann, C. 1966. The marine turtles and snakes occurring in New Zealand. *Records of the Dominion Museum, Wellington*, **5**, 201–215.

McCapra, F. 1990. The chemistry of bioluminescence: origins and mechanism. pp. 265–278. In *Light and Life in the Sea* (P.J. Herring, A.K. Campbell, M. Whitfield and L.R. Maddock, eds), Cambridge University Press, Cambridge.

McClintock, J.B. and Janssen, J. 1990. Pteropod abduction as a chemical defence in a pelagic Antarctic amphipod. *Nature*, London, **346**, 462–464.

McFall-Ngai, M.J. and Ruby, E.G. 1991. Symbiont recognition and subsequent morphogenesis as early events in an animal–bacterial mutualism. *Science*, **254**, 1491–1494.

McKenzie, R.A. and Homans, R.E.S. 1937. Rare and interesting fishes and salps in Bay of Fundy and off Nova Scotia. *Proceedings of the Nova Scotian Institute of Science*, **19**, 276–281.

McLean, N. and Nielsen, C. 1989. *Oodinium jordani* n.sp., a dinoflagellate (Dinoflagellata: Oodinidae) ectoparasitic on *Sagitta elegans* (Chaetognatha). *Diseases of Aquatic Organisms*, **7**, 61–66.

McReynolds, J.S. and Gorman, A.L.F. 1975. Hyperpolarizing photoreceptors in the eye of a primitive chordate, *Salpa democratica. Vision Research*, **15**, 1181–1186.

Ménard, F., Dallot, S. and Thomas, G. 1993. A stochastic model for ordered categorical time series. Application to planktonic abundance data. *Ecological Modelling*, **66**, 101–112.

Ménard, F., Dallot, S., Thomas, G. and Braconnot, J.-C. 1994. Temporal fluctuations of two Mediterranean salp populations from 1967 to 1990. Analysis of the influence of environmental variables using a Markov chain model. *Marine Ecology Progress Series*, **104**, 139–152.

Mensinger, A.F. and Case, J.F. 1992. Dinoflagellate luminescence increases susceptibility of zooplankton to teleost predation. *Marine Biology*, **112**, 207–210.

Merrett, N.R. and Marshall, N.B. 1981. Observations on the ecology of deep-sea bottomliving fishes collected off northwest Africa (08°–27°N). *Progress in Oceanography*, **9**, 185–244.

Mertens, H. 1830. Beschreibung der *Oikopleura*, einer neuen Molluskengattung. *Mémoires de l'Académie Impériale des Sciences de St.-Pétersbourg*, **1** (6), 205–220.

Metcalf, M.M. 1893. The eyes and subneural glands of *Salpa*. In *The Genus Salpa* (W.K. Brooks, ed.), pp. 307–371. *Memoirs of the Biological Laboratory Johns Hopkins University*, Vol. 2.

Metcalf, M.M. 1918. The Salpidae: a taxonomic study. *Bulletin of the United States National Museum*, **100**, 1–193.

Metcalf, M.M. and Hopkins, H.S. 1919. *Pyrosoma*. A taxonomic study based upon the collections of the U.S. Bureau of Fisheries and the U.S. National Museum. *Bulletin of the United States National Museum*, **100**, 195–276.

Metcalf, M.M. and Lentz-Johnston, M.E.G. 1905. The anatomy of the eyes and neural glands in the aggregated forms of *Cyclosalpa dolichosoma-virgula* and *Salpa punctata. Biological Bulletin, Woods Hole*, **9**, 195–212.

Meurice, J.C. 1970. Contribution à l'étude du genre *Ritteriella: Ritteriella amboinensis* Apstein. *Annales de la Société Royale Zoologique de Belgique*, **100**, 191–214.

Meurice, J.C. 1974. Contribution à l'étude du genre *Ritteriella: Ritteriella picteti* (Apstein, 1904). *Bulletin de la Société Royale des Sciences de Liège*, **43**, 473–492.

Mianzan, H.W., Pajaro, M., Macharandiarena, L. and Cremonte, F. 1997. Salps: possible vectors of toxic dinoflagellates. *Fisheries Research*, **29**, 193–197.

Michaels, A.F. and Silver, M.W. 1988. Primary production, sinking fluxes and the microbial food web. *Deep-Sea research*, **35**, 473–490.

Miller, C.A. and Landry, M.R. 1984. Ingestion-independent rates of ammonium excretion by the copepod *Calanus pacificus. Marine Biology*, **78**, 265–270.

Miller, R.L. 1991. Spawning and sperm, approach to the egg in the salp *Thalia democratica* Forskal 1775). *American Zoologist*, **31**, p138A.

Miller, R.L. and King, K.R. 1983. Sperm chemotaxis in *Oikopleura dioica* Fol, 1872 (Urochordata: Larvacea). *Biological Bulletin, Woods Hole*, **165**, 419–428.

Mills, C.E. and Goy, J. 1988. *In situ* observations of the behavior of mesopelagic *Solmissus* narcomedusae (Cnidaria, Hydrozoa). *Bulletin of Marine Science*, **43**, 739–751.

Mills, C.E. and Lean, N. 1991. Ectoparasitism by a dinoflagellate (Dinoflagellata: Oodinidae) on 5 ctenophores (Ctenophora) and a hydromedusa (Cnidaria.) *Diseases of Aquatic Organisms*, **10**, 211–216.

Minami, T. and Tanaka, M. 1992. Life cycles in flatfish from the northwestern Pacific, with particular reference to their early life histories. *Netherlands Journal of Sea Research*, **29**, 35–48.

Monniot, C. 1990. Diseases of Urochordata. In *Diseases of Marine Animals* (O. Kinne, ed.), Vol. 3, pp. 569–636. Biologisches Anstalt Helgoland, Hamburg.

Monniot, C. and Monniot, F. 1966. Un Pyrosome benthique: *Pyrosoma benthica* n.sp. *Comptes Rendus Hebdomodaire des Séances de l'Académie des Sciences, Paris*, **263D**, 368–370.

Monterey Bay Aquarium, 1996. Jellies and other ocean drifters. VHS video program, 35 min.

Montgomery, J.C., Macdonald, J.A. and Housley, G.D. 1988. Lateral line function in an Antarctic fish related to the signals produced by planktonic prey. *Journal of Comparative Physiology*, **163A**, 827–833.

Moreno, C.A. and Jara, H.F. 1984. Ecological studies on fish fauna associated with *Macrocystis pyrifera* belts in the south of Fueguian Islands, Chile. *Marine Ecology Progress Series*, **15**, 99–107.

Morin, J.G. 1983. Coastal bioluminescence: patterns and functions. *Bulletin of Marine Science*, **33**, 787–817.

Morris, C.C. and Deibel, D. 1993. Flow rate and particle concentration within the house of the pelagic tunicate *Oikopleura vanhoeffeni*. *Marine Biology*, **115**, 445–452.

Morris, R.J., Bone, Q., Head, R., Braconnot, J.C. and Nival, P. 1988. Role of salps in the flux of organic matter to the bottom of the Ligurian Sea. *Marine Biology*, **97**, 237–241.

Mosely, H. 1892. *Notes by a Naturalist. An Account of Observations Made During the Voyage of HMS Challenger* (see p. 505). John Murray, London.

Moss, E.L. 1870. On the anatomy of the genus Appendicularia, with the description of a new form. *Transactions of the Linnean Society, London*, **27**, 299–304.

Mullin, M.M. 1983. *In situ* measurement of filtering rates of the salp, *Thalia democratica*, on phytoplankton and bacteria. *Journal of Plankton Research*, **5**, 279–288.

Mumm, N. 1993. Composition and distribution of mesozooplankton in the Nansen Basin, Arctic Ocean, during summer. *Polar Biology*, **13**, 451–461.

Murray, J. and Hjort, J.H. 1912. *The Depths of the Ocean*. Macmillan, London.

Myers, A.A. and Giller, P.S. (eds) 1988. *Analytical Biogeography*. Chapman and Hall, London.

Nakamura, E.L. 1965. Food and feeding habits of Skipjack Tuna (*Katsuwonus pelamis*) from the Marquesas and Tuamoto Islands. *Transactions of the American Fisheries Society*, **94**, 236–242.

Nakamura, E.L. 1970. Observations on the biology of the myctophid *Diaphus garmani*. *Copeia*, 1970 (2), 374–377.

Nealson, K.H. 1979. Alternative strategies of symbiosis of marine luminous fishes harboring light-emitting bacteria. *Trends in Biochemical Science*, **4**, 105–110.

Nealson, K.H. and Hastings, J.W. 1979. Bacterial bioluminescence: its control and ecological significance. *Microbiology Reviews*, **43**, 496–518.

Nealson, K.H. and Hastings, J.W. 1980. Luminescent bacterial endosymbionts in bioluminescent tunicates (W. Schwemmler and H.E.A. Schenk, eds), pp. 461–466. In *Endocytobiology: Endosymbiosis and Cell Biology*. de Gruyter, Berlin.

Nedreaas, K. 1987. Food and feeding habits of young saithe, *Pollachius virens*, on the coast of western Norway. *Fiskeridirektoratets skrifter*, **18**, 263–301.

Nelson, G. 1986. Models and prospects of historical biogeography. In *Pelagic Biogeography* (A.C. Pierrot-Bults, S. van der Spoel, B.I. Zahuranec and R.K. Johnson, eds), pp. 214–218. *Unesco Technical Papers in Marine Science*, Vol. 49. UNESCO, Paris.

Nelson, G. and Platnick, N.I. 1981. *Systematics and Biogeography, Cladistics and Vicariance*, 567 pp. Columbia University Press, New York.

Nemoto, T. and Saijo, Y. 1968. Trace of chlorophyll pigments in stomachs of deep sea zoo-plankton. *Journal of the Oceanography Society of Japan*, **24**, 310–312.

Nesterov, A.A. 1981. The feeding of *Scomberesox saurus* (Scomberesocidae) and its trophic relationships in the epipelagic zone of the ocean. *Journal of Ichthyology*, **21** (2), 55–69.

Neumann, G. 1906. *Doliolum Wissenschaftliche Ergebnisse der Deutschen Tiefsee Expedition 1898–1899*, **12**, 97–243.

Neumann, G. 1913a. Die Pyrosomen der deutschen Tiefsee Expedition. *Wissenschaftliche Ergebnisse der Deutschen Tiefsee Expedition 1898–1899*, **12**, 293–421.

Neumann, G. 1913b. Salpae II: Cyclomyaria et Pyrosomida. *'Das Tierreich,'* **40**, 1–38.

Neumann, G. 1913c. Die Pyrosomen und Dolioliden der Deutschen Sudpolar Expedition. *Deutsche Sudpolar Expedition, 1901–1903*, **14**, 17–34.

Neumann, G. 1935. Cyclomyaria. Tunicata. In *Handbuch der Zoologie* (W. Kükenthal and T. Krumbach, eds), Vol. 5, pp. 24–400. de Gruyter, Berlin, Germany.

Nevitt, G. and Gilly, W.F. 1986. Morphological and physiological properties of non-striated muscle from the tunicate, *Ciona intestinalis*: parallels with vertebrate skeletal muscle. *Tissue and Cell*, **18**, 341–360.

Nichols, J.T. and Breder, C.M. 1927. The marine fishes of New York and southern New England. *Zoologica (New York)*, **9**, 1–192.

Nichols, J.T. and Murphy, R.C. 1922. On a collection of marine fishes from Peru. *Bulletin of the American Museum of Natural History*, **46**, 501–516.

Nicolas, G., Bassot, J.-M. and Nicolas, M.-T. 1990. The advantages of cryotechniques: application to bioluminescent cells. In *Proceedings of the 12th international Congress of Electron Microscopy, Seattle*, pp. 486–7. San Francisco Press, San Fancisco.

Nierop, M.M. van and den Hartog, J.C. 1984. A study on the gut contents of five juvenile loggerhead turtles, *Caretta caretta* (Linnaeus) (Reptilia, Cheloniidae), from the southeastern part of the North Atlantic Ocean, with emphasis on coelenterate identification. *Zoölogische Mededeelingen*, **59** (4), 35–54.

Nishikawa, J. 1995. Ecological study of pelagic tunicates, salps and doliolids, PhD thesis, pp. 1–358, University of Tokyo (in Japanese).

Nishikawa, J. and Terazaki, M. 1995. Measurement of swimming speeds and pulse rate of salps using a video equipment. *Bulletin of the Plankton Society of Japan*, **41**, 170–173.

Nishikawa, J., Nagonobu, M., Ichii, T., Ishii, H., Terazaki, M. and Kawaguchi, K. 1995. Distribution of salps near the South Shetland Islands during austral summer, 1990–1991 with special reference to krill distribution. *Polar Biology*, **15**, 31–39.

Nishimura, S. 1958. Quelques remarques sur l'ingestion du salpe, *Salpa fusiformis Cuvier*, chez les maquereaux pêchés dans la mer du Japan orientale. *Annual Reports of the Japan Sea Regional Fisheries Laboratory*, **4**, 105–112.

Nival, P. and Nival, S. 1976. Particle retention efficiency of an herbivorous copepod, *Acartia clausi* (adult and copepodite stages) effects on grazing. *Limnology and Oceanography*, **21**, 24–38.

Nival, P., Nival, S. and Palazzoli, I. 1972. Données sur la respiration de differents organismes communs dans le plancton de Villefranche-sur-Mer. *Marine Biology*, **17**, 63–76.

Nival, P., Braconnot, J.C., Andersen, V., Oberdorff, T., Choe, S.M. and Laval, Ph. 1985. Estimation de l'impact des Salpes sur le phytoplancton en mer Ligure. *Rapports et Procès-verbaux des Réunions — Commission Internationale pour l'Exploration Scientifique de la Mer Méditerranée*, **29**, 283–286.

Noji, T.T., Estep, K.W., MacIntyre, F. and Norrbin, F. 1991. Image analysis of faecal material grazed upon by three species of copepods; evidence for coprorhexy, coprophagy and coprochaly. *Journal of the Marine Biological Association of the United Kingdom*, **71**, 465–480.

O'Sullivan, D. 1983. A guide to the pelagic tunicates of the Southern Ocean and adjacent waters. *Australian national Antarctic Research Expeditions, Research Notes*, **8**, VI–98.

Ogawa, Y. and Nakahara, T. 1979. Interrelationships between pelagic fishes and plankton in the coastal fishing ground of the southwestern Japan Sea. *Marine Ecology Progress Series*, **1**, 115–122.

Ohtsuka, S. and Kubo, N. 1991. Appendicularians and their houses as important food for some pelagic copepods. *Proceedings of the 4th International Conference on Copepoda, Bulletin of the Plankton Society Special Volume,* 535–551.

Ohtsuka, S. and Onbé, T. 1989. Evidence of selective feeding on larvaceans by the pelagic copepod *Candacia bipinnata* (Calanoida: Candaciidae). *Journal of Plankton Research,* **11**, 869–872.

Ohtsuka, S., Kubo, N., Okada, M. and Gushima, K. 1993. Attachment and feeding of pelagic copepods on appendicularian houses. *Journal of the Oceanographic Society of Japan,* **49**, 115–120.

Ohtsuka, S., Ohaye, S., Tanimura, A., Fukuchi, M., Hattori, H., Sasaki, H. and Matsuda, O. 1993. Feeding ecology of copeppodite stages of *Eucalanus gingii* in the Chukchi and northern Bering Seas in October 1988. *Proceedings National Institute for Polar Research. Symposium on Polar Biology,* **6**, 27–37.

Olsson, R. 1962. Reissner's fibre apparatus in its most primitive condition. *General and Comparative Endocrinology,* **2** (6), 617–618 (abstract no. 24).

Olsson, R. 1963. Endostyles and endostylar secretions: a comparative histochemical study. *Acta Zoologica (Stockholm),* **44** (3), 299–328.

Olsson, R. 1965. The cytology of the endostyle of *Oikopleura dioica. Annals of the New York Academy of Sciences,* **118**, 24, 1038–1051.

Olsson, R. 1975. Primitive coronet cells in the brain of *Oikopleura* (Appendicularia, Tunicata). *Acta Zoologica (Stockholm),* **56**, 155–161.

Olsson, R., Holmberg, K. and Lilliemarck, Y. 1990. Fine structure of the brain and brain nerves of *Oikopleura dioica* (Urochordata, Appendicularia). *Zoomorphology,* **110** (1), 1–7.

Omori, M. and Ikeda, T. 1984. *Methods in Marine Zooplankton Ecology,* pp. 332. John Wiley, New York.

Ormières, R. 1964. Recherches sur les sporozoaires parasites des tuniciers. *Vie et Milieu,* **15**, 823–946.

Orzech, J.K. and Nealson, K.H. 1984. Bioluminescence of marine snow: its effect on the optical properties of the sea. *Proceedings of the Society of Photo-Optical Instrument Engineers,* **489**, 100–106.

Owen, R.W. 1966. Small-scale, horizontal vortices in the surface layer of the sea. *Journal of Marine Research,* **24** (1), 56–66.

Ozawa, T., Kawai, K. and Uotani, I. 1991. Stomach content analysis of chub mackerel *Scomber japonicus* larvae by quantification. I method. *Bulletin of the Japanese Society of Scientific Fishery,* **57**, 1241–1245.

Paffenhöfer, G.-A. 1973. The cultivation of an appendicularian through numerous generations. *Marine Biology,* **22**, 2, 183–185.

Paffenhöffer, G.-A. 1976. On the biology of Appendicularia of the southeastern North Sea. In *Proceedings of the 10th European Symposium on Marine Biology,* Osteno, Belgium. Sept. 17–23, 1975 (G. Persoone and E. Jaspers, eds), Vol. 2, pp. 437–455. Universal Press, Wetteren.

Paffenhöfer, G.-A. 1985. The abundance and distribution of zooplankton on the southeastern shelf of the United States. In *Oceanography of the Southeastern U.S. Continental Shelf, Coastal and Estuarine Sciences,* Vol. 2, pp. 104–114.

Paffenhöfer, G.-A. and Harris, R.P. 1976. Feeding, growth and reproduction of the marine planktonic copepod *Pseudocalanus elongatus* Boeck. *Journal of the Marine Biological Association of the United Kingdom,* **56**, 327–344.

Paffenhöfer, G.-A. and Knowles, S.C. 1978. Feeding of marine planktonic copepods on mixed phytoplankton. *Marine Biology,* **48**, 143–152.

Paffenhöfer, G.-A. and Knowles, S.C. 1979. Ecological implications of fecal pellet size, production and consumption by copepods. *Journal of Marine Research,* **37**, 35–49.

Paffenhöfer, G.-A. and Lee, T.N. 1987. Development and persistence of patches of Thaliacea. *South African Journal of Marine Science,* **5**, 305–318.

Paffenhöfer, G.-A. Sherman, B.K. and Lee, T.N. 1987. Abundance, distribution and patch formation of zooplankton. *Progress in Oceanography,* **19**, 403–436.

Paffenhöfer, G.-A., Wester, B.T. and Nicholas, W.D. 1984. Zooplankton abundance in relation to state and type of intrusions onto the southeastern United States shelf during summer. *Journal of Marine Research,* **42**, 995–1017.

Paffenhöfer, G.-A., Stewart, T.B., Youngbluth, M.J. and Bailey, T.G. 1991. High-resolution vertical profiles of pelagic tunicates. *Journal of Plankton Research,* **13**, 971–981.

Paffenhöfer, G.-A., Atkinson, L.P., Lee, T.N., Verity, P.G. and Bulluck, L.R. 1995. Distribution and abundance of Thaliaceans and copepods off the southeastern U.S.A. during winter. *Continental Shelf Research,* **15**, 255–280.

Page, R. 1990. Component analysis, a valiant failure. *Cladistics,* **6**, 119–136.

Palma, S.G. 1985. Migracion nictemeral del macroplancton gelatinoso de la bahia de Villefranche-sur-Mer, Mediterraneo Noroccidental. *Investigación Pesquera,* **49**, 261–274.

Panceri, P. 1872. Gli organi luminosi e la luce dei Pyrosoma e della foladi. *Atti della Reale Accademia delle Scienze Fisiche e Matemastiche di Napoli,* **5**, 1.51.

Parin, N.V. 1970. *Ichthyofauna of the Epipelagic Zone.* Keter Press, Jerusalem.

Parsons, T.R., Lebrasseur, R.J. and Fulton, J.D. 1967. Some observations on the dependence of zooplankton grazing on the cell size and concentration of phytoplankton blooms. *Journal of the Oceanographic Society of Japan,* **23**, 10–17.

Pavlov, Yu.P. 1991. Information on morphometrics and ecology of pomfrets of the genus *Brama* inhabiting the southeastern Pacific Ocean. *Journal of Ichthyology,* **31**, 120–124.

Pavlova, E.T., Petipa, T.S. and Sorokin, Y.I. 1971. Bacterioplankton as food for pelagic marine organisms. In *Funktsionirovanie pelagicheskikh soobshchestv tropicheskikh raionov okeana.* (ed. M.E. Vinogradov), pp. 142–151. Izdatel 'stvo "Nauka" Moscow. Published in 1973 as *Life activity of pelagic communities in the ocean tropics,* pp. 156–165. Israel Program for Scientific translations, Jerusalem.

Pavlova, T.P. 1979. Some peculiarities of the biology of *Epigonus denticulatus* (Dieuzeide, 1950) from the South Paciac elevation. *Studies on Fish Biology and Commercial Oceanography,* **10**, 92–98 (in Russian).

Pavshtiks, E.A. 1972. Seasonal variations in the number of zooplankton in the region of the North Pole. *Doklady Biological Sciences,* **196**, 1–6.

Pavshtiks, E.A. and Rudakova, V.A. 1962. On the problem of yearly changes in the development of plankton and feeding conditions of herring in the Norwegian Sea. *Trudy Polyarnogo Nauchno-issledovatel'skogo Instituta Morskogo Rybnogo Khozyiaistva i Okeanografii Murmansk,* **14**, 209–222 (in Russian).

Paxton, J.R. 1989. Synopsis of the whalefishes (family Cetomimidae) with descriptions of four new genera. *Records of the Australian Museum,* **41** (2), 135–206.

Pearcy, W.G. and Ambler, J.W. 1974. Food habits of deep-sea macrourid fishes off the Oregon coast. *Deep-Sea Research,* **21**, 745–759.

Pearcy, W., Nishiyama, T., Fujii, T. and Masuda, K. 1984. Diel variations in the feeding habits of Pacific salmon caught in gill nets during a 24-hour period in the Gulf of Alaska. *Fishery Bulletin,* **82** (2), 391–400.

Perasso, R., Baroin, A., Qu, L.H., Bachellerie, J.P. and Adoutte, A. 1989. Origin of the algae. *Nature, London,* **339**, 142–144.

Pérès, J.M. 1943. Recherches sur le sang et les organes neuraux des Tuniciers. *Annales de l'Institut Océanographique,* **21**, 229–359.

Permitin, Yu.Ye. and Tarverdiyeva, M.L. 1972. The food of some Antarctic fish in the South Georgia area. *Journal of Ichthyology*, **12**, 104–111.

Permitin, Yu.E. and Tarverdieva, M.I. 1978. Nutrition of Nototheniidae and Chaenichthyidae fishes off the South Orkney Islands *Biologiya Morya*, **2**, 75–81 (in Russian).

Perissinotto, R. and Pakhamov, E.A. 1997. Feeding association of the copepod *Rhincalanus gigas* with the tunicate salp *Salpa thompsoni* in the Southern Ocean. *Marine Biology*, **127**, 479–483.

Péron, F. 1804. Mémoire sur le nouveau genre *Pyrosoma. Annales de la Musée d'Histoire Naturelle*, **4**, 437–446.

Peterson, W.T., Brodeur, R.D., and Pearcy, W.G. 1982. Food habits of juvenile salmon in the Oregon coastal zone. *Fishery Bulletin*, **80**, 841–851.

Pfannkuche, O. and Lochte K. 1993. Open ocean pelago-benthic coupling: cyanobacteria as tracers of sedimenting salp feces. *Deep-Sea Research*, **40**, 727–737.

Pierantoni, U. 1921. Gli organi luminosi simbiotici ed il loro ciclo ereditario in *Pyrosoma giganteum. Pubblicazione della Stazione Zoologica di Napoli*, **3**, 191–222.

Pierrot-Bults, A.C. and van der Spoel, S. 1979. Speciation in macrozooplankton. In *Zoogeography and Diversity of Plankton* (S. Van der Spoel and A.C. Pierrot-Bults, eds), pp. 144–167. Bunge, Utrecht.

Pinkas, L., Oliphant, M.S. and Iverson, I.L.K 1971. Food habits of albacore, bluefin tuna and bonito in Californian waters. *California Department of Fish and Game Fishery Bulletin*. **152**, 1–105.

Platnick, N.I 1991. On areas of endemism. In *Austral Biogeography* (P.Y. Ladiges, C.J. Humphries and L.W. Martinelli, eds). *Australian Systematic Botany*, vol. 4, Comments (2 unnumbered pages).

Podrazhanskaya, S.G. 1969. Nutrition of rock grenadier (*Macrurus rupestris*) in some regions of the Northwestern Atlantic and in Iceland waters. *Trudy Molodykh Vsesoyuznyi Nauchno-issledovatel'skii Okeanografii (UNIRO) Moscow*, **1**, 54–73 (in Russian).

Pomerat, C.M. 1957. *Olfactory Epithelium of Human Foetus in Tissue Culture* (film). University of Texas.

Pomeroy, L.R. and Deibel, D. 1980. Aggregation of organic matter by pelagic tunicates. *Limnology and Oceanography*, **25**, 643–652.

Pomeroy, L.R. and Deibel, D. 1986. Temperature regulation of bacterial activity during the spring bloom in Newfoundland coastal waters. *Science*, **233**, 359–361.

Pomeroy, L.R. and Wiebe, W.J. 1993. Seasonal uncoupling of the microbial loop and its potential significance for the global cycle of carbon. In *Trends in Microbial Ecology* (R. Guerrero and C. Pedros-Alio, eds). Spanish Society for Microbiology, Barcelona.

Pomeroy, L.R., Hanson, R.B., Gillivary, P.A., Sherr, B.F., Kirchman, D. and Deibel, D. 1984. Microbiology and chemistry of fecal products of pelagic tunicates: rates and fates. *Bulletin of Marine Science*, **35**, 426–439.

Pontekorvo, T.B. 1974. Some peculiarities of distribution of the hydrobiological and hydrological characters in the regions of the banks of the Hawaiian seamount. *Izvestiya Tikhookeanskogo Nauchno-issledovatel'skogo Instituta Rybnogo Khozyaistra i Okeanografii*, **92**, 32–37 (in Russian).

Porter, K.G. and Porter, J.W. 1979. Bioluminescence in marine plankton: a coevolved antipredation system. *American Naturalist*, **114**, 458–461.

Purcell, J.E. 1981. Dietary composition and diel feeding patterns of epipelagic siphonophores. *Marine Biology*, **65**, 83–90.

Purcell, J.E. and Madin, L.P. 1991. Diel patterns of migration, feeding, and spawning by salps in the subarctic Pacific. *Marine Ecology Progress Series*, **73**, 211–217.

Quoy, J.R.C. and Gaimard, J.-P. 1824–1826. *Voyage autour du Monde, entrepris par l'Ordre du Roi....exécuté sur les Corvettes de SM. l'Uranie et la Physicienne, pendant les années 1817, 1818, 1819 et 1820, publié sous les auspices...par Louis de Freycinet, 1824. Zoologie par (Jean René Constantin) Quoy et (Jean-Paul) Gaimard*. Vols 1–4, pp712, Pillet aîné, Paris.

Quoy, J.R.C. and Gaimard, J.-P. 1827. Observations zoologiques faites à bord de l'Astrolabe, en mai, 1826, dans le détroit de Gibraltar. *Annales des Sciences Naturelles*, **10**, 1–21.

Quoy, J.R.C. and Gaimard, J.-P. 1833. Zoologie. In *Voyage de Découvertes de l'Astrolabe*, Vol. 4, pp. 390 + atlas. J. Tastu, Paris.

Quoy, J.R.C. and Gaimard, J.-P. 1836. Voyage de découvertes de l'Astrolabe, exécuté par ordre du Roi, pendant les années 1826–1829 sous le commandement de R.I. [sic] Dumont D'Urville. Zoologie. T. III. *Isis, von Oken*, **29** (2), 95–159.

Radovich, J. 1952. Food of the Pacific sardine, *Sardinops caerulea*, from central Baja California and southern California. *California Fish and Game*, **38** (4), 575–585.

Rae, B.B. 1967. The food of the dogfish, *Squalus acanthias* L. *Department of Agriculture and Fisheries Scottish Marine Research*, **1967** (4), 1–19.

Raitt, D.F.S. and Adams, J.A. 1965. The food and feeding of *Trisopterus esmarkii* (Nilsson) in the northern North Sea. *Marine Research*, **3**, 1–28.

Rajagopal, P.K. 1962. Respiration of some marine planktonic organisms. *Proceedings of the Indian Academy of Science, Series B*, **55**, 76–81.

Randall, J.E. 1967. Food habits of reef fishes of the West Indies. *Studies in Tropical Oceanography*, **5**, 665–847.

Rassoulzadegan, F. and Etienne, M. 1981. Grazing rate of the tintinnid *Stenosemella ventricosa* (Clap. & Lachm.) Jörg. on the spectrum of the naturally occuring particulate matter from a Mediterranean neritic area. *Limnology and Oceanography*, **26**, 258–270.

Rathke, H. 1835. Beschriebung der Anchinia, einer neuen Gattung der Mollusken von Eschsholtz, mitgeheir von Rathke. *Mémoires présentés à l'Académie Impériale des Sciences à St-Petersbourg par divers savants*, **2**, 177–179.

Redden, A.M. 1994. Grazer-mediated chloropigment degradation and the vertical flux of spring bloom production in Conception Bay, Newfoundland, PhD thesis, 250 pp. Department of Biology, Memorial University of Newfoundland, St Johns.

Rege, M.S. and Bal, D.V. 1963. Some observations on the food and feeding-habits of the silver pomfret *Pampus argenteus* in relation to the anatomy of its digestive system. *Journal of the University of Bombay*, **31**, 75–79.

Reilly, G.A. and Gorgy, S. 1948. Quantitative studies of summer plankton populations of the western North Atlantic. *Journal of Marine Research*, **7**, 100–118.

Reinke, M. 1987. Zur Nahrungs- und Bewegungsphysiologie von *Salpa thompsoni* und *Salpa fusiformis* On the feeding and locomotory physiology of *Salpa thompsoni* and *Salpa fusiformis. Berichte zur Polarforschung (Reports on Polar Research)*, **36**, 1–89.

Reintjes, J.M and King, I.E. 1953. Food of yellow fin tuna in Central Pacific. *Fishery Bulletin*, **54** (81), 90–110.

Riehl, M.W. 1992. Elemental analyses of oikopleurids and factors affecting house production rate of *Oikopleura vanhoeffeni* (Tunicata, Appendicularia) in coastal Newfoundland waters. MS thesis, 64 pp. Memorial University of Newfoundland, St Johns.

Riisgård, H.U. 1989. Properties and energy cost of the muscular piston pump in the suspension feeding polychaete *Chaetopterus variopedatus. Marine Ecology Progress Series*, **56**, 157–168.

Risso, A. 1816. Histoire Naturelle des Crustacés des Environs de Nice, pp.175. Librairie Grecque-Latine-Allemande, Paris.

Ritter, W.E. and Byxbee, E.S. 1905. The pelagic Tunicata. *Memoirs of the Museum of Comparative Zoölogy at Harvard College*, **26** (5), 193–216.

Robertson, J.D. 1957. Osmotic and ionic regulation in aquatic invertebrates. In *Recent Advances in Invertebrate Physiology* (B.T. Scheer, ed.), Vol. 2, pp. 29–246. University of Oregon Press, Eugene.

Roboz, Z. 1886. Adatok a gregarinak Ismeretének. *Ertekezések Termész Magyar Akademie*, **16**, 1–34. (In Hungarian)

Roe, H.S.J., James, P.T. and Thurston, M.R. 1984. The diel migrations and distributions within a mesopelagic community in the North East Atlantic. 6. Medusae, ctenophores, amphipods and euphausiids. *Progress in Oceanography*, **13**, 425–460.

Roe, H.S.J., Badcock, J., Billett, D.S.M., Chidgey, K.C., Domanski, P.A., Ellis, C.J., Fasham, M.J.R., Gooday, A.J., Hargreaves, P.M.D., Huggett, Q.J., James, P.T., Kirkpatrick, P.A., Lampitt, R.S., Merrett, N.R., Muirhead, A., Pugh, P.R., Rice, A.L., Russell, R.A., Thurston, M.H. and Tyler, P.A. 1987. Great Meteor East: a biological characterization. *Institute of Oceanographic Sciences Deacon Laboratory Report*, **248**, 322 pp.

Roger, C. 1982a. Macroplancton et micronecton de l'Atlantique tropical. I. Biomasses et composition taxonomique. *Océanographie Tropicale*, **17**, 85–96.

Roger, C. 1982b. Macroplancton et micronecton de l'Atlantique tropical. II. Cycles de l'azote et du phosphore. Remarques sur la mesure de production. *Océanographie Tropicale*, **17**, 177–185.

Roméo, M., Gnassia-Barelli, M. and Carré, C. 1992. Importance of gelatinous plankton organisms in storage and transfer of trace metals in the northwestern Mediterranean. *Marine Ecology Progress Series*, **82**, 267–274.

Roshchin, E.A. 1991. Aspects of the life cycle of *Trematomus eulepidotus* (Nototheniidae) in the Indian Ocean sector of the Antarctic. *Journal of Ichthyology*, **31** (4), 1–11.

Roskell, J. 1986a. Thaliacea in continuous plankton records during 1983. *Année Biologique*, **40**, 77–79.

Roskell, J. 1986b. Thaliacea in continuous plankton records during 1984. *Année Biologique*, **41**, 77–41.

Rowedder, U. 1979. Feeding ecology of the myctophid *Electrona antarctica* (Gunther, 1878) (Teleostei). *Meeresforschung*, **27**, 252–263.

Runnström, S. 1931. Eine Uebersicht über das Zooplankton des Herdla- und Hjelte-fjordes. *Bergens Museum Aarbok*, **7**, 1–67.

Russell, B.C. 1983. The food and feeding habits of rocky reef fish of north-eastern New Zealand. *New Zealand Journal of Marine and Freshwater Research*, **17**, 121–145.

Russell, F.S. and Hastings A.B. 1933. On the occurrence of pelagic tunicates (Thaliacea) in the waters of the English Channel off Plymouth. *Journal of the Marine Biological Association of the United Kingdom*, **18**, 635–640.

Russell, F.S. and Colman, J.S. 1935. The zooplankton IV. The occurrence and seasonal distribution of the Tunicata, Mollusca and Coelenterata (Siphonophora). *Scientific Reports of the Great Barrier Reef Expedition*, **2**, 205–234.

Salensky, W. 1883. Neue Untersuchungen ueber die embryonale Entwicklung der Salpen. *Mittheilungen aus der Zoologischen Station zu Neapel*, **4**, 90–171 and 327–402.

Salensky, W. 1903. Etudes anatomiques sur les Appendiculaires. I. *Oikopleura vanhoeffeni* Lohmann. *Mémoires de l'Académie Impériale des Sciences de St.-Pétersbourg*, **13** (7), 1–44.

Salensky, W. 1904a. Etudes anatomiques sur les Appendiculaires. II. *Oikopleura refescens* Fol. *Mémoires de l'Académie Impériale des Sciences de St.-Pétersbourg*, **15** (1), 1–54.

Salensky, W. 1904b. Etudes anatomiques sur les Appendiculaires. III. *Fritillaria pellucida* Busch. *Mémoires de l'Académie Impériale des Sciences de St.-Pétersbourg*, **15** (1), 55–90.

Salensky, W. 1904c. Etudes anatomiques sur les Appendiculaires. IV. *Fritillaria borealis* Lohmann. *Mémoires de l'Académie Impériale des Sciences de St.-Pétersbourg*, **15** (1), 91–106.

Salensky, W. 1905. Zur Morphologie der Cardialorgane der Appendicularien. In *Compte Rendus des Séances du Sixième Congrès International de Zoologie, tenu à Berne du 14 au 16 Août* 1904, pp. 381–383. Kandiget fils, Genève.

Savigny, J.C. 1816. Mémoires sur les Animaux sans Vertèbres. 2 Vols., 358 pp, Deterville ed. Paris.

Salvini-Plawen, L. von. 1972. Cnidaria as food sources for marine invertebrates. *Cahiers Biologie Marin*, **13**, 385–400.

Sardou, J., Etienne, M. and Anderson, V. 1996. Seasonal abundance and vertical distributions of macroplankton and micronekton in the Northwestern Mediterranean Sea. *Oceanologica Acta*, **19**, 645–656.

Savage, R.E. 1937. The food of North Sea herring 1930–1934. *Fishery Investigations*, **15**, 57.

Schnack, S.B. 1985. Feeding by *Euphausia superba* and copepod species in response to varying concentrations of phytoplankton. In *Antarctic Nutrient Cycles and Food Webs* (W.R. Siegfried, P.R. Condy and R.M. Laws, eds). Springer-Verlag, Berlin.

Schneider, G. 1990. A comparison of carbon based ammonia excretion rates between gelatinous and non-gelatinous zooplankton: implications and consequences. *Marine Biology*, **106**, 219–225.

Seapy, R.R. 1980. Predation by the epipelagic heteropod mollusk *Carinaria cristata forma japonica*. *Marine Biology*, **60**, 137–146.

Seeliger, O. 1895a. *Die Pyrosomen der Plankton-Expedition. Ergebnisse der Plankton-Expedition der Humboldt Stiftnung*, Vol. 2, E.b. pp. 95. Lipsius und Tischer, Kiel und Leipzig.

Seeliger, O. 1895b. Tunicata (Mantelthiere). 1. Die Appendicularien, Copelata. *Bronn's Klassen und Ordnungen des Tier-Reichs*, Vol. 3 (supplement 4–5), pp. 97–144. Leipzig.

Seguin, G. and Ibanez, F. 1974. Zooplancton provenant de radiales effectuées le long de la côte sénégalaise par le navire océanographique *'Laurent Amaro'* en mai 1968. *Bulletin de l'Institut Fondamental d'Afrique Noire, Serie A, Sciences Naturelles*, **36**, 842–879.

Seki, H. 1973. Red tide of *Oikopleura* in Saanich inlet. *La Mer (Bulletin de la Société Franco-Japonaise d'Océanographie)*, **11** (3), 153–158.

Selys-Longchamps, M. de 1901. Etude du développement de la branchie chez *Corella*, avec une note sur la fonction des protostigmates chez *Ciona* et *Ascidiella*. *Archives de Biologie*, **17**, 673–711.

Selys-Longchamps, M. de 1939. Origine des premières ébauches cardiaques chez les Tuniciers. *Travaux de la Station Zoologique de Wimereux*, **13**, 629–634.

Serventy, D.L. 1956. The southern bluefin tuna, *Thunnus thynnus maccoyii* (Castelnau) in Australian waters. *Australian Journal of Marine and Freshwater Research*, **7**, 1–43.

Seymour Sewell, R.B. 1953. The Pelagic Tunicata. *Scientific Reports of the John Murray Expedition, 1933–34*, **10**, 1–90.

Shandikov, G.A. 1986. Biological description of *Nototheniops tohih* (Balushkin) (Nototheniidae) from the Ob' and Lena Banks, Indian Ocean sector of the Southern Ocean, In *Morphology and Distribution of Fishes of the Southern Ocean*, pp. 91–109. Zoological Institute of the Academy of Sciences, Leningrad (in Russian).

Sheard, K. 1965. Species groups in the zooplankton of eastern Australian slope waters, 1938–41. *Australian Journal of Marine and Freshwater Research*, **16**, 219–254.

Shelbourne, J.E. 1953. The feeding habits of plaice post-larvae in the Southern Bight. *Journal of the Marine Biological Association of the United Kingdom*, **32** (1), 149–160.

Shelbourne, J.E. 1957. The feeding and condition of plaice larvae in good and bad plankton patches. *Journal of the Marine Biological Association of the United Kingdom*, **36** (3), 539–552.

Shelbourne, J.E. 1962. A predator–prey size relationship for plaice larvae feeding on *Oikopleura*. *Journal of the Marine Biological Association of the United Kingdom*, **42** (2), 243–252.

Sheldon, R.W., Prakash, A. and Sutcliffe, W.H. 1972. The size distribution of particles in the ocean. *Limnology and Oceanography*, **17**, 327–340.

Sheldon, R.W., Sutcliffe, W.H., Jr and Paranjape, M.A. 1977. Structure of pelagic food chain and relationship between plankton and fish production. *Journal of the Fisheries Research Board of Canada*, **43**, 2344–2353.

Shiga, N. 1976. Maturity stages and relative growth of *Oikopleura labradoriensis* Lohmann (Tunicata, Appendicularia). *Bulletin of the Plankton Society of Japan*, **23**, 81–95.

Shiga, N. 1982. Regional and annual variations in abundance of an appendicularian, *Oikopleura labradoriensis* in the Bering Sea and the northern North Pacific Ocean during summer. *Bulletin of the Plankton Society of Japan*, **29**, 119–128.

Shiga, N. 1985. Seasonal and vertical distributions of Appendicularia in Volcano Bay, Hokkaido, Japan. *Bulletin of Marine Science*, **37** (2), 425–439.

Shiga, N. 1993a. First Record of the Appendicularian, *Oikopleura vanhoeffeni* in the Northern Bering Sea. *Bulletin of the Plankton Society of Japan*, **39**, 107–115.

Shiga, N. 1993b. Regional and vertical distributions of *Oikopleura vanhoeffeni* on the northern Bering sea shelf in summer. *Bulletin of the Plankton Society of Japan*, **39**, 117–126.

Shih, C.-T. 1986. Biogeography of oceanic zooplankton. In *Pelagic Biogeography* (A.C. Pierrot-Bults, S. van der Spoel, B.J. Zahuranec and R.K. Johnson, eds), *UNESCO Technical Papers in Marine Science*, Vol. 49, pp. 250–253. UNESCO, Paris.

Shimamoto, N. and Watanabe, J. 1994. Seasonal changes in feeding habit of red sea bream *Pagrus major* in the eastern Seto Inland sea, Japan. *Bulletin of the Japanese Society for Scientific Fisheries*, **60**, 65–71.

Shpak, V.M. 1976. Nutrition of southern poutassu *Micromesistius australis* Norman 1937 in the southwestem Pacific Ocean. *Studies on Fish Biology and Commercial Oceanography*, **7**, 37–45 (in Russian).

Shuntov, V.P. 1979. *Ichthyofauna of the Southwestern Pacific Ocean.* Pishchevaya Promyshlennost'O-vo, Moscow (in Russian).

Shust, KV. 1978. On the distribution and biology of members of the genus *Micromesistius* (Family Gadidae). *Journal of Ichthyology*, **18** (3), 490–493.

Shust, K.V. and Pinskaya, I.A. 1978. Age and rate of growth of six species of notothenid fish (Family Nototheniidae). *Journal of Ichthyology*, **18** (5), 743–749.

Siegel, V. and Harm, U. 1996. The composition, abundance, biomass and diversity of the epipelagic zooplankton communities of the southern Bellinghausen Sea (Antarctic) with special reference to krill and salps. *Archive of Fishery and Marine Research*, **44**, 115–139.

Sigl, A. 1912. Adriatische Thaliaceen Fauna. *Sitzungsberichte der Akademie der Wissenschaften in Wien, Mathematik-Naturwissenschaftliche Klasse*, **12**, 463–508.

Sigl, A. 1913. Die Thaliaceen und Pyrosomen des Mittelmeeres und der Adriatik. Gesammelt während der fünf Expeditionen S.M. Schiff "Pola" 1890–1894. *Denkschrift für Mathematik-Naturwissenschaftlichen Klasse*, **88**, 260–289.

Silver, M.W. 1971. The habitat of *Salpa fusiformis* in the California Current as defined by stomach content studies and the effect of salp swarms on the food supply of the plankton community, pp. 1–20. PhD thesis, University of California, San Diego, CA.

Silver, M.W. 1975. The habitat of *Salpa fusiformis* in the California Current as defined by indicator assemblages. *Limnology and Oceanography*, **20**, 230–237.

Silver, M.W. and Alldredge, A.L. 1981. Bathypelagic marine snow: deep-sea algal and detrital community. *Journal of Marine Research*, **39**, 501–530.

Silver, M.W. and Bruland, K.W. 1981a. Differential feeding and fecal pellet composition of salps and pteropods, and the possible large-scale particulate organic matter transport to the deep sea. *Marine Biology*, **53**, 249–255.

Silver, M.W. and Bruland, K.W. 1981b. Differential feeding and fecal pellet composition of salps andpteropods and the possible origin of the deep-water flora and olive green "cells". *Marine Biology*, **62**, 263–273.

Silver, M.W. and Gowing, M.M. 1991. The 'particle' flux: origins and biological components. *Progress in Oceanography*, **26**, 75–113.

Sivaprakasam, T.E. 1963. Observations on the food and feeding habits of *Parastromateus niger* (Bloch) of the Saurashtra coast. *Indian Journal of Fisheries*, **10** (1), 140–147.

Skaramuca, B. 1980. Kvalitativno i kvantitativno rasprostranjenje populacija apendikularija u Jadranskom moru (The qualitative and quantitative distribution of the appendicularian population in the Adriatic Sea), Doktorska disertacija, prirodoslovno-matematickom fakultet sveucilista u Zagrebu, 202 pp. (in Serbo-Croat, English abstract).

Skaramuca, B. 1983. Kvantitativno u kvalitativno rasprostranjenje populacija apendikularija u otvorenim vodama Jadranskog mora. *Acta Adriatica*, **24**, 133–177. (in Serbo-Croat, English summary.)

Skinner, H.A. 1961. *The Origin of Medical Terms*, 2nd edn, 437 pp. Williams & Wilkins, Baltimore.

Sleigh, M.A. editor. 1974. *Cilia and Flagella*, 500 pp. Academic Press London.

Small, L.F., Fowler, S.W. and Unlu, M.Y. 1979. Sinking rates of natural copepod fecal pellets. *Marine Biology*, **51**, 233–241.

Small, L.F., Fowler, S.W., Moore, S.A. and La Rosa, J. 1983. Dissolved and fecal pellet carbon and nitrogen release by zooplankton in tropical waters. *Deep-Sea Research*, **30**, 1199–1220.

Smayda, T.J. 1969. Some measurements of the sinking rate of fecal pellets. *Limnology and Oceanography*, **14**, 621–625.

Smith, A.B., Lafay, B. and Christen, R. 1992. Comparative variation of morphological and molecular evolution through geologic time: 28S ribosomal RNA versus morphology in echinoids. *Philosophical Transactions of the Royal Society of London B*, **338**, 365–382.

Smith, J.M. 1978. *The evolution of sex*, pp. 1–222. Cambridge University Press, Cambridge.

Smith, M.M. and Heemstra, P.C. 1986. *Smith's Sea Fishes.* Springer-Verlag, Berlin.

Soest, R.W.M. Van 1972. Latitudinal variation in Atlantic *Salpa fusiformis* Cuvier, 1804 (Tunicata, Thaliacea). *Beaufortia*, **20** (262), 59–68.

Soest, R.W.M. Van 1973a. The genus *Thalia* Blumenbach, 1798 (Tunicata, Thaliacea) with descriptions of two new species. *Beaufortia*, **20** (271), 193–213.

Soest, R.W.M. Van 1973b. A new species in the genus *Salpa* Forskal, 1775 (Tunicata, Thaliacea). *Beaufortia*, **21** (273), 9–15.

Soest, R.W.M. Van 1974a. Taxonomy of the subfamily Cyclosalpinae Yount, 1954 (Tunicata, Thaliacea), with descriptions of two new species. *Beaufortia*, **22** (288), 17–55.

Soest, R.W.M. Van 1974b. A revision of the genera *Salpa* (Forskal, 1775), *Pegea* Savigny, 1816, and *Ritteriella* Metcalf, 1919 (Tunicata, Thaliacea). *Beaufortia*, **22** (293), 153–191.

Soest, R.W.M. Van 1975. Zoogeography and speciation in the Salpidae (Tunicata, Thaliacea). *Beaufortia*, **23** (307), 181–215.

Soest, RW.M. Van 1979a. Revised classification of the order Pyrosomatida (Tunicata, Thaliacea) with the description of a new genus. *Steenstrupia*, **5**, 197–217.

Soest, RW.M. Van 1979b. North–South diversity. In *Zoogeography and Diversity of Plankton* (S. Van der Spoel and AC. Pierrot-Butts, eds), pp. 103–111. Bunge, Utrecht.

Soest, R.W.M. Van 1981. A monograph of the order Pyrosomatida (Tunicata, Thaliacea). *Journal of Plankton Research*, **3** (4), 603–631.

Sogin, M.L., Elwood, H.J. and Gunderson, J.H. 1986. Evolutionary diversity of eukaryotic small subunit rRNA genes. *Proceedings of the National Academy of Sciences USA*, **83**, 1383–1387.

Sorokin, Y.I. 1973. Quantitative evaluation of the role of bacterioplankton in the biological productivity of tropical Pacific waters. In *Life Activity of Pelagic Communities in the Ocean Tropics* (M.E.V. Nogradov, ed.), pp. 98–135. Israel Scientific Translations.

Staples, R.F. 1966. The distribution and characteristics of surface bioluminescence in the oceans. *United States Navy Oceanographic Office, Technical Reports*, **184**, i–vi and 1–48.

Steinberg, K.D., Silver, M.W., Pilskaln, H.C., Coale, L.S. and Paduan, B.J. 1994. Midwater communities on pelagic detritus (giant larvacean houses) in Monterey Bay, California. *Limnology and Oceanography*, **39**, 1606–1620.

Stier, A. 1937. Beiträge zur Embryonalentwicklung der *Salpa pinnata*. *Zeitschrift für Morphologie und Ökologie der Tiere*, **33**, 582–633.

Steuer, A. 1908. Die Sapphirinen und Copilien der Adriatica. *Bollettino della Societa adriatica di Scienze Naturali in Trieste*, **24**, 157–166.

Streiff, R. 1908. Über die Muskulatur der Salpen und ihre systematische Bedeutung. *Zoologische Jahrbucher, Abteilung Systematik*, **27**, 1–82.

Sutton, M.F. 1960. The sexual development of *Salpa fusiformis* (Cuvier). *Journal of Embryology and Experimental Morphology*, **8**, 268–290.

Suyehiro, Y. 1942. A study of the digestive system and feeding habits of fish. *Japanese Journal of Zoology*, **10** (1), 1–303.

Swanberg, N.R. and Anderson, O.R. 1985. The nutrition of radiolarians: trophic activity of some solitary *Spumellaria*. *Limnology and Oceanography*, **30** (3), 646–652.

Swift, E., Biggley, W.H. and Napora, T.A. 1977. The bioluminescence emission spectra of *Pyrosoma atlanticum, P. spinosum* (Tunicata) *Euphausia tenera* (Crustacea) and *Gonostoma* sp. (Pisces). *Journal of the Marine Biological Association of the United Kingdom*, **57**, 217–213.

Swift, E., Lessard, E.J. and Biggley, W.H. 1985. Organisms associated with stimulated epipelagic bioluminescence in the Sargasso Sea and the Gulf Stream. *Journal of Plankton Research*, **7**, 831–848.

Swift, E., Biggley, W.H., Verity, P.G. and Brown, D.T. 1983. Zooplankton are major sources of epipelagic bioluminescence in the southern Sargasso Sea. *Bulletin of Marine Science*, **33**, 855–863.

Swofford, D.L. 1988. PAUP, phylogenetic analysis using parsimony, Version 3.0. Distributed by the Illinois Natural History Survey, Champaign, Illinois.

Sykes, P.F. 1982. The ultrastructure of the luminescent granules of *Oikopleura* (Urochordata: Larvacea), M.S. thesis, 57 pp. California State University, Long Beach, CA.

Taggart, C.T. and Frank, K.T. 1987. Coastal upwelling and *Oikopleura* occurence ('Slub'): a model and potential application to inshore fisheries. *Canadian Journal of Fisheries and Aquatic Sciences*, **44** (10), 1729–1736.

Taguchi, S. 1982. Seasonal study of fecal pellets and discarded houses of Appendicularia in a subtropical inlet, Kaneohe Bay, Hawaii. *Estuarine and Coastal Shelf Science*, **14** (5), 545–555.

Takano, H. 1954. Food of the mackerel taken near Oshima Island in 1953. *Bulletin of the Japanese Society of Scientific Fishery*, **20** (8), 694–697.

Talbot, F.H. 1960. Notes on the biology of the Lutjanidae (Pisces) of the East African coast, with special reference to *L. bohar* (Forskal). *Annals of the South African Museum*, **45**, 549–573.

Tarasov, N.I. 1956. *Luminescence of the Sea*, 203 pp. Academy of Sciences of the USSR, Moscow.

Tarverdieva, M.I. 1972. Daily food consumption and feeding pattern of the Georgian cod (*Notothenia marmorata* Fischer) and the Patagonian toothfish (*Dissostichus eleginoides* Smitt) (Fam. Nototheniidae) in the South Georgia area. *Journal of Ichthyology*, **12** (4), 684–692.

Tarverdieva, M.I. and Pinskaya, I.A. 1980. The feeding of fishes of the families Nototheniidae and Chaenichthyidae on the shelves of the Antarctic Peninsula and the South Shetlands. *Journal of Ichthyology*, **20** (4), 50–60.

Tebeau, C.M. and Madin, L.P. 1994. Grazing rates for three life history stages of the doliolid *Dolioletta gegenbauri* Uljanin (Tunicata, Thaliacea). *Journal of Plankton Research*, **16**, 1075–1081.

Terry, R.M. 1960. Investigations of inner continental shelf waters off lower Chesapeake Bay. Part III. The phorozooid stage of the tunicate *Doliolum nationalis*. *Chesapeake Science*, **2**, 60–64.

Théodoridès, J. 1989. Parasitology of marine zooplankton. *Advances in Marine Biology*, **25**, 117–177.

Théodoridès, J. and Desportes, I. 1968. Sur trois grégarines parasites d'invertébrés marins. *Bulletin de l'Institut Océanographique Monaco*, **67** (1387), 1–11.

Theroud, S. and Gorsky, G. 1996 Efficient removal of POM by Larvaceans during the 1995 spring bloom in the N.W. Mediterranean. *EOS, Transactions of the American Geophysical Union*, **76**, 116 (abstract).

Thompson, H. 1948. *Pelagic Tunicates of Australia*, pp. 196. Commonwealth Council for Scientific and Industrial Research, Melbourne.

Todaro, F. 1902. Sur les organes excréteurs des Salpidés (Salpidae Forbes). *Archives Italiennes de Biologie*, **38**, 33–48.

Tokioka, T. 1955. General consideration on Japanese Appendicularian fauna. *Publications of the Seto Marine Biological Laboratory*, **4**, 251–261.

Tokioka, T. 1956a. On chaetognaths and appendicularians collected in the central part of the Indian Ocean. *Publications of the Seto Marine Biological laboratory*, **5** (2), 197–202.

Tokioka, T. 1956b. *Fritillaria arafoera* n. sp., a form of the sibling species: *Fritillaria haplostoma*-complex (Appendicularia: Chordata). *Pacific Science*, **10** (4), 403–406.

Tokioka, T. 1957. Two new appendicularians from the eastern Pacific, with notes on the morphology of *Fritillaria aequatorialis* and *Tectillaria fertilis*. *Transactions of the American Microscopical Society*, **76** (4), 359–365.

Tokioka, T. 1958. Further notes on some appendicularians from the eastern Pacific. *Publications of the Seto Marine Biological Laboratory*, **7** (1), 1–17.

Tokioka, T. 1960. Studies on the distribution of appendicularians and some thaliaceans of the North Pacific, with some morphological notes. *Publications of the Seto Marine Biological Laboratory*, **8** (2), 351–443.

Tokioka, T. 1961. Appendicularians of the Japanese Antarctic research expedition. *Bulletin of the Marine Biological Station of Asamushi*, **10** (4), 241–245.

Tokioka, T. 1964. Taxonomic studies of appendicularians collected by the Japanese Antarctic research expedition 1957. *Scientific Reports — Japanese Antarctic Research Expedition 1956–1962*, **21E**, 1–16.

Tokioka, T. 1979. *Neritic and oceanic zooplankton*. In *Zoogeography and Diversity of Plankton* (S. Van der Spoel and A.C. Pierrot-Bults, eds), pp. 126–143. Bunge, Utrecht, The Netherlands.

Tokioka, T. and Berner, L. 1958a. Two new Doliolids from the eastern Pacific Ocean. *Pacific Science*, **12**, 135–138.

Tokioka, T. and Berner, L. 1958b. On certain Thaliacea (Tunicata) from the Pacific Ocean, with descriptions of two new species of Doliolids. *Pacific Science*, **12**, 317–326.

Tokioka, T. and Bhavanarayana, P.B. 1979. Note on the occurrence of a swarm of *Salpa cylindrica* Cuvier (Tunicata, Salpidae) in Sardinera Lagoon, Puerto Rico. *Proceedings of the Biological Society of Washington*, **92** (3), 572–576.

Tokioka, T. and Caabro, J.A.S. 1956. Apendicularias de los mares cubanos. *Memorias de la Sociedad Cubana de Historia Natural 'Felipe Poey'*, **23** (1), 37–95.

Tomo, A.P. 1971. *Aves y Mamiferos Antarticos*. Dirección Nacional Antartica Divulgar 1. Instituto Antarticos Argentina, Buenos Aires.

Torres, J.J. and Childress, J.J. 1983. Relationship of oxygen consumption to swimming in *Euphausia pacifica*. I. Effects of temperature and pressure. *Marine Biology*, **74**, 79–86.

Toselli, P.A. and Harbison, G.R. 1977. The fine structure of developing locomotor muscles of the pelagic tunicate, *Cyclosalpa affinis* (Thaliacea: Salpidae). *Tissue and Cell*, **9**, 137–156.

Tournier, H. 1968. Note préliminaire sur *Gadus poutassou* de Méditerranée. *Rapports et procès-verbaux des Réunions du Commission Internationale de l'Exploration de la Mer Méditerranée*, **19** (2), 317–319.

Trégouboff, G. 1916. Sur quelques protistes parasites rencontrés à Villefranche-sur-Mer. *Archives de Zoologie Experimentale et Génerale*, **55** (3), 35–47.

Trégouboff, G. 1956. Prospection biologique sous-marine dans la région de Villefranche-sur-mer, en juin 1956. *Bulletin de l'Institut Océanographique de Monaco*, **53** (1085), 1–24.

Trégouboff, G. 1965. La distribution verticale des Doliolides au large de Villefranche-sur-mer. *Bulletin de l'Institut Océanographique de Monaco*, **64** (1333), 1–47.

Trégouboff, G. and M. Rose. 1957. *Manuel de Planctonologie Méditerranéenne*, 589 pp. Centre National de la Recherche Scientifique, Paris.

Trepat I. 1983. Thaliaceos de la plataforma gallega (NO de España). *Resultados expedicíones Cientificas. Suplemento Investigacíon Resquera*, **11**, 139–147.

Trueman, E.R., Bone, Q. and Braconnot J.-C. 1984. Oxygen consumption in swimming salps (Tunicata: Thaliacea). *Journal of Experimental Biology*, **110**, 323–327.

Tsuda, A. and Nemoto, T. 1992. Distribution and growth of salps in Kuroshio warm-core ring during summer 1987. *Deep-Sea Research*, **39** (Supplement 1) 219–229.

Turley, C.M. 1993. The effect of pressure on leucine and thymidine incorporation by free living bacteria and by bacteria attached to sinking oceanic particles. *Deep-Sea Research*, **40**, 2193–2206.

Turley, C.M. and Mackie, P.J. 1994. Biogeochemical significance of attached and free-living bacteria and the flux of particles in the NE Atlantic Ocean. *Marine Ecology Progress Series*, **115**, 191–203.

Turner, J.T. 1977. Sinking rates of fecal pellets from the marine copepod *Pontella meadii*. *Marine Biology*, **40**, 249–259.

Turner, J.T., Tester, P.A. and Fergusson, R.L. 1988. The marine cladoceran *Penilia avirostris* and the 'microbial loop' of pelagic food webs. *Limnology and Oceanography*, **33**, 245–255.

Tyler, H.R. and Pearcy, W.G. 1975. The feeding habits of three species of lanternfishes (Family Myctophidae) off Oregon, U.S.A. *Marine Biology*, **32**, 7–11.

Udvardy, M.D.F. 1954. Distribution of appendicularians in relation to the Strait of Belle Isle. *Journal of the Fisheries Research Board of Canada*, **11** (4), 431–453.

Uljanin, B. 1884. Die Arten der Gattung Doliolum im Golfe von Neapel und der angrenzenden Meeresabschnitten. *Fauna und Flora des Golfes von Neapel*, **10**, 1–140.

Urban, J.L. McKenzie, C.H. and Deibel, D. 1992. Seasonal differences in the content of *Oikopleura vanhoeffeni* and *Calanus finmarchicus* fecal pellets: illustrations of plankton food web shifts in coastal Newfoundland waters. *Marine Ecology Progress Series*, **84**, 255–264.

Urban, J.L., McKenzie, C.H. and Deibel, D. 1993a. Nanoplankton found in fecal pellets of macrozooplankton in coastal Newfoundland waters. *Botanica Marina*, **36**, 267–281.

Urban, J.L., Deibel, D. and Schwinghamer, P. 1993b. Seasonal variations in the densities of fecal pellets produced by *Oikopleura vanhoeffeni* (C. Larvacea) and *Calanus finmarchicus* (C. Copepoda). *Marine Biology*, **117**, 607–613.

Uye, S.I. and Ichino, S. 1995. Seasonal variations in abundance, size composition, biomass and production rate of *Oikopleura dioica* (Fol) (Tunicata: Appendicularia) in a temperate eutrophic inlet. *Journal of Experimental Marine Biology and Ecology*, **189**, 1–11.

Vaillant, M.L. 1886. Remarques sur l'appareil digestif et la mode d'alimentation de la tortue luth. *Comptes Rendus Hebdomadaires des Seances de l'Académie des Sciences Paris*, **123**, 654–656.

Valiani, S. 1940. Contributo all biometria e alla studio dell'alimentazione dei pesci (*Argentina sphyraena* L). *Bolletino di Pesca Piscicoltura ed Idrobiologia*, **16** (4), 125–129.

Van Daele, Y. and Goffinet, G. 1987. Composition chimique et organisation de la tunique de deux ascidies *Phallusia mammillata* et *Halocynthia papillosa*. *Annales de la Société Royale Zoologique de Belgique*, **117**, 181–199.

Van der Spoel, S. and Heyman, R.P. 1983. *A Comparative Atlas of Zooplankton. Biological patterns in the Oceans*, pp. 1–186. Bunge, Utrecht. The Netherlands.

Van der Spoel, S. Pierrot-Bults, A.C. and Schalk, P.I-T. 1990. Probable Mesozoic vicariance in the biogeography of Euphausiacea. *Bijdragen Dier Kunde*, **60** (314), 155–162.

Van Zyl, R.P. 1959. A preliminary study of the salps and doliolids off the west and south coasts of South Africa. *Division of Fisheries of South Africa Investigations Report*, **40**, 3–31.

Verighina, I.A. and Golovan, G.A. 1978. Peculiarities of the structure of the alimentary canal of fish of the family Alepocephalidae. *Journal of Ichthyology*, **18** (2), 251–261.

Vernières, P. 1933. Essai sur l'histoire naturelle des Appendiculaires de Banyuls et de Sète (thèse pharmacie, Montpellier). *Bulletin de l'Institut Océanographique Monaco*, **30**, 617, 60p.

Vernières, P. 1934. Les Appendiculaires de la mer de Villefranche: *Pelagopleura haranti* n. sp. *Bulletin de la Société Zoologique de France*, **59** (2), 160–163.

Visser, J. de 1986. Transition zones and salp speciation. In *Pelagic Biogeography* (A.C. Pierrot-Bults, S. Van der Spoel, B.J. Zahuranec and R.K. Johnson, eds), *UNESCO Technical Papers in Marine Science*, Vol. 49, 266–269. UNESCO, Paris.

Visser, J. de and Van Soest, R.W.M. 1987. *Salpa fusiformis* populations of the North Atlantic. *Biological Oceanography*, **4** (2), 193–209.

Vogt, C. 1854. Recherches sur les animaux inférieurs de la Méditerranée. Second mémoire. Sur les Tuniciers nageants de la mer de Nice. *Mémoires de l'Institut Genèvois, Section des Sciences Naturelles et Mathématiques*, **2**, 1–102.

Wada, H., Makabe, K.W., Nakauchi, M. and Satoh, N. 1992. Phylogenetic relationships between solitary and colonial ascidians, as inferred from the sequence of the central region of their respective 18S rDNAs. *Biological Bulletin, Woods Hole*, **183**, 448–455.

Wada, H. and Satoh, N. 1994. Details of the evolutionary history from invertebrates to vertebrates, as deduced from the sequences of 18S rDNA. *Proceedings of the National Academy of Sciences USA*, **91**, 1801–1804.

Wallace, B.J. and Malas, D. 1976. The significance of the elongate, rectangular mesh found in capture nets of fine particle filter feeding trichoptera larvae. *Archiv für Hydrobiologie*, **77**, 205–212.

Wallace, G.T.Jr., Mahoney, O.M., Dulmage, R., Storti, F. and Dudek, N. 1981. First-order removal of particulate aluminium in oceanic surface layers. *Nature, London*, **293**, 729–731.

Watson, G. 1966. *Seabirds of the Tropical Atlantic Ocean*. Smithsonian Identification Manual, Smithsonian Institution, Washington.

Weigmann-Haass, R. 1983. Zur Taxonomie und Verbreitung der Gattung *Cyllopus* Dana 1853 (Amphipoda: Hyperiidea) im antarktischen Teil des Atlantik. *'Meteor' Forschungs Ergebnisse Reihe D*, **36**, 1–11.

Weihs, D. 1977. Periodic jet propulsion of aquatic creatures. *Fortschritte der Zoologie*, **24**, 171–175.

Welsch, U. and Storch, V. 1969. Zur Feinstruktur der Chorda dorsalis niederer Chordaten [*Dendrodoa grossularia* (v. Beneden) und *Oikopleura dioica* Fol]. *Zeitschrift für Zellforschung*, **93**, 547–559.

Wenner, C. 1979. Notes on fishes of the genus *Paraliparis* (Cyclopteridae) on the middle Atlantic continental slope. *Copeia*, **1979**, 145–146.

Werner, E. and Werner, B. 1954. Über den Mechanismus der nährungserwerbs der Tunicaten, speziell der Ascidien. *Helgoländer Wissenschaftliche Meeresuntersuchungen*, **5**, 57–92.

Wesenberg-Lund, E. 1926. Appendiculater og Salper. *Meddelelser om Grønland*, **23** (supplement 19), 117–132.

Wheeler, A. 1969. *The Fishes of the British Isles and North-west Europe*. Michigan State University Press, East Lansing.

Wheeler, A. 1975. *Fishes of the World*. Macmillan, New York.

Wheeler, A. 1978. *Key to the Fishes of Northern Europe*, 380 pp. Frederick Warne, London.

White, H.H. 1979. Effects of dinoflagellate bioluminescence on the ingestion rate of herbivorous zooplankton. *Journal of Experimental Marine Biology and Ecology*, **36**, 217–224.

Widder, E.A., Latz, M.I. and Case, J.F. 1983. Marine bioluminescence spectra measured with an optical multichannel detection system. *Biological Bulletin, Woods Hole*, **165**, 791–810.

Widder, E.A., Bernstein, S.A., Bracher, D.F., Case, J.F., Reisenbichler, K.R., Torres, J.J. and Robison, B.H. 1989. Bioluminescence in the Monterey Submarine Canyon: image analysis of video recordings from a midwater submersible. *Marine Biology*, **100**, 541–551.

Wiebe, P.H., Madin, L.P., Haury, L.R., Harbison, G.R. and Philbin, L.M. 1979. Diel vertical migration by *Salpa aspera* and its potential for large-scale particulate organic matter transport to the deep-sea. *Marine Biology*, **53**, 249–255.

Wiley, E.O. 1981. *Phylogenetics. The Theory and Practice of Phylogenetic Systematics*, 439 pp. John Wiley, New York.

Wiley, E.O. 1988. Parsimony analysis and vicariance biogeography. *Systematic Zoology*, **37**, 271–290.

Willey, A. 1893. Studies on the Protochordata. On the origin of the branchial stigmata, praeoral lobe, endostyle, atrial cavities in *Ciona intestinalis*, with remarks on *Clavelina lepadiformis*. *Quarterly Journal of Microscopical Science*, **34**, 317–360.

Williams, D.M. and Hatcher, A. 1983. Structure of fish communities on outer slopes of inshore, mid-shelf and outer shelf reefs of the Great Barrier reef. *Marine Ecology Progress Series*, **10**, 239–250.

Williams, G.C. and Mitton, J.B. 1973. Why reproduce sexually? *Journal of Theoretical Biology*, **39**, 545–554.

Williams, R. 1983. The inshore fishes of Heard and Donald Islands, Southern Indian Ocean. *Journal of Fish Biology*, **23**, 283–292.

Winkler, J. 1975. Variability of the oral musculature in the genera *Salpa* Forskål 1775 and *Weelia* Yount 1954 (Tunicata, Thaliacea). *Bulletin of the Zoological Museum of the University of Amsterdam*, **4** (18), 149–163.

Winstanley, R.H. 1978. Food of the trevalla *Hyperoglyphe porosa* (Richardson) off southeastern Australia. *New Zealand Journal of Marine and Freshwater Research*, **12** (1), 77–79.

Withers, P.C. 1992. *Comparative Animal Physiology*, 949 pp. Saunders College Publishing, Philadelphia.

Witzell, W.N. 1983. Synopsis of biological data on the Hawksbill Turtle, *Eretmochelys imbricata* (Linnaeus, 1766). *FAO Fisheries synopsis* (137), 78p.

Wolfenden, R.N. 1911. Die marinen Copepoden der deutschen Sudpolar-Expedition 1901–1903. II. Die pelagischen Copepoden der Westwinddrift und des sudlichen Eismeeres. *Deutsche Sudpolar Expedition*, 1901–1903, **12**, 181–380.

Wood, R.J. and Raitt, D.F.S. 1968. Some observations of the biology of the greater silver smolt, particularly in the northeastern Atlantic Ocean. *Rapports et Procés-verbaux des Réunions — Commission Internationale pour l'Exploration de la Mer Méditerranée*, **158**, 64–73.

Wyatt, T. 1973. The biology of *Oikopleura dioica* and *Fritillaria borealis* in the Southern Bight. *Marine Biology*, **22**, 137–158.

Wyatt, T. 1973. Production dynamics of *Oikopleura dioica* in the southern North Sea, and the role of fish larvae which prey on them. *Thalassia Jugoslavica*, **7**, 435–444.

Yasuda, F. 1960. The types of food habits of fishes assured by stomach contents examination. *Bulletin of the Japanese Society of Scientific Fisheries*, **26**, 653–662.

Yoder, J.A., Atkinson, L.P., Bishop, S.S., Hofmann, E.E. and Lee, T.N. 1983. Effect of upwelling on phytoplankton productivity of the outer southeastern United States continental shelf. *Continental Shelf Research*, **4**, 385–404.

Yoon, W.-D. 1995. Contribution des organismes macrozooplanctoniques au flux de la matière organique. Thèse de Doctorat de l'Université Pierre et Marie Curie Paris VI., pp. 137+7 annexes.

Yoon, W.-D., Marty, J.C., Sylvain, D., and Nival, P. 1996. Degradation of faecal pellets in *Pegea confoederata* (Salpidae, Thaliacea) and its implication in the vertical flux of organic matter. *Journal of Experimental Marine Biology and Ecology*, **203**, 147–177.

Young, J.W. and Blaber, S.J.M. 1986. Feeding ecology of three species of midwater fishes associated with the continental slope of eastern Tasmania, Australia. *Marine Biology*, **93**, 147–156.

Young, J.W. and Davis, T.L.O. 1990. Feeding ecology of larvae of southern bluefin, albacore and skipjack tuna (Pisces: Scombridae) in the eastern Indian Ocean. *Marine Ecology Progress Series*, **61**, 17–29.

Young, R.E. 1981. Color of bioluminescence in pelagic organisms. In *Bioluminescence: Current Perspectives* (K.H. Nealson, ed.), pp. 72–81. Burgess Publishing Co., Minneapolis.

Young, R.E. 1983. Oceanic bioluminescence, an overview of general functions. *Bulletin of Marine Science*, **33**, 829–845.

Young, R.E., Kampa, E.M., Maynard, S.D., Mencher, F.M. and Roper, C.F.E. 1980. Counterillumination and the upper depth limits of midwater animals. *Deep-Sea Research*, **27A**, 671–691.

Youngbluth, M.J. 1984. Manned submersibles and sophisticated instrumentation: tools for oceanographic research. In *Subtech '83, the Design and Operation of Underwater Vehicles*, Proceedings of the International Conference 15–17 November 1983, London, pp. 335–344. Society for Underwater Ecology, London.

Youngbluth, M.J., Bailey, T.G., Jacoby, C.A., Davoll, P.J. and Blandes-Eckelbarger, P.I. 1986. Submersible-based measurements of the sources, densities, distributions and sinking rates of marine snow aggregates and euphausiid fecal pellets. In *Aggregate Dynamics in the Sea* (A.L. Alldredge and E.O. Hartwig, eds), *Workshop Report, Office of Naval Research*, pp. 191–192. American Institute of Biological Science, Washington.

Youngbluth, M.J., Bailey, T.G. and Jacoby, C.A. 1990. Biological explorations in the mid-ocean realm: food webs, particle flux and technology advancements. In *Man in the Sea* (Y.C. Lin and K.K. Shida, eds), Vol. 2, pp. 191–208. Best Publishing, San Pedro.

Yount, J.L. 1954. The taxonomy of the Salpidae (Tunicata) of the central Pacific Ocean. *Pacific Science*, **8** (3), 276–330.

Yount, J.L. 1958. Distribution and ecologic aspects of central Pacific Salpidae (Tunicata). *Pacific Science*, **12** (2), 111–130.

Zagami, G., Badalmenti, F., Guglielmo, L. and Manganaro, A. 1996. Short term variations of the zooplankton community near the straits of Messina (North-eastern Sicily): relationships with the hydrodynamic regime. *Estuarine, Coastal and Shelf Science*, **42**, 667–681.

Zandee, M. and Roos, M. 1987. Component-compatibility in historical biogeography. *Cladistics*, **3**, 305–332.

Zeldis, J.R., Davis, C.S., James, M.R., Ballara, S.L., Booth, W.E. and Chang, F.H. 1995. Salp grazing: effects on phytoplankton abundance, vertical distribution and taxonomic composition in a coastal habitat. *Marine Ecology Progress Series*, **126**, 267–283.

Zeller, U., Peeken, I. and Bodungen, B. von 1994. Vertical fluxes and lateral advection of biogenic particles at the Barents Sea continental margin (Northern North Atlantic). *EOS, Transactions of the American Geophysics Union*, **75** (3), 127 (abstract).

Zernov, S.A. 1913. K voprosu ob izuchenii zhizni Chernogo morya. (Contribution à l'étude de la vie dans la mer Noire). *Mémoires de l'Académie Impériale des Sciences de St. Pétersbourg*, **32** (1), 1–299 (in Russian).

Systematic index

Note: only the names of pelagic tunicates are included. In the tables of Chapter 12, references are only given to those identified to species.

Order Appendicularia

Family 1. Oikopleuridae

Family 2. Fritillaridae

Family 3. Kowalewskiidae

Subject index

Note: entries are grouped under each of the four classes of pelagic tunicates; to look up all references to a particular topic, it is necessary to look separately under each group.

Pyrosomas

Doliolids

Salps

Appendicularians